AF835662

ESTATES GAZETTE LAW REPORTS

Estates Gazette Law Reports

1993

Volume 1

Edited by
Barry Denyer-Green LLM PhD FRICS
of the Middle Temple, barrister

Audrey Boyle BA
Publisher's editor

Estates Gazette

A member of Reed Business Publishing

THE ESTATES GAZETTE LIMITED
151 WARDOUR STREET, LONDON W1V 4BN

First published 1993
ISBN for complete set of 2 volumes: 0 7282 0190 9
ISBN for this volume: 0 7282 0191 7

ISSN 0951-9289

©The Estates Gazette Limited 1993

Typesetting by Digital Graphics, 20-26 Brunswick Place, London N1 6DZ
Printed and bound at the Bath Press, Avon

CONTENTS

SUPPLEMENT

Cases not reported in "Estates Gazette"

TABLE OF CASES

Titles of cases shown in bold type
Names of parties reversed shown in ordinary type

INDEX OF SUBJECT-MATTER

AGRICULTURE

A Queen's Bench Division

July 10 1992
(Before Mr Justice OWEN)

HAMMOND v ALLEN AND OTHERS

Estates Gazette February 27 1993
[1993] 08 EG 122

Agricultural holdings — Whether court has jurisdiction to order specific performance of repairing liability against* B *defendant landlords — Whether plaintiff tenant entitled to damages representing cost of carrying out works falling within a notice to repair — Whether defendant landlords counternotice was valid

The plaintiff holds a tenancy of Little Eaves Farm, Cheadle, Staffordshire. Holdings under an oral agreement with the defendant landlords in respect of which the Fixed Equipment Model Clauses set out in Schedule 1 to the Agriculture (Maintenance, Repair and Insurance of Fixed Equipment) Regulations 1973 are deemed incorporated. The farmhouse had fallen into serious disrepair and the costs of doing essential repairs and replacements were in the order of C £35,000. By clause 1(1) of the model clauses, the landlord has the right and liability for all repairs and replacements to certain defined parts of the farmhouse. By a notice dated January 12 1990 the plaintiff requested the defendants to make the specified necessary repairs or replacements. No works were done by the defendants, but by a notice dated November 1 1990, issued after the commencement of this action, the defendants claimed that the works required by the tenant would constitute improvements the cost of which the defendants were not liable for and requiring the question of liability to be determined by arbitration. Contrary to model D clause 12(3)(*a*) the defendants failed to serve any counternotice in time. By model clause 12(2) if a landlord fails to execute any repairs or replacements within three months of a notice the tenent may execute such repairs and recover the cost from the landlord. Not being able to afford the cost of repairs and replacements, the plaintiff commenced the present action seeking specific performance and/or the payment of damages representing the cost of such works. The defendants argued that the notice served in November 1990 was an effective counternotice time not being of the essence.

***Held*: The counternotice of November 1 1990 was invalid. The landlord is given, from receipt of the tenant's notice, one** E **month only to serve his counternotice, and is given three months only to do the necessary works himself. It cannot have been Parliament's intention that a recalcitrant landlord could have divested the tenant of his vested right by permitting a late counternotice. The landlords did not exercise their right to contest liability and could not do so now. There no longer remained any claim, question or difference between the parties under the model clauses and accordingly the court had jurisdiction to enforce the plaintiff's right: see *Grayless* v *Watkinson*. The plaintiff was entitled to pursue his claim for damages. The court may, but will not necessarily, make an** F **order for specific performance.**

The following cases are referred to in this report.

Goldsack v *Shore* [1950] 1 KB 708; [1950] 1 All ER 276; (1950) 66 TLR (Pt 1) 636
Grayless v *Watkinson* [1990] 1 EGLR 6; [1990] 21 EG 163
Jeune v *Queens Cross Properties Ltd* [1974] 1 Ch 97; [1973] 3 WLR 378; [1973] 3 All ER 97; (1973) 26 P&CR 98; [1973] EGD 976; 228 EG 143
United Scientific Holdings Ltd v *Burnley Borough Council* [1978] AC 904; [1977] 2 WLR 806; [1977] 2 All ER 62; (1977) 33 P&CR 220; [1977] EGD 195; 243 EG 43 & 127, HL G

This was a claim by the plaintiff, Robert Keith Hammond, for the determination of certain preliminary questions relating to a notice served on the defendants, George Henry Allen, Graham Thomas Allen, George Allen and Barbara Mary Allen, his landlords, under the model clauses in Schedule 1 to the Agriculture (Maintenance, Repair and Insurance of Fixed Equipment) Regulations 1973.

David Stockhill (instructed by Cowlishaw & Mountford, of Uttoxeter) appeared for the plaintiff; Joanne Moss (instructed by Hacking Ashton Jervis & Co, of Newcastle under Lyme) represented the defendants.

Giving judgment, **OWEN J** said: The plaintiff is, and has been for H over 30 years, the tenant of Little Eaves Farm, Cheadle, Staffordshire. The defendants are his landlords. The farm is an agricultural holding of some 23 acres. In addition, the plaintiff farms another 40 acres. The plaintiff took and holds the tenancy by an oral agreement. He lives in the farm house, which he holds under the tenancy. Over the years the farm house has fallen into serious disrepair. It is estimated that the cost of doing essential repairs and making essential replacements will be about £35,000, that being the lowest tender obtained by the tenant. The plaintiff does not have such a sum and cannot borrow that sum. He contends that the repairs and replacements are the defendants' obligation.

It is common ground between the plaintiff and the defendants that by virtue of section 7 of the Agricultural Holdings Act 1986 the Fixed J Equipment Model Clauses set out in Schedule 1 to the Agriculture (Maintenance, Repair and Insurance of Fixed Equipment) Regulations 1973 (SI 1973, No 1473) are "deemed to be incorporated" in the plaintiff's contract of tenancy.

By clause 1(1) of that schedule, the landlord has the right and liability for "all repairs and replacements to the under mentioned parts of the farm house . . .". The undermentioned parts are extensive and are extended still further by subparas (2) and (3). There is a proviso to subpara (1):

> . . . That in the case of repair and replacements to floor boards, interior staircases and fixed ladders (including bannisters or handrails) doors and windows and opening skylights (including frames) eaves — guttering and downpipes, the landlord may recover one half of the reasonable cost thereof K from the tenant.

In accordance with his contention, and in accordance with the model clauses, the plaintiff served a notice dated January 12 1990, requesting the defendants to make the specified necessary repairs or replacements. The defendants have not in any way complied with this notice. The only responses by the defendants have been to issue a notice requiring the tenant to comply with his obligations relating to hedges and fences — with which, I am told, the tenant has complied — and eventually to issue a notice dated November 1 1990 (ie after the issue and service of the writ in this action claiming that the works required by the tenant would constitute improvements for the cost of which the defendants are not liable and requiring "the question of

A our liability to execute such items of repair or replacement to be determined by arbitration under the provisions of the above Acts".

Clause 12(1) of the regulations provides that:

> If the landlord fails to execute repairs . . . which are his liability within three months of receiving from the tenant a written Notice specifying the necessary repairs and calling on him to execute them the tenant may execute such repairs and, except to the extent to which under the terms of Part I hereof the tenant is liable to bear the cost, recover . . . the reasonable cost from the landlord forthwith.

Clause 12(2) provides that:

> If the landlord fails to execute any replacements which are his liability within three months of receiving from the tenant a written notice specifying the necessary replacements and calling on him to execute them, the tenant may B execute such replacements and, except to the extent to which under the terms of Part I hereof the tenant is liable to bear the cost, recover . . . the reasonable cost from the landlord forthwith; provided that the tenant shall not be entitled to recover in respect of the aggregate of the replacements so executed by him in any year of the tenancy any sum in excess of whichever is the smaller of the two following sums, that is to say, a sum equal to the rent of the holding for that year or £500.

This clause has been somewhat amended by later regulations. However, the amendment is of no materiality for present purposes. Clause 12(3)(*a*) provides that:

> If the landlord wishes to contest his liability to execute any repairs or replacements specified in a notice served upon him by the tenant under either of the last foregoing subparagraphs he shall within one month serve a C counternotice in writing upon the tenant specifying the grounds on which and the items of repairs or replacement in respect of which he denies liability and requiring the question of liability in respect thereof to be determined by arbitration under the Act.

Clause 12(3)(*b*) provides that:

> Upon service of the counternotice on the tenant, the operation of the notice (including the running of time thereunder) shall be suspended, in so far as it relates to the items specified in the counternotice, until the termination of an arbitration determining the question of liability in respect of those items.

Clause 12(3)(*c*) provides that:

> In this subparagraph, "termination" in relation to an arbitration, means the date on which the arbitrator's award is delivered to the landlord.

D This must mean the date on which the determination of liability is made.

As has been stated, the landlords did not serve a counternotice in time, nor did they carry out the repairs replacements. Accordingly, the tenant, after the three-month period, became entitled to do the specified repairs and recover the reasonable costs from the landlord *forthwith*.

Failing payment forthwith, the tenant's remedy would have been by an action in these courts for the amount expended by him.

Similarly, in respect of replacements, the tenant, after the three-month period, became entitled to make the replacements and recover in these courts in accordance with the model clauses which had E become a part of the contract of tenancy.

Not being able to afford the cost of repairs and replacements, the plaintiff has brought this action. Before me there are three issues to be determined: (l) whether this court has any jurisdiction to order specific performance against the defendant; (2) whether the court has any jurisdiction to order the defendants to pay damages, representing the cost of such works as the plaintiff may prove to fall within the notice dated January 12 1990 and also within the defendants' repairing and replacement obligations contained within the Agriculture (Maintenance, Repair and Insurance of Fixed Equipment) Regulations; and (3) whether the counternotice, served on behalf of the defendant, dated November 1 1990, is valid. It will, I think, be convenient to deal with the last of these questions first.

F The procedure contemplated by clause 12(3)(*a*) is that if the landlord wishes to contest his liability to execute any of the repairs or replacements specified in the tenant's notice, he "shall" within one month serve a counternotice asking that that *liability* should be determined. This was not done. However, the defendants argue that their notice, served in correct form but well out of time in November 1990, was an effective counternotice. As I understand their argument, they contend that time is not of the essence despite the use of the imperative "shall". They argue that the status of the time-limit is *quasi-contractual*.

I have been referred to the decision of the House of Lords in *United Scientific Holdings Ltd* v *Burnley Borough Council* [1978] AC 904. The House considered two appeals concerning leases containing rent review clauses. These rent review clauses indicated a timetable. Their G lordships were unanimous in commending the modern law in the cases of contracts of all types, as correctly summarised in *Halsbury's Laws of England*, 4th ed, vol 9, para 481, at p338:

> Time will not be considered to be of the essence unless: (1) the parties expressly stipulate that conditions as to time must be strictly complied with; or (2) the nature of the subject matter of the contract or the surrounding circumstances show that time should be considered to be of the essence.

Although it would be misleading to consider the problem in the *United Scientific* case to be that of this case, nevertheless, in my judgment, some guidance may be found. The surrounding circumstances, and the purpose and wording of the Act, are all of some importance in considering whether time was of the essence. H

The model clauses place clear obligations on the landlord for the benefit of the tenant. It is the tenant who will be — to use the words of the defendants' counsel — "producing the nation's food". It is the tenant who needs protection. The model clauses place clear liabilities on the landlord to repair and replace. Such repairs and replacements should take place without any formal notice. If they do not, the procedure provided is, in my judgment, intended and designed to protect the tenant by avoiding the law's delays and by providing a speedier process. As I read and understand the procedure, the landlord is given, from receipt of the tenant's notice, one month and one month only to serve his counternotice, and is given three months and three months only to do the necessary works himself. At the end of the one-month period his liability is conclusively presumed or J found against him. At the end of the three-month period, the landlord, who might be able to do the work more cheaply, nevertheless has no right to do the work himself. The tenant's entitlement to do the work himself and recover the cost forthwith then comes into existence. His right becomes vested. It is conceded by the defendants that here the tenant became so entitled but, it is said, the defendants were entitled to terminate that entitlement or vested right. In my judgment, this cannot be so. The tenant might, and had he had the means would, have incurred substantial contractual liabilities at the end of the three-month period. In my judgment, Parliament cannot have intended that a recalcitrant landlord could have divested the tenant of his vested right even less, as could well have been the case, could have involved the tenant in further liability. K Accordingly, I answer the third preliminary issue by saying, "No".

Next, I turn to the first and second preliminary issues, which may be taken together. The defendants, having taken no steps to comply with their obligations, even after receipt of the tenant's notice, argue that, even assuming the counternotice to be invalid, although the tenant may do the repairs and then recoup from the landlord, in the event that the tenant is impecunious and unable to borrow the money to finance the repairs or replacements so that he is unable to do them, as is the case here, the tenant is without remedy and cannot enforce his rights. The only remedy available to the tenant, argues Miss Moss for the defendants, is for the plaintiff to do the work himself and then claim the cost. This court, she argues, has no jurisdiction to award L any remedy now claimed by the plaintiff, although presumably she would concede that if the plaintiff had done the works, claimed the cost of doing them and had not been paid "forthwith" he could then have sued for that cost in this court.

Principally the defendants say that their argument in favour of this somewhat surprising contention — even more surprising in view of the legal truism expressed by Sir Raymond Evershed MR in *Goldsack* v *Shore* [1950] 1 All ER 27 — that the jurisdiction of the King's courts must not be taken to be excluded unless there is clear language in the statute which is alleged to have that effect — is based on the decision and reasoning in *Grayless* v *Watkinson* [1990] 1 EGLR 6*. I see no reason to refer to the facts of this case great detail.

There, at first instance, the recorder purported to hold that an M agricultural tenant had remedies at common law in addition to his options conferred by para 12, and so, the recorder held, the tenant having himself done the work, he could recover the reasonable cost from the landlord at common law, without having to submit to the restrictions on the landlord's liability in the final proviso in para 12(2). This approach was rejected. The reasoning for the immediate decision in that case is apparent. Further help for this case may be obtained from a consideration of the judgments. At p9, Dillon LJ said:

> The recorder gave judgment for the respondent for the full amount of

---

*Editor's note: Also reported at [1990] 21 EG 163.

£7,992.50 claimed, together with interest. He accepted an argument from counsel for the respondent — foreshadowed, I apprehend in the reply — that, if the landlord has a liability to the tenant under the 1973 regulations to replace the roof of the barn, the tenant has remedies at common law in addition to his option under para 12(2) and so, having himself done the work, can recover the reasonable cost from the landlord at common law, without having to submit to the restriction on the landlord's liability in the final proviso in para 12(2). One of the troubles about that argument, however, is that it would deprive the final proviso in para 12(2) of all effect, except where the tenant, being ill-advised, chose to agree to the landlord's liability being limited by the proviso.

The true position, as I see it, is that para 12(2) is the sole source of the tenant's right to recover the cost of replacing the barn roof from the landlord, and the tenant must accept para 12(2) as a whole, including the proviso. The only liability of the appellant as landlord which is available for the respondent as tenant to enforce where the respondent has carried out works after the appellant had failed to do so is a liability qualified by the proviso.

At p8H, Ralph Gibson LJ said:

I agree that that part of the recorder's judgment cannot be sustained. The Act and the regulations made under it did not imply into the contract of tenancy a separate or independent term that the landlord should replace the roof, or any other part of the premises. The effect of the regulations is as stated in regulation 3:

"The provisions set forth in the Schedule . . . shall be deemed to be incorporated in every contract of tenancy . . . except in so far as they would impose on one of the parties to an agreement in writing a liability which under the agreement is imposed on the other."

The tenant, therefore, acquires no contractual right and the landlord is subjected to no contractual liability, by reason of incorporation of those provisions, save as contained in the provisions in the Schedule. To the extent that the rights and liabilities there described are qualified or limited, the tenant and the landlord can neither acquire nor be subjected to any greater rights or liabilities by reason of incorporation of the provisions. The provisions of the Schedule do not, of course, reduce any rights or liabilities given or imposed by the terms of the tenancy agreement itself. If the tenancy agreement had contained an obligation under the landlord to replace the roof, the provisions in the Schedule would not, as I understand the position, reduce the tenant's rights in that regard: both rights and obligations, those under the tenancy agreement itself and those incorporated therein by the regulations, would coexist.

At p9 Stuart-Smith LJ said:

Three questions arise for determination in this appeal.

The first is whether the recorder was right in giving judgment for the whole amount expended upon the replacement of the barn roof by the tenant, namely £7,992.50, without regard to the limitation or restriction of liability imposed by the proviso to para 12(2) of the Schedule to the 1973 regulations. In my judgment he was not. I agree entirely with the reasons given by Dillon LJ for holding that the tenant's sole right derived from para 12(2), and he is bound by the restriction imposed by the proviso.

It seems that the Court of Appeal at least held that, when a tenant relies on the model clauses, he must take those clauses for better or for worse. He must accept all the clauses, since all of them, in the absence of agreement in writing to the contrary, are deemed to be incorporated into the contract. Here, there was no written agreement and the tenant had to rely on the model clauses. Accordingly, in my judgment, he was obliged to follow the procedure set out in para 12 and implied into his contract of tenancy. In other words, it was incumbent upon him to serve his notice and, had the landlords served a counternotice in time, it would have been incumbent upon him to go to arbitration to determine liability. The position here, however, is that the landlords did not exercise their right to contest liability and cannot now do so. It is nevertheless instructive to see what would have happened had the counternotice been served in time. The purported counternotice is at p19 of the bundle of documents. That notice may be seen as a notice the question in which has been determined against the landlords by their failure to serve a counternotice. In other words, the question of their liability to execute such repairs or replacements has been found against them.

Subject to one further argument, I see no reason why that having been decided, as it happens by default, the tenant should not now bring proceedings in these courts for the consequences of that liability, ie the cost of doing the works for which the landlords are liable. To do so would seem to be similar to the procedure in *Grayless* and in accordance with the decision and reasoning of Knox J in *Tustian* v *Johnstone*.

The one argument which remains to be considered is based upon clause 15 of the model clauses which provides that:

If any claim, question or difference shall arise between the landlord and the tenant under the foregoing provisions hereof, not being a matter which, otherwise than under the provisions of this paragraph, is required by or by virtue of the Act or section 19 of the Agriculture (Miscellaneous Provisions) Act 1963 (notice to remedy Breach of Tenancy Agreement) or Regulations or Orders made thereunder or the foregoing provisions hereafter be determined by arbitration under the Act, such claim, question or difference shall be determined, in default of agreement, by arbitration under the Act.

Miss Moss argues that there is here a claim, question or difference between the landlord and the tenant under the "foregoing provisions". Clearly, she says, that is so, since it is only as a result of those provisions and under the terms of them that this claim is made. In these circumstances, she argues, there must be an arbitration.

The plaintiff, on the other hand, argues that his claim is in contract, the model clauses having been incorporated into that contract and forming the basis of his claim. Liability having been decided, he contends that he has the right to sue for the reasonable cost of the repairs, for a declaration that he is entitled to the reasonable cost and even for an order for specific performance.

The defendants argue that the sole source of the tenant's right to recover the costs of replacing or repairing is clause 12 and not the contract. For myself, I do not see the force of this argument, since clearly the whole of clause 12 must be deemed to be incorporated into the contract, as indeed must be the whole of regulation 15. As I see it, the question now is whether this present claim, presented to this court by writ and statement of claim, is made under "the foregoing provisions" and, if so, what effect does regulation 15 have?

In my judgment, no "claim, question or difference" now arises between the landlords and the tenant under "the foregoing provisions". There could have been a "claim, question or difference" as to liability — but no more, but that has been decided by the defendants' failure to serve a counternotice. The plaintiff now has a right, the existence of which was to be decided either by arbitration or by the defendants' failure to act. The defendants not having sought arbitration, that right has been established by default. Now, the plaintiff is merely seeking to enforce that right in a manner similar to that in which Mr Grayless sought to enforce and was held entitled to enforce his right in *Grayless* v *Watkinson*. As I understand the purport of para 5 of the rejoinder, the defendants do not argue to the contrary. Accordingly, and especially bearing in mind the words quoted from Sir Raymond Evershed, I reject the defendants' argument that this court has no jurisdiction to right a patent wrong.

Since the contrary was argued before me with some enthusiasm, I add that, had it been necessary for me to consider whether the defendants had conferred jurisdiction upon the courts by taking steps in the action, I would not have found in favour of the tenant's contention.

In the circumstances, in my judgment, the plaintiff tenant, relying upon the terms of section 97 of the Act, in so far as it is necessary for him to do so, *is entitled* to pursue his remedies in these courts.

As to specific performance, no separate argument has been addressed to me. However, I note that Knox J in *Tustian* v *Johnstone* cited *Jeune* v *Queens Cross Properties Ltd* [1974] 1 Ch 97, a decision of Pennycuick V-C with approval. In these circumstances, I answer the first question "Yes". The court "may" grant such relief. I do not say it will, since it is said to be "a jurisdiction which should be carefully exercised".

In my judgment, the plaintiff is also entitled to pursue his claim for damages. The plaintiff shall have an order for costs and there will be leave to appeal.

Declaration accordingly. Leave to appeal.

Chancery Division

December 21 1992

(Before Mr Justice FERRIS)

LUTTENBERGER v NORTH THORESBY FARMS LTD AND OTHERS

Estates Gazette May 1 1993

[1993] 17 EG 102

Agricultural holdings — Failure to pay rent — Whether notice to quit under Case D valid — Whether notice to quit fraudulently served

The freehold of Eastfield House, North Thoresby, Lincolnshire, was owned by the plaintiff. By an agreement dated June 30 1982 the plaintiff's predecessors in title granted a yearly tenancy of the farm to the first defendant company, the shares of which were owned by the second and third defendants, Mr and Mrs Popoff. By the agreement the rent was payable half-yearly on April 6 and October 10 each year. On October 10 1989 rent amounting to £15,264 became payable. On November 17 1989, the rent not having been paid, the plaintiff's solicitor, Miss Prince, served a notice to pay requiring payment of rent within two months and a statutory demand on the first defendant. On October 19 1989 a cheque drawn in the sum of £14,805.60 on the company's account, signed only by the third defendant, was sent to the defendants' then solicitor Mr Nutt. The defendants were entitled to deduce drainage rates of £450 (not the £458.40 in fact deducted) from the rent. On December 22 1989 Mr Nutt sent the cheque to Miss Prince and, by reason of a dispute between Miss Prince and the Post Office about the delivery of recorded mail and her refusal to accept an item of post on December 30 1989, the cheque was not received by her until the afternoon of January 19 1990. Miss Prince caused the service on the company on January 19 1990 of the notice to quit under Case D before she was aware of the receipt of the cheque on that date. The cheque was sent to Miss Prince's bank with instructions for special clearance and to be returned and not re-presented if payment was not met at once on presentation. Because the mandate to the paying bank required two signatures on the cheque, and it had only the third defendant's, at the paying bank's request, the second defendant added his signature to the cheque before it was paid on January 24 1990. On February 5 1990 Mr Nutt sent a cheque for £8.40 representing the overdeduction of drainage rates.

The plaintiff sought an order for possession against the defendants following the expiration of the notice to quit. On behalf of the second and third defendants it was contended that the statement in the notice to quit concerning non-payment of rent was false and that that false statement was made fraudulently.

***Held*: Order for possession was granted. (1) The proposition in *Beevers* v *Mason* that payment of rent by cheque relates back to the date of posting of the cheque, and not the date when the cheque is honoured, does not apply where the cheque is one which the paying bank, as between itself and its client, was not bound to honour, because, for example, it lacked any signature. In the present case the cheque lacked the necessary two signatures, and when the second was added, and payment was made, that payment did not relate back to the date of posting, but only to the date when the second signature was given. (2) The second defendant's willingness to add his signature constituted a fresh mandate to pay on the cheque, this was given on January 24 1990; in these circumstances the cheque posted on December 22 1989 was not "being honoured" for the purposes of what was said in *Beevers* v *Mason* and the payment of the cheque did not relate back to the date of posting under the proposition in that case and the rent was paid only on January 24. (3) The *de minimis* rule applies to the payment of rent and a notice to quit under case D. Accordingly, the shortfall of £8.40 was, on the facts, *de minimis* and would have been ignored had payment otherwise been made on time. Having regard to the evidence, Miss Prince had reasonable grounds for believing that the statement concerning non-payment of rent, which was contained in the notice to quit, was a true statement when the notice was served. That was enough to rebut the allegation of fraud.**

The following cases are referred to in this report.

Angus v *Clifford* [1891] 2 Ch 449
Beevers v *Mason* (1978) 37 P&CR 452; 248 EG 781, CA
Derry v *Peek* (1889) 14 App Cas 337
Magdalen College, Oxford v *Heritage* [1974] 1 WLR 441; [1974] 1 All ER 1065; (1974) 27 P&CR 169; [1974] EGD 1; 230 EG 219, CA
Price v *Romilly* [1960] 1 WLR 1360; [1960] 3 All ER 429
Rous v *Mitchell* [1991] 1 WLR 469; [1991] 1 All ER 676; *sub nom Stradbroke (Earl of)* v *Mitchell* [1991] 1 EGLR 1; [1991] 03 EG 128 & 04 EG 132

This was a claim by the plaintiff against the first defendant, North Thoresby Farms Ltd, and the second and third defendants, Mr and Mrs Michael Popoff, for possession of Eastfield House Farm, North Thoresby, Lincolnshire, following the expiration of a notice to quit under Case D for non-payment of rent.

Derek Wood QC and Olav Ernstzen (instructed by Prince & Co, of Sherburn) appeared for the plaintiff; Robert Reid QC and John McGhee (instructed by H V Beckett & Co, of Market Rasen) represented the second and third defendants; the first defendant did not appear and was not represented.

Giving judgment, **FERRIS J** said: Eastfield House Farm is a substantial farm, some 630 acres in extent, at North Thoresby, Lincolnshire. In the events which have happened the freehold is vested in the plaintiff, Mrs Alexandra Helen Luttenberger, in her capacity as personal representative of two great aunts — Mrs Helena Johanna Butt and Miss Eva Motley Robinson — who were sisters. In this action Mrs Luttenberger seeks an order for possession of the farm against North Thoresby Farms Ltd and against Mr Michael Dimitry Popoff and his wife Mrs Lesley Popoff. The company was, and according to the Popoffs still is, the tenant of the farm. Mr and Mrs Popoff between them hold 10 shares of £1 each in the company, which represent its only issued share capital. At all material times they have been the only directors of the company and Mrs Popoff has been its secretary. Mr and Mrs Popoff are in occupation of the farm, pursuant to arrangements with the company, the precise nature of which is unclear. The company has taken no part in this action. It is, however, common ground that the Popoffs are in no worse and, correspondingly, in no better position than the company itself. In effect, therefore they have presented the case which the company could have presented on its own behalf if it had defended this action.

The farm came into the ownership of Mrs Butt and Miss Robinson many years ago. It appears that some parts of it belonged exclusively to Mrs Butt and other parts of it exclusively to Miss Robinson, the remainder belonging to them jointly. Some time in or before the year 1979 they granted an oral tenancy of the farm as a whole to Mr Popoff who is the brother of Mrs Luttenberger and thus himself a great-nephew of the two ladies.

On November 15 1979 Mr Popoff assigned this tenancy to the company. Subsequently, by an agreement in writing dated June 30 1982, Mrs Butt and Miss Robinson granted a new tenancy to the company for one year from April 6 1982 until April 6 1983, continuing from year to year thereafter until determined at the end of any year of the tenancy by not less than 12 months' notice in writing given by either party to the other. The initial rent under this tenancy was £17,000 pa. By the time of the events with which I am concerned this rent had been duly increased, under the machinery applicable to agricultural tenancies, to £30,528 pa. This rent was payable half-yearly in arrears on April 6 and October 10 in each year.

Miss Robinson died on January 7 1986 and Mrs Butt died on April 28 1987. Mrs Luttenberger is the sole executor of the will of each of them. She has not yet completed the administration of either estate. In particular, I understand that some pecuniary legacies under one or both wills remain unpaid. Subject to the due completion of the administration of each estate, Mrs Luttenberger is the sole residuary legatee under each will. For present purposes, however, it suffices to say that the legal estate in the entirety of the farm is vested in Mrs Luttenberger as personal representative of one or other of her two great-aunts.

Unfortunately, Mrs Luttenberger and her brother Mr Popoff have been estranged for some years and a state of considerable hostility has existed and continues to exist between them. It seems that they are in litigation against each other in at least four pending actions, in three of which Mrs Luttenberger is the plaintiff. It was said on behalf of Mrs Luttenberger that the company has been an irregular payer of the rent due under the tenancy. It does in fact appear that the company has not paid its rent strictly on time, but a schedule of payments which was not disputed shows that the delay in payment has been of the order of two to six weeks. All the rent becoming due under the tenancy down to October 10 1989 was paid by the company to Mrs Luttenberger, the last payment in respect of the rent due on April 6 1989 having been made by cheque dated April 21 1989. It is convenient to note at this stage that the actual payments made by the company were always for a sum rather less than the full amount of the rent payable. This was because the company deducted from the rent, as it was entitled to do under section 72 of the Land Drainage Act 1976, certain payments which it had made in satisfaction of drainage rates payable by the landlord.

This action is brought as a consequence of the company's failure to pay in due time the instalment of rent, amounting to £15,264, which became payable on October 10 1989. There is no doubt that this rent was not paid on the due date. Equally, there is no doubt that Mrs Luttenberger received the greater part of this instalment of rent on January 22 1990 and the balance, amounting to a mere £8.40, on February 5 1990. The issue between the parties is whether the delay in payment, coupled with the steps taken by or on behalf of Mrs Luttenberger, which I have yet to mention, have the result that a notice to quit served on behalf of Mrs Luttenberger on January 19 1990 was effective to determine the company's tenancy on April 5 1991. There is no doubt that, if that notice to quit was effective, the termination of the company's tenancy would have effect on the latter date.

Before I explore the facts of this case in greater detail, I propose to say something about the impact on this tenancy of the Agricultural Holdings Act 1986 and certain decisions made under it or its predecessor. This was not in dispute, so I can take it quite shortly. The position may, I think, be summarised as follows:

(1) By section 25(1) of the Act a notice to quit an agricultural holding is invalid if it purports to terminate the tenancy before the expiration of 12 months from the end of the then current year of the tenancy. The contractual provisions of the company's tenancy agreement which I have referred to are consistent with this statutory requirement.

(2) By section 26(1) the ability of a landlord of an agricultural holding to terminate a tenancy of that holding is severely curtailed. Generally, the tenant has the right to serve a counternotice in response to any notice to quit and, if he does so, the notice to quit is not to have effect unless the Agricultural Land Tribunal consents to its operation which it may do only in certain limited circumstances.

(3) However, by section 26(2), this general curtailment does not apply in certain cases which are specified in Part I of Schedule 3. One of these cases is Case D which, so far as material, is applicable where:

> at the date of the giving of the notice to quit the tenant had failed to comply with a notice in writing served on him by the landlord, being either:
>
> (a) A notice requiring him within two months from the service of the notice to pay any rent due in respect of the agricultural holding to which the notice to quit relates.

Then there is an alternative in (b), which does not matter. Then the case concludes:

> and it is stated in the notice to quit that it is given by reason of the said matter.

(4) Where a landlord serves a notice to quit, any question whether it was within the exception to the general restriction on the effect of notice to quit which is provided by section 26(2) is to be determined by arbitration pursuant to para 1 of Schedule 4 and the regulations made pursuant to section 29.

(5) The relevant regulations are those contained in the Agricultural Holdings (Arbitration on Notices) Order 1987 (SI 1987 No 710). By article 9, where a tenant wishes to contest any question arising under section 26(2), he must within one month after the service of the notice to quit on him serve on the landlord a notice in writing requiring the question to be determined by arbitration under the Act. By article 10 a notice requiring arbitration ceases to be effective three months after it is given unless an arbitrator has been appointed or an application has been duly made for the appointment of an arbitrator.

(6) Arbitration is the only means by which a tenant can challenge a landlord's contention that a notice to quit falls within the exception provided by section 26(2). Accordingly, if the tenant does not duly require arbitration or if a notice requiring arbitration is rendered of no effect by the failure in due course to obtain or apply for the appointment of an arbitrator, the landlord's claim that the case falls within section 26(2) cannot be disputed by the tenant unless the notice on which the landlord relies is tainted by fraud. This is, in substance, the effect of the decisions of the Court of Appeal in *Magdalen College* v *Heritage* [1974] 1 WLR 441* and *Rous* v *Mitchell* [1991] 1 WLR 469†. In the latter of these cases it was held, on the facts, that the landlord's notice was tainted by fraud.

As I have already indicated, the underlying question in this case is whether a notice to quit served on behalf of Mrs Luttenberger on January 19 1990 was effective to terminate the company's tenancy. That notice was expressed to be given on the ground that the circumstances were within Case D. The company wished to challenge that assertion and, in due time, it gave notice requiring the matter to be determined by arbitration. However, it failed to apply for the appointment of a arbitrator within the three-month period, with the result that its notice requiring arbitration ceased to have effect and the company cannot now seek arbitration. Mrs Luttenberger seeks an order for possession on the ground that the company's tenancy terminated on April 5 1991 by virtue of the notice to quit. Mr and Mrs Popoff resist that claim on the ground that the claim in the notice to quit that the requirements of Case D were satisfied was false and that this claim was made fraudulently.

In order to evaluate this contention it is necessary to explore the facts with some care. The story which unfolds is a remarkable one.

I take it up in October 1989. At that time, as I have said, all rent due from the company under its tenancy had been received. But Mrs Luttenberger and Mr Popoff were on the worst possible terms and were in litigation with each other on several matters. Unfortunately, their personal hostility seems to have been mirrored by a similar hostility on the part of their respective solicitors. These were, on behalf of Mrs Luttenberger, Miss Prince, sole principal of Prince & Co of Sherburn, near Malton, North Yorkshire, and, on behalf of Mr and Mrs Popoff, Mr Kevin Nutt, a partner in the firm of Epton & Co of Lincoln. Such was the hostility between these two solicitors that they were not willing even to speak to each other over the telephone.

As the time for payment of the rent due on October 10 1989 approached, Mrs Luttenberger was very anxious to receive payment so as to enable her to pay money due to the Inland Revenue and also to pay counsel's fees for litigation between her and Mr Popoff. On October 5 1989 Miss Prince, on behalf of Mrs Luttenberger, wrote to the secretary of the company drawing attention to the fact that the half-yearly instalment of rent amounting to £15,264 would fall due for payment on October 10 1989 and requesting that payment be made by cheque payable to Prince & Co clients' account. This reminder and the request for payment by cheque to Prince & Co were, it seems, in accordance with the practice which Mrs Luttenberger, through Miss Prince, had adopted for some time. The somewhat pedantic reservation of position which Miss Prince inserted in the letter, which I will not read, seems also to have been typical of Miss Prince's general style.

On October 25, the rent not having been received by Miss Prince or Mrs Luttenberger, Miss Prince sent a reminder to the company by recorded delivery post. Still the rent was not received and, on November 17 1989, Miss Prince took two further steps. First, she served on the company a formal notice of the kind referred to in Case D in Part I of Schedule 3 to the Agricultural Holdings Act 1986, requiring payment of the rent within two months. The notice was given in the prescribed form. It referred to the relevant statutory provisions and drew the attention of the tenant to the fact that non-compliance with the notice might be relied on as a reason for service of a notice to quit under Case D. Second, Miss Prince prepared and caused to be served on the company a statutory demand in respect of the unpaid rent of £15,264. Miss Prince gave evidence, which I accept, that she did this after obtaining Mrs Luttenberger's express instructions and because, in her experience, corporate debtors frequently responded to a statutory demand when they failed to respond to other pressures and because the 21-day period, after which a winding-up petition could be presented, was shorter than the two-month period which must elapse before a notice to quit could be served. When the rent remained unpaid after the expiration of the 21-day period, Miss Prince prepared on behalf of Mrs Luttenberger a petition for the winding up of the company founded upon the outstanding rent and the company's failure to satisfy the statutory demand. The petition was presented to the High Court on December 19 1989 and was, in the usual way, fixed to be heard on Wednesday, February 7 1990. The petition was served on the company on December 20 1989, or thereabouts, and it immediately came to the attention of Mr Popoff.

All these steps had, so far as Mrs Luttenberger and Miss Prince were concerned, brought forth no reaction of any kind from the company or Mr Popoff. In fact, as is now known from the evidence of Mr Popoff, a number of relevant steps had been taken by him. The company's only significant asset appears to have been its tenancy of

* Also reported at [1974] 230 EG 219.
† Also reported at [1991] 1 EGLR 1.

A the farm. But it also had a bank account at Midland Bank, Louth, where Mr Popoff kept other accounts of his own. The company's account was used only for the purpose of paying rent due under the tenancy. In the middle of October 1989 the balance standing to the credit of this account was only some £12. Mr Popoff gave evidence that on or about October 10 1989 he wrote out a cheque in the company's cheque book for the sum of £14,805.60 payable to Prince & Co. The cheque is dated October 10 1989 and was signed by Mrs Popoff. I accept the evidence of Mr Popoff that this cheque was written and signed on October 10, or thereabouts, and that it was intended to be in respect of the rent due from the company on that date. Mr Popoff said, and I accept, that the cheque, when it had been signed, was sent to Mr Nutt of Epton & Co. On October 18 1989, B which was about the earliest date on which the cheque could be expected to have been presented for payment, Mr Popoff caused to be transferred from another account of his into the company's account at Midland Bank, Louth, an amount exactly equal to the amount of the cheque. The company was thereafter in funds to an extent sufficient to enable the cheque to be paid.

The reason why the cheque was for the sum of £14,805.60 and not for the full amount of £15,264, which was the rent due, was that Mr Popoff deducted from the latter figure the sum of £458.40 which he calculated to be the amount of the landlord's drainage rate, which the company was entitled to deduct from the rent.

Although the cheque was, as I have said, sent to Mr Nutt, he did not C immediately send it on to Prince & Co. According to Mr Popoff, Mr Nutt advised that he should hold on to the cheque for a while. His reasons for this advice are unclear. It seems that Mr Nutt suffered a breakdown in April 1991. He and his firm have ceased to act for the Popoffs and the company and he was not called to give evidence. This piece of advice, and other advice which he is said to have given, and actions which he took between this time and February 1990, seem inexplicable and, to say the least, imprudent. But it is not necessary for me to make findings about that. It will suffice to say that Epton & Co did, in fact, retain the company's cheque. This continued to be the case, notwithstanding Miss Prince's letter of October 25, the notice to pay dated November 19 and the statutory demand of the same date, all of which, I accept, were passed on by Mr Popoff to Mr Nutt D soon after they were received.

The position changed only on December 22 1990, after the winding-up petition had been served. When the petition was drawn to Mr Nutt's attention he decided to send the cheque to Prince & Co. This he did by enclosing it in a plain white envelope together with a covering letter dated December 22 1989. The envelope was addressed to Prince & Co. It was stamped with ordinary adhesive postage stamps. There was no franking or other wording on the envelope to indicate that it had been sent by Epton & Co. Most important of all, for present purposes, the letter was sent by recorded delivery post, not ordinary post. The date of posting was December 22 1989.

It so happens that, at this time and, indeed, since about the end of E September 1989, Miss Prince had been conducting an intemperate correspondence with the Post Office about her experience of the recorded delivery service. Her initial complaint was that the Post Office was not delivering letters sent by her firm by recorded delivery service until such time as the addressee, or someone on his behalf, was prepared to sign a receipt. This, Miss Prince complained, was leading to delays which meant that in some cases vital dates for service were passed and, so she claimed, unscrupulous recipients were enabled to avoid service. Miss Prince's assertion was that the obligation of the Post Office was to effect delivery in due course, with or without a receipt, and, in cases where no receipt was obtainable, to certify that delivery had been effected. In successive letters Miss Prince expressed her complaint in unrestrained language which did not shrink from F alleging that a particular addressee was "on too friendly terms with the Post Office official" and to assert that the Post Office had on one occasion "conspired to enable the addressee to evade completely service of the Notice by Post". In the same letter as that in which the last assertion was made Miss Prince asked the District Head Postmaster at York:

> Can you adequately ensure that Post Office officials carry out their duties honestly and intelligently, or ought my Member of Parliament to take these matters up with the Minister?

Although Miss Prince's initial complaints were made in respect of recorded delivery letters sent by her firm, she extended it to apply to recorded delivery items addressed to her. Her reasoning was that, if the Post Office was not delivering items sent by her when a receipt was unobtainable, it might not be delivering items addressed to her if, G at the time of delivery, there was no one to give a receipt on her behalf. As it would be inconsistent with her argument to accept that recorded delivery mail should always be signed for by the recipient, Miss Prince adopted the stance that she would not sign for recorded delivery mail addressed to her or her firm.

Miss Prince lives in a flat above her office in Sherburn, near Malton. Sherburn is served by a subpost office, which is situated within a shop carried on by a company named Sherburn Stores Ltd. The proprietor of the company, a Mr Holmes, is also the subpostmaster.

On Saturday December 30 1989 Miss Prince went out jogging which, she said, was her usual custom on a Saturday. When she returned to her premises she found a note from the postman to the H effect that he had a recorded delivery letter addressed to her firm, but had not been able to deliver it. The note stated that it was available for collection by her at the post office in Sherburn. Miss Prince went immediately to the post office and asked for the letter. The subpostmaster, Mr Holmes, showed her the letter, but, as Miss Prince does not wear her glasses when she goes jogging and the letter was shown to her only very briefly, she could not discern anything special about it. Mr Holmes asked Miss Prince to sign a receipt for the letter, but she declined to do so as a matter of principle. When Miss Prince refused to sign a receipt Mr Holmes refused to hand over the letter. It seems that, very shortly thereafter, the letter was returned by Mr Holmes to his head office or to another department within the J Post Office organisation.

On January 2 1990 Miss Prince wrote letters concerning her complaints about recorded delivery mail to Mr Holmes, the subpostmaster at Sherburn, and to her Member of Parliament. The letter to Mr Holmes was delivered by hand and he, it seems, at once passed it on to his head office in York. On January 11 Miss Prince wrote to the District Head Postmaster at York complaining, in her usual intemperate way, about the lack of response to earlier letters which she had written to him and to the letter to Mr Holmes.

On Thursday January 11 1990 the petition to wind up the company was advertised in a local newspaper, *Grimsby Evening Telegraph*. On the next day, Friday January 12, Mrs Luttenberger had a conversation with Mr Hewson who was the manager of the farm on K behalf of the company or on behalf of Mr and Mrs Popoff. Mr Hewson told Mrs Luttenberger that he had seen the advertisement of the petition and, being worried about his own position, had spoken to Mr Popoff about it, saying that he understood that the petition was based on non-payment of rent for the farm. Mr Popoff then told Mr Hewson that there was no need to worry about it because the rent had in fact been sent by recorded delivery post. Mr Hewson passed this response on to Mrs Luttenberger. Mr Hewson did not himself give evidence at the trial. This account of what he told Mrs Luttenberger comes from the evidence of Mrs Luttenberger. It is, however, consistent with Mr Popoff's own evidence and the accuracy of Mrs Luttenberger's account of what she has told was not challenged. L

Mrs Luttenberger passed on to Miss Prince by telephone on the evening of Friday January 12 the information which she had received from Mr Hewson. Miss Prince then mentioned to Mrs Luttenberger that she had been having difficulties with the Post Office concerning recorded delivery mail. In my judgment, this must have been because the thought crossed Miss Prince's mind that the recorded delivery letter for which she had declined to give a receipt nearly a fortnight before might have contained a cheque for the rent of the farm. Miss Prince said in evidence that this was the only recorded delivery letter addressed to her which she had known about during the winter months of 1989-90. After discussion with Mrs Luttenburger, Miss Prince prepared three letters during the weekend. These are all dated January 15 1990, which was the Monday after the weekend, but the M evidence, which I accept, was that they were typed by the end of Sunday January 14 and were brought by hand to Mrs Luttenberger and Mr Hewson by Mr Boswell, Miss Prince's part-time assistant, during the evening of Sunday January 14.

The three letters were addressed to Mr Hewson, the company (being stated to be for the attention of Mr or Mrs Popoff) and to Epton & Co. The letter to Mr Hewson does not matter much for present purposes. It relates only to arrangements for service of the letter addressed to the company on Mr Popoff. Each side attaches importance of the other two letters and I must deal with them in some detail.

The letter addressed to the company was in the following terms:

A Dear Sir and/or Madam

Tenancy Agreement, Eastfield House Farm, North Thoresby Petition to wind up North Thoresby Farms Ltd

As you know, we represent Mrs Alexandra Helen Luttenberger in her capacity of personal representative of both Mrs Helen Johanna Butt (deceased) and of Miss Eva Motley Robinson (deceased) and thereby as landlord of Eastfield House Farm, North Thoresby.

As you know, North Thoresby Farms Ltd was served on 24th November 1989 with a statutory demand for the amount of £15,264. The period in which that money was to be paid in order to comply with the statutory demand was 21 days. There can be no doubt that the said sum was not paid within 21 days. Accordingly, on grounds arising because of the failure of North Thoresby Farms Ltd to comply with the said statutory demand and also on other grounds, a petition to wind up North Thoresby Farms Ltd was presented on
B 19th December 1989 and was served upon North Thoresby Farms Ltd on 20th December 1989, the said petition being served at the registered office of North Thoresby Farms Ltd upon Mr M D Popoff as a director of North Thoresby Farms Ltd.

Mrs Luttenburger has been informed by Mr John Hewson of North Thoresby that, because Mr Hewson saw the said winding up petition advertised last week, Mr Hewson asked Mr M D Popoff what Mr Popoff was doing about it and that Mr Popoff replied to Mr Hewson that the rent on Eastfield House Farm has recently been paid up to date by recorded delivery. This letter serves to confirm that no payment of such rent or such amount in any form has been received by our client or by us, whether by recorded delivery or by any other means. No part of the £15,264 mentioned in the said statutory demand has ever been paid by North Thoresby Farms Ltd or by anyone else.

C This letter is being delivered by you today by hand in order to ensure that at the hearing at the Royal Courts of Justice on 7th February 1990 of the petition to wind up North Thoresby Farms Ltd you will not be able truthfully to deny that you received this letter by way of notice that, as of this date, all of the amount of £15,264 demanded in the above mentioned statutory demand remains unpaid. Our client does not waive any of her rights nor any of the rules.

Yours faithfully Prince & Co

The letter addressed to Epton & Co enclosed a copy of the letter addressed to the company which I have just read. It referred to the statutory demand, the petition and Mr Popoff's assertion to Mr Hewson that the company had recently paid the rent. It then continued:

D That claim by Mr Popoff is untrue. However, because of that claim by Mr Popoff it seems possible that there is an intention on the part of North Thoresby Farms Ltd to oppose the petition to wind up North Thoresby Farms Ltd at the hearing on 7th February 1990.

Because you have represented North Thoresby Farms Ltd in the past, in case North Thoresby Farms Ltd does intend to oppose the petition and to instruct your firm to act, even though we have never heard to that effect from you as of this date, we deem it advisable to ensure that you receive now this letter and the enclosed photocopy of the letter to North Thoresby Farms Ltd, dated 15th January 1990.

The letter concluded with a request for acknowledgement expressed in not very polite terms.

There is no doubt that the letter to the company was handed by Mr
E Hewson to Mr Popoff at about 8.15am or 8.30am on the morning of January 15. The letter to Epton & Co was delivered to that firm's offices in Lincoln by Mrs Luttenberger herself at about 10.35am on the same day. Soon after he received the letter addressed to the company, Mr Popoff telephoned Mr Nutt about it and, on the afternoon of January 15, he took the letter to Mr Nutt at his office and left it there for Mr Nutt to deal with. Mr Nutt and Mr Popoff checked with Midland Bank at Louth and found that the cheque for the rent had not been paid. Mr Popoff said that they wondered where the cheque was. Neither of them made any attempt to communicate with Miss Prince or Mrs Luttenberger.

Shortly before this time, Mr Popoff had discovered that he had made a mistake in calculating the amount of the owner's drainage
F rates which the company was entitled to deduce from the rent. He had erroneously included in the owner's rates to be deducted an amount of £8.40 in respect of land owned by an entity known as Mapletoff Foundation, which was not the liability of Mrs Luttenberger. As a result, Mr Popoff had deducted £458.40 from the rent instead of the correct amount of £450. Having discovered this mistake, he asked the clerk to Louth Drainage Board to confirm what was the correct amount of the owner's drainage rate. On January 16 1990 he collected from the clerk's office a letter confirming that the amount was £450. He mentioned this to Mr Nutt who, according to Mr Popoff, advised that the amount was so small that it could be dealt with later.

On January 16 1990 the expiry of the two-month period, referred to in the notice to pay served on November 17, was imminent. Miss G
Prince, who had heard nothing in response to her letters of January 15, prepared a notice to quit in proper form relying on Case D in the following terms:

This notice is given on and for the following ground and reason and pursuant to the appropriate paragraphs set out in the Agricultural Holdings Act 1986, Schedule 3, Part I: Case D

At the date of the giving of this notice you have failed to comply with a written notice, dated the 17th day of November 1989, served on you by your landlord which required you, within two months of service of the notice, to pay rent due in respect of the whole.

On the same day Miss Prince sent this notice to a process server with instructions to serve it on Mr or Mrs Popoff on behalf of the
company on Friday January 19 or so soon thereafter as he was able to H
do so. Miss Prince heard nothing further about the payment of rent before January 19 and did not countermand these instructions. The notice to quit was duly served on the company on the morning of January 19.

During the week beginning January 15 1990, probably some time on January 18, representatives of the Post Office arranged to meet Miss Prince concerning her complaints about recorded delivery mail. A meeting was arranged for the afternoon of January 19. Before it took place Miss Prince had indicated to the Post Office officials that it would serve no useful purpose unless they brought with them the letter for which she had refused to sign on December 30, but,
according to Miss Prince, the Post Office officials indicated that that J
letter could not then be found. When the officials arrived at Miss Prince's office on January 19, however, they did, in fact, bring the letter with them. I accept that Miss Prince did not know, until it was produced to her at about 2.48pm on that day, that the letter would be produced to her by these officials. The envelope could then be seen to have been postmarked as having been posted in Lincoln on December 22 1989. It had attached to it a Post Office sticker indicating that delivery had been refused on December 30 1989. When the letter was opened it was found to contain the letter from Epton & Co dated December 22 and the company's cheque for rent, which had already been drawn as I have described.

Miss Prince decided that the cheque should be specially presented
for payment. She made a photocopy of it and caused there to be typed K
on the page containing the photocopy instructions to her bank, the Yorkshire Bank, Bridlington, in the following terms:

The cheque photocopied below is for special clearance at once. However, please note that, if payment is not made upon the said cheque at once upon presentation, but the cheque is returned to Yorkshire Bank for any reason, do not re-present this cheque anywhere at any time. Instead, return it to us at once.

Those instructions were signed by Miss Prince on behalf of her firm.

The Yorkshire Bank at Bridlington stays open late on a Friday afternoon when Miss Prince herself took the cheque, together with
her typed instructions on the photocopy, to the bank and at about L
4.45 pm she personally gave it to the manager, Mr Naylor.

There then followed another strange series of events. I heard a number of banking witnesses, all of whom agreed that, when a cheque is specially presented, the collecting bank sends it by first-class post to the paying bank together with a note to the effect that the collecting bank will telephone the next day to ascertain whether the cheque has been paid or dishonoured. This process is described as ascertaining the fate of the cheque.

In the present case the company's bank, which was the paying bank, was Midland Bank, Louth, but the company's bank account was not controlled within that branch, but by the agricultural division of the corporate banking section of Midland Bank, Lincoln.
The officer responsible for the company's account within the M
corporate banking section was Mr Edwin John Rigby. Before the cheque was presented for payment he already knew about it. This was because, on hearing about the winding-up petition, he had telephoned Mr Nutt to ascertain what the position was. Mr Nutt then informed Mr Rigby about the cheque having been drawn and he emphasised to Mr Rigby that it was important that the cheque should be paid immediately it was presented for payment. After checking the balance on the company's account and, presumably, satisfying himself about the impact of section 127 of the Companies Act 1985, Mr Rigby informed the branch at Louth that the cheque was to be paid when it was presented.

It seems unlikely that the cheque was put into the post by Yorkshire

Bank at Bridlington until Monday January 22, for it did not arrive at Midland Bank, Louth, until the morning of January 24. Mr Purdie, who was then the manager of the Louth branch, was involved in the opening the post at the beginning of that day when he saw the cheque. He had previously received Mr Rigby's message and, shortly after 9 am on that day, the cheque was released by him into the general office of the branch with instructions that it could be met. A clerk in the general office then carried out certain routine checks in the course of which it was discovered that the company's mandate to the bank required that cheques be signed by both Mr and Mrs Popoff. This cheque had been signed by Mrs Popoff alone, as I have already mentioned. The same thing had happened with the cheque for the rent paid in October 1988 and almost certainly with other rent cheques as well and the bank had paid these cheques, probably without realising the departure from the mandate. But when the absence of a second signature was discovered within the bank on January 24 1990 it was decided that the best thing to do would be to obtain the missing signature. Either the Louth branch or the corporate banking section in Lincoln then telephoned Mr Popoff and asked him to call in at the Louth branch in order to sign the cheque.

A little later on the morning of January 24, probably at about 10.30 am, the Yorkshire Bank telephoned the Midland Bank, Louth, to discover the fate of the cheque. This was done on behalf of Yorkshire Bank by a clerk named Michelle Smith. She spoke to a clerk in the Midland Bank named Doreen Smith. The evidence of both of them was to the effect that Doreen Smith told Michelle Smith that the cheque was not paid at that stage. Michelle Smith said that she was told that it was not paid because it required a second signature. Doreen Smith could not remember saying this, but did not disagree that she might have said it, that being the reason why the cheque was held up at that stage.

Michelle Smith's evidence was that, on receiving this information, she telephoned Miss Prince who told her that if the cheque was not paid it must be returned immediately. Michelle Smith said that within a few minutes she telephoned Midland Bank, Louth, again and passed on this request. She could not remember whether she spoke to the same person as before or precisely what reply she received. Doreen Smith remembered being party to a second telephone conversation. Her recollection was that the Midland Bank was being pressed for an answer about the fate of the cheque, but she said that she could not give an answer at that time. Doreen Smith said that she would not have replied that the cheque was dishonoured because she knew that it was going to be paid. She could not remember being asked for the cheque to be returned.

I have no doubt that both Michelle Smith and Doreen Smith were, like the other bank witnesses, witnesses of complete truth, doing their best to assist the court on matters which happened nearly three years ago. On the only point of any significance on which their evidence differed, namely whether Michelle Smith asked that the cheque be returned, I prefer the evidence of Michelle Smith. I accept that, by January 24, Miss Prince was not prepared to wait if the cheque was not paid immediately upon being received by Midland Bank. I have no doubt that she told Michelle Smith to ask for the cheque back if it was not paid and I am sure that Michelle Smith passed on this request. It is equally clear, however, that this request was not appreciated within Midland Bank or, if it was appreciated, it was misunderstood or simply ignored. At about lunchtime on January 24 Mr Popoff called in at the Midland Bank, Louth, and added his signature to the cheque. Midland Bank then paid the cheque and informed Yorkshire Bank by telephone that it had done so.

The subsequent events which are relevant to this action can be shortly stated. On January 29 1990 Miss Prince wrote another long letter to Epton & Co and on February 5 1990 Epton & Co replied to that letter and to the earlier letter of January 15. In the last paragraph of their letter of February 5 Epton & Co referred to the miscalculation of the owner's drainage rate. They enclosed a copy of the confirmatory letter from Louth Drainage Board and a cheque drawn on the company's bank account for £8.40, the amount remaining due in respect of the November 1989 rent. This cheque was dated January 16 1990, which was the date on which Mr Popoff received confirmation that he had made a mistake. In his evidence, however, Mr Popoff frankly stated that the cheque must have been drawn on or after January 24 1990 because it is signed by both himself and Mrs Popoff and he said that it was only on that date that they were reminded that two signatures were required. This cheque was paid by Midland Bank, Louth, on February 13 1990.

The only other relevant events are ones which I have already mentioned, namely the service of a notice requiring arbitration and the subsequent failure of the company to apply in due time for the appointment of an arbitrator.

It is accepted that, on the facts, the company can resist Mrs Luttenberger's claim for possession only if it can establish, first, that the statement in the notice to quit concerning non-payment of rent was false; and, second, that this false statement was made fraudulently. On behalf of Mrs Luttenberger it is denied that the company can establish either of these things, but it is emphasised that the company must establish both of them if it is to succeed.

The statement which is said to have been false is the part of the notice to quit which I have already read, notably the assertion that:

> At the date of giving of this notice you have failed to comply with a written notice dated the 17th day of November 1989 . . . which required you within 2 months of service of the notice to pay rent due in respect of the whole.

The truth or falsity of this statement has to be assessed as at January 19 1990 when the notice to quit was served, that is to say, at a time before the recorded delivery letter containing the company's cheque was actually received by Prince & Co. At that time the rent was certainly unpaid in the literal sense, because money had not changed hands between the tenant and the landlord. But the company relies on the reasoning contained in *Beevers* v *Mason* (1978) 37 P&CR 452 as establishing that the rent is, in law, to be regarded as having been paid before that time. *Beevers* v *Mason* was another case concerning the validity of a notice to quit based on Case D. On the facts it was found that, two days before the expiration of the two-month period prescribed by the notice to pay, the tenant had posted a letter addressed to the landlord containing a cheque for the unpaid rent. By a strange coincidence the date on which the cheque was paid was December 22, the same day of the same month as the posting of the cheque in this case, and by an equally strange coincidence postage appears to have been effected in Sherburn, which I take to be the very village in which Miss Prince has her office, but neither of these facts is of any relevance. The Court of Appeal held — I read from p458:

> On general principles, the landlord should have the rent in cash in his hands by the due date. This requirement may, however, be waived by express arrangement, or by necessary implication where the facts are sufficiently strong to establish that the landlord has shown that he is content to accept payment by cheque posted by the due date of payment. Inferences of this nature are not to be too readily drawn, but, where the facts support them clearly and emphatically, they are not to be dismissed.

Such an inference was made in *Beevers* v *Mason* and Mr Derek Wood QC, on behalf of Mrs Luttenberger, accepts that it ought to be made here. On this footing the Court of Appeal held at p460:

> When the cheque was put in the post, then, subject only to its being honoured, the rent was paid.

In the present case the Popoffs contend that the cheque was honoured when it was first presented on January 24 1990 and that, accordingly, the rent must be regarded as having been paid when the recorded delivery letter was posted on December 22. Accordingly, it is argued that it was false to assert on January 19 1990 that the rent had not been paid within the two-month period.

On behalf of Mrs Luttenberger Mr Wood challenges this analysis on three grounds. First, he contends that the analysis made in *Beevers* v *Mason* is applicable only where the cheque is one which, as between the tenant and his bank, the bank is bound to meet. Second, he says that on the facts the cheque was not honoured on first presentation and was therefore not honoured within the statement of the Court of Appeal which I have quoted. Third, he says that the rent was not paid in due time because, on any view, £8.40 of it remained unpaid until February 5 1990 (or, indeed, February 13 if actual payment of the second cheque is the relevant event).

As to the first of these points the argument is that Midland Bank was not bound to meet, and in fact declined to meet, the company's cheque, because the only signature that it bore was that of Mrs Popoff. In so far as Mr Wood's argument was as I have stated it, namely that a cheque constitutes payment within the *Beevers* v *Mason* analysis only when the bank is bound to honour it, I do not accept the proposition. A bank is not bound to honour a cheque if its customer will thereby exceed a permitted overdraft facility. If the bank does nevertheless honour the customer's cheque, the payee of the cheque cannot, in my view, object on the ground that it was not bound to do so. In this case the rent due in October 1988 was, it seems, paid by a cheque drawn on the company's account and signed by Mrs Popoff alone. On that occasion the Midland Bank honoured the cheque when it was presented, probably because it had itself omitted to confirm the requirements of the mandate. If, as was the usual practice, the cheque was sent by post, it would not, in my judgment, have been open to Mrs Luttenberger afterwards to say that the *Beevers* v *Mason* analysis was inapplicable and that payment of this instalment of rent was made otherwise than on the date on which the cheque was posted.

Nevertheless, I consider that Mr Wood's contention would be correct if it were advanced in a modified form. It seems to me that in saying "subject only to [the cheque] being honoured" the Court of Appeal in *Beevers* v *Mason* must have been envisaging that the cheque would be honoured without further intervention on the part of the tenant. If, for example, a tenant sends by post an unsigned cheque which the landlord presents for payment, not noticing the absence of a signature, it seems to me that that does not constitute any kind of payment on any date. If, when asked to pay the cheque, the tenant's bank notices the absence of a signature and, instead of returning the cheque, contacts its customer and gets him to sign the cheque, which it then pays, the resulting transfer of money is undoubtedly a payment, but in my view it is a payment made only when the tenant signs the cheque or when the cheque is then paid.

In the present case the cheque was not wholly unsigned, but it lacked a signature without which, on January 24 1990, Midland Bank was not prepared to pay the cheque. That is how I interpret the evidence of Mr Purdie and, so far as material, Mr Rigby in this case. It is possible that, if Mr Popoff had not called in to sign the cheque during the day, the bank would have reconsidered the matter and agreed to pay it at or just before the close of business. But I find that no decision to this effect was ever made. The only decision which was made was that the cheque would be paid if Mr Popoff came in and signed it. If he did not, the matter might be reconsidered, but the bank would not be bound to pay. In my judgment there is no valid distinction between these circumstances and the case which I have postulated of a cheque which contains no signature at all when presented for payment.

As to Mr Wood's second argument, there is no doubt, in my view, that when Michelle Smith first telephoned to discover the fate of the cheque on January 24 the cheque was not paid. The same remained the case when she telephoned a second time after speaking to Miss Prince. In my judgment the cheque remained unpaid until Mr Popoff signed it at about the middle of the day.

The question of whether this amounts to the cheque "being honoured" for the purpose of what was said by the Court of Appeal in *Beevers* v *Mason* depends upon what is banking practice in relation to special presentation. On behalf of the Popoffs it was contended that, when a cheque is specially presented, the paying bank has until the end of the day on which it is received in which to decide whether or not to pay it. If that be right, the cheque was honoured in due time. On behalf of Mrs Luttenberger it was said that when the collecting bank telephones to discover the fate of the specially presented cheque the paying bank must decide immediately whether or not to pay it. If it does not decide to pay it the cheque is not honoured. The requirement for an immediate decision would, of course, entitle the paying bank to carry out the normal checks, for example, as to correspondence of words and figures, date, signatures and customer's balance, if it has not already done this when telephoned. But, subject to this, the paying bank cannot have further time unless the collecting bank agrees.

Each party called an expert witness in support of its contention, Mr Oblein being called on behalf of Mrs Luttenberger and Mr Luckhurst on behalf of the Popoffs. I do not propose to review their evidence in any detail. It will suffice to say that I found the evidence of Mr Oblein more convincing than that of Mr Luckhurst. In his report Mr Luckhurst accepted that the paying bank would normally advise the collecting bank as to the fate of the cheque upon receipt of the collecting bank's phone call, but he maintained that there were circumstances in which it would be common for the paying bank to ask the collecting bank to call again later. I found the particular circumstances put forward by Mr Luckhurst improbable, at least in so far as Mr Luckhurst put them forward as giving the paying bank a right to defer its decision. I can readily understand that in some of these circumstances a paying bank might ask the collecting bank for more time to consider the matter and I can also understand that in some circumstances the collecting bank may agree to give extra time. But if the collecting bank does not agree I do not see anything which enables the paying bank to defer its decision. The position would, in my view, be straightforward. If it is not prepared to pay the cheque when asked, it will dishonour it. Mr Oblein's view was, to my mind, more logically consistent than that of Mr Luckhurst and it was in accordance with a statement on the very point contained in a publication called *Questions on Banking Practice*, 1978 ed, issued by the Institute of Bankers under the authority of its council. I do not accept that the banking practice has changed since 1978, although, as I have indicated, I do accept that it is not uncommon for the collecting bank to agree to a request from the paying bank for more time in which to make its decision. However, that, in my view, is entirely a matter for the collecting bank to decide.

In the present case Midland Bank was not prepared to pay the cheque on January 24 1990 at the time of Michelle Smith's first telephone call on that day or at the time of her second telephone call, made a short time after the first and after Miss Prince had instructed her to ask for the return of the cheque. The Yorkshire Bank did not agree to Midland Bank's expressed or implied request to be allowed to postpone its decision. On the contrary, as I find, it requested the return of the cheque in the second telephone call. Midland Bank's retention and subsequent payment of the cheque was justified as between itself and its customer, the company, but that justification was not attributable to the mandate constituted by the cheque in its original form. It was attributable to Mr Popoff's expressed willingness to call in and sign the cheque, followed by his implementation of that intention. That, in my view, constituted a fresh mandate given on January 24. As between Mrs Luttenberger and the company, Mrs Luttenberger was entitled to retain the money transferred to her on January 24 as a result of Midland Bank's performance of this new mandate. But that transfer of money was, in my judgment, a payment made on January 24 and not one which dates back of December 22 1989 in accordance with the analysis made and applied in *Beevers* v *Mason*.

As to Mr Wood's third ground of challenge of the applicability of the *Beevers* v *Mason* analysis, Mr Wood relied on the fact that Case D applies where, at the date of the giving of the notice to quit, the tenant has failed to comply with a notice requiring him to pay "any rent due" in respect of the relevant holding. He argued that a notice to pay a specific sum representing all the rent due is not complied with by payment of a lesser sum when there is no dispute that the shortfall represents rent due. Mr Wood referred me to *Price* v *Romilly* [1960] 1 WLR 1360, where Diplock J (as he then was) held, in relation to a notice to quit based on a failure by the tenant to comply with a notice requiring him to remedy breaches of covenant relating to matters other than payment of rent, that substantial compliance with the notice was not enough for the tenant. Diplock J said (at p1,362):

> [Counsel] contended that it was sufficient if there was a "substantial compliance" with the terms of the notice. Those words are not in the Act, and I see no reason for reading them into the Act. No doubt, of course, the *de minimis* rule would apply, but the Act seems to me to be plain in its terms, and, if a tenant fails to comply fully with the requirements set out in the notice, being requirements for remedying breaches of terms or conditions of his tenancy which are not inconsistent with the fulfilment of his responsibilities to farm in accordance with the rules of good husbandry, the provisions of Section 24(2)(*d*) of the Act are fulfilled.

On behalf of the Popoffs, Mr Robert Reid QC did not question, in this court, that substantial compliance would not be enough, although he reserved his position on this point in case the matter should go further. But he relied upon the dictum of Diplock J concerning the *de minimis* rule. Mr Wood pointed out against this that it was only a dictum, made in relation to a different case within what is now Part 1 of Schedule 3 to the 1986 Act. He argued that there is no room for the *de minimis* rule in relation to failure to comply with an obligation to pay rent, where what is required for compliance is known with precision.

In my judgment, the *de minimis* rule is applicable to Case D. In an Act which is designed to give protection to agricultural tenants, I do not think that Parliament can have intended to exclude this general rule and to entitle a landlord to recover possession when a tenant has made only an insignificant underpayment of rent.

There remains the question of what underpayment will fall to be ignored under the *de minimis* rule. In my judgment, this will depend upon the circumstances as a whole, and not exclusively on the amount of the underpayment. An underpayment of a tiny sum may not fall to be disregarded if it arises from a course of conduct by a tenant deliberately designed to irritate the landlord. In this case, however, I accept that the underpayment was made as a result of a *bona fide* mistake on the part of Mr Popoff which he made good soon after discovering it. The amount involved is 0.055% of the rent. In my judgment, this is, in all the circumstances, to be disregarded as being *de minimis*.

Nevertheless, it follows from what I have said concerning the first two grounds of the attack on the application of the *Beevers* v *Mason* analysis that I find that the statement, in the notice to quit, to the effect that the company had failed to comply with the two-month notice to pay rent was not a false statement. That is enough to determine the outcome of this action, but it would be quite wrong for me not to state my conclusion on the second main issue, namely that the statement, if false, was made fraudulently. That issue involved an allegation of fraud made against a solicitor which, although presented by Mr Reid with proper detachment, was made with equally proper resolution.

On the issue of what constitutes fraud I was referred to the well-known passages in the judgments in *Derry* v *Peek* (1889) 14 App Cas 337 and *Angus* v *Clifford* [1891] 2 Ch 449 which were cited in *Rous* v *Mitchell* [1991] 1 WLR 469, a case where, as I have remarked, a statement was held to have been made fraudulently in a context akin to that which I have to consider. In *Derry* v *Peek* Lord Hershell said at p374:

> I think the authorities established the following propositions: First, in order to sustain an action of deceit, there must be proof of fraud and nothing short of that will suffice. Secondly, fraud is proved when it is shown that a false representation has been made (1) knowingly, or (2) without belief in its truth, or (3) recklessly, careless whether it be true or false. Although I have treated the second and third as distinct cases, I think the third is but an instance of the second for one who makes a statement under such circumstances can have no real belief in the truth of what he states. To prevent a false statement being fraudulent, there must, I think, always be an honest belief in its truth and this probably covers the whole ground, for one who knowingly alleges that which is false has obviously no such honest belief. Thirdly, if fraud be proved, the motive of the person guilty of it is immaterial. It matters not that there was no intention to cheat or injure the person to whom the statement was made.

Lord Hershell went on to state (at p375):

> At the same time I desire to say distinctly that when a false statement has been made the questions whether there were reasonable grounds for believing it, and what were the means of knowledge in the possession of the person making it, are most weighty matters for consideration. The ground upon which an alleged belief was founded is a most important test of its reality. I can conceive many cases where the fact that an alleged belief was destitute of all reasonable foundation would suffice of itself to convince the Court that it was not really entertained, and that the representation was a fraudulent one. So, too, although means of knowledge are, as was pointed out by Lord Blackburn in *Brownlie* v *Campbell*, a very different thing from knowledge, if I thought that a person making a false statement had shut his eyes to the facts, or purposely abstained from inquiring into them, I should hold that honest belief was absent, and that he was just as fraudulent as if he had knowingly stated that which was false.

In *Angus* v *Clifford* Bowen LJ said at pp471-2:

> Now, whether you take the inquiry in the one order, or in the other, whether you regard it from the point of view which is indicated in the expressions which are well known to the common law, and which have been used by Lord Hershell and Lord Bramwell, that a man is bound to have some honest belief in a statement if he makes it, or whether you treat the matter in the inverse order, with regard to the necessity of finding at least some recklessness to truth, that is to say, some indifference to truth which amounts to dishonesty, in either view it seems to me the result is the same. A man ought to have a belief that what he is saying is true.

Mr Reid submitted that fraud, within these expositions of the expression, was made out in this case by the following route. First, he said, Mrs Luttenberger clearly has animosity towards Mr Popoff. This regretfully is the case, although it appears to me that the animosity is mutual and thankfully I have not been concerned as to its origin or justification. Second, it was said that Mrs Luttenberger was and is on close terms with her solicitor, Miss Prince, with whom she communicated up to three times a week during the relevant period. This, too, I would accept although it is, I think, largely immaterial because if Miss Prince had the requisite state of mind in relation to the statement in the notice to quit there can be no doubt that her fraud affected Mrs Luttenberger, however close or distant their relationship.

Third, it was said that, by the evening of January 12 1990, when Mrs Luttenberger told her what she had heard from Mr Hewson, Miss Prince must have had at least a strong suspicion that the recorded delivery letter which she had refused to sign for on December 30 and which she had not therefore received contained a cheque in payment of the rent for the farm. If she did not have this suspicion then she was closing her eyes to an obvious possibility. This, too, I would accept.

If, therefore, Miss Prince had taken no further steps before causing the notice to quit to be served, it might have followed that she would have been reckless to the extent necessary to constitute fraud in making the statement contained in the notice to quit without further inquiry.

However, matters did not rest there. Miss Prince wrote the letters dated January 15 1990 to which I have referred and she went to some pains to ensure that these letters reached Mr Popoff on behalf of the company first thing on January 15 and Epton & Co in good time the same morning. Mr Reid said that these letters made matters worse, not better. He said they were verbose letters written in Miss Prince's provocative style. By the time that they were written Mrs Luttenberger and Miss Prince had decided to seek to recover possession of the farm if they could. They were no longer concerned merely to obtain payment of the rent. Yet, complained Mr Reid, the letters, while reciting at some length the statutory demand and the winding-up petition and anticipating an attempt to resist the petition on the basis of a denial, made no reference at all to the imminent expiry of the two-month period referred to in the notice to pay, which, on Miss Prince's own evidence, was the reason why she was anxious to be sure that the letters were received as early as possible on January 15. Miss Prince, it was suggested, was deliberately setting a trap by concealing her real intentions. Reference was made to some of the allegations made, it was said without justification, in Miss Prince's intemperate correspondence with the Post Office. It was suggested that Miss Prince is given to making reckless allegations and that, in all the circumstances, the allegation that rent remained unpaid was made recklessly without belief in its truth.

In my judgment, it would not be right to bring the correspondence with the Post Office into consideration of Miss Prince's state of mind at the time when the notice to quit was served. The allegations made in that correspondence have not been pleaded as particulars of a propensity to recklessness and the existence or non-existence of a justification for them has rightly not been explored in any detail at the trial of this action. I entirely accept that Miss Prince has a tendency, at least on some occasions, to write letters couched in intemperate terms. These appear to me to be, at worst, somewhat unprofessional and probably unproductive, but for the most part they are more aptly described as merely silly. I do not regard them a sign of recklessness in the sense in which that term is used in relation to fraud.

As to the letters of January 15, I accept that by this date Mrs Luttenberger and Miss Prince had decided to try to obtain possession of the farm if this was possible. I do not think that they had any real belief that possession might be obtained until this time or perhaps a few days earlier, because the dire consequences make it extremely unlikely that an agricultural tenant will fail to pay rent after a formal notice to pay has been served unless it is hopelessly insolvent. When the notice to pay and the statutory demand were served and the winding-up petition was presented, I consider that Mrs Luttenberger and Miss Prince were concerned only to exert the maximum amount of pressure in order to obtain payment of rent. As the two-month period referred to in the notice to pay drew towards a close, I think that their attitude changed and that they realised that, contrary to their earlier expectation, possession might indeed be obtainable. I do not doubt that the omission of a reference to the notice to pay and to the two-month period in the letters of January 15 was deliberate, but I cannot regard this as recklessness. To hold that it ought to have been referred to would be close to holding that Miss Prince owed a duty of care to the company to warn of the possible consequences of the action she was taking and to advise how these consequences could be avoided. This would be manifestly absurd. Both letters of January 15 did make it quite clear that, at that date, the rent had not been received. Both letters frankly refer to the information which Mrs Luttenberger had received from Mr Hewson to the effect that the rent had been sent by recorded delivery post. The letter sent to the company stated:

> This letter serves to confirm that no payment of such rent or of such amount in any form has been received by our client or by us, whether by Recorded Delivery or by any other means.

The letter to Epton & Co repeated this in that it enclosed a copy of the letter to the company and it stated in terms that Mr Popoff's assertion to Mr Hewson was untrue. Up to a point, the letters had the effect that Miss Prince must have intended. Mr Popoff took the letter to the company to Mr Nutt, who presumably had received the letter to

A Epton & Co. Mr Popoff and Mr Nutt were sufficiently concerned about them to check the state of the company's bank account and to ascertain that the cheque for rent had not been presented for payment. What Miss Prince could not have been expected to anticipate was that, having received the letters and correctly appreciated their impact to this extent, neither Mr Popoff nor Mr Nutt would do anything to contact her and at least point out that a cheque had been sent but appeared to have been lost in the post. They would, indeed, at that time have had sufficient time to stop the apparently lost cheque and send the money by telegraphic transfer, banker's draft or, at the very least, a fresh cheque. They took none of these steps. I am satisfied that, if they had done any of these things before the two-month period expired, Miss Prince could and would B have arranged for the planned service of the notice to quit to be cancelled.

I accept Miss Prince's evidence that the reason why she went to some length to ensure that her letters of December 15 were received early on that day was that she wanted there to be as much time as possible between the time when the company and its solicitors would know that the rent had not been received and the time when the two-month notice to pay expired. When these letters brought forth no reaction at all so far as Miss Prince was concerned, she was, in my judgment, entitled to set at rest whatever doubts she might have had as to the contents of the undelivered recorded delivery letter. In my judgment, she had reasonable ground to believe that the statement C concerning non-payment of rent which was contained in the notice to quit was a true statement at the time when that notice was served. That is more than enough to rebut the allegation of fraud by recklessness. Accordingly, this branch of the Popoffs' case fails as well.

I will hear counsel as to the precise order which I ought to make.

After hearing submissions as to costs, **FERRIS J** said: I will make an order for the legal aid taxation of the costs of both parties for the periods that they were covered by a legal aid certificate. As between the parties, it seems to me that the plaintiff, Mrs Luttenberger, is entitled to an order for her costs against all the defendants, that order not to be enforced against Mr and Mrs Popoff in relation to costs D attributable to any period when they were legally aided without further order of the court.

As to the scale of costs, I take the view that this was something of a desperate throw by the defendants in which they, for reasons for which neither Mrs Luttenberger nor her solicitor were in any way to blame, were in a position where they could succeed only by alleging fraud. They did not shrink from making that allegation which was an essential ingredient of their case. The impact of it is not reduced by the fact that they also needed to establish matters other than fraud in order to succeed. They played what I say I regard as rather a desperate throw for high stakes. It involved serious allegations against Mrs Luttenberger and her solicitor, which have failed. In my judgment, it is a case for indemnity costs, so that the order which I make against E the company and Mr and Mrs Popoff, subject to the restriction that I have indicated in respect of periods when they were legally aided, is that they shall pay Mrs Luttenberger's costs to be taxed on an indemnity basis.

Chancery Division

November 10 1992

(Before Mr J SUMPTION QC, sitting as a deputy judge of the division)

F BROWN v TIERNAN

Estates Gazette May 8 1993

[1993] 18 EG 134

Agricultural holdings — Grazing agreement — Whether restriction to specified period of year — Whether simultaneous grazing of ponies for pleasure outside meaning of agricultural land

Until he died on December 26 1987, Sir Douglas Howard lived at Clophill House, Clophill, Bedfordshire, a property which included a paddock of just under 2 acres. In about 1973 Sir Douglas made an agreement with the plaintiff for the use of the paddock for grazing. The defendant, who purchased G the property from Sir Douglas' executors, claimed that the agreement was in ôntemplation of the use of the paddock for grazing only during the period of approximately 10 months from May to March each year when the daffodils were not in bloom, and was therefore within the proviso in section 2(3)(*a*) of the Agricultural Holdings Act 1986; the plaintiff therefore did not have the protection of the Act. The evidence was that the plaintiff did not graze his cattle between March and May each year, although he grazed a mare on the property throughout the year which was kept for the enjoyment of his children until she died in 1982. The plaintiff contended that H the agreement was to graze for the whole of each year.

***Held*: The application for a declaration that the plaintiff had a tenancy protected by the Agricultural Holdings Act 1986 was granted. It was a term of the agreement that the plaintiff would not graze cattle during the daffodil season and this restriction did not extend to ponies. Although the mare was used for pleasure and not for any trade or business, the paddock was let to the plaintiff for the purposes of a business and its simultaneous pleasure use did not take the agreement outside the definition of agricultural land in section 1(4)(*a*) of the 1986 Act.** J

The following cases are referred to in this report.

Reid v *Dawson* [1955] 1 QB 214; [1954] 3 WLR 810; [1954] 3 All ER 498; (1954) 53 LGR 24, CA
Scene Estate Ltd v *Amos* [1957] 2 QB 205; [1957] 2 WLR 1017; [1957] 2 All ER 325; (1957) 56 LGR 14, CA
Watts v *Yeend* [1987] 1 WLR 323; [1987] 1 All ER 744, CA

This was the hearing of an originating summons of the plaintiff, Colin Brown, for a declaration that he held an agricultural tenancy of land at Clophill House, Clophill, Bedfordshire, belonging to the defendant, Mr Tiernan.

Peter Cranfield (instructed by Park Woodfine, of Bedford) appeared for the plaintiff; Richard Bray (instructed by E T Ray & Co, K of Bletchley) represented the defendant.

Giving judgment, **MR J SUMPTION QC** said: Sir Douglas Howard, a retired diplomat, lived for many years at Clophill House, Clophill, Bedfordshire, until he died on Boxing Day 1987. Sir Douglas' property included, in addition to the house and its garden, a paddock of just under 2 acres, which lay beyond the garden separated from it by an iron grill and railings. Mr Colin Brown is a local farmer. He claims that in the spring of 1973 Sir Douglas let the paddock to him from year to year for keeping livestock and that he thereby became the tenant of an agricultural holding for the purposes of the Agricultural Holdings Act 1986. The question in this action is whether he is right. If he is, then it is not disputed that his rights bind L the defendant, Mr Tiernan, who bought the house and its adjoining land from Sir Douglas' executors in July 1988.

Disregarding immaterial provisions of the Act, what Mr Brown has to show in order to make good his claim is that he was the tenant under "a contract for an agricultural tenancy": section 1(1). For that purpose two relevant conditions must be satisfied. First, the land must be let for use as "agricultural land": section 1(2). This means "land used for agriculture which is so used for the purposes of a trade or business": section 1(4)(*a*). "Agriculture" includes grazing: section 96(1). Second, the letting must be "for a term of years or from year to year": section 1(5). The tenant is assisted in satisfying the second of these requirements by section 2 of the Act, which provides — in substance — that a licence or letting for less than a year shall take M effect as if it were an agreement to let the land for a tenancy from year to year. There is, however, an important proviso to section 2, which is contained in subsection 2(3)(*a*). That reads:

> This section does not apply to an agreement for the letting of land, or the granting of a licence to occupy land . . . made (whether or not it expressly so provides) in contemplation of the use of the land only for grazing or mowing (or both) during some specified period of the year.

The effect of the proviso is that, if a licence or tenancy is granted in contemplation of the use of the land for grazing for a period of less than a year or for successive periods of less than a year, then the agreement is not a "contract of tenancy" because on that footing it will not be an agreement for letting land for a term of years or from year to year and will not by virtue of section 2(1) and (2) be deemed to be such an agreement: see *Reid* v *Dawson* [1955] 1 QB 214 and *Scene Estate Ltd* v *Amos* [1957] 2 QB 205.

The proviso in section 2(3)(*a*) refers to the contemplated use of the land. In *Scene Estate* v *Amos* [1957] 2 QB 205 at p 211, Denning LJ (as he then was) explained this as follows:

> . . . In my opinion the object of the word "contemplation" in the proviso is to protect a landlord who has not expressly inserted a provision that it is for grazing only, or for mowing only, or that it is for a specified part of a year; but, nevertheless, both parties know that that is what is contemplated. Often a landlord may let a field to a man by word of mouth, saying: "You can have the field this year the same as you had it last year". Both sides mean it to be for grazing only and mean it to be only for a few weeks of the spring, but they do not say so expressly. In such circumstances, even though nothing is expressed in the agreement, nevertheless the landlord can still take advantage of the proviso. That seems to me to be the real object of introducing the "contemplation" of the parties.

It has not been disputed, and is in any event plain from the evidence, that Mr Brown did make an agreement with Sir Douglas Howard in about 1973, which allowed him the use of the paddock for grazing in consideration of a small rent, initially £10 per year, subsequently increased to £20 and then to £25. The question at issue is whether that agreement was made in contemplation of the use of the paddock for a period of less than a year or successive periods of less than a year.

The defendant's case is that the agreement between Sir Douglas and Mr Brown contemplated the use of the land for grazing only during the period of approximately 10 months from May to March of each year when the daffodils in the paddock were not in bloom. Mr Brown denies that there was any limitation on his use of the paddock in the daffodil season.

The primary submission of Mr Bray, who appeared for the defendant, was that any agreement which contemplated the use of land for grazing necessarily contemplated that use during a period less than a year because grazing is of its nature a seasonal activity. This was put to me as a proposition of law, based on the decision of the Court of Appeal in *Watts* v *Yeend* [1987] 1 WLR 323. However, all that I can extract from that case is that where it is proved on the evidence that land has been let for a season or for some period less than a year, the proviso applies notwithstanding that the parties have not agreed precisely when the season or period is to begin and end. What is contemplated by parties to an agreement of this kind seems to me to be a question of fact. The evidence which I have heard suggests that land let for grazing may be grazed at any and all times of the year, depending on the hardiness of the animals in the winter months and on their numbers and appetite in relation to the area of grass. This dispute, therefore, turns on what Sir Douglas Howard and Mr Brown agreed or contemplated in 1973 and during the following years when Mr Brown used the paddock.

Since Sir Douglas and Mr Brown were the only persons present when the original agreement was made and Sir Douglas is dead, the only evidence of it was Mr Brown's. I approach his evidence with some caution. This is partly because it is always right to test rather carefully evidence which is given many years after the event when one party is not available to give his side of the story, and it is partly because Mr Brown struck me as a witness who was apt to state his case more dogmatically and with a greater wealth of circumstantial detail than his recollection really permitted. This was particularly evident when he was dealing with matters which he regarded — rightly or wrongly — as weak points in his case.

I find that what happened was as follows. Sir Douglas Howard was, as all the evidence has shown, an extremely keen gardener. Because of the position of the paddock, the view of it was an important feature of his garden. One of the high points of the gardening year at Clophill House was the daffodil season. The daffodils, which were densely planted at the near end of the paddock, were a spectacular sight in March and April, and friends and relations of Sir Douglas would come specially at that time of year in order to see them. They were Sir Douglas' "pride and joy" as his cousin, Mrs Allen, told me. Grazing cattle during the flowering season is liable to damage the plant, for although the cattle do not eat daffodils, they tend to trample over them. Grazing horses or ponies is apparently less damaging. It was therefore important to Sir Douglas that if livestock were put into the paddock during the daffodil season, they should not be cattle.

Until 1971, the paddock was used for grazing cattle by a Mr Palmer. I do not know what the precise terms were of his agreement with Sir Douglas — they are both dead — but Mr Palmer's son, Mr Donald Palmer, who assisted his father told me that in those days cattle were never put into the paddock during the daffodil season because Sir Douglas would not have permitted it. In 1971, Mr Palmer senior gave up beef production and no longer needed the paddock. So Mr Palmer junior spoke to Mr Brown and asked him whether he would like it. When Mr Brown replied that he would, Mr Palmer took him to see Sir Douglas' caretaker and gardener, Mrs Izzard. There was then a long delay of at least a year, after which Mr Brown received a summons to go and see Sir Douglas at Clophill House to discuss the matter.

According to the oral evidence of Mr Palmer, but not his written statement, when he first spoke to Mr Brown about the paddock he told Mr Brown that he would not be allowed to put cattle into the paddock during the daffodil season. Mr Brown, although he remembered discussing the paddock with Mr Palmer, denied that he had been told anything about the daffodil season or about any restrictions on his rights to graze the land when the daffodils were in bloom. He said that at the time when he approached Sir Douglas about the paddock, he did not yet know that the daffodils existed. He had never seen the paddock and knew nothing about it; he had received no information about it from Mr Palmer. Moreover, although it was spring, he believed that he saw no daffodils in flower when he visited Sir Douglas and that he would have remembered seeing them if they had been out. I do not accept that evidence. I find that when Mr Brown went to visit to Sir Douglas in the spring of 1973 to discuss taking the paddock, he knew that he would not be allowed to graze cattle there during the daffodil season because Mr Palmer junior had told him so.

Mr Brown's evidence of his agreement with Sir Douglas was that he told Sir Douglas that he wanted to use the paddock for grazing ponies and some small cattle, and that Sir Douglas agreed to let him have it for that purpose at a rent of £10 per year. I accept that much of Mr Brown's evidence about the exchange. It is consistent with all that subsequently happened. Mr Brown also told me that there was no mention of daffodils nor of any restriction on his right to use the paddock at any time of year. I do not accept that part of his evidence. I can summarise my reasons quite shortly. Mr Brown acknowledged that Sir Douglas' main interest in letting or licensing the use of the paddock was that the arrangement would make his garden more attractive. The paddock had become very overgrown since Mr Palmer had left and an attempt had been made to mow it by machine which had failed. Sir Douglas wanted the long grass kept down. He also liked to look out from his windows at the pastoral sight of cattle and horses on the land beyond his garden. This is apparent from Mr Brown's own evidence.

An important part of the amenities of Sir Douglas' garden was the view of the daffodils in spring. Mr Palmer had at some stage accepted restrictions on the grazing of cattle during the daffodil season, and I find it extremely improbable that similar conditions were not imposed from the outset on Mr Brown. Moreover, when Mr Brown began to use the paddock he never, in practice, did graze cattle there during the daffodil season. He told me that this was because his ordinary practice and the common practice of farmers of the district was not to put his cattle out for grazing until the first week of May, when the daffodils were over anyway. But he did accept that even if he had for some reason needed to graze cattle in the paddock during the daffodil season, he would not have done so because he realised that Sir Douglas would not have wanted him to.

Mr Brown suggested that he drew this conclusion for himself, entirely unprompted by Sir Douglas. I find that he drew it because Sir Douglas had told him so at their first meeting. I accept that that conclusion is at variance with the only direct evidence which I have heard on the subject. However, I do not regard Mr Brown's recollection as sufficiently reliable to outweigh the circumstantial evidence and the strong inherent probabilities. Moreover, I am satisfied that even if it had not been discussed in terms, it was contemplated by both parties that whatever use was made of the paddock would be consistent with the wellbeing of the daffodils and that meant keeping cattle out of the paddock during their flowering season.

There remains the question of the ponies, which is important for a reason that I shall come to. Ponies seem to have been a lesser threat to daffodils. Moreover, it is plain from the evidence of all the witnesses — except one — that Mr Brown did, in fact, keep a mare and, from time to time, her foals in the paddock from 1973 until the mare died in

1982. The mare was in the paddock throughout the year, including the daffodil season, except for short periods when she was taken out for some special reason. Sir Douglas was perfectly capable of protesting when he was displeased at the use which was being made of his paddock. But it does not appear that he ever did protest at this particular use. I think that it is probable that Sir Douglas' reservation about the presence of animals in the paddock during the daffodil season was in terms confined to cattle and did not extend to ponies. However, even if that were not the position, then there can be no doubt that within a year or two Sir Douglas had accepted that the use of the paddock by the mare would be allowed throughout the year.

In the result, I find that the use which Sir Douglas and Mr Brown contemplated for the paddock was that livestock might be grazed there throughout the year, except that during the period when the daffodils were in flower the livestock should not include cattle or other animals liable to damage the flowers. Nothing was said about the term of the arrangement. But it was implicit in Sir Douglas' acceptance of an annual rent that the arrangement, whether it was technically a tenancy or a licence, would continue from year to year.

I now return to the Act. If the agreement between Sir Douglas and Mr Brown had been confined to the grazing of cattle, it would, in my judgment, have been excluded from the Agricultural Holdings Act 1986 by the proviso in section 2(3). On that footing the only permitted user would have been grazing for a specified period of the year — namely the period of 10 months or so when the daffodils were not in flower. However, the agreement was not limited in that way, as I have found. Mr Brown was permitted to use the paddock for grazing at any time of the year, provided that cattle were not in the paddock during the daffodil season.

The defendant's riposte to this point is to say that the use of the paddock by the mare and her foals is irrelevant. The reason is that the cattle, which were part of Mr Brown's farming stock, were used for the purposes of his business as a farmer, whereas the mare and her foals were not used for any trade or business purpose, but for Mr Brown's children. Section 1(4) of the Act provides, it was suggested, that the only relevant user is for trade and for business purposes. Mr Brown does not dispute the relevant facts. He accepts that the mare and her foals were used for pleasure and not for business. But those facts are, in my judgment, irrelevant in law.

The requirement of the Act that the land should be used for the tenant's trade or business is part of the definition of "agricultural" land. By section 1(2) an agricultural holding must relate to "agricultural land". However, this requirement is satisfied if the land is let for the purposes of a trade or business, notwithstanding that it is simultaneously used for another purpose, such as keeping a mare for pleasure, or, for that matter, growing wild flowers for pleasure. The fact that there is a period during the term of the letting or licence when the land — although still let or its use still licensed — will not be used for both business and pleasure, but used only for pleasure, makes no difference unless either: (i) the business purpose is relatively so minor that it can no longer be regarded as the purpose of the letting or licence; or (ii) the agreement can be regarded as two distinct agreements, one relating to business use and the other relating to other use. Neither of these points could be made in the circumstances of this case. Mr Brown was let into the land to use it for the purposes of his farming business albeit that he might use it for other purposes also. The paddock was, in fact, so used thereafter. It was, accordingly, in Mr Brown's hands "agricultural land".

Once that point is reached, the limitation in section 1(4) to use for the purposes of a trade or business has no further relevance. The question posed by section 2(3)(*a*) is a different question which arises only once it is shown that the land was "agricultural land" and therefore capable of being the subject of an agricultural holding. It is then necessary to ask whether the contemplated use of the land was "only" grazing or mowing or both during some specified period of the year. If it was more than grazing or mowing, or if it was grazing or mowing for a year or more, the proviso does not apply. Trivial or insignificant use may be disregarded.

This is not a case — such as those which have given rise to the reported decisions — in which the tenant was allowed into possession simply for a limited period of grazing at a time and then went out of possession until the next season. Mr Brown was in occupation of the paddock throughout the year. The mare and her foals were visible signs of his continuing occupation, notwithstanding that the daffodils were in bloom and the cattle were elsewhere. If a tenant is permitted to use the land for the whole year, and is permitted to use it for other purposes when it is not being used for grazing, it can no longer be said that the contemplated user of the land is "only" for grazing during a part of the year. Where, as in this case, some grazing continues throughout the year, albeit by different animals some of which were not used in the tenant's business, it must be clearer still that the proviso has no application.

I do not imagine that Sir Douglas Howard intended to create an agricultural holding on land which was, in a sense, an extension of his garden and, indeed, had to be reached through it. However, Sir Douglas was not a lawyer and he does not appear on this occasion to have taken legal advice. He no doubt relied on his personal relationship with the users of his paddock in order to solve whatever problems might arise. In my judgment, he reached an agreement with Mr Brown the legal result of which was to create an agricultural holding, whether he realised it or not. Subject to counsel's submissions on the form of the order, I shall therefore make the declaration which the plaintiff seeks.

Declaration accordingly.

Queen's Bench Division

January 26 1993

(Before Lord Justice KENNEDY)

R v DAIRY PRODUCE QUOTA TRIBUNAL AND ANOTHER, EX PARTE P A COOPER & SONS

Estates Gazette May 15 1993

[1993] 19 EG 138

Milk quota — Applicant for secondary quota under the Dairy Produce Quota Regulations 1991 — Whether Dairy Produce Quota Tribunal misunderstood evidence before it and whether tribunal should have given applicants an opportunity to deal with evidence of which it had any doubt

The applicants are a family partnership consisting of Mr and Mrs David Cooper and their two sons engaged in the business of dairy farming at Matts Hill Farm, Sittingbourne, Kent. In 1989 the applicants purchased plant for the production and sale of ice cream outside the new milk quota system. By reason of the death of Mr David Cooper in November 1989, and problems with his estate, the proposed ice cream business had not got off the ground by February 1991 when the UK Government had to extend the quota system to additional milk products, such as ice cream, with effect from October 29 1991. The Dairy Produce Regulations 1991 provide for the award of quota to producers making sales of produce such as ice cream by giving a primary quota to established producers and a secondary quota to be awarded to producers who could show that by March 1 1991 they were committed to developing their production. The applicants' application to the Dairy Produce Quota Tribunal was heard on November 4 1991. In accordance with para 6 of Schedule 9 to the Dairy Produce Quota Regulations 1991 the tribunal had to determine: (1) the quantity of ice cream made from milk produced on the applicants' farm that might reasonably be expected to be produced annually using the facilities acquired by the applicants as a result of their investment; and (2) the quantity of ice cream that might reasonably be expected to be sold annually by direct sale by the applicants, in that regard the tribunal had to have regard to the applicants' established pattern of sales of ice cream and to the applicants' commitment to supply that product. The hearing, which lasted about one and a half hours, was attended by Mr Mark Cooper, one of the partners in the applicant firm, with his solicitor. The determination of the tribunal noted that the applicants' statement of case merely contained estimated figures for the quantities of milk applied in the manufacture of ice cream for 1989, 1990 and 1991. The estimate being 4,050

A ltrs of ice cream made from 6,236 ltrs of milk. The determination also noted that the applicants estimated in the quota year 1991-1992 that it would manufacture 150,000 ltrs of ice cream from 330,000 ltrs of milk and that it claimed a direct sales primary quota of 6,237 ltrs and a secondary quota of 425,000 ltrs. The tribunal concluded that in the light of estimates for each year of 4,050 ltrs of ice cream from 6,263 ltrs of milk, it was totally unable to reach an estimation of the production annually of 150,000 ltrs of ice cream from 330,000 ltrs of milk, which it considered could be described as reasonable as no production or sales anywhere approaching such quantities had been achieved. The evidence, it was said,
B was wholly insufficient to reach any reasonable determination in excess of the estimated figure of 6,237 ltrs for 1991. It was submitted on behalf of the applicants that the tribunal made a fundamental error of law in that it so misconstrued para 6 of Schedule 9 to the 1991 regulations as to regard itself as limited to historically achieved sales of ice-cream as distinct from evidence of reasonably anticipated sales. Further, that if the tribunal correctly understood its function, it must have misunderstood the evidence, because the evidence was all one way.

Held: **The application was allowed and the tribunal's**
C **determination set aside for the matter to be remitted to make a fresh determination. It is clear from reading the determination as a whole that the tribunal was well aware that it was dealing with an application for secondary quota and, therefore in the end, with forecasts rather that past performance, but it could not ignore past or present performance, not only because forecasts are usually based upon such performance but also because para 6(2) of Schedule 9 positively required the tribunal to have regard to it and to any commitment to supply. Where a tribunal gives reasons justice requires that the reasons should have explained, however**
D **briefly, why it was that all or part of the evidence which was all one way, as it was here, was found to be unacceptable, especially if that was something which did not emerge during the course of the hearing. By the end of this hearing more was required of the tribunal than a simple assertion that the applicants had failed to discharge the onus of proof. There were doubts whether the tribunal had arrived at its decision by a proper route and without regard to irrelevant considerations or considerations which should not have been regarded as relevant unless the applicants or their solicitors were given an opportunity to deal with them. The question posed by para 6(1)(a) of the Schedule ought to be answered without regard to**
E **sales potential with the focus, at this stage, being simply on productive capacity. The determination had to be set aside because it failed to tell the applicants in broad outline and in simple terms by what process of reasoning the tribunal arrived at the conclusion that 6,237 ltrs was the appropriate figure to adopt.**

The following case is referred to in this report.

Fox v *(PG) Wellfair Ltd* [1981] 2 Lloyd's Rep 514; (1982) 263 EG 589, CA

This was an application by way of judicial review of a decision of the Dairy Produce Quota Tribunal which, on December 21 1991
F determined the applicants' application for secondary quota under Schedule 9 to the Dairy Produce Quota Regulations 1991. The respondents to the application being the Dairy Produce Quota Tribunal and the Ministry for Agriculture, Fisheries and Food.

Joanne Moss (instructed by Brachers, of Maidstone) appeared for the applicants; Nicholas Paines (instructed by the solicitor to the Ministry of Agriculture, Fisheries and Food) represented the respondents.

Giving judgment, **KENNEDY LJ** said: This is an application for judicial review of a determination of the Dairy Produce Quota Tribunal for England and Wales dated December 20 1991 relating to an application by the applicants for a secondary quota under Schedule 9 to the Dairy Produce Quota Regulations 1991.

The applicant firm is a family partnership which, at various times, G
consisted of Mr and Mrs David Cooper and their two sons. It has been engaged in the business of dairy farming at Matts Hill Farm near Sittingbourne, Kent, for many years.

In 1988, in response to the advent of milk quotas, Mr David Cooper decided to diversify into the production and sale of ice cream outside the new quota system.

In 1989 plant was purchased for the new enterprise, the purchase being financed in part by a farm diversification grant and in part by the sale of part of the milking herd and the wholesale quota.

In November 1989 Mr David Cooper died. Soon afterwards Mrs Cooper was seriously ill. Retail greengrocer shops, run by David Cooper, had to be sold and there were problems with his estate so that the proposed ice cream business had not got off the ground by H
February 1991 when the UK Government had to bow to European pressure and extend the quota system to additional milk products, such as ice cream, with effect from October 29 1991.

The 1991 regulations were made to provide for the award of quota to producers making sales of produce such as ice cream. Those regulations provided for primary quota to be awarded to established producers and for secondary quota to be awarded to producers who could show that by March 1 1991 they were committed to developing their production. Those seeking a secondary quota award had to apply to the Dairy Produce Quota Tribunal, which then had to make an award in accordance with para 6 of Schedule 9 to the 1991 regulations, which provides: J

> (1) For the purposes of an award of secondary quota under paragraph 5, the Tribunal shall determine in respect of each producer
> (a) The quantity of additional milk products made from milk produced on his holding that might reasonably be expected to be produced annually using
> (i) the facilities available to the producer on 28th February 1991, and
> (ii) the additional facilities expected to be available to the producer on 31st March 1992 by virtue of money expended or a contract entered into as mentioned in paragraph 5(a) and (b) before 1st March 1991;
> (b) the quantity of the additional milk products referred to in subparagraph (a) that might reasonably be expected to be sold by direct sale by the producer annually.
>
> (2) In making the determination referred to in subparagraph 1(b) the Tribunal shall have regard to the producer's established pattern of sales of K
> additional milk products and to the producer's commitment to supply such products.
>
> (3) For the purposes of an award of secondary quota, the Tribunal may also determine in respect of each producer —
> the quantity, or a part of the quantity, by which the sum of the producer's wholesale and direct sales quota (other than any primary quota awarded under paragraph 2), taking account of any wholesale and direct sales quota permanently or temporarily transferred by or to him since 28th February 1991, exceeds the quantity of the dairy produce other than additional milk products which that producer may be expected to sell by direct sale or deliver to a purchaser in the quota year ending on 31st March 1992,
>
> if in the Tribunal's opinion it is fair and reasonable that such a quantity or a part of a quantity should be taken into account in making the award. L

These applicants made their application to the Dairy Produce Quota Tribunal in the prescribed form and the matter came before the tribunal for consideration on November 4 1991. In accordance with para 6 of the Schedule, the tribunal then had to determine:

(a) The quantity of ice cream made from milk produced on the applicants' farm that might reasonably be expected to be produced annually using, in this case, the facilities acquired by the applicants as a result of their investment: see para 6(1)(a).

(b) The quantity of ice cream that might reasonably be expected to be sold annually by direct sale by the applicants: see para 6(1)(b).

In dealing with question (b) the tribunal had to have regard to the applicants' established pattern of sales of ice cream and to the M
applicants' commitment to supply that product: see para 6(2).

Mr Mark Cooper, one of the partners in the applicant firm, attended at the hearing with his solicitor. The tribunal was presided over by Lord Grantchester, QC and the hearing, according to Mr Cooper, lasted about one and a half hours. It is clear from the affidavit of Mrs Sayer, who was secretary to the tribunal — though not sitting with the tribunal on that day — that the length of the hearing was average. Para 5 of her affidavit reads:

> The hearings were conducted in a fairly informal way and applicants could bring with them their own agents to advise them or speak on their behalf. During the course of a hearing the Tribunal would ask the applicant to describe his general dairy farm circumstances, would go on to examine his production investments and theoretical capacity, the production actually

A likely to be achieved and the applicant's prospects for selling his product, giving the applicant every opportunity to present written or oral evidence on these points. His quota position, which might affect the award, was also discussed. The Tribunal applicant had been briefed on the substance and purpose of the Tribunal by the paper he had already received. In particular the Notes for Guidance go through the questions set down in Schedule 9 for the Tribunal's determination, and point out that the onus is on the applicant to satisfy the Tribunal on these matters.

As is clear from that explanation, the proceedings were not, in any sense, adversarial. Mrs Sayer, having interviewed the acting secretary to the tribunal, Mr Crowther, who was at the relevant hearing and, having read the material which was before the tribunal, asserts, in para 6 of her affidavit, that in dealing with the Cooper
B application the tribunal took its normal course. Part of her affidavit reads:

> The applicant and his adviser were given opportunities to explain, and did so explain, why production and sales had not been developing. They were questioned on his sales prospects specifically, and were given the opportunity to explain, and did so explain, the basis for his expectations. After the hearing (which lasted approximately one and a half to two hours) the Tribunal discussed the case prior to the chairman drafting the decision. Lord Grantchester has confirmed that the evidence submitted by the applicant was carefully looked at again before he finally circulated the draft decision to the other two members for their consideration. The decision was agreed by all three members.

Mr Cooper, for his part, says, in para 10 of his affidavit, that he
C was "not very happy" with the hearing. He says that:

> Part of the time was occupied with investigating matters related to milk production and ice cream manufacturing capacity. The assessment of sales potential of the product was not ignored but was dealt with in a low key way with no indication at all that the Tribunal would be unable to make a determination without further evidence. However, the most unusual feature of the hearing was that a large part of the time was spent disputing the chairman's incorrect interpretation of para 6(3) of Schedule 9 to the 1991 regulations.

I turn now to the determination itself, the first three paragraphs of which give the background to the application. In para 4 the tribunal accepts that the plant and machinery for making ice cream, which was purchased in 1989, cost £54,179 and that its capacity was 425,000
D ltrs of milk producing 276,500 ltrs of ice cream per annum.

Para 5 of the determination states:

> No figures are contained in the Application Form for the quantity of milk applied in the manufacture of ice cream in 1990 or in any previous year, or for the quantities of ice cream produced. The Statement of Case merely contains estimated figures for 1989, 1990 and 1991, (being unable to estimate at all for 1988). In each year the estimate is 4,050 litres of ice cream made from 6,236 litres of milk. These figures are most unsatisfactory.

Para 6 then begins:

> According to the Application Form the Applicant Firm estimates that, in the quota year 1991/92, it will manufacture 150,000 litres of ice cream from 330,000 litres of milk produced on its holding and that it will sell 50,000 litres of such ice cream from a farm shop and the balance of 100,000 litres to other
> E local shops. According to the Statement of Case the Applicant Firm claims a Direct Sales primary quota of 6,237 litres and a secondary quota of 425,000 litres.

Para 6 then goes on to state that in para 2.38 of the statement of case the applicant firm seeks to justify its claim for secondary quota and it accurately quotes part of para 2.39. In fact, as I understand it, para 2.39 is not seeking to justify the claim for secondary quota, but simply setting out the projected sales growth on which the business plan is based and explaining why the applicants seek a quota equivalent to full production from the outset.

In para 7 of the determination the tribunal turns to what it describes as "The first question for our determination" as set out in
F para 6(1)(a) of Schedule 9 to the 1991 regulations; that is the quantity of ice cream that might reasonably be expected to be produced annually with the facilities available.

Having expressed sympathy with the applicants, the tribunal continues:

> But they are not going to manufacture ice cream without outlets for the sale thereof, and the only figure which we have for sales in 1990 and 1991 are an estimate for each year of the sale of 4,050 litres of ice cream from 6,263 litres of milk from their herd. In the light thereof we are totally unable to reach an estimation of the production annually of 150,000 litres of ice cream from 330,000 litres of milk which we consider could be described as reasonable. Even less could we reasonably determine the higher figures for ice cream and milk contained in the Statement of Claim. No production or sales anywhere approaching such quantities has been achieved. In our opinion, the evidence before us is wholly insufficient to reach any reasonable determination in this G
> case in excess of the estimated figure of 6,237 litres for 1991.

Para 8 of the determination I can omit because, as the tribunal says, it deals with a situation which does not arise in this case. There is no para 9 but in para 10 the tribunal states:

> In the light of our determination under subparagraph 6(1)(a) of the Ninth Schedule as set out above, our determination under subparagraph 6(1)(b) should also, in our view, be 6,237 litres. The evidence before us is wholly insufficient to enable us to make any greater determination of sales for ice cream annually which can reasonably be expected to be achieved by the Applicant Firm.

In para 11 the tribunal expresses its conclusion as to excess of quota. There is no challenge to that aspect of the determination and I say no more about it. H

At the start of the hearing before me Miss Moss, for the applicants, asked me to exclude the affidavits filed on behalf of the respondents many months ago on the basis that they contained hearsay, were tendentious and raised issues not raised by her notice of motion. Miss Moss indicated that if I refused to exclude the evidence — which I did — she would apply to amend the grounds on which she sought relief. Eventually, at the end of the hearing she did so. That application to amend was opposed by Mr Paines and I deal with it later in this judgment.

There are a number of grounds set out in the notice of motion on which relief is sought, but having regard to the submissions made by Miss Moss, it seems to me that they can be summarised thus. J

(1) Either the tribunal misunderstood its function in law or it misunderstood the evidence before it because there was unchallenged evidence of projected production and sales which the tribunal, for no apparent reason, did not accept: grounds 1 and 2.

(2) If the tribunal had any serious doubt about any aspect of the evidence it should, before giving weight to the doubt, have given the applicants an opportunity to deal with it: ground 5.

(3) The way in which the tribunal expressed its decision was such as to indicate that it must have erred in one or other of the ways alleged: ground 6.

The amendment which Miss Moss sought leave to make was to add a ground which would allege that the tribunal

> failed to intimate to the applicants for their comments any specific evidence or K
> materials or reasons relied on by the Tribunal to justify rejection of the applicants' sales projections, and also gave no, or no sufficient reasons for rejecting them.

It will be clear from the way in which I have summarised the original grounds on which relief is sought that the proposed amendment adds nothing to them. I therefore refuse leave to make that amendment.

Miss Moss' first submission was that the tribunal made a fundamental error of law in that it so misconstrued para 6 of Schedule 9 to the 1991 regulations as to regard itself as limited to historically achieved sales of ice cream as distinct from evidence of reasonably anticipated sales. That I cannot accept because, as is clear L
from Schedule 9 itself, the concept of a secondary quota came into existence in order to do justice to those who had already financially committed themselves to enter or to expand into the market for additional milk products, such as ice cream, at the moment when the quota system was introduced. It is clear from reading the determination as a whole that the tribunal was well aware that it was dealing with an application for secondary quota and, therefore in the end, with forecasts rather than past performance, but it could not ignore past or present performance, not only because forecasts are usually based upon such performance but also because para 6(2) of Schedule 9 positively required the tribunal to have regard to it and to any commitment to supply.

Miss Moss then goes on to submit that if the tribunal correctly M
understood its function it must have misunderstood the evidence because the evidence was all one way. The short answer to that submission seems to me to be, as Mr Paines submitted, that the evidence was bound to be all one way because of the nature of the proceedings and that no adjudicating body is bound to accept all of the evidence placed before it — otherwise there can be no need for the adjudication.

But that may not be a complete answer because, at least where a tribunal does give reasons, as this tribunal does, it seems to me that justice requires that the reasons should have explained, however briefly, why it was that all or part of the evidence which was all one way was found to be unacceptable, especially if that was something

A
which did not emerge during the course of the hearing. I do not go so far as to accept Miss Moss' proposition that it is a breach of natural justice for a tribunal to receive evidence without challenge and then to take an applicant by surprise by rejecting it. She invited me to consider the decision of the Court of Appeal in *Fox* v *(PG) Wellfair Ltd* [1981] 2 Lloyd's Rep 514, but the problem there was that the arbitrator introduced his own expertise to contradict the evidence of the applicants' expert witnesses without giving those witnesses any opportunity to deal with his proposals.

I do not find anything in that decision to support Miss Moss' wide general proposition that a tribunal such as that with which I am concerned must, even in general terms, indicate what difficulty it
B
finds with evidence so that the applicant or the applicants' legal adviser can deal with it.

As Mr Paines pointed out, the tribunal may simply find the evidence inadequate or insufficient to establish that which the applicant seeks to prove, whether it be an anticipated level of production or a projected level of sales, in which case the tribunal can say just that when it comes to make its determination. It seems from para 16 of the affidavit of Mr Thomas, a senior executive officer at the Ministry of Agriculture, Fisheries and Food, that there was a danger of such an outcome in the present case because, contrary to the submission of Miss Moss that the Cooper application was well presented, he says:

C
> Part of my responsibility was preliminary scrutiny of all applications for quota. In contrast with the vast majority of applications received, Mr Cooper's application seemed to me imprecise and poorly supported: no records were provided, no evidence was supplied concerning contracts or agreements to supply existing or new outlets and only some of the documents relating to a claim under the Farm Diversification Grant Scheme were included. Question 20 of the application form MSQ 61 asked the applicant to indicate if a business plan had been prepared, and, if so, to supply a copy of the plan. Mr Cooper did not indicate that such a plan existed, nor was a copy provided. This would suggest that no business plan had been prepared at the time the application for quota was made.

However, as I shall explain in a moment, it seems to me that by the end of the hearing more was required of the tribunal than a simple assertion that the applicants had failed to discharge the onus of
D
proof. Parts of the contents of the two affidavits now put in evidence by the respondents do cause me to wonder whether the tribunal did arrive at its decision by a proper route and without regard to irrelevant considerations or considerations which should not have been regarded as relevant unless the applicants or their solicitors were given an opportunity to deal with them. For example, in para 17 of the affidavit of Mr Thomas there is a reference to documents which came into existence when the applicants sought a farm diversification grant (ie the grant which provided part of the finance for the plant obtained in 1989). Mr Thomas detects some inconsistency between a letter dated December 18 1989 and the letter dated April 20 1991 which was submitted with the quota application. I was told during the hearing before me that the 1989 documents were not before the
E
tribunal and were only referred to by Mr Thomas so that I should "have the full picture". If they were not before the tribunal I do not see why they have been brought to my attention. If they were before the tribunal and the tribunal detected the inconsistency to which Mr Thomas refers, that was just the sort of matter which, to my mind, should have been drawn to the attention of the applicants or to the applicants' solicitor before the tribunal acted upon it to the applicants' detriment.

Similarly, in para 8 of her affidavit, Mrs Sayer questions the capacity of the applicants' plant. She says:

> The Tribunal's decision records that they accepted "the capacity of the plant and machinery as set out in the Statement of Case at 425,000 litres of milk
> F
> producing 276,500 litres of ice cream per annum". The Tribunal thereby accepted the theoretical capacity of the plant and machinery for processing, but this by no means necessarily equates to reasonably expected production. This capacity assumes 8 hours a day 5 days a week, 52 weeks a year running of the plant with no seasonal fluctuation, no technical problems or slowing for any reason. Even if no doubts existed on the producer's business prospects, therefore, this quantity would be unlikely to have been the Tribunal's final determination of production (let alone of sales).

In fact, that extract highlights my principal criticism of this determination, namely that it is not possible to know from it what the tribunal did decide, but if the tribunal doubted whether the actual capacity of the plant would have been the same as its theoretical capacity (a capacity which, as Mrs Sayer points out, was not based upon a 24-hour day or a seven-day week) then, as it seems to me, those
G
are the sort of doubts which the tribunal should have given the applicants an opportunity to dispel.

At the end of her affidavit Mrs Sayer says that "the Chairman, Lord Grantchester, has seen this affidavit and is in agreement with it". What exactly that observation is meant to convey, I am not sure. It tends to suggest that what might otherwise be regarded as personal speculation on the part of Mrs Sayer was in fact the approach adopted by the tribunal and if that is the position it does not surprise me that the applicants are concerned.

That brings me to the way in which the tribunal expressed its decision. The evidence, to my mind, does not support the submission that the hearing was perfunctory and Mr Cooper accepts that production capacity and sales potential were both canvassed,
H
although he asserts that the latter was dealt with in a low key way. It was then incumbent on the tribunal to address itself to the problem posed by para 6(1)(a) of Schedule 9 to the regulations; in other words, to ask itself what quantity of ice cream made from milk produced on the applicants' farm might reasonably be expected to be produced annually using the available plant.

Because the next subparagraph in the Schedule deals with sales, it seems to me that the question posed by para 6(1)(a) ought to be answered without regard to sales potential with the focus, at this stage, being simply on productive capacity. This segregation of productive capacity and sales potential seems to me to be of some importance, not only because it accords with the Schedule, but also because it enables an unsuccessful or only partially successful
J
applicant to know why he has failed.

In the present case the tribunal did not adopt that two-stage approach, but instead moved straight from what Mrs Sayer described as the theoretical capacity of the plant to the question of sales. Actual capacity was therefore never properly addressed, but, having regard to the next paragraph of the Schedule, that should not have undermined the determination if there was a proper approach to the question of potential sales. However, in my judgment, that approach also was flawed. For the purposes of para 6(1)(b), what the tribunal had to decide was the quantity of ice cream that might reasonably be expected to be sold by direct sale annually. Inevitably, therefore, the focus had to be on the future even though, by virtue of para 6(2), the
K
tribunal had to have regard to any established pattern of sales and to any commitment to supply. No doubt that provision would be of great assistance to producers who could show a steadily rising production rate year by year over several years and contractual commitments based on the assumption that the rise would continue. But that was not this case. In order to justify their assertion that over a relatively short period they could build up sales to match the capacity of their plant, these applicants relied on six factors, namely:

(1) Their investment, the argument being that people do not spend money buying plant to make something unless they really believe they can sell it.

(2) Their retail business experience in the greengrocery trade.

(3) The evidence of a nationwide potential market for ice cream: L
see, for example, the Cream Report 1990 and so forth.

(4) Their pilot scheme: see especially para 2.42 of their statement of case.

(5) The letter which they had obtained from Billy Boy Frozen Foods confirming that less than a week before the tribunal hearing primary negotiations were continuing with a view to that company distributing the applicants' product nationwide.

(6) The accountants' business plan, which showed the ice cream project to be viable in business terms.

Yet when the tribunal came to address itself to potential sales it says nothing about any of those factors and mentions only past production before going on to make a determination in line with past
M
production. Even if the evidence was not very persuasive that, on the face of it, is a surprising conclusion, namely that the applicants cannot reasonably be expected to sell more than they had been selling thus far. But it is not simply because it is a surprising conclusion that I have come to the conclusion that it cannot stand. In my judgment, the determination must be set aside because it fails to tell the applicants in broad outline and in simple terms by what process of reasoning the tribunal arrived at the conclusion that 6,237 ltrs was the appropriate figure to adopt.

In the circumstances of this case that determination could not properly be justified by the tribunal saying simply, as it does in para 10, that:

> The evidence before us is wholly insufficient to enable us to make any greater

A determination of sales for ice cream annually which can reasonably be expected to be achieved by the Applicant Firm.

In para 9 of her affidavit Mrs Sayer puts forward some reasons why the tribunal might not have regarded the applicants' evidence as adequate to support their claim. She points to what she considers to be shortcomings of the pilot scheme and says that the letter from Billy Boy Frozen Foods might have been viewed with caution as evidence of sales prospects. Maybe, but as the tribunal said nothing about either the pilot scheme or the letter, her comments cannot be said actually to represent the reasoning processes of the tribunal, even though the chairman of the tribunal saw and approved her affidavit.

Accordingly, in my judgment, the determination of the tribunal and the award of milk quota made by the ministry upon the basis of B the tribunal's determination must both be set aside and the matter must be remitted to the tribunal so that it can make a fresh determination. It is not for this court to pre-empt the jurisdiction of the tribunal by making the declarations sought and it is for the minister to decide what quota should be awarded when he receives a valid and effective tribunal determination.

Queen's Bench Division

February 12 1993

C (Before Mr Justice POPPLEWELL)

R v MINISTRY OF AGRICULTURE, FISHERIES AND FOOD, EX PARTE COX

Estates Gazette June 5 1993

[1993] 22 EG 111

Milk quota — Whether there was an effective transfer of a milk quota sold to the applicant — Whether the Ministry of Agriculture, Fisheries and Food had power to rescind the registration of the milk quota in the name of the applicant

D By the agreement dated September 10 1986 Hamps Valley Farms Ltd ("HVF") agreed to grant the applicant, Mrs Cox, a grazing licence in respect of 9.39 acres of their farm and to transfer a milk quota of 65,000 ltrs. The licence was a personal agreement with a right to graze beef cattle or sheep only for a period from September 1 1986 to July 31 1987 in consideration of a fee of £300. A further £1,300 was paid for the transfer of milk quota. The applicant did not exercise her grazing rights. The Milk Marketing Board registered the transfer of quota in the applicant's name. By a letter of May 12 1988 HVF claimed that they were entitled to the milk quota because it had not been validly transferred to the applicant and they required the E respondent, the Ministry of Agriculture, Fisheries and Food, to adjust the register. On July 11 1988, the ministry declined to do so. By an award of December 14 1990 an arbitrator concluded that occupation of the land was necessary to entitle the applicant to acquire the milk quota. On April 25 1991 the respondent informed the applicant that in the light of the arbitrator's award the minister intended to rescind the wholesale quota register entry in respect of the intended transfer of 65,000 ltrs of quota from HVF to the applicant. The applicant brought proceedings by way of judicial review of the respondent's decision to alter the register.

F *Held*: **The application was granted together with an appropriate declaration that the respondent is not entitled to rescind the wholesale quota register entry in respect of the 65,000 ltrs of quota acquired by the applicant in September 1986. Commission regulation 1371/84, article 5(2), is to be read as requiring the transferee to be operating the holding before he or she is entitled to the transfer of the milk quota. The applicant did not have such occupation. It would have been open to the respondent to alter the registration of the milk quota made in October 1986. However, the respondent did not have power to vary his decision of July 11 1988 that it was not prepared then to rescind the milk quota registered in favour of the applicant. It would be inequitable in the light of** G **the common practice that an agreement to graze with exclusive right of occupation, but without actual occupation, was accepted by the respondent in relation to the transfer of dairy quota that the applicant should now be deprived of her quota.**

The following cases are referred to in this report.

Customs & Excise Commissioners v *APS Samex* [1983] 1 All ER 1042
Denton Road (No 56) Twickenham Re [1953] Ch 51; [1952] 2 TLR 676
Kuhn v *Landwirtschaftskammer Wesems* [1992] 2 CMLR 242
Posthumus v *Oosterwoud* [1992] 2 CMLR 336
Wachauf v *Bunderesamt für Ernährung und Forstwirtschaft (Case 5/88)* [1991] 1 CMLR 328; [1989] ECR 2609 H

This was an application by the applicant, Mrs Margaret Cox, of a decision of the respondent, the Ministry of Agriculture, Fisheries and Food, of its decision in April 1991 to revoke the applicant's registration of milk quota, pursuant to leave granted by Hutchison J on May 23 1991.

Paul Morgan QC (instructed by Burges Salmon) appeared for the applicant; Peter Roth (instructed by the solicitor to the Ministry of Agriculture, Fisheries and Food) represented the respondent.

Giving judgment, **POPPLEWELL J** said: This is an application for judicial review pursuant to leave granted by Hutchison J on May 23 1991.

J

The parties

The applicant, Mrs Cox, is a farmer at Malpas, Cheshire. Hamps Valley Farms Ltd are the owners of a farm at Onecote, nr Leek, Staffordshire. Though they are an interested party and have been notified of these proceedings they have not taken part in them. The respondents are the Ministry of Agriculture, Fisheries and Food, who are responsible for the milk quota regime.

The issues

Hamps Valley Farm Ltd ("HVF") sold to the applicant a milk quota of 65,000 ltrs. The first issue is whether there was an effective transfer of that quota. In about October 1986 the Milk Marketing Board, acting as agent for the respondent had registered that quota in the name of Mrs Cox and sought in April/May 1991 to revoke that K registration. The second issue is whether they had the power to rescind that entry.

The facts

These are not substantially in dispute. By an agreement made on September 10 1986 the applicant and HVF entered into a written agreement whereby HVF agreed to grant the applicant a grazing licence in respect of some 9.39 acres of their farm and to transfer a milk quota of 65,000 ltrs to the applicant so that, on the expiration of the licence to graze, the quota would remain part of the applicant's milk quota. It was a personal agreement with the applicant. She had the right to graze with beef cattle or sheep only, for a period from September 1 1986 to July 31 1987. She was required to keep all the L fences, gates, ditches and water courses in condition and to use only cattle or sheep for the purpose of grazing the land. The applicant was to pay £300 in respect of the grazing rights and a further £1,300 in respect of the transfer of the milk quota. There were provisions as to compensation in the event of a number of eventualities. There was an arbitration clause in relation to disputes and the parties acknowledged that they had each been advised to seek separate independent legal advice before entering into the agreement.

Having entered into that agreement the applicant did not exercise her right and the land was vacant from the commencement of the agreement until January 1987 when HVF entered into physical occupation of the land.

In about October 1986 the Milk Marketing Board, pursuant to the M Dairy Produce Quotas Regulations 1986, registered the transfer of the milk quota to the applicant from HVF. The applicant believed that the temporary transfer of part of a holding was sufficient to enable the transfer of the milk quota from HVF to her and that it was not necessary for her to enter into physical occupation of the land.

On July 11 1988 the respondents wrote to HVF in reply to their letter of May 12 1988 (a copy of which has not been found). However, it is clear from the letter of July 11 that HVF were claiming that they were entitled to the milk quota because it had not been validly transferred to Mrs Cox and they required the ministry to adjust the register.

The letter of July 11 1988 reads:

Your letter of 12th May addressed to the MMB about the validity of a number of quota transfers has been passed to me for reply. Please accept my apologies for the delay in replying to you.

I have considered the case carefully in the light of the allegations made in your letter but I am of the view that it would not be correct for the Ministry to adjust the quota register as you have requested. In reaching this conclusion, I would emphasise the following points.

(a) The operative dates of the quota transfers to which you refer were 24 months and 22 months ago but you did not question their validity near that time. Neither did you submit quota transfer forms claiming quota when the grazing licences had expired.

The reference to 24 months and 22 months indicates that there was an agreement with some other person of a similar nature with which this case is not concerned.

The letter went on:

(b) The Dairy Produce Quotas Regulations 1986 apply to the transfers in this case. A mechanism for dealing with disputes concerning transfers of milk quota, is laid down in Schedule 4 Part I of the Regulations. Under these procedures, if agreement cannot be reached between the transferor and the transferee(s) then the matter must be decided by arbitration.

If you had informed the Ministry that the land transferred did not, in fact, take place, then the provisions of the Regulations would have helped you resolve any dispute over the quota, with the transferees concerned.

The matter did come before an arbitrator and on December 14 1990 he made his award. He came to the conclusion that occupation of the land was necessary to entitle the applicant to acquire the milk quota, he heard arguments about the compensation and made an order as to costs.

The decision of the arbitrator as to matters of law are not binding on me. It is not suggested that they should be. Nor is it suggested that they have any legal effect in relation to the register. On April 25 1991 the ministry wrote to the applicant as follows:

I write further to your letters of 4th and 20th February to give notice of the Minister's intention to rescind the wholesale quota register entry in respect of the intended transfer in September 1986 of 65,000 litres of quota from Hamps Valley Farms to your client Mrs M E Cox As requested in your letter of 4th February I confirm that no change in the registration of the Milk Quota will be made within 21 days of the date of this letter.

In reaching a decision on this case the Minister has had regard to the arbitrator's award dated 14th November 1990, and to the representations made in your letters (particularly those of 20th December and 4th February) together with the papers which you have provided. In view of the potentially serious effects of this decision for your client, my clients have reviewed this case as a whole with the benefit of further legal advice and still remain of the view that physical occupation is required in order to effect a transfer of quota. In view of the clear evidence that there was no physical occupation in this case, the Minister cannot accept that a valid transfer of quota has taken place, though in view of your client circumstances it is not without regret that this decision has been reached.

The respondents undertook not to alter the register pending the hearing of these proceedings. The respondents' position is that they have a duty to keep a proper register and if the agreement in September 1986 did not have the effect of transferring the milk quota to Mrs Cox then the register must be put right. If they had declined in 1991, as they did in July 1988, to adjust the register at the instance of HVF a challenge to that decision by HVF would almost certainly fail whatever the legal position for the reasons set out in that letter, namely delay. However, the effect of the respondents' proposal now to adjust the register will be to transfer the quota back to HVF from Mrs Cox arising from the agreement over six years ago.

First issue

Where there is a temporary transfer of part of a holding is it necessary that there should be physical occupation by the transferee to enable her to receive the benefit from the transferor of a milk quota?

This depends on the interpretation of European directives and domestic regulations.

The quota system was introduced by regulation 856/84. It was done by introducing a levy payable by producers or purchasers of cows' milk and its object was to curb the increase in milk production. Member states were to introduce a levy system in accordance with one of two formulas. The formula adopted by the UK was formula B. A levy was to be paid on milk produced over a certain quantity. Each country was given a guaranteed total quantity and in the UK the Milk Marketing Board allocated to producers a fixed quota. Above that quota a levy which is 100% of the price of milk was to be paid. A quota therefore is treated as an amount to represent the exemption from levy and the quota can move from one producer to another. It is not in dispute that you can have a permanent transfer of a quota on the back of a temporary transfer of land, but there must be a relevant land transaction. The issue in this case is whether there has been. There is no dispute that there is a substantial market in the transfer of quotas.

By article 5c(6) of Council Regulation 804/68:

The Council, acting on a proposal from the Commission in accordance with the voting procedure laid down in Article 43(2) of the Treaty, shall lay down the general rules for the application of this Article, and in particular those relating to the determination of the reference quantities and the amount of the levies referred to . . .

7. Detailed rules for the application of this Article shall be adopted in accordance with the procedure laid down in Article 30.

Thus it is said on behalf of the applicant that the rights are laid down by Council Regulations and the procedure by Commission Regulations.

The first regulation to consider is article 7 of Council Regulation 857/84. It reads:

Where an undertaking is sold, leased or transferred by inheritance, all or part of the corresponding reference quantity shall be transferred to the purchaser, tenant or heir according to procedures to be determined.

By article 12 producer is defined as:

(c) producer: a natural or legal person or group of natural or legal persons farming a holding located within the geographical territory of the Community:
— selling milk or other milk products directly to the consumer, and/or
— supplying the purchaser;

A holding is defined:

(d). . . all the production units operated by the producer and located within the geographical territory of the Community.

It has been accepted that a holding may include a mixed farm.

Article 7 was amended by Council Regulation 590/85. The reason is set out in the preamble which reads:

Whereas the application of Article 7 of Regulation (EEC) No 857/84, concerning the transfer of reference quantities in the event of the sale, leasing or inheritance of a holding, may in certain cases result in difficult situations at an economic and social level; whereas it is therefore appropriate, in order to enable a lessee whose lease on a holding is due to expire to continue milk production elsewhere, that Member States be authorized to place at the disposal of such lessee all or part of the reference quantity corresponding to the holding which he is leaving; whereas, with the same purpose in mind, comparable provisions should be introduced for the benefit of the departing producer where land is transferred to public authorities or for purposes of public use.

The new article 7 reads as follows:

1. Where a holding is sold, leased or transferred by inheritance, all or part of the corresponding reference quantity shall be transferred to the purchaser, tenant or heir according to procedures to be determined.

. . .

4. In the case of rural leases due to expire, where the lessee is not entitled to an extension of the lease on similar terms, Member States may provide that all or part of the reference quantity corresponding to the holding which forms the subject of the lease shall be put at the disposal of the departing lessee if he intends to continue milk production.

It is submitted on the applicant's behalf that those regulations make no reference at all to a lessee going into possession and they are to be contrasted with the provisions in new article 7 relating to a departing lessee if he "intends to continue milk production". Thus it is said the general effect of the Council Regulations is to transfer the quota whenever there is a sale, a lease or a will.

The relevant Commission Regulations are, first, no 1371/84. The preamble reads:

Having regard to Council Regulation (EEC) No 857/84 of 31 March 1984 adopting general rules for the application of the levy referred to in Article 5c of Regulation (EEC) No 804/68 . . . and in particular . . . Article 7(1).

Article 5 reads:

For the purposes of applying Article 7(1) of Regulation (EEC) No 857/84 and without prejudice to Article 7(3) of the same Regulation, the following rules shall apply to the transfer of reference quantities granted to producers and purchasers in application of formulas A and B and of reference quantities granted to producers selling for direct consumption:

1. Where an entire holding is sold, leased or transferred by inheritance, the corresponding reference quantity shall be transferred in full to the producer *who takes over the holding* [emphasis supplied].

2. Where one or several parts of a holding is sold, leased or transferred by inheritance, the corresponding reference quantity shall be distributed among

A the producers *operating the holding* [emphasis supplied] in proportion to the areas used for milk production . . .

3. The provisions of subparagraphs 1 and 2 above shall also be applicable in other cases of transfer which, under the various national rules, have comparable legal effects as far as producers are concerned.

It is submitted on behalf of the applicant that the phrase "who takes over the holding" in para 1 and "operating the holding" in para 2 are descriptive of the producer and not a necessary element in the transfer of a quota. The effect of para 3 is not in dispute, it is that the phrase "comparable legal effect" is intended to treat an agreement of the sort under review as if it had the same legal effect as a lease thus avoiding the difficulties inherent in different concepts of property law in different countries.

B Article 5 has been amended by the Commission Regulation 1043/85. The preamble to this amendment reads as follows:

Whereas pursuant to the second subparagraph of paragraph 1 and paragraph 4 of Article 7 of Regulation (EEC) No 857/84 Member States may, . . . where a non-renewable lease expires and cannot be extended on similar terms, provide all or part of the reference quantity corresponding to the holding or to the part of the holding concerned is to be put at the disposal of the departing producer; whereas that right constitutes a derogation from the principle laid down in the first subparagraph of Article 7(1) whereby a reference quantity may not be transferred without the land; whereas the said derogation was introduced in order to overcome certain economic and social problems and allow producers to continue milk production; whereas the extent of that derogation should therefore be specified.

C The following is then added to Article 5:

In the event of the application of . . . Article 7(4) of the said Regulation, concerning rural leases which are due to expire and which cannot be extended on similar terms, all or part of the reference quantity corresponding to the holding or to the part of the holding which is the subject of the transfer or of the said lease shall be put at the disposal of the producer concerned if he intends to continue milk production, provided that the sum of the reference quantity thus made available to him and the quantity corresponding to the holding which he takes over on which he continues milk production does not exceed the reference quantity which was available to him before the land was transferred or before the lease expired.

It is submitted on behalf of the respondent that the preamble makes clear that a quota may not be transferred without the land. But

D what is meant by the land transaction? Is it that the quota is transferred when the relevant legal rights are transferred or created or is the quota only transferred when the legal rights are exercised?

No explanation has been given for the different wording in paras 1 and 2 of article 5. Para 3 of article 5 says nothing about the exercise of rights.

The relation between Council Regulations and Commission Regulations is governed, first, by article 189 of the Treaty of Rome:

A regulation shall have general application. It shall be binding in its entirety and directly applicable in all Member States.

It was also referred to by the Advocate-General Monsieur Jean Mischo in *Kuhn* v *Landwirtschaftskammer Wesems* [1992] 2 CMLR

E 242 at p256 where he said:

Having regard to the Court's interpretation, in paragraph 13 of *Wachauf*, of Article 7 of Regulation 857/84, which takes precedence over the provisions of Commission Regulations 1371/84 and 1546/88, there can be no objection on the ground that the last paragraph of Article 5 of Regulation 1371/84 or Article 7(2) of Regulation 1546/88 provides that . . .

The court did not have to consider whether that was a correct interpretation of the law. However, they ruled at p262:

It is contrary to Council Regulations 857/84 . . . adopting *general* rules for the application of the levy referred to Article 5c of Regulation 804/68 . . . as supplemented by Commission Regulation 1371/84 laying down *detailed* rules for the application of the additional levy.

[Emphasis supplied]

F In *Wachauf* v *Bunderesamt für Ernährung und Forstwirthschaft (Case 5/88)* [1991] 1 CMLR 328, the Advocate-General described the Council Regulations as being general rules for the application of the levy system and the Commission Regulations as setting out detailed rules.

Mr Wachau was a tenant of a farm under a tenancy agreement originally made in 1959 between his parents and the owner of the farm, the princess. His tenancy agreement expired on January 31 1983 and he vacated the farm early in 1985. He sought compensation under Council Regulations and it was decided that by article 7 of regulation 857/84 the community legislature intended that at the end of an agricultural tenancy the milk quota pertinent to the lease land should in principle return to the lessor who retakes possession of the holding. It was said that the legal effect of the surrender of a lease is G essentially the same as that of its grant, namely the transfer of the leased property from one party to the other. He observed at p340 that it was a principle that the quota follows the land on transfer and that article 54 of regulation 1371/84 was by way of derogation from the principle, namely to permit member states to provide a narrowly defined category of cases that outgoing tenants may retain all or part of the quota if they wish to continue milk production. The court in giving judgment at p347 said:

In fact, as the United Kingdom and the Commission rightly points out, Article 12(d) is intended to define the scope of the rules relating to transfers of reference quantities following a *change of ownership or occupancy of the holding*.

[Emphasis supplied] H

The applicant submits that somebody may well be an owner without at the same time being an occupier and that that does not suggest physical occupation of the holding. On the same page:

It is apparent from the provisions quoted, considered as a whole, that the Community legislature intended that at the end of the lease the reference quantity should in principle return to the lessor who retakes possession of the holding, subject, however, to the member-states' power to allocate all or part of the reference quantity to the departing lessee.

At p348 the judgment goes on:

Article 5 of Commission Regulation 1371/84 laid down the detailed rules governing the transfer of reference quantities following a change in the ownership or occupancy of a holding. J

After the expiry of Mr Wachauf's lease the land was let to six different tenants and the corresponding quota was divided between them. There is nothing in the decision to indicate that the princess ever occupied the farm, indeed the references by the court to ownership rather suggests otherwise. One of the strong arguments submitted by the minister is that there should not be a trade in milk quotas and that if there is to be a transfer of milk quotas it must be as a result of transfer of land. At first sight this has very great sense and a good deal of logic. It seems strange if milk quotas can change hands by simply signing an agreement giving somebody the right to graze when they never intend to exercise that right. It is no more than an agreement to transfer the quota. K

However, when the grazing agreement is analysed it does not suggest that the land is to be used in any way for milk production. Indeed the opposite is true. The right which the applicant has does not include a right to graze with a diary herd; it is a right to graze with beef cattle or sheep only. Thus in order to acquire the milk quota, even according to the ministry's argument, all that the applicant needs to do is to graze the land with some sheep. I inquired during the course of argument of counsel whether putting one sheep on the land for a day would be sufficient. After initial acceptance of that proposition counsel was, on instruction, more circumspect and contended that the use of the land would have to be more than *de minimis*.

Accepting that argument for the moment, nevertheless if a milk quota can be legally transferred to someone using the land for grazing L sheep the argument that these regulations are designed to prevent the ready transfer of milk quotas seems to me to be much weakened.

There is of course power in the Milk Marketing Board to prevent a transfer if it is to be "a sham". But if grazing by sheep for a sufficiently substantial period will enable a milk quota to be transferred that agreement could scarcely be described as a sham.

My attention was also drawn to the domestic regulations which have to be read subject to the European regulations. The Dairy Produce Quotas Regulations 1986 (DPWR) were made pursuant to section 2(2) of the European Communities Act 1972 to implement the relevant EEC legislation. Regulation 8 reads as follows:

(1) Where there is a change of occupation of the entirety of or a part of a M holding, the transferee shall within 21 days of that change of occupation notify the Minister in writing of the date of that change of occupation and the holding or part of a holding occupation of which has changed, and shall submit to the Minister such evidence relating to that change of occupation, and within such time, as the Minister may reasonably require . . .

(3) Where there is a change of occupation of part of a holding:

(a) the apportionment of quota shall be carried out as laid down in Schedule 4, Parts I and III . . .

(6) For the purposes of this regulation a change of occupation of a holding or part of a holding shall be treated as a case to which the first sentence of Article 5(3) of the Commission Regulation applies . . .

By section 2(1) Commission Regulation means the Commission Regulation 1371/84D. "Occupier" includes, in relation to land in

A respect of which there is no occupier, the person entitled to grant occupation of that land to another person. That definition, indicates that "occupier" means somebody in physical occupation otherwise it would be unnecessary to define "occupier" additionally as somebody who is entitled to grant occupation. In view of the definition provided in the Act itself it is not, in my judgment, very fruitful to consider how "occupation" has been interpreted in other statutes or situations. "Occupation" includes possession as its primary element but also includes something more. I find it difficult to hold that an intention to occupy which may never be performed is equivalent to "occupation". In the instant case it required the physical use of the land to some degree. In ordinary English parlance occupation and a right to occupy are two totally different concepts.

B Mr Paul Morgan QC, for the applicant, has mounted a formidable argument on the construction of the Council and Commission Regulations to which I hope I have done full justice. Like all matters of construction there is a limited scope for argument. I have come to the conclusion, though not without some hesitation, that the Commission Regulation 1371/84 article 5(2) is to be read as requiring the transferee to be operating the holding before he or she is entitled to the transfer of the milk quota. I find nothing inconsistent between the Commission Regulations and the Council Regulations. They are to be read together. The Commission Regulations are laying down detailed rules to supplement the general rules provided by the Council Regulations. Looked at in that light it does not seem to me
C there is indeed any conflict.

Both parties urged upon me that I should adopt a purposive approach to the interpretation of the regulations. My attention was drawn to the observations of the Advocate-General in *Posthumus* v *Oosterwoud* [1992] 2 CMLR 336 at p346 where he said:

> Accordingly, while it may well be natural for the parties in the first instance to seek to reach agreement on the amount of quota to be transferred, any such agreement must be based on the criterion laid down by the Community legislation or on such other criteria as have been laid down by member-States. Moreover, member-States must in my view, take appropriate measures to ensure that such agreements are consistent with those criteria. If member-States do not take such measures, then it is impossible to see how the requirements of the Community legislation can be satisfied. Such measures might also include the scrutiny of the terms of compensation agreed, in order
> D to ensure that such agreements do not amount to a disguised form of trading in quotas.

It seems to me that, whether the purposive approach is adopted or the more traditional approach of the English courts, nevertheless the same result is reached.

Finally, on this matter I deal with one further argument. It was submitted that if I were in doubt I should submit the matter to the European court under article 177 of the Treaty of Rome, more particularly if there were or appeared to be regulations in contradiction. To this end my attention was drawn to the well-known decision of *Customs & Excise Commissioners* v *Samex* [1983] 1 All ER 1042 where Bingham J, as he then was, at p1054 gave guidance as
E to the proper approach which courts of first instance should adopt in relation to community law. I have considered that decision and the arguments. In the light of the conclusion to which I have come, namely that the Community Regulations are sufficiently clear, I do not propose to seek the assistance of the European Court. In any event, if a different view is taken by another court there is no bar to the matter being sent to the European Court subsequently.

That is not however an end of the matter. The second question which arises is whether the minister either had the power to alter the register or should be allowed to alter the register.

This question is entirely governed by the Dairy Produce Quotas Regulations 1986. By regulation 8(1), to which reference has already been made, the transferee has to notify the minister in writing of the
F date of the change of the occupation and the holding to which the occupation relates and:

> . . . shall submit to the Minister such evidence relating to that change of occupation, and within such time, as the Minister may reasonably require.

Under regulation 8(7) the minister may provide such forms as he reasonably considers to be necessary for the purposes of this regulation.

Under regulation 30(2):

> For each region the Minister shall —
> (*a*) prepare a wholesale register entry in respect of each producer in that region. . .
> (*b*) maintain: —
> (i) wholesale . . . register for each region.

And by regulation 30(5): G

> The Minister shall amend the registers which he is required by this regulation to maintain to record any allocations or adjustments made under or by virtue of these regulations, and shall inform any person to whom an amendment relates and any purchaser affected by an amendment of that amendment.

The penalties are contained in regulation 36:

> Subject to paragraph (2), any person who:
> (a) fails without reasonable excuse to comply with a requirement imposed on him by or under these regulations or the Community legislation, or
> (b) in connection with these regulations or the Community legislation, makes a statement or uses a document which he knows to be false in a material particular or recklessly makes a statement or uses a document which is false in a material particular —
> shall be guilty of an offence and liable . . . H

Para (3):

> The Minister may, following any conviction under paragraph 1(b) against which there is no successful appeal, by notice served (within the period of 12 months . . .) . . . reduce his quota to such an extent as may reasonably be regarded by the Minister as being attributable to the falsehood on which the conviction was founded.

By Schedule 4, para 1:

> Subject to paragraphs 6 and 8 where there is a change of occupation of part of a holding the apportionment of quota relating to that holding shall be carried out —
> (a) where within 21 days of the change of occupation the transferee submits to the minister — J
> . . .
> (ii) a statement, signed by the transferor and the transferee, that they have agreed that the quota shall be apportioned according to areas used for milk production as specified in the statement.

Para 8 reads:

> Where the Minister reasonably considers that the areas used for milk production on a holding are not as specified in a statement made for the purpose of paragraphs 1(a)(ii) . . . he shall give notice of this fact in writing to the person who made the statement and the apportionment or prospective apportionment of that quota shall then be made by arbitration.

Both the applicant and HVF made applications on Form T2. The applicant's document is dated November 5 1986 and contained the following declaration: K

> 1. I/We confirm that I/we have/will occupied/occupy all/part of the holding stated in Part 1 of this form with effect from the date shown in Part 1.
> 2. I/We request that all/part of the quota registered in the name of the producer referred to in Part 1 of this form be transferred to me/us at the above address(es). . .
> 4. I/We understand that I/we are liable for supplementary levy which may become payable after I/we have taken possession of all/part of the holding stated in Part 1 of the Form.

There is an explanatory note which indicates that the Milk Marketing Board will be responsible for processing transfers.

It is suggested on behalf of the minister that the Milk Marketing Board was just going through a rubber-stamping exercise and because of the volume of work a decision to register was not a matter L which should be treated as "decision". I do not take such a view of the regulations. Regulation 8 indicates that the minister is to look at the matter in the light of the evidence submitted to him and may require further evidence before the registration.

It is submitted on behalf of the minister that where the parties notify the Secretary of State that there is a change of occupation an entry to that effect on the register is not a determinational decision that a change of occupation has been properly notified or indeed taken place. Further that as the minister has a duty under regulation 30 to maintain the register it is by necessary implication a continuing duty to maintain an accurate register. Alternatively, as he was informed by the parties that there was change of occupation when he discovers that information is incorrect he is entitled to revoke a M registration made on that basis.

The applicant makes these submissions. First, there is no power in the minister under the regulations to revoke the registration and in the absence of express statutory power it cannot be altered. Reliance was placed on the decision of *Re 56 Denton Road Twickenham* [1953] 1 Ch 51 at p56 where Vaisey J said:

> . . . where Parliament confers upon a body such as the War Damage Commission the duty of deciding or determining any question, the deciding or determining of which affects the rights of the subject, such decision or determination made and communicated in terms which are not expressly preliminary or provisional is final and conclusive, and cannot in the absence of express statutory power or the consent of the person or persons affected be altered or withdrawn by that body.

A I accept that proposition as well founded. How does it apply to the decision not to alter the register in 1988, but to alter it in 1991?

However, while it is true that it was an administrative decision by a minister, it is also clear the decision which has to be made at an early stage in the transaction may not reflect actuality and is necessarily provisional. Thus a transferee fills in the T2 saying:

I will occupy.

At that stage that may be the genuine intention. Based on that the Milk Marketing Board registers the transferee. Suppose the transferee falls ill and never does occupy the holding. It cannot be that the minister is debarred thereafter from adjusting the register to take account of the fact that the holding has never been occupied.
B Accordingly, I have no difficulty in deciding that it was open to the minister to alter the registration of October 1986.

However, the matter does not rest there. The letter from the ministry purporting to rescind the register entry is dated April 25 1991. I have set out the contents earlier in this judgment.

By their letter of July 11 1988, to which I have already referred, the ministry was making a decision that in the light of the information that there had been no occupation by the applicant they were nevertheless not going to alter the register because of the delay of HVF. That decision seems to me to come fairly and squarely within the principle laid down by Vaisey J and in other cases to which I was referred in *Administrative Law* by Professor Sir William Wade at pp253 to 257.

C Although the letter of July 11 1988 from the minister suggests that the HVF should seek arbitration initially, nevertheless it has not been argued before me that there has been some new fact, which emerged between July 11 1988 and April 1991, which entitled the minister to go back on a decision which had been made. It was always open to HVF at the end of the grazing licence to seek alteration of the register and thereafter if they were dissatisfied by the decision of July 11 1988 to challenge that decision. I am wholly unpersuaded either that the minister had the power to vary his decision of July 11 1988 or if he had the power it was a proper exercise of that power.

There is yet an additional matter which confirms me in this view. In the papers at p185, though not argued before me, are guidelines
D issued by the ministry which preceded the coming into force of the Dairy Produce Quotas Amendment Regulations 1988 under which the minimum interest in land was now to be a tenancy for at least 10 months. Para 9 at p190 reads:

The Dairy Produce Quotas Regulations say that the transfer of quota occurs when there is a change of occupation. The Community Regulations themselves are not completely clear on the extent to which a transferee needs to exercise his rights over the land transferred to him, but the Government takes the view that the transferee should have at least an exclusive right of occupation. Where no physical change of occupation takes place, it is likely that a full explanation of the circumstances will need to be provided to the Ministry. The more clearly it can be shown that a transferee has exercised his rights over the land acquired, the easier it is to justify the transfer of quota. This has always been the case, and is not changed by the amendments to the
E Regulations.

It is clear that what happened in this case, namely an agreement to graze with exclusive right of occupation, but without actual occupation, was common practice among those with milk quotas thereby transferring the quotas without apparent demur by the ministry. Both the applicant and HVF proceeded on the basis that that was sufficient as a matter of law. In the light of those guidelines I am not surprised.

It would be inequitable in any event in the light of that that this applicant should now be deprived of her quota with the consequence of having to find a suitable sum to cover the additional levy for which she has now become liable over the years.

F In the light of this judgment I shall invite the parties to make submissions on the form of the order I shall make.

County Court G

January 3 1992

(Before Mr Recorder BURGESS, sitting at Bodmin County Court)

CARSON v CORNWALL COUNTY COUNCIL AND ANOTHER

Estates Gazette January 23 1993

[1993] 03 EG 119

Agricultural holdings — Claim for compensation in respect of milk quota on termination of tenancy — Whether arbitrator erred in valuation method H

The following case is referred to in this report.

Wachauf v *Bunderesamt für Ernährung und Forstwirtschaft* (Case 5/88) [1991] 1 CMLR 328; [1989] ECR 2609

Paul Morgan (instructed by Horwood & James, of Aylesbury) appeared for the applicant tenant, Cyril David Carson (deceased), acting by his executor Midland Bank Trust Co Ltd; Nicholas Green (instructed by Burges Salmon, of Bristol) represented the first respondents, Cornwall County Council; the second respondent, John Edgar Pallett, the arbitrator, did not appear and was not represented. J

The applicant, Cyril David Carson, farmed an agricultural holding on an annual tenancy from the first respondents, Cornwall County Council, which ended on September 29 1987 as a result of a notice given by the applicant to the first respondents. The holding was then relet by the first respondents to another tenant. The applicant made a claim for compensation in respect of his part of the milk quota which he gave up. The claim was referred to arbitration and the second respondent, John Edgar Pallett, was appointed as arbitrator. The second respondent sat with a legal adviser and the parties agreed that the relevant quota for the purpose of Schedule 1 to the Agriculture Act 1986 was 210,040 litres; the standard quota was 158,792 litres and the tenant's fraction of the standard quota was K
103,849 litres . On December 10 1990 the second respondent made an interim award in the total sum of £26,469.40 based upon "method B", this method involves the capitalisation of the value of annual quota leasing, the second respondent preferring this method to the value which is obtained by "method C", which is based on sales with or without quota of freehold land with vacant possession. Method C produced a figure of £23,978.73, but the second respondent found the evidence given for this method to be rather light, whereas the evidence produced for method B had seemed to him to be stronger. Both parties appealed against the arbitrator's award.

After dismissing the landlord's appeal, **Mr Recorder Burgess** said: The tenant's appeal alleges that the arbitrator: (1) erred in law in L
declining to value the compensatable litreage of 103,849 litres in accordance with Part A (method A) of para 14 of the reasons for the award; (2) erred in law in accepting the legal opinion of Professor Usher as to the matters relevant to method A of para 14; (3) erred in law in declining to hold that any doubt as to the efficacy of the method A kind of transfers of quotas was already taken into account by persons bidding for quota in the open market and so the value of the quota was the price paid for it in the open market; or (4), alternative to (3), the arbitrator erred in law in failing to resolve the alleged doubt as to the efficacy of method A transfers of quotas and which should have been resolved by finding that those transfers or those pursuant to a *Gladstone* v *Bower*'s type of agreement were effective to transfer quota; and, finally, by (5) alternative to the M
foregoing that he erred in law in failing to give proper and adequate reasons for the award.

Ground 5 has found no favour whatsoever. The arbitrator was asked to state his reasons when he published his award and did so. The reasons and the legal opinion which accompanied it are contained in some 31 pages. It may be said that he could have said more about one aspect of matters rather than another (depending upon whom is reading the document), but it is clear that a great deal of thought has gone into the reasons and, upon any view, they are proper reasons and they are adequate reasons.

The essence of the tenant's appeal is that the arbitrator should have accepted method A because valuation by that method based on

A market forces produces a greater figure and thus the tenant is benefited. Method A is valuation by reference to sales or transfer of quota from one producer to another by transactions involving a grazing tenancy or a grazing licence. At para 13 of his reasons the arbitrator finds that the value he has to settle is to be assessed by reference to the extent to which the landlord is better off as a result of the extra quota (in this case 103,849 litres) contributed by the tenant's efforts at the date when the tenancy terminated. Mr Pallett there reminds himself that at that date the landlord has the option of vacant possession if so desired.

At para 14 he deals with method A valuation. The parties had agreed that the "open-market value" of the quota was 33p per litre at termination. His reasons discuss the deductions which either party
B had contended for and in the result, deducting 2.7p per litre for rent, he values the quota at 30.3p net, but then says that having considered the legal opinion of Professor Usher as to this method of valuation he has decided to disallow such a method. It is therefore necessary now to consider the advice given by the professor.

The advice contains a detailed summary of the background to milk quotas, which are entirely a creature of European Community law introduced in 1984 with the aim of reducing the then excessive output of milk and milk products in the countries of the European Economic Community. It imposed a punitive levy on producers (Mr Carson) or purchasers (the Milk Marketing Board in the United Kingdom), who exceed defined reference quantities. "Producer" is defined as a
C person farming a holding selling milk and a "holding" is defined as all the production units operated by the producer and located within the geographical territory of the community. Article 7 of para 1 of regulation 857/84 provides that where a holding is sold, leased or transferred by inheritance all or part of the corresponding reference quantity is to be transferred according to procedures to be determined. Eventually in *Wachauf* v *Bunderesamt für Ernährung und Forstwirtschaft* [1989] ECR 2609 it was held that the concept of a transfer "with comparable legal effects" to a lease to be taken to include an operation whereby a leased holding reverts to the lessor on expiry of the lease, as happened between the tenant and the landlord in this case and, further, that where the allocation of the quota was the result of the tenant's own labours, it would be a breach of the
D tenant's fundamental rights to deprive him of the fruits of his labour without compensation. But *Wachauf* was not concerned with valuation; that subject was considered, on regulations which came into effect at a later date, by the European Court of Justice in Case C121/90 *Posthumus* v *Oostewoud*. In his submission to the court Advocate-General Jacobs says at the conclusion of para 22:

It will then be for the National Court

— *viz* the county court in England and Wales —

to determine what amount of compensation for loss of quota is appropriate in all the circumstances

— and at the end of para 23 —

E Community law is concerned with the allocation of quota and the possible requirement of compensation for loss of quota, but not with compensation for other forms of loss or for unjust enrichment.

The Act lays down no bar to a smallholdings authority selling their land with vacant possession if they considers it meet to do so. Indeed, section 49(4) specifically reserves the authority's powers under section 165 of the Local Government Act 1933 to sell land and I, therefore, agree with Professor Usher that any increase in the land's value as the result of the tenant's efforts must be treated as being the same as the increase of value in the hands of any other owner of land. Thereafter, he reviewed the operation of markets in quota in the light of the arguments raised by both parties concerning transactions relating to the sale or transfer of grazing tenancies and grazing
F licences and what are now known as the *Gladstone* v *Bower*'s type of tenancy. In the result he doubted (a) whether grazing licences or G
tenancies had the effect of transferring quota by virtue of Article 7, para 1 of regulation 857/84 or (b) whether a licence or a tenancy for less than one marketing year could have the effect of transferring quota and (c) reminded the arbitrator that land which the producer could not use for milk production, in the way many of the short-term licences or tenancies prohibit, does not constitute part of the holding under regulation 857/84. The nub of the doubts expressed as (a) and (b) above might have been resolved by general evidence of entries in the register (or the absence of such entries) maintained by the board or the ministry. Evidence as to whether short-term quota transactions are or are not recognised for registration purposes will no doubt be produced on some other occasion, but that is not a H
matter for the court on this appeal.

The reasons for this legislation are well summarised in *Wachauf*'s case, where, at para 17 of the judgment, it is said:

Fundamental rights form an integral part of the general principles of the law, the observance of which is ensured by the Court . . .

(18) The . . . rights recognised by the Court are not absolute, however, but must be considered in relation to their social function. Consequently restrictions may be imposed on the exercise of those rights . . . so that they do not constitute, with regard to the aim pursued, a disproportionate and intolerable interference, impairing the very substance of those rights. Having regard to those criteria, it must be observed that Community rules which, upon the expiry of the lease, had the effect of depriving the lessee, without compensation, of the fruits of his labour and his investments in the tenant J
holding would be incompatible with the requirements of the protection of fundamental rights in the Community legal order.

At para 21 it is said:

However that conclusion does not preclude the possibility for a departing lessee to obtain compensation calculated on the basis of all or part of the relevant reference quantity when that is justified by the extent of the lessee's contribution to the building-up of milk production on the holding.

From this it is quite clear that the compensation one is considering is not concerned with sums paid by farmers or graziers, possibly in urgent need of additional grazing or pasture, who may be prepared to pay very much "above the odds" for short-term licences or grazing agreements or *Gladstone* v *Bower*'s type tenancies. The K
compensation is to go to those tenants who have been concerned with the building-up of milk production on the holding. In the light of the plain guidance there given by the court, it is clear that the arbitrator is not to be concerned with valuations which have been found by the method A type of valuation. He was therefore right to give little weight to the calculations he himself had ascertained and to "disallow this method of valuation" as he expressed himself in para 1 on p16 of the reasons for the award.

In those circumstances, reviewing para 2 of the tenant's grounds for appeal set out in his notice, I am satisfied that there were no errors of law as contended at subparas (1), (2), (3) or (4). As to (1) and (3), the arbitrator did not err in declining to value in accordance with method A. As to (2), the doubts that Professor Usher expressed were L
properly expressed, but, in my view, there is no doubt that the law is there to compensate for what is being surrendered "the building-up of the holding" — not to give, as Advocate-General Jacobs was to exemplify in the *Posthumus* case "unjust enrichment". Subpara 3 says that the value of quota was the price paid for it in the open market — that is true for that type of quota — but the type of quota here being valued is concerned with compensation for the "building up of milk production on the holding" — a different thing entirely. It is for that reason that the alternative ground in (4) cannot be sustained.

In all the circumstances, I find myself unable to accept the arguments put forward on the late Mr Carson's behalf and his appeal will therefore be dismissed. M

ARBITRATION

A Chancery Division

July 30 1992

(Before Mr Justice KNOX)

ARNOLD AND OTHERS v NATIONAL WESTMINSTER BANK PLC

Estates Gazette January 9 1993

[1993] 01 EG 94

Arbitration — Whether award of arbitrator should be
B ***remitted under section 22 of the Arbitration Act 1950 where later court decisions disclose that there had been an error of law — Whether time for application for remission should be extended***

In *National Westminster Bank plc* v *Arthur Young McClelland Moores & Co* (1985) 273 EG 402 Walton J held that under a lease of 7 Rolls Buildings, Fetter Lane, London EC4, the notional lease postulated by the rent review clauses did not itself contain a rent review provision; he declined to certify that a question of law was one which should be considered by the Court of Appeal. That decision was subsequently considered to have been wrong by the Court of
C Appeal and the House of Lords: see *Arnold* v *National Westminster Bank plc* [1990] 1 EGLR 137 and [1991] 2 EGLR 109. In the present proceedings the plaintiff tenants sought an order remitting the award of March 19 1984 by the arbitrator, as varied by Walton J on November 26 1984, pursuant to section 22 of the Arbitration Act 1950 on the grounds that it contained an error of law, and an order extending time for making such an application.

***Held*: The plaintiffs' applications were dismissed. The delay in making the applications was excessive. The path for the plaintiffs' remedy under section 22 of the 1950 Act was**
D **clear at the latest with the report of the decision in *Basingstoke and Deane Borough Council* v *Host Group Ltd* or in *Equity & Law Life Assurance Society plc* v *Bodfield Ltd*. There was a delay of three and three-quarter years before the issue of the originating motion, which could not be abridged. There is no suggestion in the authorities of any special provision for, or possibility of, remission on the discovery of an error of law. The error of law was not appealable and it would not be right to find a way around a limitation on appeals imposed by Parliament. Estoppel *per rem judicatam* applied. The changes made to appealing an arbitrators' award by the Arbitration**
E **Act 1979 did not impliedly enlarge the power of the court to remit under section 22 of the 1950 Act.**

The following cases are referred to in this report.

Basingstoke and Deane Borough Council v *Host Group Ltd* [1988] 1 WLR 348; [1988] 1 All ER 824; (1987) 56 P&CR 31; [1987] 2 EGLR 147; 284 EG 1587, CA
British Gas Corporation v *Universities Superannuation Scheme Ltd* [1986] 1 WLR 398; [1986] 1 All ER 978; (1986) 52 P&CR 111; [1986] 1 EGLR 120; 277 EG 980
Burnard v *Wainwright* (1850) 19 LJQB 423
Carl Zeiss Stiftung v *Rayner and Keeler Ltd (No 2)* [1967] 1 AC 853; [1966] 3 WLR 125; [1966] 2 All ER 536, HL
Elizabeth v *Motor Insurers' Bureau* [1981] RTR 405, CA **F**
Equity & Law Life Assurance Society plc v *Bodfield Ltd* (1987) 54 P&CR 290; [1987] 1 EGLR 124; (1987) 281 EG 1448, CA; *affirming* [1985] 2 EGLR 144; (1985) 276 EG 1157
GKN Centrax Gears Ltd v *Matbro Ltd* [1976] 2 Lloyd's Rep 555
Keighley, Maxsted & Co v *Bryan Durant & Co* [1893] 1 QB 405
Ladd v *Marshall* [1954] 1 WLR 1489; [1954] 3 All ER 745, CA
King v *Thomas McKenna Ltd* [1991] 2 QB 480; [1991] 2 WLR 1234; [1991] 1 All ER 653, CA
Montgomery Jones & Co Ltd v *Liebenthal & Co Ltd* (1898) 78 LT 406
Mutual Shipping Corporation v *Bayshore Shipping Co Ltd* [1985] 1 WLR 625; [1985] 1 All ER 520; [1985] 1 Lloyd's Rep 189, CA
National Westminster Bank plc v *Arthur Young McCelland Moores & Co* [1985] 1 WLR 1123; [1985] 2 All ER 817; [1985] 2 EGLR 13; (1985) 272 EG 171; CA; *affirming* [1985] 1 EGLR 61; (1985) 273 EG 402

G

This was an application by originating motion to remit an arbitrator's award under section 22 of the Arbitration Act 1950 determining a rent review under a lease held by the plaintiff trustees, Michael John Arnold, Peter Edwards and Simon Francis Phillips, from the defendant, National Westminster Bank plc, as trustees for British Rail Pension Trustee Co Ltd.

James Munby QC and Jonathan Gaunt QC (instructed by Freshfields) appeared for the plaintiffs; Hazel Williamson QC (instructed by Stephenson Harwood) represented the defendants.

Giving judgment, **KNOX J** said: This is the latest, and perhaps not the last, chapter in the history of attempts by the applicants, a firm of chartered accountants, whom I will call "the tenants", under what **H**
was originally a subunderlease, but which I will call "the lease", of premises at 7 Rolls Buildings, Fetter Lane, London EC4, to undo the effects of a decision of Walton J on November 26 1984, whereby he varied the award of an arbitrator, acting under a rent review clause in the lease, and held, first, that the notional lease postulated by the rent review clause should not itself be assumed to contain a rent review provision and, second, refused to certify that a question of general public importance was involved or that the question of law raised was one which, for some other special reason, should be considered by the Court of Appeal.

That decision has now been shown to have been wrong on both counts by decisions of the highest authority in the Court of Appeal **J**
and in the House of Lords.

The tenants since then have been understandably trying, with varying degrees of success, to undo its effects.

I take the facts from some of the fairly numerous judgments which have been given in the past by courts of first instance, the Court of Appeal and the House of Lords. The term granted by the lease ran from October 28 1976 to June 24 2008, and the rent was initially one of £800,000 pa. There were provisions for review, to take effect from the midsummer quarter day, in the years 1983, 1988, 1993, 1998 and 2004. The review procedure adopted involved what was called the "fair market rent", and defined as follows:

> . . . such amount as shall represent a yearly rent at which the demised premises might reasonably be expected to be let at the Relevant Review Date in the open **K**
> market by a landlord to a tenant without a premium with vacant possession and subject to the provisions of this subunderlease other than "the rent hereby reserved there being disregarded any effect on rent of any of the matters set out in paragraphs (a) (b) or (c) of section 34 of the Landlord and Tenant Act 1954 (as amended).

There was provision for arbitration if the parties failed to agree, as they did in relation to the rent payable from the midsummer quarter day in 1983.

There were a number of disputes involved in the arbitration, but what has proved to be the main subject of dispute was whether, under that definition of "fair market rent", which I read, the rent under the

A hypothetical lease had to be fixed on the basis that it contained the same provisions for rent reviews as were contained in the lease itself or, on the other hand, contained no provision for review.

The arbitrator to whom the matter was referred, by an award published on March 19 1984, decided that the hypothetical lease should be treated as containing review provisions and, on that basis, he fixed the rent at £1,003,000 pa. But he also found what the rent would be based on the other view, namely that there was no rent review clause in the hypothetical lease and that rent he found at a figure some 20% higher, that is to say £1,209,000 pa. So there was a difference, obviously, of £206,000 pa during the remainder of that review period of five years from midsummer 1983.

At that stage the landlords were the National Westminster Bank B plc, as trustees for British Rail Pension Trustee Co Ltd and PosTel Properties Ltd, which, as their names indicate, were pensions organisations for British Rail and the Post Office, respectively. They were beneficially interested in equal shares.

The matter proceeded on appeal before Walton J, who, on November 26 1984, gave a judgment varying the arbitrator's award. The order reads:

> It is ordered that the said interim award dated 19th March 1984 be varied by substituting therefor an award in the sum of £21,209,000 per annum, as set out in his alternative award at paragraph 10(8) of the said award, and the court does not think fit

— I pass over (1) —

> C (2) to grant a certificate, pursuant to section 1(7)(b) of the aforementioned Act, that this decision is one of general public importance

The aforementioned Act, of course, is the Arbitration Act of 1979.

In the course of his judgment, Walton J said (it is reported at (1984) 273 EG 402, at p402):

> The first and I think probably the main question which arises here is whether when the arbitrator is deciding what is to be the fair market rent for the next rent period, he is to do that upon the basis that the lease contains (as, of course, we know it does in fact contain) a rent review clause or whether, on the other hand, he is to fix the fair market rent on the basis that there is no such clause in the lease. I do not think that there is a presumption one way or the other. I think that in every case that must depend upon the precise terms of the lease, because it must be very much borne in mind that there is no such thing as D a fair market rent of any premises in the abstract. There is only a fair market rent upon a set of abstractions which may be actual, may be hypothetical and in most cases under rent reviews are a mixture of the one and the other. For example, here we have that the premises are expected to be let with vacant possession, which is, of course, something which we know as a fact just is not the case because the tenants are actually in possession and are certainly not just going to move out for the purpose of a rent review.
>
> So the attempt by Mr Morritt

— the tenant's counsel —

> to poison my mind in advance to achieve the lower of the two values by defining the fair market rent as something which favours the tenant, as in fact he was attempting to do, I do not think impresses me very much.

Walton J, having dealt with the question of whether there was a E presumption one way or the other, went on to find that the hypothetical lease should not be treated as containing a rent review provision and, as already mentioned, he refused a certificate under section 1(7) of the Arbitration Act 1979.

Before the order of Walton J that I have read was passed and entered, a second attempt to obtain a certificate under section 1(7)(*b*) was made to, and rejected by, Walton J.

On June 13 1985 the Court of Appeal dismissed an appeal from his decision on the ground that it had no jurisdiction to hear it.

There then ensued some correspondence, in the course of which Mr Arnold, on behalf of the tenants, wrote two letters, the first dated January 6 1986 and addressed to a gentleman at Folkard & Hayward, the estate agents and chartered surveyors who were acting for the F landlord, in the course of which Mr Arnold said:

> in the circumstances we (the tenants) have been looking for an alternative way of dealing with "the problem and, in particular, are contemplating an action for rectification. Our review shows good evidence to support this and I anticipate the writ will be issued during the course of January.

And 11 days later he wrote to a gentleman at Maxwell Batley, the landlords' solicitors, a letter in the course of which he said:

> I really would emphasise that, because of the long-term consequences to us, we do not feel able to allow any stone to be left unturned in reversing the effect of Mr Justice Walton's judgment. As previously indicated, a route of rectification is our most likely next course. However, you were kind enough to indicate that further expensive litigation might not be necessary and that you would discuss the matter further with your clients to see whether a sensible compromise might be worked out. I remain willing to discuss matters directly G with your clients, but obviously would like to do so sooner rather than later so that I may turn off some expensive legal meters.

On February 6 1986 the vice-chancellor, Sir Nicolas Browne-Wilkinson, gave his decision in *British Gas Corporation* v *Universities Superannuation Scheme Ltd* [1986] 1 WLR 398*, and set out his views regarding the correct approach to questions of construction of rent review clauses. He said at p403A:

> In these circumstances, there are in my judgment conflicting decisions as to the correct approach to the construction of these clauses. I am accordingly free to adopt the approach I prefer. In my judgment the correct approach is as follows: (a) words in a rent exclusion provision which require *all* provisions as to rent to be disregarded produce a result so manifestly contrary to H commercial common sense that they cannot be given literal effect; (b) other clear words which require the rent review provision (as opposed to all provisions as to rent) to be disregarded (such as those in the *Pugh* case, 264 EG 823) must be given effect to, however wayward the result; (c) subject to (b), in the absence of special circumstances it is proper to give effect to the underlying commercial purpose of a rent review clause and to construe the words so as to give effect to that purpose by requiring future rent reviews to be taken into account in fixing the open market rental under the hypothetical letting.
>
> I am conscious that such approach is perilously close to seeking to lay down mechanistic rules of construction as opposed to principles of construction. But there is an urgent need to produce certainty in this field.

Very nearly one year later two events occurred. First, the British Rail Pension Trustee Co Ltd, on March 5 1987, contracted to buy out the J half-share in the reversion vested in the other tenant in common. That contract contains a covenant, which operates to put the burden of any refund that eventually might be ordered in respect of the rents paid after the first rent review, should Walton J's decision be upset, upon the purchaser, British Rail Pension Trustee Co Ltd, by way of indemnity to PosTel Properties Ltd.

At almost the same time the Court of Appeal approved what Sir Nicolas Browne-Wilkinson V-C had said, which I have quoted just above, in the *British Gas* case, in a decision called *Equity & Law Life Assurance Society plc* v *Bodfield Ltd* (1985) 276 EG 1157†. Dillon LJ described them as "rough guidelines", but subject to that description he adopted them.

In those circumstances, the second lot of proceedings, entitled K "1987-A-No 3648", were instituted by the tenants. The specially endorsed writ contained a statement of claim which in the first 19 paragraphs set out the rectification claim, and it is not necessary for me to read that. Para 20 went on as follows:

> Further or in the alternative to their claim for rectification, the Plaintiffs seek a determination as to the basis on which those reviews are to be conducted and contend that on the true construction of the definition of "the fair market rent" set out in paragraph 5 above, the rent is to be reviewed on the basis of a letting for a term equal to the residue of the term unexpired at the relevant review date with provision for a rent review on 24th June 1993, 1998 and 2004.

There was inserted a "1988" in manuscript: nothing turns on this aspect of the matter.

L The plaintiffs', that is to say the tenants', claim for relief, besides rectification sought under para 3:

> In so far as may be necessary as a consequence of any such order for rectification, an order remitting the question of the rent payable with effect from 24th June 1983, alternatively the award of the Arbitrator as varied by the Judgment referred to in paragraph 7 above, to the Arbitrator, pursuant to section 22 of the Arbitration Act 1950.

The reference to "the judgment referred to in paragraph 7 above" is, of course, Walton J's judgment. The defence raised, among other things, laches as a basis of defence to the rectification claim. The landlords then applied to strike out para 20, by a summons dated October 1 1987, on the ground that it disclosed no reasonable cause of action.

M The matter came before Sir Nicolas Browne-Wilkinson V-C and he observed that:

> During the hearing, the parties agreed that I should treat as being before me a preliminary issue whether in the circumstances I have recited, the Plaintiffs are estopped by the judgment of Walton J from putting forward the case pleaded in para 20 of the statement of claim.

His decision on the subject, after a discussion of the authorities regarding issue estoppel, was as follows ([1989] 1 Ch 63, at p70):

> In my judgment a change in the law subsequent to the first decision is capable of bringing the case within the exception to issue estoppel. If, as I think, the

*Editor's note: Also reported at [1986] 1 EGLR 120.

†Editor's note: Also reported at [1985] 2 EGLR 144.

A yardstick of whether issue estoppel should be held to apply is the justice to the parties, injustice can flow as much from a subsequent change in the law as from the subsequent discovery of new facts. In both cases the injustice lies in a successful party to the first action being held to have rights which in fact he does not possess. I can therefore see no reason for holding that a subsequent change in the law can *never* be sufficient to bring the case within the exception. Whether or not such change does or does not bring the case within the exception must depend on the exact circumstances of each case.

Turning again to the circumstances of the present case, I have no doubt that justice does require the matter to be relitigated. The following factors are in my judgment relevant.

1. There is a continuing contractual relationship of landlord and tenant under which, if there is an issue estoppel, the decision of Walton J will regulate four further rent reviews and thereby affect the rent payable until the end of

B the term.

2. Because of the peculiarities of the procedure applicable to appeals from arbitrators, unlike the ordinary case of a prior decision by a judge, the decision of Walton J was not subject to appeal. Therefore a matter of a very great financial importance involving millions of pounds will, if an issue estoppel applies, be decided on a point of law which the lessees have never had the opportunity to test in the higher courts.

3. The decision whether or not to permit an appeal was the decision of Walton J himself and there was no right of appeal against his refusal to certify the matter fit for appeal. The lessees took every possible step to test the decision in the earlier case in the higher courts but without success.

4. Subsequent decisions, in particular that of the Court of Appeal in the *Equity & Law Life* case, 276 EG 1157, make it, at the lowest, strongly arguable that the decision of Walton J was wrong.

C And for those reasons he dismissed the application to strike out para 20 of the statement of claim.

The next significant event was that on November 3 1987 the Court of Appeal gave its judgment in *Basingstoke and Deane Borough Council* v *Host Group Ltd* [1988] 1 WLR 348*, in terms which supported the views as to the principles of construction in rent reviews that had been expressed by the vice-chancellor in the *British Gas* case and endorsed by the Court of Appeal itself in the *Equity & Law Life* v *Bodfield* case.

On October 31 1988 Gibson J refused to allow the construction issue to be taken as a second preliminary issue. During the course of that decision he said:

D The plaintiffs accept

— that, of course, is the tenants —

that the rectification point will in any event have to be tried if they are to recover the additional rent paid between 1983 and 1988. The total sum in dispute in respect of that past rent is a little over £1 million. So that while it is rightly said by Mr Gaunt that four-fifths of what is at stake in the litigation (in the sense of four out of the five rent reviews) would be conclusively determined if the plaintiffs won on both the issue estoppel point and the construction point, on any footing there is a substantial amount remaining in dispute.

The landlords appealed the decision of the vice-chancellor to the Court of Appeal, which in November 1989 dismissed the appeal. Lord Justice Dillon, in the course of the judgment, said at [1990] Ch

E 573 at p587 at letter F:

The simplest and oldest form of estoppel *per rem judiscatam* is what is now called "cause of action estoppel". That is referred to by Lord Reid in *Carl Zeiss Stiftung* v *Rayner & Keeler Ltd (No 2)* . . . where he said:

"There is a vast amount of authority on estoppel per rem judicatam. 'The object of the rule of res judicata is always put upon two grounds — the one public policy, that it is the interest of the state that there should be an end of litigation, and the other, the hardship on the individual, that he should be vexed twice for the same cause, (*per* Lord Blackburn in Lockyer v *Ferryman*) . . . And the general principle is clear that the earlier judgement relied on must have been a final judgment, and that there must be identity of parties and of subject matter in the former and in the present litigation."

One instance of cause of action estoppel . . . is in my view, pace the Vice-

F Chancellor, the decision of Jenkins J in *In Re Waring (No 2)* [1948] Ch 221. In that case a testator had by his will made in February 1939 bequeathed two annuities payable free of income tax. He had given directions to his trustees to set up an annuity fund and he died in 1940. The trustees took out a summons asking whether, in setting aside the fund, they ought to provide for the annuities at their face value, or whether they fell to be reduced under the wartime provisions of section 25 of the Finance Act 1941. Farwell J held in *In Re Waring* [1942] Ch 309 that section 25 did not apply, but his decision was reversed by the Court of Appeal, who held [1942] Ch 426 that section 25 did apply, notwithstanding that the testator had not died until 1940. The trustees accordingly appropriated a fund sufficient to pay 2029ths of the gross amount of the tax free annuities in accordance with the section. Only one of the anuitants had been a party to that decision and no representation order had been made. Subsequently the House of Lords, in a case between other G litigants, *Berkeley* v *Berkeley* [1946] AC 555, had held that *In Re Waring*

— in the Court of Appeal —

had been wrongly decided and therefore it followed that the section which replaced section 25 did not apply where the testator had died after the commencement of the war, which was the crucial date. Jenkins J held that in the case of the annuitant who was a party to the previous proceedings, the principle of res judicata applied and he was bound by the decision of the Court of Appeal that section 25 of the Act of 1941 applied to reduce his tax free annuity to the appropriate fraction of the full amount. The second annuitant, however, who had not been a party to the previous decision, and was not bound by any representation order, was entitled to claim retrospectively the full amount of her annuity. That is a case where the earlier decision had decided the precise point. There was no further point to litigate, although it H was suggested that the change in the statutory provision produced a further point.

Lord Justice Dillon's conclusion is found at p595E of the report, where he said:

If therefore the binding nature of an issue estoppel, where a point has been litigated in earlier proceedings, may be qualified on new facts of sufficient cogency being brought forward in a second action, why as a matter of common sense should it not equally be qualified as a result of new developments in law? If one has in mind Lord Upjohn's overriding consideration of working justice and not injustice, there can be no reason why not.

The Vice-Chancellor said in his judgment under appeal . . . :

"If, as I think, the yardstick of whether issue estoppel should be held to apply J is the justice to the parties, injustice can flow as much from a subsequent change in the law as from the subsequent discovery of new facts."

Respectfully, I entirely agree with him. I pray in aid also the observation of Cumining-Bruce LJ in *Property & Reversionary Investment Corporation Ltd* v *Templar*.

It is not enough for a litigant who has argued a point of law and lost, to say, in order to avoid issue estoppel, merely that it is arguable in the light of other decisions at first instance that the previous decision at first instance might have been wrong. He must go further. But the tenants here do go further. The Vice-Chancellor said, at p67:

"To put it no higher, there is a very substantial chance that if the rent review clause in this sub-underlease were now to be construed in the light of the Court of Appeal decision, the opposite result would be achieved."

K

And Lord Justice Dillon goes on:

That seems to be an understatement, but it is not necessary for us to reach a final conclusion. I merely draw attention to three points. (1) It is admitted that Walton J's approach was not in accordance with the guidelines since laid down and approved by this court. (2) There was nothing in this lease, properly construed, to constrain him to the extreme (and to my mind ridiculous) position that he seems to have taken of assuming a hypothetical lease with no covenant for payment of rent and no power of re-entry for non-payment. (3) His decision is contrary to commercial sense and the underlying commercial purpose of a rent review clause in that he has saddled the tenants with the rent for a term they do not have, that is to say a term at a fixed rent without reviews.

Staughton LJ said this on the issue of flexibility, having observed in passing that he regarded the words "the rent hereby reserved" in the L rent review clause as referring to the basic rent of £800,000 pa and to nothing else and that he had considered Walton J's reasons for reaching a different conclusion, but could not agree with him (see p596G):

Nevertheless that decision remains binding between the parties so far as concerns the rent payable between June 1983 and June 1988. The question is whether it is still binding from June 1988 onwards.

On flexibility in relation to issue estoppel and cause of action estoppel he said (at p597E):

So far as I recollect it was not argued that there is any decision which compels us to one conclusion or the other on this dispute. But a good many cases have touched upon it. As Miss Williamson, also for the defendants, pointed out, there are two lines of authority, and two views expressed by judges of great M distinction. Thus Lord Devlin in *Connally* v *Director of Public Prosecutions* . . . said: "Res judicata imposes a rigid bar and Wigram V-C's principle a flexible one."

And Lord Justice Staughton says:

I take this to mean that there is no exception in a case of cause of action estoppel or issue estoppel in the strict sense, but that there is flexibility in the case of non-issue estoppel.

And he went on to deal with the other line of authority which indicated that there might be flexibility in the second and third categories of what he called non-issue estoppel and issue estoppel in the strict sense. But there is no suggestion that there was any flexibility in cause of action estoppel.

*Editor's note: Also reported at [1987] 2 EGLR 147.

A His conclusion is to be found at p598D of the report in these terms:

> There then arises the question, which is equally difficult, whether such circumstances exist in the present case.

That is to say, circumstances where a matter can be reopened notwithstanding issue estoppel.

> As was conceded before us, it cannot by itself be enough that the previous decision was arguably wrong in law; nor, in my judgment, is it by itself enough that the previous decision was plainly wrong in law. The remedy for such errors is that provided by the appellate process. Where the legislature has chosen to limit the right of appeal, as in section 1(7) of the Arbitration Act 1979, it is not for the courts to circumvent that policy: see *Bynoe* v *Governor and Company of the Bank of England* [1902] 1 KB 467, 471, where Sir Henn Collins MR said:
> B "If we yielded to his argument, we should be constituting ourselves a Court of Appeal in a matter in which the legislature has thought fit to declare that there shall be no appeal."
>
> However, the remedy which the the defendants seek in this action is not wholly equivalent to what might have been granted in an appeal from Walton J if one had been permitted. The rent for 1983 to 1988 remains as Walton J determined it.

And he concluded that with the continuing contractual relationship the further elucidation of the law since Walton J's decision was sufficient to justify the decision to allow the plaintiffs to argue the construction of the lease anew. He said at p599A:

> But I do not say that the result would have been different if there had been no further elucidation of the law. I have difficulty in seeing why the plaintiffs C would have failed if the original decision had merely been plainly wrong (as I think it was), but should succeed now that subsequent courts have shown that to be the case. That difficulty may have to be resolved on another occasion.

Finally, Mann LJ expressed his conclusion at p600E in the report:

> I would not disagree with the Vice-Chancellor's finding of special circumstances in this case. I regret to say that Walton J's decision was in my view plainly wrong. If it were arguably wrong, I might have taken a different view as to whether there was a special circumstance.
>
> I wish to add only this. I was at one time during the course of argument troubled by *In Re Waring (No 2)* [1948] Ch 221. However, that case is distinguishable in my view, because it was a decision of the court which it was sought to reopen. Here there is no attempt to re-open the rent review for 1983 to 1988. The present question is as to a totally independent rent review and the D respondents to this appeal have expressly disclaimed any intent to upset the previous determination. Therefore *In Re Waring (No 2)* no longer troubles me.

It was pointed out, no doubt principally as a result of that observation which I have read from Mann LJ's judgment, by Mr Munby QC, who appeared for the tenants, that the rent review for 1983 to 1988 was indeed being challenged by the rectification claim in those very proceedings, and that elucidation was recorded as having been made clear to the Court of Appeal. In the result, it seems to me that the three judges of the Court of Appeal in that decision were unanimous in holding, first, that Walton J was wrong on construction — and they said that with varying degrees of emphasis, but they were all agreed on that point. Second, they were all agreed E that there was a need for exceptional circumstances for it to be possible to relieve a litigant from the consequences of issue estoppel. They all recognised the distinction between cause of action estoppel, which was inflexible, and recorded the fact that it was not sought to disturb the rent between 1983 and 1988 as a result of the issue estoppel argument as being a significant factor in their decision.

The next event chronologically was that a further application was made, by coincidence to me, for a certificate under section 1(7) of the Arbitration Act 1979, and I rejected that application in July of 1990.

Next there was an appeal to the House of Lords against the decision of the Court of Appeal, which on April 25 1991 was dismissed. Lord Keith gave the leading judgment and he said at p103 of the report, which starts at [1991] 2 AC 93*, having reviewed what had happened F in the past:

> It therefore appears that there are powerful grounds for the view that Walton J wrongly construed the rent review clause in the parties' lease and that he did so by virtue of an approach to the question of construction which was wholly incorrect in law.

In fact, he went somewhat further later in his speech, when at p11 he said:

> I am satisfied, in agreement with both courts below, that the instant case presents special circumstances such as to require the plaintiffs to be permitted to reopen the question of construction decided against them by Walton J, that being a decision which I regard as plainly wrong.

*Editor's note: *Arnold* v *National Westminster Bank plc* also reported at [1991] 2 EGLR 109.

He then, at p104D of the report, dealt with the difference between cause of action estoppel and issue estoppel. He said: G

> It is appropriate to commence by noticing the distinction between cause of action estoppel and issue estoppel. Cause of action estoppel arises where the cause of action in the later proceedings is identical to that in the earlier proceedings, the latter having been between the same parties or their privies and having involved the same subject matter. In such a case the bar is absolute in relation to all points decided unless fraud or collusion is alleged, such as to justify setting aside the earlier judgment. The discovery of new factual matter which could not have been found out by reasonable diligence for use in the earlier proceedings does not, according to the law of England, permit the latter to be re-opened. The rule in Scotland, which recognises the doctrine of res noviter veniens ad notitiam, is different: see *Phosphate Sewage Co Ltd* v *Molleson*. There is no authority there, however, for the view that a change in the law can constitute res noviter. H The principles upon which cause of action estoppel is based are expressed in the maxims nemo debet bis vexari pro una et eadem causa and interest rei publicae ut finis sit litium. Cause of action estoppel extends also to points which might have been but were not raised and decided in the earlier proceedings for the purpose of establishing or negativing the existence of a cause of action.

He then went on to consider the substance of the matter, and I can pass over that and come to his conclusion, after reading, first of all, the quotation from what Lord Reid had said in *Carl Zeiss Stiftung* v *Rayner & Keeler Ltd (No2)* [1967] 1 AC 853, at p917C. This is what Lord Reid said:

> The difficulty which I see about issue estoppel is a practical one. Suppose the first case is one of trifling importance but it involves for one party proof of facts which would be expensive and troublesome and that party can see the J possibility that the same point may arise if his opponent later raises a much more important claim. What is he to do? The second case may never be brought. Must he go to great trouble and expense to forestall a possible plea of issue estoppel if the second case is brought? This does not arise in cause of action estoppel: if the cause of action is important, he will incur the expense: if it is not, he will take a chance of winning on some other point. It seems to me that there is room for a good deal more thought before we settle the limits of issue estoppel. But I have no doubt that issue estoppel does exist in the law of England. And, if it does, it would apply in the present case, if the earlier judgment had been a final judgment of an English court.

Lord Keith stated the issue for their lordships in the following terms at p109E of the report:

> Your Lordships should appropriately, in my opinion, regard the matter as K entire and approach it from the point of view of principle. If a judge has made a mistake, perhaps a very egregious mistake, as is said of Walton J's judgment here, and a later judgment of a higher court overrules his decision in another case, do considerations of justice require that the party who suffered from the mistake should be shut out, when the same issue arises in later proceedings with a different subject-matter, from reopening that issue?

And his preferred solution was in the following terms, at p110D of the report:

> In the instant case there was no right of appeal against the judgment of Walton J because he refused to grant a certificate that the case included a question of law of general public importance. There can be little doubt that he was wrong in this refusal, as is shown by the large volume of litigation on the construction of rent review clauses and the decisions in that field which I have mentioned L earlier. I consider that anyone not possessed of a strictly legalistic turn of mind would think it most unjust that a tenant should be faced with a succession of rent reviews over a period of over 20 years all proceeding upon a construction of his lease which is wholly unfavourable to him and is generally regarded as erroneous. It is true that the chosen method of settling disputes about rent reviews, namely arbitration, involves that the right of appeal from the judge of first instance is subject to limitations. But these limitations were not present when the lease was entered into in 1976, having been introduced by the Arbitration Act 1979. There is much force also in the view that the landlord, if the issue cannot be reopened, would most unfairly be receiving a very much higher rent than he would be entitled to on a proper construction of the lease. The public interest in seeing an end to litigation is of little weight in circumstances under which, failing agreement, there must in any event be arbitration at each successive review date. Estoppel per rem judicatam, M whether cause of action estoppel or issue estoppel, is essentially concerned with preventing abuse of process. In the present case I consider that abuse of process would be favoured rather than prevented by refusing the plaintiffs permission to reopen the disputed issue.

And he agreed with what the vice-chancellor had said at first instance and the appeal was dismissed.

That was an end to the second litigation so far.

The third litigation is what is before me by an originating motion. The tenants ask for an order remitting the award of March 19 1984 by the arbitrator, as varied by the order of Walton J on November 26 1984, to the arbitrator, pursuant to section 22 of the Arbitration Act 1950, and, second, for an order extending the applicants' time for making this application.

A Finally, the last event that I should recite is that on March 24 1992 the construction issue was finally decided by Master Monroe in favour of the tenants, who now have the benefit of cause of action estoppel in relation to all rent reviews later than 1983 in favour of the contention for which they have so long struggled.

In my judgment, there are two categories of reasons why the tenants' application should fail in this case. One is short and the other significantly longer. The short one is that the delay seems to me excessive and such as I should not extend time for. I do this for the following reasons. In my judgment, the path for the tenants' remedy, if they have one by way of section 22, was clear, at latest, when the Court of Appeal gave its decision in *Basingstoke and Deane* v *Host* B *Group Ltd* or, at latest, when it was reported. The decision was on November 3 1987. *Weekly Law Reports* carried a report of it on March 4 1988 and *All England Reports* carried a report on April 1 1988. If that is not right, the decision of the Court of Appeal in *Equity & Law Life* v *Bodfield* in February 1987, reported in 1987 in *Estates Gazette*, provided the path to that remedy, if it exists.

The remedy sought, of course, was by the originating motion to which I have just referred, which was issued on December 5 1991. The time-limit that the rules lay down under Ord 73, r5, is 21 days. Subrule (1) reads:

> An application to the court —
> (*a*) to remit an award under section 22 of the Arbitration Act 1950

— I pass over (*b*) and (*c*) —

C
> . . . must be made, and the summons or notice must be served, within 21 days after the award has been made and published to the parties.

Even making all due allowances for the time it took to publicise the Court of Appeal's successive decisions, there has been very nearly three and three-quarter years' gap between — I take Lady Day 1988 as being the day by which the remedy was crystal clear, until the originating motion was actually issued. In my view, it is not an answer that the tenants were concentrating on the issue estoppel litigation. In a sense, it was incumbent on the tenants to inform the courts dealing with the issue estoppel of other possible remedies as potentially relevant circumstances. Indeed, Mr Munby very properly did just that when what fell from Mann LJ in the Court of Appeal suggested D that he did not perhaps have in mind that the rectification claims in the second set of litigation would, if successful, have the result of reversing the decision regarding the rent payable between 1983 and 1988. I should make it clear that I am not suggesting that at any stage the tenants, in saying what they have said in correspondence and in various courts regarding the situation concerning the rents payable in respect of the first review period from 1983 to 1988, were doing more than state accurately what they saw the ambit of the proceedings as then constituted. What they said has in every respect, so far as I am aware, been strictly accurate. But what I am saying is that the existence of the remedy by way of remission under section 22 of the Arbitration Act was a relevant factor which, if it had been perceived, would, I feel sure, have been referred to by the tenants in making their E applications to the successive courts, and in those circumstances it does seem to me that the three and three-quarter years' delay is greater than it would be right for me to bridge over by extending time, although undoubtedly I have jurisdiction, as I see it, to do so.

The longer reasons are as follows. Remission under section 22 of the 1950 Act derives from section 8 of the Common Law Procedure Act 1854. That read:

> In any Case where Reference shall be made to Arbitration as aforesaid the Court or a Judge shall have Power at any Time, and from Time to Time, to remit the Matters referred, or any or either of them, to the Re-consideration and Re-determination of the said Arbitrator, upon such Terms, as to Costs and otherwise, as to the said Court or Judge may seem proper.

F The remedies that the court had available for correcting errors before that 1854 Act was passed were as follows. As regards error of law, there was in general no power to correct, but if an arbitrator admitted a mistake, there was a power to set aside an award *in toto*, and, if an error on the face of the award occurred, there was also power to set aside the award *in toto*, but it was an all-or-nothing remedy and in practice proved to be somewhat too wholesale. So the practice grew up which is described in *Mustill and Boyd's Commercial Arbitration*, 2nd ed, at p438, in the following terms:

> During the first half of the nineteenth century it became the practice to remedy this gap in the jurisdiction of the Court by including in those submissions to arbitration which were made a rule of court an express provision that in the event of either of the parties disputing the validity of the award, or moving to set it aside, the Court should have power to remit the matters referred, or any of them, to the reconsideration of the arbitrator. The wording of this clause G made it plain that the purpose was to mitigate the rigour of the existing law by giving the Court an alternative remedy in cases where it would otherwise have set the award aside; and the courts did not as a rule interpret the clause as conferring a general discretion to remit on grounds other than those which would have sustained an application to set aside.

The 1854 Act, by section 8, which I have read, gave statutory force to that existing practice.

Section 5 of the same Act empowered arbitrators to state a special case for the opinion of the court.

The Arbitration Act 1889, by section 10, re-enacted part of section 8 of the Common Law Procedure Act 1854. Subsection (1) reads:

> In all cases of reference to arbitration the Court or a judge may from time to time remit the matters referred, or any of them, to the reconsideration of the H arbitrators or umpire.

That is a slightly condensed version of the first part of section 8 of the 1854 Act, and section 20 reproduced the latter part of that last section. Section 20 reads:

> Any order made under this Act may be made on such terms as to costs, or otherwise, as the authority making the order thinks just.

The power to state a case was preserved and there was introduced a power to state a consultative case by section 19, which read:

> Any referee, arbitrator or umpire may at any stage of the proceedings under a reference, and shall, if so directed by the Court or a judge, state in the form of a special case for the opinion of the Court any question of law arising in the course of the reference. J

The 1950 Act repealed the 1889 Act. Section 22 re-enacts section 10 of the Arbitration Act 1889, the power to remit, and section 28 re-enacts section 20 of the 1889 Act, both of which I have just read. The procedure by case stated was retained. As it stands now, section 22(1) reads:

> In all cases of reference to arbitration the High Court or a judge thereof may from time to time remit the matters referred, or any of them, to the reconsideration of the arbitrator or umpire.

Subsection (2) is not material.

The statutory power to remit is thus quite general in its terms and always has been since its creation in 1854 under the Common Law Procedure Act. K

In 1979 the Arbitration Act, by section 1, abolished the earlier systems for correcting errors of law. What it provided was this, in subsection (1):

> In the Arbitration Act 1950 . . . section 21 (statement of case for a decision of the High Court) shall cease to have effect and, without prejudice to the right of appeal conferred by subsection (2) below, the High Court shall not have jurisdiction to set aside or remit an award on an arbitration agreement on the ground of errors of fact or law on the face of the award.

Then subsection (2):

> Subject to subsection (3) below, an appeal shall lie to the High Court on any question of law arising out of an award made on an arbitration agreement; and on the determination of such an appeal the High Court may by order —
> (*a*) confirm, vary or set aside the award; or L
> (*b*) remit the award to the reconsideration of the arbitrator or umpire together with the court's opinion on the quest of law which was the subject of the appeal;
> and where the appeal is remitted under paragraph (*b*) above the arbitrator or umpire shall, unless the order otherwise directs, make his award within three months after the date of the order.

There was no amendment to section 22 of the Arbitration Act 1950, which remains as it was when first passed. Judicial interpretation over the years and practice has provided limitations on the very general power to remit, which appears to have become fairly restricted towards the end of the last century but has subsequently become somewhat enlarged.

One established basis for remission since even before the Common M Law Procedure Act 1854 has been the discovery of fresh evidence. At any rate, if it satisfies the quite stringent tests on the lines of (but not identical to) those laid down in *Ladd* v *Marshall* [1954] 1 WLR 1489, for the introduction of fresh evidence in appellate courts, it is very long established that there may be a remission on the basis of the discovery of fresh evidence. This was decided in *Burnard* v *Wainwright* (1850) 19 LJQB 423, which is, of course, before the 1854 Act, and that decision was followed and given effect to by the Court of Appeal in the *Keighley Maxsted & Co* v *Bryan Durant & Co* arbitration [1893] 1 QB 405, where Lord Esher MR said at p409:

> There have been many decisions upon the provisions relating to arbitration in the Common Law Procedure Act, 1854, and especially as to the right

A construction of s8, which gave power to the Court to remit the matters referred to the reconsideration of the arbitrator. It has been held that one effect of the Act was that it continued the ordinary law as to decisions by arbitrators, that is to say, that they are final and conclusive both of the law and of the facts, and that whether there has been a mistake either of law or fact the parties cannot by themselves set it up. Where, however, the submission contains a power to refer back to the arbitrator, if the party alleges that there has been a mistake on the arbitrator's part either of law or fact, the Court gives this effect to the power: that upon such a state of facts alone the decision cannot be questioned either on the law or the facts, the parties having chosen their arbitrator for better or worse; but that if the arbitrator himself informs the Court that he thinks he has made a mistake either of law or fact, and both he and the party approach the court, the court would send the matter back to him for reconsideration, although such a course will not be taken on the mere
B allegation of one of the parties. That was the law under the Act of 1854, and that is the view adopted in *Dinn* v *Blake* . . . , in which case it is said in terms that an award cannot be sent back to the arbitrator on the mere ground of mistake, the exceptions being where there has been corruption or fraud, where there is a mistake of law or fact apparent on the face of the award, and where the arbitrator himself admits that he has made a mistake. That law or rule of the Court, therefore, applies only where there is an allegation by the arbitrator that he has made a mistake of law or fact; and the case of *Dinn* v *Blake* . . . is no authority for anything beyond.

In the present case there is no allegation of any mistake by the arbitrators during the progress of the arbitration; but it is said that all the case was not [then] before them, further material evidence having since been discovered. Our decision to-day will apply only to cases where evidence has been discovered since the award, and we do not determine whether in other cases
C and under other circumstances the Court would be justified in sending back the award, as, for example, where the evidence was known to one of the parties, but kept back by him in order to deceive the arbitrator. The only questions here are, Has new evidence been discovered since the award was made, and is that a good ground for sending the award back, though the arbitrator has not asked us to interfere? Now, before the Act of 1854, this point was decided in *Burnard* v *Wainwright* . . . , and it was held that in such a case the Court had power to refer the award back to the arbitrator for reconsideration. In that case the submission contained a clause empowering the Court to remit the matters to the arbitrator, and it was therefore the same as all submissions are at the present day, for the effect of the Act of 1889 is that every submission must be taken to contain such a clause. Wightman J decided that if evidence were discovered after the decision of the arbitrator the Court might (though, of course, it would not necessarily in every case) remit the
D matter for reconsideration, although in that case there had been no request of the arbitrator to that effect. That being the law at that time, we ought to adopt it now.

And they did.

Other established categories where remission was allowed were cases of corruption or fraud or misconduct on the part of the arbitrator, error on the face of the award, an arbitrator requesting the correction of an admitted mistake of fact and law and procedural mishaps. There is no suggestion in the authorities of any special provision for, or possibility of, remission on the discovery by a subsequent decision of an error of law.

The decision in *Montgomery, Jones & Co Ltd* v *Liebenthal & Co (UK) Ltd* (1898) 78 LT 406 was described by Lord Donaldson MR in
E *King* v *Thomas McKenna Ltd* [1991] 2 QB 480, at p488, in the following terms:

In *In re Montgomery, Jones & Co* and *Liebenthal & Co* . . . , Mr Sanderson of counsel, appearing for the respondents, put forward the argument that the jurisdiction was limited to remission on four defined grounds. The applicants' case was based upon a complaint that the arbitrators had gone wrong on a point of law. Smith LJ accepted the argument of the respondents. But Chitty LJ, having referred to the *Keighley Maxsted* case, said, . . .

"It is not, however, now necessary to limit the operation of section 10 to those four grounds, for we are asked to remit the matter to the arbitrators solely upon the ground that they have gone wrong on a point of law. That, if it came within any part of the Abritration Act, would come within section 19 [the special case section] and [not] section 10."

F Collins LJ agreed with both judgments. This again . . . [said Lord Donaldson] . . . provides powerful persuasive authority, but, in the light of the fact that Collins LJ whilst agreeing with the view of Smith LJ also agreed with Chitty LJ that it was not necessary to decide the point, I do not regard the decision as in any way tying our hands.

And that enabled the Court of Appeal in *King* v *Thomas McKenna* to decide the case in the way in which they did.

The case was concerned with a misunderstanding by counsel as to the effect of her submissions in thinking, erroneously, that she had effectively asked for costs to be held over by the arbitrator, where in fact she had not effectively done so and, in those circumstances, there was an application under section 22 of the 1950 Act for the award to be remitted on the ground of a procedural mishap. At first instance the application was granted and the appeal before the Court of

G Appeal was dismissed. The case contains a valuable analysis by Lord Donaldson of the jurisdiction to remit. At p491C he said this:

In my judgment the remission jurisdiction extends beyond the four traditional grounds to any cases where, nothwithstanding that the arbitrators have acted with complete propriety, due to mishap or misunderstanding, some aspect of the dispute which has been the subject of the reference has not been considered and adjudicated upon as fully or in a manner which the parties were entitled to expect *and* it would be inequitable to allow any award to take effect without some further consideration by the arbitrator. In so expressing myself I am not seeking to define or limit the jurisdiction or the way in which it should be exercised in particular cases, subject to the vital qualification that it is designed to remedy deviations from the route which the reference should have taken towards its destination (the award) and *not* to remedy a situation in which, despite having followed an unimpeachable route, the arbitrators have
H made errors of fact or law and as a result have reached a destination which was not that which the court would have reached. This essential qualification is usually underlined by saying that the jurisdiction to remit is to be invoked, if at all, in relation to procedural mishaps or misunderstandings. This is, however, too narrow a view since the traditional grounds do not necessarily involve procedural errors. The qualification is however of fundamental importance. Parties to arbitration, like parties to litigation, *are* entitled to expect that the arbitration will be conducted without mishap or misunderstanding and that, subject to the wide discretion enjoyed by the arbitrator, the procedure adopted will be fair and appropriate. What they are *not* entitled to expect of an arbitrator any more than of a judge is that he will necessarily and in all circumstances arrive at the "right" answer as a matter of fact or law. That is why there are rights of appeal in litigation and no doubt would be in arbitration were it not for the fact that in English law it is left to the
J parties, if they so wish, to build a system of appeal into their arbitration agreements and few wish to do so, preferring "finality" to "legality", to adopt Lord Diplock's terminology.

The tenants' case for remission is that remission is (as that passage I have just read) not limited to procedural mishaps or misconduct but, the tenants submit, is available wherever justice cannot otherwise be done, and I was referred to what Lord Justice Stephenson said in *GKN Centrax Gears Ltd* v *Matbro Ltd* [1976] 2 Lloyd's Rep 555, at p576, in the second column, where the learned lord justice said:

I accept that this Court has statutory jurisdiction to remit an award (in the form of a special case) not only for error on the face of the award, mistake admitted by the arbitrator, misconduct by the arbitrator or additional evidence, but, in the words of Lord Justice Parker in *Universal Cargo Carriers*
K *Corporation* v *Citati* . . . "if justice cannot otherwise be done". But I would not go so far as Mr Justice Donaldson did and regard it as a discretion to remit any award "if it considers that there is a possibility of injustice being done by inadvertence": *Centrals Morska Importowa Eksportowa* v *Companhia Nacional de Navegacao.*

To send an award back to enable justice which has certainly not been done to be done (as Mr Justice Donaldson did in *The Aristides Xilas*) . . . is one thing; to send it back to make sure that justice which may possibly not have been done is done is quite another, and, as it seems to me, would, in the words of Lord Justice Willmer in *Tersons Ltd* v *Stevenage Development Corporation*:

". . . cut at the root of the whole purpose of arbitration, the basic idea of which is that the arbitrator's decision shall be final".

L Similarly, in *Mutual Shipping Corporation* v *Bayshore Shipping Co Ltd* [1985] 1 WLR 625, the headnote reads:

A dispute between the owners and charterers of the *Montan* was referred to a sole arbitrator. The arbitrator made an award under which the charterers were to pay the owners US $62,402.13. The arbitrator was not asked to give reasons for his award but, in accordance with the practice of London maritime arbitrators, he provided the parties with written reasons on the basis that they were not to form part of or to be used in any way in connection with his award. The document revealed to the charterers that the arbitrator had mistakenly attributed the evidence of the owners' expert witness to that of the charterers and vice versa, that it was the evidence of the charterers' expert witness which he had preferred and that he should have ordered the owners to pay the charterers US $27,527.87. The matter was brought to the attention of the arbitrator who wrote to the parties admitting the error. The charterers
M applied to the High Court for an order pursuant to section 22 of the Arbitration Act 1950 that the award be remitted to the arbitrator for reconsideration.

And Mr Justice Hobhouse granted the application and the Court of Appeal dismissed an appeal against that decision.

Sir John Donaldson MR (as he then was), at p632A, said:

The principal supervisory review powers of the English courts are contained in sections 22 and 23 of the Arbitration Act 1950. Section 23 empowers the court to set an award aside if the arbitrator has misconducted himself or the reference. Section 22 empowers the court to remit an award to an arbitrator for reconsideration. It provides the ultimate safety net whereby injustice can be prevented, but it is subject to the consideration that it cannot be used merely to enable the arbitrator to correct errors of judgment, whether on fact or law, or to have second thoughts, even if they would be better thoughts.

A And, finally, there is the decision in *Thomas* v *McKenna*, on which the tenants rely, of which I have already read an extract from the Master of the Rolls' judgment. The principal submission for the tenants was that an analogy between issue estoppel cases can be drawn with remission cases in a variety of aspects and that the latest development in the law of issue estoppel in the House of Lords in *Arnold* v *Westminster Bank plc* (this case) should be applied by analogy for the purpose of remission here, in that it can be said, under both issue estoppel and remission, that the principle of finality should, if its application involves a sufficient degree of injustice, bow to the demands of justice. Fresh evidence is capable of producing such a state of affairs if the stringent conditions regarding the admissibility of fresh evidence are met. So, it is argued, should the B discovery of new law. "New law" is an expression that is used in the special sense of a change in the generally received perception of what the law is as the result of a new decision, rather than a positive change in statutory law. It is just such new law that sufficed in *Arnold* v *Westminster Bank plc* to persuade the House of Lords to uphold the decision that issue estoppel should not, in the circumstances of this case, be treated as binding in regard to the rent reviews after the first. The tenants accept that mere error of law is not enough to grant an application for remission. There was here, they say, much more than mere error of law and indeed a double error of law, in that there has been a subsequent discovery of new law, in the sense in which I have described a moment ago, which establishes the critical error of law, C that is to say the error of construction.

There are, in my judgment, three reasons why the tenants' application should not succeed.

First, applying what was said by Lord Donaldson in *King* v *Thomas McKenna* in the passage which I have quoted, it seems to me that their application is indeed based not on an argument that the route which the reference took was incorrect, but that the destination that it reached was the wrong one as a matter of law. Furthermore, the error of law was that of Walton J and is one which is not appealable and, in my judgment, it would not be right, as was said in the *Bynoe* v *Bank of England** case, for the court to be astute to find a way round the limitation which has expressly been imposed by Parliament on the extent to which appeals can be made against errors D of law. Even on the wide interpretation by Lord Donaldson in *King* v *Thomas McKenna* of the jurisdiction to remit as not limited to the four classic categories, it seems to me that the tenants' case falls outside the ambit as formulated by Lord Donaldson.

Second, if the analogy between estoppel and remission is to be taken to its logical conclusion, it involves also importing the very clear distinction which, in my view, exists between estoppel *per rem judicatam* and issue estoppel. The latter is flexible in special circumstances, as the House of Lords' decision in *Arnold* v *National Westminster Bank* shows. The former is not, as has been emphasised in several passages (which I have quoted and need not repeat) in the decisions in the Court of Appeal and in the House of Lords in *Arnold* E v *National Westminster Bank*.

Third, before the 1979 Act was passed and removed the jurisdiction to remit on the grounds of error on the face of the award, I do not consider that the court would have gone outside the very well-established boundaries for the court's intervention where an error of law was relied upon, that is to say error on the face of the award or under the procedure by way of case stated and I do not consider that the removal of both those remedies by the 1979 Act and the substitution of a system of appeals impliedly enlarged the power of the court to remit under section 22 of the 1950 Act.

For those three reasons, it seems to me that, as a matter of substance, this application should fail.

Before parting with this case I should mention the decision in F *Elizabeth* v *Motor Insurers' Bureau* [1981] RTR 405. The headnote reads:

> At about 5.15 pm on the 2 December 1976 the appellant was riding his motor cycle along the A2 road in the direction of Strood when a van in front of him braked suddenly. The appellant also braked but his motor cycle ran into the back of the van injuring the appellant. The van driver drove away and remained untraced. On December 7 1976 the police took a statement from the driver of a car which was travelling behind the van and the motor cycle in which it was stated that the witness saw that "the van in front of the motor cycle appeared to brake suddenly". The appellant applied for compensation to MIB who rejected the claim stating that the evidence including that of the independent following car driver (who in October 1977 stated that he could not after nine months remember details of the accident and did not wish to be G bothered any further) indicated that if the appellant had "kept a correct distance and a proper lookout the accident would never have occurred". He appealed to an arbitrator under clause 11 of the Agreement. The arbitrator said that he found it "impossible to say that the [appellant] had proved on the balance of probabilities that the van driver did anything negligent which caused his accident" and that he could not allow the appeal or make an award. Stocker J refused to remit the case to the arbitrator under section 22 of the Arbitration Act 1950.
>
> The Court of Appeal, allowing the appeal, held:
>
> . . . that the arbitrator had erred in putting the burden of proof on the appellant to show that the untraced van driver had braked without good reason whereas the burden of proof was on the untraced van driver to prove that he had good reason for braking suddenly . . .; that in the absence of evidence from the van driver the appellant's version of the accident should H have been accepted . . .; and that, accordingly, the case should be remitted to another arbitrator for reconsideration under section 22 of the Arbitration Act 1950.

No cases on section 22 of the 1950 Act or its predecessors in the 1854 and 1889 Act were cited to the Court of Appeal.

That decision is, in my view, explicable on the basis that the Court of Appeal took the view that what happened was a procedural mishap or its equivalent. But, even if that is wrong, I do not regard that decision as sufficient to deter me from following the decisions of the Court of Appeal after full argument on the authorities on section 22, such as *King* v *Thomas McKenna*, which do lead to the conclusions that I have sought to express earlier in this judgment.

For those reasons, I propose to dismiss this application. J

*Editor's note: Reported at [1902] 1 KB 467.

Chancery Division

March 11 1993

(Before Mr Gavin LIGHTMAN QC, sitting as a deputy judge of the division)

TECHNO LTD v ALLIED DUNBAR ASSURANCE PLC

Estates Gazette June 5 1993

[1993] 22 EG 109 K

Arbitration — Landlord and tenant agreeing statement of facts including assumptions as basis for valuation — Whether arbitrator entitled to depart from agreed assumption without further reference to the parties

By a lease dated September 26 1991 the applicant holds a term of 25 years expiring in 2006 at an initial rent of £450,000 pa subject to review in 1991, 1996 and 2001. The demised premises consist of a warehouse of 99,000 sq ft consisting of eight bays, each of some 12,000 sq ft. The rent review clause provided for the determination of the annual rent per sq ft to L be multiplied by 99,032 and that the relevant rent per sq ft should be the annual rent at which a self-contained building in the immediate locality of the demised premises comprising approximately 11,000 sq ft of accommodation of similar standard of construction and of similar age at the relevant review date might reasonably be expected to be let in the open market. Following a reference to arbitration in relation to the rent review in 1991, on March 26 1992 the arbitrator gave directions and a timetable for the making of submissions and counter-submissions. The arbitrator invited the parties to clarify the main assumption to be made in assessing the rent at review. The parties' surveyors agreed a statement of facts M including an assumption that: "The property to be valued is situated in the immediate vicinity and is of a similar age, height, construction and layout to the existing property with a gross internal floor area of 11,000 sq ft . . .". In his award dated July 16 1992 the arbitrator, in dealing with the agreed assumption, noted that the lease was entirely silent as to "height, construction and layout", was non-specific as to such physical features requiring only the characterisation of the hypothetical premises and buildings not their slavish reconstruction to a gross internal area of 11,000 sq ft. The arbitrator formed the view that the tenant's surveyor had misdirected himself in limiting the characteristics of the hypothetical premises further than the realities of the formula in the lease required. The arbitrator awarded the relevant rent per sq ft at £7 pa. The applicant tenant contended that the arbitrator had misconducted himself in discarding the assumption agreed to by the parties' surveyors, the conduct being either an excess of jurisdiction or a breach of the principles of natural justice or procedural fairness, alternatively there has been a procedural mishap or misunderstanding which rendered it inequitable to allow the award to stand.

***Held*: The application was allowed and the award remitted for reconsideration as a whole by the arbitrator. The facts and assumptions agreed by the parties' surveyors were intended to be a legally binding contract, but it was a contract of a special character. On principle and as a matter of common sense the very nature of the contract requires as a special incident of such contract that the arbitrator should be entitled either to enforce it or to release the parties from it as the demands of justice require. If, for example, the statement of facts proves ambiguous or unworkable or leads to problems or consequences unforseen by the parties when entered into, or contains mistakes which subsequently come to light, good cause may exist for an application to the arbitrator to release the parties from it. The release, however, must be (if both parties do not agree in effect to the deletion of one provision only) a release from the agreement as a whole. The agreement is one which the tribunal may refuse to enforce if it is inequitable to do so applying the analogy of principles governing the enforcement of in- terlocutory consent orders. Before exercising his discretion not to enforce the agreement the arbitrator must give an opportunity to both parties to be heard on the issue. The arbitrator was not entitled, without any notice to either party, to treat one of the concessions as ineffective and to be ignored. The award was remitted in its entirety to the arbitrator with the direction, that after giving the parties an opportunity to adduce evidence or make submissions in respect thereof, the arbitrator consider and determine whether to require the parties to abide by the agreed statement of facts, and in particular the concession in issue, and to receive such further or revised evidence or submissions as each party to the arbitration may wish to present to him in the light of such determination.**

The following cases are referred to in this report.

Bahamas International Trust Co Ltd v *Threadgold* [1974] 1 WLR 1514; [1974] 3 All ER 881; [1975] EGD 1; (1974) 233 EG 47, HL
Boots the Chemists Ltd v *Pinkland Ltd* [1992] 2 EGLR 176; 28 EG 118
Clark (H) (Doncaster) Ltd v *Wilkinson* [1965] Ch 694; [1965] 2 WLR 751; [1965] 1 All ER 934, CA
Purcell v *Trigell* [1971] AC 358; [1970] 3 WLR 884
King v *Thomas McKenna Ltd* [1991] 2 QB 480; [1991] 2 WLR 1234; [1991] 1 All ER 653, CA

This was an application by the applicant, Techno Ltd, under sections 22 and 23 of the Arbitration Act 1950 to set aside, alternatively remit, an award of an arbitrator in rent review proceedings under a lease dated September 26 1991 the reversion to which was owned by the respondents, Allied Dunbar Assurance plc and others.

Hazel Williamson QC (instructed by Rabin Leacock & Lipman) appeared on behalf of the applicant; Paul Morgan QC (instructed by Nabarro Nathanson) represented the respondents.

Giving judgment, **Mr Gavin Lightman QC** said: This is an application by Techno Ltd ("the tenant") for an order: (1) that the interim award ("the award") made by Mr John Phillips [FRICS, ACIArb] as arbitrator ("the arbitrator") in an arbitration pursuant to a rent review clause in a lease dated September 26 1991 ("the lease") be set aside pursuant to section 23 of the Arbitration Act 1950; and/or (2) that the matters referred to in the arbitration (namely the determination of the relevant rent per sq ft of the demised premises) be remitted for reconsideration to the arbitrator pursuant section 22 of the 1950 Act. The tenant is content with an order for remission to the same arbitrator and, since any grounds entitling the tenant to an order to set aside the award must equally entitle the tenant to an order for remission, the tenant is content to limit the relief sought to the second head.

The lease is for the term of 25 years expiring in 2006 at the initial rent of £450,000 subject to review in 1991, 1996 and 2001. It is the initial review which is in progress — a review agreed to take place almost immediately after the lease was granted for reasons unnecessary to explain. The demised premises consist of a warehouse of some 99,000 sq ft consisting of some eight bays, each of some 12,000 sq ft. The relevant provisions of the rent review clause are contained in clause 4 of the lease which (so far as material) reads as follows:

> (A) . . . The annual rent payable from a review date shall be the higher of
> (i) a sum equal to the relevant rent per square foot (as that term is hereinafter defined) at that review date (fixed by agreement or by the surveyor as hereinafter mentioned) multiplied by 99032 being the floor area in square feet of the accommodation provided by the buildings standing upon the demised premises and
> (ii) the annual rent payable immediately before that review date
> (B) The relevant rent per square foot shall be the annual rent per square foot at which a self-contained building in the immediate locality of the demised premises comprising approximately 11,000 square feet floor area of accommodation . . . of similar standard of construction to the buildings now situate on the demised premises . . . and of similar age to such buildings but well maintained in all respects . . . at the review date might at the relevant review date reasonably be expected to be let in the open market between a willing tenant and a willing Landlord for a term of years equivalent to the residue then remaining unexpired of the said term or a term of Ten years (with upward only review at the expiration of every fifth year) whichever is the greater with vacant possession under a lease containing terms similar (*mutatis mutandis*) to the terms of this Lease.

Provision is made for the appointment of an arbitrator in default of agreement of the relevant rent per sq ft; and also for payment of interest on any difference between the existing and reviewed rent between the review date and the date of determination of the reviewed rent and its payment.

Mr Phillips was duly appointed arbitrator. On March 26 1992 at a meeting with the parties' surveyors he gave directions. As well as setting a time table for submissions and counter-submissions, following a reading of the rent review provisions he stated that it would assist him if the surveyors would clarify the main assumptions to be made in assessing the rent at review, and thus eliminate as many areas of dispute or uncertainty as possible and to clarify the basis of the valuation to be adopted by the arbitrator.

The parties' surveyors thereupon agreed a statement of facts and this statement of facts included in section 5 the following:

> 5.0 Assumptions to be made in assessing the rent at review
> . . . (iv) The property to be valued is situated in the immediate vicinity and is of a similar age, height, construction and layout to the existing property with a gross internal floor area of 11,000 sq ft and with all necessary parking, loading and unloading facilities.

The tenant's surveyor, Mr Andrew J Ilsley [ARICS], in his affidavit dated February 10 1993 deposes to the circumstances and manner in which the statement of facts come into being:

> 4. The Statement of Facts was agreed between Mr Waud [the landlord's surveyor] and me and Section 5 was agreed on the basis of what was appropriate: for example Mr Waud wished to include in Section 5 a reference to the height of the building which he considered to be a positive factor in favour of the [landlord]. Thus the Statement of Facts as signed by Mr Waud and me was a reasoned and considered document acceptable to both of us.

Both parties' surveyors on April 23 1992 sent their submissions to the arbitrator and on May 6 and 7 1992 their counter-submissions. In Mr Ilsley's submissions on behalf of the tenant, he placed assumption (iv) in section 5 of the statement of facts in the forefront of his case and in particular laid weight on the assumption regarding layout, which (on his construction of it) required the arbitrator to assume that the notional 11,000 sq ft building was a shrunken version of the demised premises. Mr Christopher Waud [ARICS], in his submissions, made no reference to this assumption and he merely invited the arbitrator to proceed on the basis that the notional building had a good clear working area, a characteristic of the demised premises. In his counter-submissions Mr Ilsley pointed out the failure of Mr Waud to refer to the assumption and reaffirmed its importance. In his counter-submissions, Mr Waud contented himself with saying that it was wrong to make this assumption as to layout. He did not expressly say the statement of facts or assumption

A (iv) did not or should not bind the parties or invite the arbitrator to proceed on this basis. Indeed it is possible to read his submissions as expressing his construction of the term "layout" in the concession and accordingly as adopting and giving effect to the concession.

After counter-submissions had been exchanged, the parties had the opportunity by letter to correct matters of fact, and this opportunity was taken advantage of, but no references made in their letters to this difference as to assumption as to layout, and neither the arbitrator nor either party raised this question or sought to resolve it at an oral hearing or by any other means.

In his award dated July 16 1992, the arbitrator dealt with this question as follows:

B *12 Description*

I fear that a good deal of eloquence has been wasted under this heading, in some of the submissions put forward having regard to the need to restrict all argument and opinion to the constraint of the Rent Review Clause and it may therefore be helpful if I dissect it hereunder, with my comments thereon in the light of the evidence . . . (f) ". . . of similar standard of construction to the Buildings now situate on the demised Premises . . ." Surprisingly, the evidence which has been submitted to me in the Statement of Agreed Facts, signed by both parties' Surveyors on the 23 April this year, attempts to vary the Rent Review Clause in Paragraph 5(iv) on page 4 thereof, in the following way — "The Property . . . is of a similar age, height, construction and layout to the existing Property . . ." The Lease is quite clear as to similar *standard* of construction and to similar age . . . but is entirely silent as to "height, construction and layout" and non-specific as to such physical features, C requiring only the characterisation of the hypothetical Premises and Buildings, not their slavish reconstruction to a gross internal area of 11,000 square feet. I have therefore formed the view that any argument or opinion evidence suggesting otherwise is not well founded and that the hypothetical Premises are required to be in context with their size and the other criteria of the Rent Review Clause.

In para 16 of his award the arbitrator stated that he had formed the view that Mr Ilsley had misdirected himself in limiting the characteristics of the hypothetical premises further than the realities of the formula in the lease required and had thereby persuaded himself towards a level of value below anything capable of evidential substantiation.

He then went on in section 17 to award the relevant rent per sq ft as D defined as at the rent review date of September 29 1991 at £7 pa.

The tenant's complaint against the award is that the arbitrator has misconducted himself in discarding the assumption agreed to in the statement of facts. It is said that this conduct either constitutes an excess of jurisdiction, his jurisdiction being restricted by the statement of facts, or a breach of the principles of natural justice or procedural fairness, in so far as the tenant was at no time given notice by the arbitrator that he had in mind discarding that assumption. Alternatively, it is contended that the course of events in the arbitration regarding the assumption constituted a procedural mishap or misunderstanding which rendered it inequitable to allow the award to stand and such a state of affairs (irrespective of any misconduct on the part of the arbitrator) was grounds for ordering a E remission: see *King* v *Thomas McKenna Ltd* [1991] 2 QB 480.

Miss Hazel Williamson QC, for the tenant, has argued that the statement of facts constitutes a package of terms agreed between the parties and constitutes a contract binding on them (and the arbitrator) like any other contract. She says that the arbitrator was not free to depart from them, still less without affording the parties an opportunity to address on this question; and that even if this procedure had been followed, after making a determination (in effect) to override the assumption, the arbitrator should have given both parties the opportunity to present their cases in the light of this ruling, since the existence of the concession certainly underlay the presentation of the tenant's case. Mr Paul Morgan QC, for the landlord, submitted that the statement of facts was no more than a F series of concessions which it was open to either party to withdraw, and that Mr Waud's submissions made clear that he did not regard the concession as operative. In any event Mr Morgan says that this is all a storm in a teacup and relief should be refused since in any event the arbitrator would have reached the same result.

Since underlying the tenant's complaint lies the disregard by the arbitrator of the assumption in the statement of facts and its status is in issue, I think that I am obliged to examine the legal nature and effect of such statement of facts. I am further encouraged to do by the knowledge that, if I direct remission, the arbitrator will be faced with the same question and that (for the saving of costs and perhaps a further application to the High Court) I ought to assist as best I can.

The parties have been unable to find any authority or text book affording guidance on this question, and I am left to approach it on G the basis of principle, analogy and common sense.

The statement of facts on the evidence before me was a package deal: both parties negotiated and agreed its terms in an endeavour to secure the more expeditious, economic and (no doubt as each party saw it) advantageous determination of the arbitration. Each party agreed the facts and assumptions in consideration of the like agreement by the other party and, as I see it (for I see nothing to indicate to the contrary), intended it to be legally binding. The agreement accordingly does constitute a binding contract. But since it is a contract intended to be only a tool in the conduct and determination of the existing arbitration, it is a contract of a special character. It is a contract which can be enforced or given effect to only in the arbitration, and it seems to me on principle and as a matter H of common sense that the very nature of the contract requires as a special incident of such contract that the arbitrator should be entitled either to enforce it or to release the parties from it as the demands of justice require. If, for example, the statement of facts proves ambiguous or unworkable or leads to problems or consequences unforeseen by the parties when entered into, or contains mistakes which subsequently come to light, good cause may exist for an application to the arbitrator to release the parties from it. So in this case if the arbitrator concludes that a concession agreed as part of a statement of facts intended to ease the task of the arbitrator has the opposite effect and creates unnecessary and unintended complications and difficulties, he may direct that the arbitration J proceeds untrammelled by it. Principles and common sense further require that in exercising his discretion the arbitrator must have regard to the reasons relied on by the party seeking its exercise: good reason must be required if effect is not to be given to a bargain deliberately made. Likewise attention must be given to the stage which the proceedings have reached: the later the stage, the greater the potential injustice to one party of allowing the other to be freed from the fetters in the litigation which he has voluntarily assumed. The release, however, must be (if both parties do not agree in effect to deletion of one provision only from the agreed statement of facts) a release from the agreement as a whole: otherwise the arbitrator would be asserting jurisdiction to substitute a new package deal for the old and to rewrite the parties' contract. K

There are two possible sources of the special incident. The first is to imply a term into the agreement that the obligations assumed are to continue in force only until the arbitrator determines otherwise.

The analogy which appears closest to this is the approach adopted by the Rules of the Supreme Court to the amendment or withdrawal of formal admissions (see Ord 27, r 2(2)) and to the withdrawal of concessions made in course of proceedings: see eg *H Clark (Doncaster) Ltd* v *Wilkinson* [1965] Ch 694 at p702. The leave of the court is required, the criterion is what is just, and what is just may depend on the extent to which the other party has proceeded to his prejudice in reliance on the concession or admission: see *Boots the Chemist Ltd* v *Pinkland Ltd* [1992] 28 EG 118. Leave will more L readily be given if the concession is one of law, and in particular if it is a question of construction of a document: see *Bahamas International Trust Co Ltd* v *Threadgold* [1974] 1 WLR 1514; but even as a question of construction for the purposes of pending litigation can be the subject of a binding agreement or compromise (eg in consideration of one of the parties not claiming rectification), so a concession on, or related to, an issue of construction can be the subject of a legally binding agreement.

The alternative route is to hold that the agreement is a subject to no such implication, but none the less is one which the tribunal may refuse to enforce if it is inequitable to do so. This would involve applying as an analogy the principles governing the enforcement of interlocutory consent orders: see *Thwaite* v *Thwaite* [1982] Fam 1 M and *Purcell* v *Trigell* [1971] AC 358.

I favour the latter approach and I have no doubt that the tribunal has a discretion whether to enforce the agreement on analogy with the like discretion whether to enforce interlocutory consent orders. In these circumstances it is unnecessary to decide whether there is to be implied a term in the agreement to the effect that it will continue in force only until the arbitrator decides otherwise. Indeed in view of the existence of the discretion vested in the arbitrator whether to enforce the agreement, it must be doubtful whether any such term need or can be implied in the agreement.

It therefore seems to me that the agreement is a binding contract, but that the arbitrator has a discretion not to enforce it. Before exercising this discretion, the arbitrator must give an opportunity to both parties to be heard on the issue and, after exercising his discretion (and in particular after exercising his discretion not to enforce it), must ensure that in the light of his decision both parties are given the opportunity to present their cases fully. In exercising this discretion, he must have regard to the fact that, unless both parties agree, he can refuse only to enforce the agreement in whole: he cannot select not to enforce only part, ie assumption (iv) or still less omit the word "layout". He should have regard to the prejudice to be occasioned to the tenants by a decision to jettison the assumption and perhaps statement of facts, but in this regard he may conclude that any prejudice can sufficiently be cured by allowing the tenant to represent its case fully on the new basis, ie in the light of the ruling and by appropriate orders as to costs.

In the present case the arbitrator, while generally treating the statement of facts as operative and effective, without any notice to either party treated assumption (iv) alone as ineffective and to be ignored. In my view, this he was not permitted to do. He was entitled to form the *prima facie* view that the concession involved and led to an untenable view as to the scope and effect of the rent review clauses; he was entitled to invite them to address him whether he should ignore that provision and, indeed, if objection was then taken to his overriding that provision alone, the statement of facts as a whole; and, in the light of the parties' submissions, he should have ruled on the question and given notice of his ruling; and then invited the parties to address him on what further opportunities should be given to adduce evidence and argument on the substantive issue in the arbitration in the light of his ruling. I have no doubt that in exercising his discretion he was entitled to have regard to any problems, whether in form of ambiguity or unintended results or otherwise, occasioned by the language of the concession. In the light of the evident weight and reliance given and placed by the tenant on the concession by the tenant's surveyor, natural justice required that Mr Ilsley be given the opportunity: (1) to make submissions to the arbitrator as to the course to be taken by him in respect of the concession and statement of facts as a whole; and (2) after any ruling precluding reliance on the concession to supplement his case to meet this critical change of circumstance.

Since the arbitrator did not follow this course, as it seems to me, the arbitrator was guilty of misconduct, and since I cannot be certain whether further evidence or submissions occasioned by the change in the ground rules might not have impacted on the award, I must remit the award to the arbitrator with appropriate directions.

It will be seen that I have decided this case on the basis of procedural irregularity by the arbitrator. In the case of a breach of the ground rules in the form of an agreed statement of facts, I prefer this analysis to that of an excess of jurisdiction involved in failing to give effect to a substantive provision in the arbitration agreement itself: *cf Couvela* v *Volkart* (1888) 4 TLR 209. It is implicit in my decision that in any event there was at the least a procedural mishap requiring a remission.

I shall order that the interim award be remitted in its entirety to the arbitrator and direct that, after giving the parties the opportunity to adduce evidence or make submissions in respect thereof, he consider and determine whether to require the parties to abide by the statement of facts and in particular section 5 assumption (iv); and to receive such further or revised evidence or submissions as each party to the arbitration may wish to present to him in the light of such determination; and that in making his award upon the "relevant rent per sq ft" he take account of such further or revised evidence or submissions. I shall further direct pursuant to section 13(2) of the Arbitration Act 1950 that the arbitrator may make his award later than three months after the date of this judgment.

Orders accordingly.

COMPULSORY PURCHASE AND COMPENSATION

A
Court of Appeal

October 26 1992

(Before Lord Justice LLOYD, Lord Justice Ralph GIBSON and Sir David CROOM-JOHNSON)

AQUILINA AND ANOTHER v HAVERING LONDON BOROUGH COUNCIL

Estates Gazette February 6 1993

[1993] 05 EG 139

B *Compulsory purchase compensation — Appeal from the Lands Tribunal — Whether decision of tribunal perverse as being made upon a basis of valuation not specifically relied upon by the claimants, alternatively upon a basis of valuation not put to the acquiring authority's expert valuer — Whether tribunal acted in breach of natural justice*

Following a reference to the Lands Tribunal by the claimants, PJ Aquilina and LJ Carberry, to have determined the proper compensation payable for the compulsory acquisition of a two-storey house used for a dental surgery at 71 Western Road, Romford, the tribunal awarded the sum of £717,000. Before the tribunal the claimants' valuer advanced
C three valuations of which one, valuation A, in the sum of £720,000, assumed office use of the existing building and the construction of new offices. The acquiring authority's valuer spoke to a primary valuation on the same basis as valuation A in the sum of £365,000. The tribunal in rejecting the claimants' valuer's alternative valuations B and C adopted a further valuation, referred to in the appeal as valuation D, based upon a free-standing scheme of development on the subject land and rates per sq ft of site area derived from the sale of a site for such redevelopment in the same road. The tribunal decided that it would have been reasonable to expect planning permission on a similar basis to be forthcoming for the subject
D site. On appeal by way of a case stated, the acquiring authority contended that the tribunal acted in breach of natural justice by making the award upon a basis of valuation not relied upon by the claimants and failing to put the basis of valuation to the acquiring authority's expert valuer. They were not given an opportunity to deal with valuation D as a separate and distinct way of putting the case.

Held: **Valuation D was not distinctly or separately put forward by the claimants; there was no direct reference to it in any part of the claimants' valuer's evidence at any stage as a separate basis of valuation, and nothing with reference to it
E was put to the acquiring authority's expert witness. The tribunal, in its expert judgment, was free to prefer valuation D as a better approach for the doing of justice between the parties but, if the tribunal thought it right to apply valuation D, it was necessary to ensure that the parties had a proper opportunity to deal with it. The acquiring authority was entitled, until given reason to do otherwise, to limit their questioning, evidence and submissions to the case put forward by the claimants based as it was upon elaborate expert evidence and legal submissions by counsel. If the tribunal had wished to proceed on valuation D the member should have** F
raised the matter at the hearing or called the parties back for further consideration. The appeal was allowed and the matter remitted to the tribunal for rehearing before another member.

The following cases are referred to in this report.

Fairmount Investments Ltd v *Secretary of State for the Environment* [1976] 1 WLR 1255; [1976] 2 All ER 865; (1976) 75 LGR 33, HL

R v *Paddington and St Marylebone Rent Tribunal, ex parte Bell London & Provincial Properties Ltd* [1949] 1 KB 666; [1949] 1 All ER 720; 65 TLR 200; 47 LGR 306

Sabey (H) & Co v *Secretary of State for the Environment* [1978] 1 All ER 586; (1977) 245 EG 397

Westminster Renslade Ltd v *Secretary of State for the Environment* (1983) 48 P&CR 255; [1983] JPL 454 G

This was an appeal by case stated from the decision made by Mr T Hoyes FRICS in the Lands Tribunal on June 7 1991 ([1991] 2 EGLR 209; [1991] 41 EG 143 and 42 EG 139) upon a reference to determine the compensation payable on the compulsory acquisition by the acquiring authority, Havering London Borough Council, of a building owned by the claimants, PJ Aquilina and LJ Carberry.

Philip Walter (instructed by Clifford Chance, for the solicitor to Havering London Borough Council) appeared for the appellants; Timothy Straker (instructed by Hunt & Hunt & Houghtons, of Romford) represented the respondents.

H
Giving the first judgment at the invitation of Lloyd LJ, **RALPH GIBSON LJ** said: This is an appeal by case stated from the decision made by Mr T Hoyes FRICS, a member of the Lands Tribunal, on June 7 1991 upon a reference to determine the compensation payable on the compulsory acquisition by Havering London Borough Council of a building used for dental surgeries, at 71 Western Road, Romford. The award was the sum of £717,000 together with an appropriate surveyor's fee and costs.

The case stated, under section 3(4) of the Lands Tribunal Act 1949, is dated June 15 1992, and the facts proved or admitted in the contentions put forward by the parties are set out in the written decision. The question of law for the decision of this court turns upon one point, namely that in the submission of the council, the tribunal J
acted in breach of natural justice by making the award upon a basis of valuation not relied upon by the claimants and to deal with which the council had no proper opportunity.

The land had upon it a two-storey house. The gross-internal floor area was 1,999 sq ft. To the rear was a single-storey annex with a gross-internal area of 401 sq ft. Planning permission was likely to be granted for a change of use from surgeries to offices and for a new single-storey office building of 2,490 sq ft gross-internal area, with associated car parking upon the rear part of the site. The valuation date was November 21 1988.

Certain matters were agreed. Those which need to be mentioned were (this can be seen at pp 8-9 of the decision): K

> There existed in November 1988 in Romford a market both for occupational leases of offices and for the freehold purchase of modest areas of office space. Both markets were at their highest, or peak, in late 1988. The 2,400 square feet of existing space on the subject site [which was] 12,565 square feet [in all] left latent potential for the construction of a new single-storey office building to the rear of 2,493 square feet; in effect 50 per cent of the site was available for new development for which planning permission could have been reasonably expected.

Matters upon which the valuers differed were noted by the tribunal as follows: first, the capital value per sq ft to attribute to the existing buildings at the material date so as to reflect physical, locational and other attributes; next, the capital value to attribute to any new offices assumed for valuation purposes to be constructed on the rear of the site as part of the process of arriving at site value; and, last, whether it was appropriate, in all the circumstances prevailing in Romford in November 1988, to consider the value of the subject premises for office redevelopment in conjunction with adjoining sites and attribute a proportionate part of the value of a larger site to the site of 71 Western Road.

Mr Parish, FRICS, who had advised and prepared the case for the claimants, had submitted a proof of evidence in which he advanced three valuations. They were briefly described by the tribunal as follows at pp 9-10 [[1991] 2 EGLR 209 at p211G]:

> Mr Parish spoke to three valuations of the subject premises.
>
> *Valuation A* . . . in the sum of £720,000, assumed office use of the existing buildings (about £520,000) and the construction of new offices on the rear half of the site (about £200,000 being £31.83 per sq ft of site area);
>
> *Valuation B*, in the sum of £840,000 . . . assumed redevelopment for offices of the total site in conjunction with other sites . . .
>
> *Valuation C*, in the sum of £lm . . . assumed redevelopment likewise but in conjunction with a local-authority-owned car park to the south-west.

It is to be noted that valuation A was claimed in addition to the sum claimed for disturbance, the largest sum claimed being £161,336, as to part of which there was agreement.

Mr Burton FRICS, who was called on behalf of the council, did not accept valuations B or C as valid bases of valuation. The tribunal agreed with Mr Burton and neither need be considered in further detail. Mr Burton agreed with the approach of Mr Parish under valuation A, but reached a smaller sum for compensation. The tribunal said of that as follows:

> Mr Burton spoke to a primary valuation . . . on the same basis as valuation A by Mr Parish, in the sum of £365,000; £300,000 . . . in respect of the existing building and 50% of the site and £65,000 . . . for the rear half of the site.

After a passage in which the tribunal considered the evidence relating to 87 Western Road, which had appeared in the papers as a comparable, and then concluded that evidence set the upper limit of value for the freehold interest and possibly total compensation (and to that passage I shall return shortly), the tribunal next considered the two valuations based upon the method applied by Mr Parish for valuation A. The tribunal said:

> . . . these valuations proceeded upon a common approach and embraced two elements: (a) the value of the existing buildings and the front part of the site (for all practical purposes 50% of the total) and (b) the value of the residue for the construction of 2,493 sq ft of new single-storey offices. I take the latter element first, for which the valuations were: Mr Parish £200,000 . . . and Mr Burton £65,000 . . . both promulgated on the residual basis but involving considerable diversity in the data employed and the detailed mode of calculation adopted.

The tribunal then reviewed the differences between the two experts and concluded with reference to (b) (at pp 12-13) [[1991] 2 EGLR 209 at p 212L]:

> Valuation A by Mr Parish falls to be reduced by £56,092 . . . £199,692 minus £56,092 = £143,600 . . . This figure, which I round up to £145,000 I regard as realistic when broadly tested against the two transactions at 89 Western Road.

The member then turned to (a), the first part of valuation A, the value of the existing surgery building as offices. He rejected Mr Parish's valuation of £520,800. He accepted Mr Burton's valuation of £300,000, but increased it by 10% because of his view of the progress of the market. He concluded (this is at pp 14-15) [1991] 2 EGLR 209 at p 213C]:

> The value of the land taken is therefore determined in the alternative at:

(a) The existing buildings and their site	£330,000
(b) The backland (50% of the site area)	£145,000
	£475,000
Less the agreed payment for title insurance	£1,000
Total	£474,000

Although the tribunal had thus fixed the proper amount of the valuation A claim at £474,000, and had rejected valuations B and C, the sum awarded was different. The sum of £474,000 with the full amount of £161,336 claimed for disturbance (of which £54,628 was in dispute) amounted to £635,336. The tribunal did not consider the outstanding issue with reference to compensation because in its view it did not arise: the alternative award for the land taken in the sum of £717,000 exceeded the valuation A claim together with the compensation.

I return now to the passage in the decision in which the tribunal determined the "alternative award". It was in effect in a passage which was a continuation of the reasons for rejecting valuations B and C. It reads (at pp 10-11 of the decision) [[1991] 2 EGLR 209 at p 212E] as follows:

> It was accepted by Mr Burton that in terms of a location for offices there was no difference material to value between 71 and 87 Western Road. The evidence of Mr Heady was clear that no 87 had been sold for redevelopment by offices and that planning permission had been granted for 6,000 sq ft on the site in isolation, namely as a free-standing scheme. In my judgment, it would also have been reasonable to expect . . . planning permission on a similar basis to be forthcoming for the site of 71 Western Road. Mr Parish included in his submitted documents two such schemes and in answer to my question he indicated that were the site to be redeveloped for offices the expected density would be equal to that approved at no 87.
>
> It is agreed that 87 Western Road realised, in late 1988, £57.14 per sq ft of site area for a free-standing development by offices. I prefer this direct evidence of site value in the locality close to the valuation date to any figure derived from residual valuations promulgated upon the assumption of implementing a scheme of redevelopment jointly with neighbouring owners. A purchaser of 71 Western Road would not put a joint project entirely out of mind, but would take comfort from the ability to undertake a free-standing scheme and would accordingly prefer to base any offer on the less risky and more immediate proposition.
>
> In re-examination, Mr Parish adopted the value for a free-standing office development, using the same unit price as 87 Western Road, as a check upon his valuation A, namely 12,565 sq ft at £57.14 = £717,964, say £718,000. I consider Mr Parish was misguided in seeking to substantiate his valuation A at £720,000 by adopting the achieved development value. In my judgment £718,000, less £1,000 agreed in respect of title insurance, sets the upper limit of value for the freehold interest and possibly total compensation, subject to what follows in relation to valuation A (Mr Parish) and . . . the claim for disturbance.

That basis of claim has been called, in argument, "Valuation D".

The point of law raised in the case is at p 20 of the bundle, and set out at some extent, but I now need read only para 3, which is as follows:

The questions upon which the decision of the . . . court is desired are:
(3) Whether the decision was perverse as being made upon:
(a) the general evidence of value before the tribunal;
(b) a basis of valuation not specifically relied upon by the claimants;
(c) a basis of valuation not put to the expert valuer called to give evidence by the acquiring authority.

In opening the matter, it was said on behalf of the council that reliance was placed only upon (b) and (c). Reference has been made to the decision of the House of Lords in *Fairmount Investments Ltd* v *Secretary of State for the Environment* [1976] 1 WLR 1255. In that case it was held that it was contrary to natural justice for the Secretary of State to confirm the order on a basis of facts as to the inadequacy of the foundations, which the owners had no opportunity of showing was erroneous, and of an opinion, with which they had no opportunity to deal, that the state of the foundations made rehabilitation impractical. I quote from the speech of Lord Russell of Killowen at p 1265 at E:

> For the Secretary of State it was in substance submitted that the question of stability and settlement had been raised, and therefore Fairmount should reasonably have anticipated all possible causes by leading evidence directly as to the state of the foundations: consequently it was not right to say that Fairmount had been deprived of the opportunity, as would have been the case had the inspector at the hearing cast doubt on the foundations but refused to hear evidence designed to correct his doubt.
>
> This submission for the appellant appears to me to be at the core of this appeal, and, my Lords, I do not, in the circumstances of the case that I have rehearsed, accept it. I entirely accept that such an inspector, in a case such as this, is not merely trying an issue or issues between the local authority and the objector owner, and may from his professional experience supply deficiencies in the case as presented by the local authority. I equally accept that he is not bound to accept as established a contention in evidence for the objector owner simply because it is not, or is not adequately, challenged or contested on the part of the acquiring authority at the hearing. Part of his function lies in his own knowledge of the subject. Nor would I wish to introduce into procedures such as this — which include, prior to report, his inspection of the site — a rigidity more appropriate to a private issue to be decided by a judge; and in that connection I do not believe that a "view" by a judge is to be equated in any way with such a site inspection by such an inspector, a possibility which (it was said) might have been in the mind of the Court of Appeal in the instant case.
>
> But in this case I am unable, consonant with the essential principles of fairness in a dispute, to uphold this compulsory purchase order. All cases in which principles of natural justice are invoked must depend on the particular circumstances of the case. I am unable, my Lords, in the instant case, to generalise. I can only say that in my opinion, in the circumstances I have outlined, Fairmount has not had — in a phrase whose derivation neither I nor

your Lordships could trace — a fair crack of the whip. A passage at p 682 of *Rex* v *Paddington and St Marylebone Rent Tribunal, ex parte Bell London & Provincial Properties Ltd* [1949] 1 KB 666 on being taken by surprise is of relevance here.

I would only add two points. The first is the suggestion that the inspector should perhaps have kept silent about his views or inference as to the foundations. With this I wholly disagree: it was not suggested by Fairmount. What he should have done was either to reconvene the hearing, or to invite the department to do so: or, in a relatively straightforward case such as this, have in writing invited views on his provisional conclusions as to the foundations and financial feasibility. There is nothing either in the statute or in any rules to prevent this.

It should be noted that at the commencement of the hearing there was reference to an application to call fresh evidence before the court, directed, as I understand it, to the question of prejudice to the council. In the event, the argument proceeded in a way which rendered it unnecessary for that application to be pursued, and I say nothing further about it. Reference was also made to *R* v *Paddington and St Marylebone Rent Tribunal, ex parte Bell London & Provincial Properties Ltd* [1949] 1 KB 666 and *H Sabey & Co Ltd* v *Secretary of State for the Environment* [1978] 1 All ER 586.

The case for the appellant council was that they were not given an opportunity to deal with valuation D as a separate and distinct way of putting the case. They acknowledge that it was open to the tribunal to proceed on the basis of valuation D, if, in the member's expert view, it had been necessary so to do, but in that case they contend that they should have been given an opportunity by the member to call any evidence with reference to it and to make submissions upon it.

Mr Straker, for the claimants, in responding to the appeal, has acknowledged the applicability of the principles stated by Lord Russell in *Fairmount Ltd*, referred to above, to these proceedings before the Lands Tribunal, but has submitted that there had been no breach of natural justice in what occurred. The member made the following points, among others. First, that there was clear reference before the hearing to development of the site as a whole by itself, not, as in valuations B or C, with the property of others. In the proof of Mr Parish, the member pointed out, there had been reference to two schemes, and there was mention of that in the passage in the decision to which I have referred, at pp 10-11, as there having been two such schemes included in the submitted documents. Mr Straker pointed out that those were not variants of valuations B or C. They were part of the contention, apparent at p 28 of the bundle, in the proof of Mr Parish, that valuation A was not only a sound basis for valuation but should be accepted in the figure put forward by him. In particular that contention included reference to pp 61 and 62 of the bundle, on which can be seen drawings, the two on the right-hand side of p 62 showing a proposed development of the entire site by itself including the demolition of the existing buildings, that being, of course, the basis of valuation D.

Mr Straker contended that the member of the tribunal, as an expert, was entitled to act upon a basis of valuation put forward by neither side at the hearing, and that there was before him sufficient evidence to justify the decision which he reached. There was reference to *Westminster Renslade Ltd* v *Secretary of State for the Environment*, a decision of Forbes J of May 9 1983. Next Mr Straker argued that the council in fact had sufficient opportunity to deal with valuation D, particularly since, in re-examination, Mr Parish had referred to it as a check valuation for affirming the validity of his approach to valuation A. In proceedings of this nature, it was said, it is very commonly the case that the decision of the tribunal is on ground not contended for by either side. Lastly it was said that the onus of proof, with reference to an alleged breach of natural justice, lies upon the party asserting that it has occurred.

For my part, I have no doubt that the main contention for the council has been made out. Mr Straker acknowledged that the case put forward for the claimants based on valuations A, B and C was that advanced at the hearing. Valuation D was not distinctly or separately put forward. There was no direct reference to it in any part of Mr Parish's evidence at any stage as a separate basis of valuation, and nothing with reference to it was put to Mr Burton, the council's expert witness. It is clear that the tribunal, in its expert judgment, was free to prefer valuation D as a better approach for the doing of justice between the parties, but, if the tribunal thought it right to apply valuation D, it was necessary to ensure that the parties had proper opportunity to deal with it. The council did not have such opportunity merely because they could have asked questions about the references at pp 60 and 61, and could have directed evidence to destroy or modify any conclusions which might otherwise be based upon that material. The council was, I think, entitled, until given reason to do otherwise, to limit their questioning, evidence and submissions to the case put forward by the claimants based as it was upon elaborate expert evidence and legal submissions by counsel. As was pointed out in argument by my lords, the passage at p 28 in the proof of Mr Parish was introduced and controlled by the sentence, "The value of the existing building coupled with the development potential of the extensive backland", which was, of course, as has been made clear, the basis of valuation A.

If the tribunal had wished to proceed on valuation D, the member should have raised the matter at the hearing or called the parties back for further consideration. Further, the reference by Mr Parish to valuation D as a check valuation in re-examination did not, in my judgment, provide any proper opportunity to the council to deal with it as a separate basis of claim. They were dealing with valuation A and on that they largely succeeded.

For my part, I would therefore allow the appeal and remit the matter to the tribunal for rehearing. It clearly should be heard by another member. That is no reflection whatever upon Mr Hoyes. I am sure he would prefer not to be required to deal further with the matter. If the claimants wish to put forward valuation D on the rehearing, it is common ground that on the facts of this case they should be free to do so, subject to complying with the requirements of the rules, and any directions of the tribunal. The undecided issue as to the sum due for disturbance, if relevant, will also require to be considered.

Lloyd LJ and **Sir David Croom-Johnson** agreed and did not add anything.

Appeal allowed with costs; matter remitted to the Lands Tribunal for expedited rehearing by a different member; application for leave to appeal to the House of Lords refused.

DAMAGES

Court of Appeal

April 7 1993

(Before Lord Justice DILLON, Lord Justice STEYN and Lord Justice ROSE)

SURREY COUNTY COUNCIL AND ANOTHER v BREDERO HOMES LTD

Estates Gazette June 26 1993

[1993] 25 EG 141

Damages — Breach of covenant — Whether covenantees entitled to damages related to additional profits made by covenantor following breach

By a contract dated November 26 1980 the defendant agreed to purchase two parcels of land of some 12.33 acres from the plaintiffs, Surrey County Council and Mole Valley District Council, at a price of £1.5m. The contract was completed by two transfers dated January 2 1981, each containing covenants in a similar form that the defendant would commence development of the land in accordance with a planning permission for the erection of 72 houses issued by the second plaintiff on December 11 1980 and would diligently pursue the development complying with the planning permission. On June 29 1983 the second plaintiffs granted the defendant a further planning permission related to 3.64 acres of the land, which had the effect of increasing the number of permitted houses from 72 to 77. The defendant completed the development in accordance with the later planning permission and not in accordance with the first permission. The plaintiffs, who did not seek an interlocutory injunction, or an injunction at trial, restraining breach of covenant, claimed damages for breach of covenant related to the additional profits made by the defendant. That claim was rejected by Ferris J who awarded nominal damages. The plaintiffs appealed.

Held: **The appeal was dismissed. The remedy at common law for a breach of contract is an award of damages, and damages at common law are intended to compensate the victim for his loss, not to transfer to the victim, if he has suffered no loss, the benefit which the wrong-doer has gained by his breach of contract. Accordingly, the plaintiffs were only entitled to nominal damages and not to damages related to the additional profits earned by the defendant as a result of the breach of the covenant. The decision in *Wrotham Park Estates Co Ltd* v *Parkside Homes Ltd* was not applied because in that case Brightman J refused to grant an injunction requiring the removal of houses erected in breach of a covenant and damages were awarded in equity under Lord Cairns' Act.**

The following cases are referred to in this report.

Anglia Television Ltd v *Reed* [1972] 1 QB 60; [1971] 3 WLR 528; [1971] 3 All ER 690, CA

Bracewell v *Appleby* [1975] Ch 408; [1975] 2 WLR 282; [1975] 1 All ER 993; [1976] EGD 190; (1974) 237 EG 731

British Westinghouse Electric & Manufacturing Co Ltd v *Underground Electric Railways Co of London Ltd* [1912] AC 673; (1912) 81 LJKB 1132; 107 LT 325

Carr-Saunders v *Dick McNeil Associates Ltd* [1986] 1 WLR 922; [1986] 2 All ER 888; (1986) 53 P&CR 14; [1986] 2 EGLR 181; 279 EG 1359

Griffiths v *Kingsley-Stubbs* Unreported June 3 1986

Hadley v *Baxendale* (1854) 9 Exch 341

Jegon v *Vivian* (1871) LR 6 Ch App 742

Johnson v *Agnew* [1980] AC 367; [1979] 2 WLR 487; [1979] 1 All ER 883; (1979) 38 P&CR 424; [1979] EGD 969; 251 EG 1167, HL

Lever v *Goodwin* (1887) 36 ChD 1

Penarth Dock Engineering Co Properties Ltd v *Pounds* [1963] 1 Lloyd's Rep 359

Phillips v *Homfray* (1871) LR 6 Ch App 770

Robinson v *Harman* (1848) 1 Exch 850; [1843-60] All ER Rep 383; 18 LJ Ex 202; 13 LTOS 141

Stoke-on-Trent City Council v *W&J Wass Ltd* [1988] 1 WLR 1406; [1988] 3 All ER 394; (1988) 87 LGR 129, CA

Strand Electric & Engineering Co Ltd v *Brisford Entertainments Ltd* [1952] 2 QB 246; [1952] 1 All ER 796, CA

Tito v *Waddell (No 2)* [1977] Ch 106; [1977] 2 WLR 496; [1977] 3 All ER 129

Walford v *Miles* [1992] 2 AC 128

Wallington v *Townsend* [1939] Ch 588; [1939] 2 All ER 225

Watson, Laidlaw & Co Ltd v *Pott, Cassels, Williamson* 1914 SC HL 18; [1914] 31 RPC 104

Whitwham v *Westminster Brymbo Coal & Coke Co* [1896] 2 Ch 538

Wrotham Park Estate Co Ltd v *Parkside Homes Ltd* [1974] 1 WLR 798; [1974] 2 All ER 321; (1973) 27 P&CR 296

This was an appeal from the decision of Ferris J, who on November 21 1991 awarded the plaintiffs, Surrey County Council and Mole Valley District Council, nominal damages in a claim against the defendant, Bredero Homes Ltd, for breach of covenants in transfers of land.

Sir William Goodhart QC and Bernard Weatherill (instructed by Howell-Jones & Partners) appeared for the appellants; Colin Rimer QC and Nicholas Peacock (instructed by Turner Kenneth Brown) represented the defendant.

Giving the first judgment, **DILLON LJ** said: This is an appeal by the plaintiffs, Surrey County Council and Mole Valley District Council, against a decision of Ferris J given on November 21 1991, after the hearing of issues directed by an earlier order. By his decision the learned judge awarded the plaintiffs nominal damages against only the defendant, Bredero Homes Ltd, for breaches of virtually identical positive covenants contained in transfers by the plaintiffs to the defendant of certain land in Surrey in 1981.

The facts are straightforward and not in dispute. I can take them from the judgment of the learned judge.

In 1980, Surrey County Council and Mole Valley District Council were respectively the registered proprietors with absolute title of two adjoining parcels of land lying to the west of the Ridgeway, Fetcham, Leatherhead, Surrey. The total area of the two parcels was some 12.33 acres. The land had originally been acquired by the councils or their predecessors for road purposes but, by 1980, it was no longer required for these purposes and the two councils decided to act together in offering the entire site for development as a housing estate.

The defendant made an offer which included a purchase price of £1.5m. This led to a contract, dated November 26 1980, between the two plaintiffs as vendors and the defendant as purchaser. This contract was completed by two transfers both dated January 2 1981 — one being by Mole Valley District Council and the other by Surrey County Council. The transfer made by Mole Valley District Council contained a covenant by the defendant with that council:

> To commence the development of the land hereby transferred in accordance with the planning permission issued by Mole Valley District Council,

A reference MO/80/1214 and dated 11th December, 1980, within six months from the date hereof and thereafter to diligently pursue the development of the land hereby transferred to its completion complying with the said planning permission.

The transfer by Surrey County Council was in substantially the same terms. The planning permission MO/80/1214, there referred to ("the first planning permission") was granted on the application of the defendant. It provided for the development of the land, ie both parcels taken together, by the erection of 72 detached bungalows and houses in six different designs according to a layout as shown in certain drawings.

The defendants started work on the development in accordance
B with the first planning permission. There were certain minor alterations by way of modification of that planning permission, but these do not matter.

A bit later, however, the defendant applied for and obtained, on June 29 1983, from Mole Valley District Council as the planning authority, a planning permission which has the effect of raising the total number of dwellings to be built on the estate from 72 to 77, the main changes being a reduction in the number of four-bedroom houses and an increase in the number of three-bedroom houses — the size and arrangement of plots being amended, but the layout of roads and verges remaining unchanged. This planning permission (the "later planning permission") was given the reference number MO/83/0368 and it related to 3.64 acres of the land.

C The defendant then completed the development of those final 3.64 acres in accordance with the later planning permission and not, as had been covenanted, in accordance with the first planning permission.

The plaintiffs object to the development of the final 3.64 acres, in accordance with the later planning permission, not on planning grounds, but on the grounds that the later permission enabled the defendant to build 77 rather than 72 houses and bungalows on the two adjoining parcels. There is no objection on planning grounds, not surprisingly, since it was Mole Valley District Council which granted the later planning permission.

The objection is on the ground that it is a more profitable planning permission for the defendant than the first planning permission,
D because more houses can be built. As a legal basis it is said by the plaintiffs and conceded by the defendant that in building 77 houses in all under the later planning permission, rather than 72 under the first planning permission, the defendant has acted in breach of the covenants in the transfers.

The plaintiffs therefore seek damages. They have never sought an interim injunction to restrain the defendant from developing the land otherwise than in accordance with the first planning permission. They never sought an injunction at the trial requiring the defendant to pull down the completed houses. They recognised that there was never any practical possibility of such an injunction being granted. There was a formal amendment of the relief sought at the trial to raise
E in form a claim for an injunction, but that was not pursued. The plaintiffs have merely sought damages which have been described as "damages at common law", as opposed to "damages in equity under Lord Cairns' Act". The plaintiffs accept that they have not suffered any damage at all of the nature of damage to adjoining property owned or occupied by them. What they claim as damages is essentially the profit made by the defendant by breaking the covenants and building 77 houses and not just 72 — or, since the plaintiffs wish to be modest in their demands in putting forward a somewhat revolutionary development of the law of damages, such a part of the profit as would reflect the reasonable premium that the defendant should have paid them for contractual permission by way of relaxation of the covenants to build the 77 houses rather than 72.

F Indeed, the plaintiffs say, and I have no reason to doubt, that their sole purpose in imposing the covenants at all — to commence and pursue the development to its completion in accordance with the first planning permission — was that the defendant would have to apply for and pay for a relaxation if it wanted to build anything more.

It is, of course, clear that had the contracts been worded otherwise there could have been provision for the payment by the defendant of an additional price of a specified amount or fixed by an appropriate formula for each extra house or bungalow, if they or their successors in title built more than 72 houses or bungalows on the land within a specified period, but that is not the contract that was made.

In putting forward the claim for damages with which we are concerned, the plaintiffs rely very strongly on the decision of Brightman J in *Wrotham Park Estate Co Ltd* v *Parkside Homes Ltd* G
[1974] 1 WLR 798, to which I shall have to come.

The starting point, however, in my judgment is that the remedy at common law for a breach of contract is an award of damages and damages at common law are intended to compensate the victim for his loss, not to transfer to the victim, if he has suffered no loss, the benefit which the wrong-doer has gained by his breach of contract. Thus it is stated in *Chitty on Contracts* 26 ed, vol 1, para 1771:

Damages for a breach of contract committed by the defendant are a compensation to the plaintiff for the damage, loss or injury he has suffered through that breach.

Similarly Viscount Haldane LC, said in *British Westinghouse Electric & Manufacturing Co Ltd* v *Underground Electric Railways* H
Co of London Ltd [1912] AC 673 at p689:

The fundamental basis is thus compensation for pecuniary loss naturally flowing from the breach.

Lord Wilberforce said, in *Johnson* v *Agnew* [1980] AC 367* at p400H:

The general principle for the assessment of damages is compensatory.

Each of these three statements is accompanied by a statement to the effect that the innocent party is to be placed, so far as money can do so, in the same position as if the contract had been performed. That follows the wording of the statement of the rule of the common law by Baron Parke in *Robinson* v *Harman* (1848) 1 Ex 850 at p855. J
That rule has been referred to in argument in the present case as the "conventional" rule.

Sir William Goodhart QC, for the plaintiffs, has pointed out that the conventional rule is not of universal application in that there are cases in which the plaintiff is awarded not what is required to place him in the same situation as if the contract had been performed, but what is required to recoup to him the expenditure which he has incurred which has been wasted because the contract has not been performed: see, for instance, *Wallington* v *Townsend* [1939] Ch 588 and *Anglia Television Ltd* v *Reed* [1972] 1 QB 60.

The principle is still compensation for loss. The difference is merely that there are cases where the contract has so palpably not been performed at all that it would be unreal to assume that it had K
been performed and impossible to calculate damages on such an unreal assumption.

Every student is taught that the basis of assessing damages for breach of contract is the rule in *Hadley* v *Baxendale* (1854) 9 Ex Ch 341, which is wholly concerned with the losses which can be compensated by damages. Such damages may, in an appropriate case, cover profit which the injured plaintiff has lost, but they do not cover an award to a plaintiff who has himself suffered no loss, of the profit which the defendant has gained for himself by his breach of contract.

In the field of tort there are areas where the law is different and the plaintiff can recover in respect of the defendant's gain. Thus in the field of trespass it is well established that if one person has, without L
leave of another, been using that other's land for his own purposes he ought to pay for such user. Thus, even if he had done no actual harm to the land he was charged for the user of the land. This was applied originally in wayleave cases where a person had without authority used his neighbour's land for passage: see, for instance, *Jegon* v *Vivian* (1871) LR 6 Ch App 742 and *Phillips* v *Homfray* (1871) LR 6 Ch App 770.

The same principle was applied where the defendant had trespassed by tipping spoil on the plaintiff's land: *Whitwham* v *Westminster Brymbo Coal & Coke Co* [1896] 2 Ch 538.

The same principle was applied to patent infringement by the House of Lords in *Watson Laidlaw & Co Ltd* v *Pott Cassels* M
Williamson [1914] 31 RPC 104. The infringer was ordered to pay by way of damages a royalty for every infringing article because the infringement damaged the plaintiff's property right, that is to say his patent monopoly. So in a case of detinue the defendant was ordered to pay a hire for chattels he had detained: *Strand Electric & Engineering Co Ltd* v *Brisford Entertainments Ltd* [1952] 2 QB 246.

Those cases do not apply in the present case as the defendant has made no use of any property of either plaintiff.

The cases have been taken still further in some fields of tort, particularly concerned with intellectual property, where it is well established that the plaintiff can choose to have either damages or an

* Editor's note: Also reported at (1979) 251 EG 1167.

account of profits made by the defendant by his wrongful acts: see, for instance, *Lever* v *Goodwin* (1887) 36 ChD 1 *per* Cotton LJ at p7. This is in line with the long-established common law doctrine of waiving the tort.

The liability in the present case is solely in contract and not in tort.

I come then to the *Wrotham Park* case. In that case the predecessor in title of the plaintiffs had, in 1935, sold some land to a predecessor in title of the defendants, subject to a restrictive covenant restricting building to a particular layout. That covenant was duly registered under the Land Charges Act 1925. In 1971 the land was sold to the defendant, who had no actual knowledge of the restrictive covenant and proceeded to build 14 houses on the land in breach of the covenant. In early 1972 the plaintiffs, as successors in title to the benefit of the covenant, issued their writ against the defendant claiming an injunction to restrain building in breach of the covenant, and demolition of anything built in breach. The plaintiffs made no application for an interim injunction. By the time the action came on for trial in July 1973, the 14 houses had all been completed and sold to purchasers with the benefit of indemnity insurance policies. At the trial Brightman J held that the plaintiffs were indeed entitled to the benefit of the covenant and the defendant was bound by it.

For obvious reasons however — that he could not shut his eyes to the fact that the houses existed and it would be an unpardonable waste of much needed houses to direct that they be pulled down — he refused to grant a mandatory injunction. He commented that no damage of a financial nature had been done to the plaintiffs by the breach of the covenant and proceeded to consider what damages, if any, he should award under the jurisdiction which had originated under Lord Cairn's Act to award damages in substitution for an injunction.

It was submitted to him that the damages should be nil or purely nominal because the value of the Wrotham Park estate was not diminished by one farthing in consequence of the breach of covenant. But the judge concluded that such a result would be of questionable fairness.

He said at p812H of the report:

> If, for social and economic reasons, the court does not see fit in the exercise of its discretion, to order demolition of the 14 houses, is it just that the plaintiffs should receive no compensation and that the defendants should be left in undisturbed possession of the fruits of their wrongdoing?

He then referred to the wayleave cases and *Whitwham* v *Westminster Brymbo Coal & Coke Co* and the other cases which I have mentioned where the same principle has been applied.

He concluded that the appropriate course was that the defendant should pay by way of damages the sum which the plaintiffs might hypothetically have been willing to pay — though actually they would never have been willing to relax the covenant to permit the defendant to do what it wanted to do on the land. He fixed that at a small percentage of the defendant's anticipated profit from building the 14 houses on the land.

The difficulty about the decision in the *Wrotham Park* case is that in *Johnson* v *Agnew*, Lord Wilberforce, after citing certain decisions on the scope and basis of Lord Cairn's Act, which were not cited to Brightman J stated in the clearest terms, at p400G of the report, that on the balance of those authorities, and on principle, he found in the Act no warrant for the court awarding damages differently from common law damages.

Sir William Goodhart submits that it follows from that analysis by Lord Wilberforce, in *Johnson* v *Agnew*, that the damages awarded by Brightman J in the *Wrotham Park* case were indeed damages assessed on recognised common law principles which should, so he says, be applied in the present case.

I doubt, however, whether that does follow from Lord Wilberforce's analysis in *Johnson* v *Agnew*. As I read his judgment Brightman J was not seeking to analyse the scope or basis of the court's jurisdiction under Lord Cairn's Act. He merely concluded that, as Parliament had expressly empowered the court to grant damages in lieu of an injunction, Parliament must have intended that in every case the court must be able to award such damages as would achieve a fair result between the parties and would not be limited to awarding nominal damages only. He sought to apply that conclusion.

That involves a conclusion by the judge that Lord Cairn's Act affected a substantive change in the law of damages and was not a merely procedural statute as *Johnson* v *Agnew* has held. It is unnecessary to refer further in this judgment to the *Wrotham Park* since that was under Lord Cairn's Act and stands or falls by that; whereas, the present case is not, and makes no pretence of being, under that Act. I should however mention in passing that we were referred to a number of cases where the measure of damages chosen by Brightman J in the *Wrotham Park* case was applied by other judges. For instance *Bracewell* v *Appleby* [1975] Ch 408*; *Carr-Saunders* v *Dick McNeil Associates Ltd* [1986] 1 WLR 922† and *Griffiths* v *Kingsley-Stubbs*, decided by this court on June 3 1986, but unreported. All those were cases where the plaintiff's cause of action lay in tort, either trespass or nuisance, where the defendant had interfered with the plaintiffs property rights. The decisions and awards of damages are amply justified by the common law principles in tort of the wayleave cases and *Whitwham* v *Westminster Brymbo*, already mentioned.

Given that the established basis of an award of damages in contract is compensation for the plaintiff's loss, as indicated above, I have difficulty in seeing how Sir William Goodhart's suggested common law principle of awarding the plaintiff, who has suffered no loss, the gain which the defendant has made by the breach of contract, is intended to go. Is it to apply, for instance, to shipping contracts or contracts of employment or contracts for building works?

Sir William suggested, in his and Mr Weatherill's skeleton argument, that the conventional measure fails to do justice and a different measure should be applied where the following conditions are satisfied: (a) the breach is deliberate, in the sense that the defendant is deliberately doing an act which he knows or should know is plainly or arguably in breach of contract; (b) the defendant, as a result of the breach, has profited by making a gain or reducing a loss; (c) at the date of the breach it is clear or probable that damages under the conventional measure will either be nominal or much smaller than the profit to the defendant from the breach; and (d) if the profit results from the avoidance of expenditure, the expenditure would not have been economically wasteful or grossly disproportionate to the benefit which would have resulted from it.

He suggested in that paragraph in the skeleton argument that the underlying principle might be that the conventional measure of damages might be overridden "in certain circumstances" by the rule that no one should benefit from his deliberate wrongdoing. In the course of his submissions Sir William limited his formulation and, while retaining conditions (a), (b) and (c), substituted for condition (d) the following:

> Damages for loss of bargaining power can be awarded if, but only if, the party in breach could have been restrained by injunction from committing the breach of contract or compelled by specific performance to perform the contract. Where no such possibility existed there was no bargaining power in reality and no right to damages for loss of it. Hence damages for loss of bargaining power cannot be awarded where there is for example a contract for the sale of goods or generally a contract of employment.

I find difficulty with that because, in theory, every time there is a breach of contract the injured party is deprived of his "bargaining power" to negotiate for a financial consideration a variation of the contract which would enable the party who wants to depart from its terms to do what he wants to do. In addition it has been held in *Walford* v *Miles* [1992] 2 AC 128 that an agreement to negotiate is not an animal known to the law and a duty to negotiate in good faith is unworkable in practice — and so I find it difficult to see why loss of bargaining or negotiating power should become an established factor in the assessment of damages for breach of contract.

Beyond that, since we are looking for the measure of damages at common law for breach of contract, apart from Lord Cairn's Act, I do not see why that should vary depending on whether the party in breach could or could not have been restrained by injunction from committing the breach or compelled by specific performance to perform the contract. Injunctions and specific performance were not remedies in the common law courts and were granted by the Court of Chancery, which, before Lord Cairn's Act, had no power to award damages, just because the common law remedy of damages was not an adequate remedy.

We were referred, in the course of Sir William's argument, to a number of other cases and, in particular, to passages in the judgment of Sir Robert Megarry VC, in *Tito* v *Waddell (No 2)* [1977] Ch 106 and to the decision of this court in *Stoke-on-Trent City Council* v *W&J Wass Ltd* [1988] 1 WLR 1406. In the latter case this court upheld the general principle that in tort a plaintiff recovered damages

* Editor's note: Also reported at (1974) 237 EG 731.
† Editor's note: Also reported at [1986] 2 EGLR 181.

equivalent to the loss he had suffered and held also that the "user" principle in the wayleave cases and *Whitwham* v *Westminster Brymbo* should not be extended to cover infringement of a market right of the plaintiff council by the holding by the defendant of an unauthorised market where the plaintiffs could not show they had suffered any actual loss by the infringement. What was sought in that case was damages calculated by reference to a notional licence fee that the plaintiff council might have charged for permitting the defendants' infringement but that was refused. I need not refer to the cases further.

As I see it, therefore, there never was in the present case, even before the writ was issued, any possibility of the court granting an injunction to restrain the defendant from implementing the later planning permission. The plaintiff's only possible claim from the outset was for damages only, damages at common law.

The plaintiffs have suffered no damage. Therefore, on basic principles, as damages are awarded to compensate loss, the damages must be merely nominal.

For these reasons, which substantially accord with those of Ferris J, I would dismiss this appeal.

Agreeing, **STEYN LJ** said: The issue in this appeal was defined by Sir William Goodhart QC, appearing for the appellants, as the correct measure of damages in a case where the following three circumstances are satisfied: (a) there has been a deliberate breach of contract; (b) the party in breach has made a profit from that breach; and (c) the innocent party is in financial terms in the same position as if the contract had been fully performed. It is an important issue, with considerable implications for the shape of our law of obligations, and I therefore add a few remarks of my own.

Dillon LJ has reviewed the relevant case law. It would not be a useful exercise for me to try to navigate through those much-travelled waters again. Instead, it seems to me that it may possibly be useful to consider the question from the point of view of the application at first principles. An award of compensation for breach of contract serves to protect three separate interests. The starting principle is that the aggrieved party ought to be compensated for loss of his positive or expectation interests. In other words, the object is to put the aggrieved party in the same financial position as if the contract had been fully performed. But the law also protects the negative interest of the aggrieved party. If the aggrieved party is unable to establish the value of a loss of bargain he may seek compensation in respect of his reliance losses. The object of such an award is to compensate the aggrieved party for expenses incurred and losses suffered in reliance of the contract. These two complementary principles share one feature. Both are pure compensatory principles. If the aggrieved party has suffered no loss he is not entitled to be compensated by invoking these principles. The application of these principles to the present case would result in an award of nominal damages only.

There is, however, a third principle which protects the aggrieved party's restitutionary interest. The object of such an award is not to compensate the plaintiff for a loss, but to deprive the defendant of the benefit he gained by the breach of contract. The classic illustration is a claim for the return of goods sold and delivered where the buyer has repudiated his obligation to pay the price. It is not traditional to describe a claim for restitution following a breach of contract as damages. What matters is that a coherent law of obligations must inevitably extend its protection to cover certain restitutionary interests. How far that protection should extend is the essence of the problem before us. In my view *Wrotham Park* v *Parkside Homes* is only defensible on the basis of the third or restitutionary principle: see *MacGregor on Damages* 15th ed, para 18, and Professor P B H Berks, *Civil Wrongs: A New World Butterworths Lectures, [1990-1991]* 55 at p71. The appellants' argument that *Wrotham Park* can be justified on the basis of a loss of bargaining opportunity is a fiction. The object of the award in *Wrotham Park* was not to compensate the plaintiff for financial injury, but to deprive the defendants of an unjustly acquired gain. While it must be acknowledged that *Wrotham Park* represented a new development, it seems to me that it is based on a principle of legal theory, justice and sound policy. In the respondent's skeleton argument some doubt was cast, by way of alternative submission, on the correctness of the award of damages for breach of covenant in *Wrotham Park*. In my respectful view, it was rightly decided and represents a useful development in our law. In *Tito* v *Waddell (No 2)* [1977] 1 Ch 107 at pp335C to 336C, the Vice-Chancellor interpreted *Wrotham Park* and the decision in *Bracewell* v *Appleby* [1975] Ch 408, which followed *Wrotham Park*, as cases of the invasion of property rights. I respectfully agree. *Wrotham Park* is analogous to cases where a defendant has made use of the aggrieved party's property and thereby saved expenses: see *Penarth Dock Engineering Co Properties Ltd* v *Pounds* [1963] 1 Lloyd's Rep 359. I readily accept that the word "property" in this context must be interpreted in a wide sense. I would also not suggest that there is no scope for further development in this branch of the law.

But in the present case we are asked to extend the availability of restitutionary remedies for breach of contract considerably. I question the desirability of any such development. The acceptance of the appellants' primary or alternative submissions, as outlined by Dillon LJ, will have a wide-ranging impact on our commercial law. Even the alternative and narrower submission will, for example, cover charter parties and contracts of affreightment where the remedy of a negative injunction may be available. Moreover, so far as the narrower submission restricts the principle to cases where the remedies of specific performance and injunction would have been available, I must confess that that seems to me a bromide formula without any rationale in logic or common sense. Given a breach of contract, why should the availability of a restitutionary remedy, as a matter of legal entitlement, be dependent on the availability of the wholly different and discretionary remedies of injunctions and specific performance? If there is merit in the argument, I cannot see any sense in restricting a compensatory remedy, which serves to protect the restitutionary interests to cases where separate remedies of specific performance or injunction, designed directly and indirectly to enforce payment, are available.

For my part, I would hold that if Sir William's wider proposition fails then the narrower one must equally fail. Both submissions hinge on the defendant's breach being deliberate. Sir William invoked the principle that a party is not entitled to take advantage of his own wrongdoing. Despite Sir William's disclaimer it seems to me that the acceptance of the propositions formulated by him will inevitably mean that the focus will be on the motive of the party who committed the breach of contract. That is contrary to the general approach of our law of contract and, in particular, to rules governing the assessment of damages. In my view, there are also other policy reasons which militate against adopting either Sir William's primary or narrower submission. The introduction of restitutionary remedies to deprive cynical contract breakers of the fruits of their breaches of contract will lead to greater uncertainty in the assessment of damages in commercial and consumer disputes. It is of paramount importance that the way in which disputes are likely to be resolved by the courts must be readily predictable. Given the premise that the aggrieved party has suffered no loss, is such a dramatic extension of restitutionary remedies justified in order confer a windfall in each case on the aggrieved parry? I think not. In any event such a widespread availability of restitutionary remedies will have a tendency to discourage economic activity in relevant situations. In a range of cases such liability would fall on underwriters who have insured relevant liability risks. Inevitably underwriters would have to be compensated for the new species of potential claims. Insurance premiums would have to go up. That, too, is a consequence which militates against the proposed extension. The recognition of the proposed extension will, in my view, not serve the public interest. It is sound policy to guard against extending the protection of the law of obligations too widely. For these substantive and policy reasons I regard it as undesirable that the range of restitutionary remedies should be extended in the way in which we have been invited to do so.

The present case involves no breach of fiduciary obligations. It is a case of breach of contract. The principles governing expectation or reliance losses cannot be invoked. Given the fact of the breach of contract the only question is whether restitution is an appropriate remedy for this wrong. The case does not involve any invasion of the plaintiff property interests even in the broadest sense of that word, nor is it closely analogous to the *Wrotham Park* position. I would, therefore, rule that no restitutionary remedy is available and there is certainly no other remedy available.

I would dismiss the appeal.

Also agreeing **ROSE LJ** said: I also agree with the reasons given in both the previous judgments.

A feature which to my mind is capable of distinguishing the present case from *Wrotham Park Estates Co Ltd* v *Parkside Homes* [1974] 1 WLR 798 is not just that damages were sought there in equity and here in contract at common law but the different conduct of the

A respective plaintiffs in response to the breach of covenant.

In *Wrotham Park* the plaintiffs objected to building works in breach of covenant as soon as they learnt of them and, within a month, issued a writ seeking restraining and mandatory injunctions. They would not have granted any relaxation of the covenants even if this had been sought. From first to last they objected to what the defendants did. I find it unsurprising that, in consequence, Brightman J awarded them substantial damages in equity as an alternative to a mandatory injunction. He did so in order to achieve a just result (see p815D).

In the present case, from first to last, the plaintiffs have neither objected, nor wished to object, to what the defendants have done.
B The fact that the plaintiffs granted planning permission for development in breach of the covenant is immaterial, for they had to comply with their statutory duties as planning authority. Having written to the defendants in December, 1982, drawing attention to the covenant and indicating a willingness to discuss its variations, the second plaintiffs did nothing to seek to enforce the covenant for the very good reason that they did not wish to do so, there being no harm to their adjoining land. They issued their writ seeking damages more C
than five years later in March, 1988. It seems to me that that history of events is fatal to the plaintiffs' attempted reliance on *Wrotham Park Estate*, where, in my respectful view, Brightman J correctly answered the particular question which confronted him. It is, in my judgment, wholly unlikely that he would have given the same answer if, in that case, the plaintiff's response to the defendant's breach of covenant had been that of the plaintiffs in the present case. Equitable relief, whether by way of injunction or damages in lieu, would, as it seems to me, have been inconceivable.

In the light of what Lord Wilberforce said, in *Johnson* v *Agnew* [1980] AC 367, at p400B to G, I see no reason why in the present case the plaintiffs should be in any stronger position as to damages because they are suing the original covenantor, rather than a D
successor who bought with notice of the covenant.

Accordingly, and for the reasons given by my lord, I, too, would dismiss this appeal.

Appeal dismissed with costs.

ESTATE AGENTS

A Queen's Bench Division

November 4 1992

(Before Judge ZUCKER QC, sitting as a judge of the division)

MULDOON v MAYS OF LILLIPUT LTD

Estates Gazette April 10 1993

[1993] 14 EG 100

Negligence — Estate agents — Whether agents negligent in advising the vendor as to the price to accept in the open market

B On July 1 1988 the plaintiff, Brian Christopher Muldoon, entered into a contract to sell his freehold property, 5 Chaddesley Wood Road, Canford Cliffs, Poole, Dorset, to a Mr and Mrs Scanlon at a price of £125,000. From about May 1988 the defendants, Mays of Lilliput Ltd, acted for the plaintiff and advised him to put the property on the market at £125,000. In the summer of 1987 Mr Croker, of the defendants, valued the property at £94,000, a figure confirmed by another agent. In May 1988 Mr Croker valued the property at £125,000. On May 27 1988 Mr and Mrs Scanlon's offer of £125,000 was accepted by the plaintiff. Under the contract, completion was to be on July 11 1988, but C after contracts were exchanged, the plaintiff was informed by property owners in the area of a possible development, which might give the properties a considerably higher value. The plaintiff alleges that the defendants were negligent and that the true minimum market value of his property at which he should have been advised to sell was £173,000.

***Held*: The action was dismissed. There was no evidence that the property had been included in any development scheme. Evidence of offers made to purchase the plaintiff's property at a time when he was contractually bound to sell to Mr and Mrs Scanlon was thoroughly unsatisfactory and as not accepted as D genuine. It was accepted that there was a rising market through 1988, and that of the comparables relied on by the plaintiff's expert, only one was truly comparable. If property values increased by 35%-40% between 1987 and 1988, in relation to the 1987 value of £94,000, such a rise would suggest a valuation in May to July 1988 of £126,900 to £131,600. On the evidence there was no material available during the period upon which a competent surveyor, exercising reasonable care and skill, would have arrived at a conclusion that the sum of £125,000 was outside the bracket into which the proper valuation should have fallen.**

E The following case is referred to in this report.

Singer & Friedlander Ltd v *John D Wood & Co* [1977] EGD 569; (1977) 243 EG 212 & 295

This was a claim by the plaintiff, Brian Christopher Muldoon, for damages against the defendants, Mays of Lilliput Ltd, in relation to advice as to the sale price of 5 Chaddesley Wood Road, Canford Cliffs, Poole, Dorset, which the plaintiff contracted to sell on July 1 1988.

Patrick McCahill (instructed by Edge & Ellison Calow Easton, of Birmingham) appeared for the plaintiff; the defendants were present in court by two of its directors, Michael Halpern and John Croker, F but were not otherwise represented.

Giving judgment, **JUDGE ZUCKER** said: On July 1 1988 the plaintiff, Mr Brian Christopher Muldoon, entered into a contract to sell his freehold property, 5 Chaddesley Wood Road, Canford Cliffs, Poole, Dorset, to a Mr and Mrs Scanlon. The contract price was £125,000. The defendants, Mays of Lilliput Ltd, are estate agents who carry on business at a number of addresses, including an address at Poole. They acted for Mr Muldoon from about May 1988 in the sale of his property. They had advised him to put the property on the market at £125,000. Mr Muldoon's case is that in so doing they were negligent; the true minimum market value of which he should have G been advised was £173,000. By this action Mr Muldoon claims damages for the loss he has suffered as a result of that alleged negligence.

I should say at the outset that, although a defence was filed and the defendants have, until recently, been represented by solicitors and counsel, they parted company with their solicitors shortly before the trial. The result was that neither solicitors nor counsel have been present on the defendants' behalf during the hearing before me. The defendants' directors, Mr John Croker [FNAEA] and Mr Michael Halpern [ACIS], have been in court. At the outset they applied to represent the defendants. I ruled that they were not entitled to do so. They did not wish to accept my indication that I would adjourn the H case so that the defendants could be properly represented. The result is that they have not been represented. It follows that I have heard no evidence or argument on behalf of the defendants. I regard that as highly regrettable in the circumstances of this case.

Mr Muldoon lives in Birmingham. He has a wife and four children, who are now aged between 15 and 22. In 1978 he bought 5 Chaddesley Wood Road as a holiday home. That property was built about 60 years ago. It is a semi-detached chalet-style house with a living room, kitchen and one bedroom on the ground floor and two bedrooms on the first floor. It has no garage but a hard standing on which three to four cars can be parked.

In the summer of 1987, because some of his children (who were by then in their teens) no longer wished to go to the property for J holidays, Mr Muldoon decided to place the property on the market and to sell it by way of an auction. He sought a valuation from Mr Croker. Mr Croker valued the property at £94,000. Mr Muldoon went to another estate agent, who confirmed that figure. Prior to the auction he received an offer in the sum of £94,000. Mr Muldoon, however, withdrew the property as he believed he could obtain £100,000. Neither Mr Muldoon nor anyone on his behalf has suggested that £94,000 was an unreasonable valuation of the property in the summer of 1987.

In October 1987 Mr Muldoon let the property to a Mrs Carika. In about March 1988 she asked Mr Muldoon if he would sell her the property. Mr Muldoon considered that £115,000 would be a fair price, but he put forward a figure of £125,000. He met Mr Carika, K who initially said that price was acceptable, but within a fortnight Mr Carika decided not to go ahead.

On May 21 1988 Mr Muldoon instructed Mr Croker that he wanted to sell 5 Chaddesley Wood Road and to buy instead a larger and more expensive property nearby, 26 Shore Road, called "Channel Breeze". He asked Mr Croker to value his property. Mr Croker came back with a figure of £125,000. Mr Muldoon felt he could confidently rely on Mr Croker because his valuation of £94,000 the previous summer had been matched by an offer in that sum, and because £125,000 was the figure he had had in mind.

On May 26 1988 Mr and Mrs Scanlon, who lived locally, offered £120,000. The following day they increased their offer to £125,000.

A Mr Muldoon accepted. On July 1 1988 contracts were exchanged with Mr and Mrs Scanlon. Completion was to be on July 11 1988. Mrs Muldoon had all along been unhappy with the sale price of £125,000. She thought it was worth more. To make matters worse, within a day or two of contracts being exchanged Mr Muldoon had a telephone conversation with Mr Percival, the owner of 5A Chaddesley Wood Road. Mr Percival told him he had been offered £135,000 for his house. It was smaller than no 5. Mr Muldoon also learned from Mr Percival that developers were interested in developing a site, which included a number of houses in the vicinity of which his property was one. I add in parenthesis that there is no evidence that a sale of number 5A Chaddesley Wood Road at the price of £135,000, or any price, in fact materialised.

B Mr Muldoon then spoke to Mr Fagg who owned 24 Shore Road nearby, which would also have been included in the proposed development. Indeed, Mr Fagg was one of the moving forces seeking to bring about the development. He was also employed by another local firm of estate agents, Rebbeck.

On July 5 1988 Mr Fagg provided Mr Muldoon with a written valuation. It was in these terms:

Valuation as private residence £220,000 to £235,000. Valuation as development site £265,000 to £275,000.

These figures led to a serious disagreement between Mr and Mrs Muldoon. She left the matrimonial home in Birmingham with her children and went to live at 5 Chaddesley Wood Road. She caused a C caution to be placed on the property. The result was that Mr Muldoon could not complete on July 11 1988. On July 11 1988 Mr and Mrs Scanlon's solicitors served a notice to complete. On August 25 1988 they issued a writ in the Chancery Division seeking specific performance of the contract of sale. Mr Muldoon resisted specific performance on the ground that there was collusion between Mr Croker and Mr and Mrs Scanlon. He reported the matter to the fraud squad of the Dorset Constabulary. That action was settled. Mr Muldoon completed the sale on February 23 1989 and paid Mr and Mrs Scanlon the sum of £5,000 in full and final settlement of their claim for damages for non-completion.

In para 16 of the amended statement of claim there is, *inter alia*, a claim to recover the sum of £5,000 paid to Mr and Mrs Scanlon, and D £5,520, the fees of Mr Muldoon's solicitors in that action. That part of the claim has been abandoned. Mr Muldoon has not pursued before me allegations of connivance or of fraud against the defendants.

Before coming to the expert evidence, I have to refer to a number of documents which are in the trial bundle and which came into existence between July 14 1988 and October 6 1988 at a time when Mr Muldoon was resisting specific performance by Mr and Mrs Scanlon. On p130 there is a letter from a company called Manway Homes Ltd. It is addressed to another local firm of estate agents, Andrew Key & Partners:

Dear Sirs,

E *Re: 5 Chaddesley Wood Road, Sandbanks, Poole*

We are prepared, as principals, to offer £210,000 (Two Hundred and Ten Thousand Pounds), for the freehold interest in the above property, subject to contract and vacant possession.

According to Mr Muldoon, he knew nothing of Manway Homes Ltd. He had not instructed Andrew Key & Partners to act on his behalf. This document was pushed through his door. It is an odd document because, if Andrew Key & Partners were the agents for the developers, it might be queried why the offer should be made in those terms to their own agent.

At p132 is a letter dated July 29 1988 simply headed "Strathavon" to Andrew Key & Partners:

Dear Sirs,

F *Re: 5 Chaddesley Wood Road, Sandbanks, Poole*

I write to confirm my offer, subject to contract, of £215,000 (Two hundred and fifteen thousand pounds).

Precisely the same comments apply to that document as to the previous one. On p133 dated August 2 1988 on Andrew Key's notepaper and addressed to Mr Muldoon, there is this letter:

Re: 5 Chaddesley Wood Road, Sandbanks

I write to confirm that after careful consideration and taking the present market conditions into account we would value the above property at a figure of £210,000 (Two hundred and ten thousand pounds) subject to contract.

Mr Muldoon did in fact obtain that valuation from Andrew Key & Partners. He exhibited it to his affidavit in the Chancery action.

On p134 is a letter dated September 29 1988 from Andrew Key & Partners to Mr Muldoon at his Birmingham address: G

Re: 5 Chaddesley Wood Road

We have pleasure in enclosing Details of Sale in connection with the above proposed sale, Subject to Contract.

Details have also been sent to all other parties concerned and no doubt a draft contract will be prepared as soon as possible.

If you have any queries, please do not hesitate to contact us.

That is clearly written as Mr Muldoon's agent. There follows the details of the sale of 5 Chaddesley Wood Road by Mr Muldoon to MJL Associates at a price of £200,000.

Mr Muldoon told me that he really knew nothing about these matters. Andrew Key & Partners were not acting on his behalf.

On p136 also dated September 29 there is a copy of the letter to Mr Muldoon which is on p134, though this time addressed to his H solicitors. On p137 there is a letter dated October 6 1988 from MJL Associates to Mr Muldoon:

I write to advise you that following negotiations through Andrew Key Estate Agents in respect of my Company's proposed purchase of your property I have instructed my Lawyers to request your Solicitors' draft Contract and ancillary documentation, requesting my Lawyers to deal with the matter with all expedition.

I therefore hope to be in a position to exchange Contracts very shortly although I would mention as a matter of courtesy that it is my intention to exchange Contracts simultaneously with exchange of Contracts in the purchases of other property so as early exchange of Contracts will very much depend as to how soon all the solicitors involved can forward to my Lawyers the draft Contracts and ancillary documentation. J

I will of course do all I can to bring an early exchange of Contracts to fruition and until then matters of course remain subject to Contract.

Mr Muldoon repeated that he had not instructed Andrew Key to negotiate on his behalf. P138 is a letter from Town & County Estate Agents in Bournemouth dated October 6 1988. It is addressed to Mr Muldoon's solicitors:

Re: 5 Chaddesley Wood Road, Sandbanks, Poole, Dorset

We understand that you act on behalf of Mr B Muldoon, Vendors of the above property and would like to advise you that we have arranged a sale, Subject to Contract on their behalf.

As the Vendors' Agents, we have pleasure in enclosing copies of the Sales Particulars, printed details and confirmation letter to the Vendor, so that you may proceed accordingly. K

It would be appreciated if you would acknowledge receipt of these details.

There follows particulars of a proposed sale in which the vendor is Mr Muldoon, the purchaser is a company called Sky High Ltd and the agreed price is £225,000.

On p140, also dated October 6 1988, there is a letter from Town & County, confirming the matter to Mr Muldoon personally. Mr Muldoon said in effect that he knew nothing about any of these offers.

Mr Keeling, a partner in Andrew Key, said that his firm was retained by the development companies referred to in his firm's letters. He said Mr Muldoon was actively engaged in selling his property through Andrew Key. He had spoken to Mr Muldoon on the L telephone. The documents that went through his firm's hands were all genuine documents. There is a complete conflict of evidence between Mr Muldoon and his own witness, Mr Keeling. Counsel for Mr Muldoon sought to place some reliance on the figures contained in those documents as showing what the value of the property would have been to a developer and seeking to apply a discount to show its value without development potential. The documents are also referred to in the report of the plaintiff's expert, Mr Nigel Holtom [ARICS], to which I shall come.

The fact is that the property has not been included in any development scheme. I have been given no proper explanation of any of the offers contained in any of the documents I have referred to. All the offers were made at a time when Mr Muldoon was contractually M bound to sell the property to Mr and Mrs Scanlon. The evidence in relation to them is thoroughly unsatisfactory. I am not satisfied that any of them were genuine.

Mr Fagg gave evidence to justify the valuation set out in his letter of July 5 1988, to which I have referred. He was not relied on by the plaintiff as an expert witness in this case. Even his lower valuation of the property as a private residence, that is the bracket of £220,000 to £235,000, is far above the figure now put forward on behalf of the plaintiff. As Mr Fagg hoped to involve his own property in any development he had an interest in placing as high a value as possible on Mr Muldoon's property. I gain no assistance from Mr Fagg's evidence or from his valuation of July 5 1988.

A The expert evidence the plaintiff did rely upon was that of Mr Holtom of Black Horse Agencies, Rumsey & Rumsey, who had premises in Poole. Mr Holtom qualified as a chartered surveyor in September 1985, but he has had practical experience since 1976. He inspected the property on September 20 1988 and his report values it as at May 27 1988. The substance of his report is as follows:

> The property has been neglected over a period of time and is in need of modernisation and refurbishment. The general condition is largely consistent with its type and age, however, we would comment as follows:
>
> The plain tile roof cladding is in minor disrepair and will warrant a degree of continual attention. Furthermore, isolated beetle infestation is evident to the timber roof framework.
>
> The chimney stack has a slight lean and warrants minor repair.
>
> Outer main walls appear superficially in satisfactory order, however, the B sub floor ventilation is inadequate. We noted deflection, and loose boards to the ground floor bedroom floor. Full investigation of all the sub floor timbers is necessary. Inadequate ventilation may lead to decay which is costly to eradicate.
>
> Internally, as mentioned previously modernisation of all services, appliances, fixtures and fittings is ideally required, together with internal and external redecoration.
>
> GENERAL REMARKS AND VALUATION
>
> The Sandbanks peninsula and the neighbouring area of Canford Cliffs are regarded as one of the foremost residential locations within the Poole/Bournemouth conurbation. The subject property whilst in a somewhat indifferent and tired condition is located in a prime spot within easy walking distance of Poole Bay beaches, together with several well established Yacht Clubs. To realise its full potential, the property requires complete C modernisation. It is ideally suited for use as a retirement or holiday home or a weekend retreat.
>
> We are of the opinion that the property with vacant possession freehold tenure has an open market value as at 27 May 1988 in the region of £173,000 (one hundred and seventy three thousand pounds).
>
> The value of the property as per the date of our inspection 20 September 1988

— Mr Holtom here refers to an earlier inspection by another member of his firm —

> was we believe in the region of £180,000 (one hundred and eighty thousand pounds). The difference between the two figures representing the market trend during this time. This was calculated by analysing the sales during the relevant period of both second hand and new build properties. The difference D equating generally to between 3-5%. For the purposes of this valuation we adopted approximately 4%.
>
> Our opinion of value is subject to the following qualifications:—
>
> We understand that a Building/Structural Survey of the property was not required and therefore our valuation assumes that the property is structurally sound and free from defects, other than those mentioned in the report.
>
> We have not been supplied with any documents relating to the freehold title in the above property and our valuation is on the assumption that a good freehold title can be shown and that it is not subject to any easements, wayleaves, restricted covenants or encumbrances.
>
> Our opinion as to the value of the property reflects a figure that it might have been reasonably expected to be sold at by private treaty as at 27 May 1988 assuming:—
>
> a. A willing vendor.
>
> E b. A reasonable period of time in which to negotiate sale.
>
> c. That values remained static during that period.
>
> d. The property would have been freely exposed to the market, and
>
> e. That no account would have been taken of any higher price which have been paid by a purchaser with a special interest.

Mr Holtom then turned to comparable evidence under three heads: address, sale price and completion:

> 1. Portside, 14 Shore Road, Sandbanks — Major structural problems to include attention to a floor slab. £170,000. 23.08.88.
>
> 2. Flat 1 Beach Walk, Banks Road, Sandbanks. Ground Floor Flat. £175,000. 23.09.88.
>
> 3. Flat 6, Sandhills, 40 Banks Road, Sandbanks — First Floor Flat. £137,000. 13.09.88.
>
> 4. Flat 5, Fourwinds, 32 Banks Road. Second Floor Flat. £142,500. F 6.05.88.
>
> 5. Flat 6, Seapoint, Martello Pk, Canford Cliffs. £196,000. 8.03.88.
>
> The Sandbanks area comprises a large diverse mixture of properties many of which are used solely for second homes, investments and/or holiday lettings.
>
> Sandbanks, the peninsula and the adjoining area of Canford Cliffs are regarded as one of the foremost residential locations along Britain's South Cost. Within a short distance the beaches in Poole Harbour provide extensive sail boarding and yachting facilities. The subject property is centrally located for all these amenities. In addition, Chaddesley Wood Road is private thereby residents enjoy convenient parking in the height of the summer season.
>
> Given the unique location and popularity, numerous older properties with large gardens or a number of adjoining properties are being converted in the majority of instances to flat developments. No property has therefore the preferred comparables and given the select nature of the area, the number of sales are small in number. The above comparables reflect a general trend in G values whereupon purpose built flats are predominant.
>
> We understand that a price was agreed for the subject property with a then tenant of £125,000. The period of the time being February/March 1988. This therefore indicates that given the buoyant market, this value would be significantly higher in the May of that year. Given increases of 35 to 40% during the period January-September, this would realise £168,750—£175,000.

I can leave out paragraphs dealing with sales advice and development potential as no reliance on development potential has been made in this case. I come to market conditions during 1988:

> The residential market during February — August 1988 experienced significant increases in value of some 35-40%. Properties were being sold very rapidly and in countless instances gaz-zumping was experienced together with H Dutch Auctions arising. The rapid pace of the increase made valuation of property difficult. This we believe increased the margin of error in valuation and was further compounded by the individual nature of the Sandbanks area.
>
> From the information provided negotiations to purchase the property commenced at £120,000 before finally agreeing the asking price of £125,000. I do not feel that once the asking price had been secured there should be any delay in the disposal of the property. Mr Muldoon had I assume agreed the marketing price at which the agent had produced the purchaser at the price required. There would be no reason to delay. If any interested party had arrived prior to the exchange of contracts with a higher offer, then the vendor must be notified. Furthermore, the property should be actively marketed until that time.
>
> I find it very hard to believe that Mays had no knowledge of the development potential of the area in question. J

Then he goes on to deal with the question of development. He goes on:

> The valuation letter provided by Mr Fagg of Messrs Rebbecks does not give the impression of a formal valuation. It does give in our opinion a market appraisal whereupon, if the property were to be actively marketed one would be talking at the submitted figures of:—
>
> Valuation as private residence £220,000—£235,000, Valuation as development site £265,000—£275,000.
>
> In our opinion these figures would appear to be very bullish but not unusual given the prevailing market conditions.
>
> Andrew Keys would appear to have genuinely arranged a sale on this property and have interest from a number of developers who are prepared to offer £200,000 plus. K
>
> CONCLUSION
>
> Valuation of residential properties is not a precise science and is very much an art. One cannot therefore be in a position to specifically calculate a set price and a degree of flexibility must be adopted. Matters as explained previously with regard to the individual nature of the Sandbanks area makes valuation more onerous.
>
> In our opinion of value a wider than normal variation must be adopted on any valuation figure. This in this instant and applicable to both the May and July dates should range from 15-20%.
>
> From this, we believe it is evident that the sale price of £125,000 is significantly below the market value at that time. The impression is that the Defendants were negligent in valuing the property at the said sale price.

In evidence Mr Holtom expanded that part of his report, which said that between February and August 1988 a significant increase in L value in the residential market of between 35% and 40% was achieved. He said that the market started to take off at the end of February and by May an increase of 30% had been achieved.

Before I turn to my conclusions in this case it may be convenient to recite the principles of law which I must apply. The onus of proof is dealt with in *Jackson and Powell on Professional Negligence*, 3rd ed, para 3-352 at p219:

> The onus of proof rests "fairly and squarely" upon the plaintiff in an action for professional negligence against a surveyor. It is not enough to prove that he was wrong. Negligence must be distinctly proved. It may even be difficult to prove he was wrong by showing that the defendant's valuation was significantly higher or lower than another surveyor's valuation or the price actually achieved or that his estimate of individual items making up the M calculation was on the high side. It must be shown that his valuation and calculations were such that they could not have been arrived at by the exercise of reasonable care and skill. Although it is sometimes stated that the onus of proof in a professional negligence action is a heavy one, the standard is no more than that applicable to civil actions generally, namely proof on a balance of probabilities.

At para 3-42 on p211 the standard of care and skill required is dealt with:

> In common with other professional persons and in the absence of an express term to the contrary the standard required of a surveyor is that of the ordinary skilled man exercising the same skill as himself. He is variously described in the cases as the "reasonably skilled", "competent", "prudent" or "average" surveyor.

At para 3-43 at p212 the text reads as follows:

> The standard is not relaxed if the required task is carried out in unusual circumstances, for example a valuation in a market showing signs of deep depression or of unusual buoyancy or volatility. Whether the standard is complied with in a particular case, however, must be determined having regard to the circumstances and knowledge current at the time when the surveyor was required to carry out his instructions. The court must be careful "not to be wise after the event". Yet guarding against hindsight may be difficult where a long time has elapsed between the allegedly negligent act and trial or where the state of knowledge at a particular time is in issue.

There are very helpful dicta in the judgment of Watkins J in *Singer & Friedlander Ltd* v *John D Wood & Co* (1977) 243 EG 212. At p213 the learned judge said:

> The valuation of land by trained, competent and careful professional men is a task which rarely, if ever, admits of precise conclusion. Often beyond certain well-founded facts so many imponderables confront the valuer that he is obliged to proceed on the basis of assumptions. Therefore, he cannot be faulted for achieving a result which does not admit of some degree of error. Thus, two able and experienced men, each confronted with the same task, might come to different conclusions without any one being justified in saying that either of them has lacked competence and reasonable care, still less integrity, in doing his work. The permissible margin of error is said by Mr Dean [one of the experts in that case], and agreed by Mr Ross [the other expert], to be generally 10 per cent either side of a figure which can be said to be the right figure, ie so I am informed, not a figure which later, with hindsight, proves to be right but which at the time of valuation is the figure which a competent, careful and experienced valuer arrives at after making all the necessary inquiries and paying proper regard to the then state of the market. In exceptional circumstances the permissible margin, they say, could be extended to about 15 per cent, or a little more, either way. Any valuation falling outside what I shall call the "bracket" brings into question the competence of the valuer and the sort of care he gave to the task of valuation.

On the same page, but further on in the judgment Watkins J said:

> If a valuation is sought at times when the property market is plainly showing signs of deep depression or of unusual buoyance or volatility, the valuer's task is made more difficult than usual. But it is not in such unusual circumstances an impossible one. As Mr Ross said, valuation is an art, not a science. Pinpoint accuracy in the result is not, therefore, to be expected by he who requests the valuation. There is, as I have said, a permissible margin of error, the "bracket" as I have called it. What can properly be expected from a competent valuer using reasonable skill and care is that his valuation falls within this bracket.

The only way the plaintiff's claim is now put is that the minimum market value which a competent surveyor should have placed on the property was £173,000. As Mr Croker valued it at £125,000, the plaintiff's loss is £48,000. This is not a case where any specific allegation of omission or commission is now made against Mr Croker. The plaintiff's case rests principally, if not entirely, on a consideration of sale prices obtained in the case of properties which Mr Holtom says are comparable.

Of Mr Holtom's five comparables only two were sold prior to the exchange of contracts on July 1 1988. I take that date rather than May 27 1988, when Mr Muldoon accepted Mr and Mrs Scanlon's offer of £125,000, because, if the market had risen before July 1 1988, it would of course have been Mr Croker's duty to inform Mr Muldoon of that rise. I am quite certain that, if the value of his property had increased before July 1 1988, Mr Muldoon would have withdrawn.

Comparable 5, Flat 6, Seapoint, Martello Park, was sold for £196,000 on March 8 1988. The estate agent's particulars prior to the sale describe it as a luxury second-floor flat built eight years earlier, two of the three bedrooms having luxury en-suite bathrooms. I note that at the beginning of the 1988 season, before the market had taken off, it was put on the market at £205,000. All those factors indicate to me that this was a superior property and not a true comparable. Comparable 4, Flat 5, Fourwinds, 32 Banks Road: completion of its sale was on May 6 1988 at £142,500. It is described as a spacious, three bedroomed, second-floor flat with extensive views over Poole Harbour. That had been marketed at £140,000. Mr Croker should have known of this sale. Comparable 1, Portside, 14 Shore Road, Sandbanks, was sold by auction on July 26 1988 for £170,000. Comparable 3, a two-double-bedroomed flat at 6 Sandhills, 40 Banks Road, Sandbanks, achieved a sale price of £137,000. Completion was on September 13 1988. It was marketed on a date about which there has been no evidence for £139,000. Finally, comparable 2, a flat at 1 Beach Walk, Banks Road, Sandbanks, described as a three-bedroomed, ground-floor, luxury apartment built in 1987, was marketed on a date unknown for £190,000. The sale was completed on September 23 1988 for £175,000.

Of course, the fact that higher prices were subsequently achieved for other properties may be evidence that a surveyor's valuation was such that it could not have been arrived at by the exercise of reasonable care and skill. But, in a rising market, what actual indication would Mr Croker have had that a value of £125,000 for a property which Mr Holtom described as neglected and requiring complete modernisation was outside a reasonable bracket? On the evidence, he would only have had comparables 4 and 5, of which no 5, I have held, is not truly comparable.

There are two other relevant facts which have emerged from the evidence. The first is the offer price prior to auction in the summer of 1987 of £94,000. It has not been suggested that that was not a reasonable value to put on the property at the time it was made. If one takes that figure as the base for an increase of 35% to 40% for the following year, it suggests a valuation in May-July 1988 of £126,900 to £131,600. £173,000 would represent an 84% increase.

The second is the fact that Mr Muldoon was willing to accept £125,000, despite his wife's misgivings that it was too low. The strong impression I received from Mr Muldoon's evidence is that he was very alive to the state of the property market in this area: see, for example, his withdrawal of the property from auction in the summer of 1987 because he thought the property was worth £100,000 and not £94,000. I bear in mind that Mr Muldoon is not an expert and would not have had all the information that would have been available to Mr Croker. But, if Mr Muldoon had any reason to believe that £125,000 could in any way have been bettered, I am sure he would have acted upon if.

Mr Holtom states in his report that, because of the state of the market in 1988, a wider than normal variation must be adopted on any valuation figure. He suggests that the range is 15% to 20%. If one takes £125,000 as being at the bottom of the range, the upper bracket would be £143,650 to £150,000. I am not satisfied on the evidence I have heard that this property was worth even £143,750 in May to July 1988. Nor am I satisfied on the evidence that during that period, and not by way of hindsight, there was any material then available upon which a competent surveyor, exercising reasonable care and skill, would have arrived at a conclusion that the sum of £125,000 was outside the bracket into which any proper valuation should have fallen.

In the circumstances the plaintiff has not made out his claim on the evidence that I have heard and I dismiss it.

Mr McCahill, I propose subject to what you say, to dismiss the claim and award the defendants its costs.

Claim dismissed with costs.

FEES AND COMMISSIONS

A
Court of Appeal

December 17 1992
(Before Sir Donald NICHOLLS V-C and Lord Justice STAUGHTON)

CHASEN RYDER & CO v HEDGES

Estates Gazette February 27 1993
[1993] 08 EG 119

Estate agents — Commission — Two agents — Whether B ***introduction by first agent was effective cause of purchase***

By a letter dated July 29 1988 the plaintiff firm of estate agents accepted the defendant's instructions to find a purchaser for Forest Hill House Nursing Home, Dorset (which had the benefit of planning permission for bungalow), and stating that the commission would be 2%; there was a further letter in the same terms on August 2 1988, save that the commission was stated as $1\frac{1}{2}$%. A number of potential purchasers were introduced by the plaintiffs, including a Mr Russell Wilson, but he made no offer and no sale resulted. In June 1989 the defendant instructed a second firm of estates C agents, Goadsby & Harding, as sole agents at a commission of 2%. On June 22 1989 the officials of the local planning authority indicated that they would support a plan to increase the floor area through an extension. This information was given to Mr Russell Wilson by Goadsby & Harding and on July 31 1989 he completed the purchase of the property. Both agents claimed commission and the defendant paid Goadsby & Harding. Judge Bates QC in Bournemouth County Court gave judgment to the plaintiff for £16,000 plus interest. The defendant appealed, contending that the plaintiff's introduction of Mr Wilson was not the effective cause of the sale.

D *Held*: **The appeal was allowed. The burden is on the plaintiff to show that his introduction in any case was the effective cause of the purchase. If, however, he shows that he was the first to introduce the purchaser, and that a purchase followed, and if no other facts are established, then it may well be that the judge will infer that the plaintiff was the effective cause. The evidential burden in such a case then passes to the defendant. The first introduction by the plaintiff was the effective cause of nothing. It produced no offer from Mr Wilson. The second introduction did not depend on the first and it was the second introduction which was the cause of the** E **purchase.**

The following cases are referred to in this report.

Chesterfield & Co Ltd v *Zahid* [1989] 2 EGLR 24; [1989] 29 EG 75
Wood (John D) & Co v *Dantata* [1987] 2 EGLR 23; (1987) 283 EG 314, CA

This was an appeal by the defendant, David Hedges, from the decision of Judge Bates QC, in Bournemouth County Court, where he gave judgment to the plaintiff, Chasen Ryder & Co, for commission on the sale of the defendant's property.

Michael Norman (instructed by Aldridge & Brownlee, of Bournemouth) appeared for the appellant; Roger Shawcross (instructed by Morris Scott & Co, of Christchurch) appeared for the F respondent.

Giving the first judgment at the invitation of Sir Donald Nicholls V-C, **Staughton LJ** said: In July 1988 Mr Hedges, the defendant, was the owner of Forest Hill House, a large building in Dorset, which was used as a residential nursing home. Mr Hedges wanted to sell it. He instructed the plaintiff, Chasen Ryder & Co, which I gather is the trading name of Mr Rhone, an estate agent specialising in that type of property. He wrote a letter on July 29 to Mr Hedges in these terms:

> Forest Hill House Nursing Home
> Thank you for your instructions to find a purchaser for your property: Forest G Hill House Nursing Home, Rushall Lane, Corfe Mullen, Wimborne, Dorset, BH21 3RT
> Our commission will be 2% of the sale price plus VAT at prevailing rate.
> We enclose a copy of our sales details Ref: N333 and a copy of this letter with a stamped addressed envelope. Would you kindly sign the copy and return to confirm our sales details are correct.

The sales details enclosed with that letter said that the price was £750,000, and the land included 2.79 acres. It was said that there was planning permission for a bungalow in the grounds, which would enable the number of residents in the home to be increased from 20 to 29.

There was a further letter on August 2 in substantially the same terms as that of July 29, except that the commission was now to be H $1\frac{1}{2}$% plus VAT. Mr Rhone introduced a number of possible buyers. Almost straightaway there was a Mr Ronald Lay and his brother. They agreed to buy, subject to contract, for the asking price of £750,000. Mr Hedges thought that he must be selling too cheaply, so he took the property off the market for a time and then put it back on different terms. The asking price was now £925,000 and the amount of land offered, 15 acres.

By November 1988 Mr Rhone had sent particulars to Mr Russell Wilson as a possible purchaser and he arranged for Mr Wilson to visit the house. Mr Wilson did that on November 22, but he made no offer and expressed no interest. There is evidence that instead in April 1989 he made an offer for another property. J

In January 1989, meanwhile, another person had agreed to buy for £835,000, subject to contract, but that sale also fell through. The same happened with a Mr Hewitt between February and April 1989, at a price of £875,000. It happened again with a Mr Wilkes, at a price of £895,000, in May and June 1989. Meanwhile, on June 17 or thereabouts, Mr Hedges instructed another firm of estate agents, Goadsby & Harding. He told them to market the property with 15 acres at a price to be advised. They were to be sole agents and receive a commission of 2%. In their letter of confirmation, they said that commission was to be payable "in the event of our being instrumental in the introduction of a person, ready, able and willing who proceeds to an exchange of contracts to purchase at a price acceptable to you". The planning permission situation was much the K same as it had been in the previous year.

But then, on June 22 1989, there was a meeting between Mr Hedges' architect and officials of the local planning authority. It would seem that at that meeting the officials were prepared to lend their support to a plan to increase the floor area of the building through an extension by an amount of 50%. This would have enabled the number of residents to be increased either to 31 or 34. Of course, the officials of the local authority do not have power to grant planning permission. The authority and their councillors have that power. But these officials were prepared to recommend it, and Mr Hedges' architect regarded the outcome of that meeting as very satisfactory. Mrs Quinn of Goadsby & Harding had met, or at any

A rate been in contact with, Mr Russell Wilson back in 1986, and she knew him to be interested in buying a nursing home. She telephoned him. The evidence of that telephone conversation is to be found in what Mrs Quinn says in her examination-in-chief:

> I spoke to Mr Wilson. He was in the market to buy a nursing home/rest home. I came across him in 1986. He had a rest home in Bournemouth known as Balhome Grange. He sold Balhome Grange in 1988.
>
> I spoke to Mr Wilson on the telephone. I felt Forest Hill House suited his requirements — development possibilities — that is what prompted me to telephone him. We made an appointment for Mr Wilson to view the property. I handled the transaction.

Mrs Quinn does not there say expressly what she said on the telephone, but one must bear in mind that she did not know that Mr B Rhone had already introduced Mr Wilson to this property some seven months earlier; so no doubt she told him a bit about it, possibly that she felt it suited his requirements and that it had development possibilities.

That must have been on or about June 23, because on that day Mrs Quinn made an appointment for Mr Wilson to view the house. Of course he had seen it seven months earlier, but he went to view it again. Presumably, or perhaps it is inference, he was told of the outcome of the planning meeting on the previous day. What Mr Hedges says in evidence is:

> He was shaking with excitement. He wanted to see the plans and the letter from the architect. He was excited about the development possibilities. He C made an offer there and then. £935,000 offer.

That, of course, was subject to contract, but this time the property was actually sold. Completion took place on July 31. Both agents claimed commission. Goadsby & Harding were paid by Mr Hedges, although it is said that that was only after they had issued proceedings. Mr Rhone, the plaintiff in this action, was not paid, so he started this action, which was tried in Bournemouth County Court before Judge Bates QC. On February 14 1992 judgment was given for Mr Rhone for some £16,000 plus interest. So Mr Hedges would have to pay twice. But he now appeals.

There is not a great deal of dispute about the law applicable to this situation. It is to be found in two cases. The first is a decision of this court in *John D Wood & Co* v *Dantata* [1987] 2 EGLR 23*. In that D case, too, there were two rival estate agents seeking commission. Of course they might, on the facts, both be entitled to it. But that does not seem to have arisen in that case. In his judgment, Nourse LJ quoted a passage from the judgment of Forbes J in the court below, which I also quote.

> I do not consider, however, that the question here turns on whether either agent was instrumental or assisted in the process which led finally to the sale. The question is whether either of them has demonstrated that it was his introduction which was the effective cause of the sale. Although the question of who was first in the field is not conclusive, I do not consider that an agent who effects a second introduction to the property (if that is not a contradiction in terms) can succeed in demonstrating that such an introduction was the effective cause of the sale, unless he can show that the interest aroused in the E purchase by the first introduction has evaporated by the time of the second.

Nourse LJ, at p25, said:

> As I have said, the learned judge recorded an acceptance by all three counsel that in order to succeed one or other of the two firms had to show that they introduced the ultimate purchaser and that such introduction was the effective cause of the purchase. That would seem to suggest that there are two questions to be answered, and it would certainly explain the importance which the learned judge attached to the chief's retention of a lively interest in the property when he went there again on September 9. In truth I think that there is but a single question to be answered: which of the two firms introduced the chief to the sale? Both language and authority established that that question must be answered by answering this further question: which of the two firms was the effective cause of the sale? Here I would gratefully adopt the F following statement of the law in *Bowstead on Agency* . . . to which the learned judge referred:
>
> "... the fact that one agent introduces a person who ultimately purchases after a later introduction by another agent will not necessarily entitle the first agent to commission. In such a case the court must determine which of the two agents was the effective cause of the transaction taking place."

Then Nourse LJ went on to say that, so far as introduction was concerned, what mattered was the introduction of the purchaser to the purchase, not to the property. So there is no dispute here that what we have to consider is: has it been shown that Mr Rhone's introduction was the effective cause of the sale? The second case was *Chesterfield & Co Ltd* v *Zahid* [1989] 2 EGLR 24†, a decision of Garland J. The judge there again quoted the passage of Forbes J at G first instance in *John D Wood* v *Dantata* and said at p25 that Nourse LJ had adopted that passage in that case. He said the same again at p26.

I do not think it right to say that Nourse LJ adopted that passage from Forbes J. He merely cited it. Nor do I think that the law is as Forbes J stated it, if indeed he was intending to state a rule of law. The burden is on the plaintiff to show that his introduction in any case was the effective cause of the purchase. If, however, he shows that he was the first to introduce the purchaser, and that a purchase followed, and if no other facts are established, then it may well be that the judge will infer that the plaintiff was the effective cause. It can therefore be said that the evidential burden in such a case passes to the defendant, H whether the other agent or the vendor, to prove more facts which displace that inference. But, even in such a case, I do not think that the further facts which the defendant then has to prove must be such as to show that interest aroused by the first introduction has evaporated, that is to say, entirely disappeared. It will be a matter for consideration in each case as to how far the defendant has to go before he has displaced an inference which might arise from the mere fact of the introduction followed by the purchase.

The judge in the present case did say that there was an evidential burden on the defendant. He did not spell that out, although he went on to say that the defendant must show that the interest had evaporated. As I have said, if the judge meant that the fact of the first introduction followed by the fact of the purchase placed some J evidential burden on the defendant, I would not have disagreed with that. But I would not have gone so far as to say that the defendant then had to show that Mr Wilson's interest had evaporated before the second introduction.

Looking at the matter afresh, it seems to me, on the evidence, that the first introduction in this case, by Mr Rhone, in the event was the effective cause of nothing. It produced no offer from Mr Wilson. It produced no interest from Mr Wilson. So far as we know, Mr Wilson disappeared for a period of seven months. What is more, during that period he was making an offer for some other property.

The second introduction by Mrs Quinn did not in any way depend upon the first. She was not aware that Mr Wilson had been shown the property as a result of the first introduction. It is true that Mr Wilson K may have carried in his head information that he had obtained through the first introduction and his first view of the house. But on this second occasion there was a different bargain on offer. Mr Shawcross says not very different: instead of the prospect of 29 residential places, either 31 or 35. There is then the fact that, according to Mr Hedges, Mr Wilson was shaking with excitement and was very interested in the possibilities of development on this second occasion. What brought that about, one asks, if it was not the second introduction? He was not shaking with excitement, so far as we know, at any time before that.

Mr Hedges' credibility is criticised by the judge, but he did not reject everything that he said. Neither side called Mr Wilson as a L witness, although he was available. So we have to make do with the evidence that we do have. That evidence shows that, following the second introduction, there was immediate interest and almost immediately a purchase. It does not always follow that, as the latin tag has it, because one thing occurs after another thing it is therefore caused by the other thing. But, to my mind, by far the most probable inference on the facts of this case is that the introduction by Mrs Quinn was the effective cause of the purchase and that the earlier introduction by Mr Rhone was not.

I would therefore allow this appeal, set aside the judgment in favour of Mr Rhone and dismiss the claim.

Sir Donald Nicholls V-C agreed and did not add anything. M

Appeal allowed with costs.

---

* Editor's note: Also reported at (1987) 283 EG 314.

† Editor's note: Also reported at [1989] 29 EG 75.

HOUSING ACTS

A
Court of Appeal

October 22 1992

(Before Lord Justice NOURSE, Lord Justice FARQUHARSON and Lord Justice EVANS)

HARROW LONDON BOROUGH COUNCIL v TONGE

Estates Gazette March 6 1993

[1993] 09 EG 109

Housing Act 1985 — Right to buy — Tenant holding secure
B *tenancy and qualified to exercise right to buy dwelling-house — Pursuant to section 123(1) and 2(*a*) of the Act, written notice claiming the right to buy required tenant's daughter should share that right — Secure tenant died — Daughter not entitled to secure tenancy — Whether daughter entitled to exercise right to buy*

On December 24 1973 William Robert Tonge was granted a tenancy of 9 Hutton Gardens, the freehold of which is now vested in the appellants, Harrow London Borough Council. Upon his death on October 9 1976 his widow, Mrs Tonge, qualified to succeed the secure tenancy. On September 6 1988 Mrs Tonge served a written notice on the council claiming to
C exercise the right to buy the house and, pursuant to section 123(1) and 2(*a*), in her notice she required that Miss Tonge, the respondent, should share the right to buy with her. By a notice dated October 12 1988, under section 124(1), the council admitted the right to buy of Mrs Tonge and Miss Tonge. The right to defer completion pursuant to section 142 until September 7 1991 was claimed and acknowledged by the council. On November 7 1990 Mrs Tonge died intestate. There having been already one succession to the secure tenancy under section 87 of the Housing Act 1985, Miss Tonge was not entitled to succeed to a secure tenancy. The council contended
D that not being herself a secure tenant she no longer had any right to buy the house. On April 15 1991 Miss Tonge's solicitors informed the council of her wish to complete conveyance. The council served a notice to quit on January 10 1991. On May 23 1991 Miss Tonge issued proceedings in the county court claiming an injunction, pursuant to section 138(3), requiring the council to complete the purchase. On July 4 1991 the council served a counterclaim in those proceedings claiming possession of the house and damages for trespass. On the hearing of those proceedings and separate proceedings for possession under Ord 24 of the County Court Rules, Judge Quarren Evans in Willesden County Court
E ordered the council to convey the house to Miss Tonge. The council appealed.

Held: **The appeal was dismissed. By reason of section 123(3) of the 1985 Act, the right to buy belonged to Mrs Tonge and Miss Tonge jointly and that they should be treated for the purposes of Part V of the Act as joint tenants. The concluding words in the subsection that "he and they shall be treated for the purposes of this Part as joint tenants" and the words "a secure tenant" in the section, meant that they were to be deemed to be joint *secure* tenants. Accordingly, Mrs Tonge and Miss Tonge were deemed to be joint secure tenants for the** F **purposes of section 138 of the Act. If they had been joint secure tenants in the real world, Miss Tonge would, by her survivorship of Mrs Tonge, have become the sole secure tenant on November 7 1990. Accordingly, the effect of section 123(3) is that Miss Tonge was deemed to have been the sole secure tenant for the purposes of section 138 from that date and, on that footing, the council's duty to convey the house to her became enforceable by injunction on their receipt of her solicitor's letters of April 15 and 17 1991. It was implicit in section 123 that the deemed secure tenant must still occupy the** G **house as her only or principal home.**

The following cases are referred to in this report.

Dance v *Welwyn Hatfield District Council* [1990] 1 WLR 1097; [1990] 3 All ER 572; (1990) 22 HLR 339, CA
East End Dwellings Co v *Finsbury Borough Council* [1952] AC 109; [1951] 2 All ER 587; (1951) 2 P&CR 135; 49 LGR 669, HL
Jennings v *Epping Forest District Council* unreported July 9 1992
Sutton London Borough Council v *Swann* (1985) 18 HLR 140, CA

This was an appeal by Harrow London Borough Council from an order made by Judge Quarren Evans in Willesden County Court, whereby he ordered the council to convey to the respondent, Miss H Patricia Jean Tonge, the freehold of 9 Hutton Gardens, Harrow Weald, Middlesex.

Roger McCarthy (instructed by the solicitor to Harrow London Borough Council) appeared for the appellants; Edward Denehan (instructed by David Gouldman & Co) represented the respondent.

Giving judgment **Nourse LJ** said: This is an appeal by Harrow London Borough Council against an order made by Judge Quarren Evans in Willesden County Court, whereby he ordered them to convey to the respondent, Miss Patricia Jean Tonge, pursuant to the right to buy provisions of the Housing Act 1985, the freehold of 9 Hutton Gardens, Harrow Weald, Middlesex. He also dismissed the council's claim for possession of the house. The arguments on the J appeal have centred mainly on the true construction and effect of sections 123 and 138 of the 1985 Act. All references to sections are to sections of that Act.

The facts are not in dispute and can be shortly stated. On December 24 1973 Miss Tonge's father, Mr William Robert Tonge, was granted a tenancy of 9 Hutton Gardens, the freehold of which is vested in the council. He lived there with his wife, Mrs Winifred Tonge, and their daughter, Miss Patricia Tonge. It was, in the case of Miss Tonge it still is, their only or principal home. Accordingly, while Mr Tonge was alive his tenancy was a secure tenancy within sections 79, 80 and 81.

Mr Tonge died on October 9 1976, when Mrs Tonge became qualified to succeed to the secure tenancy by virtue of section 87. On K September 6 1988, pursuant to section 122(1), Mrs Tonge served a written notice on the council claiming to exercise the right to buy the house. Pursuant to section 123(1) and 2(*a*), Mrs Tonge, in her notice, required that Miss Tonge should share the right to buy with her. That brought into operation section 123(3), to which careful consideration must presently be given. Mrs Tonge's notice was received by the council on September 7 1988. On October 12 1988, pursuant to section 124(1), the council duly served a written notice admitting the right to buy of Mrs and Miss Tonge.

It is agreed that by June 6 1989 all matters relating to the grant and the amount to be left outstanding or advanced on the security of the house had been agreed or determined within section 138(1). Shortly stated, the position was that, after applying the maximum statutory discount of 50%, the price had been fixed at £41,200 (£41,722 including costs and stamp duty) and the amount to be left outstanding on mortgage at £20,175. The right to defer completion pursuant to section 142 until September 7 1991 (three years after the council's receipt of Mrs Tonge's notice under section 122(1)) had been claimed by the Tonges and acknowledged by the council. A holding deposit of £150 had been duly paid.

On November 7 1990, when the three-year period had 10 months still to run, Mrs Tonge died intestate. There having already been one succession to the secure tenancy under section 87, it is agreed that the final words of that section disallowed another. The council thought, and have throughout contended, that that meant that Miss Tonge, not being herself a secure tenant, no longer had any right to buy the house. They returned the deposit and served notices to quit both on her and on the President of the Family Division. However, on April 15 1991 solicitors acting for Miss Tonge wrote to the council stating that she wished to complete the purchase by the transfer to her of the freehold interest in the house. Two days later, on April 17, they sent the council an engrossed conveyance which had been signed as a deed by Miss Tonge. On April 22 the council replied, contending that she had no right to buy under the 1985 Act and that, although a notice to quit had been served on January 10 1991, she remained in unauthorised occupation of the property.

On May 23 1991 Miss Tonge issued proceedings in the county court claiming an injunction pursuant to section 138(3) requiring the council to complete the purchase. On July 4 1991 the council served a counterclaim in those proceedings claiming possession of the house and damages for trespass. On the same day they issued separate proceedings for possession under Ord 24 of the County Court Rules. Both proceedings came before Judge Quarren Evans on December 5 1991, when he gave judgment for Miss Tonge and made the order to which I have referred.

Mr McCarthy, for the council, and Mr Denehan, for Miss Tonge, have each advanced arguments in relation to sections 123 and 138. Both here and below Mr Denehan has based himself primarily on section 138, which, so far as material, is in these terms:

> (1) Where a secure tenant has claimed to exercise the right to buy and that right has been established, then, as soon as all matters relating to the grant and to the amount to be left outstanding or advanced on the security of the dwelling-house have been agreed or determined, the landlord shall make to the tenant —
>
> (*a*) if the dwelling-house is a house and the landlord owns the freehold, a grant of the dwelling-house for an estate in fee simple absolute . . .
>
> in accordance with the following provisions of this Part.
>
> (2) . . .
>
> (3) The duty imposed on the landlord by subsection (1) is enforceable by injunction.

It is established by two decisions of this court and, I would think, by common sense that the opening words of section 138(1), "Where a secure tenant has claimed to exercise the right to buy," require that the person exercising the right shall continue to be a secure tenant up to the moment of completion: see *Sutton London Borough Council* v *Swann* (1985) 18 HLR 140 and *Jennings* v *Epping Forest District Council* July 9 1992 (unreported). If that were not so, the right to buy could be enforced by someone who had ceased to occupy the property as his only or principal home. Here Mr McCarthy submits that Miss Tonge has never been the secure tenant of the house. On the other side, Mr Denehan, relying on another decision of this court — *Dance* v *Welwyn Hatfield District Council* [1990] 1 WLR 1097 — has argued that, in the circumstances I have stated, Mrs Tonge and Miss Tonge became jointly entitled to the equitable interest in the house on June 6 1989; that Miss Tonge became solely entitled to it by her survivorship of Mrs Tonge on November 7 1990; and that she thereupon became and thereafter remained the secure tenant within section 138(1).

In my opinion, Mr Denehan's primary argument is faced with formidable objections. Although to do so goes further than anything which was decided in *Dance*, I am prepared to assume that the entitlement to the equitable interest in the property can arise under section 138(1) before there has been a request for completion. But even on that assumption it is difficult to see how someone who is entitled in equity under the equivalent of an uncompleted contract for the purchase of property can be said to be a tenant of that property. His occupation has no contractual origin; he has no legal estate in the property; and in any event it must be very doubtful whether his status can be more than that of a licensee.

In the circumstances I very much doubt whether Mr Denehan's primary argument could succeed. But, since the objections to it have not been fully explored in argument and, since I am satisfied that our decision in this case must be governed not by section 138, but by section 123, I need say no more about it. I would add a word of caution only against the introduction of the analysis attempted in the *Dance* case into other cases where it is unhelpful and may, as it seems, cause confusion. The analysis was helpful in the *Dance* case, where the essential question was whether the right to buy had already been "exercised" for the purposes of section 121(1). The Vice-Chancellor and I thought that if a stage had been reached at which the equitable interest in the property had passed to the secure tenant, the right must have been exercised. That is not a question which arises in the present case.

Section 123 is in these terms:

> (1) A secure tenant may in his notice under section 122 require that not more than three members of his family who are not joint tenants but occupy the dwelling-house as their only or principal home should share the right to buy with him.
>
> (2) He may validly do so in the case of any such member only if —
>
> (*a*) that member is his spouse or has been residing with him throughout the period of 12 months ending with the giving of the notice, or
>
> (*b*) the landlord consents.
>
> (3) Where by such a notice any members of the tenant's family are validly required to share the right to buy with the tenant, the right to buy belongs to the tenant and those members jointly and he and they shall be treated for the purposes of this Part as joint tenants.

I have already explained how it was that section 123(3) came into operation here. The consequences it prescribed were, first, that the right to buy belonged to Mrs Tonge and Miss Tonge jointly and, second, that they should be treated for the purposes of Part V of the 1985 Act (sections 118 to 188) as joint tenants. In my view, it is necessary thus to divide the consequences prescribed, as being the only way in which due effect can be given to all the words of the provision.

Mr McCarthy submits that the right belonged to Mrs Tonge and Miss Tonge together and not to one of them alone. if section 123(3) had ended at the word "jointly", I would have seen great force in that submission. But we cannot disregard the succeeding words "and he and they shall be treated for the purposes of this Part as joint tenants". Those words are a deeming provision. It is a provision which tells us not only what is to be deemed, but for what purposes it is to be deemed.

What was to be deemed here was that Mrs Tonge and Miss Tonge were joint tenants. It is clear from the words of subsections (1) and (3) of section 123, "A secure tenant" and "Where by such a notice any members of the tenant's family" etc, supported by the terms of section 118(2), that they were to be deemed to be joint *secure* tenants. Indeed, as Mr Denehan has pointed out, it is difficult to see how someone who satisfies the tenant condition in section 81 could become a joint tenant with a secure tenant without herself becoming a secure tenant.

Then we must ask for what purposes Mrs Tonge and Miss Tonge were to be deemed to be joint secure tenants. The answer is for the purposes of Part V of the 1985 Act. Part V includes section 138. So they were deemed to be joint secure tenants for the purposes of that section. What was the effect of that? Mr McCarthy accepts, correctly, that if Mrs Tonge and Miss Tonge had been joint secure tenants in the real world, Miss Tonge would, by her survivorship of Mrs Tonge, have become the sole secure tenant on November 7 1990. So the effect of section 123(3) is that Miss Tonge is deemed to have been the sole secure tenant for the purposes of section 138 from that date. On that footing there can be no doubt that the council's duty to convey the house to her became enforceable by injunction on their receipt of her solicitors' letters of April 15 and 17 1991 at the latest.

On this short ground I think that Miss Tonge is entitled to succeed on this appeal. I emphasise that we are dealing only with a case where the deemed secure tenant still occupies the house as her only or principal home. It is, I think, implicit in section 123(1) that that state of affairs must continue up to the moment of completion. That again is only a matter of common sense.

It is no disrespect to Mr McCarthy's very clear argument to say no more than that it was unable to persuade me that the effect of section 123(3) in the circumstances of this case is not very plain. At every point I was reminded of some well-known observations of Lord Asquith of Bishopstone in *East End Dwellings Co* v *Finsbury Borough Council* [1952] AC 109 at p132:

A If you are bidden to treat an imaginary state of affairs as real, you must surely, unless prohibited from doing so, also imagine as real the consequences and incidents which, if the putative state of affairs had in fact existed, must inevitably have flowed from or accompanied it. . . . The statute says that you must imagine a certain state of affairs; it does not say that having done so, you must cause or permit your imagination to boggle when it comes to the inevitable corollaries of that state of affairs.

Although Judge Quarren Evans accepted Mr Denehan's primary argument as the basis of his decision, there is a passage in his judgment which shows, I think, that he would have been prepared to base it on section 123(3) alone. He said:

The plaintiff accepts that she is not in her own right a secure tenant, but I decide that she can rely upon the wording of section 123(3) to defeat the B contentions of the council. She and her mother are to be treated in relation to the right to buy as joint tenants.

The judge arrived at a correct decision. I would affirm it and dismiss this appeal accordingly.

Farquharson and **Evans LJJ** agreed and did not add anything.

Appeal dismissed with costs; legal aid taxation of the respondent's costs; application for leave to appeal to the House of Lords refused.

C

Court of Appeal

October 8 1992

(Before Lord Justice GLIDEWELL, Lord Justice MANN and Lord Justice LEGGATT)

MUIR GROUP HOUSING ASSOCIATION LTD v THORNLEY AND ANOTHER

Estates Gazette March 13 1993

[1993] 10 EG 144

D ***Housing Act 1985 — Right to buy — Whether the loss by a secure tenant of his status as such entitles the landlord to refuse to convey the dwelling-house notwithstanding that all matters relating to the grant having been agreed between the parties***

On June 20 1980, the appellant, Muir Group Housing Association Ltd, granted the respondents, Wilbraham George Thornley and his wife Anita May Thornley, as joints tenants, a monthly tenancy of 23 Parrymead, Woodley, Stockport, the respondents being secure tenants. On May 23 1989 the respondents served upon the appellant a written notice claiming to exercise the right to buy the freehold of the E house under Part V of the 1985 Act. On June 23 the appellant admitted by written notice their right to buy and in December 1989 the respondents received a written notice of the appellant's proposed terms of sale. By July 9 1990 all matters relating to the grant of the freehold had been agreed and completion was to take place on August 31. Towards the end of July the appellant discovered that on August 28 1989 the respondents had moved out of the house on taking up a three-year appointment as resident wardens of a girl-guide establishment in Derbyshire. The appellant further discovered that on October 12 1989 the respondents had sublet the house on an assured shorthold tenancy in breach of a F condition of the tenancy agreement. By reason of section 93(2) of the 1985 Act, the independent effect of the subletting was that the respondents ceased to be secure tenants. On August 3 1990 the appellant informed the respondents that because of the subletting the sale would not proceed and on November 27 served a notice to quit, which expired on December 31. In the appellant's proceedings for possession, the respondents counterclaimed that the appellant became under a statutory duty to grant to them a freehold estate in the house. In Stockport County Court, Judge Blackburn dismissed the appellant's claim for possession and ordered it to grant the respondents the freehold.

Held: **The appeal was allowed. A process in four stages can** G **be identified in the provisions of Part V of the Housing Act 1985: the first stage is the claim to exercise the right to buy, the fourth and consummating stage that of grant. Between those stages are the intermediate second and third stages of the establishment of the right and the agreement or determination on, or of, the terms of the grant. A tenant who has ceased to be a secure tenant has no right which he can enforce under section 138(3) of the Act. The court was bound by its own decision in *Sutton London Borough Council* v *Swann*. The decisions in *Enfield London Borough Council* v *McKeon* and *Dance* v *Welwyn Hatfield District Council* were concerned with a** H **different question, namely at what stage in the process towards the conveyance does the right to buy of a person, who is at all relevant times a secure tenant, become indefeasible. In section 125(1) the word "tenant" must be an abbreviation of the phrase "secure tenant" which precedes it; section 139(2) does not refer to "*a* secure tenancy" but uses the definite article thereby presupposing that there *is* a secure tenancy immediately before the grant. Section 155(1) presupposes that the grantee is "the secure tenant". Section 118(1) provides that it is a secure tenant who is given the right to buy; throughout the stated circumstances there is a person who is at first "*a* secure tenant" and matures into "*the* secure tenant",** J **the physical identity of that person may change, for example on a succession, that is immaterial.**

The following cases are referred to in this report.

Dance v *Welwyn Hatfield Districtg Council* [1990] 1 WLR 1097; [1990] 3 All ER 572; (1990) 22 HLR 339, CA

Enfield London Borough Council v *McKeon* [1986] 1 WLR 1007; [1986] 2 All ER 730; (1986) 18 HLR 330; (1987) 85 LGR 24, CA

Opoku-Forfieh v *Haringey London Borough Council* October 17 1988 unreported

Sutton London Borough Council v *Swann* (1985) 18 HLR 140, CA

This was an appeal by Muir Group Housing Association Ltd from K a decision of Judge Blackburn given in Stockport County Court on January 14 1992 dismissing the appellant's claim for possession and on the respondents' counterclaim, ordering the appellant to grant the respondents a freehold estate in 23 Parrymead, Woodley, Stockport.

John Martin QC (instructed by Mason & Moore Dutton, of Chester) appeared for the appellant; Jeffrey Terry (instructed by Chronnell Fitzpatrick & Jones, of Hyde) represented the respondents.

Giving the first judgment at the invitation of Glidewell LJ, **Mann LJ** said: This is an appeal by Muir Group Housing Association Ltd against a decision of Judge Blackburn, which was given in Stockport County Court sitting at Oldham on January 14 1992. By his decision L the judge dismissed the appellant's claim for possession of the house known as 23 Parrymead, Woodley, Stockport, and, on the defendants' counterclaim, ordered the appellant to grant the defendants an estate in fee simple absolute in the house. The defendants in the action were Mr and Mrs W G Thornley, who are the respondents to the appeal.

The appeal raised a single question under the "right to buy" provisions, which are now consolidated in Part V of the Housing Act 1985 and which were first introduced by the Housing Act 1980. The single question is whether the loss by a secure tenant of his status as such entitles the landlord to refuse to convey an estate in the dwellinghouse even although all matters relating to the grant have been agreed between the parties. The appellant contends that it does. M The judge held that it does not and the respondents support his conclusion.

The circumstances of the case commence on June 20 1980 when the appellant as freeholder granted to the respondents as joint tenants a monthly tenancy of 23 Parrymead. It is common ground that for the purposes of the 1985 Act the tenancy was a secure tenancy and the respondents were secure tenants. On May 23 1989 the respondents served upon the appellant a written notice claiming to exercise the right to buy the freehold of the house. On June 23 the appellant served upon the respondents a written notice admitting their right to buy and, on a date in December, the respondents received a written notice of the appellant's proposed terms of sale. The notice was dated

A October 6 and the reason for its late receipt is neither known nor material. By July 9 1990 all matters relating to the grant of the freehold had been agreed and completion was to take place on August 31. It did not take place on that or any subsequent date. It did not take place because towards the end of July the appellant discovered that on August 28 1989 the respondents had moved out of the house on taking up a three-year appointment as resident wardens of a girl-guide establishment known as "The Children's Inn" at Rowarth, Derbyshire. The appellant further discovered that on October 12 1989 the respondents had sublet the whole of the house on an assured shorthold tenancy at a profit rent. The subletting was a breach of a condition of the 1980 tenancy agreement, whereby the
B respondents agreed not to sublet, and it had the independent statutory effects of the tenancy ceasing to be a secure tenancy and consequentially, of the respondents ceasing to be secure tenants (1985 Act, section 93(2)). The judge found that the respondents had not "fully realised" that subletting was a breach of the tenancy agreement or might affect the exercise of their right to buy. However, there is no suggestion that the respondents were guilty of a fraudulent concealment and the judge found them to be "entirely straightforward people".

On August 3 1990 the appellant's solicitors informed those acting for the respondents that because of the subletting the sale would not proceed. On November 27 the appellant sent to the respondents a notice to quit which expired on December 31. The house was not
C vacated and proceedings for possession were commenced on the grounds that the secure tenancy had ceased in October 1989 and that the contractual tenancy had been effectively determined by the notice to quit. The defence, so far as now material, was in effect that on July 9 1990 the appellant became under a statutory duty to grant to the respondents a fee simple absolute in the house and that the respondents' correlative right to a grant defeated the claim to possession. By a counterclaim the respondents sought the injunctive enforcement of the alleged duty. In a reply and defence to counterclaim the appellant denied the duty on the ground that the respondents had ceased to be secure tenants in October 1989. The rival contentions were argued before the judge who, in a short extempore judgment given at the end of a long day, concluded: "I am
D satisfied that the defendants' right by July 9 1990 had crystallised in a way under section 138 which entitled them . . . to an injunction". He accordingly made the dispositions to which I have referred. The significance of his reference to section 138 will hereafter appear.

I turn to the provisions of Part V of the 1985 Act, which of necessity are elaborate. The whole is introduced by section 118, which provides in subsection (1) that:

> A secure tenant has the right to buy, that is to say, the right, in the circumstances and subject to the conditions and exceptions stated in the following provisions of this Part —
>
> . . . to acquire the freehold of the dwelling-house.

Subsection (2) makes provision for the case (as here) of joint
E tenants and provides for the right to belong to all of them. Section 120 provides exceptions to the right to buy but none was here applicable. Section 121 provides circumstances in which the right to buy cannot be exercised. Subsection (1) provides that it cannot be exercised ". . . if the tenant is obliged to give up possession of the dwelling-house in pursuance of an order of the court or will be so obliged at a date specified in the order". Subsection (2) provides that the right cannot be exercised if it belongs to a person who is (broadly put) insolvent.

Section 122(1) provides: "A secure tenant claims to exercise the right to buy by written notice to that effect served on the landlord". It was such a notice that the respondents gave on May 23 1989 and it is common ground that they were entitled to do so. Section 124 provides for a landlord's notice admitting or denying the right to buy
F and it was such a notice that the appellant gave on June 23 1989. Section 125(1) provides: "Where a secure tenant has claimed to exercise the right to buy and that right has been established (. . . by the landlord's admission . . .), the landlord shall . . . serve on the tenant a notice" of his opinion as to the purchase price and of other specified matters. The appellant's notice dated October 6 1989 was such a notice. Section 125 and its alphabetical additions are followed by nine sections dealing between them with the amount and determination of the purchase price (including discount) and with the right to a mortgage. Section 138 is important and is the section to which the judge referred. It deals with the duty of the landlord to convey the freehold and (omitting what is presently immaterial) it provides:

> (1) Where a secure tenant has claimed to exercise the right to buy and that right has been established, then, as soon as all matters relating to the grant and to the amount to be left outstanding or advanced on the security of the dwelling-house have been agreed or determined, the landlord shall make to the tenant — G
>
> (*a*) if the dwelling-house is a house and the landlord owns the freehold, a grant of the dwelling-house for an estate in fee simple absolute, . . .
>
> in accordance with the following provisions of . . . [the Act].
>
> (3) The duty imposed on the landlord by subsection (1) is enforceable by injunction.

The date by which agreement was reached in this case was July 9 1990 (by when the respondents had ceased to be secure tenants) and it was by the exercise of subsection (3) that the injunction was granted. Section 139 provides for the terms and effect of a conveyance of the H
freehold and subsection (2), particularly, provides: "The secure tenancy comes to an end on the grant to the tenant of an estate in fee simple, . . . in pursuance of the provisions of this Part relating to the right to buy . . .". Section 155(1) further provides in relation to the terms of a grant that where the sale is at a discount (as usually it must be: see section 129), then the conveyance "shall contain . . . a covenant binding on the secure tenant and his successors in title" to the effect that in certain circumstances the discount is to be repayable.

Mr John Martin QC, who appeared for the appellant, attached importance to this provision. I conclude my summary visit to Part V by observing that section 141(3) (as amended) and section 142 entitle J
a tenant to defer completion in certain circumstances for (now) three years after the service of the notice under section 122.

A process in four stages can be identified in the sections to which I have referred. The first stage is the claim to exercise the right to buy. The fourth and consummating stage is that of grant. The intermediate second and third stages are those of the establishment of the right and the agreement or determination on or of the terms of the grant. In *Enfield London Borough Council* v *McKeon* [1986] 1 WLR 1007, a decision of this court, it was important to determine at what stage in that process the right to buy is exercised. The case arose under the predecessor provisions of the 1980 Act as amended by the Housing and Building Control Act 1984, but it will be convenient if I refer to the provisions of the 1985 Act. The defendant in an action for K
possession brought by the appellant authority was by succession the secure tenant of a house owned by the authority. She had on August 30 1984 served a notice under section 122 and on September 26 the plaintiffs had served a notice under section 124 admitting the defendant's right to buy. On November 29 the plaintiffs served a section 125 notice but on October 10 they had served a notice of their intention to seek possession of the house on ground 16 in Part III of Schedule 2, namely that the accommodation was more extensive than was reasonably required by the tenant. In January 1985 the plaintiffs commenced proceedings for possession. The defence was on the ground that the right to buy the house had been exercised and established, and that the plaintiffs' resultant duty to convey the freehold overrode any right they might have to possession. The L
defendant's argument in particular was that her right had been "exercised" by the service of a notice under section 122 and that section 121 therefore could no longer bite. The argument was rejected by Slade LJ who said (at p 1015G):

> In my judgment, the short answer to the defendant's point of law is that, for the purpose of applying section 2(4)(b) . . . [21] (on the true construction of which her case depends) the Act of 1980 [1985] treats a tenant as purporting to exercise his right to buy at any time and from time to time when he takes steps towards implementation of that right, up to and including completion of the purchase. If, therefore, any of the circumstances set out in Part II of Sched 1 [section 121] subsist at any time between the time when he serves his section [122] notice and completion, his right to buy ceases to be exercisable.

Eastham J, who was the other member of the court, agreed with M
that conclusion. In reaching it, Slade LJ expressed himself as influenced by three considerations (see p1015A-G). The first was the linguistic contrast between a claim to exercise the right and the exercise of the right which is found in the opening words of sections 125(1) and 138(1). The second was that, especially where completion is deferred, it can hardly have been the intention of Parliament that a landlord should be deprived before completion of his ordinary powers including those of obtaining possession on the statutory grounds. The third consideration was that if the defendant had been correct none of the events in section 121, including insolvency, could prevent completion.

The decision in the *McKeon* case was considered by this court in

A *Dance* v *Welwyn Hatfield District Council* [1990] 1 WLR 1097. In the later case matters had proceeded to the third stage, that is to say, the stage of the agreement contemplated by section 138 of the 1985 Act, but the landlord refused to complete and sought possession of the house on ground 10 in Part II of Schedule 2, namely that the landlord intended on obtaining possession to demolish the building. The plaintiff tenants sought an injunction pursuant to section 138(3). The defendant landlord asserted its right to possession and contended that under section 121(1) the right to buy is not further exercisable if an order for possession is made. It asked that the claim to possession be heard first. The county court judge thought *McKeon's* case constrained him to decide that the tenants' right to buy was still in the
B process of being exercised and that a further exercise in the form of a claim to a grant would be inhibited by the making of an order for possession. The tenants' appeal was allowed by this court. The leading judgment was given by Nourse LJ, who pointed out that the facts in *McKeon's* case differed from those before the court in "an important respect", that is to say the stage of agreement on terms had not there been reached ([1990] 1 WLR 1097 at p1103A).

He said (the same at G):

> Mr Haines, for the defendants, submitted that the decision in *Enfield London Borough Council* v *McKeon* [1986] 1 WLR 1007 is binding on this court and, moreover, that it is decisive of this case. He relied particularly on the view of Slade LJ that the Housing Act 1985 treats a tenant as purporting to exercise his right to buy at any time and from time to time when he takes steps towards
> C implementation of that right, "up to and including completion of the purchase". Although others might have preferred the simple view of section 121(1) which had been urged on Slade LJ and Eastham J by the tenant, we must certainly accept that their decision is binding on us. I do not, however, think that it is decisive of this case. Although the words of Slade LJ on their face apply here, the decision is only a binding authority for cases where the facts are the same. As I have already pointed out, the facts of this case are different. Indeed, the facts of *Enfield London Borough Council* v *McKeon* [1986] 1 WLR 1007 were such that it never became necessary to consider section 138, to whose effect on the facts of this case I now turn.

Nourse LJ analysed the effect of the coming into operation of section 138(3) and concluded (at p1105A) that as from the date of the agreement contemplated by section 138(1):

> D . . . the plaintiffs, being entitled to enforce the defendants' duty by injunction, became the equitable owners of the freehold in the property. At that stage, if not beforehand, they must, on any natural use of language, have "exercised" their right to buy for the purposes of section 121(1) of the Act of 1985.

Accordingly, section 121(1) could no longer bite.

McCowan LJ said (at p1106G) that the plaintiffs were by the date of the hearing in the county court indefeasibly entitled to completion and that this circumstance distinguished *McKeon's* case. Sir Nicolas Browne-Wilkinson V-C (as he then was) agreed with both judgments.

The decision in *Dance's* case compels some modification of Slade LJ's phrase "up to and including completion of the purchase" because the right must now be regarded as exercised at the conclusion of the third rather than of the fourth of the four stages. The necessity
E for this modification was perceived by Judge Blackburn when he put *McKeon's* case to one side (as he expressly did) and regarded the respondents' right as "crystallising" on July 9 1990. However, necessary as the modification is, it reduces (perhaps eliminates) the weight of the second and third considerations to which Slade LJ attached importance in coming to his conclusion. What Nourse LJ referred to as the simple view which "others might have preferred" could seem the more attractive in consequence. However, *McKeon's* case remains binding on this court for precisely what it decided, which is that the achievement of neither stage one nor two constitutes an exercise of the right to buy.

I have examined the cases of *McKeon* and *Dance* because they were discussed in the court below and before this court. However, it must
F be observed that in both cases the tenant throughout remained a secure tenant and accordingly neither addresses the situation where the secure tenant has lost his status.

Mr Martin submitted that the loss of status prevents there being an enforceable right in favour of the erstwhile secure tenant because the statute premises that the grant of an estate under Part V is a grant to a secure tenant. If this argument is correct then the loss of status must have the same effect whether it occurs before or after stage three, although the landlord will doubtless refuse to agree terms if he knows of the loss. In the present case he agreed terms in ignorance.

There are two cases in this court where a secure tenant had lost his status after a claim under section 122 and an admission under section 124. In *Sutton London Borough Council* v *Swann* (1985) 18 HLR 140 the defendant sought to resist a reference to an exercise of his right to G buy. The plaintiff council had followed their admission of the right to buy by a notice under section 125, but the defendant allowed the matter to lapse and he moved out of the premises thereby ceasing to be a secure tenant. Ackner LJ (as he then was) delivered a judgment with which Parker LJ and Sir David Cairns agreed. In the quotation from that judgment which follows I have substituted references to the 1985 Act for those in the 1980 Act, but I have retained the reference to section 16(11) of the 1980 Act for it is in somewhat different terms to the successor section 139(2). Ackner LJ said ((1985) 18 HLR 140 at p144):

> In my judgment this case is simply dealt with by the following propositions.
>
> In March 1982 Mr Swann was entitled to serve a notice under section 5 of the Housing Act 1980, claiming the right to buy, because he was a secure tenant, H and this was admitted by the notice in reply served by the respondents. The respondents provided Mr Swann with a perfectly straightforward offer for him to accept so the matter could proceed. He took no further step. The matter did not proceed. Accordingly, the offer which the respondents were obliged to make under the statute lapsed by efflux of time. When Mr Swann returned to the charge a year or more later, that offer, as I have indicated, having lapsed, he was no longer a secure tenant and therefore he no longer was entitled to apply under section [122] of the Act to buy the premises.
>
> Mr Swann takes a point, which in fact is dealt with by the learned judge, and in my view properly dealt with by the learned judge. He submits that, given that he was a secure tenant and as such made a claim as a secure tenant in March 1982 to buy, thereafter he has a right to purchase which cannot be defeated by any change in his status; ie by his thereafter ceasing to be a secure tenant. J
>
> In regard to this, the learned judge said: "Section[s 138 to 142] set out the steps to completion of the right to buy. Section [138(1)] uses the term "secure tenant". Section[s 138(2) to 142 use] the term "landlord" and, or "tenant". She then came to section 16(11), and I set that out in full. It reads as follows:
>
> "On the grant to 'a secure tenant' of an estate in fee simple or of a lease in pursuance of this Chapter the secure tenancy of a dwelling-house shall come to an end and, if there is then a sub-tenancy, section 139 of the Law of Property Act, 1925 shall apply as on a merger or surrender"
>
> That section proceeds in my judgment on the assumption that, at the end of the procedure, when the various investigations have been made, the various terms spelt out and accepted, and the time for the grant comes, that grant has to be made to a secure tenant. Accordingly the status of "secure tenant" has to exist, not only at the time when the claim to buy is made, but also at the time K when the grant comes to be made. If during the period between claim and grant the tenant has ceased to be a secure tenant, he is not entitled to that grant.

Mr Martin relied upon this passage and submitted that the change in the language of what was section 16(11), and is now section 139(2), during the process of consolidation, did not affect the assumption upon which the provision proceeded. Mr Jeffrey Terry, who appeared for the respondents, submitted that the observations of Ackner LJ in regard to the assumption of the statute were, first, *obiter* in that the basis of the decision was the lapse of the procedure and, second, were inconsistent with the decision in *Dance*'s case. I cannot dismiss the observations as *obiter* (even as such they would be highly persuasive) because they were directed to what was apparently L an independent ground of defence which required a separate disposition. Nor can I accept that the observations are inconsistent with the decision in *Dance*'s case. *Swann*'s case was cited in the court in *Dance* although it was not referred to in the judgments. The two cases are different in the most material respect that Mr and Mrs Dance had remained secure tenants throughout.

The second case where a secure tenant had lost his status after a claim under section 122 and an admission under section 124, is *Opoku-Forfieh* v *Haringey London Borough Council* unreported (1988), but of which we were given a *Lexis* transcript. In that case the defendant sought to defend possession proceedings by reference to a right to buy which had been established prior to the termination of his secure tenancy, but which does not appear to have proceeded to the M stage of an agreement as to terms. The leading judgment was given by Ralph Gibson LJ with whom Croom-Johnson LJ agreed. Ralph Gibson LJ treated *McKeon*'s case as determinant, but said that the same result could be achieved by the application of *Swann*'s case. The judgments in *McKeon* and *Swann* he regarded as "entirely consistent in their analysis of the relevant provisions of the Housing Acts". Mr Terry submitted that the observations in regard to *Swann*'s case were *obiter*. I agree, but they obviously have persuasive force.

It was Mr Terry's argument that the status of *secure* tenant had to exist at the time of claim and of admission but that its loss thereafter was irrelevant save that it exposed the tenant to the risk of a possession order which, by reason of section 121, would prevent the achievement of stage three. He drew our attention to the use in the provisions of Part V of both the phrase "secure tenant" and the word "tenant". Mr Terry submitted that the phrase and the word were employed deliberately and not interchangeably. He sought to demonstrate that the phrase "secure tenant" was employed in importing a requirement only to the stage of establishment and that thereafter it is used either as descriptive of what has occurred before and at that stage or as dealing with a contingency. For example, in section 138(1) the prefatory words "where a secure tenant has" are narrative but the only requirement is that the grantee must be a "tenant". An example of the suggested contingent usage is section 139(2), which Mr Terry suggests is to provide for the case where there happens to be a secure tenancy.

In my judgment, and with respect to him, Mr Terry's argument is ingenious but unconvincing. My reasons are five in number. First, in for example, section 125(1) the word "tenant" must be an abbreviation of the phrase "secure tenant" which precedes it and the draftsman must be expected to have a consistent usage. Second, section 139(2) does not refer to "*a* secure tenancy" but uses the definite article, thereby presupposing more strongly than the provision relied upon by Ackner LJ that there *is* a secure tenancy immediately before the grant. Third, section 155(1) presupposes that the grantee is "the secure tenant". The fourth, under section 118(1) it is a secure tenant who is given the right to buy and the right is described as being "the right, in the circumstances . . . stated in the following provisions of this Part . . . to acquire the freehold of the dwelling-house". These words seem to me to enact the assumption that throughout the stated circumstances there is a person who is at first "*a* secure tenant" and who matures into "*the* secure tenant". The physical identity of that person may change, for example on a succession (see section 136), but that is immaterial. My fifth reason, and of itself decisive in this court, is that the decision in *Swann*'s case actually decided in favour of the assumption which I have identified.

I, accordingly, conclude that a tenant who has ceased to be a secure tenant has no right which he can enforce under section 138(3). The conclusion accords with my belief that most people would find it surprising if Parliament had enacted right-to-buy provisions which enabled a person to acquire a house at a discount when he no longer occupied it as his only, or principal, home. I emphasise that the question which my conclusion answers is quite different from the question as to when the right to buy is exercised by a secure tenant. Upon that question, which primarily arises under section 121, the law is stated in the cases of *McKeon* and *Dance*. They have a reconcilable relationship, but the reasoning in the earlier case is damaged by the conclusion in the latter.

For the reasons which I have now formulated, we decided, at the conclusion of argument, that this appeal should be allowed. The judge found that the notice to quit sent on November 27 was properly served upon the respondent before November 30 1990 and accordingly the appropriate order is one dismissing the respondents' counterclaim and ordering possession as claimed by the appellant.

Agreeing, **Leggatt LJ** said: Since the judge did not develop his reasoning, I need not say in words of my own why I agreed that for the reasons given by Mann LJ this appeal should be allowed, the counterclaim dismissed and an order for possession made in favour of the appellant.

Also agreeing, **Glidewell LJ** said: I agree with the reasons given in his judgment by Mann LJ for our decision, announced at the conclusion of the argument, that this appeal should be allowed. I add some short observations of my own, because we are differing from the learned judge.

My main reason for concluding that a tenant who has ceased to be a secure tenant has no right which he can enforce under section 138(3) is that, in my view, the relevant sections of the statute clearly have this effect. Thus under section 118(1) it is "a secure tenant" who is given the right to buy, ie in this case, "to acquire the freehold of the dwelling-house". It follows that a person who ceases to be a secure tenant, ceases to have that right. In section 125, in this context, the words "the tenant" clearly mean "the secure tenant", the full phrase with which the section starts.

As Mann LJ points out, in section 139(2) the phraseology used, "The secure tenancy comes to an end on the grant to the tenant," etc can only mean that there is to be a secure tenancy until the grant. I note also the point that section 155(1) presupposes that the conveyance was to a secure tenant. For those reasons, I agree that the meaning of section 138 is that the landlord shall make a grant to the secure tenant, a right which only a secure tenant can enforce by injunction.

If, however, I had any doubt about the meaning of the statutory provisions, I should take into account the fact that the statutory regime gives the secure tenant considerable benefits — not merely to require his landlord to sell him the freehold, but also to sell it to him at a discount — which result in corresponding disadvantages for the landlord. It would follow, therefore, that the provisions of the statute should be strictly construed so as to preclude a person who has, for whatever reason, ceased to be a secure tenant from being able to claim the advantages.

Finally, I agree that we are bound by the decision of this court in *Sutton London Borough Council* v *Swann* (1985) 18 HLR 140 to conclude that a person who was no longer a secure tenant was no longer entitled to the right to buy. The decisions in *Enfield London Borough Council* v *McKeon* [1986] 1 WLR 1007 and *Dance* v *Welwyn Hatfield District Council* [1990] 1 WLR 1097 were concerned with a different question, namely at what stage in the process towards a conveyance does the right to buy of a person who is at all relevant times a secure tenant become indefeasible? There is therefore nothing in those two decisions which prevents us from arriving at the decision we have announced.

For those reasons, which do no more than emphasise some of the reasoning of Mann LJ, with all of which I agree, I conclude that this appeal should be allowed. It follows that the respondents' counterclaim is dismissed and an order that they should give up possession of the dwelling-house must follow.

Appeal allowed with costs, application for leave to appeal to the House of Lords refused.

Court of Appeal

January 27 1993

(Before Sir Thomas BINGHAM MR, Lord Justice McCOWAN and Lord Justice HIRST)

TAYLOR v NEWHAM LONDON BOROUGH COUNCIL

Estates Gazette June 12 1993

[1993] 23 EG 126

Landlord and tenant — Housing Act 1985 — Whether secure tenant entitled to enforce right to buy by injunction where landlord housing authority issues possession proceedings

By a tenancy agreement dated August 6 1984 the plaintiff is a secure tenant of 96 Carlyle Road, Manor Park, London E12, within the meaning of section 79 of the Housing Act 1985. In October 1990 the plaintiff exercised her right to buy the house by written notice to the defendant landlords under section 122 of the Act; the defendants duly admitted the plaintiff's right to buy and ultimately a purchase price of £32,860 was agreed. On June 24 1991, following receipt by the plaintiff of a notice from the landlords of their intention to seek possession, the defendants issued proceedings in Bow County Court claiming possession of the house on the ground that the plaintiff was in breach of the terms of her tenancy being guilty of conduct which harassed and annoyed her neighbours, particulars being given of a number of acts committed between April 1988 and August 1989, being acts of an aggressive, violent, insulting and offensive nature and being the product of hostility borne of racial enmity. The victims of the plaintiff's conduct were almost all, if not entirely, drawn from the local Asian community. On July 12 1991 the plaintiff put in a defence to that claim denying the allegations. On March 1992 the plaintiff issued the present proceedings in Bow County Court claiming an injunction to enforce her right to buy under section 138(3) of the 1985 Act; on April 3 1992 the defendants put in a defence denying the plaintiff's right and pleading the acts of annoyance, discrimination and harassment, as set out

A in the possession action, and relying on section 121(1) of the 1985 Act to contend that the plaintiff was not entitled to exercise effectively her right to buy. On the hearing of a preliminary issue as to whether the county court had discretion to refuse the grant of an injunction to the plaintiff, Judge Goldstein held that the defendants' allegations did not give grounds upon which they could effectively resist the injunction; he had no discretion but to grant the injunction and ordered that the defence be struck out, that the sale of the house be completed and that the possession proceedings be stayed pending appeal. The defendants appealed, relying

B upon *Dance* v *Welwyn Hatfield District Council* for authority that any ground which would enable a defendant to resist an order for specific performance, in particular hardship to third parties, or any ground which might induce a court properly to refuse the grant of an injunction in the exercise of its discretion, will enable a local authority to resist the grant of an injunction under section 138(3).

***Held*: The appeal was dismissed. The effect of the judgment of McCowan LJ in *Dance* v *Welwyn Hatfield District Council* was that, once the conditions in section 138(1) are satisfied, the tenant is entitled as of right to an injunction to enforce the landlord's duty to convey the property to him on the terms**

C agreed or determined and the court has no discretion to withhold such an injunction. In the present case, like the *Dance* case, the claim for possession had not been decided when the conditions of section 138(1) were satisfied and therefore did not provide a ground for withholding the injunction. It is plain, as a matter of construction, that Parliament did not intend that general considerations of hardship should afford any ground for refusing the grant of an injunction under section 138(3) where there is a right to buy under section 118.

The following case is referred to in this report.

D

Dance v *Welwyn Hatfield District Council* [1990] 1 WLR 1097; [1990] 3 All ER 572; (1990) 22 HLR 339, CA

This was an appeal by the defendants, Newham London Borough Council, from an order striking out their defence at a hearing of a preliminary issue into an application by the plaintiff, Cathryn Taylor, for an injunction under section 138 of the Housing Act 1985 to enforce her right to buy 96 Carlyle Road, Manor Park, London E12.

David Watkinson (instructed by the solicitor to Newham London Borough Council) appeared for the appellants; David Brook (instructed by Kenneth Elliott & Rowe, of Romford) represented the

E respondent.

Giving the first judgment, **SIR THOMAS BINGHAM MR** said: The plaintiff in this action is the tenant of a house owned by the defendant local authority. She is entitled to buy the house under the provisions of Part V of the Housing Act 1985. The issue in the appeal is whether, on her application to enforce that right to buy by injunction, the court has a discretion to refuse that relief. There is further a subsidiary question as to whether on the facts here, if it does have such a discretion, it should exercise it to deny her an injunction.

The facts are very short and very simple. The plaintiff lives at 96 Carlyle Road, Manor Park, London E12. The defendant local authority are her landlords and own the house. By virtue of a tenancy

F agreement dated August 6 1984 the plaintiff is a secure tenant of the house within the meaning of section 79 of the Housing Act 1985. She is a person entitled to buy the house under section 118 of the 1985 Act.

In October 1990 she claimed to exercise her right to buy the house by written notice to the landlords under section 122 of the Act. The local authority duly admitted the plaintiff's right to buy under section 124. In due course, but after a certain amount of negotiation and pressure, a purchase price of £32,860 was agreed.

Just before that stage was reached, on January 18 1991, the tenant received from the landlords notice of the landlords' intention to seek possession. That was some months before the particulars of the offer to buy were finally agreed with the landlords. The outcome was that on June 24 1991 the local authority issued proceedings in Bow County Court, claiming possession of the house. The grounds upon G which that claim was made was that the plaintiff was in breach of the terms of her tenancy. Putting the matter very shortly, it was contended that the plaintiff herself and her family had been guilty of conduct which harassed and annoyed her neighbours. Details were given in the pleading of a number of acts said to have been committed between April 1988 and August 1989, being acts of an aggressive, violent, insulting and offensive nature and being, as it was alleged, the product of hostility born of racial enmity. Certainly, the pleading was to the effect that the victims of the plaintiff's conduct were almost, if not entirely, drawn from the local Asian community.

A defence was put in dated July 12 1991 to that claim for possession in which the right to possession and also the facts pleaded were H roundly denied. It is right to emphasise that those allegations have not been investigated and they should be regarded as no more than allegations which are contradicted. It is, however, relevant to point out that at the date of the claim the events were some two to three years old and that, at the time when the plaintiff exercised her right to buy the house by giving notice of her intention to do so, there was no claim for possession made against her, let alone any possession order.

The next stage of the proceedings came on March 9 1992 when these proceedings were issued in Bow County Court and the plaintiff in those proceedings claimed an injunction to enforce her right to buy under section 138(3) of the Act. The defendants put in a defence on April 3 1992 in which, in para 12, they denied the plaintiff's right to J an injunction. They pleaded the acts of annoyance, discrimination and harassment set out in the particulars of claim in the possession action and they relied on section 121(1) of the Housing Act 1985 to contend that the plaintiff was not entitled to exercise effectively her right to buy. The county court ordered that the claim for an injunction should be heard on May 5 1992 when the possession action was also due to be heard.

Thus it was that the parties went to Bow County Court on May 5 1992 and then, by agreement, there was argument on a preliminary point whether the county court had discretion to refuse the grant of an injunction to the plaintiff. Judge Goldstein, who heard the argument, gave his judgment on May 12 1992 and it was plainly a carefully considered judgment. I would take this opportunity not K only to acknowledge our indebtedness to the learned judge for giving such a clear and careful judgment, but also to observe how helpful it now is when county court judges' judgments reach us in the form of transcripts taken from the tape and give us a very much better idea of what the learned judge actually said than was available in days of yore.

In his judgment the learned judge very plainly indicated the issue which he had to decide. He said, referring to the plaintiff on p2B:

> She has now become entitled, under Part V of the 1985 Housing Act to exercise her right to buy that house from her landlord. Three important matters are not in dispute in this case: first, that she is entitled in law to make her application to buy; second, that she is not a disqualified person under L section 121 of the Act from exercising that right; last despite some delay in the matter all the formalities for completing the transfer of the property from the landlords to the secure tenant are now in place.
>
> The defendants, however, do not wish to go through with the sale. They say it would not be equitable to force them to complete the sale of this house to this plaintiff. Thus it is that this action has been brought on behalf of the plaintiff who seeks to enforce the sale under section 138(3) of the same Act by asking this court to grant her an injunction.
>
> To that claim the defendants have put in and put forward the defence, discussion of which has formed the preliminary issue in this trial. If I could just summarise the defence — I hope doing it no injustice — what is being said is to grant an injunction is of course to exercise a discretionary remedy which the court may or may not so exercise. That the court should exercise the discretion in the defendant's favour and not enforce the sale for the following M reasons: first, it would be a hardship to the defendants if the injunction were granted; second, the defendants owe a duty to third parties and in all the circumstances of this case it would be unreasonable if the court were to act in such a way as to prevent the defendant landlords from discharging its duty to those third parties.
>
> The real basis for putting forward such a claim arises in this way. In the defence, and indeed in another separate obviously linked matter, the defence outline an alleged claim for possession which they say they have against the defendant based upon what is basically one act of alleged racial harassment in 1989, proceedings of which were only commenced in June 1991.

The learned judge then reviewed the Act and in particular the authority *Dance* v *Welwyn Hatfield District Council* [1990] 1 WLR 1097 (to which I shall come) and then at p10B of his judgment he said:

A It seems to me here that what these defendants are seeking to do is one of two things. They are either seeking to circumvent the principle of the right to buy, which the court obviously cannot and will not condone, or they are seeking to impose some form of condition that any proposed purchaser must be a person of whom the landlords approve and nowhere in the legislation can I find support for that proposition.

He accordingly held that the landlords' allegations did not give grounds upon which the landlords could effectively resist the injunction and, that being so, he did not go into the allegations. He did, however, add that, even if the complaints were made out and even if he did have a discretion to withhold an injunction, he thought it very doubtful if he could make a possession order in the local authority's favour in all the circumstances. That question, however,
B did not arise for decision since he ruled that the defence of April 3 1992 disclosed no reasonable defence. He accordingly ordered that the defence be struck out, that the sale of the house be completed and that the possession proceedings be stayed pending appeal. He also gave leave to appeal.

It is not necessary for the purposes of this judgment to go at great length into the provisions of the Housing Act 1985 Part V. It suffices, I think, to refer to section 118(l)(*a*) which provides:

> A secure tenant has the right to buy, that is to say, the right, in the circumstances and subject to the conditions and exceptions stated in the following provisions of this Part —
> (*a*) if the dwelling-house is a house and the landlord owns the freehold, to acquire the freehold of the dwelling-house;

C Section 120 sets out exceptions to the right to buy by reference to cases specified in Schedule 5, and section 121 defines the circumstances in which the right to buy cannot be exercised. So far as material it reads:

> (1) The right to buy cannot be exercised if the tenant is obliged to give up possession of the dwelling-house in pursuance of an order of the court or will be so obliged at a date specified in the order.
> (2) The right to buy cannot be exercised if the person, or one of the persons, to whom the right to buy belongs —
> (*a*) has a bankruptcy petition pending against him . . .
> (*c*) is an undischarged bankrupt, or
> (*d*) has made a composition or arrangement with his creditors the terms of which remain to be fulfilled.

D There follow a series of sections outlining the procedure to be followed when the right to buy is exercised and dealing with the right to a mortgage and ascertainment of the purchase price. Then one comes to section 138, which is the relevant section for present purposes. It provides:

> (1) Where a secure tenant has claimed to exercise the right to buy and that right has been established, then, as soon as all matters relating to the grant and to the amount to be left outstanding or advanced on the security of the dwelling-house have been agreed or determined, the landlord shall make to the tenant —
> (*a*) if the dwelling-house is a house and the landlord owns the freehold, a grant of the dwelling-house for an estate in fee simple . . .
> in accordance with the following provisions of this Part.
> E (2) If the tenant has failed to pay the rent or any other payment due from him as a tenant for a period of four weeks after it has been lawfully demanded from him, the landlord is not bound to comply with subsection (1) while the whole or part of that payment remains outstanding.
> (3) The duty imposed on the landlord by subsection (1) is enforceable by injunction.

It is that last provision which founds the plaintiff's claim in this action.

In submitting in this court that the learned judge below erred in holding that he had no discretion, Mr Watkinson, who argues the case on behalf of the defendant local authority, relies principally on a recent authority of this court in *Dance* v *Welwyn Hatfield District Council* [1990] 1 WLR 1097, to which I have already referred. In that
F case the tenants sought to exercise their right to buy, but they wanted to leave part of the purchase price outstanding and accordingly made arrangements to defer their purchase for a period of three years. The council acknowledged that the tenants had chosen to defer the purchase. The council also accepted a deposit towards the purchase price, and the council confirmed that there was no compulsory purchase scheme which affected houses in the area of that which the plaintiffs wished to buy. Unhappily, however, the housing committee of the council had resolved to demolish houses including the plaintiffs'. Accordingly, eight months later the council informed the tenants that they proposed to demolish the house and offered to return the deposit. The tenants, however, wrote a letter on September 2 1988 in which they required the council to complete the sale of the property to them and specified a date by which completion was to G
take place. The council did not comply with that notice and instead served a notice of their own, seeking possession on the grounds that they proposed to demolish the house. At that stage the tenants issued proceedings, claiming an injunction under section 138(3) of the 1985 Act. The council for their part counterclaimed for possession and said that the right to buy was defeated by section 121(1), which is the section already referred to, which makes the existence of a possession order a ground of defence. The council also issued separate possession proceedings. The learned judge in that case acceded to a request by the council that the tenants' claim for an injunction should be adjourned until the merits of the claim for possession had been investigated. Against that decision the tenants appealed and the appeal in this court was allowed. H

There were observations in the course of the first judgment given by Nourse LJ in this court on which Mr Watkinson principally founds his submissions now. He points out that in the opening sentences of his judgment Nourse LJ made reference to the similarity between the right to buy provided by the Act and an option to purchase. He also draws attention to Nourse LJ's statement on p1104 of the report below E that:

> The right to an injunction in a case where there is no contract cannot be any different in its incidents from the right to an order for specific performance in a case where there is a contract;

And he draws attention to the passage on p1105 below F where Nourse LJ says: J

> The comparison may or may not be exact, but the injunction cannot be *less* readily available than an order for specific performance, which, although it is a discretionary remedy, is refused only on well settled grounds, for example hardship to the party against whom it is granted.

In reliance on those observations in a judgment with which the Vice- Chancellor agreed, Mr Watkinson submits that any ground which would enable a defendant to resist an order for specific performance, in particular hardship to third parties, or any ground which might induce a court properly to refuse the grant of an injunction in the exercise of its discretion, will enable a local authority to resist the grant of an injunction under section 138(3). Accordingly, he submits that the injury to third-party neighbours of the plaintiff's is a legitimate consideration to which attention should K
be given in considering whether the statutory remedy of an injunction should be granted.

Approaching the matter at this stage purely on the basis of this authority, I find myself wholly unpersuaded by that argument. In drawing attention to the possible analogy with an option to purchase or with specific performance Nourse LJ was doing what any lawyer tends to do when confronted by a new statutory animal, which is to consider and reflect on what analogies there are to that statutory animal in the existing law. Thus it was natural for Nourse LJ to consider options to purchase and specific performance, both of which clearly have some resemblance to the statutory animal created by the 1985 Act. L

What, however, I feel sure that Nourse LJ did not intend was that the analogy in the general law should wag the statutory dog. He was dealing with a case in which the right to an injunction was quite clear. There was no question of hardship and I find no reason whatever to suppose that Nourse LJ was intending to introduce into this straightforward statutory remedy doctrines which have grown up around both the grant of specific performance and the grant of an injunction. I would, however, observe, in case there should be any doubt about it, that various grounds which do enable a court properly to resist an order for specific performance such as mistake, misrepresentation or fraud would unquestionably afford a reason for not making an order under section 138(3) because they would be grounds for saying that the conditions in section 138(1) had not been M
fulfilled.

It is not, however, necessary to confine one's consideration of the reasoning in *Dance* to the judgment of Nourse LJ alone since my lord, McCowan LJ, was also a member of that court and gave a judgment with which the Vice-Chancellor also agreed. I draw attention in particular to what McCowan LJ said at p1106C:

> From receipt of the letter of 2 September 1988 the defendants were under a duty to convey the house to the plaintiffs on or before 26 September 1988. It follows that from that date the defendants were in breach of their statutory duty to convey the house, and the plaintiffs were entitled to an injunction to enforce that duty. It would, in my view, be very regrettable if the defendants could, by refusing or failing to carry out their statutory duty, improve their position. Therefore, unless I am obliged by the words of the statute or by

A authority to decide otherwise, I would hold that, certainly by the time 26 September 1988 had passed, it was too late for the defendants to take steps to defeat the plaintiffs' entitlement to completion.

Section 121(1) of the Housing Act 1985 could not assist the defendants at the hearing before the judge because at that date it was not possible to say that "the tenant is obliged to give up possession of the dwelling-house in pursuance of an order of the court or will be so obliged at a date specified in the order".

McCowan LJ then distinguished an earlier authority of this court which had been considered in argument and continued at H:

Before us, Mr Haines for the defendants put the matter in this way: "I accept the judge had a discretion to grant the plaintiffs the injunction they sought. Had he done so, I could not have complained, but he chose to exercise his B discretion by adjourning the application in order to hear at the same time the defendants' claim for possession. Unless it can be said that that exercise of discretion was wholly wrong, this court should not interfere." In my judgment, the judge had, in the circumstances of this case, no discretion to refuse the plaintiffs an injunction, and no question therefore arose, as the judge thought, of a "balancing exercise required to exercise the discretion either to grant or refuse an injunction and consequently to grant or refuse an order for possession." Accordingly, I do not consider that he had a discretion to adjourn the matter to allow the defendants to get their claim for possession before the court.

This was, I repeat, a judgment with which the Vice-Chancellor agreed and plainly McCowan LJ was holding in robust terms that there was no discretion and that the learned judge was obliged to C make the order which was asked in favour of the tenants.

I understand the upshot of that judgment as being that, once the conditions in section 138(1) are satisfied, the tenant is entitled as of right to an injunction to enforce the landlord's duty to convey the property to him on the terms agreed or determined and the court has no discretion to withhold such an injunction.

In this case, like *Dance's* case, the claim for possession had not been decided when the conditions of section 138(1) were satisfied and therefore did not provide a ground for withholding the injunction. It may be going too far, and unnecessarily far, to hold that there is no conceivable situation in which there could be any discretionary ground for withholding an injunction but, for my part, I find it almost impossible to imagine circumstances in which such a D discretion would arise or be properly exercisable.

As a second string to his bow Mr Watkinson relied on the general law, and in particular a passage drawn from *Spry on Equitable Remedies* (4th ed 1990) at pp435 to 436. The passages to which he particularly drew our attention were these. First, a sentence to this E effect:

And even if the subject matter is new in the sense that rights are being created that had no prior equivalents that were protected by injunctions ordinary equitable considerations may be expected to apply unless a statutory intention to the contrary is established.

The second passage was:

However all these matters involve questions of construction, and care should be taken to ensure both that preconceptions by reference to the general practices of courts of equity do not cause legislation conferring special powers to be misapplied and also that in the exercise of widely expressed powers a desire to assist an inferred legislative policy does not cause discretionary considerations such as hardship to be overlooked.

Those observations occur in a passage dealing with statutory F injunctions and I do not for a moment question that it is the duty of any court, considering and construing a provision providing for the grant of an injunction, to construe the provisions in question and endeavour to give them their correct effect. What, however, seems to be crucial in the present situation is that section 118 creates what is described as a right to buy, and section 138, setting out the landlords' duty to convey to the tenant, expresses it in equivocal and unqualified mandatory terms. It seems to be quite plain as a matter of construction that Parliament did not intend in this instance that general considerations of hardship should afford any ground for refusing the grant of an injunction under section 138(3).

That consideration is in my mind greatly strengthened when one G recalls the history of this particular right. It was introduced for the first time in 1980 at a time when a number of local authorities strongly resisted parting with the ownership of publicly-owned accommodation to those who then lived in it, and it seems to me that we should be doing great violence to the obvious intention of Parliament if we did not recognise that it was Parliament's intention to block to the maximum the opportunities open to reluctant landlords to obstruct the acquisition of title by their tenants.

For all those reasons I consider that the learned judge in the court below reached an entirely correct decision. There was no discretion to refuse the injunction for which the plaintiff asked and, accordingly, I would dismiss the appeal.

McCowan and **Hirst LJJ** agreed and did not add anything. H

Appeal dismissed with costs; leave to appeal to the House of Lords refused; application for a stay of the injunction refused.

LANDLORD AND TENANT

GENERAL

A
Court of Appeal

July 6 1992

(Before Lord Justice NOURSE, Lord Justice McCOWAN and Lord Justice BELDAM)

BRENT LONDON BOROUGH COUNCIL v O'BRYAN AND OTHERS

Estates Gazette January 16 1993

[1993] 02 EG 113

B *Landlord and tenant — Ord 113 proceedings — Whether agreement to grant tenancy*

In or about 1988 the respondent, Michelle O'Bryan, who runs an infant school, occupied an ex-scout hut at 161 Willesden Lane, Brent, London NW6, owned by the plaintiff council, for the purposes of her school; the site adjoined Vernon House Day School belonging to the appellants. On August 29 1988 the first respondent wrote to the appellants' department of development expressing an interest in purchasing the site and on September 29 1988 she sent a cheque for £54.50 to the appellants for what she described as ground rent. On September 19 1988 the appellants' schools
C subcommittee resolved that, on the assumption that Vernon House Day School was to move shortly and the school and the adjoining site would be sold, the monthly letting of the site could be agreed by the council's valuers. On November 28 1988 the appellants' education officer wrote to the first respondent stating that "the Education Committee agreed to the monthly letting of the premises to you on the terms laid down by the Valuers. Accordingly, I am informing the Valuers of this decision so that they can draw up the relevant contract". In March 1989 the appellants' land and building subcommittee resolved that the respondent be offered a
D monthly tenancy until February 1990. A letter headed "subject to contract without prejudice" and offering terms for a monthly tenancy was sent to the respondent on February 13 1990. Although the respondent's solicitors replied seeking clarification of the offer, no acceptance was received and by a letter of June 18 1990 the appellants sought possession of the site. The appellants issued an originating summons under Ord 113 of the RSC seeking possession of the site and an adjoining area of land also occupied by the respondent. Because the respondent claimed that she had a tenancy the master remitted the matter to the county court, where Mr Assistant Recorder
E Morris held that the letter of November 28 1988 amounted to an agreement to grant a tenancy. The appellants appealed.

Held: **The appeal was allowed. The letter of November 28 1988 amounted to no more than an expression of the willingness of the appellants to make an offer to the respondent to let the premises to her on a monthly basis when their valuer had decided what the appropriate terms were. If it were an offer the assistant recorder had not considered whether that offer had ever been accepted. The assistant recorder did not dismiss the appellants' application on the basis merely that he found an arguable case for the respondent and the appellants could take further proceedings if they so** F
wished. At the trial the parties embarked on a hearing of the issues between them. There was no evidence from which the court could draw an inference that the appellants accepted the respondent as a tenant at will, at sufferance or on any other basis.

The following cases are referred to in this report.

Javad v *Aqil* [1991] 1 All ER 243; [1990] 2 EGLR 82; [1990] 41 EG 61
Sudbrook Trading Estate Ltd v *Eggleton* [1983] 1 AC 444; [1982] 3 WLR 315; [1982] 3 All ER 1; (1982) 44 P&CR 153; [1982] EGD 392; 265 EG 215, HL G

This was an appeal from a decision of Mr Assistant Recorder Morris, sitting at the Wood Green Trial Centre, who had dismissed the application of the appellants, Brent London Borough Council, under Ord 113 of the RSC, remitted by the High Court to the county court, for possession of a site at 161 Willesden Lane, Brent, occupied by the respondent, Mrs Michelle O'Bryan.

Terence Gallivan (instructed by the solicitor to Brent London Borough Council) appeared for the appellants; Shane Dougall (instructed by Peter Kolbe, of Barnet) represented the respondents, Michelle O'Bryan and other teachers at L'Isle aux Enfants.

Giving the first judgment at the invitation of Nourse LJ, **BELDAM** H
LJ said: In the London Borough of Brent, on the site of 161 Willesden Lane, there was for a number of years a scout hut. It adjoined Vernon House Day School. The 27th Willesden troop of scouts ceased to use the hut in about 1986.

The first defendant runs an infant school for young children, many of foreign background, under the style of L'Isle aux Enfants. Initially she ran the school from her home, but the demand gradually exceeded the number for which the borough were prepared to grant her permission to run it from her home. Apparently two departments of the appellants were concerned — the social services department, whose function it was to issue registration certificates for nursery groups or schools, and the planning department.

Coming across the site of the now disused scout hut in about 1988, J
the first defendant, in her own words, through sheer frustration decided to occupy the premises at some time between May and August 1988. On August 29 of that year she wrote to the appellants' department of development, saying that she understood the site was no longer required and expressing an interest in purchasing it, and on September 29 she sent them £54.50 in cash in payment for what she described as ground rent for the hut site. Her occupation thus became known to the council and the director of private and voluntary play groups told the respondent that the nursery could not be re-registered at that address until the education department gave permission for the use of the hut for that purpose.

The situation of the scout hut and its future were considered at a meeting of the council's education subcommittee on September 19 K
1988. The background to this site was that no 161, being next door to Vernon House Day School at no 163, had in fact been earmarked for sale by the borough's schools subcommittee in either 1989 or 1990. Those plans, of course, were subject to the Department of Education's consent for relocating Vernon House Day School. Thus obviously no long-term arrangement could be made, but the schools subcommittee on September 26 resolved that "on the assumption that Vernon House will move mid-term", which was October, "the Committee agreed to a monthly letting of the premises on terms to be agreed by the valuers". The council's valuers were, of course, the normal avenue for negotiation for either the letting or sale of premises.

A On October 7 the acting director of education wrote a memorandum to the director of development referring to the arrangements which the education subcommittee had sanctioned as "a licence to use the building, but at an economic rent". On November 28 the education officer asked the council's valuation officers to "arrange for an assessment of what it would take to rent the premises to Mrs O'Bryan on a monthly basis". These departmental exchanges, entirely within the council's normal organisation, were not at that time known to the respondent. But on November 28 the education officer wrote to the respondent a letter in these terms. He said:

> I am writing to inform you that the Education Committee agreed to the B monthly letting of the premises to you on the terms laid down by the Valuers. Accordingly, I am informing the Valuers of this decision so that they can draw up the relevant contract.

By the end of 1988 it had become clear that the Department of Education would take longer than expected to give approval for the movement of Vernon House Day School and in March 1989 at a meeting of the appellants' land and building subcommittee it was resolved that the respondent be offered a monthly tenancy until the end of February 1990. However, it was not until February 13 1990 that any offer of a tenancy was made to the respondent. When it was made it was made in these terms. The letter was headed "Subject to Contract Without Prejudice" and addressed to the first respondent. It was in regard to 161 Willesden Lane, NW6:

> C As you are aware, the Education Committee has agreed to a monthly letting of the above property you occupy.
>
> I am proposing to grant you a monthly licence under the following terms.
>
> 1. Monthly rental of £400 per month retrospectively from 1.5.89.
>
> 2. Fees to be paid 1st day of the month in advance and by Standing Order Mandate.
>
> 3. 14 days Notice of termination of agreement by either side.
>
> May I please have confirmation of your acceptance in order to seek the Committee's approval.

That was signed by the Assistant Director of Development (Employment). That letter was answered by solicitors instructed on behalf of the respondent on March 30 1990. It read:

> 161 Willesden Lane NW6 — Subject to Contract
>
> D We have been instructed to act on behalf of Michelle O'Bryan with respect to this matter.
>
> We refer to your letter of 13th February, and have the following observations to make:
>
> 1. Does the monthly rent include payment with respect to rates?
>
> 2. Kindly advise us who is responsible for all structural and external repairs to the building?
>
> 3. Kindly note that our client carried out repairs to the roof on 4th March 1990, in addition our client has spent a considerable amount of money decorating the premises insofar as, when she occupied the premises originally in September 1988, there were no windows, no toilets, no kitchen, and the premises were infested with rats.
>
> As you will note our client uses the premises as a school with 40 children attending between the ages of 2 and 5 years of age. Our client would ideally E like to purchase this building together with house number 163 Willesden Lane. We understand from our client that she has spoken to Mr Benhaim of the Education Department, who does not object to our client's proposal. We await to hear from you soon.

By April 1990, apparently, arrangements had been able to be made for the removal of Vernon House Day School, and the council, having had no acceptance of their offer, instructed solicitors, who wrote to the respondent on June 18 1990 requiring possession on July 20. The respondent through her solicitors then sought to negotiate a purchase, suggesting in the course of those negotiations that she could offer a substantially lower sum than the market value as she was entitled to claim to be in occupation pursuant to a tenancy to which Part II of the Landlord and Tenant Act 1954 applied. F Eventually the appellants issued the proceedings which give rise to this appeal. They claimed possession from the respondent under the provisions of Ord 113, of no 161 and 163 Willesden Lane, because it appears that the respondent had been making use of the playground at the back of the premises at no 163, initially when they were not required for the purposes of the school and subsequently more generally.

The appellants' application came before the master in the usual way. There was an affidavit in support from the appellants and one in response by the respondent, who claimed in the course of that affidavit that she had the right to remain in the premises. In those circumstances, as is customary when issues of fact have to be determined in order to establish whether the owner of premises can make good his claim to occupation under those provisions, the G master remitted the application to the county court and it was heard at the Wood Green Centre by Mr Assistant Recorder Morris. Both parties embarked on the trial on the basis of the issues raised by the affidavits. There had been considerable discovery of documents principally given by the appellants, who had produced the various minutes of meetings and notes for briefing and inter-departmental memoranda to which I have already made reference. In support of their claim the appellants called Mrs Jadhav, chief planner, projects and resources for development, to give evidence. She completed her evidence-in-chief and, at the end of the first day of the hearing, Mr Dougal, who appeared on behalf of the defendants, reserved further cross-examination. When the court reconvened on the following day, the learned recorder embarked upon lengthy legal submissions, H in the course of which he eventually said:

> Well, Miss Hutt,

— Miss Hutt was then appearing for the appellant —

> I'm against you at the moment. I find that this letter of the 28th November 1988 is constituting sufficient memorandum in writing to constitute a tenancy, a monthly tenancy.

Miss Hutt asked him whether that related both to the land at 161 and 163 Willesden Lane and the assistant recorder said: "That we haven't addressed our minds to at this stage." Later, after hearing further discussion, he said:

> At the moment I'm making a ruling that it was a tenancy granted. All I'm now J determining is what the extent of the property was. If you'd like to hear Mrs O'Bryan, and if necessary you can then recall Mrs Jadhav.

That was the course which was then adopted. So, in the course of cross-examination of Mrs Jadhav, the assistant recorder had concluded, on the basis of the letter which I have quoted, that a tenancy was created though he had not determined the nature or extent of the property demised. He then proceeded to hear further evidence. Although he heard further evidence about the extent of the letting, he made no findings on that or any other terms of the tenancy and, in due course, he gave a judgment which I can describe only as being deficient in either the relevant findings of fact or proper reasons for his decision. After reciting the terms of the letter of November 28 1988 and going on to record that nothing further had K then happened, he said:

> That in my view is an agreement to grant a tenancy of the premises, a monthly tenancy of the premises to Mrs O'Bryan and that fully defeats the application by the council to remove Mrs O'Bryan as a squatter.

He then went on to consider that part of the appellants' application in which they claimed to be entitled to possession of no 163. That, so far as I can discern, is the extent of the reasons he gave for his decision.

From the facts which I have recited it is clear that, apart from the payment of £54 odd said to be ground rent for the use of the scout hut in September, no further payments had been made. In my view, the assistant recorder, who was referred in the course of argument to the L case of *Javad* v *Aqil* [1991] 1 All ER 243*, would have done well to have heeded the advice of the Vice-Chancellor giving the first judgment in that case. At p 248F, he said:

> Of course, when one party permits another to enter or remain on his land on payment of a sum of money, and that other has no statutory entitlement to be there, almost inevitably there will be some consensual relationship between them. It may be no more than a licence determinable at any time, or a tenancy at will. But when and so long as such parties are in the throes of negotiating larger terms, caution must be exercised before inferring or imputing to the parties an intention to give to the occupant more than a very limited interest, be it licence or tenancy. Otherwise the court would be in danger of inferring or imputing from conduct, such as payment of rent and the carrying out of repairs, whose explanation lies in the parties' expectation that they will be able to reach agreement on the larger terms, an intention to grant a lesser interest, M such as a periodic tenancy, which the parties never had in contemplation at all.

Those words of caution, it seems to me, apply with even greater force to the facts in the present case where the appellants were faced with someone who, without their permission or licence, had, in fact, taken possession of a part of the property — that is the scout hut — and then extended it to a larger portion and had paid nothing thereafter. It would have been wise for the assistant recorder to have considered all the factors which were present in the documents to which I have referred and to have considered, in the light of those documents, what, in fact, was the meaning of the letter of November

---

*Editor's note: Also reported at [1990] 2 EGLR 82.

28 1988. Moreover, he would then have had to have considered, if this amounted to an offer to let the premises to the respondent, whether it was ever accepted. How it could be thought or even argued that the letter itself created a tenancy I confess I have failed to understand. It seems to me that the submission of the appellants is unanswerable, that it amounts to no more than an expression of the willingness of the appellants to make an offer to the respondent to let the premises to her on a monthly basis when their valuer has decided what the appropriate terms are to be.

Faced with this position, Mr Dougall sought, though without a respondent's notice, to support the decision of the assistant recorder with an argument that, under Ord 113 and the corresponding rule of the county court in Ord 24, a defendant, faced with a claim for possession in summary form such as this, has to show only an arguable case and that the assistant recorder could have decided the issues between the parties merely upon the basis that the appellants had failed to show that they were entitled to possession. The defendant had established an arguable case and he could have dismissed the application, leaving it to the appellants to take further proceedings for possession if they wished.

The first answer to that submission is that the assistant recorder did not make such a finding. Mr Dougall accepts that Ord 113 and County Court Ord 24 would apply in the case of a licensee whose licence to occupy the premises had been determined or to a tenant at sufferance or will whose right had similarly been determined. In my judgment, here the parties having embarked upon the hearing of the issues between them, it was incumbent upon the assistant recorder to decide that issue. The mere fact that it might involve an examination of documents and the hearing of evidence is beside the point. That is the purpose for which the claim was remitted to the county court. Two further points were taken by Mr Dougall. First, he contended that the letter of November 28 itself amounted to an offer by the appellants to let the premises to the respondent on terms that were to be agreed by the valuer. He contended that that offer was in fact accepted by the respondent remaining in possession of the premises.

There are, of course, cases when, having entered under an agreement for a lease, a party remains in possession under that agreement and it is a reasonable inference that the relationship of landlord and tenant was intended to continue between them, but in cases in which reliance is placed on an inference of that kind it has been stressed that there is no other reasonable inference to be drawn. That is far from the case here where, as I have said, the respondent initially took occupation without permission or licence of any kind. The appellants, faced with her presence in the property, clearly did not then take steps to evict her, but from their failure to do so I do not see how an inference can be drawn that they had accepted that she would be their tenant. Even if the letter were capable of amounting to an offer, which I do not think it was, her remaining in possession was simply the result of their acquiescence to her presence there. It was not referable to any grant of a tenancy. The second argument advanced was that the offer itself, in the letter of November 28, was capable of acceptance because it amounted to the offer of a tenancy at a rent to be agreed between the appellants and the respondent. Mr Dougall made reference to, as he had before the assistant recorder, the well-known case of *Sudbrook Trading Estate Ltd* v *Eggleton* [1983] 1 AC 444*, a case of a very different character which involved the question of an option to purchase. At p 479A the difference between the kind of case with which the House of Lords was there dealing and the case before us was clearly pointed out by Lord Diplock. He said:

> It may be that where upon the true construction of the contract the price to be paid is not to be a fair and reasonable one assessed by applying objective standards used by valuers in the exercise of their professional task but a price fixed by a named individual applying such objective standards as he personally thinks fit, and that individual, without being instigated by either party to the contract of sale, refuses to fix the price or is unable through death or disability to do so, the contract of sale is thereupon determined by frustration. But such is not the present case. In the first place the contract upon its true construction is in my view a contract for sale at a fair and reasonable price assessed by applying objective standards.

That observation is sufficient to point the difference between this case in which there was no question of a letting at a fair and reasonable rent. All that was said by the appellants was that the education committee were prepared to let the premises to the respondent on terms to be fixed by their valuer.

I can see no evidence from which the court could draw the inference that the appellants accepted the respondent as a tenant at will, at sufferance or any other basis. Faced with the situation that she had already taken possession of the premises without their permission or licence, they considered what they were to do. They were clearly prepared, on terms laid down by their valuer, to let the premises to her on a monthly basis, but that is a long way from saying that they accepted her as a tenant on a monthly basis with terms to be agreed. From her continued occupation and the appellants' forbearance, the very most that could be inferred is that they acquiesced in her possession and, at the highest, there was an implied licence for her to continue to occupy the premises. That licence was validly determined when the time which she was given came to an end on July 20 1990, and, in my judgment, the appellants fully establish their right to possession and quite properly claimed possession under the summary procedure laid down.

I would therefore allow the appeal.

McCowan LJ agreed and did not add anything.

Also agreeing, **Nourse LJ** said: I also agree that this appeal must be allowed for the reasons given by Beldam LJ. Since we are reversing the decision of the assistant recorder, I will state the essence of my reasoning in my own words.

The assistant recorder held that the letter of November 28 1988 was an agreement to grant a monthly tenancy of the premises to Mrs O'Bryan. Mr Dougall accepts that that cannot be literally correct and that the highest it can be put is that the letter was an unconditional offer by the council, capable of acceptance by Mrs O'Bryan, to let the premises to her on a monthly tenancy at a rent to be laid down by the valuers. Even if it is put no higher than that, the submission assumes, first, that the letter was intended to be such an offer as opposed to one subject to a contract to be drawn up in the first instance by the valuers and thereafter agreed and exchanged between the parties; second, that there would be no difficulty over the identification of "the premises"; third, that "the terms" to be laid down by the valuers referred only to rent; fourth, that an offer to let premises at a rent to be prescribed without limit by servants of the landlords is, on acceptance by the tenant, capable in law of constituting an enforceable agreement for a letting. Although to make all these assumptions in favour of Mrs O'Bryan is no doubt highly unrealistic, I am nevertheless prepared, for the purposes of meeting the remainder of Mr Dougall's argument, to make them.

Where, then, does that leave him? He has to say that, by remaining in possession and, I will assume, by carrying out repairs and decorations to the premises and perhaps by making some improvements to them, Mrs O'Bryan accepted the offer made to her by the council and, moreover, accepted it whatever rent might be laid down by the valuers. Here my credulity is tested to breaking point. It is impossible to infer that Mrs O'Bryan agreed to pay whatever rent the valuers might lay down. The only reasonable inference is that she was awaiting intelligence of a figure before she decided whether to accept it. So there was no acceptance of the offer at that stage. Moreover, it is clear from the council's letter to Mrs O'Bryan of February 13 1990 and from her solicitors' reply of March 30 1990 that when a rent of £400 was laid down it was not accepted.

In the circumstances, making every credible assumption in favour of Mrs O'Bryan, I am satisfied that there never was an agreement between her and the council and therefore no tenancy as defined by section 69(1) of the Landlord and Tenant Act 1954. For the further reasons given by Beldam LJ that means that the council are entitled to possession of the premises.

Appeal allowed with costs here and below; application for leave to appeal to the House of Lords refused.

---

*Editor's note: Also reported at (1982) 265 EG 215.

A

Court of Appeal

July 31 1992

(Before Lord Justice NEILL, Lord Justice STEYN and Sir Christopher SLADE)

DARLINGTON BOROUGH COUNCIL v DENMARK CHEMISTS LTD

Estates Gazette January 16 1993

[1993] 02 EG 117

B *Landlord and tenant — Forfeiture of lease — Whether discretion to refuse relief for forfeiture properly exercised — Whether Court of Appeal can exercise discretion to grant relief — Whether tenant entitled to an inquiry in respect of unjust enrichment by landlord*

By a lease dated May 19 1987 the respondent council demised a site to the appellant company for a term of 125 years at a yearly rent of one peppercorn for a premium of £12,000. By clause 3(iii) of the lease the appellant covenanted at its cost within a defined timetable to erect a medical practitioner's surgery building. The lease contained negative covenants by the appellant: (i) not to use or permit the use of the premises
C for the purposes of retail sales save in connection with a retail pharmacy; and (ii) not to use or permit the use of the premises otherwise than as a surgery for qualified medical practitioners for use in connection with the treatment of National Health Service patients together with a retail pharmacy. Following the grant the appellant erected a pharmacy, which opened for business on August 11 1987, but no surgery has ever been built on the land. On January 23 1989 the respondents served a notice on the appellant under section 146 of the Law of Property Act 1925 alleging breach of covenants in that no surgery was erected nor were the premises used for that
D purpose. No remedial work was carried out and on September 1 1989 the respondents issued the writ in the present proceedings claiming possession following forfeiture. Judge Bryant, in Darlington County Court, refused the appellants relief from forfeiture. The appellants appealed.

***Held*: The appeal was dismissed. On the evidence before him, that the appellant was losing money and its directors had no intention of building the surgery at once, the trial judge was entitled to conclude that, if he granted relief on condition that the surgery was erected immediately, the appellant would be unwilling to comply with the condition. Although in an**
E **appropriate case an appellate court can exercise a fresh discretion under section 146(2) of the 1925 Act where circumstances have changed and new evidence is available, it is important to distinguish between a change of circumstances and a change of heart. The affidavit evidence filed with the Court of Appeal at the close of, and after, the hearing and directed to the willingness and financial ability of the appellant to erect the surgery did not show a change of circumstances. Any enrichment by the respondents in recovering possession with the benefit of the pharmacy building erected by the appellants was not unjust.**

F The following cases are referred to in this report.

Hyman v *Rose* [1912] AC 623

Shiloh Spinners Ltd v *Harding* [1973] AC 691; [1973] 2 WLR 28; [1973] 1 All ER 90; (1973) 25 P&CR 48, HL

Southern Depot Co Ltd v *British Railways Board* [1990] 2 EGLR 39; [1990] 33 EG 45

This was an appeal by Denmark Chemists Ltd from a decision of Judge Bryant, sitting in Darlington County Court, who made an order for possession of premises in favour of the respondents, Darlington Borough Council, against the appellants and dismissed the appellants' claim for relief from forfeiture.

Nicholas Patten QC and John Fryer-Spedding (instructed by Grunhut Makepeace & Duffy, of Tyne & Wear) appeared for the appellants; Christopher Parker (instructed by the solicitor to G Darlington Borough Council) represented the respondents.

Giving judgment, **NEILL LJ** said: This is an appeal by Denmark Chemists Ltd ("Denmark") from the order of Judge Bryant dated July 10 1991, whereby it was ordered that Denmark do give Darlington Borough Council ("Darlington") possession of the land and premises in Denmark Street, Darlington, identified on a plan annexed to a lease between Darlington and Denmark dated May 19 1987. The order for possession was made in forfeiture proceedings brought by Darlington against Denmark, in which Denmark unsuccessfully sought relief from forfeiture under section 146(2) of the Law of Property Act 1925 ("the 1925 Act").

By the lease made on May 19 1987 Darlington demised to Denmark H (then called Plansweep Ltd) the piece of land identified on the plan annexed thereto for the term of 125 years at the yearly rent of one peppercorn. The lease was expressed to be made in consideration of the sum of £12,000 and of the expense to be incurred by Denmark under the lease and in consideration of the covenants on the part of Denmark contained in the lease.

Clause 3 of the lease contained no less than 35 covenants by Denmark. Five of these covenants are relevant in these proceedings. In addition, the lease contained a proviso for re-entry if Denmark should "fail to comply with the covenants contained in clause 3(iii) and clause 3(iv)" or "if any other covenant agreement condition or stipulation" on Denmark's part should not "be performed or J observed". It is not necessary to refer further to clause 3(iv) which related to the construction of a road. By clause 3(iii) Denmark covenanted:

> At their own cost within six months from the date hereof to commence and within either the period of two years from the date hereof or twelve months of the first occupation of any part of the Demised Premises whichever shall be the sooner to complete the erection upon the Demised Premises of a building or buildings and works (together hereinafter called "the said building") to be used as a surgery for qualified medical practitioners together with a retail pharmacy and for no other purpose in a thoroughly sound and workmanlike manner and in conformity in every respect with plans designs layouts and specifications to be approved by the Council in its capacity as Landlord and in addition to any approval required of the Council in any other capacity Provided that if the Tenants are unable for reasons beyond their control to K complete the said buildings within the said period of two years the Council on application being made to it in writing by the Tenants before the expiry of the said period may extend the period by such further period or periods of time as may appear to the Council to be reasonable.

Clause 3(xxxi) contained a further positive covenant by Denmark:

> Within the period of twelve months from commencing to use the Demised Premises as a retail pharmacy and in any case within two years from the date hereof to commence to use the Demised Premises as a surgery for qualified general medical practitioners as referred to in subclause (xxx) hereof.

In addition, the lease contained three relevant negative covenants. Clause 3(xi):

> Not to use or permit the Demised Premises to be used for the purpose of retail L sales other than for retail sales made in connection with the use of the Demised Premises as a retail pharmacy as permitted by sub-clause (xxx) hereof.

Clause 3(xxx):

> Not to use the Demised Premises or permit or suffer the same to be used otherwise than as a surgery for qualified general medical practitioners for use in connection with the treatment of National Health patients only together with a retail pharmacy and not to carry on or permit to be carried on upon the Demised Premises any other profession trade or business whatsoever.

Clause 3(xxxii):

> Except as permitted by sub-clause (xxxi) hereof not to use the Demised Premises or any part thereof as a retail pharmacy without simultaneously using the greater proportion of the floor area contained in the said building as a surgery for qualified general medical practitioners as referred to in sub- M clause (xxx) hereof.

Following the execution of the lease Denmark erected a pharmacy, which opened for business on August 11 1987, but no surgery has ever been built on the land.

On January 23 1989 Darlington served a notice under section 146 of the 1925 Act setting out the following alleged breaches of covenant:

> (a) The tenants have failed to commence within six months of the date of the Lease and complete within twelve months of the llth August, 1987 being the first occupation of the Demised Premises the erection upon the Demised Premises of a surgery for qualified general medical practitioners together with a retail pharmacy.

A (b) The tenants have failed to commence within twelve months of commencing to use the Demised Premises as a retail pharmacy the use of the Demised Premises as a surgery for qualified general medical practitioners.

(c) The tenants have used the Demised Premises as a retail pharmacy without simultaneously using the greater proportion of the floor area contained in the building as a surgery for qualified general medical practitioners.

It will be seen that the notice alleged breaches of Clause 3, subclauses (iii), (xxxi) and (xxxii) of the lease. The notice required Denmark to remedy the breaches within a reasonable time and to pay compensation and costs. No remedial work was carried out, however, and on September 1 1989 Darlington issued the writ in the present proceedings. The proceedings were subsequently remitted B for trial to Darlington County Court.

At the hearing before Judge Bryant in May 1991 three principal arguments were advanced on behalf of Denmark:

(a) That the covenants in the lease had been frustrated and should be treated by the court as suspended because after the execution of the lease they had discovered that the Darlington area had previously been declared an "intermediate" area by the local family practitioner committee ("the FPC") and that it was therefore unlikely that any general medical practitioner would be permitted by the FPC to practise at the premises.

(b) That they should be relieved from forfeiture because they were ready, able and willing to carry out their obligations under the lease C whenever there was a use for such a surgery.

(c) That if Darlington recovered possession of the premises Darlington would be unjustly and inequitably enriched and that therefore Denmark should receive restitution from Darlington in respect of such unjust enrichment.

The judge rejected the argument based on "frustration" on the simple ground that the alleged "frustrating" event had occurred before the date of the execution of the lease. The judge said:

> Frustration essentially depended on the occurrence of a fundamentally different state of affairs to that which prevailed at the date the contract was made. It cannot apply to a case where the state of affairs before the contract was made is exactly the same as the state of affairs afterwards even if one party was mistaken about that state of affairs and subsequently discovers his mistake. I have no doubt that the doctrine of frustration has no application to D the circumstances of this case and that the defendants' defence fails. It may be that the doctrine of mistake would have some relevance but the defendants have declined, no doubt because of the consequences if it were successful, to pursue such a defence.

It seems to me that the judge was plainly right on this issue and no argument based on frustration has been pressed in this court.

The judge then turned to consider the claim for relief from forfeiture. He took account of the following circumstances:

(1) That the correspondence leading up to the grant of the lease showed clearly that Darlington were concerned that Denmark should build a surgery and a pharmacy which would operate together and that Denmark should not build or operate a pharmacy alone; in an E earlier part of his judgment the judge referred to the letter from Darlington's chief estates officer who had written on March 30 1987:

> If a retail pharmacy is to be attached to the Surgery and operates on a smaller scale than the treatment part of the premises, then this is acceptable to the Council, provided that the pharmacy does not operate alone.

(2) That if he granted relief from forfeiture without imposing a condition that the surgery must be erected immediately, the situation which Darlington had been at such pains to prevent, "namely the erection and operation of a pharmacy without an accompanying surgery would be allowed to continue".

(3) That Denmark's ignorance of the fact that the FPC was unlikely to give permission for a new practice to be opened at the premises was not the fault of Darlington, but resulted from F Denmark's "own failure to make simple and straightforward inquiries which any prudent person would have made immediately before agreeing to the terms of the lease".

(4) That if he granted relief from forfeiture with a condition that Denmark erected a surgery immediately, he would be imposing a condition with which Denmark "would certainly be unwilling and probably financially unable to comply". The judge therefore concluded that in all the circumstances it was not a proper case for granting relief from forfeiture.

Finally, the judge considered the claim for unjust enrichment, which was based on the contention that if Darlington recovered possession of the land they would obtain a pharmacy built by Denmark at a cost of £90,000. This was a benefit for Darlington, it was argued, for which it was equitable that Denmark should receive G compensation.

The judge rejected this claim also, holding that there was no reason to suppose that Darlington would be "enriched" by having the pharmacy; indeed they might wish to have it removed. He further held that even if Darlington were enriched by the pharmacy such enrichment was neither unjust nor inequitable. "I am unable to see," said the judge, "how the council's failure to appreciate that the defendant had not done its basic homework or to do it for the defendant can properly be stigmatised as unconscionable."

Denmark's case in the Court of Appeal

In this court counsel for Denmark made five main submissions:

(1) That in exercising his discretion not to grant relief from H forfeiture the judge had failed to consider

(a) the effect of forfeiture on Denmark and on Mr and Mrs Hussain, the owners of the shares in Denmark, and

(b) the fact that the failure to build a surgery *immediately* would not cause any hardship to Darlington.

(2) That the judge had erred in not considering the alternative solution whereby Denmark would have been required to erect the surgery forthwith, but on terms that the operation of the covenants as to user would be suspended for the time being.

(3) That in any event there had been a change of circumstances; there was now evidence before the court which showed that Denmark had sufficient financial resources to build the surgery.

(4) That the discretion given to the court under section 146(2) of the J 1925 Act was very wide. The court should look at all the circumstances.

(5) That though the judge rejected the evidence that Denmark had spent £90,000 on the pharmacy he was satisfied that about half that amount had been spent. Accordingly, if relief from forfeiture were refused and Darlington recovered possession, there should be an inquiry to determine the extent to which Darlington would be "enriched" by the work carried out by Denmark on the land.

Towards the close of his submissions counsel put forward a draft order which, he submitted, contained suitable conditions on which relief from forfeiture could be granted. The conditions in the draft order were in these terms: K

> 1. That within four weeks of the date of this order the Defendant shall apply to the Plaintiff for building regulation approval and for landlord's approval (under sub-clause 3(iii) of the Lease dated 19th May 1987 ("the Lease") of the plans designs layouts and specifications in respect of the surgery to be built at Denmark Street, Darlington ("the Surgery").
>
> 2. That within three months of the date of this order or within six weeks (if later) of receiving the last of the building regulations approval and the landlord's approval as aforesaid the Defendant shall commence the construction of the Surgery.
>
> 3. That the Defendant shall construct the Surgery in accordance with its obligations under clause 3(iii) of the Lease and shall complete the construction of the Surgery within nine months of commencing the construction thereof or such further period as the County Court may allow.
>
> 4. That within the period of two years commencing from the date of this L order or such further period as the County Court may allow (and for this purpose and for the purpose of condition 3 above the Defendant is to be at liberty to apply to the Darlington County Court) the Defendant shall commence to use the demised premises in accordance with the covenants contained in sub-clauses 3(xxx)-(xxxii) inclusive of the Lease.

I propose to consider Denmark's arguments under three headings:

(a) Whether the judge erred in refusing to grant relief from forfeiture.

(b) Whether this court is entitled to exercise a fresh discretion and, if so, whether we should grant relief from forfeiture.

(c) Whether Denmark are entitled to compensation in accordance with the principle that, in appropriate circumstances, compensation may be recovered from another party where (i) that party has been M enriched by the efforts of the claimant; and (ii) it is unjust that that party should retain the benefit of his enrichment.

The judge's refusal to grant relief

The statutory power to grant relief from forfeiture is conferred by section 146(2) of the 1925 Act, which, so far as is material, provides as follows:

> . . . the court may grant or refuse relief, as the court, having regard to the proceedings and conduct of the parties under the foregoing circumstances, thinks fit; and in case of relief may grant it on such terms, if any, as to costs, expenses, damages, compensation, penalty, or otherwise, including the granting of an injunction to restrain any like breach in the future, as the court, in the circumstances of each case, thinks fit.

A The discretion to grant relief is a wide one, as Earl Loreburn LC explained in relation to section 14(2) of the Conveyancing Act 1881, the forerunner of section 146(2), in *Hyman* v *Rose* [1912] AC 623 at p 631:

> I desire in the first instance to point out that the discretion given by the section is very wide. The Court is to consider all the circumstances and the conduct of the parties. Now it seems to me that when the Act is so express to provide a wide discretion, meaning, no doubt, to prevent one man from forfeiting what in fair dealing belongs to some one else, by taking advantage of a breach from which he is not commensurately and irreparably damaged, it is not advisable to lay down any rigid rules for guiding that discretion.

Counsel for Denmark relied on this passage in relation to the judge's decision and also in support of his submission that this court B should exercise a fresh discretion and grant relief. In addition he relied on the questions posed by Morritt J in *Southern Depot Co Ltd* v *British Railways Board* [1990] 2 EGLR 39* at p 44, where he said:

> I return to the words of the Lord Chancellor in *Hyman* v *Rose* . . ., which describe the purpose of section 146(2). I would paraphrase it by posing the questions whether the damage sustained by British Rail is proportionate to the advantage it will obtain if no relief is granted and, if not, whether in all the circumstances it is just that British Rail shall retain that advantage. I answer both questions in the negative.

It was submitted that in the present case the advantage to Darlington of regaining possession of the site on which the pharmacy had already been erected was quite disproportionate to the minimal damage which they would suffer if the erection, or at any rate the C occupation, of the surgery was delayed for the time being. It was further submitted that the judge should have considered the alternative solution whereby the surgery would have been constructed as soon as possible, but the covenant as to user would be suspended until some medical practitioners could be found to occupy it.

For my part, I have no doubt whatever that the judge was entitled to refuse relief on the basis of the evidence before him. What has been called "the alternative solution" in this court was not an issue before the judge. Mr Hussain told the judge in the course of his evidence that his business was losing money and that the surgery would cost within £150,000 to construct. He made it clear that he had no intention of D constructing the surgery at once. In cross-examination he said:

> As long as area is classified [as] intermediate I do not have to build the surgery. When they change classification I will fulfil my obligation.

In the light of this answer, which was consistent with Mr Hussain's contention before the judge that the obligation to build a surgery had been "frustrated", the judge was fully justified in concluding that if he granted relief on condition that the surgery was erected immediately "the defendant would certainly be unwilling . . . to comply". The judge also concluded that Denmark would probably be "financially unable" to build the surgery immediately. I see no ground whatever for disturbing this conclusion.

Whether this court can and should grant relief

E It was submitted on behalf of Denmark that even if the judge was entitled to refuse relief on the basis of the evidence then before him this court was entitled to and should exercise a fresh discretion and grant relief because the circumstances had changed. It was said that Denmark were now prepared to proceed with the construction of the surgery and that finance was now available.

I am satisfied that in an appropriate case an appellate court can exercise a fresh discretion under section 146(2) where circumstances have changed and new evidence is available. But it is important to distinguish between a change of circumstances and a change of heart by the person seeking relief. It is also important to bear in mind a passage in the speech of Lord Wilberforce in *Shiloh Spinners Ltd* v *Harding* [1973] AC 691, at p 723:

F
> . . . it remains true today that equity expects men to carry out their bargains and will not let them buy their way out by uncovenanted payment. But it is consistent with these principles that we should reaffirm the right of courts of equity in appropriate and limited cases to relieve against forfeiture for breach of covenant or condition where the primary object of the bargain is to secure a stated result which can effectively be attained when the matter comes before the court, and where the forfeiture provision is added by way of security for the production of that result.

It is true that Lord Wilberforce was speaking of the general equitable jurisdiction to grant relief from forfeiture but, in my view, it is equally necessary in applying section 146(2) to remember the "stated result" which the bargain between the parties was intended to secure. In the present case the purpose of the bargain was to secure G the construction of a pharmacy *and a surgery*. The council made it quite clear that they intended to avoid having a pharmacy alone on the land. Towards the close of the hearing of the appeal counsel for Denmark produced affidavits from Mr Hussain and Mr William Duffy dated June 24 1992. Mr Duffy is the solicitor who is acting for Denmark in these proceedings. In addition, since the hearing ended, further affidavits have been sent to the Registrar of Civil Appeals by both parties. This evidence is directed to the financial capacity of Denmark to construct a surgery on the land.

I propose to look at the evidence put forward on behalf of Denmark without taking account, at this stage, of the criticisms made of it by Mr Derek Bainbridge [ARICS], a quantity surveyor team leader with Darlington, in his affidavit dated July 3 1992. H

It appears from the evidence filed by Denmark:

(a) That Mr and Mrs Hussain are the joint owners of their matrimonial home at East Boldon in the County of Tyne & Wear.

(b) That this property has a value of £190,000 and is charged to a building society to secure a loan of £50,000.

(c) That Mr and Mrs Hussain are prepared further to charge the property to secure funding for the construction of the surgery.

(d) That Mr Hussain is prepared to undertake to apply within four weeks for building regulations approval for the construction of the surgery in accordance with the lease.

(e) That a company called Littlehaven Developments (South Shields) Ltd ("Littlehaven") have written to Mr Duffy by letter J dated June 24 1992 to confirm that they are willing to lend Denmark a sufficient sum to enable the surgery to be constructed.

(f) That the cost of construction of the surgery has been estimated by a firm of architects in Darlington at £177,000, excluding VAT.

(g) That the ability of Littlehaven to undertake a contract of the value of £177,000 has been confirmed by Northern Rock Housing Trust Ltd, with whom Littlehaven are engaged in a large joint venture at South Shields having a sales value of £6,900,000.

It was argued on behalf of Denmark that in the light of this evidence this court should grant relief on the basis of a change of circumstances. To refuse relief would confer a benefit on the council which would be quite disproportionate to the damage to the council if relief were granted. K

I am satisfied that relief should be refused. The pharmacy opened for business in August 1987. For a long time the attitude of Mr Hussain and Denmark was that they were under no obligation to build a surgery until the area was reclassified. That was the position at the trial. There has now been a change of heart and Denmark say they are willing to construct the surgery. They have also put before us at the last minute some proposals as to how the construction will be financed. In my view, however, where a tenant had been in clear breach of covenant for years and the judge at the original hearing was not satisfied that the tenant was in a financial position to remedy the breach, it is necessary for the tenant, if he wishes to rely on a change of circumstances, to put forward precise evidence: L

(a) of the sources of finance which are available and of the capacity of the tenant to meet any interest payments; and

(b) to explain why the finance was not available when the matter was before the judge.

In the present case it seems clear that the pharmacy is not profitable and that the equity value of the matrimonial home is less than the estimated cost of building the surgery. The evidence of possible help from Littlehaven has come very late; it was not available before the judge because at that stage the primary contention of Denmark was that the relevant covenants had been frustrated.

One must look at all the circumstances and the conduct of Denmark in the past. I consider that it would be wrong to grant relief M on this evidence.

(c) *Unjust enrichment*

It was submitted that without evidence it was impossible to decide whether the council would be enriched by the presence of the pharmacy on the land. The question of enrichment, it was said, should be referred for inquiry. I see the force of this submission. On the other hand, I can see no reason to interfere with the judge's conclusion that any enrichment would not be unjust. Denmark should have known the status of the area when they signed the lease. Furthermore, counsel for Denmark was unable to formulate any recognised principle of unjust enrichment which would allow Denmark to recover compensation in circumstances such as these

*Editor's note: Also reported at [1990] 33 EG 45.

A following the refusal of relief from forfeiture. It is unnecessary, however, to consider this point further as any enrichment would plainly not be unjust.

I would dismiss the appeal.

Steyn LJ and **Sir Christopher Slade** agreed and did not add anything.

Appeal dismissed with costs.

B
Court of Session

Outer House

July 9 1992

(Before Lord SUTHERLAND)

POSTEL PROPERTIES LTD v MILLER & SANTHOUSE PLC

Estates Gazette February 13 1993

[1993] 06 EG 107

C ***Landlord and tenant — Covenant to occupy premises as shop — Whether pursuers' conclusion for implement competent — Whether landlords entitled to an order compelling a tenant to occupy and use a shop for retail purposes***

Clause 5 of Part IV of a lease of shop premises in the Sauchiehall Centre, Glasgow, provided that the defenders, the lessees, "shall occupy the leased premises and shall not without the written consent of the landlords leave the leased premises unoccupied for a period exceeding 21 days . . .". Clause 13 provided that the defenders were to "keep and use the . . . premises solely as retail premises". On the pursuers' claim the defenders admitted that they had closed the shop, D but when the case came on procedure roll the defenders' first plea-in-law was that the pursuers' conclusion for implement was incompetent for lack of precision, lack of enforceability and exclusion of remedy in terms of the lease.

***Held*: The defenders' first plea-in-law was upheld and the action dismissed. What was sought by the pursuers lacked the necessary precision for a decree *ad factum praestandum*: what was required to be done to comply with clauses 5 and 13 of the lease was not a clear and specific act to be performed at a clear and specific time but a general duty to be performed over a long period when circumstances might change. However, on E defenders' alternative arguments, there was no reason why decree for specific implement should not be made against the defenders notwithstanding that they were a limited company. The fact that the lease provided that the tenants were to indemnify the landlords for a breach of clause 5 did not exclude the remedy of specific implement. Furthermore, the pursuers' second conclusion for damages, being ancillary to the first, failed.**

The following cases are referred to in this report.

Fleming & Ferguson v *Paisley Magistrates* 1948 SC 547
Grahame v *Magistrates of Kirkcaldy* (1882) 9 R (HL) 91
F *Grosvenor Developments (Scotland) plc* v *Argyll Stores Ltd* 1987 SLT 738
Harvey v *Smith* 1904 6 F 511
Hendry v *Marshall* 1878 5 R 687
Lochgelly Iron & Coal Co v *NB Railway Co* 1913 1 SLT 405
MacLeod v *Alexander Sutherland Ltd* 1977 SLT (Notes) 44
McArthur v *Lawson* 1877 4 R 1134
Middleton v *Leslie* 1892 19 R 801
Salaried Staff London Loan Co Ltd v *Swears and Wells Ltd* 1985 SLT 326
White & Carter (Councils) Ltd v *McGregor* [1962] AC 413; [1962] 2 WLR 17; [1961] 3 All ER 1178, 1962 SC (HL) 1

The case was heard on procedure roll on the first and second pleas of the defenders, Miller & Santhouse plc, as to competency and relevancy of part of the pursuers' action seeking decree ordaining the defenders to comply with the terms of a lease.

C J M Sutherland QC and R Skinner (instructed by Tods Murray WS, of Edinburgh) appeared for the pursuers; A C Hamilton QC and R Ivey (instructed by Archibald Campbell & Harley WS, of Edinburgh) represented the defenders. G

Giving his opinion, **Lord Sunderland** said: In this action the pursuers, as landlords under a lease of shop premises, seek decree ordaining the defenders as tenants under the lease to occupy the premises and trade therefrom, with a further conclusion for damages. The leased premises are a shop unit in the Sauchiehall Centre, Glasgow. The lease was for a period from April 17 1987 to April 16 2012. Clause 5 of Part IV of the lease provides:

> Within ten weeks of the date of entry under this lease to install in the leased premises suitable and necessary fixtures, fitments and fittings of good quality required to make the leased premises suitable in all respects for the permitted use thereof, any necessary planning consent and/or building warrant being obtained by the tenants at their expense and the drawings and specifications in relation to any fitting out works by the tenants as aforesaid shall be submitted for the landlords approval, such approval not to be unreasonably withheld or delayed and the tenant shall occupy the leased premises and shall not without the written consent of the landlords leave the leased premises unoccupied for a period exceeding 21 days, the tenants indemnifying the landlords against all loss or damage suffered by the landlords in consequence of the failure by the tenants to occupy the leased premises including without prejudice to the foregoing generality any loss or damage to or depreciation in the capital or rental value of the leased premises attributable to the tenant's failure to occupy or in consequence of any other breach by the tenants of the tenant's obligations. H J

Clause 13 of Part IV provides:

> To keep and use the leased premises solely as retail premises within Class 1 of the Town and Country Planning (Use Classes) (Scotland) Order 1973 . . .

followed by certain other restrictions on use.

Clauses 14 and 15 which deal with prohibition of acts causing a nuisance or endangering the insurance of the centre commence respectively with the words:

> not to do or permit to be done upon or in connection with the leased premises or any part thereof any act or thing . . .

and

> not to do or permit to be done or omit to do within or upon the leased premises or any part of the centre any act or thing . . . K

The pursuers aver that the defenders used the premises for a period, but then closed the shop and ceased trading there. The defenders admit that they closed the shop and explained that the reason they did so was because their activities were producing a loss. They aver that they endeavoured to find another tenant for the premises to whom their rights and obligations under the lease could be assigned in terms of clause 30(b), but they have been unable to do so.

The case came on procedure roll on the first and second pleas of the defenders, which are pleas to competency and relevancy, and the fourth plea for the pursuers which seeks decree *de plano*. Under their L first plea the defenders argued that the pursuer's conclusion for implement was incompetent on three grounds, namely lack of precision, lack of enforceability and exclusion of that remedy in terms of the lease. On the first ground it was argued that an order *ad factum praestandum* must be precise, a proposition vouched in general by *Walker on Civil Remedies*, p 269, *Burn Murdoch on Interdict,* p 158, *Middleton* v *Leslie* 1892 19 R 801, *Fleming & Ferguson* v *Paisley Magistrates* 1948 SC 547 and *Hendry* v *Marshall* 1878 5 R 687.

In connection with leases the defenders referred to *Grosvenor Developments (Scotland) plc* v *Argyll Stores Ltd* 1987 SLT 738. In *Grosvenor Developments* the defenders were the tenants of a supermarket within a development owned by the pursuers and the M lease ran from 1974 to 2016. The lease provided that the subjects were let for the retail sale of all foodstuffs and all hardware, electrical goods and non-foods "as commonly sold in supermarkets and discount stores" and the use of part of the premises as an off-licence. It was provided that:

> the tenants shall take possession of and use and occupy the premises for the foregoing purpose(s) within 3 calendar months from the date of entry hereunder and shall thereafter continue to use and occupy the same for said purpose(s) and during all normal business hours shall keep the premises opened for business throughout the whole period of this lease.

The defenders intimated their intention to discontinue trading and the pursuers sought interdict against the defenders from ceasing to continue to occupy and use the supermarket premises. It was held that it was not competent to obtain an interdict which in terms compelled the performance of an act and accordingly the action was dismissed. In addition, however, it was held that the obligation sought to be implemented was too general to be enforced by specific implement. The obligation to occupy and use premises and carry on a business therein involved continuous acts of management in which multifarious actions are required and it also required decisions over a period of 42 years as to what at any one time is commonly sold in supermarkets. Accordingly, the obligation was one where the defenders would have difficulty in deciding at any given time whether they were acting in breach of it or not. An order from the court must be precise and specific so that the defenders know throughout the period when the order is enforced exactly what they are required to do and what they are prohibited from doing. A further problem raised in the opinion of Lord Jauncey was that the order sought was that the defenders continue to occupy and use the premises without any limitation, whereas certain provisions in the lease contemplated an interruption of occupancy and use in the event, for example, of the premises being damaged or destroyed or the lease being irritated. The terms of the interdict sought therefore required the defenders to do more than they were required to do in terms of the lease.

Counsel for the defenders in the present case founded strongly on *Grosvenor Developments* as showing that an obligation to carry on commercial activity will not be enforced by specific implement because the order would inevitably lack the necessary precision. Furthermore, the terms of clause 13 in the present lease do not even make it clear that there is a positive obligation to trade from the premises. All that is said is that the premises must be kept and used solely for the purpose of retail trade within class 1 of the relevant regulations. This should be construed as meaning that clause 13 was purely a restrictive clause and did not provide any positive obligation. In any event even if there were an obligation to use the premises, does that necessarily mean that the defenders require to continue trading in a particular fashion? Would it be sufficient to open the shop for half an hour per week with one employee selling matches stationed at a counter just inside the door? If that would not be enough, what would be sufficient to comply and how can the defenders know what it is that they must do? In these circumstances there is a total lack of precision in the pursuers' first conclusion and accordingly that conclusion is incompetent.

In reply, counsel for the pursuers argued that specific implement is a normal remedy which should be granted unless there is some special hardship or exceptional circumstances or implement is impossible. It is for the defenders to aver and prove what such exceptional circumstances would be and if they are unable to do so they have no answer to a claim for implement. In *McArthur* v *Lawson* 1877 4 R 1134, it was said by Lord President Inglis that a contract which cannot be enforced by specific implement in so far as regards its form and substance is no contract at all and cannot form the ground of an action of damages.

It was not argued by the defenders in this case that there was anything wrong with the contract of lease, or that it was unenforceable and, if that be so, then its terms must be enforceable by specific implement. Where a contract has been repudiated the innocent party has an option. He may either accept the repudiation and sue for damages for breach of contract, or if he chooses, he may refuse to accept the repudiation and then the contract remains in full effect: *White & Carter (Councils) Ltd* v *McGregor* 1962 SC (HL) 1; *Salaried Staff London Loan Co Ltd* v *Swears & Wells Ltd* 1985 SLT 326. All that the pursuers are seeking to do here is to enforce the terms of a contract which the defenders do not attack as being in any way defective in form or substance. *Grosvenor Developments* is not in point as the issue there was whether interdict was an appropriate remedy and the other observations in the case were *obiter*. In any event the decision amounted to saying only that the terms of the clause in that particular lease were too vague to be enforceable.

There is no difficulty in understanding the concept of occupying premises, nor is there any difficulty in understanding the concept of using the premises for the purposes of retail trade. The obligation in clause 13 is to keep and use the premises and, if this was intended only to be a negative obligation, it could easily have been phrased in precisely the same way as clauses 14 and 15. The difference in wording shows that it was intended to be a positive obligation. This is perfectly comprehensible in the context of a commercial lease of a unit in a shopping development because lack of use detracts from the viability of a shopping centre. The point taken by Lord Jauncey in *Grosvenor Developments* about other clauses in the lease giving rise to circumstances where occupancy and use could not be enforced is not a valid one, as if decree is granted in the present case this would not abolish all the other clauses in the lease, but would merely be tantamount to saying that the whole terms of the lease should remain in full force and effect. Counsel for the pursuers went on to argue that the defenders' averments relating to the loss which they were making in their trading and their difficulties in finding an alternative tenant were irrelevant. The circumstances required to deprive the pursuers of their right to a decree *ad factum praestandum* would have to be exceptional: *Grahame* v *Magistrates of Kirkcaldy* (1882) 9 R (HL) 91, *White & Carter (Councils) Ltd (supra)* and *Salaried Staff London Loan Co Ltd (supra)*.

It is no answer for a party in breach to say that it appears that the contract into which he entered has become unduly onerous. There is nothing exceptional in a company trading at a loss in times of recession and equally there is nothing unusual in a company finding difficulty in obtaining alternative tenants for a development which is, in any event, substantially unlet. It is nowhere suggested by the defenders that the pursuers do not have an interest to enforce the contract rather than merely claim the alternative remedy of damages. That being so the defenders' averments, which purport to show that there are some exceptional circumstances which would justify the court in exercising its equitable jurisdiction to refuse a decree for specific implement, are irrelevant and therefore decree should be granted *de plano*.

In reply to this latter point, counsel for the defenders pointed out that in the case of *Salaried Staff London Loan Co*, where what was sought was payment of rent for the period up to the date when the action was raised, the Lord President said that, if the pursuers continued to maintain the contract and continued to sue for payment of unpaid rent in subsequent actions, it may well be that different considerations would arise. Lord Ross observed that it was significant that the defenders made no averments which might justify the view that it would be manifestly unjust and unreasonable to allow the pursuers the remedy which they sought in that action, namely payment of the rent. The defenders offered no explanation for purporting to renounce the lease. They did not say that it had become uneconomic for them to continue as tenants, nor that it would be uneconomic for any other tenant and they did not aver that any attempt had been made by them to find a substitute tenant or that it had been impossible to find such a substitute tenant.

Lord Ross went on to observe that the lack of any such averments appeared to be fatal to the defenders' contention, as there was no material in that case which would justify the court in concluding that it would be manifestly unjust and unreasonable to allow the pursuers the remedy which they were seeking.

In the present case the defenders had made appropriate averments. They gave an explanation for ceasing to trade, namely that it had become uneconomic for them to do so. They averred that the centre was not a popular development and they aver that they have made substantial efforts to find a substitute tenant, even to the extent of offering a substantial premium to any incoming tenant, but that they have been unable to find anyone to take on the obligations. It was argued that it was a reasonable inference from Lord Ross' opinion that, had the averments made in the present case been made in that case, he would have been prepared to accept that there was sufficient at least to go to proof on the question of whether it was unjust and unreasonable to allow the pursuers even the limited remedy which they were seeking in that case. *A fortiori* these averments are very relevant when what is sought is a decree ordaining the defenders to continue uneconomic trading for the next 20 years, or until such time as they can find a substitute tenant which, at present, appears to be a virtual impossibility.

The first matter to be considered is what is meant by the words "to keep and use the leased premises solely as retail premises" within the meaning of clause 13. In my opinion, as a matter of construction this is intended to be a positive obligation. It is expressed as a positive obligation even though all the rest of the clause produces further restrictions upon what use can be made of the premises. Had it been intended to be a negative restriction, it would have been perfectly possible to phrase it in exactly the same way as clauses 14 and 15, in other words "not to keep and use the leased premises other than as retail premises" etc. *Prima facie*, therefore, there is a duty on the defenders to occupy the premises and to use the premises as retail

A premises within the terms of clauses 5 and 13. I have no doubt that a perfectly valid contract can be framed in terms such as this and such a contract could not be said to be void for uncertainty. It does not, however, in my opinion, necessarily follow that every term in a valid contract can necessarily be made the subject of an order for specific implement. The obvious example is a term providing for the payment of money. Another example would be where a contract provides that one party should use their best endeavours to do certain things. This is not an uncommon form of words in a perfectly valid contract, but plainly could not be made the subject of an order for specific implement.

Finally, there is the position as appears in *Grosvenor* B *Developments* where what was said to be a perfectly valid lease was nevertheless regarded as being unsuitable for pronouncing an order for specific implement. Where what is required to be done by one party is not a clear and specific act to be performed at a clear and specific time, but is simply a general duty to be performed over a long period during which circumstances might change, I consider that the inevitable aura of vagueness which must surround any such proposition is fatal to a party seeking an order *ad factum praestandum*. It will not, in my opinion, do for the pursuers simply to say that everybody knows what is meant by occupying a shop and everybody knows what is meant by using a shop as retail premises. If that argument were valid, all that would be necessary in the pursuers' first conclusion would be an order ordaining the defenders to carry C out their obligations under the lease without giving any further specification and it is, in my opinion, clear that the court would never pronounce such an order.

While *Salaried Staff London Loan Co* was a case involving only payment of rent, the observations made by the court in that case would fortify the view that the court would be slow to grant a decree of specific implement which would involve active participation by the defenders in trading from premises for the next 20 years. No case was put before me where such an onerous obligation was imposed by the court. The only case which comes anything near it is *Grosvenor Developments* and in that case the proposition was firmly rejected. I see no material ground for distinguishing this case from *Grosvenor Developments* and, accordingly, I take the view that what is sought D by the pursuers in this case lacks the necessary precision for a decree *ad factum praestandum*.

The second ground upon which it was said that the action was incompetent was that the defenders, being a limited company, cannot have a decree *ad factum praestandum* pronounced against them. The historic sanction for breach of such an order was imprisonment and a corporate body cannot be imprisoned. In *Lochgelly Iron & Coal Co* v *NB Railway Co* 1913 1 SLT 405, Lord Kinnear expressed doubts as to whether a company could be subject to a decree *ad factum praestandum* because it could not be imprisoned. In *Gloag on Contract* p 659 it is said that such a conclusion cannot be brought against a company without an E alternative conclusion for damages. There have been some cases in which an order *ad factum praestandum* appears to have been pronounced against a corporate body, but it also appears that in those cases no point was taken that such an order was incompetent.

Whatever may be the historic basis for the idea that the only sanction for breach of an order *ad factum praestandum* is imprisonment, it does not appear to me that such a proposition in modern times could any longer be sustainable. The Law Reform (Miscellaneous Provisions) (Scotland) Act 1940, section 1, provides that no person shall be apprehended or imprisoned on account of his failure to comply with a decree *ad factum praestandum* except in accordance with certain conditions. These conditions include a provision that the person in right of such a decree has to make an F application to the court, which will not grant warrant for imprisonment unless satisfied that the defender has wilfully refused to comply with the decree. Under subsection 2 it is provided that, on any application in pursuance of subsection 1, the court may, in lieu of granting warrant for imprisonment, recall the decree on which the application proceeds and make an order for payment by the respondent to the applicant of a specified sum or make such other order as appears to the court to be just and equitable in the circumstances.

It may well be that this Act was not intended to cover the situation where a company or corporate body was the defender as it proceeds on the basis that an application can be made for imprisonment of the defender. It is, however, clear that the court is empowered on any view to substitute for the decree *ad factum praestandum* an order for G payment or such other order as appears appropriate. The pursuers contended that the idea that the only sanction was imprisonment was based historically on the proposition that a person in right of a decree was entitled at his own hand to enforce the decree by imprisonment and that the 1940 Act had removed that absolute right.

The position was, however, different where the matter was in the control of the court. If a party is in breach of a court order then that party can be brought before the court and will be treated as being in contempt of court. The penalty for contempt of court may be imprisonment but may, of course, also be any such other order as the court may think appropriate. Accordingly, if a decree *ad factum praestandum* is passed against a corporate body and that body wilfully refuses to comply with the order, it can be brought before the H court as being in contempt and the court can pronounce such remedy as appears to be appropriate in the circumstances.

Counsel for the defenders argued that it is clear from what was said by Lord Kinnear that the historical basis undoubtedly was that imprisonment was the only proper remedy for failure to obtemper a decree *ad factum praestandum* and that there was no good reason for having any change in that position. Counsel accepted that it was perhaps somewhat illogical that interdicts can be and are daily pronounced against corporate bodies whereas a decree ordaining a corporate body to perform a particular act should for technical and historical reasons be regarded as incompetent.

In my view, it is indeed illogical that the court cannot pronounce a J decree to which an innocent party would *prima facie* be entitled just because there was thought to be historically only one penalty for breach of that order. I see no reason whatsoever why a party in breach of an order ordaining performance of a specific act should not be treated in exactly the same way as a party who is in breach of interdict and, if that be so, there is no reason why a decree *ad factum praestandum* should not pass against a corporate body. As was said in *Gloag on Contract*, p 657, the object of a decree is to enable the pursuer to secure his right, not to punish the defender. If that is correct, it would appear strange that difficulties in punishing the defender should prevent an innocent pursuer securing his right. In *MacLeod* v *Alexander Sutherland Ltd* 1977 SLT (Notes) 44, the pursuers sought implement of a clause in missives providing that the K defenders should perform certain specified building works. The defenders were a company which was in receivership. Lord Stott refused decree of implement on the ground that, as a receiver had been appointed, the company was not in a position to obtemper any decree and might find itself in the position of being in breach of an order if the receiver chose not to act. He added some observations on the subject of enforcement in the following terms:

> The only sanction suggested by counsel for the pursuer for failure to obtemper such a decree against a limited company would be the imposition of a monetary penalty, and counsel was unable to refer to any report of such having been done. That, I think, is why the learned author of *Gloag on Contract* at p 659 suggests that an action against a corporate body concluding for the performance of some act which can be performed vicariously is L incompetent without an alternative conclusion for damages, and may afford the explanation also for Lord Kinnear's observations in *Lochgelly Iron & Coal Co Ltd* v *North British Railway Co* where he had expressed doubts as to whether a contract with the Railway Company would support a decree *ad factum praestandum*. For the purposes of the present case, it is unnecessary to affirm so wide a proposition, and counsel for the defenders did not invite me to do so . . . It would, I think, be inequitable to grant a decree *ad factum praestandum* against a company which can act only through the agency of a receiver and which, if the receiver should decline to act, would inevitably be in contempt of Court in respect of an omission which they would be powerless to remedy. To pronounce such an order in a case in which the sanction of a monetary penalty would have to be imposed on defenders for a contempt which they have no means of purging except through the intervention of an agent whose actings they have no power to control, would not only be M ineffective, but would as it seems to me, be unfair to the defenders and no doubt also to their creditors.

It would appear from these observations that his lordship would have accepted that, if decree *ad factum praestandum* could have been pronounced against the defenders, he would have been prepared to do so and that the sanction of a monetary penalty appeared to him to be an appropriate sanction in the event of there being a breach of the order. For these reasons, in my opinion, the defenders attack on the competency on this ground must fail.

The defenders' third ground of attack was that the remedy of specific implement was excluded by the terms of the lease. Clause 5 provides for the situation where the tenants failed to occupy the

A premises for a period exceeding 21 days and in that situation it is provided that the tenants should indemnify the landlords against all loss and damage suffered by them. The lease has accordingly provided the appropriate sanction for a failure to occupy the premises and it would not be appropriate for the court to order specific implement.

In my opinion, the complete answer to this proposition can be found in *Gloag on Contract*, p 720, where the learned author says that when an obligation is fortified by a penalty it is still open to enforce the obligation and the debtor cannot escape simply by paying the penalty.

Finally, on competency, the defenders argued that if the first
B conclusion fell to be dismissed then the second conclusion for damages should also be dismissed. It was apparent from the second plea-in-law for the pursuers that damages were intended to be subsidiary to the pronouncement of specific implement and would be payable only in the event of the defenders failing to implement the contract. The second conclusion is somewhat ambiguous in its terms, but the interpretation of it must be governed by the plea-in-law. Counsel for the defenders referred to *Harvey* v *Smith* 1904 6 F 511 where an action concluded for implement of missives of sale and failing implement for damages. The defender argued that the action was premature in respect that the time for implement provided by the missives had not arrived. The action was dismissed as premature so far as the first conclusion is concerned, but the Lord Ordinary
C allowed a proof on damages.

On appeal the opinion was expressed that the second conclusion was ancillary to the first and that, if the first conclusion were dismissed, damages could not be awarded under the second. Counsel for the pursuers argued that there was still a right to damages even if the court felt that there was a breach of contract, but for exceptional reasons would not enforce that contract. The second conclusion and the second plea simply constituted a claim for damages for breach of contract. In my opinion, having regard to the terms of the second plea-in-law, it is clear that the claim made in the second conclusion must be ancillary to decree of specific implement being pronounced under the first conclusion. The second plea-in-law makes it clear in terms that damages are sought in the event of the defenders' failure to
D implement the contract. This, in my opinion, is not the same thing as saying that the pursuers accept that the defenders have repudiated the contract and seek damages in respect of that breach. As I have held that the first conclusion must fail, it follows that the second conclusion must fail also. This does not, of course, prevent the pursuers from raising an appropriate action for damages for breach of contract, but both the conclusion and the plea-in-law in such an action would be in different terms from those contained in the present conclusion and plea-in-law.

In conclusion the defenders argued that the pursuers' averments in condescendence 5 relating to damages were irrelevant. The basis on which the pursuers seek damages is a capitalisation of the yield drop
E of the centre as a whole as a result of the defenders' failure to occupy the leased premises. The defenders argued that what is contemplated in the lease is not a loss to the centre as a whole, but a loss in respect of the leased premises alone. I have already referred to the provisions of clause 5 of Part IV of the lease, which provides in relation to loss and damage that there is included any loss or damage to or depreciation in the capital or rental value of the leased premises attributable to the tenant's failure to occupy. The defenders' argument is that the way in which damages are calculated in condescendence 5 is not in relation to depreciation of the capital or rental value of the leased premises, but relates to the whole of the centre instead. In reply the pursuers argued that, while there was specific reference to depreciation in the value of the leased premises, the clause in fact provided that the
F landlords should be indemnified against all loss or damage suffered by them in consequence of the failure by the tenants to occupy the leased premises.

Counsel referred to the averments of the effect of the closure of this particular shop and argued that these averments were relevant to show that there was damage to the capital value of the centre as a whole arising naturally and directly from the defenders' breach and that such loss and damage fell squarely within the general provisions of clause 5. In my opinion, the way in which the damages are calculated is indeed somewhat unusual. However, the pursuers have made averments offering to prove what is a consequence of the defenders' failure to occupy the leased premises and, if they can establish these averments, I consider that the method of calculation of loss is one which could fall within clause 5. I would not be prepared G
at this stage to say that the method of calculation was so hopelessly irrelevant that it should not have been remitted to probation.

On the whole matter I shall sustain the defenders' first plea-in-law and dismiss the action.

Court of Appeal

September 16 1992

(Before Lord Justice GLIDEWELL, Lord Justice MANN and Lord Justice LEGGATT) H

TOWER HAMLETS LONDON BOROUGH COUNCIL v ABDI

Estates Gazette February 13 1993

[1993] 06 EG 102

Landlord and tenant — Secure tenancy — Whether terms of licence between housing authority and private landlord made grant of tenancy by the local authority a secure tenancy — Whether defendant tenant entitled to rely upon matters susceptible to judicial review in defence to possession proceedings J

In discharging their statutory responsibility under Part III of the Housing Act 1985, relating to "housing the homeless", the appellant council provided short-term accommodation to the respondent, Mrs Lul Hassam Abdi and her family from October 30 1990 in Flat B, 366 Green Lanes, London N4. On March 21 1991 the council offered the respondent permanent accommodation at 32 Sleaford House, Lincoln Estate, London E3, which she considered unsuitable for her medical condition; she rejected the offer. The council having informed the respondent on May 21 1991 that their duty under Part III of the 1985 Act had been performed, they served a notice to K
quit on her on June 20 1991. Upon the respondent failing to give up possession the council commenced possession proceedings. Flat B was owned by a private landlord and held by the council under a licence dated March 9 1988. In the proceedings the council made applications: (1) for the trial of a preliminary issue as to whether the respondent had a secure tenancy by reason of the requirements at para 6 of Schedule 1 to the 1985 Act; and (2) that the respondent's defence and counterclaim be struck out as either disclosing no reasonable defence or cause of action or as being abusive. Giving no reasons Judge Medawar QC dismissed both applications. The L
council appealed.

Held: **The appeal was allowed; the preliminary question determined in favour of the council, the defence and counterclaim was struck out and an order for possession was made. The judge's omission of reasons for the exercise of his discretion was regrettable: it is the duty of a judge to give reasons, albeit brief, for his decision. (1) Conditions (*a*) and (*b*) in para 6 of Schedule 1 to the Housing Act 1985 were satisfied. It was inherent in the agreement by the private landlord that vacant possession was granted to the council; the purpose of occupancy by homeless persons was unachievable save by the grant of vacant possession confirmed by the** M
provision in the agreement for giving vacant possession on its termination. The agreement between the private landlord and the council contained a provision for the licensor to obtain possession when he required, namely upon the giving of a seven-day notice. (2) It was difficult to understand how an asserted breach of duty could be an answer to a claim for possession of Flat B. The assertion of a breach of duty has as its necessary corollary an allegation of infringement of right. If the infringed right has no existence in private law but exists solely in public law, then the complaint is apt for judicial review and to raise it in another way is abusive. The only

private right acquired by the respondent was one to the accommodation of which she rejected. She had, and has, no other private right although she would of course have had a sufficient *locus* to challenge timeously the legality of the council's performance of their public law functions.

The following cases are referred to in this report.

Ali v *Tower Hamlets London Borough Council* [1992] 3 All ER 512,
Avon County Council v *Buscott* [1988] QB 656; [1988] 2 WLR 788; [1988] 1 All ER 841; (1988) 86 LGR 569, CA
Cocks v *Thanet District Council* [1983] 2 AC 286; [1982] 3 WLR 1121; [1982] 3 All ER 1135; (1982) 81 LGR 81, HL
Eagil Trust Co Ltd v *Pigott-Brown* [1985] 3 All ER 119, CA
O'Reilly v *Mackman* [1983] 2 AC 237; [1982] 3 WLR 1096; [1982] 3 All ER 1124, HL
Tower Hamlets London Borough Council v *Miah*[1992] 2 WLR 761; [1992] 2 All ER 667; (1991) 90 LGR 151; [1992] 1 EGLR 39; [1992] 18 EG 147
Wandsworth London Borough Council v *Winder* [1985] AC 461; [1984] 3 WLR 1254; [1984] 3 All ER 976; (1984) 83 LGR 143, HL

This was an appeal by Tower Hamlets London Borough Council from a decision of Judge Medawar QC, sitting in Edmonton County Court on June 17 1992, whereby he rejected two applications, an application for a question to be tried as a preliminary issue and one for the defence and counterclaim of the respondent, Mrs Lul Hassam Abdi, to be stuck out in possession proceedings brought by the council against the respondent in relation to Flat B, 366 Green Lanes, London N4.

Ashley Underwood and Lisa Giovannetti (instructed by the solicitor, Tower Hamlets London Borough Council) appeared for the appellant; Alan Tyrrell QC and Mark Loveday (instructed by Timmis Desai) represented the respondent.

Giving the first judgment at the invitation of Glidewell LJ, **MANN LJ** said: There is before the court an appeal against a decision of Judge Medawar QC given in Edmonton County Court on June 17 1992. On that day the learned judge had before him two applications by Tower Hamlets London Borough Council, which had been made by them in an action in which they were the plaintiffs and in which the defendant was Mrs Lul Hassam Abdi. The applications were: (1) an application under the County Court Rules, Ord 13, r 2, for a question to be tried as a preliminary issue; (2) an application under Ord 13, r 5, for the defence and counterclaim to be struck out either as disclosing no reasonable defence or cause of action or as being abusive.

The judge dismissed both of those applications and the council now appeal against his decision.

Each application was one which called for an exercise of judicial discretion and, I at once observe, that the judge gave no reasons for exercising his discretion as he did. This was a regrettable omission. Recognised exceptions apart, it is the duty of a judge to give reasons, albeit brief reasons, for his decision. That duty exists so that the parties and, where necessary, this court can know the basis on which the judge has acted. I draw attention to the observations made in this court in *Eagil Trust Co Ltd* v *Pigott-Brown* [1985] 3 All ER 119. As it is, in this case we are deprived of the benefit of the knowledge to which I have referred. Mr Tyrrell QC, for the respondent defendant, did suggest that we should remit for the judge to exercise his discretion and to give reasons for that exercise. That would be a most unfortunate course, for the matter has now been fully argued before this court and I see no necessity for adopting it. That course would arguably be necessary only if the judge had declined to exercise his discretion at all and it is not suggested, nor is it to be inferred, that that was the case.

Background

The council have, as a housing authority, a statutory responsibility under Part III of the Housing Act 1985 which relates to "housing the homeless". In the exercise of powers conferred under that part, the council determined in June 1990 that Mrs Abdi and her family were unintentionally homeless persons and, accordingly, the council accepted a statutory obligation to provide the family with suitable accommodation: see sections 65 and 69, as amended, of the 1985 Act. The council discharged their duty in the first instance by providing accommodation as from October 30 1990 in Flat B, 366 Green Lanes, N4. There was a written agreement between the council and Mrs Abdi relating to that accommodation dated October 29, which was prefaced in this way:

> Dear Mrs Abdi,
> HOUSING ACT 1985, (PART III) — PROVISIONS OF TEMPORARY ACCOMMODATION
> You will now be moving into your new temporary home. This is not your final accommodation but we hope you will be happy there until a permanent dwelling is found for you.

There were then terms and conditions as to the occupancy. Mrs Abdi and her family moved in.

On March 21 1991 the council offered Mrs Abdi permanent accommodation at 32 Sleaford House, Lincoln Estate, E3. Mrs Abdi regarded that offer as being an offer of unsuitable accommodation, the unsuitability being that it did not cater for her medical condition. The council considered Mrs Abdi's rejection to be unreasonable and, on May 21, they informed her that their duty under Part III of the Act had been performed by the making of the offer which she had rejected. On June 20 1991 they served a notice to quit Flat B. Mrs Abdi did not comply with that notice and the council commenced possession proceedings, in the course of which the two applications now before the court came to be made.

Flat B is what is called private-leased accommodation which, we were told, has to a great extent supplanted bed-and-breakfast accommodation as temporary accommodation for the homeless. It is a means by which local housing authorities can obtain the temporary use of private-sector accommodation for the homeless. It is a means recognised by provisions in the 1985 Act, which prevent tenancies or licences of such temporary accommodation from becoming secure tenancies or licences. Ordinarily, a letting or a licence by a housing authority of a separate dwelling to an individual becomes a secure tenancy or licence with the accompanying restrictions on recovery of possession. But among the exceptions to the ordinary rule is one found at para 6 of Schedule 1 to the 1985 Act:

> *Short-term arrangements*
> 6. A tenancy is not a secure tenancy if —
> (*a*) the dwelling-house has been leased to the landlord with vacant possession for use as temporary housing accommodation,
> (*b*) the terms on which it has been leased include provision for the lessor to obtain vacant possession from the landlord on the expiry of a specified period or when required by the lessor,
> (*c*) the lessor is not a body which is capable of granting secure tenancies, and
> (*d*) the landlord has no interest in the dwelling-house other than under the lease in question or as a mortgagee.

The terms under which the council held Flat B are in a licence dated March 9 1988 and made between the owner of the premises and the council. Clause 1 provides:

> The Grantor permits the [Licensee] to use the premises at

— which can then be identified from another page —

> for the sole purpose of enabling the [Licensee] to provide temporary housing accommod ation for the homeless persons as hereinafter defined but not so as to create a relationship of landlord and tenant between the Grantor and Licensee or to confer on the Licensee such rights as would amount in law to a tenancy of the premises or to create any estate or proprietory interest therein or to create any form of partnership or joint venture between Grantor and Licensee.

That the instrument is a licence rather than a lease is not material for the purposes of para 6. This court so held in *Tower Hamlets London Borough Council* v *Miah* [1992] 2 WLR 761*. The conclusion of the court was that the word "leased" in the paragraph is apt in context to include "licence". We were not invited to depart from that decision and we could not do so even if we so wished.

It is the council's case that the conditions in para 6 were here satisfied, that, therefore, Mrs Abdi acquired no statutory interest in Flat B, that the notice to quit was sufficient to terminate her contractual rights under the agreement of October 29 and that the claim for possession is, accordingly, unanswerable.

The defence filed in answer to that case has two components. First, it is alleged that two of the conditions in para 6 of Schedule 1, that is to say, (*a*) and (*b*), are not satisfied, with the consequence that the exception to the creation of a "secure licence" is not satisfied and a secure licence was created as from October 1990. The second component is a denial of an averment in the particulars of claim that the rejection of permanent accommodation was unreasonable, together with a counterclaim for a declaration that the council had not discharged their duty under the Act to the defendant. The components of that defence prompted the applications for, first, a

*Editor's note: Also reported at [1992] 1 ELGR 39.

A preliminary issue as to whether there was a secure tenancy on the grounds that that would be a determinate question and, second, for the defence and counterclaim to be struck out as abusive, on the grounds that issues as to the performance of the statutory obligations could be raised only in proceedings for a judicial review.

As to the first application, I entertain no doubt that the preliminary issue suggested is one good to be such, for it is short, requires no evidence save the licence agreement and notice to quit and, subject to the defence concerning non-performance of a statutory obligation, determinate. We have in effect been asked to resolve the preliminary issue and I pass to consider it.

The issue raised, secure licence or not, is narrowed as to whether
B subparas 6(*a*) and (*b*) are satisfied in the circumstances. As to (*a*), Mr Tyrrell submitted that this subparagraph was not satisfied because the agreement, which was the only material before the court, did not in terms grant vacant possession and, indeed, gave the licensor a right to enter, and inferentially to remain, for any purpose. The material clauses of the licence are clauses 4 and 6:

> 4. Throughout the subsistence hereof the Licensee shall have the right to nominate persons being homeless persons as defined by the provisions Part 3 of the Housing Act 1985 . . . to occupy the Premises on the terms hereinafter appearing and the Grantor shall provide for the use of the Nominees all such services as may be reasonably requested by the Licensee . . .
>
> 6. The terms on which the Licensee shall permit the nominees to be temporarily accommodated at the Premises shall reserve and acknowledge the right of the Grantor to enter the Premises for any reason at any time
> C without having to give any previous notice to the Licensee or their Nominees and further shall ensure that the Licensee can give vacant possession of the Premises to the Grantor on termination of this Licence.

Mr Underwood, for the plaintiff council, submitted that it was inherent in the agreement, particularly having regard to clause 4, that vacant possession was granted. I agree: the purpose of occupancy by homeless persons is unachievable save by the conferment of vacant possession on the council, and that this was done is shown, in my judgment, by the provision for giving up of vacant possession on termination. One cannot give up what one has not acquired. As to the right of entry, I cannot regard that as being adverse to the acquisition of vacant possession. Indeed, the contrary. A possessor does not require the reservation of a right to enter on that which he possesses.

D As to subpara (*b*), Mr Tyrrell submitted that that was not satisfied because neither is there a specified period in the licence nor is there a provision for vacant possession when required by the licensor. Mr Underwood drew our attention to clauses 2, 6 and 8. Clause 6 I have already read. Clauses 2 and 8 are in these terms:

> 2. This Licence shall subsist until revoked as hereinafter provided.
>
> 8. This Licence may be terminated by the Grantor giving to the Licensee not less than 7 days notice in writing.

Mr Underwood submits that, taken together, the three clauses provide for the licensor to obtain possession on the expiry of a specified period, that is to say, seven days, or alternatively when required. The first submission is plainly arguable. The second is, in
E my judgment, unanswerable. There was provision for the licensor to obtain possession when he required, that is to say, by the giving of a seven-day notice. Accordingly, I reject Mr Tyrrell's submissions in regard to subparas (*a*) and (*b*) and I should remark that he accepted they were not conspicuously encumbered by merit. I, accordingly, conclude that there was no secure licence and, the notice to quit being accepted as sufficient to end the contractual licence, I can turn to the defence and counterclaim concerning the council's alleged failure to provide suitable accommodation.

The pleadings upon the point are short. Paras 9 and 10 of the particulars of claim state by way of narrative as follows:

> 9. On 4th April 1991 the Plaintiffs made an offer to the Defendant of a
> F secure tenancy at 32 Sleaford House, London E3 which offer the Defendant unreasonably declined.
>
> 10. By a letter dated the 21st May 1991 the Plaintiffs notified the Defendant that they had discharged their duty to the Defendant under Part III of the 1985 Act.

I said that those paragraphs were narrative because, in my view, they are not averments material to the claim for possession. However that may be, they attracted these paragraphs in the defence, that is to say, paras 4 and 5:

> 4. Paragraph 9 is denied insofar as the Defendant avers that her rejection of the said accommodation upon medical grounds, the details of which have been supplied to the Plaintiff was in all the circumstances of this case reasonable.
>
> 5. For the matters hereinbefore pleaded, paragraph 10 is denied.

G The counterclaim repeated, among others, paras 4 and 5, and claimed "A declaration that the Plaintiffs have not discharged their duty under Part III of the Housing Act 1985".

Mr Underwood submits that the allegations in the defence and counterclaim are allegations entertainable only in judicial review proceedings for they are not referenced to any private law right to remain in Flat B. Mr Underwood relied upon *Avon County Council* v *Buscott* [1988] QB 656. In that case, gypsies sought to resist proceedings under Ord 113, not by reference to any private right to remain on the council's land, but by reference to the council's failure to provide accommodation for them in alleged breach of the duty imposed by the Caravan Sites Act 1968. After referring to the speech of Lord Fraser in *Winder*, to which I shall come, Lord Donaldson of
H Lymington MR said at p663D:

> There is a fundamental difference between this case and *Winder's* case. Mr Winder was seeking to raise a true defence. He was saying that he had a valid tenancy, that he did not owe any rent and accordingly was not liable to eviction. It was a defence on the merits. In the present case the defendants do not allege any right to occupy the land and accordingly do not deny that they are liable to be evicted. They do not suggest that they have any defence on the merits. What they say is quite different, namely, that the council is not entitled to enforce its rights. It is not entitled to come to the court to seek an eviction order. If one can imagine a private company whose memorandum and articles limited its powers to acting in a way which was *Wednesbury* reasonable and which sought to evict the defendants from its land in similar circumstances, the defendants would be seeking to strike out the action for want of authority
> J on the part of the plaintiff to bring the action. When a defendant is seeking, in effect, to strike out an action on the basis of a public law right, he should, in my judgment, proceed by way of an application for judicial review, thus ensuring that the matter is dealt with speedily as a preliminary point and in a manner which gives the public authority and the public which it serves the protections enshrined in the judicial review procedure.

Mr Underwood says that those observations are compelling in his favour. He says that, by analogy, once the contractual arrangement between the council and Mrs Abdi was terminated by the notice to quit, there was no private right to remain in the flat.

Mr Tyrrell relies upon passages in the speech of Lord Fraser in *Wandsworth London Borough Council* v *Winder* [1985] AC 461. In that case, a defendant was allowed to raise as a defence to a claim for rent due that the amount of the rent was unlawful, even though the
K decision as to the amount was a decision in the public domain which was susceptible to judicial review. The landlord housing authority had argued that susceptibility to judicial review meant a challenge in another way was abusive and they relied upon *O'Reilly* v *Mackman* [1983] 2 AC 237. Lord Fraser said at p509E:

> It would in my opinion be a very strange use of language to describe the respondent's behaviour in relation to this litigation as an abuse or misuse by him of the process of the court. He did not select the procedure to be adopted. He is merely seeking to defend proceedings brought against him by the appellants. In so doing he is seeking only to exercise the ordinary right of any individual to defend an action against him on the ground that he is not liable for the whole sum claimed by the plaintiff.

L Mr Tyrrell points out that here the defendant did not choose the forum and that she raises the asserted breach of duty by way of defence. It is true that the defendant did not choose the forum, but I find it difficult to understand how the asserted breach of duty could be an answer to a claim for possession of Flat B. As Mr Underwood put it, the discharge of duty is not pinned to the provision of particular premises.

However that may be, there is, in my judgment, a more fundamental objection. The assertion of breach of duty has as its necessary corollary an allegation of infringement of right. If the infringed right has no existence in private law but exists solely in public law, then the complaint is apt for judicial review and to raise it in another way is abusive: see *Cocks* v *Thanet District Council* [1983]
M 2 AC 286. Mrs Abdi is complaining that she has a right to have suitable accommodation. Her complaint is indistinguishable from that of the plaintiff in *Ali* v *Tower Hamlets London Borough Council* [1992] 3 All ER 512. In that case, the homeless plaintiff had refused an offer of accomodation because he considered it unsuitable and brought an action against the housing authority alleging breach of statutory duty and claiming an injunction to secure performance of the duty. This court held that county court had no jurisdiction to entertain the action. The reasoning was that although the performance of public law duties under Part III of the Act could give rise to a private right, that right does not arise until performance of the duty is completed by the allocation of accomodation adjudged suitable by the local authority. The private right is then one simply to

that accommodation. Nolan LJ, with whose judgment Parker and Farquharson LJJ agreed, said at p519e:

> It then follows that in my judgment the public law duties of Tower Hamlets were not discharged until they had completed the process of deciding upon the suitable accommodation which they were obliged to secure for Mr Ali. If this process was properly carried out as a matter of public law, then the consequential private law right of Mr Ali was simply a right to the accomodation which Tower Hamlets had decided to be suitable.

Those observations are binding upon this court and from them it follows that the only private right acquired by Mrs Abdi was one to the accommodation which she rejected. She had, and has, no other private right although she would of course have had a sufficient *locus* to challenge timeously the legality of the council's performance of their public law function. Accordingly, in my judgment, the arguments sought to be deployed by her would be abusive and the defence and counterclaim should accordingly be struck out.

I would allow the appeal in regard to the second application. I reach that conclusion with no regret. Under the law as it at present stands, questions of suitability are eminently for local authorities and not, subject to the supervisory jurisdiction of the High Court, for the ordinary courts.

If the allegations concerning breach of duty are struck out, as I think they should be, then the preliminary point which I have already discussed is actually determinate of this action. I would accordingly allow the appeal on the first application as well as on the second, determine the preliminary point in favour of the plaintiff council and accordingly make an order for possession, but would invite comment on the terms of that order.

LEGGATT LJ and **GLIDEWELL LJ** agreed and did not add anything.

Appeal allowed; order for possession granted; application for leave to appeal to the House of Lords refused; no order as to costs save for legal aid taxation.

Chancery Division

October 2 1992

(Before Mr E C EVANS LOMBE QC, sitting as a deputy judge of the division)

TENBRY INVESTMENTS LTD v PEUGEOT TALBOT MOTOR CO LTD

Estates Gazette February 13 1993

[1993] 06 EG 104

Landlord and tenant — Rent — Where landlord a company resident outside the UK and tenant obliged to deduct tax from rent — Whether tenant entitled to deduct from future rents tax omitted to be deducted from past rents

By a lease dated March 19 1990 the plaintiff, Tenbry Investments Ltd, a company resident in Jersey, let to the defendant, Peugeot Talbot Motor Co Ltd, premises at Mile End Road, London E1. By clause 2 of the lease the yearly rent was payable "without deduction except such as the tenant may be required by law to deduct". From the commencement of the lease until December 25 1990 the defendant paid the yearly rent at the rate of £50,000 per quarter, without deduction, directly into the plaintiff's bank account in Jersey. On March 11 1991 the defendant's managers realised the tax should have been deducted from the rent payments pursuant to the provisions of the Income and Corporation Taxes Act 1988 and such deductions should have amounted to rather more than £50,000 by that date. The gale of rent due on March 25 1991 was not paid by the defendant on the grounds that it was entitled to set off against that sum the amount which the defendant ought to have deducted from earlier instalments of rent and for which it was liable to account to the Inland Revenue: see sections 4, 43, 349, 3502 of, and Schedule 16 to, the Income and Corporation Taxes Act 1988. On March 19 1991 the plaintiff appointed an agent in the UK and directed the defendant to make all further rent payments to that agent. The defendant failed to take advantage of section 821 of the 1988 Act in making adjusting deductions in respect of instalments of rent paid after the commencement of a tax year, but before the rates of tax are finally fixed by the coming into force of a relevant finance Act. The plaintiff claimed the rent due for March 25 1991. The defendant relied on section 349(1) of the 1988 Act contending that it was entitled to make the appropriate deductions.

***Held*: Judgment for the plaintiff. The words "shall on making the payment deduct" used in section 349(1) are plain and to be compared with the words which were being construed by the Court of Appeal in *Taylor* v *Taylor*. Once a relevant payment is complete the right to deduct tax in respect of that payment is lost. The power conferred by the section on the payer is to be able on making any payment to deduct out of it a sum representing the amount of the tax on that payment. It is not possible to construe the yearly rent as one payment thus permitting a deduction in respect of the whole year against a particular quarter's rent. The legislature plainly contemplated deductions from each separate payment as it is made.**

The following cases are referred to in this report.

Johnson v *Johnson*
Taylor v *Taylor* XV1 ATC 218

This was a claim for arrears of rent by the plaintiff, Tenbry Investments Ltd, against the defendant, Peugeot Talbot Motor Co Ltd, due under the lease dated March 19 1990 in respect of premises at Mile End Road, London E1.

Fay Stockton (instructed by Taylor Joynson Garrett) appeared for the plaintiff; Jonathan Peacock (instructed by Wragge & Co, of Birmingham) represented the defendant.

Giving judgment, **Mr E C EVANS LOMBE QC** said: In this case the plaintiff, Tenbry Investments Ltd, a company resident outside the UK, in Jersey, claims against the defendant, Peugeot Talbot Motor Co Ltd, £50,000 with interest, being a quarter's instalment of rent becoming due on March 25 1991, pursuant to a lease dated March 19 1990 whereby the plaintiff let to the defendant certain premises at Mile End Road, London E1.

The defendant was the assignee from the original lessee. By clause 2 of the lease, the plaintiff let the premises at Mile End Road to the defendant for a term of 25 years from March 19 1990. The tenant:

> . . . paying therefore unto the landlord yearly during the term and so in proportion for any period less than a year and whether demanded or not but without deduction except such as the tenant may be required by law to deduct, notwithstanding any stipulation to the contrary. Basic Rent, the Basic Rent which shall be in advance and by means of a banker's standing order if required by the landlord by equal quarterly payments in each of the rent days except the first payment which should be made on the date hereof and be in respect of the period from and including the 19th day of March 1990 to the rent day next following the date hereof.

There then followed provisions for the payment of insurance, rent and further rent, being VAT and other payments becoming due under the provisions of the lease.

On the day the lease was entered into the defendant paid £2,287.67 advance rent covering the period until the next rent day provided for in the lease. That was March 22 1990, when a sum of £50,000 was paid by the defendant to the plaintiff, being a full quarter's rent in advance without deduction.

On July 6, September 29 and December 25 1990 similar payments of £50,000, being a quarter's rent without deduction, were paid by the defendant to the plaintiff. All the payments of rent which I have described were paid directly into the plaintiff's bank account in Jersey.

On March 11 1991 the defendant's managers realised that they should have been deducting tax from the rent payments which they had made pursuant to the provisions of the Income and Corporation Taxes Act 1988, which I will set out later in this judgment. In respect of the instalments of rent already paid, such deductions should have amounted to rather more than £50,000.

On March 25 1991 a further quarter's instalment of rent became due but was not paid by the defendant on the grounds that the defendant was entitled under the provisions of the 1988 Act to set off

A against that sum the amount which the defendant ought to have deducted from the earlier instalments of rent and for which it was liable to account to the revenue. The plaintiff did not accept that the defendant was entitled to act in this way and these proceedings were commenced by a writ issued by the plaintiff on June 14 1991.

In the meantime, on March 19 1991, the plaintiff had appointed an agent in the UK for the management of its properties and had directed the defendant to make all further rent payments to that agent.

It is common ground that by the combined effect of sections 4, 43, 349, 350 of, and Schedule 16 to, the Income and Corporation Taxes Act 1988, the defendant was obliged, on making payments of rent, to deduct a sum equal to the basic rate of income tax applicable at the B time and to account for that tax to the revenue. Section 43(1), so far as material, provides as follows:

> Section 78 of the Management Act (taxation of non-residents in name of agent) shall not apply to tax on profits or gains chargeable to tax under Schedule A, or on any of the profits or gains chargeable under Case VI of Schedule D — . . .
>
> (*b*) which arise under the terms of a lease, but to a person other than the landlord, or otherwise arise out of any disposition or contract such that if they arose to the person making it they would be chargeable under Schedule A, where payment is made (whether in the United Kingdom or elsewhere) directly to a person whose usual place of abode is outside the United Kingdom, but Sections 349(1) and 350 shall apply in relation to the payment as they apply to annual payments charged with tax under Case III or IV of Schedule D and not payable out of profits or gains brought in to charge to income tax.

C Accordingly, section 43 applies the provisions of sections 349(1) and 350 to the payments of rent made by the defendant to the plaintiff direct to its bank account in Jersey. Section 349(1) provides materially as follows:

> Where . . . [(*c*) any rent . . .] is not payable or not wholly payable out of profits or gains brought into charge to income tax, the person by or through whom any payment thereof is made shall, on making the payment, deduct out of it a sum representing the amount of income tax thereon.

Section 350(1) provides:

> Where any payment within section 349 is made by or through any person, that person shall forthwith deliver to the inspector an account of the payment, and shall be assessable and chargeable with income tax at the basic rate on the D payment, or on so much thereof as is not made out of profits or gains brought in to charge to income tax.

Section 350(4) provides:

> Section 349 and this section has effect subject to the provisions of Schedule 16 which has effect for the purposes of regulating the time and manner in which companies resident in the United Kingdom —
>
> (*a*) are to account for and pay income tax in respect of payments from which tax is deductible under Section 349, and . . .

Para 2(1) of Schedule 16 provides:

> A company shall for each of its accounting periods make, in accordance with this Schedule, returns to the collector of the relevant payments made by it in that period and of the income tax for which it is accountable in respect of those E payments.

There then follow provisions requiring such company to make quarterly returns by prescribed dates in respect of such payments and deductions.

Section 821 of the 1988 Act permits a payer of rent in the position of the defendant to make adjusting deductions in respect of instalments of rent paid after the commencement of a tax year but before the rates of tax are finally fixed by the coming into force of the relevant Finance Act. The Finance Act 1990 came into force on July 26 1990. However, the defendant accepts that it failed to take advantage of the provisions of section 821 to make deductions from the next instalment of rent becoming due as provided for in section 821 and, accordingly, can base no part of its defence to the plaintiff F on the provisions of that section.

The defendant also concedes that the failure to make deductions from instalments of rent constitutes voluntary payments by the defendant to the plaintiff, under a mistake of law and, accordingly, the defendant is not in a position to claim restitution by the plaintiff of the sums it failed to deduct and for which it may become assessable to tax by the revenue.

Clause 2 of the lease entitles the plaintiff to receive payment of the rent by quarterly payments without deduction except such as the tenant may be required by law to deduct notwithstanding any stipulation to the contrary. It follows that the defendant can resist the plaintiff's claim only on the basis that a construction of section 349(1) permits the defendant now to deduct from instalments of rent becoming due the tax which should have been deducted from the G earlier instalments. It was not suggested in the course of the defendant's submissions that the defendant had any other defence.

The important words which fall to be construed in section 349(1) are "where any rent is not payable out of profits the person by or through whom any payment thereof is made shall on making the payment deduct out of it a sum representing the amount of Income Tax thereon". In *Taylor* v *Taylor*, XVI ATC 218 the Court of Appeal was construing the then equivalent of what is now section 348 of the 1988 Act, namely r 19(1) of the All Schedules Rules to the Income Tax Act 1918 which provided:

> Where any annual payment (whether payable as a personal debt or obligation by virtue of any contract or whether payable half-yearly or at any shorter or H more distant periods) is payable wholly out of profits or gains brought into charge to tax, no assessment shall be made upon the person entitled to such annual payment, but the whole of the profits or gains shall be assessed and charged with tax on the person liable to the annual payment without distinguishing the same; and the person liable to make such payment shall be entitled on making such payment to deduct and retain thereout a sum representing the amount of the tax thereon at the rate or rates of tax in force during the period in which the said payment was accruing due. The person to whom such payment is made shall allow the deduction upon receipt of the residue of the same, and the person making the deduction shall be acquitted and discharged of so much money as is represented by the deduction, as if that sum had been actually paid.

The Court of Appeal was construing r19(1) in a case where a J husband had been making payments to his wife, provided for under a separation agreement, from his taxable income, gross, without deduction of the tax thereon. He unsuccessfully attempted to resist a claim, brought under the provisions of the deed of separation, for unpaid instalments, on the ground that he was entitled to set off against those instalments the amount of the tax he had failed to deduct previously. In his judgment at p220, Slesser LJ says:

> By August 1932 the whole of the weekly payments up to that date had been made. The right under Rule 19 in [the All Schedule Rules] the husband is to be entitled, on making such payments, to deduct and retain the amount of the tax. In my opinion, up to August 1932, he had made such payments, and he had omitted to make any deductions, and it is not open to him now, therefore, under the wording of the rule, to make a deduction on making such payments; K because, in August 1932, there were no payments in arrear, and the only payments had all been completely made, and his right of deduction had gone. But the position thereafter is, in my opinion, different; after August 1932 he was at all times in arrear with his payments, and at no time after that date can it be said that he had made his payments so as to prevent his availing himself of the right which he has of making the deductions under Rule 19.

Later, at p221, Slesser LJ continues:

> I would add, in answer to the argument addressed to us by Mr Jacob, only that, in my view, it is not right, in dealing with Rule 19, to have regard to the fact that the payments were made actually week by week. The basis upon which this could be taken to be an annual payment is to be found in the authorities to the effect that it is an annuity, because it provides for payment for more than one year, and the mere fact that it is paid in periods of less than a year seems to me to be expressly reserved in the language of Rule 19 itself L where, giving the right of deduction, the rule says: "whether payable half-yearly or at any shorter time or more distant periods".
>
> In my opinion, whenever the payment for the whole year under consideration has not been made, it is still open to the person making such payment to deduct and retain thereout a sum representing the amount of tax thereon, notwithstanding that, for some particular weeks during that period, the payments had been made in full. The rule points, in my opinion, to the year as the standard to be considered, and not the week, the half-year, or any shorter period . . .

The Court of Appeal therefore concluded that once a payment falling within the rule had been completely made without deduction the right to deduct was lost. What constituted a payment for the purposes of the rule was to be derived from the agreement pursuant to M which the payment was to be made. In that case the payment was the annual sum becoming due under the provisions of the agreement between the August of each succeeding year. Where, however, payments in respect of any year were in arrear it remained open to the payer to deduct the tax becoming due in that year from the remaining payments referable to that year. It mattered not that the agreement provided that the yearly payments were to be made by weekly or other instalments, some of which might have been paid in full.

Mr Peacock for the defendant contends that the decision in *Taylor* v *Taylor* is inapplicable to the issue before me. It is certainly not a decision construing section 349 or its then equivalent under the All Schedules Rules, namely r21(1). Mr Peacock contended that the essential difference between sections 348 and 349 is that, whereas

A deductions under the former section are permissive and the revenue had no interest whether deductions were made or not, because they would get their tax anyway, under section 349 deduction is compulsory and if the payer fails to deduct he can be assessed for the sums not accounted for (see Schedule 16), interest runs in respect of sums unpaid (section 87 of the Taxes Management Act), and there are statutory penalties for failure to account for tax under section 98 of that Act. It follows, so he submits, that section 349(1) should be construed more liberally in favour of the payer so that the right to deduct is not lost when payments are complete.

The position adopted by the plaintiff in this case lacks merit, but I am reluctantly compelled to the conclusion that I cannot accept Mr Peacock's contention. The words "shall on making the payment
B deduct" used by the subsection are plain and are to be compared with the words "shall be entitled on making such payment to deduct" which were the words being construed by the Court of Appeal in *Taylor* v *Taylor*.

In my judgment, I cannot place on the words used in section 349 any different construction than that placed by the Court of Appeal on the equivalent words in r19(1) and must hold that once a relevant payment is complete the right to deduct tax in respect of that payment is lost.

The next question to be considered is what is the payment for the purposes of the present case and can any relevant payment be said to remain open so as to entitle the defendant to deduct tax becoming due
C in respect of that payment? To contend that the payment in question here was a yearly payment, albeit payable by quarterly instalments and so, if in arrears, the payer would be entitled to deduct from the arrears of tax not deducted from earlier instalments would not, even if correct, help the defendant. If the approach of the court in *Taylor* v *Taylor* is adopted, the year in question must be the year from the first rental date provided for by the lease. Four quarterly payments having subsequently been made, payment in respect of that year is complete and, following, the right to deduct is lost. There are no other relevant payments.

Mr Peacock, however, drew my attention to the case of *Johnson* v *Johnson*, another decision of the Court of Appeal applying r19 of the All Schedules Rules to payments being made under a separation
D agreement. By contrast with *Taylor* v *Taylor* and for a reason which does not emerge from the report, the Court of Appeal in that case appear to have adopted tax years and the amounts becoming due during each tax year as the relevant payment. Mr Peacock, as an alternative submission, urged that there should be a similar approach in the present case. On that basis when the March 1991 quarterly rent payment became due the tax year 1990 to 1991 remained open and deductions could be made from the instalment of rent then due in respect of the amounts of tax which should have been deducted from the three rental payments made in that tax year.

I am unable to accept that submission also. On this issue the words used by section 349(1) differ materially from the words used by
E section 348 and 19(1) of the All Schedules Rules. In particular, in section 349 only the word "any" appears in the phrase "the person by whom any payment thereof is made". "Any payment" in that phrase plainly refers to any payment, *inter alia*, of rent. The power as conferred by the section on the payer is to be able on making any payment to deduct out of it, ie the payment, a sum representing the amount of the tax thereon, ie on that payment.

Furthermore, the method of accounting for sums deducted prescribed for deductions made under section 349 by section 354, which I have already partly quoted above, make plain that the legislature was contemplating deductions from each separate payment as it is made.

In my judgment, therefore, the right to deduct under section 349(1)
F by contrast with section 348 arises when any relevant payment is made and is lost in respect of that payment if it is made gross. For these reasons, in my judgment, the plaintiff's action succeeds.

Judgment for the plaintiff with costs.

Queen's Bench Division G

Liverpool District Registry
November 7 1991
(Before Mr Justice STEYN)

FAMOUS ARMY STORES v MEEHAN AND ANOTHER

Estates Gazette March 6 1993
[1993] 09 EG 111

Landlord and tenant — Arrears of rent — Whether tenants entitled to set off against arrears of rent building costs they incurred to remedy an alleged breach of a duty by the landlord H
to obtain a fire certificate

By a lease dated November 14 1989 the defendants were granted a term of premises by the plaintiff with an initial rent-free period of three months. On the defendants failing to pay rent in two sums of £3,993.82 and £812.50, the plaintiff issued proceedings and obtained summary judgment under RSC Ord 14. On appeal from the district registrar's decision the defendants contended they were entitled to set off against the rent building costs and other losses arising from an alleged breach of duty by the plaintiff to obtain a fire certificate. The plaintiff argued that by clause 1 of the lease the defendants J
covenanted to pay all rent "without any deductions".

***Held*: The appeal was dismissed. It was a question of construction of the new lease. The ordinary meaning of "without deductions" is that the rent must be paid in full and there must be no deductions of any kind which otherwise might be permissible by law. The ordinary meaning of the words is therefore wide enough to exclude any right of setoff under the lease.**

The following cases are referred to in this report.

British Anzani (Felixstowe) Ltd v *International Marine Management (UK) Ltd* [1980] QB 137; [1979] 3 WLR 451; [1979] 2 All ER 1063; (1978) 39 P&CR 189; [1979] EGD 414; 250 EG 1183 K
Gilbert-Ash (Northern) v *Modern Engineerging (Bristol)* [1974] AC 689; [1973] 3 WLR 421; [1973] 3 All ER 195, HL
Reardon Smith Line Ltd v *Yngvar Hansen-Tangen* [1976] 1 WLR 989; [1976] 3 All ER 570; [1976] 2 Lloyd's Rep 621, HL

This was an appeal from the decision of the district registrar giving judgment on a claim for arrears of rent under Ord 14 to the plaintiff, Famous Army Stores, against the defendants, Christopher Francis Meehan and Monica Mary Jones.

Bartley Jones (instructed by Davies Wallis Foyster, of Liverpool) appeared for the plaintiff; Anthony Elleray (instructed by Taylors Bridge, of Wigan) represented the defendants. L

Giving judgment, STEYN J said: This is an appeal against two decisions dated August 9 1991, in terms of which a district registrar gave judgment under Ord 14 in favour of the plaintiff, the landlord, against the defendants, the tenants of the leased premises, in sums in respect of rent of respectively £3,993.82 and £812.50 together with interest and costs. The sums claimed represented rent, which admittedly became due under a lease dated November 14 1989. The only defence pleaded is one of set off. It is pleaded in the same way in both actions. The defence, in so far as it is material, reads as follows:

(a) At the commencement of this suit the Plaintiffs were and still are indebted to the Defendants in an amount which by far exceeds the Plaintiffs' claim. M

(b) That the Plaintiffs were responsible for the obtaining of a Fire Certificate for the premises referred to in para 1 of the Statement of Claim and for the carrying out of any work necessary for that Certificate to be obtained; this they failed to do within a reasonable time, despite being aware from a period prior to the granting of the lease, of the necessity of such a certificate and the defects which required rectification before a certificate would be given. It has therefore not been possible for the Defendants to safely and legally occupy the premises with the exception of a small part of the floor area.

(c) That as a consequence of the requirements referred to in para 6(b) above the Defendants have incurred building costs in the sum of £5,620 in order that they could legally and safely occupy a small part of the premises only.

(d) That the failure of the Plaintiffs to obtain the necessary Fire Certificate within a reasonable time has caused damage to the Defendants' business in that they have been compelled to conduct it from one room in the premises,

A forgo the appointment of staff necessary for the proper development and advancement of the business and suffer all those matters which are commensurate with the circumstances in which the Defendants have found themselves through the failure of the Plaintiffs to remedy or abate the nuisance complained of.

(e) That they are entitled to set off against the Plaintiffs' claim in accordance with the Supreme Court Act 1981 section 49(2)(a):

(i) the sum of £5,620 referred to in (c) above;

(ii) a compensatory sum equal to the Defendants' loss of use of the floor area amounting to £50 per week from the date of signing of the lease, which at the date of suit amounted to £3,200;

(iii) a sum in respect of damages to be assessed by the Court.

The plaintiff's submission in answer to the defence is twofold.
B First, the plaintiff submitted that the defendants had failed to establish any arguable breach of duty, whether arising by virtue of the terms of the lease, common law or statute. Accordingly, the plaintiff submits that the defence is misconceived. It seems that the district registrar did not hear full argument on this aspect and I heard no submissions on it at all.

It was agreed that I should concentrate in the first place on the second point, namely the plaintiff's assertion that the parties had agreed that there would be no right of set off. It is axiomatic that the right of set off may be excluded by agreement. It is also clear that clear express words are required to rebut the presumption that neither party intends to abandon any remedies for breach of contract which may arise by operation of law: see *Gilbert Ashe (Northern)* v
C *Modern Engineering (Bristol)* [1974] AC 689 at p717H.

The tenants were granted a rent-free period of three months, thereafter rent instalments were payable. The lease provided in clause (1), and repeated in clause (2)(i), that the payment of rent was to be made "without any deductions". It was agreed that the words "except as aforesaid" which were inserted in clause (2)(i) were inserted by mistake and can safely be ignored. It has not been pleaded or suggested in argument that in the context the words "without any deductions" have been impressed with any special meaning by custom or usage. No implication of any term has been put forward. The question for decision is simply a point of construction. It is, of course, trite that a judge, who is faced with a question of construction, should consider all classes of evidence which arguably
D forms part of the objective setting in which a contract is to be construed: see *Reardon Smith Line Ltd* v *Yngvar Hansen-Tangen* [1976] 1 WLR 989 at p997C. But here there is no evidence of surrounding circumstances advanced by either side which can assist on the question of construction. The question is simply what the words "without any deductions" in the context of the rent payable under this lease would convey to the ordinary speaker of English. It seems to me that the ordinary meaning of "without any deductions" is that the rent must be paid in full and that there must be no deductions of any kind which otherwise might be permissible by law. The ordinary meaning of the words is therefore wide enough to exclude any right of set off under this lease.

E On behalf of the defendants it was submitted that the words "without any deductions" or kindred expressions, first crept into leases at a time when the only matters capable of set off were rates and taxes. The next step is then to say that, without any evidence of usage or custom, this meant that as a matter of construction those words acquired a crystallised meaning and that the words "without any deductions" are therefore a reference simply to deductions for rates and taxes only. Let me assume without further examination that the defendants are right in these assertions of historical fact and legal development, but then there is still a fundamental flaw in the legal reasoning. The purpose of construction is to determine the meaning of the words at the time of the making of the contract. The concept of the "original intent" of an instrument, which features in the current
F debate about the United States constitution, has no place in the construction of contracts in English Law. To ask what words meant in other contracts at an earlier stage of legal development or to ask even what words in a standard form of contract meant when the form was first published is to pose entirely the wrong question. The sole question is: what the words meant when the contract was made, and it is common ground that when this contract was made in November 1989 the law of set off was already fully developed. The words "without any deductions" quite naturally embraced any deduction by way of any set off.

On behalf of the defendants one further matter was urged on me. It was said that if I resolved this issue in favour of the plaintiff it would come as a surprise to conveyancers. It is not possible for the court to poll conveyancers on this issue. In truth their answers may not be G
harmonious: some may say "Yes," others "No" and yet a larger body possibly "I do not know." There is no decision on the point and there are no *obiter dicta* to provide guidance. The point was apparently argued in *British Anzani (Felixstowe) Ltd* v *International Marine Management (UK) Ltd* [1980] 1 QB 137. In that case there is no reasoned decision on the point.

In the absence of a reasoned decision *Anzani* can be of no assistance. One is driven back to the starting point of the process of construction, namely the ordinary meaning of the relevant words in the lease.

In my judgment, the words in question exclude a right of set off: there was, therefore, no defence to the claim for judgment under Ord 14 in respect of the rent. It follows that the appeal against the district H
registrar's decision must be dismissed.

For the avoidance of doubt I must make clear that there was a further claim of the order of £300 in respect of "liquidated" damages to which the provision about excluding the right of set off did not apply. By agreement that aspect was not argued before me and by agreement the order of the district registrar was varied to make clear that there would be unconditional leave to defend on that aspect.

Finally, I simply add that, when announcing my decision, I made clear that I grant leave to appeal in this case since it is a point of some importance to conveyancers and it would be of benefit to conveyancers to have the judgment of the Court of Appeal on the matter. J

Chancery Division

October 28 1992

(Before Sir Donald NICHOLLS V-C)

RPH LTD v MIRROR GROUP (HOLDINGS) LTD AND OTHERS

Estates Gazette April 3 1993 K
[1993] 13 EG 113

Landlord and tenant — Tenant and assignee of original tenant insolvent — Whether original tenant entitled to orders against administrator of insolvent assignee requiring assignment or enforcement of covenant by solvent assignee to payment

Robert Maxwell Estates Ltd is the owner of the reversion to an underlease of Orbit House, New Fetter Lane, London EC4, granted on August 6 1970 to the plaintiff, RPH Ltd. In February 1972 RPH Ltd assigned the underlease to Mirror Group (Holdings) Ltd, a company which is now insolvent and subject to an administration order. In March 1979 Mirror L
Group (Holdings) Ltd assigned the underlease to Mirror Group Newspapers plc, a solvent company. Finally, in June 1987 Mirror Group Newspapers plc assigned the underlease to Maxwell Communication Corporation plc, a company which is insolvent and the subject of an administration order. The annual rent payable under the underlease is £2,650,000; Maxwell Communication Corporation plc failed to pay the quarter's rent due at Christmas 1991 and has paid no rent since. Robert Maxwell Estates Ltd issued a writ claiming against RPH Ltd, as original tenant, arrears of rent for the three quarters up to June 1992, and interest. The underlease M
being registered land under the Land Registration Act 1925, section 24(1)(*b*) of that Act implied a covenant upon the transferee at each assignment of the underlease, *inter alia*, to pay the rent. In this application, RPH Ltd sought an order that the administrators of Mirror Group (Holdings) Ltd assign to RPH Ltd the benefit of Mirror Group Newspapers plc's covenant with Mirror Group (Holdings) Ltd to pay the rent, alternatively, it sought an order directing the administrators of Mirror Group (Holdings) Ltd to bring proceedings against Mirror Group Newspapers plc requiring it to pay the rent to the landlord, Robert Maxwell Estates Ltd,

with RPH Ltd indemnifying Mirror Group (Holdings) Ltd against the costs of the proceedings. As a further alternative, RPH Ltd sought leave under section 11 of the Insolvency Act 1986 to commence proceedings against Mirror Group (Holdings) Ltd claiming as relief an order that it institute against Mirror Group Newspapers plc such proceedings.

***Held*: The application was dismissed. The implied covenant contained in section 24(1)(*b*) of the 1925 Act is a covenant by the transferee with the transferor that during the residue of the term the transferee will pay the rent and will indemnify the transferor against all claims on account of non-payment. That covenant cannot be taken to envisage that if the transferee further assigns the lease, thereafter quarter by quarter the transferee himself will solemnly continue to pay the rent. However, the obligation is not one which would entitle RPH Ltd to compel Mirror Group (Holdings) Ltd to enforce the benefit of its covenant with Mirror Group Newspapers plc. There was no direct legal nexus between the two covenants. If the landlord has recourse to RPH Ltd for payment of the rent, and RPH Ltd turns to Mirror Group (Holdings) Ltd, Mirror Group Newspaper plc's covenant. But as a matter of law Mirror Group (Holdings) Ltd cannot be compelled to do so; it could release Mirror Group Newspapers plc from its covenant without being in breach of its (Mirror Group (Holdings) Ltd's) obligations to RPH Ltd under the covenant it has entered into with RPH Ltd. Nor could the court order Mirror Group (Holdings) Ltd to assign to RPH Ltd the benefit of Mirror Group Newspapers plc's covenant.**

The benefit of Mirror Group Newspapers plc's covenant is an asset of Mirror Group (Holdings) Ltd. Although Mirror Group (Holdings) Ltd could not be compelled to assign the benefit of that covenant to RPH Ltd, the administrators of Mirror Group (Holdings) Ltd are at liberty to make such an assignment if in their view that would be for the benefit of Mirror Group (Holdings) Ltd and its creditors.

No cases are referred to in this report.

This was an application by the applicant, RPH Ltd, for relief against the first respondent, Mirror Group (Holdings) Ltd and the administrators, in relation to a lease of Orbit House, New Fetter Lane, London EC4; the first respondent having the benefit of implied covenants on the part of the sixth respondent, Mirror Group Newspapers plc.

Robert Dicker (instructed by Freshfields) appeared for the applicants; Susan Prevezer (instructed by Allen & Overy) represented Mirror Group (Holdings) Ltd and the administrators; Anthony Trace (instructed by Lovell White Durrant) represented Mirror Group Newspapers plc.

Giving judgment, **Sir Donald Nicholls V-C** said: This application arises out of a type of claim which has been much in the news recently: a claim made by a landlord for rent from the original tenant because the present tenant has become insolvent. In this case the insolvent present tenant is one of the Maxwell group of companies, Maxwell Communication Corporation plc ("MCC"). This company is the subject of an administration order. For simplicity I shall refer to this and all the other companies involved by their present names. Ironically the present landlord is also a company in the Maxwell Group, Robert Maxwell Estates Ltd ("RMEL"). RMEL is in administrative receivership. It is the present landlord of a long lease (strictly an underlease) of Orbit House, New Fetter Lane, London EC4, granted by the then landlord on August 6 1970 to RPH Ltd. RPH Ltd is a non-trading subsidiary of Reed International plc. It will be convenient to refer to RPH Ltd as "Reed". Reed is the applicant making this application.

The lease has been the subject of three successive assignments. First, in February 1972 or thereabouts the lease was assigned by Reed to Mirror Group (Holdings) Ltd ("Holdings"). Holdings is insolvent and an administration order is in force. Second, in March 1979 or thereabouts Holdings in turn assigned the lease to Mirror Group Newspapers plc ("Newspapers"). Newspapers is solvent. Finally, in June 1987 Newspapers assigned the lease to MCC.

The annual rent currently payable under the lease is £2,065,000. MCC failed to pay the quarter's rent due at Christmas 1991 and it has paid no rent since then. RMEL has therefore turned to Reed as the original tenant. As the original tenant Reed covenanted to pay the rent throughout the whole term of the lease. In law it is liable on this covenant, notwithstanding it parted with all interest in the lease over 20 years ago. A writ was issued by RMEL on the September 21 1992 claiming £1,548,750 as the rent due for the three-quarters up to June 1992, and interest.

In the ordinary way when a lessee assigns the lease he obtains a covenant from the assignee that for the future the assignee will pay the rent and carry out the other covenants in the lease and will indemnify the assignor against any claims arising from future breaches. Such a covenant is implied under section 77(1)(*c*) of the Law of Property Act 1925 or, in the case of registered land, under section 24(1)(*b*) of the Land Registration Act 1925. So far as material, section 24(1) provides:

> On the transfer, . . . of any leasehold interest in land under this Act, unless there be an entry on the register negativing such implication, there shall be implied . . .
>
> (*b*) on the part of the transferee, a covenant with the transferor, that during the residue of the term the transferee and the persons deriving title under him will pay, perform, and observe the rent, covenants, and conditions by and in the registered lease reserved and contained, and on the part of the lessee to be paid, performed, and observed, and will keep the transferor and the persons deriving title under him indemnified against all actions, expenses, and claims on account of the non-payment of the said rent or any part thereof, or the breach of the said covenants or conditions, or any of them.

The title to the lease with which this application is concerned was registered, and there was no entry on the register negativing the implication of the covenant set out in section 24. Thus, by virtue of the covenant implied by section 24 Reed is entitled to look to Holdings for an indemnity, Holdings in turn (subject to a point I shall mention later) is entitled to look to Newspapers for an indemnity and so on down the chain to the insolvent present tenant. Unhappily for Reed, Holdings is itself in financial difficulty as I have said. Thus, a claim by Reed against Holdings under this covenant is likely to yield little if anything for Reed as an unsecured creditor. If the buck has to stop there and Reed can look only to Holdings, the loss flowing from MMC's inability to pay the rent will fall on Reed. The landlord will not suffer because it can look to Reed, but Reed's own rights of indemnity are against a company of doubtful value. On the other hand, if by one means or another Reed can reach Newspapers, Reed will not be out of pocket. In that event the loss will fall on Newspapers because its right of indemnity was given by the insolvent MMC.

Who then is to bear the loss: Reed or Newspapers? This application represents an attempt by Reed to find a way around the difficulty and pass the loss on to Newspapers. Reed seeks an order that the administrators of Holdings assign to Reed the benefit of Newspapers' covenant with Holdings. Alternatively, Reed seeks an order directing the administrators of Holdings to bring proceedings against Newspapers requiring it to pay the rent to RMEL, with Reed indemnifying Holdings against the costs of the proceedings. As a further alternative, and as an ultimate fall back, Reed seeks leave under section 11 of the Insolvency Act 1986 to commence proceedings against Holdings claiming as relief an order that Holdings institute against Newspapers the proceedings I have just mentioned.

Newspapers has sought leave to intervene on this application. Tn the event I heard full argument from Newspapers, without opposition from Reed. Reed, nevertheless, continued to maintain that Newspapers had no standing on this application and should not be permitted to intervene. The participation of the administrators of Holdings was confined to a statement of their attitude to the relief sought.

The argument before me ranged far and wide. I heard erudite submissions on a number of points. In the end there is, in my view, one insuperable difficulty confronting Reed. This can be stated quite shortly. Reed's difficulty is that even on the most favourable construction of section 24(1)(*b*) for which Reed contended, Holdings' obligation to Reed was no more than a positive obligation to pay the rent for the rest of the term. I have to say I do not find this a tenable construction in respect of the rent falling due *after* Holdings had assigned the lease. The implied covenant extends to the whole of the residue of the term. In its simplest form it is a covenant by the transferee with the transferor that during the residue of the term the transferee will pay the rent and will indemnify the transferor against all claims on account of non-payment. That covenant cannot be taken to envisage that if the transferee further assigns the lease, thereafter quarter by quarter the transferee himself will solemnly continue to pay the rent. The obligation must be an obligation to see that the rent is paid. Due payment of the rent is being guaranteed by the transferee for the residue of the term.

However, for present purposes the distinction matters not. Either way, the obligation is not one which would entitle Reed to compel Holdings to enforce the benefit of *its* covenant with Newspapers. There is no direct legal nexus between the two covenants. Of course, if the landlord has recourse to Reed for payment of the rent and Reed turns to Holdings, Holdings in its own interest can be expected to enforce Newspapers' covenant. But as a matter of law Holdings cannot be compelled to do so. Holdings could release Newspapers from Newspapers' covenant without being in breach of its (Holdings') obligations to Reed under the covenant it has entered into with Reed. (I pause to interpose that if the administrators of Holdings were to do this, of course the creditors of Holdings could have legitimate grounds for complaint.) That being so, there is no basis on which the court as part and parcel of enforcing Reed's own rights against Holdings under Holdings' covenant can compel Holdings to sue Newspapers. Nor can the court order Holdings to assign to Reed the benefit of Newspapers' covenant. Reed has unsecured claims against Holdings and nothing more. It cannot, as an unsecured creditor, lay claim on its own account to the benefit of Newspapers' covenant.

Mr Dicker laid emphasis on the established principle that a person entitled to an indemnity does not have to pay the debt and then sue the surety to recover the amount he has paid. In a suitable case the court will direct the surety to pay the amount due directly to the creditor, here the landlord. The present case, he submitted, is a stronger one. The covenant under section 24 created more than a quasi-surety relationship. The first limb of the covenant was not merely a covenant to indemnify; here there was a positive obligation to pay the rent and the court should grant an injunction compelling Holdings specifically to perform its obligation. Holdings admittedly cannot be subjected to an injunction compelling it to find money, not least because it is now in the course of administration initiated for para (*d*) in section 8(3) of the Insolvency Act 1986. There is no prospect of Holdings surviving as a going concern. Reed is not entitled to preference over other unsecured creditors. But, it is submitted, Holdings can and should be made specifically to perform its obligation so far as it can by being required to take such steps as are open to it to compel Newspapers to pay the rent to the landlord. Thus Holdings should be compelled to sue Newspapers or, preferably, the benefit of Newspapers' covenant should be assigned to Reed thus enabling Reed to sue Newspapers directly.

This line of argument is not without its attractiveness. However, the fundamental difficulty remains. The court is being asked to make an order requiring Holdings to take steps vis-a-vis Newspapers which Holdings *vis-à-vis* Reed is under no legal obligation to take. It therefore cannot be right, under the guise of specific performance of Holdings' covenant to pay the rent and to indemnify Reed against non-payment, to grant an injunction compelling Holdings to take that step.

That, however, does not mean that Reed has necessarily reached the end of the road. The attitude of the administrators is that it would be unconscionable for Holdings to enforce its right of indemnity against Newspapers and then hold on to the money recovered to swell the general assets of Holdings rather than use the money to repay Reed. So, left to themselves, the administrators would have been minded to consent to the relief sought by Reed. They regard the benefit of Newspapers' covenant as having no commercial value as an asset of Holdings. Newspapers, however, is understandably anxious not to find itself at risk of being sued by Reed. So it told the administrators that if they did consent that would be a misfeasance on their part. Faced with this the administrators altered their stance. They are not prepared to consent to the relief sought, although they themselves have not presented arguments in opposition to it.

In my view, the benefit of Newspapers' covenant is an asset of Holdings for whatever it may be worth. Although Holdings cannot be *compelled* to assign the benefit of that covenant to Reed, the administrators are at liberty to make such an assignment if, in their view, that would be for the benefit of Holdings and its creditors. The administrators have not themselves sought directions from the court on whether they should do so, or regarding the terms on which such an assignment could properly be made. That being so, it is not for me to interfere with the exercise by them of discretions entrusted to them. However, that is a course open to them.

I must mention one other matter. Newspapers claims that in any event it is not liable to indemnify Holdings as claimed. When Newspapers took its assignment it entered into a direct covenant with the then landlord thenceforth to pay the rent for so long as the term granted by the lease should be vested in Newspapers. That covenant was contained in the licence to assign to which Holdings itself was a party. It is therefore, so it is submitted, not now open to Holdings to rely on section 24(1)(*b*) as imposing on Newspapers a greater obligation in respect of rent. Further, no tears should be shed for Holdings. Newspapers took steps to limit its liability for future rent; Holdings did not. In my view, this is not a point I should decide on this application. Newspapers can seek to raise this matter as a defence if any proceedings are brought against it on the covenant.

Finally, I must turn to the question of Newspapers' application to intervene on this application for directions. *Prima facie* Newspapers has no standing to be heard. Newspapers is, or claims to be, a creditor of Holdings. However, on this application by Reed, Newspapers is not concerned to protect its position as a creditor of Holdings. It is not concerned with the effect which granting or withholding the relief sought might have upon it as a creditor interested in Holdings' assets. Every argument of substance addressed to me on behalf of Newspapers was concerned, not with its position as a creditor of Holdings or with any loss it might suffer as a creditor if the directions sought were granted, but with other reasons why Reed was not entitled to obtain against Holdings or the administrators the relief Reed was seeking.

However, as I have indicated, Newspapers presented all the arguments it would have wished to advance in any event. The arguments included the point on which I have decided that Reed's application fails. Newspapers has done so in the circumstance that the administrators' counsel did not consent to the application, did not present any contrary arguments and was content to leave Newspapers to present the contrary argument. In that unusual circumstance it really would be bizarre if I were to refuse leave for Newspapers to intervene. The fact is it has done so and with effect. I shall therefore formally direct that it be joined as an additional respondent to this application.

As I have indicated, many more matters were canvassed before me which, having regard to my conclusion, I need not pursue and I therefore shall not do so. The only footnote I add is that Mr Trace contended that in any event the court has no jurisdiction to give any directions as sought because the Insolvency Act 1986 makes no provision for an application for directions such as this. The administrators themselves may apply for directions under section 14(3); creditors may present petitions under section 27(1); applications may be made for relief under section 11; but that is all. I do not accept this. As noted in *Re Atlantic Computer Systems* [1991] BCLC 606 at p634A, the court may exercise control over the administrators as officers of the court and may give directions to that end. So the court has jurisdiction. However, for the reasons I have given I shall not give any of the directions sought by Reed.

Application dismissed with costs.

Chancery Division

July 2 1992

(Before Judge MICKLEM, sitting as a judge of the High Court)

HOCKLEY ENGINEERING CO LTD v V&P MIDLANDS LTD AND ANOTHER

Estates Gazette May 8 1993

[1993] 18 EG 129

Landlord and tenant — Alleged forfeiture — Whether landlord entitled to forfeit without notice under section 146(9)(e) of the Law of Property Act 1925 — Whether landlord in breach of covenant for quiet enjoyment — Whether landlord trespassed upon demised premises

By a lease dated May 4 1988 the plaintiff, Hockley Engineering Co Ltd, held the term of units F and G at 100

Dudley Road East, Oldbury, Warley, West Midlands, the reversion was held by the first defendant, V&P Midlands Ltd. On June 6 1989 the plaintiff went into creditor's voluntary liquidation and Mr B J Ward FCA was appointed the liquidator. Shortly before going into liquidation the plaintiff had entered into negotiations with the second defendant, KJD Reinforcements Ltd, to sublet part of unit F; permission had been given to the second defendant by the plaintiff to store stock-in-trade in part of that unit and to lock the doors. That permission was given without the consent of the first defendant and was a breach of covenant. On June 9 1989, following a request by a director of the first defendant, a representative of the liquidator handed over the keys to the premises. After that date, by way of an arrangement to allow the liquidator to sell the plaintiff's equipment, the liquidator was granted access to the premises only on request and with the assistance of the first defendant. At the time when the first defendant re-entered the premises, it had not served any notice on the plaintiff under section 146 of the Law of Property Act 1925. The plaintiff claimed against the first defendant damages for trespass and for breach of the covenant for quiet enjoyment; the plaintiff contending that, by reason of the first defendant's actions, it had lost the opportunity of selling the lease at a premium. On behalf of the first defendant it was contended that its re-entry was lawful having regard to the exceptions to the need for a notice before entry under section 146(9) of the 1925 Act, in particular that the personal qualifications of the plaintiff were important to the first defendant, the premises being close to its headquarters and registered office; accordingly para (*e*) of section 146(9) was satisfied.

***Held*: Judgment was given to the plaintiff. Section 146(9)(*e*) of the 1925 Act did not apply. The exceptions in subsection (9) are concerned with the land which is both the subject of a lease to the lessee and also land of a particular type, or which has some special quality. In the context of the subsection one would expect para (*e*) to identify another type of land or land having a special quality which can be objectively determined by the court. Para (*e*) must be read as a whole and so read the two phrases "personal qualities of the tenant" and "on the ground of neighbourhood" would appear at first sight to colour each other. The inquiry which the court must undertake is as follows: is the lessee, whoever he may be, of the property a neighbour of the lessor or someone claiming through him? If yes, is there something about the property comprised in the lessee's lease, having regard to the close relationship involved in being neighbours, which makes it of particular importance to the lessor, or someone claiming through him, who lives next door that the tenant of that land has certain personal qualities rather than others? There was absolutely nothing special about the land, the subject of the plaintiff's lease, which could possibly bring it within either category identified by para (*e*). The first defendant acted unlawfully in re-entering the premises on June 1 1989. On the evidence the proper sum by way of damages was £12,000, being the loss of the premium to the liquidator and allowing for the risks in obtaining possession of the premises from the second defendant.**

The following case is referred to in this report.

Horsey Estate Ltd v *Steiger* [1899] 2 QB 79; [1895-9] All ER Rep 515; 68 LJQB 743; 80 LT 887; 47 WR 644; 15 TLR 367, CA

This was a claim for damages for trespass and for breach of covenant of quiet enjoyment by the plaintiff, Hockley Engineering Co Ltd, against the first defendant, V&P Midlands Ltd, in relation to a lease dated May 4 1988 of units F and G at 100 Dudley Road East, Oldbury, Warley, West Midlands.

Timothy Fancourt (instructed by Alsop Wilkinson) appeared for the plaintiff; Simon Clegg (instructed by Norman Carless & Co, of Stourbridge) represented the first defendant; the second defendant did not appear and was not represented.

Giving judgment, **JUDGE MICKLEM** said: The plaintiff company in this action, Hockley Engineering Co Ltd, went into creditor's voluntary liquidation on June 6 1989. Immediately before going into liquidation the plaintiff was a tenant under a lease, dated May 4 1988, of units F and G at 100 Dudley Road East, Oldbury, Warley, West Midlands. The first defendant, V&P Midlands Ltd, was its landlord.

Shortly before going into liquidation, the plaintiff had sought to improve its financial position by entering into negotiations with the second defendant, KJD Reinforcements Ltd, to sublet part of unit F to it. Those negotiations had come to nothing, but not before the plaintiff had allowed the second defendant to store stock-in-trade in part of unit F and to lock the doors against the plaintiff as well as the rest of the world. That permission was given without the consent of the first defendant. By giving it the plaintiff incurred the displeasure of the first defendant and giving it was a breach of covenant.

Mr B J Ward FCA, a partner in Rabjohns Insolvency Services of 1 and 2 College Yard, Worcester, and a licensed insolvency practitioner, was appointed the liquidator of the plaintiff on June 6 1989. On June 8 1989 one of the liquidator's staff, Mr Graham John Dowsell, attended the plaintiff's premises at 100 Dudley Road East to tab the company's equipment with a view to its sale. Mr Donald Barry Richardson, the joint managing director of the first defendant, came in with two others while Mr Dowsell was there, asked Mr Dowsell what he was doing and requested Mr Dowsell to hand over the keys of the plaintiff's premises. Mr Dowsell did not hand over the keys, but in his presence Mr Richardson arranged for the back door, through which Mr Dowsell had obtained access, to be boarded up.

On the following day, June 9 1989, another of the liquidator's staff, Mr Andrew Richard Williams, called at the premises to collect papers. Again Mr Richardson came in with another man and, having established that Mr Williams was an authorised agent of the liquidator, again demanded the keys to the premises. Mr Williams handed over the keys. Before he left Mr Williams heard Mr Richardson again discussing the boarding up of the premises and the changing of the locks.

Also on June 9 1989 Mr Dowsell had a meeting with Mr Richardson. There is a dispute as to whether, as Mr Dowsell says, he offered on behalf of the liquidator to surrender the premises for a premium of £20,000 or whether, as Mr Richardson's recollection is, no such subject-matter as a surrender and much less any figure was put to him.

I prefer on this the evidence of Mr Dowsell. It seems to me that the evidence of Mr Richardson in his original witness statement was considerably slanted. It attempted to show that all that the first defendant was doing was securing the premises against outside thieves and intruders and that what took place was a surrender by agents of the liquidator. That was simply not the fact and, faced with the contemporary correspondence, Mr Richardson ultimately accepted that what he was purporting to do was to re-enter and exercise the right of forfeiture in the lease.

I hold that the proposal was made on behalf of the liquidator to surrender the premises for £20,000, but it was not, of course, accepted.

What was arranged at that meeting on June 9 between Mr Dowsell and Mr Richardson was a method whereby the liquidator should have access to enable him to sell the plaintiff's equipment. After that date access to the liquidator was granted by the first defendant on request and, as I understand it, an agent of the first defendant unlocked the padlocks, unboarded what was necessary and allowed the liquidator's staff and prospective purchasers to inspect the various items of equipment.

Mr Richardson accepted in cross-examination that what the first defendant had done was to exercise its right of re-entry under the plaintiff's lease. I find that the first defendant re-entered on June 9 1989.

Rent under the plaintiff's lease was payable six months in advance and at the date when the plaintiff company went into liquidation was paid up until the end of June 1989. At the time it re-entered the first defendant had not served any notice on the plaintiff under section 146 of the Law of Property Act 1925.

By this action the plaintiff claims against the first defendant damages for trespass and damages for breach of covenant for quiet enjoyment. The substance of the claim is that because of the first defendant's actions, which the plaintiff claims were wrongful, the plaintiff company lost the opportunity of selling the lease at a premium on a market which, it is common ground, was buoyant.

By this action the plaintiff claims against the second defendant damages for trespass. The second defendant acknowledged service by solicitors and put in a defence long since, but its solicitors were removed from the record in about October 1991 and the second defendant has, in the course of these proceedings, itself gone into liquidation. I am satisfied that the liquidator of the second defendant was aware of the date and place of this trial, but did not choose to attend and it has not been represented at the trial.

There is a certain amount of common ground between the plaintiff and the first defendant. First, it is common ground that the first defendant was precluded by section 146 of the 1925 Act from re-entering in reliance on the plaintiff's breach of covenant, in allowing the second defendant into occupation of part of the demised premises without the written consent of the first defendant, without first serving a notice under that section, which it did not do.

The plaintiff's lease was for a term of 10 years from November 1 1987 at an initial rent of £18,000 pa, payable six months in advance, with rent reviews every three years. It is common ground that, having regard to the rent review clause in the plaintiff's lease and the market conditions since June 6 1989, there was no premium value in the lease after November 1 1990, the date of the first rent review. I am satisfied that the practical effect of what the first defendant did was to disable the plaintiff from selling its lease.

Unless the first defendant can establish that the plaintiff's lease is a lease of land and falls within one of the exceptions identified in section 146(9) of the 1925 Act, the first defendant's re-entry was unlawful having regard to the terms of subsection (10) of that section and the decision of the Court of Appeal in *Horsey Estate Ltd* v *Steiger* [1899] 2 QB 79. Counsel for the first defendant did not argue to the contrary.

Section 146(9) is in these terms:

> This section does not apply to a condition for forfeiture on the bankruptcy of the lessee or on taking in execution of the lessee's interest if contained in a lease of
>
> (*a*) agricultural or pastoral land;
> (*b*) mines or minerals;
> (*c*) a house used or intended to be used as a public house or beershop;
> (*d*) a house let as a dwelling house, with the use of any furniture, books, works of art or other chattels not being in the nature of fixtures;
> (*e*) any property with respect to which the personal qualifications of the tenant are of importance for the preservation of the value or character of the property, or on the ground of neighbourhood to the lessor, or to any person holding under him.

This subsection reproduces section 2(3) of the Conveyancing and Law of Property Act 1892. Although it has been around for a long time the researches of counsel have not revealed any authority on para (*e*) on the second limb of which the first defendant relies in this case.

In his submissions on the subsection counsel for the first defendant first drew attention to the provisions of section 205 (1)(*i*) of the Law of Property Act 1925 and established that "bankruptcy" in this Act in relation to a corporation means the winding up thereof. As to the phrase in para (*e*) "personal qualifications" he drew attention to the fact that "personal" necessarily involved a relationship to the tenant and said it made no difference that here the tenant was a corporation rather than an individual person.

As to "qualifications" he referred me to the *Oxford English Dictionary* which identifies no less than eight separate meanings for the word "qualification" and opted, if I may so put it, for the second meaning identified in the *Oxford English Dictionary*, namely:

> 2. The distinctive quality of a person or thing; condition, character, nature.

He submitted that it was the distinctive quality of the plaintiff company in this case, to which the lease was originally granted, and it was to the company that attention must be given. I do not accept that for reasons which I shall come to.

As to "importance" and "neighbourhood" counsel for the defendant drew attention to the fact that the demised premises are very close — across a narrow court yard — to the headquarters and registered office of the first defendant and the registered office is also that of many other companies, which are associated with it. He drew attention to the evidence of substantial expenditure of money on advertising those companies and submitted that the evidence was that nearly all of them emphasised the address, 100 Dudley Road East. He said that both the fact that the place was the headquarters of the first defendant and the expenditure advertising that fact were matters of importance and submitted that the court should be satisfied that the particular condition and surroundings of the headquarters were of importance to the first defendant.

He then submitted that the closeness of the headquarters premises to the demised premises meant that they were in effect neighbours and that I should conclude that this was a case which fell within the paragraph or, as he put it, that the first defendant had been acting within the exception in para (*e*).

Counsel for the plaintiff had in the forefront of his submissions one to this effect; that it was impossible for a landlord to assert that a case fell within para (*e*) of subsection (9) save in a case of which it could be said that there was an absolute prohibition against assignment or parting with possession. He drew attention to the fact that section 146 imposed a restraint on a lessor in two quite separate ways. First, by imposing the requirement of a notice before taking any action; and, second, by providing for relief against forfeiture. He submitted that both those protections for the tenant amounted in effect to impediments to the landlord and pointed out that where it applies, subsection (9), excludes both protections and both impediments.

He submitted next that the importance of the exceptions in subsection (9) was that a landlord in such a case was able to take immediate action and he submitted that it could be seen precisely why the various categories were set out in paras (*a*) to (*e*).

With regard, for example, to agricultural land in para (*a*) he pointed out that such land requires continuous attention at certain times of the year and that it was of particular importance that a landlord should be able to re-enter quickly without the necessity of serving a notice, and the delay entailed by that, in regard to agricultural land.

Similarly, with regard to mines and minerals, he pointed to the difficulty of a landlord who had let a mine and was concerned for safety and so forth. Similar considerations applied, and I accept, to the beer shop in para (*c*) and to the house let with works of art and so forth in para (*d*). Counsel then submitted that the reference to "bankruptcy" at the beginning of subsection (9) pointed generally in the direction of the draughtsman having an individual person in mind, though he accepted that, having regard to the definition of "bankruptcy", the subsection could apply to a tenant who was a corporation which had gone into liquidation. His submission — which I reject — was that para (*e*) applied only where it could be shown by the landlord that he was really prepared only to let to that individual tenant and that the way the landlord could demonstrate that was by prohibiting any assignment or parting with possession at all in his lease.

In support of that counsel for the plaintiff drew attention to the terms of section 19 of the Landlord and Tenant Act 1927 and pointed out that, at the time when the 1925 Act passed into law, in a case where the covenant was not absolute, it was perfectly possible to refuse unreasonably as a landlord. He said that it was only really since the passing of the Landlord and Tenant Act 1927 that the ambit of para (*e*) has been restricted and therefore the court ought not to take the view that because it had only had a very restricted ambit that could not be the true construction.

With some prompting from the court, counsel for the plaintiff made an alternative submission to the effect that para (*e*) contemplated the sort of property in respect of which only a particular tenant would be suitable on one of the two stated grounds. So that, if the actual lessee loses control of his interest in one of the two stated ways the landlord should be entitled to forfeit immediately. He went on to submit that the evidence here fell far short of establishing that the original tenant, that is, the plaintiff here, fell within any such category and he submitted that it was extremely difficult for any landlord to establish that the case came within this paragraph if, by the terms of his lease, any assignment whatever was permitted.

I do not accept the submission of counsel for the first defendant. It did not seem to me that they really answered the immediate problem which presents itself, as it seems to me, on the first consideration of subsection (9)(*e*). That problem can be identified by the following example.

A landlord owns two terraced houses, which are next door to each other. He lets one to A and the other to B. A and B each derive title from the landlord and they are neighbours. By virtue of the very fact that they are neighbours the personal qualities of the other are of importance to each because, as we all know, the quality of one's life

A can be ruined by an incompatible neighbour. Clearly subsection (9)(*e*) does not apply to that situation.

Put another way the words "on the ground of neighbourhood" cannot mean that. If they did, the exception would be so large that a great part of the force of the rule would be removed. Moreover, if para (*e*) applied to that situation it would be absolutely astonishing that no one had taken the point over the last 100 years since the wording first appeared in section 2(3) of the Conveyancing and Law of Property Act 1892.

I start with two general observations on the question of construction. Section 146(9)(*e*) of the Act must of course be construed in context. Section 146 is itself part of a group of sections in Part V of the 1925 Act, which part is headed "Leases and tenancies". B With two exceptions however the sections in Part V are couched in the terminology of "lessor" and "lessee". The term "landlord" is not used and the word "tenant" appears only twice. The reference to a "tenant" in subsection (9)(*e*) was the only reference to a tenant in Part V that I have been able to find apart from that in the side note to section 151, which appears in the printers' copy as "Provisions as to attornment by tenants".

This latter reference is easily explicable. The expression to "attorn tenant" was embedded in the law on which the 1925 Act was imposed. It was important that the side note be in that form in order to identify the familiar situation which is the subject-matter of section 151. No one spoke naturally then nor speaks now of an C "attorning lessee", but it is noticeable than when, having identified the subject-matter of the section by the side note, one comes to the operative part of the section, section 151(2) speaks of "an attornment by the lessee". In that it is entirely consistent with the rest of the terminology in this part. No construction of section 146(9)(*e*) will be acceptable which does not give full force to the exceptional use of the word "tenant" in it.

The other general observation is as follows — and here I accept a point made by counsel for the plaintiff — the notice required to be served by a lessor under section 146 operates as a protection for the lessee, but from the point of view of the lessor it is an impediment to swift action.

Turning to subsection (9) itself, on consideration of the D paragraphs of the subsection other than para (*e*) it is apparent that each of them is concerned with land, which is both the subject of a lease to the lessee and also land of a particular type, or which has some special quality. It is, I think, further apparent that the type of quality is perceived as justifying the removal of the protection afforded generally to lessees of land by requiring a notice to be served. Justification for that is perceived to be the special features of the types of land identified in the several paragraphs. It is clear that the special types of land identified in paras (*a*) to (*d*) of subsection (9) are types which can be identified objectively as having the characteristics which determine whether the lessee's land in question falls or does not fall within the terms of any one of these paragraphs. In its context I would then expect para (*e*) to identify another type of E land or land having a special quality which can be objectively determined by the court. Further, I would expect para (*e*) in its context to identify another situation in which the interests of the lessor in getting the lessee out of possession quickly on his bankruptcy or liquidation can be seen to justify denial to the lessee of the protection of the notice which section 146 gives to other lessees.

With these preliminary observations I turn to the terms of subsection (9). The phrase "a condition of forfeiture" is clearly a reference to a proviso for re-entry by another name. The reference to "forfeiture on the bankruptcy of the lessee" in context — as counsel for the first defendant pointed out — involves a reference to a corporation being wound up, having regard to the definition in F section 205(1) of the Act.

Section 146(5) reads as follows:

> (5) For the purposes of this section —
> (*a*) "Lease" includes an original or derivative under-lease; also an agreement for a lease where the lessee has become entitled to have his lease granted; also a grant at a fee farm rent, or securing a rent by a condition;
> (*b*) "Lessee" includes an original or derivative under-lessee, and the persons deriving title under a lessee; also a grantee under any such grant as aforesaid and the persons deriving title under him;
> (*c*) "Lessor" includes an original or derivative under-lessor, and the persons deriving title under a lessor; also a person making such grant as aforesaid and the persons deriving title under him . . .

Having regard to the provisions of paras (*a*), (*b*) and (*c*) of section 146 (5), in those definitions the word "under-lease" (by virtue of G para (*d*)):

> . . . includes an agreement for an under-lease where the under-lessee has become entitled to have his underlease granted.

I emphasise the wording of subsection (5) to make it clear that it is in contemplation that subsection (9) shall have potentially a very wide ambit having regard to the words used and the definition of those words for the purposes of the section which I have quoted.

Whereas under paras (*c*) and (*d*) the property is identified as a narrow class of property, namely a house, potentially para (*e*) is much wider. It applies to a lease of any property. But of course it only applies if the property has the characteristics defined in para (*e*) itself. There are two alternative situations identified in para (*e*). The property will fall into the exception if it can objectively be said of it H either that the personal qualifications of the tenant are of importance for the preservation of the value or character of the property or it can be said of it that the personal qualifications of the tenant are of importance on the ground of neighbourhood to the lessor or to any person holding under him.

The Act does not set out to identify further any particular situation in which the tenant's personal qualifications are of importance of either of the alternative grounds mentioned. That appears to be left to the court.

In my judgment, para (*e*) must be read as a whole and so read the two phrases "personal qualities of the tenant" and "on the ground of neighbourhood" would appear at first sight to colour each other. J The root meaning of the word "neighbour" is someone, an individual, who lives next door. "Neighbourhood" in the paragraph is clearly not a reference to the word "neighbourhood" where it means the area in which premises or property are situated. "On the ground of neighbourhood" means, as it seems to me, "because they are neighbours". It is not a natural or usual use of the English language to refer to a corporation as having personal qualities, nor is it natural to refer to a corporation as one's neighbour.

Of course one can say that one lives "next door" to a company just as one can, if one lives next to a church, speak in conversation of those lying in the graveyard as being "quiet neighbours". But in each case there is an element of conscious artificiality in the expression. K

Reading para (*e*) with that in mind I conclude that the inquiry which the court must undertake is first of all this. Is the lessee, whoever he may be, of the lessor's property a neighbour of the lessor or someone claiming through him? If yes, is there something about the property comprised in the lessee's lease, having regard to the close relation involved in being neighbours, which makes it of particular importance to the lessor, or someone claiming through him, who lives next door that the tenant of that land has certain personal qualities rather than others?

The "importance" referred to in the paragraph has to be an importance which the court can see objectively and the importance must arise out of the special qualities of the land, the subject of the lessee's lease, and must also be such as to justify the view that it would L be right to deprive the lessee of his right to a notice on going bankrupt in the interest of the lessor in getting anyone who is on the land, and has become bankrupt, out quickly.

In my judgment, the word "tenant" is used in para (*e*) because attention is focused at that point not on the status of the individual holding the land — that is whether he is the lessee or a sublessee — but on the quality of the land itself and the tenant of it, whoever he may be, from time to time. The use of the word "tenant" in a group of sections which otherwise uses "lessor" and "lessee" seems to me to draw attention to and emphasise the fact that the subject-matter of the paragraph is, as I have indicated, the land comprised in the lessee's lease.

It follows from what I have said above that I regard it as wholly M immaterial whether the lessor did or did not himself subjectively regard the personal qualities of the tenant as of importance to him and wholly immaterial, therefore, whether he did or did not make that view clear to the prospective tenant, either before he took the lease or by the terms of the lease, or by his conduct.

If a lessor wishes to have a person with particular personal qualities as his lessee in a particular property he can achieve this, first of all, by choosing the tenant carefully and, second, by making sure that the tenant has no power whatever to assign or part possession with the property.

I proceed on the footing that notwithstanding the emphasis in personal terminology on "personal qualities" and

A “neighbourhood”, to which I have referred, para (*e*), having regard to all the other indications, applies where the lessee is a corporation.

In my judgment, there is in this case absolutely nothing special about the land, the subject of the plaintiff's lease, which could possibly bring it within either category identified by section 146(9)(*e*). The evidence was not directed to, much less did it establish, any such case.

The land on which the first defendant's development as a whole is built was formerly a tip beside a canal. The first defendant and its associated companies have developed it piecemeal over some 40 years; originally, as I understand it, to house its various activities in relation to the motor trade and a haulage business. All that can be B said in this case is that the plaintiff and the first defendant are neighbours. Therefore, the first defendant fails to establish that its re-entry was a lawful re-entry. I am satisfied that the first defendant acted, in other words, unlawfully in re-entering the premises on June 9 1989 and from that date was in breach of its covenant of quiet enjoyment in the plaintiff's lease which is in the common form:

> The landlord hereby covenants with the tenant . . . that the tenant paying the rents hereby reserved and observing and performing the tenants covenants here and before contained shall and may peaceably hold and enjoy the property during the term hereby granted without any interruption or disturbance from or by the landlord or any person lawfully claiming through under or in trust for it.

Given that the exception to para (*e*) does not apply, the contrary C was not argued by the first defendant.

I turn to quantum. I heard expert witnesses as to the value of the plaintiff's lease as at June 6 1989. The plaintiff's expert valued it at £26,000. The view of the first defendant's expert was that the plaintiff would have had to *pay* someone a sum of the order of £28,458 to take the lease off its hands. They were cross-examined about their mystery or art.

The plaintiff's expert was Mr Martin Harry Bloomer, a BSc in estate management and a fellow of the Royal Institution of Chartered Surveyors. He is a partner in Woolley Pritchard & Co, chartered surveyors, of 330 High Street, West Bromwich, and has extensive experience of valuing industrial property in this area. His was the firm which was employed by the liquidator of the plaintiff to D market its assets in June 1989 and he was the partner under whose control Mr Dowsell, whom I have mentioned, worked. He discussed with Mr Dowsell the value of the plaintiff's lease in June 1989 before it was advertised for sale and I accept his explanation that the lease was advertised inviting offers over £10,000 in order to attract the maximum amount of interest in a property which was being marketed by a liquidator. As he put it, the advertisement was not a statement of value but a marketing tool.

Mr Bloomer was, it seemed to me, an excellent witness and sustained his opinion as to value under cross-examination. He gave persuasive reasons for thinking that the defendant's expert's method of approaching the problem was defective and I accept Mr Bloomer's E evidence as to value. I thought his criticism of some of the comparables mentioned by the defendant's valuer was valid and in particular as to the three on which the first defendant's valuer particularly relied, that is, those on the immediate estate. I thought Mr Bloomer's reasons for rejecting them were valid reasons.

Mr Christopher White [FNAEA], the first defendant's expert, despite his considerable experience, did not appear to me to be a persuasive witness. He rated his comparables in order of merit and put at the top of the list the letting effected by the first defendant of the premises which had formerly been let to the plaintiff — the first defendant's letting to the second defendant.

On the evidence of Mr Richardson himself — and quite apart from that evidence — it was perfectly obvious that that was not an open F market letting and it cannot be a sensible comparable in the present case, notwithstanding that it was a letting on the same estate and indeed of the same property.

With regard to the two other properties which were let on the estate, one in 1988 and one in 1990, which Mr White rated as his next-best comparables, I think Mr Bloomer would be prepared to accept that they are so close that normally one would have regarded those as the best comparables. But when one looks at them and adjusts the rents as appropriate (and for this purpose it seems to me that Mr Bloomer's adjustments, which are slightly different from Mr White's, are appropriate) one finds not only that the rents are much lower than rents elsewhere but also that contrary to the entire market trend the rent in 1990 was lower than the rent in 1988. I accept the evidence of Mr Bloomer that that is so extraordinary and such a G bizarre result that one really cannot rely on these comparables as of any importance at all. Mr Bloomer's phrase was that he would “discount them”, but on examination what he meant by that was that he would disregard them altogether. I need not go into the matter further.

I accept Mr Bloomer's evidence that having examined the matter in depth in recent times he believes — and will advise the liquidator now — that the plaintiff's lease was at the time of the liquidation worth £26,000. That being Mr Bloomer's present considered opinion, after going into the matter in detail, there are powerful reasons for thinking that on the more limited material available to him at the time of the creditor's meeting on June 19 1989 Mr Bloomer thought that the market value of the property was substantially less than £26,000. H There are good reasons for thinking also that the liquidator, properly advised by Mr Bloomer, would have accepted substantially less than £26,000 if the property had been sold reasonably shortly after the plaintiff company went into liquidation.

The situation shortly before the creditor's meeting on June 19 1989 is conveniently summarised in a letter written by Mr Ward, the liquidator, to the solicitors then acting for him as liquidator on June 9 1989, in which he enclosed a copy of the resolution whereby he was appointed liquidator of the plaintiff some three days previously. He says:

> As you are aware the landlord has occupied the premises although the rent has been paid up to 24th June 1989. It would be of assistance therefore if you J could write to the landlord and advise him of his rights in this matter and that although there may be a small outstanding debt due to him for electricity supplied to the unit and/or insurance premiums these debts cannot be used for distraint proceedings.
>
> Messrs Woolley Pritchard & Company [Mr Dowsell's firm] are acting for me in this matter and it is hoped that the landlord may prove to be more reasonable with them and allow the removal of certain third party goods. But, he is aware of my intended action in asking you to forcibly explain my rights to access the premises. A creditor's meeting is fixed for 19th June and notice of such meeting will now be sent to the landlords as they claim to be creditors in these proceedings. There is also another problem with regard to this matter as a firm known as KJD Reinforcements [that is the second defendant] have illegally occupied the unit and I would be pleased if you will issue proceedings to evict them as a matter or urgency.
>
> K In this respect I enclose a copy of correspondence found amongst the company's books and papers wherein Messrs Shakespeare Duggan Lea & Company have initiated some action. It may be necessary therefore for you to contact that firm to obtain the up to date position and I agree that should they have reasonable outstanding costs with regard to this matter then you may give an undertaking to them for payment provided that we are able to take over their files.
>
> If you require any further information please do not hesitate to contact Andy Williams or Graham Dowsell at Woolley Pritchard.

The correspondence referred to as having come from Shakespeare Duggan Lea & Co must have included a letter of March 1 1989 headed “Subject to contract and formal lease” and addressed to the second defendant from Shakespeares acting for the plaintiff. They say:

> We act for the Hockley Engineering Co Limited of units F and G, 100 Dudley Road East and we understand Mr Tubbs of our clients has been speaking to L you regarding terms of a proposed underletting of part of unit F. We therefore write to set out the proposed terms as follows . . .

They then identify the area to be let by reference to a plan and in the second paragraph they state:

> The terms will be for two years from the date of possession under the lease and the parties will make a joint application to the County Court for an order excluding security of tenure . . .

Then they say:

> The rent will be £7,600 pa plus rates and the cost of usual services and a proportion of the insurance premium for the buildings insurance payable by the landlords. M
>
> Otherwise the lease will be in the same form as our clients' present lease of units F and G and you will be responsible for our reasonable costs and disbursements in connection with the underletting and those of the landlord's solicitors in connection with the grant of a licence to underlet.
>
> On the assumption that you are agreeable to proceeding on these terms could you either please pass a copy of this letter to your solicitors and ask them to respond to us direct or give us their details and we will contact them.

Then Shakespeares refer to the fact that the second defendant is already a tenant of the first defendant and they hope to keep references to the minimum. The letter continues:

> We understand that you may be storing certain raw materials in these premises at the moment. [That is on the premises which are demised to the plaintiff by the first defendant.] Our clients have no objection to this provided you

> A acknowledge that these are stored entirely at your own risk and our clients reserve the right to ask you to remove them forthwith if necessary, pending completion of the lease. However, our clients are not prepared to allow machinery to be moved into the premises pending completion of the lease.

The correspondence referred to by the liquidator in the letter of June 9, to which I have referred, must also have included a copy of a letter written by Shakespeare Duggan Lea & Co, as solicitors for the plaintiff, to the second defendant on March 20 1989, the first paragraph of which makes it clear that the plaintiff's position is that the second defendant was not permitted to move any machinery in or to start trading and the second paragraph of which states:

> In view of this our clients have resolved not to proceed with the proposed underletting and so require you to vacate the premises and leave them clean B and tidy no later than Thursday of this week. We have copied this letter to Challinor & Roberts who we trust will confirm to you that you have no legal right to occupy the premises. We have asked our clients to contact us on Wednesday with confirmation that the unit has been or is being vacated in accordance with this demand.

That was a demand to leave at the latest by the end of March 23 1989.

Mr Ward's letter to his solicitors of June 9 1989 makes it clear that Mr Ward has not yet ascertained whether Shakespeares have actively started proceedings or taken any further steps to recover possession of the part of the demised premises which were being used for storage by the second defendant and shows that at that time the liquidator contemplated and indeed in that letter gave instructions for the issue of proceedings to evict the second defendant. The situation recorded C in that letter is consistent with Mr Dowsell's recollection that when the property was advertised on June 16 1989 the full position with regard to the second defendant was not known. But I must assume that the property was advertised at a time when Mr Ward knew that he had to take proceedings to evict the second defendant, and that the instructions to advertise were given after he had given instructions to start proceedings against the second defendant. I infer that Mr Ward was proposing to sell, having recovered possession. On June 16 the liquidator, on the advice of Mr Dowsell, who had consulted with Mr Bloomer, advertised the plaintiff's lease for sale in a newspaper in terms, as I have indicated, of "offers in excess of £10,000". I accept the evidence of Mr Dowsell and Mr Bloomer that this is a deliberately low-starting threshold established with a view to getting people in to D bid against each other and I certainly accept that they both expected to receive offers greater than £10,000 in what was a buoyant market.

But, as counsel for the plaintiff rightly accepted, there comes a point in such circumstances when the seller raises the proposed selling price beyond a certain point, at which the proposing purchaser who has been encouraged to come forward with an offer in excess of £10,000 says words to the effect of "Well, wait now, you were inviting offers in excess of £10,000 and you are now asking [let us say] half as much again". The inference I draw from that advertisement is that even in a buoyant market Mr Bloomer and Mr Dowsell, on the information available to them, did not expect to actually sell the property for much over £15,000.

E Second, the clear evidence as to the figure which is stated in the statement of affairs presented to the creditor's meeting is this. Mr Tubbs, as director of the plaintiff, signed that statement of affairs on June 19 1989, but he accepted figures which were put before him by Mr Williams. Mr Williams was working with Mr Dowsell. Mr Williams was well aware that the property was being advertised and offers were being invited in excess of £10,000 and the figure he put in to the statement of affairs, and to which Mr Tubbs gave the sanction of his name as a director, as a figure of £7,500 for this lease.

Mr Williams' evidence was that he liked, in such circumstances, to value items of property low so that creditors of the company thereafter would not feel disappointed and would indeed, if the property was sold for more than £7,500, on the face of it be happy.

F I infer from the inclusion in the statement of affairs of the lease at a value of £7,500 in these circumstances that the liquidator, Mr Ward and his advisers, took the view that at the very least they would have in hand at the end of the day, after the sale of this property with all the potential difficulties of which they were aware, the sum of £7,500.

What then is the loss which the plaintiff has suffered? There are a substantial number of factors. There was a buoyant market. The sale was by a liquidator and in those circumstances Mr Dowsell was of the view that it was wise to start the bidding low as the important thing was to get people interested in the case of sales by liquidators, which were perhaps more difficult than other sales.

It is not clear of course, whatever the true value of the property and whatever the value with the liquidator was prepared to accept, that G either of those figures would actually be achieved on the sale. What has to be valued is the chance of selling something worth so much. Here, before ever the property could be marketed, in the contemplation of the liquidator, the position of the second defendant had to be regularised. That was likely to cost some money even if the costs were recovered against the second defendant. It was clearly unlikely that that sort of problem could be sorted out in a day or so and, indeed, on June 30 1989 the liquidator tendered the next six months' rent in advance of £9,000, which was rejected. He would not expect, in the end, to have to pay the full £9,000 for the pleasure of recouping himself by selling the lease because no doubt on an assignment there would be an apportionment. But he would have to face a period in which there was actually nobody in the premises, they H were not being used, but he was actually paying out rent at £4,500 a quarter.

The matter is imponderable. If it took him three months to sell he would have to pay £4,500 of rent in the meantime and would recover from the assignor only the rest of the £9,000. If it took him less time he might get less. The longer he had to think about it the more near the true value, as now estimated by Mr Bloomer, Mr Bloomer might have reached in his own mind. That is to say the later the sale the more the liquidator would have been seeking to achieve on the sale. Whatever happened there were going to be some costs of disposing of this lease.

As to the solicitors' costs the probability is that they would have been recovered from an assignee. As to the agent's costs and the J marketing costs they were going to be met by the liquidator in all probability.

Not only was there a problem about getting the second defendant out, but there was also the potential problem of the attitude of the first defendant, because the first defendant, as I well understand from Mr Richardson's evidence, was incensed. From the point of view of the first defendant the plaintiff had stopped trading, let somebody into the premises and, as Mr Richardson put it, "walked away from the problem". It is perfectly true that at the time which I am considering, June 19, the first defendant was saying "we have re-entered". But *non constat* that it would not receive advice that that course was unlawful and that it could re-enter only after serving a notice. If a notice had been served there was potential further cost to K the liquidator of selling the lease. There is also a further small potential cost in terms of payments that would have to be made to the landlord for insurance and electricity and minor matters like that.

It seems to me that if you have a property worth X and you only have a chance of selling it, the best you have got is a less than 100% chance of X. But in this case, in a buoyant market, when the evidence now is that the property was probably worth £26,000 it seems to me that there is a very high chance indeed that the liquidator would have got what I think he was hoping to get at the time in a quick sale, which is £15,000. He might actually have got more, given the information which is now available. But against that he would have had the costs I have indicated and the exact amount of the net proceeds would have L depended upon how quickly the premises were sold.

Some set-off from £15,000 must be allowed and, doing the best I can with a number of imponderables, I think that the damage which the plaintiff suffered was a loss of £12,000 in this case.

As against the second defendant the plaintiff is, as it seems to me, entitled to judgment for damages for trespass. Undoubtedly the second defendant was let into occupation and storage was allowed. Whatever the rights and wrongs about machinery it seems to me that the letter of March 20, to which I have referred, made it unequivocally clear that the plaintiff wanted the second defendant out and was determining its licence. Counsel for the plaintiff, very properly, has drawn attention to the fact that the second defendant was saying, "Well, we did not have reasonable notice". On the other M hand, having read the two letters which I have read, it is clear from the first letter of March 1 that, from receipt of that, the second defendant must have known that the plaintiff was saying "You are in on sufferance and we are entitled to get you out forthwith". In those circumstances, the quite short notice given by the letter of March 20 seems to me to be sufficient. Therefore, it seems to me that beginning on March 24 1989 the second defendant was a trespasser. The plaintiff had paid rent down to the end of June, but the second defendant was in occupation in part of unit F. The plaintiff is entitled to damages for trespass for the period beginning on March 24 1989 and ending on June 30 1989.

Counsel has drawn my attention to the figure of £7,600 pa which was put forward by Shakespeares as a figure agreed. One simply does not know if the figure was agreed as there is no answer to that letter. However, I am prepared, as counsel submitted I should, to take that as the figure that was in contemplation between the plaintiff and the second defendant at that time and as representing a proper rent and, therefore, as being an appropriate measure of what the plaintiff has lost by reason of the fact that the second defendant did not remove itself from the premises, that is the measure of damages for trespass.

Counsel assures me, and I am in no doubt that he is right, that a rent of £7,600 pa works out at £20.82 per day, that the period is 99 days and the figure which he claims for damages is therefore £2,061. As against the second defendant I award that sum.

Judgment for the plaintiff with costs.

Court of Appeal

January 18 1993

(Before Lord Justice NOURSE and Lord Justice STEYN)

RAMDATH v OSWALD DALEY (T/A D & E AUTO SPARES) AND ANOTHER

Estates Gazette May 22 1993

[1993] 20 EG 123

Landlord and tenant — Wrongful eviction of residential tenant — Whether assessment of damages could properly include exemplary damages against landlord's agent — Whether trial judge entitled to make separate orders for damages against each defendant

The first defendant was the lessee of premises at 71 Salusbury Road, London NW6; the plaintiff occupied the second-floor flat since 1982 when he moved in as licensee of a Mr Ayscough, who died in 1988. The first defendant continued to accept rent from the plaintiff, who became the first defendant's subtenant by operation of law. In July 1991 the second defendant, who had begun to manage the premises on behalf of the first defendant, demanded that the plaintiff pay a weekly rent of £100 in place of £40. The plaintiff, having refused to pay the increase, returned home on August 14 1991 to find the locks on his flat had been changed and some of his possessions outside; it was not until September 19 1991 that he was able to return to his flat. On five occasions during August 1991 the second defendant verbally abused the plaintiff in the presence of witnesses. In Willesden County Court the recorder awarded £2,000 general damages, £510 special damages and £1,000 exemplary damages against the first defendant which, after allowing for arrears of rent, made a total of £2,830 damages. He awarded £1,250 general damages (to include an element of aggravated damages), exemplary damages of £2,500 and a further sum of £2,647 for property missing from the flat, against the second defendant. The defendants appealed.

Held: **The first defendant's appeal was dismissed; the second defendant's appeal was allowed in part. The recorder had properly directed himself as to making an award of exemplary damages against the first defendant and had the evidence to award the special damages and damages for loss of property. The recorder was entitled to make two separate awards of damages against the defendants having regard to their different roles: that of the first defendant as authority and encouragement; and that of the second defendant of execution, which extended over a much longer period. However, exemplary damages should not have been awarded against the second defendant as it was not clear that he fell within Lord Devlin's second category in *Rookes* v *Barnard*.**

The following cases are referred to in this report.

Drane v *Evangelou* [1978] 1 WLR 455; [1978] 2 All ER 437; (1977) 36 P&CR 270; 246 EG 137

Rookes v *Barnard* [1964] AC 1129; [1964] 2 WLR 269; [1964] 1 All ER 367; [1964] 1 Lloyd's Rep 28, HL

This is an appeal against the decision of Mr Recorder Bridges Adams, sitting in Willesden County Court, who on January 3 1992 ordered awards of damages against the defendants, Oswald and Gary Daley, in an action by the plaintiff, Kaselal Ramdath, for trespass to land and goods.

Lorraine Irwin (instructed by Goodwin Harte, of Harrow) appeared for the appellant defendants; Martin Seaward (instructed by T V Edwards & Co) represented the respondent plaintiff.

Giving the first judgment, **NOURSE LJ** said: This is an appeal against an award of damages made by Mr Recorder Bridges Adams on January 3 1992 in an action in Willesden County Court arising out of the wrongful eviction of a tenant by the landlord and his son. The learned recorder described the manner of the eviction as not simply discourteous, but brutal. He said that the case was an extremely bad one.

The material facts, as found by the recorder, can be summarised as follows. The property in question is a second-floor flat at 71 Salusbury Road, London NW6. The first defendant, Mr Oswald Daley, is the lessee of those and other premises at that address. The plaintiff, Mr Kaselal Ramdath, has occupied the flat since early in 1982, when he moved in as the licensee of the subtenant of the flat, a Mr Ayscough, who died in 1988. The recorder found that the first defendant knew of Mr Ayscough's death shortly after it occurred, but continued to accept rent from the plaintiff. On the basis of that finding he held that the plaintiff became the first defendant's subtenant by implication of law. None of that is in issue on this appeal.

It appears that until the summer of 1991 relations between the plaintiff and the first defendant and his son, the second defendant Mr Gary Daley, were perfectly normal. The recorder evidently thought that the events I have to recount may have been caused by the fact that, in the summer of 1991, the rent payable by the first defendant under his lease of the premises as a whole was increased from £4,000 to £11,500 a year. In any event, he found that from about June 1991 onwards the second defendant began to manage the premises on behalf of the first defendant. At some time in July the second defendant demanded that the plaintiff's rent, which had been increased from £35 to £40 a week in the early part of 1989, should be increased to £100 a week.

The plaintiff, having refused to pay the increase demanded, returned home on August 14 to find that the locks on the flat had been changed and that some of his possessions had been removed into the corridor outside the flat. He immediately contacted the police and the housing department of the local authority. Notwithstanding those and other efforts made by him to gain access to the flat, it was not until September 19 that he felt it safe to return to the premises. During that period he spent only one night in the flat. So he was out of occupation for a period of something under five weeks. When he got back he found that many of his belongings had gone missing.

Those are the basic facts of the case, but it is necessary to add that there were further incidents on August 15, 16, 17, 22 and 28, when the second defendant verbally abused the plaintiff or threatened him in a serious manner, on each occasion in the presence of other people — either a local authority officer, one of two police constables or a friend of the plaintiff. The recorder, having reviewed the evidence as to all those matters with some care, stated his conclusion as to the factual position at the bottom of p7 of the note of his judgment:

> Although there were perhaps one or two minor inconsistencies in the evidence given for the plaintiff broadly speaking I accept it as correct. I find that the first defendant learnt of Mr Ayscough's death shortly after it occurred and thereafter knew perfectly well that he was dealing with the defendant. I find that the plaintiff enjoyed a tenancy by implication of law. I also accept the evidence of the plaintiff about his eviction and that of him and his witnesses about his subsequent harassment and the removal of his possessions.

The recorder then turned to questions of damages. He dealt first with damages as against the first defendant. He awarded £2,000 general damages and £510 special damages in respect of expenses related to the plaintiff's eviction. He required the plaintiff to give credit for £680 in respect of unpaid rent, that is to say rent which, although proffered, had not been accepted by the first defendant. He also awarded £1,000 by way of exemplary damages. In the result he made a total award of £2,830 by way of damages against the first defendant.

A He then turned to damages as against the second defendant. At p8 of the note he said:

> So far as the manner of the eviction is concerned, the way in which the plaintiff's possessions were treated and the threats to which he was subjected I regard these matters as being primarily the responsibility of the second defendant. The manner of the eviction was not simply discourteous, but brutal. The plaintiff was evicted from what had been his home for many years and of which, as I have found, he had been the tenant for at least two and a half years. By way of general damages I award £1,250. I include in this sum an element in respect of aggravated damages because of the obvious affront to the plaintiff's feelings. In addition, I regard this as an appropriate case for awarding substantial exemplary damages. The second defendant acted with flagrant disregard of the law as is evidenced by his remark to the police officer
> B on August 15. The case is an extremely bad one and I award exemplary damages in the sum of £2,500. I have seen a schedule of property missing from the plaintiff's flat of which the total value is put forward as £2,647. Neither the items nor the values put forward in respect of this part of the claim were seriously challenged and I accordingly direct that the second defendant pay this sum to the plaintiff.

In the result the recorder awarded total damages against the second defendant of £6,397. I will refer to a later passage in his judgment in due course.

In support of their appeal to this court the defendants have advanced seven grounds of appeal, of which only three are of any substance. I will deal with them in a somewhat different order from that in which they have been dealt with by Miss Irwin on their behalf.
C I deal first with the insubstantial grounds.

The ground advanced in para 5(ii) of the notice of appeal is that the recorder misdirected himself on the burden and standard of proof when he held that the first defendant was liable to pay exemplary damages to the plaintiff. It was established by the decision of this court in *Drane* v *Evangelou* [1978] 1 WLR 455 that an action for wrongful eviction, more accurately an action for trespass to land and goods, is capable of falling within the second category of cases in which, according to the speech of Lord Devlin in *Rookes* v *Barnard* [1964] AC 1129, at p1,227, exemplary damages may be awarded. Lord Devlin said:

> Where a defendant with a cynical disregard for a plaintiff's rights has calculated that the money to be made out of his wrongdoing will probably
> D exceed the damages at risk, it is necessary for the law to show that it cannot be broken with impunity. This category is not confined to moneymaking in the strict sense. It extends to cases in which the defendant is seeking to gain at the expense of the plaintiff some object — perhaps some property which he covets — which either he could not obtain at all or not obtain except at a price greater than he wants to put down. Exemplary damages can properly be awarded whenever it is necessary to teach a wrongdoer that tort does not pay.

In *Drane* v *Evangelou*, at p459F, Lord Denning MR, having read the second part of this passage, said:

> To my mind this category includes cases of unlawful eviction of a tenant. The landlord seeks to gain possession at the expense of the tenant — so as to keep or get a rent higher than that awarded by the rent tribunal — or to get possession from a tenant who is protected by the Rent Acts. So he resorts to
> E harassing tactics. Such conduct can be punished now by the criminal law. But it can also be punished by the civil law by an award of exemplary damages.

In support of ground 5(ii) Miss Irwin has submitted that the recorder's judgment does not sufficiently lay the basis for an award of exemplary damages against the first defendant. I do not accept that submission. In my view, it is clear that the judgment, read as a whole, contains a finding, perhaps implied rather than expressed, and perhaps not expressed because the recorder regarded the matter as so obvious, that the first defendant had authorised and encouraged the second defendant to throw the plaintiff out of the flat. On that footing this was a case where the recorder was clearly entitled to make an award of exemplary damages against the first defendant. The amount of it has not been questioned. I need say no
F more than that it seems to me to have been a perfectly proper award to make.

I go next to ground 5(iii), which claims that the recorder misdirected himself on the burden and standard of proof when he held that the claim for special damages of £510 against the first defendant was not seriously challenged and thereby proved. Once again there is nothing in this ground of appeal. Miss Irwin has complained that the £510, which was calculated by reference to the plaintiff's estimate of his living-out expenses during the period when he was out of occupation, was not sufficiently particularised and, further, that it was or may have been at variance with the £50 a week he said he had had to pay for his temporary accommodation with friends. As to that, I need say only that the plaintiff gave the evidence before the recorder, who was entitled to accept and act on it. It G
appears that the defendants, who were not represented at the trial, did not cross-examine him, or did not cross-examine him at all effectively, on that point. Accordingly, there was no reason why the recorder should not have accepted his evidence. This ground of appeal also fails.

I turn next to ground 3, which is that there was no, alternatively insufficient, evidence to support the recorder's finding that the second defendant had removed any and/or particular items of the plaintiff's property from the flat. Very similar observations apply to this ground of appeal. A schedule was put in and the plaintiff deposed to the accuracy of it when giving his evidence. There was therefore evidence on which the recorder could conclude that the losses to
which that schedule related had in fact been suffered; and Mr H
Seaward, who appeared for the plaintiff in the court below and has also appeared before us, has explained the one minor mathematical curiosity in an entirely satisfactory way. Accordingly, this ground of appeal fails. The figure of £2,647 for special damages against the second defendant stands.

Finally, among the insubstantial grounds of appeal, I refer to ground 4, which claims that there was no, alternatively insufficient, evidence to support the recorder's finding that the plaintiff had sustained ascertainable losses for which the defendants were liable to compensate the plaintiff for any special damages. That is putting in a general way only what has already been put in a particular way in
grounds 5(iii) and 3. It adds nothing to those grounds and I need say J
no more about it.

I now come to the three grounds of appeal which are of some substance. Grounds 7(i) and 7(ii) run together. Ground 7(i) claims that the recorder erred in law when he made two separate awards of damages against the defendants; ground 7(ii) claims that he erred in law when he did not pronounce judgment upon the plaintiff's overall compensatable losses. Miss Irwin's point here is that, at any rate so far as the eviction was concerned, the defendants were liable, if they were liable at all, as joint tortfeasors. If, as I have now held to have been the case, the recorder found that the first defendant authorised and encouraged the second defendant to throw the plaintiff out, then, at any rate up to that stage, the defendants were joint
tortfeasors. Miss Irwin goes on to say that since a joint award was not K
made there is a danger, she would say a probability, that there has been an element of double counting and double recovery by which the plaintiff has wrongly profited.

For a while this point has troubled me, but on the whole I think that the explanation given by Mr Seaward is correct. Mr Seaward asked the recorder to make a joint award against both defendants, but he did not do so. His reason for that is expressed on p9 of the note:

> I was asked to make a joint order against both defendants but in view of the fact that one of them is liable in contract and tort and the other in tort only, their different degrees of involvement and the difficulty of assessing whether at all times the second defendant was acting within the scope of his authority as agent for the first defendant I decline to do so.

L

Originally, Mr Seaward was disposed to explain the recorder's separate awards on the footing that he had drawn a line as at August 15, the last date on which there was any evidence of the first defendant himself having played a part in the matter. However, after an exchange with the court, his preferred explanation is that the recorder was really distinguishing between authority and encouragement — the role of the first defendant — on the one hand, and execution — the role of the second defendant — on the other. It seems to me that that is the most probable explanation of the course which the recorder took. On that footing I do not see why he should not have made separate awards, bearing in mind particularly that the second defendant's physical involvement in the matter was very
much greater than the first defendant's and extended over a much M
longer period. In the circumstances, I would reject grounds 7(i) and (ii), emphasising that the general damages awarded against the first and second defendants respectively were by no means excessive.

I come finally to ground 7(iii), which claims that the recorder erred in law when he made two distinct and separate awards of exemplary damages against the defendants. The first point taken here is that the second defendant was someone who could not and did not fall within Lord Devlin's second category. If that is correct, there is no way in which exemplary damages could properly have been awarded against the second defendant.

In order to support the second defendant's inclusion in the second category Mr Seaward has made a number of telling points. He has

A referred us to a passage in the second defendant's statement and to various passages in the oral evidence, which, he submits, clearly show that the second defendant was associating himself with the interests of the first defendant and was, as he put it, assuming the mantle of the landlord. There was evidence to show that the second defendant was making a deliberate calculation of what was in his best financial interests and was acting accordingly. I take full account of these points, but the evidence does not, in my view, show more than that the second defendant was closely concerning himself with the first defendant's interests. It does not show, on a balance of probabilities, that the second defendant himself had a sufficient interest in the matter to bring him within Lord Devlin's second category.

Before we could hold that the recorder was entitled to make an
B award of exemplary damages against the second defendant we would have to be satisfied that he had applied the right test. Mr Seaward has told us that he referred the recorder to a passage in *Megarry's Rent Acts*, 11th ed, vol 1, pp384-385 under the heading "Aggravated and Exemplary Damages". But he very properly accepts that that passage does not spell out the test in sufficient detail.

Mr Seaward next submitted that, even so, it is open to us in this court, on the authority of what was said by Lawton and Goff LJJ in *Drane* v *Evangelou* at pp461H and 463D, to treat the £2,500 as aggravated damages and to increase the award of general damages accordingly. In *Drane* v *Evangelou* this court was confronted with one global award of £1,000, which the judge in the court below had
C described as exemplary damages. Both Lawton and Goff LJJ thought that that figure would not have been excessive for aggravated damages so that, even if exemplary damages ought not to have been awarded, the figure could be upheld as aggravated damages. Basing himself on that, Mr Seaward submits that we should take the same course here.

While that submission is an attractive one, I feel unable to accept it in a case where the recorder has expressly included in the general damages of £1,250 "an element in respect of aggravated damages because of the obvious affront to the plaintiff's feelings". Aggravated damages and exemplary damages are not only different in themselves; they are awarded for two different purposes. Aggravated damages are awarded to compensate the plaintiff for
D injury to his proper feelings of dignity and pride and for aggravation generally, whereas exemplary damages are awarded in order to punish the defendant. So I do not think it permissible for us to hold that the figure which the recorder thought was sufficient for general damages, including aggravated damages, ought to be increased by adding to it the whole or a part of the sum which he awarded as exemplary damages. As a last resort, Mr Seaward asked us to increase the amount of the exemplary damages awarded against the first defendant. It seems to me to be clear that there is no way in which we could properly take that course.

I have therefore come to the conclusion, with some reluctance, that the second defendant's appeal must be allowed in relation to the sum of £2,500 exemplary damages. I express my reluctance for two
E reasons: first, because I think this may have been a case where the second defendant was fortunate not to have been represented in the court below; if he had been, his legal representative would have had to explain to the recorder how it was that he did not fall within Lord Devlin's second category. He might not have succeeded in satisfying the recorder that he did, on the facts, fall outside that category. Second, looking at the recorder's figure for damages against the second defendant as a whole, namely £6,397, I cannot say that it was at all excessive in the light of the conduct of the second defendant, which, with every justification, the recorder described as brutal.

I would dismiss the first defendant's appeal. I would allow the second defendant's appeal to the extent stated and reduce the
F damages awarded against him from £6,397 to £3,897.

STEYN LJ agreed and did not add anything.

First defendant's appeal dismissed; second defendant's appeal allowed to the extent indicated in the judgment.

Chancery Division G

February 9 1993

(Before Mr Terrence CULLEN QC, sitting as a deputy judge of the division)

NEW ENGLAND PROPERTIES v PORTSMOUTH NEW SHOPS STERLING SURVEYS v NEW ENGLAND PROPERTIES NEW ENGLAND PROPERTIES v EX-ELECTRONICS (UK) LTD

Estates Gazette June 12 1993

[1993] 23 EG 130 H

Landlord and tenant — Repairs — Service charge — Whether expenditure on reroofing recoverable under covenant by tenants to pay service charges for repairs

New England Properties is the owner of commercial premises at Grove House, Headley Road, Grayshot, Surrey. The property is a T-shaped building the ground floor being partly occupied by retail shops while the rest remains a car park for the tenants of the building. The first floor is divided into offices, the seven units having been let on leases granted in the 1980s. Following the storm in October 1987, and a
surveyor's warning that there was a real danger of roof J
collapse, three schemes were put before the tenants one of which was chosen and was carried out in 1989 at an expenditure to the landlord of £202,000. The office leases are in common form and by clause 4(a) of the fourth schedule to the common form lease the landlord covenants "to keep and maintain or procure to be kept and maintained the Scheme and in particular without prejudice to the generality of the foregoing its exterior and its structure . . . in good and substantial repair and condition and decoration and to renew or replace the same or any part or parts thereof whenever such renewal or replacement shall be necessary". Each lease was K
for a term of 25 years and the "scheme" was defined as the office and shopping development known as Grove House. The original roof was inadequate in a number of respects: the roof trusses were not thick enough and did not have sufficient lateral restraint nor longitudinal rafter ties, and the roof was built with a 22.5° pitch. The consequence of those defects, allied with the storm, was that the roof structure within the tail of the T became badly distorted and unstable; the design life of the original roof structure would not have exceeded a maximum of 16-20 years. The works carried out involved the insertion of new trusses with a 30° pitch, with more substantial
timbers forming the trusses and rafters, and a proper system L
of lateral and horizontal restrain; this produced a different roof structure, the 30° pitch meant the roof was less susceptible to water penetration and wind lift, the truss members were of proper thickness, there was a reduction in the unrestrained rafter lengths, an increase in the number of longitudinal ties, the bracing of each truss was stronger and the roof structure was strapped into the gable walls. On behalf of the tenants it was contended that the works carried out were not repairs within the meaning of clause 4(a) and the expenditure thereon was not recoverable by the landlord
under the service charge provisions. M

Held: **The landlord had carried out works of repair. The expression "renew or replace" in clause 4(a) is not confined to producing a replica of the original. Notwithstanding that the leases were for only 25 years and the tenants were commercial tenants of small office suites, the literal meaning of the words in clause 4(a) were clear. Applying the approach set out in *Holding & Management Ltd* v *Property Holding & Investment Trust plc*, it was quite plain that the roof was not an independent part of the property; it could not be said to have been altered or improved to an extent that it could be said not to fall within the normal usage of the word "repair".**

A **Although there were other options for repair considered, it did not mean that the option adopted meant that the works were not repairs. The selection of the higher-pitch option with better wind and water resistant qualities and overall strength did not mean that the landlord would get back a building that was changed or altered or added to so far as the roof was concerned. This was a borderline case and, looking at the matter broadly and taking into account in particular the nature and extent of the defects in design and construction, the fact that the comparative cost of any patching as opposed to replacing the whole roof was not a substantial difference, and**
B **that one still had a roof fulfilling precisely the same function, the new roof was not such a different roof that it must be regarded as an improvement.**

The following cases are referred to in this report.

Brew Bros Ltd v *Snax (Ross) Ltd* [1970] 1 QB 612; [1969] 3 WLR 657; [1970] 1 All ER 587; (1969) 20 P&CR 829; [1969] EGD 1012; 212 EG 281, CA
Elite Investments Ltd v *TI Bainbridge Silencers Ltd (No 2)* [1987] 2 EGLR 50; (1987) 283 EG 747
Holding & Management Ltd v *Property Holding Investment Trust plc* [1989] 1 WLR 1313; [1990] 1 All ER 938; (1989) 21 HLR 596; [1990] 1 EGLR 65; [1990] 05 EG 75, CA
Lister v *Lane and Nesham* [1893] 2 QB 212, CA
Post Office v *Aquarius Properties Ltd* [1987] 1 All ER 1055; [1987] 1 EGLR
C 40; (1987) 281 EG 798, CA; [1985] 2 EGLR 105; (1985) 276 EG 923
Norwich Union Life Insurance Society v *British Railways Board* [1987] 2 EGLR 137, (1987) 283 EG 846

This was a hearing of a preliminary issue in proceedings commenced by the plaintiff landlord, New England Properties, against all the first-floor tenants for arrears of service charges, and the hearing of an originating summons by which representative tenants, Sterling Surveys, sought a declaration as to the true meaning of the leases by which the tenants held units at Grove House, Headley Road, Grayshot, Surrey.

Vivian Chapman (instructed by Wheldon Houlsby & Scott, of South Shields) represented the landlord, New England Properties;
D Timothy Bowles (instructed by Thomas Eggar Verrall Bowles, of Worthing) appeared on behalf of the tenants, Portsmouth New Shops and others.

Giving judgment, **MR T CULLEN QC** said: This is a dispute over the cost of reroofing a property known as Grove House, Headley Road, Grayshot, Surrey. The property is a commercial one. It is a T-shaped building with the head of the T fronting on to Headley Road. The ground floor is part occupied by retail shops and the rest forms a car park for the tenants of the building.

The first floor is divided into offices, of which there are seven units which were let on leases granted in the middle of the 1980s. The great storm in October 1987 lifted tiles on the tail of the T and, as a result,
E the landlord engaged a surveyor to inspect. He found that there had been a movement in the roof so far as the tail of the T was concerned, with a cracking in the gable walls at the end of the T. There was a real danger of collapse and three schemes were put before the tenants, leaving open the question of who was ultimately going to bear the cost. It was agreed that works were necessary and a scheme was chosen which was estimated to cost £186,000, but in the end turned out to cost £202,000, because some protection for the tenants during the works was required.

It is common ground that the original roof was inadequate in a number of respects, in particular the roof trusses were not thick enough and did not have sufficient lateral restraint, nor sufficient longitudinal rafter ties. Those longitudinal rafter ties that it had were
F not all continuous, thereby weakening the roof structure. In addition, the roof was built with a 22.5° pitch, which increased the risk of water penetration and the risk of lifting tiles in strong winds.

The consequence of those defects, allied with the storm, was that the roof structure within the tail of the T became badly distorted and unstable, so that it was leaning upon the gable wall, which led to the cracking of that wall. Further, where the tail of the T met the head of the T there were severe distortions and twisting, so that the factor of safety against full collapse was minimal. And it is common ground that the design life of the original roof structure would not have exceeded a maximum of 16-20 years.

The works carried out involved the insertion of new trusses with a 30° pitch, with more substantial timbers forming the trusses and rafters, and a proper system of lateral restraint and horizontal G
restraint. The increased pitch meant a higher profile, so there had to be fundamental alterations at the junction of the tail of the T to its head. A system of purlins was replaced by "A" frames.

That work resulted in a different roof structure in a number of respects. First, the 30° pitch meant that the roof was less susceptible to water penetration and wind lift. Second, the truss members were now of a proper thickness. Further, there was a reduction in the unrestrained rafter lengths from 3,600mm to 2,700mm and there was an increase in the number of longitudinal ties. Where there had previously been five there were now 12. In addition, the bracing of each truss was considerably stronger than the original bracing and the roof structure was strapped in to the gable walls at 4m centres whereas previously no such restraint existed. H

The dispute is over who is liable for the £202,000 paid by the landlord in 1989. A number of actions were started in respect of it. First, there was an originating summons by the tenant of office units 4, 5 and 6 seeking a declaration as to the true construction. Being by the tenant, the declaration sought was, of course, that the landlord cannot recover the sum.

A representation order was obtained in respect of the other office tenants, appointing the tenant of units 4, 5 and 6 to represent them. Unfortunately the tenant of units 4, 5 and 6 went into liquidation. The claim against it has been settled. The originating summons was amended on the first day of this hearing to include the tenants of units 1, 2, 3 and 7. J

The second set of proceedings was a county court claim by the landlord against Ex-Eltronics (UK) Ltd, the tenant of unit 1, for arrears of service charge. That was transferred to the High Court and appears before me under 1990 N 6237.

There was an action in the High Court by the ground-floor retail tenants, 1990 N 11633. That was settled as to two of the three tenants. It was, when this hearing opened, unsettled in respect of the third, but it is now, I am informed, settled in principle and I am simply asked to adjourn that generally on the basis that it will be finally settled so that the action does not have to come on again.

The last action, with which I am concerned, was started in 1992 by the landlord against all the first-floor tenants for arrears of service charge. In that action a preliminary issue was ordered, which raises K
the construction point raised in the originating summons. In that action there has been a consent order between the landlord and the original tenants and guarantors of units 4, 5 and 6.

So the live issue before me is simply the construction point raised in the originating summons and in the preliminary issue in the 1992 action.

The framework of the office leases so far as they deal with the recovery of costs involved in works to the exterior or structure of the building is that the landlord carries them out and, if they are recoverable, he then recovers the costs from the tenants. The costs are apportioned between the units so that each tenant bears a fair proportion and, in the case of voids, the landlord bears the relevant L
proportion.

The crucial clause imposing the obligation upon the landlord is clause 4(a) of the fourth schedule to the lease. I should add that the leases of all seven units are in identical terms so far as the point before me is concerned. The fourth schedule contains covenants by the lessor. Clause 4 starts:

> Except insofar as such matters are the responsibility of the lessee under the provisions of this lease:
>
> (a) to keep and maintain or procure to be kept and maintained the Scheme and in particular without prejudice to the generality of the foregoing its exterior and its structure and foundations including the matters referred to in paragraphs (i) to (iv) inclusive of proviso 3 of clause 41 of the third schedule of this lease, the service area, passenger lifts (if any), lifts, lift shafts, boilers (if M
> any), central heating (if any) and all apparatus, equipment, plant and machinery serving the same in good and substantial repair and condition and decoration and to renew or replace the same or any part or parts thereof whenever such renewal or replacement shall be necessary.

I would add that the paragraphs of proviso 3 of clause 41 referred to simply exclude specifically certain structural parts of the demised premises from the tenants' obligation to repair. That obligation to repair in clause 41(a) of the third schedule included a provision upon which some argument has been addressed to me. This is the tenants' repairing obligation and in it it contains these words:

> It is hereby agreed and declared there shall be included in this covenant as repairable by the lessee (including replacement whenever this shall be

A necessary) all fixed floor coverings and lessors' fixtures and fittings together with the facias, windows, window frames . . .

And then various other fittings to the demised premises which I need not read in detail.

The only other terms in the lease which I will mention is that each lease is for 25 years and the scheme which was referred to in clause 4(a) is defined in the schedule of particulars as:

> The lessor's office and shopping development known as Grove House, Headley Road, Grayshot, in the county of Hampshire [as it then was] the boundaries of which are shown edged red on plan number two annexed hereto.

I will first take the pure point of construction raised by the last
B words of clause 4(a); those are the words "and to renew or replace the same or any part or parts thereof whenever such renewal or replacement shall be necessary". Mr Bowles for the tenants told me to look at the whole (and of course I accept I must do that) and he says if I do that then the last words simply particularise the preceding obligation to keep the scheme in good and substantial repair and condition.

As to that, if those words are given effect literally they plainly do not merely particularise the preceding obligation. As a matter of language, they form an additional obligation and would cover the replacement of the roof.

Mr Bowles, for the defendants, first asked me to construe the words so that the reference to the renewal or replacement of the
C scheme is cut down so as to refer to such renewal or replacement piece by piece so that both it and the subsequent — what seems to me would then become otiose reference to any part or parts thereof — are simply directed to repairing by the renewal of parts, and in aid of this construction he referred me to the words in the tenants' repairing obligation, which I have just cited as showing that "replacement" was included in "repair". But as to that submission I get no assistance from those words in the tenants' repairing obligation. They are simply setting out the obvious: that "repair" can include "replacement". And, in the context of that clause, someone presumably thought that it was necessary to set out the obvious. I do not see how that assists me to read these last few words in clause 4(a).

Alternatively, Mr Bowles says that the words "renew or replace"
D must be controlled by some test (perhaps akin to the test involved in construing a repairing covenant) so that the "replacement or renewal" must be a replica of the original or at any rate must not involve any substantial element of radical improvement.

On this point I was referred to one authority, *Norwich Union Life Insurance Society* v *British Railways Board* [1987] 2 EGLR 137, where Hoffmann J was unable to construe the lease there in the way that Mr Bowles asked me to do in this case. The clause in that case was in the following terms:

> To keep the demised premises in good and substantial repair and condition and when necessary to rebuild, reconstruct or replace the same and in such repair and condition to yield up the same at the expiration or sooner
> E determination of the said lease.

Hoffmann J held that the clause was quite clear and indicated that the draftsman had two separate concepts in mind, that of repair and that of rebuilding, reconstructing or replacing. In that case, Hoffmann J pointed out that a covenant to rebuild the entire premises would be unusual, but he held that since the term was for 150 years it would not be as inconceivable as it would have been in *Lister* v *Lane and Nesham* [1893] 2 QB 212 to which he had been taken.

Mr Bowles also relied upon the words in this clause 4(a) "whenever such renewal or replacement shall be necessary". He said they must refer back to the primary obligation to keep in good and substantial repair. Hoffmann J rejected a similar argument in the *Norwich*
F *Union* case and I reject the argument in this case. It seems to me that the words in this case are put in to protect the tenant and ensure that the landlord can renew or replace only when it is necessary.

Pausing there, I also reject Mr Bowles' alternative argument that "replacement" must have some implied undefined gloss on it so that an improved replacement is not within it.

As to whether there can be any replacement at all, a control is perfectly obviously there, by reason of the requirement that it must be necessary. As to whether there is a gloss on the sort of work that can be done, I need not consider it further because it seems to me, if there were such a gloss, it would not be some sort of undefined implication that Mr Bowles would seek. I think if one had some gloss on those words it would be related to the normal measure of reasonableness: the replacement must be reasonable. And here the G
evidence is that the tenants' surveyor, Mr John A Lethbridge [BSc FICE FIStructE], in cross-examination, in terms agreed that (a) replacement was necessary and (b) the actual work that was done was reasonable.

So, with that alternative argument out of the way one is left, on the literal construction, with an absolute obligation to renew or replace whenever necessary. This lease, unlike the *Norwich Union* lease, is for only 25 years and the tenants are commercial tenants of small office suites. Mr Bowles says that, in those circumstances, it is inconceivable that the parties could have intended to impose an obligation to renew the whole premises or any part of them whether substantial or not. H

From the landlord's point of view, of course, there is nothing inconceivable about it whatsoever. So far as the tenants are concerned, the words are, in my judgment, perfectly clear. I do not see how any tenants' adviser could have misread them. The tenants may have had good reasons for taking the risks involved: with a new building, the risks inherent in such a clause may well have been acceptable in the light of the rent being asked. I do not see that I can speculate about that. In any event, even if it were improbable in the extreme or even inconceivable that a tenant would have been advised to accept such a term, I have found the words to be clear, and I am simply unable, by any acceptable path of construction, to get out of them the meaning that Mr Bowles has asked me to. So that, in my judgment, the words bear their literal meaning. J

That conclusion is sufficient to impose liability on the tenants, but I should deal with the question on the footing that the last words do, in fact, add nothing to the repairing obligation. I was referred to a number of authorities on that question. The first one I would mention is *Post Office* v *Aquarius Properties Ltd* [1985] 2 EGLR 105. The nature of the works involved in that case are set out at p105 between J and K. The learned judge said:

> For most of the time between 1979 and 1984 the basement was ankle deep in water. This appears to have been the result of a rise in the level of the local water table combined with defects in construction and possibly the design of the building. In 1984 the water table subsided again and since then the basement had been dry.
>
> The tenants' lease expires in 1991, but the building has a life expectancy of K
> many years and it is therefore agreed by landlord and tenants that remedial work is necessary to make the basement waterproof in case the water table should rise again. The issue in this case is whether such work is "repair", within the meaning of the tenants' covenant.

The covenant is set out at p107B:

> keep in good and substantial repair the demised premises and every part thereof.

On p107 between letters E and F, Hoffmann J, having set out the principles between B and D, holds that the distinction is not always between repair and rebuilding or reconstruction: it can be between repair and alteration or addition or improvements. On the facts there, the learned judge held that — various schemes were proposed L
— whatever scheme was carried out there would be a substantial structural addition to the basement. He held that, looked at in all the circumstances, that would involve structural alteration and improvement and not repair.

I was taken to a case on the other side of the line, a decision of Judge Baker *Elite Investments Ltd* v *TI Bainbridge Silencers Ltd (No 2)* [1987] 2 EGLR 50. That case involved the replacement of the roof covering of an industrial building by new material. In that case, Judge Baker reviewed a considerable number of authorities and at p47M, having considered *Brew Bros* v *Snax (Ross) Ltd* [1970] 1 QB 612, he held:

> As a result of this case, the test as I see it is whether the totality of the work can properly be described as repairs, since it involves no more than renewal or M
> replacement of defective parts, or whether it is in effect renewal or replacement of substantially the whole subject-matter of the demise.

He points out that one relevant factor is the question of the cost of the repairs as compared with either the cost of putting up a new building or the value of the repaired building. He left open the question of whether value could ever be the correct comparable, but in that case he had no doubt that cost was the right comparable, and there the cost of the new roof was some 8% of the cost of a new building. He also reviewed *Post Office* v *Aquarius* and at p48M he distinguished that case in these words:

> There was an improvement in that case because at the end of the day it was a basement which could cope with a totally unexpected rise in the water table.

A In this case, although it will be an improved material in the roof, it is not in any sense giving the landlords something different.

Two recent decisions of the Court of Appeal have indicated that tests such as whether the works involve the replacement or renewal of substantially the whole subject-matter of the demise are not always the beginning and the end and one has to take a broad view, either simply by looking at every possible factor with an open mind or else in the way in which Hoffmann J has done: by considering whether one was comparing repairs with reconstruction or with improvement and the like.

This approach is set out first in *Holding & Management Ltd* v *Property Holding & Investment Trust plc* [1990] 1 EGLR 65 at p68
B where, at F, Nicholls LJ cites from Hoffmann J's decision in *Post Office* v *Aquarius*. He then cites a passage of Sachs J's judgment in *Brew Brothers* v *Snax* and goes on between G and H:

> Thus the exercise involves considering the context in which the word "repair" appears in a particular lease and also the defect and remedial works proposed. Accordingly, the circumstances to be taken into account in a particular case under one or other of these heads will include some or all of the following: the nature of the building, the terms of the lease, the state of the building at the date of the lease, the nature and extent of the defect sought to be remedied, the nature, extent, and cost of the proposed remedial works, at whose expense the proposed remedial works are to be done, the value of the building and its expected lifespan, the effect of the works on such value and lifespan, current building practice, the likelihood of a recurrence if one remedy rather than another is adopted, the comparative cost of alternative
> C remedial works and their impact on the use and enjoyment of the building by the occupants. The weight to be attached to these circumstances will vary from case to case.
>
> This is not a comprehensive list. In some cases there will be other matters properly to be taken into account.

So, on this approach, one simply looks at those factors and any other factors which one might consider relevant and then one has to decide whether the works can fairly be regarded as repair in the normal usage of that word.

That approach had been foreshadowed by a decision of the Court of Appeal in *Post Office* v *Aquarius Properties Ltd*, the Court of Appeal decision being reported in [1987] 1 All ER 1055*. In the judgment of Ralph Gibson LJ at p1064D to the end of that page one
D again gets the more general approach than by applying any formal test such as: "Is it a subordinate part or is it substantially the whole?"

Returning to the facts of this case in the light of those authorities, it seems quite plain to me that the roof is not an independent part of the scheme. It is an integral part of the building which is comprised in that scheme. The evidence is that one option available was simply to strengthen the roof. If that had been carried out, it seems to me that it would plainly have been within the repairing covenant. Looked at in the narrow test, it was not substantially the whole, even though the cost of the strengthening would be some 20% or slightly less of the cost of rebuilding the whole. The roof, in my view, could not be said to have been altered or improved to an extent that it could be said not
E to fall within the normal usage of the word "repair".

That option of merely strengthening was not adopted. It simply was not sensible. Merely to strengthen it would have left one with a roof pitch of 22.5° with the same risk of water penetration and wind lift and the probability that there would be a recurrence of the damage that had occurred once as a result of the 1987 storm. Another reason for not adopting it was that, in any event, it might not even have succeeded, regardless of the effects of future wind or rain.

Two other options, as I understand it, included a 30° pitched roof, one of which was ultimately chosen. With those two options, again costs were in the region of 20%, this time slightly more than 20%.

Whatever test or approach one uses, it is a continuing theme of the authorities that it is a matter of degree, but I do not consider that the
F option adopted meant that the works were in respect of substantially the whole or that the slightly higher costs meant that the works were not in respect of only a subsidiary part using those tests, nor on the broader approach do I consider that the fact that the option selected would be that an inadequate roof was replaced by an adequate one by itself means that it ceased to be a matter of repair. That fact would equally be the case if there had been strengthening alone. Then one would obviously hope that, if a feasible strengthening option was carried out, that would change an inadequate roof into an adequate one.

The question here, to my mind, is whether the selection of the higher-pitch option with better wind- and water-resistant qualities and stronger overall because of better restraints, rather than simply G adopting the strengthening option, meant that the landlord would get back a building that was changed or altered or added to so far as the roof was concerned or, as Hoffmann J put it, whether the result was that, as a matter of English, there was an improvement and not a repair. One result of the higher pitch was that the cost was reduced: to have replaced the roof retaining the 22.5° pitch would have cost more. That, of course, would have involved no change to the exterior of the building, but it would have continued to keep in place a roof that was likely to suffer from water penetration and wind damage.

Mr Bowles rightly says that it was not just a matter of profile: the new roof was very much better strengthened, with better tied-in trusses. That is so, but that was done, on the evidence, on the basis H that something different had to be put in place in order to prevent a recurrence of the damage and it is common ground that what was done was a reasonable and proper way of replacing the inadequate roof.

It seems to me that this is a borderline case. I think it is somewhere between *Post Office* v *Aquarius* and the *Elite* case. Looking at the matter broadly and taking into account in particular the nature and extent of the defects in design and construction of the roof and the fact that, as a result, it was in imminent danger of collapse; the fact that the comparative cost of merely patching up as opposed to replacing the whole roof was not a substantial difference; the fact that it was pointless rather than impossible to replace the roof by one of similar construction with a 22° pitch; and the fact that, at the end J of the day, one still has a roof fulfilling precisely the same function but doing so in a way which will not involve the occupiers with further disruption by recurring defects, I consider that the new roof is not such a different roof that it must be regarded as an improvement or some other class of work — alteration or addition or change, whatever one likes to call it. I do not regard it as coming under that category rather than a matter of repair. I therefore find for the landlord on this point as well.

As I have said, there are four actions and different relief will be required in respect of each action to give effect to my decision.

K

Chancery Division

December 3 1992

(Before Mr Jonathan SUMPTION QC, sitting as a deputy judge of the division)

LONSDALE & THOMPSON LTD v BLACK ARROW GROUP PLC AND ANOTHER

Estates Gazette June 26 1993

[1993] 25 EG 145

Landlord and tenant — Insurance — Where lessor insures in L ***his own name the entire interest in the property the subject of the lease, whether the insurer's obligation is limited to the injury done to the value of the lessor's reversion or extends further to the whole damage to the property***

By a lease dated December 4 1978 the first defendant, the freehold owner of a warehouse in Liverpool, demised the premises to the plaintiff for 25 years. By clause 3(2) of the lease the first defendant covenanted that it would insure the premises in a sum not less than the full reinstatement value against specified risks including fire. Clauses 1 and 2(2) provided that the tenant should pay by way of additional rent M a sum corresponding to the premium. By clause 3(2) the first defendant covenanted in case of destruction or damage to the premises by an insured risk to ensure that all moneys payable under or by virtue of any policy of insurance should be laid out and applied in rebuilding, repairing or otherwise reinstating the demised premises. It was implicit in the scheme of the lease that, although the tenant had a general obligation to reinstate, he could not be required to do so if the damage requiring reinstatement had arisen from an insured peril and the landlord had recovered the reinstatement costs from the insurer. The landlord insured the periods with the second

*Editor's note: Also reported at [1987] 1 EGLR 40.

A defendants and by clause 13 of section A of the insurance policy it was agreed that in the event of the property insured being destroyed or damaged the basis upon which the amount payable under the policy was to be calculated should be the reinstatement of the property destroyed or damaged. The tenant was not a coinsured. On December 21 1989 the landlord contracted to sell the freehold, subject to the lease in favour of the tenant, and completion took place on March 22 1990. Meanwhile on February 8 1990 the warehouse was destroyed by fire. However, completion proceeded and the price of the freehold was paid in accordance with the terms of the contract
B without regard to the fire damage. Under section 83 of the Fire Protection Metropolis Act 1774 the lessee was entitled to require the second defendant insurers to discharge such liability as they might have to their insured under the policy by reinstating the premises. A notice was duly served on the insurers under the Act, but they declined to comply with it on the ground that they did not have any contractual liability to their insured and could not therefore have any statutory liability to the lessee. They contended that they were not obliged to do more than indemnify the insured in respect of injury to his freehold interest, because it was only in respect of that interest that he could have suffered any loss. The result of
C the contract of sale was that at the time of the fire the insured had parted with its entire beneficial interest in the property, retaining no other interest than its vendor's lien for the price. When upon completion the insured received the price in full, without any abatement on account of the fire, it was completely indemnified for its loss. As for the insured's liability to lay out the insurance proceeds to reinstatement, that arose only if and when they were received. On behalf of the second defendants it was contended that nothing was payable by them.

Held: **The first defendant was entitled to recover the whole**
D **damage to the property, including that suffered by the tenant's interest, and the second defendants' liability was not limited to the injury done to the value of the first defendant's reversion. The landlord owed an obligation under the lease to insure for the reinstatement value and he therefore had no insurable interest up to that value because if he did not insure and the premises were damaged, he would be liable to the tenant. The full reinstatement value of the premises, which ought on that footing to have been recoverable from the insurer, would be the measure of his liability. Under the insurance policy the subject-matter of the insurance was**
E **damage to the premises up to their reinstatement value and under the lease it was expressly provided that upon receipt of insurance proceeds the landlord had an obligation to lay them out in reinstatement for the benefit of the tenant. Without express agreement a person who has contracted to indemnify another can never refuse on the ground that an indemnity can be obtained from someone else. If then the insurers duly paid the reinstatement cost to the insured before completion, they would not have any subrogation rights to recover an equivalent amount out of the purchase price once it was paid because the landlord, having received the insurance moneys, would have become liable to the tenant to lay them out in**
F **reinstatement of the premises; the subsequent receipt of the purchase price would not in the least diminish the loss.**

The following cases are referred to in this report.

Castellain v *Preston* (1883) 11 QBD 380
Collingridge v *Royal Exchange Assurance Corporation* (1877) 3 QBD 173; 47 LJQB 32; 42 JP 118,
London & North Western Railway Co v *Glyn* (1859) 1 EBE 652
Maurice v *Goldsborough & Co Ltd* [1939] AC 452; [1939] 3 All ER 63; 108 LJPC 75; 55 TLR 714
Powles v *Innes* (1843) 11 M&W 10; 12 LJ Ex 163
Prudential Staff Union v *Hall* [1947] KB 685; (1948) 63 TLR 392
Rayner v *Preston* (1881) 18 ChD 1
Rowlands (Mark) Ltd v *Berni Inns Ltd* [1986] QB 211; [1985] 3 WLR 964; [1985] 3 All ER 473; [1985] 2 EGLR 92; (1985) 276 EG 191, CA
Tomlinson (A) (Hauliers) v *Hepburn* [1966] AC 451; [1966] 2 WLR 453; G
[1966] 1 All ER 418; [1966] 1 Lloyd's Rep 309, HL
Waters v *Monarch Fire and Life Assurance Co* (1856) 5 E&B 870

This was the hearing of a summons under RSC Ord 14A to determine a preliminary question of law as to whether the liability of the second defendants, American International Underwriters (UK) Ltd, under a policy of insurance, was limited to the injury done to the value of the first defendant's, Black Arrow Group plc's, then interest in warehouse premises in Liverpool let to the plaintiff, Lonsdale & Thompson Ltd.

Hazel Williamson QC and John Nicholls (instructed by Simmons & Simmons) appeared for the plaintiff; Michael Hart QC and William George (instructed by Harrison Dury & Co, of Preston) H
represented the defendants.

Giving judgment, **Mr Jonathan Sumption QC** said: This is a summons under RSC Ord 14A to determine a question which is rather more difficult than at first sight appears. If a lessor insures in his own name the entire interest in the property the subject of the lease, what is the measure of the insurer's obligation? Is it limited to the injury done to the value of the lessor's reversion? Or may he recover the whole damage to the property, including that suffered by the tenant's interest, accounting to him for the balance? The point is of some wider interest, because the problem is certainly not confined to the insurance of real property by lessors. It is quite common for people to insure in their own name the property or liabilities of others J
with whom they have some business or other connection. This gives rise to no difficulty if the assured can be treated as contracting on their behalf. But what if he cannot?

The facts of this case are relatively straightforward and, in the respects which matter, undisputed. Black Arrow plc, the first defendant, was the freehold owner of a warehouse in Liverpool. On December 4 1978 it let it to the plaintiff for 25 years. By clause 3(2) of the lease the landlord covenanted that he would cause the premises to be insured in a sum not less than the full reinstatement value against specified risks including fire. Clauses 1 and 2(2) provided that the tenant should pay by way of additional rent a sum corresponding to the premium.

Several clauses dealt with the reinstatement of the premises if they K
were damaged. Clause 2(4) was a general provision requiring the tenant to "repair, maintain, and keep and where necessary rebuild and replace the demise premises". But the lease also provided, by clause 3(2), that it should be the duty of the landlord:

> . . . in case of destruction or damage to the demised premises by any insured risk . . . to ensure . . . that all moneys payable under or by virtue of any such policy of insurance as aforesaid (except for insurance moneys received in respect of loss of rent) shall with all convenient speed . . . be laid out and applied in rebuilding, repairing or otherwise reinstating the demised premises.

By clause 2(14)(d), if the premises were damaged or destroyed by an insured peril and the insurance moneys were irrecoverable on
account of an act or default of the tenant, then the tenant should be L
required to pay the landlord the full costs of rebuilding and reinstatement.

It necessarily follows from the scheme of the lease that, although the tenant has the general obligation to reinstate, he cannot be required to do so if the damage requiring reinstatement has arisen from an insured peril and the landlord has recovered the reinstatement cost from the insurer. This is implicit in: (i) clause 2(14)(d) of the lease, which makes little sense on any other basis; (ii) the fact that the landlord is required (not only entitled) to insure and to lay out loss payments in reinstatement; and (iii) the obligation of the tenant indirectly to pay the premium: *Mark Rowlands Ltd* v *Berni*
Inns Ltd [1986] QB 211*. M

The landlord insured with the second defendants, American International Underwriters (UK) Ltd. The insurance covered all risks of physical loss or damage to the premises, subject to irrelevant exclusions. Clause 13 of section A provided:

> It is hereby agreed that in the event of the property insured being destroyed or damaged the basis upon which the amount payable under . . . the policy is to be calculated shall be the reinstatement of the property destroyed or damaged . . .

There were provisions which envisaged that the premises were, or at least might be let. But it is common ground, and in any event plain,

* Editor's note: See also [1985] 2 EGLR 92.

A that the landlord and its associated and subsidiary companies were the only insured. The tenant was not a coinsured.

On December 21 1989 the landlord contracted to sell the freehold, subject to the lease in favour of the tenant. Completion occurred on March 22 1990. Between these two dates, on February 8 1990, the warehouse was destroyed by fire. However, completion proceeded and the price of the freehold was paid in accordance with the terms of the contract without regard to the fire damage.

Under section 83 of the Fire Protection Metropolis Act 1774 the lessee was entitled to require the insurers to discharge such liability as they might have to their insured under the policy by reinstating the premises. A notice has been duly served on the insurers under the Act. But the insurers have declined to comply with it, on the ground that
B they do not have any contractual liability to their insured and cannot therefore have any statutory liability to the lessee. They say that they are not obliged to do more than indemnify the insured in respect of injury to his freehold interest, because it is only in respect of that interest that he can have suffered any loss. The result of the contract of sale was that, at the time of the fire, the insured had parted with his entire beneficial interest in the property, retaining no other interest than his vendor's lien for the price. When, upon completion, the insured received the price in full, without any abatement on account of the fire, he was completely indemnified for his loss. As for the insured's liability to lay out the insurance proceeds in reinstatement, that arises only if and when they are received. It follows, the insurers
C say, that nothing is payable by them. The question on this application is whether they are right. In my judgment they are not.

The foundation of the insurer's case is the decision of the Court of Appeal in *Castellain* v *Preston* (1883) 11 QBD 380. That case had some points in common with the present one. A vendor contracted to sell property, but the buildings on it were destroyed by fire between contract and completion. The vendor was insured and the insurers paid him the full amount of the damage before completion. The vendor then received, upon completion, the full purchase price. It had been held in previous litigation between the vendor and the purchaser that the vendor was not accountable to the purchaser for the insurance proceeds, there being no provision to that effect in the contract of sale: *Rayner* v *Preston* (1881) 18 ChD 1. In these
D circumstances the insurers sued the vendor as their assured for a sum corresponding to what they had paid him on account of the loss. The ground of their claim was not that they had never been liable under the policy. It was that the subsequent recovery of an undiminished purchase price had made good the insured's loss. The question at issue was whether the insurers were subrogated to the right of the insured to receive an undiminished price under the contract of sale. The Court of Appeal held that they were, overruling Chitty J, who had held that the claims to which the insurers were subrogated were confined to those connected with the circumstances of the loss. The ground of the Court of Appeal's decision was that, because property insurance was *prima facie* a contract of indemnity, the insurer's right
E of subrogation must be extensive enough to ensure that no more than an indemnity was recovered. The insurer was therefore subrogated to every right of the insured whose exercise would diminish his loss. That included the right to receive the proceeds of his contract of sale without abatement for fire damage.

Brett LJ delivered the leading judgment. At p390 he said:

> . . . there was a right in the defendants to have the contract of sale fulfilled by the purchasers notwithstanding the loss, and it was fulfilled. The assured have had the advantage therefore of that right, and by that right, not by a gift which the purchasers could have declined to make, the assured have recovered, notwithstanding the loss, from the purchasers, the very sum of money which they were to obtain whether this building was burnt or not. In that sense I cannot conceive that a right, by virtue of which the assured has his
> F loss diminished, is not a right which, as has been said, affects the loss. This right . . . affects the loss by enabling the assured, the vendors, to get the same money which they would have got if the loss had not happened.

Cotton LJ said, at p396:

> Here the purchasers have paid the money in full, and as the property was valued between the vendors and the purchasers at £3,100, the vendors got that sum in respect of that which had been burned, but which had not been burned at the time when the contract was entered into. They had fixed that to be the value, and then any money which they get from the purchasers, and which together with £330, the sum paid by the office, exceeds the value of the property as fixed by them under the contract to sell, must diminish, and in fact entirely extinguishes the loss occasioned to the vendors of the property by the fire. Therefore, though it cannot, to my mind, be said that the insurers are entitled, because the purchase is completed, to get back the money which they have paid, yet they are entitled to take into account the money subsequently G
> received under a contract for the sale of the property existing at the time of the loss, in order to see what the ultimate loss was against which they gave their contract of indemnity.

Bowen LJ said, at pp401-2:

> What is really the interest of the vendors, the assured? . . . Their interest . . . is that at law they are the legal owners, but their beneficial interest is that of vendors with a lien for the unpaid purchase-money; they would get ultimately all the purchase-money provided the matter did not go off owing to defective title. Such persons in the first instance can obviously recover from the insurance company the entire amount of the purchase-money. That was decided in the case of *Collinaridge* v *Royal Exchange Assurance Corporation* (1877) 3 QB 173; but can they keep the whole, having lost only half? Surely it would be monstrous to say that they could keep the whole, having lost only half . . . They would be getting a windfall by the fire, their contract of H
> insurance would not be a contract against loss, it would be a speculation for gain. Then what is the principle which must be applied? It is a corollary of the great law of indemnity, and is to the following effect: That a person who wishes to recover for and is paid by the insurers as for a total loss, cannot take with both hands. If he has a means of diminishing the loss, the result of the use of those means belongs to the underwriters. If he does diminish the loss, he must account for the diminution to the underwriters.

The problem about the submission which Mr Hart, on behalf of the insurers, seeks to base on this decision is that it elides two distinct questions. The first is whether the vendor's recovery from the insurer is confined to the loss which he has suffered in respect of his own limited interest in the property at the time of the fire. This question is directed to determining what the vendor's initial loss is and what, J
therefore, would be payable to the vendor if the claim were adjusted and settled immediately after the fire and before completion. The second question is the one with which *Castellain* v *Preston* was concerned, namely whether the subsequent receipt by the vendor of the full purchase price agreed before the fire goes to diminish that loss.

The starting point in considering the first question is that the landlord unquestionably had an insurable interest in the premises up to the full reinstatement cost. I am inclined to think that it would be enough to give him that interest that he was liable to lay out the insurance proceeds on reinstatement, even though that liability would arise only after he had received them. The whole law as to what amounts to an insurable interest and when it is required, is derived K
from the statutory avoidance of wagering contracts, and it is hard to see how an insured who was obliged to spend the proceeds on reinstatement could be said to wager on the occurrence of a fire. But this is a difficult issue which I do not need to decide. The landlord owed an obligation under the lease to insure for the full reinstatement value. He therefore had an insurable interest up to that value because if he did not insure and the premises were damaged, he would be liable to the tenant. The full reinstatement value of the premises, which ought on that footing to have been recoverable from the insurer, would be the measure of his liability: see *Maurice* v *Goldsborough & Co* [1939] AC 452, at p462.

The question as to whether the measure of the insurer's liability is L
limited to the injury done to the landlord's reversion therefore depends not on any overriding principle of law but on the terms of the policy. Insurances on property are *prima facie* to be construed as contracts of indemnity. Subject to the express terms of the policy the measure of the indemnity is the diminution in the value of the thing insured as a result of the operation of the insured peril. The parties may agree that the damage suffered by the thing insured will be assessed on some agreed basis. There may, for example, be an agreed undamaged value. There may be a provision (such as the one found in this case) that the cost of reinstatement, which would otherwise be no more than evidence of the diminution in value of the property, shall be the measure of the insurer's liability. But provisions such as these do not prevent the contract from being one of indemnity. They M
merely require the value of the indemnity to be calculated on conventional facts. It remains necessary to look at the particular position of the insured to see whether in the circumstances he would, by recovering the contractual measure, obtain more than an indemnity.

If the insured has only a limited interest in the property, being, for example, a tenant or reversioner, a trustee, a mortgagee or a bailee, the value of his own interest may have diminished by much less than the value of the property or the cost of its reinstatement. But it does not necessarily follow that, if the insured recovers the whole diminution in the value of the property or the whole cost of reinstatement, he will be getting more than an indemnity. That must depend on what his legal obligations are as to the use of the insurance proceeds when he has got them. If he is accountable for the proceeds to the owners of the other interests, then he will not be receiving more than an indemnity if the insurer pays the full amount for which the property was insured. This will be so, whether the insured is accountable to the owners of the other interests as a trustee of the proceeds of the insurance or simply on the basis that he owes them a contractual obligation to pay those proceeds over to them or to employ them in reinstatement. None of this means that a party with a limited interest who insures the entire interest in the property is insuring on behalf of the others as well as for himself. All that it means is that his obligations as to the use of the insurance moneys once they have been paid are relevant in determining whether he will recover more than an indemnity by getting the measure of loss provided for in that policy.

The clearest illustration of the operation of these principles is the case of a trustee. There has never been any doubt that a trustee may insure the whole beneficial interest in property of which he holds only the legal estate, and that he may recover from the insurers the entire diminution of its value notwithstanding that the beneficial owners were not coinsureds. The reason is not that a trustee is personally liable to beneficiaries for damage done by insurable risks irrespective of whether he has received the insurance proceeds, for he will usually not be. The reason must be that he is accountable to the beneficiaries for such insurance proceeds as he may receive. Being accountable to them on that basis, the payment to him of those proceeds cannot give him more than an indemnity even though they exceed the value of his personal interest.

That this is the correct analysis of the position is demonstrated by a series of decisions about insurances effected by bailees on property belonging to their bailors. The leading case is *Waters* v *Monarch Fire & Life Assurance Co* (1856) 5 E&B 870. In that case it was held that a bailee who insured the entire interest in goods in his custody could recover their entire value, notwithstanding that he has himself suffered no loss. Lord Campbell CJ said, at p881:

> The last point that arises is, to what extent does the policy protect those goods. The Defendants say that it was only the Plaintiffs' personal interest. But the policies are in the terms: contracts to make good "all such damage and loss asa may happen by fire to the property hereinbefeore mentioned." That is a valid contract; and , as the property is wholly destroyed, the value of the whole must be made good, not merely the particular interest of the Plaintiffs. They will be entitled to apply so much to cover their own interest, and will be trustees for the owners as to the rest. The authorities are clear that any assurance made without orders may be ratified by the owners of the property, and then the assurers become trustees for them.

Wightman J said, at p882:

> Can the Plaintiffs recover their value? It seems to me that they may, unless there be something making it illegal to insure more than the Plaintiff's own interest. Mr Lush does not contend that any statute applies. It has been decided that, if no statute applies, a person insured may recover the amount contracted for: and, that being so, I think the Plaintiffs entitled to recover the whole value.

Crompton J delivered a concurring judgment.

The effect of this famous decision was summarised by Crompton J (who had sat in it) in *London & North Western Railway Company* v *Glyn* (1859) 1 E&E 652, at p663 as follows:

> Mr Lloyd said that the questions were, Who are the assured? and What is their loss? I answer, first, the assured are the Plaintiffs, both as trustees and as carriers; secondly, the loss is the loss of the property as trust property, that is to say as property in which the Plaintiffs are beneficially interested to the extent of their lien, and as to the residue to which they are trustees for the true owners.

These decisions were approved by the House of Lords in *A Tomlinson (Hauliers)* v *Hepburn* [1966] AC 451. There is a succinct statement of the principle at pp467-8 in the speech of Lord Reid:

> A bailee can if he chooses merely insure to cover his own loss or personal liability to the owner of the goods either at common law or under contract and if he does that of course he can recover no more under the policy than sufficient to make good his own personal loss or liability. But equally he can if he chooses insure up to his full insurable interest — up to the full value of the goods entrusted to him. And if he does that he can recover the value of the goods though he has suffered no personal loss at all. But in that case the law will require him to account to the owner of the goods who has suffered the loss or, as Lord Campbell says, he will be trustee for the owners. I need not consider whether this is a trust in the strict sense or precisely on what ground the owner can sue the bailee for the money which he has recovered from the insurer . . . The fact that a bailee has an insurable interest beyond his own personal loss if the goods are destroyed has never been regarded as in any way inconsistent with the overriding principle that insurance of goods is a contract of indemnity.

It is true that a bailee has a rather special status in English law, having in many respects the rights of an owner as against third parties. But the decisions in *Waters, Glyn* and *Tomlinson* do not turn on any principle peculiar to the law of bailment. Similar principles apply to insurances in quite different fields. I have already given trustees as one example. Another is the case of a trade union, which insures the property of its members against burglary. It may recover the value of the stolen property, accounting for it to its members: *Prudential Staff Union* v *Hall* [1947] KB 685. A third is the case of the shipowner who sells his ship but undertakes to keep the insurance on foot and assigns the benefit of it to the purchaser. The law might have been that if a loss subsequently occurs, the insurer is not liable because his assured has not suffered any and the assignee can have no better right than he had. But there is good authority that the insurer must pay: *Powles* v *Innes* (1843) 11 M&W 10, *per* Parke B, and *Arnould's Law of Marine Insurance and Average*, (16th ed) (1981), vol i, p173. It is implicit in *Rayner* v *Preston* (1881) 18 ChD 1 that the same would be true if real property were sold on those terms, even apart from section 47 of the Law of Property Act 1925: see, in particular, Brett LJ at p12.

The authors of these judgments regarded them as turning on two critical factors. The first was that in each case the subject-matter of the insurance was the whole interest in the property insured and not simply the assured's interest. That was treated as a question of construction: see, in particular, Lord Reid in *Tomlinson* at p469. It is usually enough that when construed on ordinary principles the policy covers the whole value of the subject-matter and not only the value of some partial interest in it. The second factor was that so far as the assured was thereby enabled to recover in excess of the value of his own interest, it had to be shown that he would be accountable for that excess, either by virtue of his own distinct legal obligations to the holders of the other interests or by virtue of a trust which the courts were (at least in some cases) prepared to construct for the occasion.

Both of these are features of the present case. By virtue of section A, clause 13, the subject-matter of the insurance is damage to the premises up to their full reinstatement value, and not simply that proportion of it which may relate to the assured's reversion. Although the assured has no general obligation under the lease to reinstate the premises after a fire irrespective of receipt of the insurance proceeds, it is expressly provided that upon receipt of insurance proceeds he has an obligation to lay them out in reinstatement, for the benefit of the tenant.

Once that point is reached, it can be seen that the decision of the Court of Appeal in *Castellain* v *Preston* is irrelevant. In *Castellain* v *Preston* the only other person, apart from the insured, who had any interest in the property was the purchaser, and it had already been held that the insured was not accountable to him. It followed that the insured would be receiving more than an indemnity if he were allowed to keep the insurance proceeds without accounting to the insurer. In the present case the insured is certainly no more accountable to the purchaser than Mr Preston was, but there is a lessee and the insured is accountable to him. Moreover, the insured continues to owe that obligation to the lessee for it cannot be affected either by the lessor's contract to sell the freehold to a third party or by its conveyance to the purchaser in due course.

Suppose that on the day after the fire the insurer had been required to meet his obligations to the insured. Even if there had been no lease, it would have been no defence for the insurer at that stage to say that in due course the insured would be entitled to receive an undiminished purchase price on completion. Without express agreement to that effect, a person who has contracted to indemnify another can never refuse on the ground that an indemnity can be obtained from someone else: *Collingridge* v *Royal Exchange Assurance Corporation* (1877) 3 QB 173. If then the insurers duly paid the reinstatement cost to the insured before completion, as was done in *Castellain* v *Preston*, would they have been entitled by virtue of their subrogation rights to recover an equivalent amount out of the purchase price once it was paid six weeks later? Of course not. The reason is that the insured, having received the insurance moneys, would have become liable to the tenant to lay them out in reinstatement of the premises. The subsequent receipt of the purchase price would not, in these circumstances, in the least diminish the loss.

A The summons asks me three questions:

1. Whether any sum is payable under or by virtue of the insurance policy in respect of such damage as the demised premises may have suffered as a result of the fire which occurred at the demised premises on 8th February 1990.

2. If the answer to 1 above is Yes whether that sum is

(a) The sum necessary fully to reinstate the demised premises in respect of such damage as may have been suffered, or

(b) Some other sum and, if so, what sum. B

3. If the answer to 1 or 2(a) is No, whether the first defendant is in breach of its obligation contained in clause 3(2) of the lease namely to insure the demised premises in the full reinstatement value thereof. The answers to questions 1 and 2(a) are "Yes" in each case. It allows that questions 2(b) and 3 do not arise.

LANDLORD AND TENANT

BUSINESS TENANCIES

Court of Appeal

November 25 1992

(Before Lord Justice NOURSE, Lord Justice EVANS and Sir Michael KERR)

CITY OF LONDON CORPORATION v FELL
HERBERT DUNCAN LTD v CLUTTONS

Estates Gazette January 30 1993

[1993] 04 EG 115

Landlord and tenant — Arrears of rent — Whether following assignment of term original tenant liable for rent during tenancy continued under Part II of the Landlord and Tenant Act 1954 — Whether liability during continuation tenancy limited to contractual rent

City of London Corporation* v *Fell

The appellant lessors granted a 10-year term to the respondent lessees from March 25 1976. The respondents assigned the term to Grovebell Group Ltd in June 1979. The yearly rent was subsequently reviewed to £38,500. Although the contractual term expired on March 25 1986, the assignee remained in possession pursuant to Part II of the Landlord and Tenant Act 1954. The assignee surrendered the tenancy to the appellant on January 23 1987. The assignee was compulsorily wound up on December 11 1986. No rent was paid after Christmas quarter day 1985. The appellant's claim for the arrears of rent against the respondents as original tenants was dismissed by Mr Desmond Perrett QC (sitting as a deputy judge of the Chancery Division). The appellants appealed.

Herbert Duncan Ltd* v *Cluttons

By an underlease dated July 9 1985 the respondent demised premises to the appellant firm for a term commencing December 1 1984 and ending March 25 1990 at a rent of £70,000 pa. The appellant assigned the term to Warringtons plc on September 27 1988. The respondent gave notice to Warringtons plc under section 25 of the Landlord and Tenant Act 1954 terminating the tenancy on March 25 1990. Following a counternotice given by Warringtons plc, and an application to the High Court for a new tenancy, under Part II of the 1954 Act, the respondent issued a summons for the determination of an interim rent pursuant to section 24A of the 1954 Act. Warringtons plc served notice discontinuing the application for a new lease and by reason of section 64 of the Act, the tenancy terminated on February 5 1991. On February 15 1991 the master made a consent order determining the interim rent at £200,000 pa. Warringtons plc continued to pay rent at the contractual rate (£60,794.24), but then went into receivership leaving a balance of £113,452.05 due at the termination of the tenancy. By clause 5(g) of the lease the expression "the said term" was stated, where the context admits, to include not only the term thereby granted but also the period of any holding over or any extension thereof whether by statute or at common law. In proceedings under RSC Ords 14A and 14 brought by the respondent against the appellant to recover the arrears of rent, Tudor Evans J gave judgment to the respondent. The appellant appealed contending that as original tenant either it was not liable for any arrears of rent arising after the expiration of the contract term or its liability was limited to the contract rent.

Held*: *City of London Corporation* v *Fell

The appeal was dismissed. Where an original tenant has assigned a tenancy before the end of the contractual term the tenancy which section 24(1) of the 1954 Act provides shall not come to an end is, and can only be, the tenancy of the assignee. Since the contractual obligations of the original tenant form no part of the legal relationship between the landlord and the assignee, and since they are not independently continued by the subsection, they are in no way affected. If the original tenant contracted to pay rent only during the contractual term, as was the case, the landlord cannot recover from him any rent payable in respect of any period after that date.

Herbert Duncan Ltd* v *Cluttons

The appeal was allowed. The covenant to pay rent was, by virtue of clause 5(g) of the lease, a covenant to pay the yearly rent of £70,000 during the contractual term and any extension of the term by section 24(1) of the 1954 Act. However, the appellant did not covenant to pay any rent other than the contractual rent of £70,000 during the continuation tenancy: they did not covenant to pay the interim rent determined pursuant to section 24A of the Act.

The following cases are referred to in this report.

City of London Corporation v *Fell* [1992] 3 All ER 224; [1992] 1 EGLR 95; [1992] 02 EG 172
Cornish v *Brook Green Laundry Ltd* [1959] 1 QB 394; [1959] 2 WLR 215; [1959] 1 All ER 373; [1959] EGD 116; (1959) 173 EG 307, CA
GMS Syndicate Ltd v *Gary Elliott Ltd* [1982] Ch 1; [1981] 2 WLR 478; [1981] 1 All ER 619; (1980) 41 P&CR 124; [1981] EGD 331; 258 EG 251
Herbert Duncan Ltd v *Cluttons* [1992] 1 EGLR 101; [1992] 22 EG 110
Junction Estates Ltd v *Cope* (1974) 27 P&CR 482; 232 EG 335
Plesser (A) & Co Ltd v *Davis* (1983) 267 EG 1039
Weinbergs Weatherproofs Ltd v *Radcliffe Paper Mill Co Ltd* [1958] Ch 437; [1958] 2 WLR 1; [1957] 3 All ER 663

In *City of London Corporation* v *Fell* this was an appeal by the City of London Corporation from a decision of Mr Desmond Perrett QC (as he then was) dismissing the appellant's claim for arrears of rent against the respondents, John Fell, John Edward James Hayward and Edward Denham Sturmer: [1992] 1 EGLR 95; [1992] 02 EG 172.

In *Herbert Duncan Ltd* v *Cluttons* this was an appeal by Cluttons from orders made by Tudor Evans J giving judgment for the respondent, Herbert Duncan Ltd, on a summons under RSC Ords 14A and 14: [1992] 1 EGLR 101; [1992] 22 EG 110.

Andrew Arden QC and Noah Weiniger (instructed by the solicitor to the City of London Corporation) appeared for the City of London Corporation; David Neuberger QC and Erica Foggin (instructed by Wilde Sapte) appeared for John Fell, John Edward James Hayward and Edward Denham Sturmer; Barry Denyer-Green (instructed by Wray Smith & Co) appeared for Cluttons; Jonathan Brock (instructed by Forsyte Kerman) appeared for Herbert Duncan Ltd.

Giving the first judgment, **Lord Justice Nourse** said: Section 24(1) of the Landlord and Tenant Act 1954 provides:

A A tenancy to which this Part of this Act applies shall not come to an end unless terminated in accordance with the provisions of this Part of this Act . . .

The well-recognised effect of that provision is to continue a tenancy of business premises after the end of the contractual term. The principal question arising on these appeals is whether it also has the effect of continuing the contractual obligations, in particular the obligation to pay rent, of an original tenant who has assigned the tenancy before that date. In *City of London Corporation* v *Fell* [1992] 3 All ER 224 Mr Desmond Perrett QC (as he then was) has answered the question in the negative. In *Herbert Duncan Ltd* v *Cluttons* [1992] 1 EGLR 101 Tudor Evans J has answered it in the affirmative. Now we have to say which of them was right.

It may be thought curious that this question, seemingly so B fundamental to the working of Part II of the 1954 Act, has not arisen earlier for decision. A possible explanation is that the answer cannot before have been doubted. More probably, it has taken the effects of a serious recession on the solvency of assignees to make the question a live one.

Although both appeals were called on together, they were argued consecutively. It is convenient to follow the course adopted in argument by dealing first with *City of London Corporation* v *Fell*, where the facts are simple and no subsidiary questions arise.

City of London Corporation v *Fell*

By a lease dated July 15 1977 and made between the plaintiffs, the City of London Corporation, as lessors, of the one part and the C defendants and another, partners in the solicitors' firm of Wilde Sapte, as lessees, of the other part, the plaintiffs demised business premises at Boston House, 63-64 New Broad Street in the City of London to the lessees for a term of 10 years from March 25 1976 (defined as "the term"), "YIELDING AND PAYING therefor during the term unto the Lessors" the yearly rent of £27,500 in advance, subject to review as at the expiration of the fifth year of the term. Clause 2(1) of the lease contained a covenant by the lessees that they:

Will pay the said yearly rents on the days and in the manner hereinbefore appointed for payment thereof.

From these provisions it is clear that the contractual liability of the lessees for rent was coterminous with the contractual term.

D On June 15 1979 the then lessees, with the licence of the plaintiffs, assigned the lease to a company called Grovebell Group Ltd ("Grovebell"). The yearly rent was subsequently reviewed and increased to £38,500 with effect from March 25 1981. After the expiration of the contractual term on March 24 1986, Grovebell continued in occupation of the premises under the 1954 Act. There can be no doubt that the effect of section 24(1) was to continue the tenancy as between the plaintiffs and Grovebell. But on December 1 1986 an order for its compulsory winding up was made in the companies court. The judge found that Grovebell remained in occupation of the premises until January 23 1987, when the lease was surrendered by its joint liquidators to the plaintiffs, who thereupon retook possession.

E Grovebell failed to pay the quarterly instalment of rent due on March 25 1986 and it paid no rent thereafter. It also failed to pay other sums payable as additional rent in respect of insurance and service charge. Having been informed by the liquidators of Grovebell that no significant assets had been recovered in the liquidation and that it was unlikely that there would be any dividend to unsecured creditors, the plaintiffs wrote to the defendants on September 7 1989 stating that they would look to them, as original lessees, for payment of the amount outstanding. The defendants denied that they were liable for any sums due in respect of any period after the expiration of the contractual term. On July 6 1990 the plaintiffs issued the writ in the action claiming payment of £33,460.64, together with interest F amounting to £4,166.54. The action came on for trial before Mr Desmond Perrett QC on June 19 and 20 1991, when he reserved judgment. On July 10 1991 he gave judgment for the defendants. The plaintiffs now appeal to this court.

Principal question

Section 24(1) of the 1954 Act provides that the "tenancy" shall not come to an end unless terminated in accordance with the statutory provisions. That word is not defined in the Act. It must therefore be given its ordinary legal meaning. To what does it refer in a case where the original tenant has assigned the tenancy before the end of the contractual term? In order that the question may be answered, some elementary propositions in the law of landlord and tenant must be restated.

G A lease of land, because it originates in a contract, gives rise to obligations enforceable between the original landlord and the original tenant in contract. But because it also gives the tenant an estate in the land, assignable, like the reversion, to others, the obligations, so far as they touch and concern the land, assume a wider influence, becoming, as it were, imprinted on the term or the reversion as the case may be, enforceable between the owners thereof for the time being as conditions of the enjoyment of their respective estates. Thus landlord and tenant stand together in one or other of two distinct legal relationships. In the first it is said that there is privity of contract between them, in the second privity of estate.

To what, in ordinary legal parlance, do we refer when we speak of a "tenancy"? I think that we refer to a particular legal relationship between tenant and landlord under which land is held by the one of H the other. A "tenant", both by derivation and by usage, is someone who "holds" land of another, for which purpose it is immaterial whether he does so by contract or by estate. Although he may remain contractually liable to the landlord, an original tenant who has assigned the tenancy, equally with an assignee who has himself assigned, cannot properly be described as the tenant. He no longer holds the land. It is the assignee who now holds the land. It is he who has the tenancy.

It follows that where an original tenant has assigned the tenancy before the end of the contractual term the tenancy which section 24(1) provides shall not come to an end is, and can only be, the tenancy of the assignee. Since the contractual obligations of the original tenant J form no part of the legal relationship between the landlord and the assignee, and since they are not independently continued by the subsection, they are in no way affected. If, as here, the original tenant has covenanted to pay rent only during the contractual term, the landlord cannot recover from him any rent payable in respect of a period after that date. Further elaboration of the principal question could only obscure the clarity of the answer. The defendants are entitled to succeed on this appeal.

The basic submission of Mr Arden QC, for the plaintiffs, was that because a tenancy can only originate in contract its continued existence is dependent on continued contractual support. Accordingly, the effect of section 24(1) must be to continue not only the obligations arising by privity of estate between the landlord and K the assignee but also those arising by privity of contract between the original landlord and the original tenant. For the reasons already stated, that submission, afflicted as it is by a confusion between the creation and the continuation of a tenancy, must be rejected. A tenancy needs a contract to create it. It does not need one to continue it. The contractual obligations which touch and concern the land having become imprinted on the estate, the tenancy is capable of existence as a species of property independently of the contract.

Mr Arden helpfully referred us to many other provisions of the 1954 Act. None of them provides any support for the view that "tenancy" in section 24(1) is to be given any other than its ordinary meaning. Indeed, by their exclusive concern with the tenant in L occupation they affirm the contrary. Mr Arden also took us to all the previous authorities on section 24(1). They establish that the effect of the subsection is to continue the common law tenancy with a statutory variation as to the mode of its determination: see, for example, *Cornish* v *Brook Green Laundry Ltd* [1959] 1 QB 394, at p409. In some of them reference has been made to the continuation, not of the tenancy, but of the "term": see, for example, *Weinbergs Weatherproofs Ltd* v *Radcliffe Paper Mill Co Ltd* [1958] Ch 437, at p446. There is no significance in this distinction in language. The term is an integral part of the tenancy and, if the one is continued, so is the other.

Unfortunately, as it now appears, the insignificance of judicial M references to the continuation of the term or the contractual term was not fully grasped in the courts below, where counsel for the respective landlords invoked them as authority for the proposition that the contractual obligations of the original tenant are also continued by section 24(1). That line of argument involves a clear *non sequitur*. Although the term is created by contract, it is continued by statute. Accordingly, while it is usual to speak of the continuation of the contractual term, it does not follow that the contractual obligations of the original tenant are continued. The true view is that none of the references in the previous authorities to the continuation of the tenancy or the term has illuminated the meaning of "tenancy" in its application to a case such as the present. A decision of my own at first instance, *GMS Syndicate Ltd* v *Gary Elliott Ltd* [1982] Ch 1, which was much relied on in the courts below, is especially unilluminating. Rather more helpful, although by analogy only, are *Junction Estates Ltd* v *Cope* (1974) 27 P&CR 482* and *A Plesser & Co Ltd* v *Davis* (1983) 267 EG 1039, in each of which it was held that a covenant guaranteeing the tenant's payment of rent and performance of his covenants was limited to the contractual term of the lease and did not extend to the period of a continuation tenancy under section 24(1).

A further point must be mentioned. Mr Neuberger QC, for the defendants, argued that even if there had been no assignment of the lease, so that the defendants remained the tenants, their contractual obligations would not have been continued by section 24(1). If the views already expressed are correct, it would seem that argument ought to be rejected. The "tenancy", the particular legal relationship between landlord and tenant, would have been that created by the original contract. It would have been that relationship which was continued by section 24(1). However, this is not a question which arises for decision in the present case. It was not fully argued. It could be of practical significance only in relation to an obligation which did not touch and concern the land. In all the circumstances it is neither necessary nor desirable to express a concluded opinion on it.

The learned deputy judge in this case came to a correct decision. I would affirm it and dismiss the appeal accordingly.

Herbert Duncan Ltd v *Cluttons*

This case has two additional features. First, the landlord claims that the covenant in the lease for payment of rent obliges the original tenants to pay it, not only during the contractual term but also during any continuation tenancy under section 24(1). Second, the landlord claims that the original tenants are liable during that period, not simply for the contractual rent, but for an interim rent determined pursuant to section 24A of the 1954 Act by a consent order made in proceedings between the assignee and the landlord.

By a lease dated July 9 1985 and made between the plaintiff, then called Herbert Duncan (Furs) Ltd, as lessor, of the one part and four partners in the chartered surveying firm of Cluttons, the defendants, as lessees, of the other part, the plaintiff demised business premises at 74 Grosvenor Street, London Wl, to the defendants for a term commencing on December 1 1984 and expiring on March 25 1990 (defined as "the said term") "YIELDING AND PAYING therefor unto the Lessor during the said term" the yearly rent of £70,000 in advance. Clause 2(l)(a) of the lease contained a covenant by the defendants that they:

> will pay the yearly rent hereinbefore reserved . . .

Clause 5(g) provided:

> The expression "the said term" shall where the context admits include not only the term hereby granted but also the period of any holding over or of any extension thereof whether by statute or at common law.

On September 27 1988 the defendants, with the licence of the plaintiff, assigned the lease to a company called Warringtons plc ("Warringtons"). On August 4 1989 the plaintiff served a notice under section 25 of the 1954 Act on Warringtons terminating its tenancy at the end of the contractual term on March 25 1990. On November 23 1989 Warringtons, having duly notified the plaintiff that it was not willing to give up possession of the premises, issued an application in the Chancery Division for a new tenancy. On December 18 1989, by summons issued in those proceedings, the plaintiff applied for an interim rent to be determined pursuant to section 24A of the Act. On November 6 1990 Warringtons served on the plaintiff notice of discontinuance of its application for a new tenancy. That meant that, by virtue of section 64 of the Act, the continuation tenancy was terminated on February 5 1991. On February 15 1991 the master made a consent order on the plaintiff's summons under section 24A determining the interim rent at £200,000 pa.

The admitted effect of the determination was that Warringtons became liable to pay rent for the period between March 25 1990 and February 5 1991 at the yearly rate of £200,000, the apportioned amount for that period being £174,246.57. However, it continued to pay at the contractual rate and no more (£6O,794.52), leaving a balance of £113,452.05. Warringtons having since gone into administrative receivership, the plaintiff may be assumed to have little hope of recovering the balance from it. In any event, it commenced the present action against the defendants by a writ issued in the Queen's Bench Division on August 9 1991. It claimed payment of £113,452.05, together with interest of £8,065.97 to date. It also claimed damages for dilapidations.

The plaintiff issued a summons for the determination of a point of construction and/or law under RSC Ord 14A and, if successful, for final judgment under Ord 14. It also issued a summons for final judgment under Ord 14 in respect of the claim for dilapidations. The summonses came on for hearing before Tudor Evans J in chambers in December 1991. On December 20 he gave judgment in open court for the plaintiff in the sum of £113,452.05, together with £14,313.08 in respect of interest. He granted the defendants leave to appeal and a stay of execution on terms which have been duly complied with. He also gave judgment for the plaintiff on its dilapidations claim for damages, interest and costs to be assessed and made an order for an interim payment of £30,000. No appeal is brought against the judge's decision so far as it related to the dilapidations claim.

The first question raised by the defendants' appeal was the principal question, on which Mr Denyer-Green, for the defendants, and Mr Brock, for the plaintiff, were content to adopt the arguments advanced by Mr Neuberger QC and Mr Arden QC respectively. For the reasons already stated, that question must be resolved in favour of the defendants. It then becomes necessary to consider two subsidiary questions.

Subsidiary questions

The first subsidiary question is whether, by virtue of the particular provisions of the lease, the defendants were liable to pay the contractual rent of £70,000 pa, not only during the contractual term but also during the continuation tenancy under section 24(1). This is a very simple question. The combined effect of the reddendum and clause 2(1)(a) of the lease was that the defendants covenanted to pay the yearly rent of £70,000 during "the said term" which, by clause 5(g), was defined to include not only the contractual term but also "the period of . . . any extension thereof whether by statute or at common law". In other words, there was a covenant to pay the rent during the period of any statutory extension of the contractual term, ie by section 24(1) of the 1954 Act. The first subsidiary question must be decided in favour of the plaintiff.

The plaintiff, by its success on that question alone, cannot acquire any financial benefit as against the defendants. That is because it has already received the full amount of the contractual rent during the continuation tenancy from Warringtons. However, its success is a necessary preliminary to the second subsidiary question, which is whether the defendants are liable, not for the contractual rent, but for the interim rent instead. This question is almost as simple as its predecessor.

So far as material, section 24A provides:

> (1) The landlord of a tenancy to which this Part of this Act applies may —
> (a) if he has given notice under section 25 of this Act to terminate the tenancy;
> . . . apply to the court to determine a rent which it would be reasonable for the tenant to pay while the tenancy continues by virtue of section 24 of this Act, and the court may determine a rent accordingly.
> (2) A rent determined in proceedings under this section shall be deemed to be the rent payable under the tenancy . . .

Mr Denyer-Green submitted, correctly, that in a case such as the present "the tenant" referred to in subsection (1) can only be the assignee. He added that Parliament cannot have intended that an original tenant should be liable for payment of a rent which, although it might be reasonable for the assignee to pay, might be unreasonable for the original tenant to pay, especially when it fell to be determined in proceedings to which he would not be a party.

Although I see great force in that submission, it is unnecessary to resort to it in this case, where everything depends on the contract between the parties. All that the defendants covenanted to pay was the yearly rent of £70,000. True it is that they covenanted to pay it during the continuation tenancy. But they did not covenant to pay any other rent. They certainly did not covenant to pay an interim rent determined pursuant to section 24A and, the principal question having been decided as it has, they cannot be fastened with any liability to do so. I cannot accept Mr Brock's submission that the deeming provision in subsection (2) in some way translated the contract into one for payment of the interim rent. For these reasons the second subsidiary question must be decided in favour of the defendants. They are entitled to succeed on this appeal.

Tudor Evans J decided the principal question and both subsidiary questions in favour of the plaintiff. He dealt very briefly with the second subsidiary question. Having referred to Mr Denyer-Green's

*Editor's note: Also reported at (1974) 232 EG 355.

submission that the liability must be limited to £70,000 and having acknowledged its attraction, the judge rejected it, saying that section 24A was in the nature of a statutory rent review. That is no doubt correct as between the landlord and the assignee. With respect to the judge, it could be correct as between the landlord and the original tenant only if the terms of their contract allowed it. They did not allow it in the present case.

In this case I would allow the appeal and discharge para 1 of Tudor Evans J's order dated December 20 1991.

Agreeing, **Evans LJ** said: In cases where the assignee is in occupation of the premises at the expiry of the contractual term, the "tenancy" referred to in section 24(1) of the Landlord and Tenant Act 1954 is, in my judgment, the legal relationship which exists between the landlord and his tenant, the assignee. The period of that tenancy was the period of the original contractual term, but it does not follow that the statute has the effect of extending the original contract between the parties to it. Where, on the other hand, the original tenant and assignor has covenanted to pay rent during the contractual term and "the period of . . . any extension thereof whether by statute or at common law", as in clause 5(g) of the lease in the second of these cases, then the landlord can enforce that covenant as a matter of contract against the original tenant during a statutory extension of the contractual term. This liability of the original tenant is not qualified, in my judgment, by the fact that the statutory extension would give the landlord no right against him, in the absence of contract.

Also agreeing, **Sir Michael Kerr** said: I agree, for the reasons stated in the judgment of Nourse LJ, that the appeal in *City of London Corporation* v *Fell* should be dismissed and that the appeal in *Herbert Duncan Ltd* v *Cluttons* should be allowed. I only wish to add a few remarks, principally with reference to the relevant provisions of the Landlord and Tenant Act 1954.

As Nourse LJ has pointed out, when a tenancy has been assigned, the assignee becomes the tenant and the assignor ceases to be the tenant. If he is the person to whom the tenancy has been granted originally, or an intermediate assignee who has concluded a direct contract with the landlord, then their contracts will continue in the absence of any contrary provision. But they will no longer have any estate or interest in the land as tenants of it. The only tenant will be the assignee.

This state of affairs is clearly reflected in the Act. Leaving aside subleases, which are irrelevant for present purposes, the references in the Act to "the tenant" designate only the tenant in occupation, not the original or any prior tenants who have parted with their interests by assignment. It is the tenant in occupation who can request a new tenancy under section 26 and it is his request which may be opposed by the landlord under section 30. It is his tenancy which is continued until the fate of his application is resolved pursuant to section 29 *et seq*, and it is in relation to his tenancy that an interim rent may meanwhile be determined pursuant to section 24A. The Act makes no reference to the continuation of any contract, only to the continuation of tenancies. Mr Arden QC submitted that there can be no tenancy without an underlying contract and that in continuing tenancies the Act must impliedly also have the effect of continuing these underlying contracts. But this is clearly incorrect. Where an assignee of a tenancy does not enter into a direct contract with his landlord the relationship between them will rest on privity of estate alone, without any privity of contract. Depending on the contractual terms, privity of contract with the landlord may continue in relation to the original tenant and any intermediate assignee who has entered into a direct contract with him. But the Act does not continue these contracts and has no effect upon them.

In this context it is interesting, and at first sight perhaps surprising, to note that the Act makes no reference to the assignment of leases in the ordinary way. The only direct reference to assignment is to be found in section 42(2)(*c*) which deals with groups of companies and provides that "an assignment of the tenancy from one member of the group to another shall not be treated as a change in the person of the tenant". It therefore follows that in that case, but in that case alone, an assignor is treated as though he continued to be the tenant, which he would not be otherwise.

Since the contract contained in the lease granted by the *City of London Corporation* contained no provision for the payment of rent after the expiry of the contractual term, and, since the Act has no effect upon this contract, it must follow that the claim for rent after its expiry must fail against the original tenant. On the other hand, clause 5(g) of the lease granted by *Herbert Duncan* contained such a provision. This was clearly directed, at least in part, to the effect of section 24 of the Act, with the result that there would be a liability in the original tenant for the original rent during any statutory continuation of the tenancy under the Act if the rent is not paid by the tenant in occupation. But nothing in the lease is directed to section 24A. The term or period of the lease is continued, together with the contractual rights and obligations for which the lease provides. These obviously include payment of the contractual rent. But they do not include payment of any increased rent which may be payable by the tenant in occupation pursuant to section 24A.

Appeal in City of London Corporation *v* Fell *dismissed with costs; appeal in* Herbert Duncan Ltd *v* Cluttons *allowed with costs; applications in both cases for leave to appeal to the House of Lords refused.*

Chancery Division

July 28 1992

(Before Mr A S GRABINER QC sitting as a deputy judge of the division)

GRAYSIM HOLDINGS LTD v P&O PROPERTY HOLDINGS LTD

Estates Gazette February 6 1993

[1993] 05 EG 141

Landlord and tenant — Landlord and Tenant Act 1954 — Whether operator of market entitled to protection of the 1954 Act — Whether landlord entitled to oppose the tenant's application for a new tenancy under section 30(1)(f)

By a lease dated April 2 1974 the plaintiff, Graysim Holdings Ltd, is the tenant of Wallasey Market, holding a term of 21 years from September 21 1970: the defendant, P&O Property Holdings Ltd, is the owner of the freehold reversion to the lease. By a notice dated November 28 1990 the defendant served the plaintiff with a notice to terminate the lease on September 29 1991 stating that it would oppose any application to the court for the grant of a new tenancy on the grounds contained in section 30(1)(*a*) and (*f*) of the 1954 Act. On December 12 1990 the plaintiff served on the defendant a counternotice stating it would not be willing to give up possession of the premises. The plaintiff acquired the premises as a shell and fitted them out as a market hall constructing stands along the walls and two rows of stalls down the centre there being 35 stalls in all. Each stall is fitted with a roller blind and a stall-holder may secure his own stall with his own padlock and key. Certain facilities and services are provided by the plaintiff for the benefit of the stall-holders. The trading hours of the market hall are from 9am to 5.50pm Monday to Saturday and the individual traders have no access either to the hall or to their stalls outside those hours. The plaintiff also provides lighting and heating and lays on electricity supplies to individual stalls and provides a market superintendent, who performs a number of services, maintains discipline among traders, and provides other supervisory duties. The stall-holders pay a set weekly charge and a service charge to the plaintiff. Four stall-holders have contracts with the plaintiff in the form of a lease; eight stall-holders have an agreement described on its face as a licence, but which has all the characteristics of a lease; 14 stall-holders have contracts on the basis of a more modern form of licence described as a "Licence Agreement" and two stall-holders have an "interim form" of licence for a period of 12 months. Before the court there were three questions for its determination: (1) whether the plaintiff's tenancy of Wallasey

A Market is protected by Part II of the Landlord and Tenant Act 1954; and, if in the affirmative, (2) what, in relation to that tenancy, is the extent of "the holding" for the purposes of the Act; and (3) whether the defendant may oppose the plaintiff's application for a new tenancy under section 30(1)(*f*) of the Act.

Held: **(1) All the stall-holders have tenancies because in the circumstances the traders undoubtedly have exclusive possession of their stalls. Under section 23 of the 1954 Act it was sufficient for the plaintiff to occupy a part only of the premises. The plaintiff did not occupy any of the stalls because**
B **as a matter of ordinary language the stalls were occupied by the individual stall-holders; accordingly the stalls do not form part of "the holding" for the purposes of section 23(3) of the Act. It could not be said that the plaintiff occupied the common parts of the market hall because they were not occupied for the purposes of any business. It followed that the plaintiff's tenancy was not protected and came to an end on September 29 1991.**

(2) On the assumption, contrary to the above judgment, that the plaintiff's tenancy is protected in whole or in part, and construing the parcels clause of the lease, the premises for the purposes of section 30(1)(*f*) includes the whole of one wall of
C **the demised premises. The ground specified in section 30(1)(*f*) was made out. The defendant's plans included the removal of part of the north wall which could be properly described as demolition work, the south wall will have to be reconstructed. The removal of the existing interior of the market hall can be described as an exercise in demolition followed by a reconstruction of a substantial part of the premises. The works proposed were very extensive resulting in a brand new market hall. It would be very impracticable for the proposed works to be carried out piecemeal and the work will in all probability take eight weeks during which it would not be**
D **possible for the plaintiff to make any use whatever of the holding. This amounts to circumstances constituting an interference to a substantial extent and for a substantial time with the use of the holding for the purposes of the business carried out by the plaintiff, and accordingly it was not entitled to the protection afforded by section 31A of the 1954 Act.**

The following cases are referred to in this report.

Ashburn Anstalt v *Arnold (No 2)* [1989] 1 Ch 1; [1988] 2 WLR 706; [1988] 2 All ER 147; (1988) 55 P&CR 137; [1988] 1 EGLR 64; [1988] 23 EG 128, CA
Bagettes Ltd v *G P Estates Ltd* [1956] Ch 290; [1956] 2 WLR 773; [1956] 1 All ER 729; (1956) 167 EG 249, CA
E *Bewlay (Tobacconists) Ltd* v *British Bata Shoe Co Ltd* [1959] 1 WLR 45; [1958] 3 All ER 652, CA
Boyer (William) & Sons Ltd v *Adams* (1975) 32 P&CR 89
Cadle (Percy E) & Co Ltd v *Jacmarch Properties Ltd* [1957] 1 QB 323; [1957] 2 WLR 80; [1957] 1 All ER 148, CA
Cerex Jewels Ltd v *Peachey Property Corporation plc* (1986) 52 P&CR 127; [1986] 2 EGLR 65; 279 EG 971, CA
Decca Navigator Co Ltd v *Greater London Council* [1974] 1 WLR 748; [1974] 1 All ER 1178, CA
Groveside Properties Ltd v *Westminster Medical School* (1983) 47 P&CR 507; 267 EG 593, CA
Hancock & Willis v *GMS Syndicate Ltd* [1983] EGD 114; (1982) 265 EG 473, CA
Heath v *Drown* [1973] AC 498; [1972] 2 WLR 1306; [1972] 2 All ER 561, HL
Joel v *Swaddle* [1957] 1 WLR 1094; [1957] 3 All ER 325, CA
F *Lee-Verhulst (Investments) Ltd* v *Harwood Trust* [1973] QB 204; [1972] 3 WLR 772; [1972] 3 All ER 619; (1972) 24 P&CR 346; [1973] EGD 467; 225 EG 793, CA
Narcissi v *Wolfe* [1960] Ch 10; [1959] 3 WLR 431; [1959] 3 All ER 71
Price v *Esso Petroleum Co Ltd* [1980] EGD 147; (1980) 255 EG 243
Redfern v *Reeves* (1978) 37 P&CR 364; 247 EG 991, CA
Ross Auto Wash Ltd v *Herbert* (1978) 250 EG 971
Street v *Mountford* [1985] AC 809; [1985] 2 WLR 877: [1985] 2 All ER 289; [1985] 1 EGLR 128; (1985) 274 EG 821, HL
Trans-Britannia Properties Ltd v *Darby Properties Ltd* [1986] 1 EGLR 151; (1986) 278 EG 1254, CA
Wandsworth London Borough Council v *Singh* (1991) 62 P&CR 219; 89 LGR 729; [1991] 2 EGLR 75; [1991] 33 EG 90, CA

This was the hearing of three preliminary questions in proceedings by the plaintiff, Graysim Holdings Ltd, against the plaintiff, P&O Property Holdings Ltd, for the grant of a new tenancy under the G
Landlord and Tenant Act 1954.

Albert Charles Sparrow QC and Patrick Powell (instructed by A Banks & Co) appeared for the plaintiff; David Neuberger QC and Nicholas Harrison (instructed by Carter Lemon) represented the defendant.

Giving judgment, **MR A S GRABINER QC** said: This case is about a covered and enclosed market hall in Merseyside. There are two issues. First, is the plaintiff entitled to any protection under the Landlord and Tenant Act 1954, Part II? Second, assuming an affirmative answer to the first issue, is the landlord entitled to oppose the tenant's application for a new tenancy of the premises on the H
ground specified in section 30(1)(*f*) of the 1954 Act? The plaintiff, Graysim Holdings Ltd ("Graysim"), carries on business as an operator of markets. It is the holding company of a group of companies with an annual turnover of £3.5m. Graysim has been in business for 25 years and operates 23 markets at various locations in England and Scotland.

The defendant, P&O Property Holdings Ltd ("P&O") is a wholly-owned subsidiary of P&O Steam Navigation Company ("the company"). P&O recently completed the redevelopment of virtually the whole of a shopping precinct, formerly known as Liscard Shopping Centres Ltd and now known as the Cherry Tree Centre. P&O Shopping Centres is also wholly owned by the company and manages its specialist covered shopping centres portfolio, which J
includes the Cherry Tree Centre. The only part of the precinct which has not been redeveloped is the market hall known as Wallasey Market.

By a lease dated April 2 1974 Wallasey Properties Ltd as lessors demised Wallasey Market to Graysim Properties Ltd as lessees for a term of 21 years from September 29 1970. Graysim is now the tenant of Wallasey Market and the freehold reversion is vested in P&O.

On November 28 1990 P&O served Graysim with a notice to terminate the lease on September 29 1991. The notice stated that P&O would oppose an application to the court for the grant of a new tenancy on the grounds contained in section 30(1)(*a*) and (*f*) of the 1954 Act. On December 12 1990, Graysim served on P&O a counternotice stating that it would not be willing to give up K
possession of the premises on the date of termination. It is common ground that the notice and counternotice comply with the requirements of the 1954 Act. It is also common ground that the relevant cutoff date is September 29 1991 so that if the 1954 Act did not apply at that date Graysim's tenancy could not continue.

There are three questions which I have to decide. They are conveniently set out in the order which was made on November 12 1991 by Master Cholmondeley Clarke:

> (i) Whether [Graysim's] tenancy of [Wallasey Market] is protected by Part II of the Landlord and Tenant Act 1954, and continues under the provisions of the said Act:
> If the answer to paragraph (i) is in the affirmative: L
> (ii) What, in relation to [Graysim's] said tenancy, is the extent of "the holding" (as defined by section 23(3) of the said Act) for the purposes of the said Act; and
> (iii) Whether [Graysim] is entitled to a new tenancy of the said premises or whether [P&O] may oppose [Graysim's] application for a new tenancy on the ground specified in section 30(1)(*f*) of the said Act.

For completeness I should say that P&O no longer relies on the ground set out in section 30(1)(*a*) of the 1954 Act.

By agreement between parties the hearing of this case was split into two self-contained parts so that questions (i) and (ii) were dealt with together in the first part and question (iii) in the second part. This judgment deals with both parts.

M

Part I (Questions (i) and (ii))

The facts

Clause 2(13) of the lease provides:

> That the demised premises shall not at any time during the term hereby granted be used or occupied otherwise than as a general market or for any purpose or in any manner inconsistent with such use or occupation nor so as to be a nuisance annoyance or damage to or in any way interfere with the quiet and comfort of the Lessors or the tenants or occupiers of any other shops offices or premises in the Shopping Precinct . . . of which the demised premises form part of any adjoining or neighbouring building and to use and occupy the demised premises during the whole of the term hereby granted and not to allow the same to remain unoccupied and to keep the same open for customers during the normal shopping hours of the neighbourhood.

Graysim acquired the premises as a shell and fitted them out as a market hall. It built stalls along the walls and two rows of stalls down the centre; there are 35 stalls. Aisles run between the stalls. The stalls are fixed and are of wooden construction. They consist of studs to which partitioning has been attached. The stalls are fitted with roller blinds which are so designed that at the end of a day's trading they can be secured by the trade padlocking the blind to the counter or some other part of the stall. Some of the security-minded stall-holders have more elaborate systems, but the essential point is that each trader secures his own stall with his own padlock and key.

Certain facilities and services are provided by Graysim for the benefit of stall-holders. There are lavatories which are intended to be exclusively for the use of stall-holders. There are bin rooms for disposal of rubbish from the market. There is also a market superintendent, Mr Jameson, who is employed by Graysim. He uses a small room in one corner of the market hall as his office. The gas heating controls for the market hall are situated in this room.

Wallasey Market was created more than 20 years ago. The installations are old and in part quite dilapidated. The floor is badly worn in places. In her evidence one of the stall-holders, Mrs Eddins, was very critical of the state of the lavatories which have been unusable for significant periods of time in recent years. In physical terms Wallasey Market has seen better times and representatives of both parties as well as stall-holders gave evidence to the effect that it is badly in need of refurbishment. In view of some of the arguments, which were advanced in relation to the second part of this judgment, I should make it plain that my use of the word "refurbishment" is merely descriptive and I have not used it as a term of art.

The trading hours of the market hall are from 9.00 am to 5.30 pm Monday to Saturday, and it is Mr Jameson's responsibility to open the hall by unlocking it in the morning and locking it at night. Individual traders have no access either to the hall or to their stalls outside trading hours unless accompanied by Mr Jameson.

According to Mr Harvey, the managing director of Graysim, an important feature of Graysim's responsibilities as operator consists of the careful selection of traders so as to ensure a good mix of goods in the market. This, in turn, encourages the public to use the market and to make it successful. Mr Harvey said that it was a matter of great importance that traders should be open for business when the market opens and that they should keep open during ordinary trading hours.

In his affidavit Mr Harvey also said:

> All the traders at Wallasey Market know that it is the custom that if they do not open by 9.30 in the morning Mr Jameson will use bolt cutters on their padlocks and will open their stalls and leave them open.

I do not accept the evidence in that passage. It was not supported by any other witness. Indeed, Mr Jameson said he had never had occasion to open the stall of any trader without authority. Under cross-examination Mr Harvey could recall no occasion when any such thing had ever happened at Wallasey Market. Moreover, none of the contractual documentation as between Graysim and the traders supports the proposition that even apart from custom Graysim reserves to itself the right to open up or enter into a stall or run it in the absence of the stall-holder. I should add that Mr Jameson has never held the keys of individual stalls.

One of Mr Jameson's tasks is to collect the rent. He recalled one incident in 1987 when a stall-holder was in arrears of rent and had been absent from his stall for some weeks. Mr Jameson broke the lock, cleared out the stall and reallocated it to another trader. This seems to me to have been an isolated example of forfeiture or distraint of goods or both.

Graysim provides lighting and heating for the market hall and lays on electricity supplies to individual stalls. Traders pay for their personal electricity consumption by a meter system. In most cases the stall meter can be read from outside the stall, but in any event meter reading is conducted by Manweb and not only Mr Jameson. There was, apparently, an occasion when Mr Jameson was required to get details of the serial numbers of the meters which were situated inside four of the stalls, but he did not suggest that he obtained entry to the stalls other than by seeking and obtaining the consent of the relevant stall-holders.

The lavatories are maintained and cleaned by Graysim. There is a burglar alarm. There is also a coin-operated telephone. Mr Jameson also performs small maintenance services in the hall, and there was some evidence that at the request of individual traders he may have done small repairs to one or two stalls. According to Graysim, Mr Jameson is also employed to maintain discipline among traders and their employees and to ensure that the aisles are kept clear so as not to impede shoppers and not to infringe fire regulations.

Mr Jameson keeps an eye on the stalls and deals with complaints about traders who are or may be in breach of their contractual arrangements with Graysim whereby they agree to supply only a specified range of goods. These arrangements are designed to ensure that the market traders between them provide a full range of choice to the public and that there is no cut-throat competition. Mr Jameson gave evidence about one or two incidents where he was obliged to intervene and help to resolve differences which had arisen between traders.

There are substantial disputes in the affidavits and in the oral testimony as to the quality and extent of the services actually provided by Graysim. Evidence given by or on behalf of P&O was to the effect that many of these so-called services were never provided and in that sense were "fictional", or, where provided, the quality of the service was very poor. Graysim, on the other hand, took the position that, although some of Mr Jameson's predecessors had not been very efficient, he had performed well and had done a good job.

Stall-holders pay a set weekly charge as well as a service charge to Graysim. The service charge is calculated as to reimburse Graysim in respect of expenses it has incurred in the provision of services. It seems that stall-holders have regularly paid the service charge, and the implication I draw is that on the whole they have been willing to pay for whatever service they have been getting. In any event, and notwithstanding the large amount of affidavit evidence which was directed to this issue, I do not think I need to resolve the debate about the quality of services provided by Graysim in order to answer the questions which have been posed.

Contractual documentation

I turn now to deal with the contractual documentation as between Graysim and the stall-holders. There are here no less than four different documents and a set of regulations.

(i) I was told that four stall-holders have contracts in terms in the form of a lease. This document looks like an ordinary lease of shop premises. The parties are described respectively as landlord and tenant and rent is payable. There is a forfeiture clause, but the landlord has no other right of entry. There are provisions which regulate opening times and there is a precise description of the goods which the tenant is permitted to display for sale.

(ii) Eight stall-holders have the "old form" licence. It is described on its face as a licence but has all the characteristics of a lease. There is a term of 21 years at a specified licence fee. The licensee agrees, *inter alia*, not to part with "possession" of a unit without the written consent of the licensor, which may not be unreasonably withheld. The licensee's covenants do not differ materially from those of the tenant under the lease I have referred to. There is a forfeiture clause but, once again, there is no other right of entry in the licensor. There is also a provision which declares that:

> nothing in this licence shall be construed as creating a tenancy.

(iii) There are 14 stall-holders who have contracted on the basis of a more modern form of licence. This is a shorter document than the lease and the old-form licence. It is described as a "Licence Agreement". The licensee pays a licence fee and a service charge. The term is for 12 months. The licensee expressly permits the licensor:

> facilities to read the electricity meter on the said unit.

As a matter of construction this provision, in my view, assumes that Graysim has no right to enter the stall. Furthermore, as I have already found, on those rare occasions when it has been necessary for Mr Jameson to enter the stalls in connection with the electricity meter or otherwise he has always sought and obtained the consent of the stall-holder.

(iv) Two stall-holders have the "interim form" of licence. This purports to be a 12-month licence and is similar to the modern-form licence.

Each of the agreements incorporates a set of rules and regulations issued by Graysim as applicable to market halls. Para 24 of this document provides:

> The licensee may have access to and use his unit and the shopping hall only at such reasonable hours as the company may determine.

There is no provision giving Graysim access to any of the units.

Mr Harvey's evidence was sensible and realistic on this point. He accepted in principle that all stall-holders were entitled to be treated in the same way regardless of any technical differences there may have been in the precise terms of their contractual arrangements with Graysim.

Strictly speaking, it is not necessary for me to express a concluded view in answer to the question did these stall-holders or any of them have tenancies with Graysim? Mr Charles Sparrow QC, who appeared on behalf of Graysim, submitted that for the purposes of his case it really did not matter whether these arrangements gave rise to tenancies or licences. It is also the case that although a few of the stall-holders gave evidence, none of them are parties to this action and, strictly speaking, they are not bound by this judgment.

My provisional view is that on the material before me, and applying the principles laid down by the House of Lords in *Street* v *Mountford* [1985] 1 AC 809*, in particular the speech of Lord Templeman (with whom the rest of their lordships agreed), especially at pp 826H-827B, all the stall-holders have tenancies. Lord Templeman emphasised that what matters in these cases is not the way in which the parties have described their legal relationship in the document, but whether their agreement has the effect in law of granting exclusive possession for a term at a rent. Broadly speaking, if it does the result will be that a tenancy has been created, and the fact that they may have described their relationship as one of licensor/licensee will be irrelevant.

In *Ashburn Anstalt* v *Arnold (No2)* [1989] 1 Ch 1† the Court of Appeal held (applying *Street* v *Mountford*) that, although no rent was payable under the relevant agreement, it did provide for exclusive possession of premises for a term certain and created a tenancy notwithstanding the fact that it was expressed in terms to be a licence. No doubt it could be argued here that stall-holders falling within categories (iii) and (iv) to which I have referred are only licensees, but in the face of Mr Harvey's evidence, it is difficult to see how some stall-holders could be tenants and other licensees.

Aside, however, from this question, the most important point is that, in my judgment and in all the circumstances which I have described, the traders in this case undoubtedly have exclusive possession of their stalls.

In order for Graysim to be protected under the 1954 Act, the occupation test must be satisfied. The relevant provisions at section 23 are as follows:

> (1) Subject to the provisions of this Act, this Part of this Act applies to any tenancy where the property comprised in the tenancy is or includes premises which are occupied by the tenant and are so occupied for the purposes of a business carried on by him or for those and other purposes . . .
>
> (3) In the following provisions of this Part of this Act the expression "the holding", in relation to a tenancy to which this Part of this Act applies, means the property comprised in the tenancy, there being excluded any part thereof which is occupied neither by the tenant nor by the person employed by the tenant and so employed for the purposes of a business by reason of which the tenancy is one to which this Part of this Act applies.

It follows that the tenant may occupy a part only of the premises and that in order to claim protection in respect of the whole or any part of the premises the tenant must "occupy" the same.

"Occupation" is an ordinary English word and the cases show that in this context it should be given its ordinary meaning and construed in a common-sense way: *Narcissi* v *Wolfe* [1960] Ch 10 at p 16 (Roxburgh J), *Lee-Verhulst (Investments) Ltd* v *Harwood Trust* [1973] QB 204 at p 213E (Sachs LJ), *Hancock & Willis* v *GMS Syndicate Ltd* (1982) 265 EG 473 at p 474 (Eveleigh LJ) and *Wandsworth London Borough Council* v *Singh* (1991) 62 P&CR 219 at p 224§ (Ralph Gibson LJ).

The cases also show that, although occupation need not be continuous, there has to be an element of actual physical occupation before the requirements of the 1954 Act can be said to be satisfied. The point is neatly summarised by Eveleigh LJ in the *Hancock & Willis* case, at p 474:

> That does not mean physical occupation every minute of the day, provided the right to occupy continues.

The same point is made by the Court of Appeal in *Trans-Brittania Properties Ltd* v *Darby Properties Ltd* [1986] 1 EGLR 151 at p 154F Slade LJ) and in the *Wandsworth* case at p 224, where Ralph Gibson LJ adopted the observations of Eveleigh LJ in the *Hancock & Willis* case.

In his submissions Mr Sparrow emphasised that this was a market and that the success or failure of any market depended on the ability of the operator, through his control of the market, to attract the public to it. Mr Sparrow drew my attention to a number of authorities in which (he submitted) the presence of a person when coupled with his ability to exercise control over premises was decisive of whether or not the occupation test had been satisfied: *Groveside Properties* v *Westminster Medical School* (1983) 47 P&CR 507* at p 509 (Fox LJ) and p 513 (Kerr LJ), the *Lee-Verhulst* case at pp 213D and 215F-G (Sachs LJ), *William Boyer & Sons Ltd* v *Adams* (1975) 32 P&CR 89 at p 92 (Templeman J) and the *Wandsworth* case at pp 225 and 226 (Ralph Gibson LJ).

By reference to the facts of this case Mr Sparrow then proceeded to emphasise those elements which showed that Graysim satisfied the presence/control test of occupation. It selected the mix of stall-holders, it imposed stringent regulations, it determined the working hours of Wallasey Market and provided advertising for the market (although the evidence in support of this was extremely thin). Through Mr Jameson it controlled the traders' access to their units outside opening hours, it maintained a vigilant watch over traders and participated in the resolution of disputes in the market place, it provided services, including on-site management and it prescribed in detail the range of goods which could be displayed by traders.

On behalf of P&O, the response of Mr Neuberger QC to this submission was that in the context of the 1954 Act there was no magic in a market and that the occupation test still had to be satisfied. He accepted (as he was bound to) that the judgments referred to by Mr Sparrow did use the words "presence" and "control", but he submitted that when properly analysed they were all cases in which, to a greater or lesser extent, there was an element of actual physical occupation on the part of the person held to have been in occupation, and that the extent of Graysim's control and presence in the sense in which those words could be applied in this context was insufficient to satisfy the requirements of the 1954 Act.

Thus in the *Lee-Verhulst* case the tenant provided room service in each of 20 furnished rooms. The director of the tenant company said that the chambermaids employed to carry out the room service had pass keys which they used: see pp 211H-212A, 213D (Sachs LJ) and p 217D-G (Stamp LJ). In the *William Boyer* case the defendant required to enter and did enter the units for the purpose of his business as landlord of the subtenants: see pp 92-93 (Templeman J). In the *Groveside* case the medical school kept the keys to the student flat and Mr Forest, the school secretary, regularly visited the flat and spent time there: see pp 510 and 513 (Fox LJ) and p 51G (Stephenson LJ). In the *Wandsworth* case Ralph Gibson LJ said at p 230:

> The council were physically present upon and exercised control over the piece of land by their servants or agents.

Finally, so far as the case law is concerned, there is *Ross Auto Wash Ltd* v *Herbert* (1978) 250 EG 971, where the tenant company, which granted licences to concession holders, was held to be in occupation of the premises. There the stalls were mobile, one of the services provided by the tenant was the running of the licensees' stalls for short periods in a day in the absence of the licensee, and Fox J found that full possession and control of the premises remained with the tenant.

In my judgment, Mr Neuberger's submissions and his analysis of the cases is correct. In view of my findings of fact and the contracts to which I have referred, it seems to me that I am bound to conclude that Graysim does not occupy any of these stalls. That result seems to me to be consistent with common sense, because as a matter of ordinary language, I think one would say that the stalls are occupied by the individual stall-holders and not by Graysim. Accordingly, they do not, in my judgment, form part of "the holding" for the purposes of section 23(3) of the 1954 Act.

In the alternative, Mr Sparrow submitted that in any event the common parts of the market hall are occupied by Graysim. He submitted that "the holding" on this hypothesis consisted of the common parts which were retained by Graysim for the purpose of carrying on a business which consisted of supplying a market enterprise to any new stall-holder. As a practical matter I would reject this argument as being artificial once it has been determined that Graysim has no protection in respect of the stalls. In any event, the point is resolved for present purposes because of the Court of Appeal decision *Bagettes Ltd* v *GP Estates Ltd* [1956] Ch 290. There the retained common parts were held to be ancillary to the business of subletting flats. It was held that the flats, when sublet, were excluded from the tenancy and it followed that the common parts could not be

*Editor's note: Also reported at [1985] 1 EGLR 128.
†Editor's note: Also reported at [1988] 1 EGLR 64.
§Editor's note: Also reported at [1991] 2 EGLR 75.

*Editor's note: Also reported at (1983) 267 EG 593.

A said to be occupied for the purposes of any business whatever. Accordingly, my conclusion is that Graysim's tenancy of Wallasey Market is not protected under the Landlord and Tenant Act 1954, Part II, and that it came to an end on September 29 1991.

Part II (Question iii)

This part of the judgment assumes (contrary to my conclusion in Part I) that Graysim's tenancy is protected in whole or in part by the 1954 Act.

Premises comprised in the holding

There is a preliminary point of construction of the lease. The relevant works P&O intends to carry out must be done to "the B premises comprised in the holding" (see section 30(1)(*f*)) and, because some of the proposed works are intended to be carried out at the boundaries of the premises, it is necessary from the outset that "the premises" should be precisely defined.

The schedule to the lease defines the premises so as

> to include one half of any walls dividing the demised premises from any adjoining premises forming part of the said Precinct TOGETHER with (a) the canopy in front of the ground floor of the demised premises.

Mr Sparrow submitted that on the true construction of the lease the outer skin of the perimeter walls of Wallasey Market does not form part of the premises. He submitted that it did not matter if the adjoining "premises" had a flank wall or not. He said that the lessor would have wanted control of the outer skin so that he could regulate C such matters as the placing of signs on the premises. He relied on the reference to the canopy in the schedule as supporting his construction. He submitted that the canopy must be a fixture to the outer skin and that it would have been unnecessary for the lease to have made express provision incorporating the canopy a part of the demise.

Mr Neuberger countered these arguments by contending that the reference to "any adjoining premises" was concerned with the case of a party-wall. Thus the area which is contiguous to the southern wall of the market hall consists of an open car park and there is no party-wall. In these circumstances Mr Neuberger submitted that the whole of the southern wall forms part of the demised premises. He D said that the reference to the canopy implicitly assumes that the outer skin of bricks forms part of the demise and that is why it was necessary for the lease to refer to the canopy only as being included.

In support of his submissions Mr Neuberger drew my attention to other provisions — namely clause 2(17) (tenant prohibited from exhibiting on the exterior of the walls), clause 2(32) (tenant prohibited from fitting any external blind to the demised premises) and clause 3(c)(iv) (tenant obliged to paint the exterior of the demised premises). These provisions presuppose that the exterior of the premises form part of the demise. Mr Neuberger submitted that "adjoining premises" simply meant "adjoining building" and it was only if there was an adjoining building that the outer skin of the boundary wall fell outside the demise; in all other cases the whole wall E formed part of the demise.

I prefer Mr Neuberger's submissions, and I agree with him that there can be no practical purpose for a wall dividing the market hall premises from the car park to be owned jointly between the lessee of the car park and the lessee of the building which houses the market hall.

Proposed works

P&O acquired Cherry Tree Centre in June 1989. It was in a dilapidated condition. Planning applications were made, consents were obtained and Bovis Construction Ltd were appointed management contractors for the redevelopment of the whole of the centre. The works, apart from the market hall, were designated as F phase 1 and were anticipated to take from February 1991 until mid-October 1991. In the event practical completion was achieved very close to target on October 21 1991. There were minor delays outside the control of the contractors. The only remaining works are those intended to be carried out to the market hall which are designated as phase 2.

If the phase 2 works are carried out, the market hall will cease to be self-contained. Under the new scheme a thoroughfare will be constructed through the market hall. This will be one of four main entrances to the centre and it will remain open when any of the shops in the centre are open.

There is no dispute between the parties as to the content of the phase 2 works, which can be summarised as follows:

(i) Most of the front (or north) brick-built wall is to be removed. It G will not be rebuilt. This will create access to a newly constructed cafeteria outside the demise and will have the practical effect of enlarging the market hall.

(ii) The whole of the interior, consisting of all the stalls, lavatories and storage areas together with the associated fixtures and fittings and all internal features, will be removed.

(iii) The existing entrance at the rear is to be blocked off. There will be a new entrance incorporating an enclosed lobby facing on to the car park. The new entrance will link up with the internal aisle creating a new main thoroughfare into the centre. One of the two new sets of double doors, which will be a feature of the new entrance, is to be erected within the demise. As part of this aspect of the works the H outer leaf or external skin of facing brickwork on part of the south wall will be removed and rebuilt.

(iv) The existing floor finish and base is to be completely removed and replaced with a new floor screed. All circulation areas are to be tiled in ceramic floor tiles.

(v) Sixteen new stalls will be erected within the market hall. They will be fully self-contained and secured. They will be brick-backed because they will be built adjacent to the existing perimeter walls. They will be constructed from metal-stud partitions and fitted with roller security shutters. It is intended that they will be constructed in different locations from the existing stalls.

(vi) New lavatories for the use of stall-holders, bin storage and access will be provided in a new location in a corner of the market J hall. Internal breeze-block dividing walls will be built in that corner area.

(vii) There will a new open-type ceiling for the hall with a solid-type ceiling above the lavatories and bin stores.

(viii) A sprinkler system will be installed and the existing electrical supply will be modified to meet current standards. Each stall will be individually metered. A fire alarm will also be installed.

As is apparent from the evidence of Mr Alan Webster [FRICS], a chartered quantity surveyor who was called as a witness on behalf of Graysim, the building in which the market hall is housed has a reinforced-concrete frame comprising structural columns and beams. It follows that the brick and block external cavity walls are in-fill between the structural columns and are not load-bearing. P&O K does not dispute this fact.

Mr Leland Jennings, a building surveyor who gave evidence on behalf of P&O, drew attention to the fact that in relation to the new entrance at what is currently the rear of the market hall two structural brick piers are to be erected which, in part at least, will be within the demise. The piers will carry steel beams which will support both the roof and fascia detail of the new lobby and the new external skin of brickwork above.

Legal test

Section 30(1)(*f*) contains the relevant ground of opposition to the grant of a new tenancy which is invoked by P&O. It provides: L

> that on the termination of the current tenancy the landlord intends to demolish or reconstruct the premises comprised in the holding or a substantial part of those premises or to carry out substantial work of construction on the holding or part thereof and that he could not reasonably do so without obtaining possession of the holding.

It is common ground that P&O intends to carry out these works and that it has the ability to do so.

The authorities establish that:

(i) "reconstruct" means "rebuild" and involves substantial interference with the structure: *Percy E Cadle & Co Ltd* v *Jacmarch Properties Ltd* [1957] 1 QB 323 and *Joel* v *Swaddle* [1957] 1 WLR 1094;

(ii) the correct approach to the question whether what is intended M will amount to reconstruction of a substantial part of the premises comprised in the holding is to look at the position as a whole and compare the effect on the premises of carrying out the proposed work with the condition and state of the premises before the work is done: see *Joel* v *Swaddle*; and

(iii) the demolition and/or construction need not be confined to load-bearing walls in order for the landlord's plans to qualify under subpara (*f*): *Bewlay (Tobacconists) Ltd* v *British Bata Shoe Co Ltd* [1959] 1 WLR 45.

In my judgment, it is abundantly clear that the ground specified in section 30(1)(*f*) is made out. P&O's plans could not be more extensive than they already are. The removal of part of the north wall is properly described as demolition work, and the same can be said of the work to be done to the outer skin of the south wall which, in addition, will have to be reconstructed. It is appropriate to describe the removal of the existing interior of the market hall as an exercise in demolition followed by a reconstruction of a substantial part of the premises. It is equally appropriate to describe all of these proposed works as the carrying out of substantial work of construction on the holding. These are very extensive works which will result in a brand new market hall.

It also seems to me to be quite unrealistic to suggest that these proposed works could be done while the market continued to function and without P&O obtaining legal possession of the holding. It is plain, in particular from the evidence of Mr Kelsall and Mr Jennings on behalf of P&O, that it would be impracticable for these proposed works to be carried out piecemeal. On their evidence, which I accept, it would be necessary for the market hall to be vacated before the work could be commenced. Clause 2(6) of the lease permits P&O, *inter alia*, to enter the demised premises for the purpose of carrying out repairs. If these proposed works could reasonably be characterised as "repairs" then it would be quite inappropriate for me to conclude that P&O required legal possession in order to do the works: see *Heath* v *Drown* [1973] AC 498. I am quite satisfied that these proposed works cannot be so characterised.

Section 31A of the 1954 Act

Graysim would be willing to have included in any new lease terms giving P&O "access and other facilities for carrying out the work intended". Graysim submits that, given access and facilities, P&O could reasonably carry out the work without obtaining possession of the holding and without interfering to a substantial extent or for a substantial time with the use of the holding for the purposes of Graysim's business. In dealing with this submission I must have regard to the physical effects of the work on the use of the holding for the purposes of Graysim's business. On the authorities I am not concerned with potential interference with the business itself or with the goodwill of the business: *Redfern* v *Reeves* (1978) 37 P&CR 364 and *Cerex Jewels Ltd* v *Peachey Property Corporation plc* (1986) 52 P&CR 127*.

Accordingly, the remaining question which I have to decide is one of fact and degree as to the time and extent of the interference during which the works are to be carried out. In *Cerex Jewels* (*supra*) and *Price* v *Esso Petroleum Co Ltd* (1980) 255 EG 243, periods of respectively two months and 16 weeks were held to fall within the statutory formula and the landlord was, in each case, entitled to possession.

So far as the facts here are concerned, P&O's evidence is that with vacant possession of the market hall the relevant works would take eight weeks. This estimate assumes that concurrent works outside the demised premises, eg outside the new entrance being constructed at the rear, would have no impact on the time expected to be taken over the work being done to the demised premises.

On behalf of Graysim, Mr Webster's evidence was that the work could be carried out during normal working hours in a reasonable five-week period, ie excluding overtime and weekend working. He went on to say that because this was retail space it had a high value, and that it was reasonable to anticipate weekend overtime which would reduce the overall contract period. In his opinion, the works should be completed in a maximum of four weeks.

On behalf of P&O, Mr Bangham, a quantity surveyor, gave evidence to the effect that significant overtime and weekend working could increase P&O's costs by as much as £33,000. He suggested that this would be an unreasonable additional cost and that it would be an uneconomical way of proceeding.

The other relevant area of difference between the witnesses related to the rescreeding of the floor. If successful in these proceedings, P&O has undertaken to rescreed the whole floor and not, as one of its plans suggests, only parts of the floor. This will require a seven-day curing process during which time it will not be possible to proceed with other works within the market hall. Graysim took the position that the rescreeding work could be done in portions so as not to inhibit other work which could be done at the same time.

I accept P&O's evidence on these matters. P&O's time estimate for phase 1 proved to be extremely accurate, and that seems to me to be a good indication of the likelihood of its estimate for phase 2 being accurate. I do not believe that P&O should be obliged to incur significant additional costs in the form of overtime and weekend working. I think this would be an unreasonable imposition. Moreover, it is not appropriate for me to conclude that there is or might be some other way in which the rescreeding works could be done. I must proceed on the basis put forward by P&O in its scheme of works: *Decca Navigator Co Ltd* v *Greater London Council* [1974] 1 WLR 748.

In these circumstances, my conclusion is that the work which it is proposed will be done to the premises will in all probability take eight weeks. During this time it will not be possible for Graysim to make any use whatever of the holding. In my view, this would in the circumstances constitute an interference to a substantial extent and for a substantial time with the use of the holding for the purposes of the business carried on by Graysim, and accordingly, Graysim is not entitled to the protection afforded by section 31A of the 1954 Act.

The answer to question (iii) is that P&0 is entitled to oppose Graysim's application for a new tenancy on the ground specified in section 30(1)(*f*) of the 1954 Act.

Declaration accordingly.

* Editor's note: Also reported at [1986] 2 EGLR 65.

Court of Appeal

October 30 1992

(Before Lord Justice DILLON, Lord Justice STEYN and Lord Justice EVANS)

SALOMON v AKIENS

Estates Gazette April 10 1993

[1993] 14 EG 97

Landlord and Tenant Act 1954 — Negotiations for new lease "subject to contract" and "subject to lease" — Tenant failing to apply to court — Whether landlord estopped from denying tenant's entitlement to new lease

The defendant held a lease of the first-floor premises at 439 Welford Road, Leicester, by assignment, the contractual term of which expired on September 29 1986; the defendant remained in possession under a tenancy continued by the 1954 Act. On December 21 1990 the plaintiff landlord served a notice under section 25 of the Landlord and Tenant Act 1954 terminating the tenancy on July 1 1991, and stating that the grant of a new tenancy would be opposed on the ground of persistent delay in paying rent. The defendant served a counternotice on January 9 1991. The parties continued to negotiate the terms of a new lease. By a letter marked "subject to lease" and dated March 22 1991 the defendant's solicitors stated that the defendant was prepared to enter into a new lease on certain terms; the plaintiff's solicitors accepted by a letter marked "subject to lease" and dated April 15. No further steps were taken until May 15 when the plaintiff's solicitors enclosed the counterpart lease for signature; the signed counterpart was not returned until June 24 1991. The plaintiff refused to complete because the defendant's cheque for the completion moneys was not accepted. The defendant's solicitors failed to make any application to the court for a new tenancy. In proceedings by the plaintiff for possession of the premises, the defendant contended that the plaintiff was estopped by her solicitors' correspondence from contending that the defendant was not entitled to a new lease, alternatively that the forbearance to issue an application to the court under the 1954 Act was consideration for a collateral contract that a lease in the terms of the final draft would be adopted. In Leicester County Court Judge Pitchers made an order for possession of the premises. The defendant appealed.

***Held*: The appeal was dismissed. There was nothing in the case to justify a departure from the well-established rule as to the meaning of "subject to lease" that either party may withdraw before exchange of lease and counterpart. There had never been a physical exchange of lease and counterpart;**

A **that indispensable condition for the formation of a contract was never satisfied. Even if the letters had not had the protective words, there was no promise by the landlord to grant a lease in consideration of the tenant's forbearance in not applying for a new tenancy. The tenant could not be said to rely on any representation by the landlord when all such representations were marked "subject to lease".**

***Per* Evans LJ dissenting: The effect of the exchange of correspondence is that the defendant gave up his right to apply for a new tenancy. There was a representation that the plaintiff no longer objected to the grant of a new tenancy in the** B **terms agreed. It was sufficient to show a representation merely that the defendant's statutory right to a new tenancy was acknowledged. Accordingly "subject to lease" would not allow a party to go back on an assumption where it would be unjust or unfair to allow him to do so.**

The following cases are referred to in this report.

Amalgamated Investment & Property Co Ltd (in liquidation) v *Texas Commerce International Bank Ltd* [1982] QB 84; [1981] 3 WLR 565; [1981] 3 All ER 577; [1981] Com LR 236, CA

Attorney-General of Hong Kong v *Humphreys Estate (Queen's Gardens) Ltd* [1987] 1 AC 114; [1987] 2 WLR 343; [1987] 2 All ER 387, PC

C *Derby & Co Ltd* v *ITC Pension Trust Ltd* [1977] 2 All ER 890; (1977) 245 EG 569

Hughes v *Metropolitan Railway Co* (1877) 2 App Cas 439

Longman v *Viscount Chelsea* (1989) 58 P&CR 189; [1989] 2 EGLR 242, CA

This was an appeal from a decision of Judge Pitchers in Leicester County Court, who had made an order for possession of premises at 439 Welford Road, Leicester, in an action by the plaintiff, Helen Salomon, for possession of the premises against the defendant, Stephen Paul Akiens.

Alison Hampton (instructed by Dixon Coles & Goddard, of Leicester) appeared for the appellant; David Pittaway (instructed by Josiah Hincks Son & Bullough, of Leicester) represented the respondent.

D

Giving the first judgment, **Dillon LJ** said: This is an appeal by the defendant in the action against a judgment of Judge Pitchers, given at the trial of the action in Leicester County Court on November 29 1991. The decision and order of the judge was that the plaintiff was entitled to possession of premises known as the first floor, 439 Welford Road, Leicester.

These were business premises within the meaning of Part II of the Landlord and Tenant Act 1954, as amended. The plaintiff was the landlord and the defendant had been the tenant by assignment. It appears that there had been a lease granted in December 1982 to a Mr Henry, who was the father-in-law of the defendant. That lease expired on September 29 1986, but the tenant remained in possession E and the tenancy was then transferred to the defendant, who was his son-in-law, and the defendant carried on business in those premises under the name of West End Taxis, a taxi business.

The landlord and tenant, through representatives, ultimately began negotiating terms for a new lease well before any question of the termination of the business tenancy, continued under the Act, had arisen. In those negotiations the documents were marked, first, "subject to contract", then later "subject to contract/lease" and then "subject to lease".

On December 21 1990 the plaintiff served a notice under the 1954 Act to terminate the lease, stating that there would be objection to the grant of a new tenancy on the ground of delay by the tenant in paying rent. The defendant duly served a counternotice on January 9 1991, F giving formal notice that he was not willing to give up possession of the property and required a new lease. This was given by his solicitors on his behalf.

It was then the defendant's task, if he wished to obtain a new lease from the court, to apply to the court for the grant of a new lease and, under the terms of the Act, the application had to be made not less than two nor more than four months after the date of service of the landlord's notice. That gave April 21 1991 as the final date, according to the timetable in the statute, for the application to the court. In the meantime the parties continued to negotiate the terms of a new lease.

On March 22 1991 the plaintiff's solicitors wrote the defendant's solicitors a letter headed "subject to lease" and, for good measure, "without prejudice" as follows:

> We are shortly coming to the time when we shall have to issue court G proceedings under The Landlord and Tenant Act to protect our client's position. Our client fears that this matter of the new lease has dragged on for far too long, which is a view you no doubt share. He is therefore prepared to sign a new Lease in the terms proposed including the new repairing obligations. The existing Counterpart Lease we hold

— that, I think, refers to the circulating draft —

> is for a period of 12 years commencing on the 29th September 1989 at a rent of £1,650 per annum. Our client is currently paying rent at the rate of £1,200 per annum and he would propose that the new 12 year period commence on the next quarter day, being 25th of this month when the new rent will become payable. This would seem a reasonable compromise as your client will have a lease drawn in the terms that she requires and also be receiving the increased rent somewhat earlier than under the terms of The Landlord and Tenant Act H notice provided of course a new Lease was granted. Our client will have the benefit of a slightly longer term if the 12 years run from 25th March next.

That was acknowledged by the plaintiff's solicitors who merely said that they were taking their client's instructions and would contact the defendant's solicitors in due course.

On April 9 the defendant's solicitors wrote again, and again marking the letter "subject to lease". They said:

> In order to protect our client's position we shall have to issue a summons under the Landlord and Tenant Act unless we hear from you by Monday, 13th April. We hope that this will not be necessary.

The reply to that from the plaintiff's solicitors, again marked "subject to lease", is dated April 15: J

> . . . we confirm that our client is prepared to accept your client's proposals to enable this matter to proceed to a prompt completion.
>
> Please return the travelling draft Lease to us duly approved so that we can arrange to engross same.
>
> Please note that we would like to complete this matter before the end of April.

That was April 15 and the defendant's solicitors replied on April 17:

> We thank you for your letter of the 15th April which arrived just as we were about to issue the Landlord and Tenant Act proceedings. On the basis of this agreement we now return the travelling draft amended in yellow.

I take the amendments to give effect to what had been put forward in the letter of March 22. That letter of April 17 was also marked by the K defendant's solicitors "subject to lease". Nothing further happened in fact until May 15, except of course that the time-limit of April 21 passed without any application to the court having been made. That however was not irrevocable because it has been held that the statutory time-limits are procedural rather than mandatory and are capable of being waived, except that any application has to be made before the existing tenancy, as continued by Part II of the Act, comes to an end because the court can have no jurisdiction to entertain an application under the Act once the landlord and tenant relationship has come to an end. The terminal date for the tenancy to expire, having regard to the notice under the Act which had been given on December 21 1990, was July 1 1991. L

The plaintiff's solicitors had said in their letter of April 15 1991, which I have read, that they would like to complete the matter before the end of April. That could, I apprehend, have been achieved with a certain amount of expedition, but I cannot suppose that anyone would have thought, around April 15 and 17, that it would not have been completed before July 1.

Be that as it may, the reply to the letter of April 17 did not come from the plaintiff's solicitors until May 15 1991. Their letter of that date is also marked, as the earlier letters were, "subject to lease". They say that they enclose the counterpart lease for signature by the defendant and apologise for the delay in forwarding the same. They also say:

> We . . . enclose a Completion Statement. Can we suggest that you telephone M the writer once your client has signed the Lease to discuss completion.

The completion statement provided for the payment to the plaintiff or her solicitors of the solicitors' charges in respect of the grant of the lease and VAT thereon — these the defendant had already agreed in the course of correspondence that he would pay on the grant of any lease — and also the difference between the new rent and the old rent under the previous tenancy, which would be payable under the new lease on the March quarter day of 1991.

There was no reply of relevance to the letter of May 15, to which I have referred, until June. On June 5 the plaintiff's solicitors wrote, and again "subject to lease", and stated that they held their "client's part of the lease duly signed and can see no reason for the delay in

A completing this matter. Please contact the writer by return to discuss completion". Nothing happened until June 24.

The reason for the delay at this stage, we have been told, was not so much inactivity on the part of the solicitors, but that the defendant was having matrimonial troubles, which raised difficulties in his use of his bank account. In any event, on June 24 the defendant's solicitors wrote to the plaintiff's solicitors enclosing the counterpart lease duly signed and they said they had sent their client a copy of the completion statement and would forward his cheque as soon as they received it.

That was followed by a letter of June 26, sent by fax, which seems to have been sent without regard to the letter of June 24, which was received by the plaintiff's solicitors on June 26. The plaintiff's
B solicitors' letter of June 26 reminded the defendant's solicitors that the defendant's lease expired under the 1954 Act notice on July 1 , and said that, if the new lease is not completed by June 28, they would advise their client "to terminate the Lease on the 1st July and . . . re-let the property on the open market".

There was an acknowledgment of that saying that the executed counterpart notice should have been received. That was on June 27. On June 28 there is another letter form the plaintiff's solicitors to the defendant's solicitors, "subject to lease", stating that they were not able to complete until they were in receipt of the moneys detailed in the completion statement of May 17 and advising that the plaintiff had received no cheque for the June quarter-day payment of rent and
C was requiring therefore the larger sum of £1,171.25 in all upon completion. That did not in fact reach the defendant's solicitors until after the suggested completion day of June 28.

There were further letters on July 1, 2 and 3, but cheque for £1,171.25 sent on July 3 was not accepted, and on July 7 notice was given by the plaintiff's solicitors of their client's termination of the licence under which he was holding because of the tenancy having expired on July 1.

In those circumstances, the primary case for the appellant is that, as a result of the correspondence in April, in particular the letters of April 9, 15 and 17, the plaintiff is estopped from contending that the defendant is not entitled to a new lease of the premises in the terms of the engrossment of the finally agreed draft. Alternatively, it is said
D that the forbearance to issue the application to the court under the 1954 Act on or before April 21 was consideration for a collateral contract that a lease in the terms of that final draft as engrossed would be adopted.

That depends on the meaning of the words, "subject to lease", which were consistently used by both parties in all the documents in the course of their correspondence before and after the service of the notice terminating the tenancy under Part II of the 1954 Act. The meaning of those words has, to my mind, been made abundantly clear in a large number of cases. There is, for instance, the decision of this court in *Longman* v *Viscount Chelsea* (1989) 58 P&CR 189*. There in the opening paragraph of his judgment at p190, Nourse LJ
E said:

It is axiomatic amongst conveyancers that negotiations for the sale of land which are expressed to be "subject to contract" cannot mature into a concluded agreement unless and until there is an exchange of contracts in accordance with ordinary conveyancing practice, before which either party can withdraw; see *Eccles* v *Bryant and Pollock* [[1948] Ch 93]. Broadly stated, the question in the present case is whether an analogous rule applies to negotiations for the grant of a lease which are expressed to be "subject to the completion of a lease", so that either party can withdraw before there has been an exchange of lease and counterpart. So stated, the question would appear to readily susceptible of an affirmative answer.

To the same effect is the decision of the Privy Council in *Attorney-General of Hong Kong* v *Humphreys Estate (Queen's Gardens) Ltd*
F [1987] AC 114 especially at pp127-128, in the opinion delivered by Lord Templeman. To the same effect is the judgment of Oliver J, as he then was, in *Derby & Co Ltd* v *ITC Pension Trust Ltd* (1977) 245 EG 569, especially at the foot of p572.

Is there anything in this case to justify a departure from the well-established rule, which indeed in *Longman* v *Viscount Chelsea* Nourse LJ at p193 held ought to be more highly regarded than the merits of individual cases? To my mind there clearly is not. The parties were negotiating throughout on the basis that neither would be bound until the terms of the lease had been embodied in formal lease and counterpart and these had been exchanged in the usual way. They had been negotiating on the basis that accepted, as each of the solicitors must have known, that either could withdraw from the G negotiations at any point before the exchange had taken place. There was no change in the stated basis of the negotiations in the subsequent documents. Moreover, the passing of the date, April 21 1991, as I have mentioned, was not irrevocable and there was no reason for supposing at that stage that it would not be possible to complete the exchange of lease and counterpart on the terms of the circulating draft speedily. It could not automatically be assumed that there would be no point taken on the final yellow amendment, to which I have referred, in the April letter that I have read.

I see nothing, therefore, in this case to distinguish it from the ordinary case of a negotiation subject to lease. It is not enough that they were negotiating against the background of a right, which could H not be indefinitely deferred, on the part of the defendant to apply to the court for the grant of a new lease to which, if he could overcome such objection as was taken on the ground of delay in payment of rent, he would be entitled. That was a negotiating factor no doubt of value. But I do not find it enough to warrant giving the words "subject to lease" in this case a different meaning from their ordinary and well-accepted meaning. I would therefore dismiss this appeal.

Agreeing, **Steyn LJ** said: Three issues arise. I will deal with them in logical order.

In a skeleton argument lodged on behalf of the appellant, the first issue raised was whether a contract in writing for the grant of a new J lease pursuant to section 2 of the Law of Property (Miscellaneous Provisions) Act 1989 came into existence. *Prima facie* the contract would only come into existence upon the physical exchange of a duly executed lease and counterpart: see *Longman* v *Viscount Chelsea* (1989) 58 P&CR 189.

The question is not whether there was a meeting of minds or a concurrence of an offer and acceptance. The question is whether an indispensable condition to the coming into existence of a contract had been fulfilled. Plainly it had not been fulfilled. There had been no physical exchange of an executed lease and counterpart of the lease. It was of course open to the parties to agree on a constructive form of exchange such as, for example, an arrangement that the contract K would come into existence when the solicitors confirm telephonically or by letter that their respective clients had signed the lease and the counterpart of the lease: see *Longman* v *Viscount Chelsea* (*supra*). Plainly there was no express agreement providing for a constructive exchange. Moreover, counsel did not suggest that there was anything in the correspondence or conduct of the parties which could conceivably warrant the implication of a term allowing a constructive exchange.

In these circumstances both sides were entitled to withdraw from the negotiations at any stage. The first argument, which counsel did not support by oral argument, must therefore be rejected.

The second issue was whether the forbearance of the tenant from issuing an application for a new tenancy under section 24 of the 1954 L Act, in consideration for the "promise" by the landlord that she would grant a new lease, created a collateral agreement pursuant to which the parties agreed to execute a new lease. Even if none of the contemporaneous exchanges had been marked "subject to lease", I would have taken the view that there was no such "promise" by the landlord. But the "subject to lease" stipulation makes it clear beyond any doubt that neither party was intended to be bound contractually to the other in any way until a formal lease had come into existence. It is therefore impossible to construct a collateral contract from the exchanges of the parties.

The third issue is whether the landlord is estopped from refusing to complete or acknowledge the validity of the lease of the premises in the form of the counterparts of the lease signed by both parties. Put in M this way the argument again founders on the "subject to lease" stipulation. The tenant cannot be heard to say that there was a representation that there was in existence a binding agreement in circumstance where every relevant communication was expressed to be "subject to lease", thereby reflecting a clear and communicated understanding that there would be no binding contract until a duly executed lease and counterpart had been exchanged: see *Derby & Co Ltd* v *ITC Pension Trust Ltd* (1977) 245 EG 569 at p572 (col 2). That only means that the particular estoppel put forward by counsel in this case is defeated. It does not mean that the stipulation "subject to lease" would by itself necessarily eliminate the operation of any estoppel. If the landlord had, for example, expressly represented in

* Editor's note: Also reported at [1989] 2 EGLR 242.

correspondence that she would not rely on the lapse of time while the parties are in negotiation and without giving reasonable notice of her intention to do so, an estoppel could have arisen. The landlord's strict legal rights could have been suspended; see *Hughes* v *Metropolitan Railway Co* (1877) 2 App Cas 439. And, in such a case, the stipulation "subject to lease" would not preclude the operation of an estoppel. It can also readily be accepted that the very fact of negotiations can sometimes give rise to an implied estoppel. That happened in *Hughes* v *Metropolitan Railway*. In that case the landlord gave his tenant notice requiring him to do repairs within six months. During that period the landlord negotiated with the tenant to buy his lease. Understandably, the tenant did not do the repairs. When the negotiations eventually broke down, the landlord tried to rely on his notice and tried to forfeit the lease. It was held that the landlord had led the tenant to believe that he would not enforce his notice without giving the tenant reasonable time to do the repairs.

Relying on this analogy, can it now be said that in the present case the tenant's solicitor was led to believe by an implied representation that the landlord would not rely on time-limits until he had given the tenant reasonable notice of his intention to rely on his strict legal rights? I fear the answer to that is "no".

Miss Hampton told us that the solicitor thought that he had negotiated a binding agreement by April 17, that is, a few days before the lapse of the statutory period for the lodging of the application for a new tenancy. That explanation is difficult to reconcile with the correspondence, including the letters of April 17 and June 24, written by the tenant's solicitor in which the words "subject to lease" appear boldly in capital letters. But I accept the solicitor's explanation as tendered by counsel. I am satisfied that the solicitor was negligent in coming to such an astonishing conclusion. As Oliver LJ (as he then was) said in the *Derby* case, at p572 (col 2), everybody knows that where parties negotiate subject to contract, there is a risk that at the end of the day either side may back out of the negotiations, up to the point where leases are exchanged. If the solicitor's conduct is judged by the objective standard of what a reasonable solicitor would have done, it is, in my judgment, a plain case where the solicitor ought to have lodged a protective application for a new tenancy within the statutory period. The cost of such an application is £30. Nothing had been said or done which could reasonably have encouraged him to conclude that the landlord was content to treat the strict time-limit as in suspense. The judge said that the solicitor was "walking towards a precipice over which he would fall" unless he took action. To that I would only add that the misfortune which befell the tenant was one entirely of the solicitor's making. In oral argument I raised with the appellant's counsel the impact of the solicitor's negligence on the alleged estoppel. In my judgment, the solicitor's negligence is inconsistent with the hypothesis that he was misled by an implied representation coming from the landlord's side. Clearly, this is not a case where the tenant's solicitor thought the strict time-limit was suspended. On analysis the analogy with *Hughes* v *Metropolitan Railway* breaks down. It follows that, in my judgment, no estoppel arises.

For these reasons, and despite the attractive argument of Miss Hampton, I would dismiss the appeal.

Dissenting, **Evans LJ** said: I have the misfortune to differ from my lords. In the circumstance I will give my reasons briefly. The essential question is whether the plaintiff was estopped from April 21 onwards from denying that the defendant was entitled to a new lease; conversely, that the defendant was permitted to stay in occupation on the premises beyond July 1 when the statutory tenancy expired.

My lord has referred to the relevan⁺ correspondence, which consists solely of the letters dated March 22 through to April 17 1991. The effect of that exchange of correspondence was that the defendant gave up his right to apply for a new tenancy within the statutory time-limit, and he did so with the full knowledge of the plaintiff that he was acting on the basis that the lease terms were agreed. Unfortunately, those letters and subsequent letters continued to be headed "subject to lease".

There was no question of extending the time for an application. There was no application because the plaintiff agreed that no application was necessary. The possibility of applying for an extension was not in anyone's mind. In fact, neither party contemplated failure to agree, precisely because the terms were agreed; unless, that is, the words "subject to lease" meant that they were not agreed so as to bind the parties. But that could not contradict what both parties said and what the defendant did by refraining from making the statutory application which he otherwise would have made.

There is some authority for the proposition that the words, "subject to contract" and, by analogy, "subject to lease" may be disregarded in certain circumstances. In the present case it seems to me that they should probably not be disregarded altogether. It was clearly intended that the terms which had been agreed should be carried forward into a formal document or documents.

Thereafter both parties proceeded on the basis that a lease would be created by agreement until, that is, the very end of June when the plaintiff imposed a deadline, at first June 28 and later July 1. In fact, as late as June 28 her solicitor demanded payment of rent due under the terms which the defendant's solicitors correctly regarded as having been agreed, subject only to the effect of the words "subject to lease". Then at that late stage the plaintiff asserted her right to withdraw on the ground that no binding contract had been entered into — that is something which may be assumed to be correct — and on the further ground that she was free to withdraw, notwithstanding the "agreement" — and I put that word in quotation marks because it appears in the defendant solicitor's letter dated April 17 — reached in April upon which the defendant had clearly and expressly relied, to his very great detriment if the plaintiff is correct.

The advice of the Board of the Judicial Committee of the Privy Council in *Attorney-General of Hong Kong* v *Humphreys Estate (Queen's Garden's) Ltd* [1987] 1 AC 114 is clear authority that no estoppel can arise except possibly in exceptional circumstance from "subject to contract" negotiations. Both parties to such negotiations are free to withdraw. Both parties accept the risk that the other party will do so. In the course of the Board's advice, Lord Templeman significantly, in my judgment, adopted the statement of the principle by Lord Denning MR in *Amalgamated Investment & Property Co Ltd (In Liquidation)* v *Texas Commerce International Bank Ltd* [1982] QB 84 at p122. Lord Templeman said:

> Lord Denning MR [in that passage] reduced the doctrine of estoppel: ". . . into one general principle shorn of limitations. When the parties to a transaction proceed on the basis of an underlying assumption — either of fact or of law — whether due to misrepresentation or mistake makes no difference — on which they have conducted the dealings between them — neither of them will be allowed to go back on that assumption when it would be unfair or unjust to allow him to do so."

In the course of Lord Templeman's judgment he refers to the judgment of Woolf J, as he then was, in what I will call the *Salvation Army* case* (1980) 41 P&CR 179. Suffice it for me to read from p127 of Lord Templeman's judgment the following passages:

> But the fact that the arrangement was "subject to contract" was irrelevant. The Salvation Army did not enter on the new site and build the new hall because they hoped that an arrangement "subject to contract" would eventually result in a concluded binding agreement. The Salvation Army took possession of the new site and built the new hall in consequence of the expressed intention on the part of the city council to acquire the old site for a purpose for which the city council held compulsory acquisition powers.

In this case, the defendant gave up his statutory right to obtain the grant of a new tenancy. It was not a case where both parties were conducting what is called, in another of the cases to which we have been referred, a normal "subject to contract" negotiation. It was a case where the defendant could obtain a new lease from the court if his negotiations with the plaintiff failed. The defendant gave up that right when the plaintiff agreed the terms which were to be embodied in a formal document. The plaintiff now says that, despite that agreement, she became free from April 21 to behave as if the defendant's right no longer existed. If the plaintiff is correct, the agreement was worth nothing to the defendant, and all this because the words "subject to lease", continue to be used in the letter headings, words which were capable of referring to the continued need, which both parties recognise and accepted, to prepare formal documents. On the other hand, if the plaintiff is correct, the agreement is worth a great deal to her because the defendant had been led to believe that a court application was not necessary in order to preserve his statutory rights, and his solicitors had said as much in the correspondence.

In my judgment these facts, on any view, come within the principle stated by Lord Denning and approved by Lord Templeman and the Board in the passage already sited. If it is useful to define the representation which, in my judgment, was clearly made, it was that the plaintiff no longer objected to the grant of a new tenancy in the

* Editor's note: *Salvation Army Trustee Co Ltd* v *West Yorkshire Metropolitan County Council.*

terms which had been agreed. It followed that the defendant could continue in possession beyond July 1 when his statutory right to possession would otherwise expire. The defendant need not show, in my judgment, a representation that there was a lease; merely that his statutory right to obtain one was acknowledged.

I append two remarks. It is not necessary, in my judgment, to consider what the position would have been if the terms of the lease had not been agreed before April 21, nor if the defendant had unreasonably failed to proceed to completion of the lease. That is a contention which the plaintiff does not raise.

Finally, if the views I have expressed are correct, it does not follow that the defendant was nevertheless free to withdraw after April 21, as Mr Pittaway in the course of his submissions has suggested. That is not necessarily a consequence of the view which, in my judgment, should be adopted here. The question of negligence by the defendant's solicitors has not been argued before us and, needless to say, I express no view whatsoever on that particular matter.

Appeal dismissed with costs.

Court of Appeal

November 24 1992

(Before Lord Justice GLIDEWELL, Lord Justice ROSE and Sir Christopher SLADE)

METHODIST SECONDARY SCHOOLS TRUST DEED TRUSTEES v O'LEARY

Estates Gazette April 24 1993

[1993] 16 EG 119

Landlord and tenant — Landlord and Tenant Act 1954 — Dwelling-house let to school trustees and occupied by school caretaker — Whether school trustees occupying dwelling-house for purposes of a business under Part II of the 1954 Act — Whether if school trustees carrying on a business in the dwelling-house they were doing so in breach of a prohibition against use for business purposes — Whether landlord consented or acquiesced in breach

The appellants are the trustees of the Methodist Secondary Schools Trust Deed 1903 and ran Kent College, a school in Canterbury. By a lease dated May 24 1974 the respondent landlord demised a dwelling-house, Greengates, Whitstable Road, Canterbury, Kent, a property immediately adjacent to the boundary of Kent College, to the trustees for a term of 15 years from June 1 1974 at an initial annual rent of £350 with provisions for rent reviews. By clause 2(X) of the lease the tenants covenanted "to use the demised premises for the purposes of a private residence in single occupation only". On July 20 1977 a Mr Lester was offered employment as a resident caretaker by the trustees and for the proper performance of his duties, he was required to live at the property on a service occupancy. Mr Lester's duties included building security at night and during holidays; he was provided with a telephone in the dwelling-house and the trustees installed an alarm system connected to the school buildings. On October 13 1988 the respondent gave notice to the trustees under section 25 of the Landlord and Tenant Act 1954 terminating the tenancy on June 1 1989. On October 26 1988 the trustees' solicitors served a counternotice under section 29 of the Act stating that they would not be willing to give up possession and on February 7 1990 the trustees issued an application in the country court for the grant of a new tenancy under Part II of the 1954 Act. At the trial of a preliminary issue, Judge James Scarlett, sitting in Canterbury County Court, found that the board of management set up the trust deed of 1903 and the local administrative governers (standing in the shoes of the trustees) were in occupation of the property at all material times by the occupation of Mr Lester, and that the occupation was partly for the purpose of the business carried on by the board of management and local administrative governers, and for other purposes. The occupation of the dwelling-house for that purpose was in breach of clause 2(X) of the lease; the respondent neither consented to nor acquiesced in the use of the dwelling-house for such purposes. Accordingly, the tenancy was not one to which Part II of the 1954 Act applies. The trustees appealed on the grounds that there was no breach of clause 2(X) of the lease, but if there had been a breach the respondent had acquiesced to it.

***Held*: The appeal was dismissed. The trustees were in breach of the covenant to use the demised premises for the purpose of a private residence in single occupation only. The insuperable obstacle facing the trustees was the judge's finding that "the property was used for the purposes of rendering Mr Lester available to hear the alarm and to receive messages and from the nearest point available to the school to react promptly". Mr Lester was using the property partly for the purpose of carrying out duties of his employment as caretaker, which had to be performed *in the property*, that is to say by continuously being on call and making himself available. The alarm and the telephone being mechanisms available to enable him to receive and respond to the calls. Accordingly, there was a breach of clause 2(X) of the lease**

Knowledge is a prerequisite of acquiescence. The respondent could not be said to have acquiesced in the continuing breach of clause 2(X) of the lease *unless he knew the facts which gave rise to the breach*. The fact which gave rise to the breach was the continuing use of the property by the school's caretaker, in part, for the purpose of being available on call during the ordinary non-working hours. The judge had found that the respondent thought that Mr Lester was one of the school's maintenance staff and that he did not know of the use to which the premises were in part being put. On those findings of fact the respondent had not acquiesced in the breach. Section 41 of the 1954 Act did not apply to make the trustees occupiers of the dwelling-house. It was the board of management, through Mr Lester, who were in occupation of the property. It was unlikely that the trustees could point to a sufficient degree of control over the property. The interposition of the board of management, which has the power to give directions to the trustees, renders this a special case. Accordingly, the trustees had not established that they were in occupation of the property for the purposes of section 23 of the 1954 Act.

The following cases are referred to in this report.

Bell v *Alfred Franks & Bartlett Co Ltd* [1980] 1 WLR 340; [1980] 1 All ER 356; (1979) 39 P&CR 591; 253 EG 903, CA
Chapman v *Freeman* [1978] 1 WLR 1298; [1978] 3 All ER 878, CA
Cheryl Investments Ltd v *Saldanha* [1978] 1 WLR 1329; [1979] 1 All ER 5; (1978) 37 P&CR 349; 248 EG 591, CA
Commissioner of Valuation for Northern Ireland v *Fermanagh Protestant Board of Education* [1969] 1 WLR 1708; [1969] 3 All ER 352; [1969] RA 475, HL
Frish Ltd v *Barclays Bank Ltd* [1955] 2 QB 541; [1955] 3 WLR 439; [1954] 1 All ER 901, CA
German v *Chapman* (1877) 7 ChD 271; 47 LJCh 250; 37 LT 685; 42 JP 358; 26 WR 149, CA
Groveside Properties Ltd v *Westminster Medical School* (1983) 47 P&CR 507; 267 EG 593, CA
Linden v *Department of Health and Social Security* [1986] 1 WLR 164; [1986] 1 All ER 691; (1985) 51 P&CR 317; [1986] 1 EGLR 108; 277 EG 543

This was an appeal by the trustees of the Methodist Secondary Schools Trust Deed from the judgment of Judge James Scarlett, sitting at Canterbury County Court, on June 12 1991, who had decided a preliminary issue as to whether the trustees held a tenancy within the meaning of Part II of the Landlord and Tenant Act 1954 from the respondent, Mr H D O'Leary.

Jonathan Ferris (instructed by Pothecary & Barratt) appeared for the appellants; Robert Jay (instructed by Furley Page Fielding & Pembrook, of Canterbury) represented the respondent.

Giving judgment, **SIR CHRISTOPHER SLADE** said: This is an appeal by the trustees of the Methodist Secondary Schools Trust Deed ("the trustees") from a judgment of Judge James Scarlett, at Canterbury County Court, delivered on June 12 1991, after the hearing of a preliminary issue in a landlord and tenant dispute. The appeal has been well argued on both sides.

The trustees are tenants of a dwelling-house known as "Greengates", Whitstable Road, Canterbury, Kent ("the property"). The caretaker of a school, Kent College, Canterbury, lives in the property. The respondent, Mr O'Leary, is the landlord. On October 13 1988, the landlord's solicitors gave notice to the trustees under section 25 of the Landlord and Tenant Act 1954 ("the Act") terminating the tenancy on June 1 1989. Para 5 of the notice stated:

> If you apply to the Court under Part II of the Landlord and Tenant Act 1954 for the grant of a new tenancy, we will not oppose your application.

At the trial of the preliminary issue, counsel for the trustees accepted that this statement did not estop the landlord from contending that Part II of the Act did not apply to the tenancy.

On October 26 1988 the trustees' solicitors served a counternotice under section 29 of the Act stating that they would not be willing to give up possession of the property on June 1 1989.

On February 7 1990 the trustees issued an application to the county court for the grant of a new tenancy under Part II of the Act. Para 4 of the application, as subsequently amended by leave at the trial, describes the nature of the business carried on at the property as being "residential accommodation for school caretaker".

Section 23 of the Act, so far as material, provides as follows:

> (1) Subject to the provisions of this Act, this Part of this Act applies to any tenancy where the property comprised in the tenancy is or includes premises which are occupied by the tenant and are so occupied for the purposes of a business carried on by him or for those and other purposes.
>
> (2) In this Part of this Act the expression "business" includes a trade, profession or employment and any activity carried on by a body of persons, whether corporate or unincorporate.
>
> . . .
>
> (4) Where the tenant is carrying on a business, in all or any part of the property comprised in a tenancy, in breach of a prohibition (however expressed) of use for business purposes which subsists under the terms of the tenancy and extends to the whole of that property, this Part of this Act shall not apply to the tenancy unless the immediate landlord or his predecessor in title has consented to the breach or the immediate landlord has acquiesced therein . . .

On October 19 1990, the landlord filed an answer to the trustees' application in which he stated that he opposed the grant of a new tenancy on the grounds that the tenancy to which the application related was not one to which Part II of the Act applied. Particulars of these grounds, drafted with section 23 of the Act in mind, were given as follows:

> (a) The Lease dated 24th May 1954 is a lease whereby the Respondent demised a dwelling-house to the Applicants and under which the permitted use of the Premises was in Clause 2(X) provided to be "for the purposes of a residence in single occupation only".
>
> (b) The Premises are neither occupied by the Applicants nor are they occupied for the purposes of a business carried on by them or for those and other purposes pursuant to section 23(1) of the Act.
>
> (c) The Premises are occupied by a caretaker employed by Kent College, Canterbury (a school).
>
> (d) If, which is denied, the Premises are occupied by the Applicants for the purposes of their business, the said use for business purposes is in breach of a prohibition of use for business purposes under the said Clause 2(X) of the Lease.

Mr Registrar Nicholson, by consent, on December 12 1990 ordered that the landlord's allegation that the tenancy was not one to which Part II of the Act applied, should be dealt with as a preliminary issue.

In the light of the landlord's answer, and of a further point raised by the trustees relating to acquiescence, the judge at the trial correctly identified four questions as requiring an answer:

(1) Was the property occupied by the trustees?

(2) If it was occupied by the trustees, was it occupied by them for the purposes of a business carried on by them or for those and other purposes?

(3) If the trustees were carrying on business in all or any part of the property comprised in the tenancy, were they doing so in breach of a prohibition however expressed for use for business purposes which subsisted under the terms of the tenancy and extended to the whole of the property?

(4) If there was such breach, has the landlord consented or acquiesced in it?

Before turning to these questions, I must summarise the relevant facts as found by the judge in his full and careful judgment.

The facts

The trustees are the trustees of the Methodist Secondary Education Trust Deed 1903. The prefatory note to this trust deed shows that Kent College, Canterbury, is one of the Methodist schools held upon the trusts set out in the deed and that the "Board of Management for Methodist Residential Schools" is the legal governing body of the school, but that, with certain exceptions, it has delegated to a body of "local administrative governors" all matters concerning the school. These matters include the appointment of staff other than the headmaster.

The recitals to the trust deed stated (*inter alia*) that the Methodist Conference proposed to apply certain moneys to the establishment of secondary schools and annually to appoint a board of management for the control and management of such schools and "to confer and impose upon such Board divers powers and duties for the control and management of such schools as the conference may think fit".

The trust deed itself contained the following provisions:

> 1. Any lands hereditaments or other property which shall be purchased leased or otherwise acquired by the Conference for the purposes of these Presents may by the direction of the Conference be conveyed to and vested in the trustees or any number of them not less than two.
>
> 2.(a) The trustees shall hold the lands or other property so vested in them as aforesaid (hereinafter called the School property) in Trust for the purposes of Secondary Schools or for other educational purposes including but without prejudice to the generality of the foregoing the provision of bursaries whether in connection with the purposes of Secondary Schools or otherwise until the Conference shall otherwise by resolution determine and thereafter for such purposes either for the advancement of Secondary Education or otherwise for such purposes in connection with the promotion of the charitable objects of the Methodist Church as the Conference shall from time to time by resolution direct.
>
> 2.(b) While the School property is held for the purposes of Secondary Schools, the Schools shall be administered and managed by the Board of Management yearly to be appointed by the Conference as the Governing Body thereof which Board shall have full control over the education given in the Schools, the School property, and all other matters relating to the Schools.
>
> 2.(c) The Board of Management may delegate to a body of local Administrative Governors all or any part of their functions. The constitution powers and duties of such Administrative Governors shall be set forth for each School in an "Instrument of Government" drawn up by the Board of Management . . .
>
> 3. The trustees shall by the direction of the Board of Management from time to time sell mortgage lease let exchange or otherwise deal with all or any part of the property for the time being subject to the trusts of these Presents . . . as the Board of Management may from time to time direct.

Clause 14 of the trust deed expressly limited the application of the trust property and its income to charitable purposes.

The property is immediately adjacent to the western boundary of Kent College. On May 24 1974 the landlord demised to the then trustees of the Methodist Secondary Schools Trust Deed of 1903 the property, together with its garage, outbuildings, yard and garden for a term of 15 years from June 1 1974, at an initial annual rent of £350 with provisions for rent reviews at the end of the third, sixth, ninth and 12th years of the term. By clause 2(X) of the lease the tenants covenanted "to use the demised premises for the purposes of a private residence in single occupation only".

The judge found that at the time of the demise the landlord understood from the then bursar of Kent College, Mr Woodroffe, that the property would be used for residential purposes only and that Kent College would use it for occupation by anyone the school wished. The first occupant was a school caterer.

There are six boarding-houses at the school accommodating 260 pupils. The total number of pupils is about 600. The classrooms are extensive and science laboratories and some of the other classrooms contain very valuable equipment, which has to be safeguarded. On July 20 1977 Mr Lester was offered by the school's assistant bursar, acting on behalf of the local administrative governors, employment as a resident caretaker upon the terms set out in a letter of that date. For the proper performance of his duties, he was required to live at the property on a service occupancy. The letter stated expressly that no tenancy would be created by the arrangement. He has occupied

A the property since that time. In about 1980 a written job description of the caretaker's duties was prepared by the then bursar of the school and agreed with Mr Lester. It set out in particular the caretaker's duties with regard to security, boiler operation and reaction to emergencies. Among the "general responsibilities" listed was "school security and emergency repair and maintenance out of normal school hours". Among the duties listed were "building security at night and during holidays ensuring doors are opened before breakfast and locked at night . . . " and "out of normal working hours or during normal working hours if a painter is not quickly available ensure that broken windows are immediately made safe . . . ".

B In order to provide for quick communication between the school authorities and the caretaker, a telephone was rented for Mr Lester by the school and installed in the property and he was provided with a pager. In order to ensure that the sounding of any alarm in the school could be heard by him a second alarm was installed in the property. These arrangements were still in operation when the present proceedings began.

The judge's judgment

In relation to the first two of the four issues which he had to determine the judge rightly pointed out that, to satisfy the provisions of section 23(1) of the Act, it is the tenant who must both "occupy" the relevant premises and "occupy them for the purposes of a C business carried on by him or for those and other purposes", but that section 23(2) contemplates the activity carried on by a corporate or unincorporated body of persons as providing a qualifying purpose within section 23(1). (I pause to say that the activity of running a school clearly constitutes a "business".)

The judge continued:

> I am satisfied first that Mr Lester, the school caretaker, is employed by the local administrative governors of Kent College, the contract of employment having been made with Mr Lester through their agent, the then bursar of the school. In making that contract the local administrative governors acted under powers delegated to them by the board of management for Methodist Residential Schools. The tenants under the lease dated May 24 1974 are the trustees of the Methodist Secondary Schools Trust Deed of 1903 as named in D the Schedule to the Lease, and not the board of management for Methodist Residential Schools or the local administrative governors. That, however, is not the end of the inquiry, because the provisions of section 41 of the Act may render occupation by the board of management and the local administrative governors, and the carrying on of a business by them, equivalent to occupation by and the carrying on a business by the tenants, the trustees.

Section 41(1) of the Act, so far as material, provides:

> Where a tenancy is held on trust, occupation by all or any of the beneficiaries under the trust, and the carrying on of a business by all or any of the beneficiaries, shall be treated for the purposes of section 23 of this Act as equivalent to occupation or the carrying on of a business by the tenant; and in relation to a tenancy to which this Part of this Act applies by virtue of the foregoing provisions of this subsection —
> E (*a*) references (however expressed) in this Part of this Act . . . to the business of, or to carrying on of business, use, occupation or enjoyment by, the tenant shall be construed as including references to the business of, or to carrying on of business, use, occupation or enjoyment by, the beneficiaries or beneficiary; . . .

The judge held that the effect of clauses 2(b) and 3 of the 1903 trust deed (together with clauses 4 and 5, which it is unnecessary to quote) was to constitute the board of management "as persons whose interest under the trust are such as to give them the right as against the trustees to occupy the school property as defined by clause 2(a) [of the trust deed] and are such that the trustees may properly let them into possession". He held that they were accordingly "beneficiaries" F within the provisions of section 41(1) of the Act in relation to the trustees' interest under the lease. Furthermore, he said, the board of management were in a position to enforce the provisions of the trust in their favour and for that reason also were "beneficiaries". He concluded:

> Accordingly, the board of management, and by virtue of the delegated powers, the local administrative governors also, stand in the shoes of the trustees under the provisions of section 41 subsection (1) for the purposes of determining whether the trustees, as tenants under the lease, occupy Greengates and occupy it for the purposes of a business carried on by them.

The judge went on to hold that the trustees were in occupation of the property because the board of management and the local administrative governors, "standing in the shoes" of the trustees, did occupy the property by their employee, Mr Lester, whom they G required to occupy the property as a term of his employment.

The circumstances in which occupation by a servant is deemed in law to be the occupation of his master are set out in the speech of Lord Diplock in *Commissioner of Valuation for Northern Ireland* v *Fermanagh Protestant Board of Education* [1969] 1 WLR 1708 at pp1721F-1731F. In the light of this guidance, the judge concluded that, in relation to Mr Lester's occupation, the test was satisfied because his obligation to reside in the property was attributable to and formed an integral part of the relationship with his employers created by the contract of employment and his residence in the property was of material assistance to his employers in the carrying out of their activities.

Having decided that the board of management and the local H administrative governors (standing in the shoes of the trustees) were in occupation of the property at all material times, the judge went on to consider whether they were in such occupation "for the purposes of a business carried on by them or for those and other purposes", within the meaning of section 23(1) of the Act.

I interpose a brief reference to two earlier authorities relating to cases where it has been said that residential property was occupied for the purposes of a business, in addition to its residential use. In this context it has been common ground between counsel on the present appeal that a distinction is to be drawn between cases in which: (1) the tenant is himself a residential occupier; and (2) others are residing in the property, but the tenant nevertheless claims to be in occupation J for the purposes of a business. The judgment of Lord Denning MR in *Cheryl Investments Ltd* v *Saldanha* [1978] 1 WLR 1329 at p1337 illustrates that in the first case business activity must be "a significant purpose" of the tenant's occupation if the case is to fall within section 23(1). In the second case, however, the test is rather different.

In *Chapman* v *Freeman* [1978] 1 WLR 1298 the tenant of a cottage used it to provide accommodation for staff employed at his nearby hotel. The Court of Appeal held that the cottage was not "occupied for the purposes of business" within the meaning of section 23(1), because the occupation of the cottage for the purpose of housing hotel staff was merely for the convenience of the business and not for a purpose necessary to the furtherance of the business. As Lord Denning MR put it (at p1300H): K

> Speaking generally, the test is whether it is *necessary* for the individual to live in the house in order to perform his own particular duties properly; or whether it is just for *convenience* that he should live there in connection with his duties.

Geoffrey Lane LJ, applying a slightly less strict test, put the matter thus (at p1301):

> It was merely a matter of convenience that the Andrews should be housed in this particular cottage. That, to my mind, is not enough. The tenant must go beyond mere convenience and show that the occupation was in furtherance of his business activities — in other words, that it was for business reasons that he was occupying the cottage and not merely for reasons of convenience.

In the present case, the judge expressed his conclusion on the L second issue before him as follows:

> It is important to determine what was, and is being done by Mr Lester in Greengates, as opposed to what work he does outside his house under the contract of employment. Plainly what Mr Lester is doing in Greengates for much of the time is residing there with his family. Save for one important area of his employment, he does not do his work in Greengates. That important area of his employment is being available as near as possible to the school to react to any emergency as I have earlier described, whether brought to his attention by the alarm in the house, or by a message received by telephone there.
> In relation to this second question, the inquiry is, is it necessary for Mr Lester to reside in Greengates to perform that part of his duties: see Lord Denning's judgment in *Chapman* v *Freeman* (1978) 3 All ER 878, and does he M occupy Greengates to further the activity of the people called Methodist in providing secondary school education. See the judgment of Lord Justice Geoffrey Lane, as he then was, in the same case. Moreover, it must be established that this area of Mr Lester's employment, that is being available to react to emergencies outside normal school hours, is a significant purpose of that employment: see *Cheryl Investments* v *Saldanha* . . .
> Having considered all the evidence, I am satisfied that it was necessary for Mr Lester to reside in Greengates to perform his duties under the contract of employment, and that he did so in order to further the activity of the people called Methodist in providing secondary education, and that being available in Greengates to react to emergencies outside normal school hours was a significant purpose of that employment.
> Accordingly, I find that the board of management and the local administrative governors, standing in the shoes of the trustees, occupied

Greengates partly for the purpose of the business carried on by them, and for other purposes.

In view of the distinction between the two classes of case mentioned above, it is common ground that the judge's references in this context to *Cheryl Investments* v *Saldanha* and to "significant purpose" were inapposite. However, in my judgment, these references do not by themselves invalidate the judge's decision on the second question.

The judge dealt with the third question succinctly as follows:

> . . . were the applicants carrying on a business in all or any part of Greengates in breach of the prohibition, however expressed, for use for business purposes, which subsisted under the terms of the tenancy and extended to the whole of Greengates.
>
> Clause 2 (X) of the lease is subsisting and extends to the whole of Greengates. The phrase "to use the demised premises for the purposes of a private residence in single occupation only", by the addition of the word "only", impliedly confines the use of the premises to use for the purposes of a private residence in single occupation. To the extent that the house was used for the purposes of rendering Mr Lester available to hear the alarm and to receive messages, and from the nearest point adjacent to the school, to react to any emergencies promptly, there was a breach of Clause 2 (X). It was a significant breach, in that the availability of Mr Lester in Greengates, was a significant area of his employment, to which his employers attached great importance.

Finally, in relation to the fourth question the judge concluded:

> I am satisfied first that the respondent was unaware that Greengates was being used by the trustees for any purpose other than as a private residence in single occupation. He did not know of the terms of Mr Lester's employment as expressed in the letter of July 20 1977, the job description of 1980, and the letter of the June 27 1988, nor of the importance placed by Mr Lester's employers on his availability in Greengates to react to emergencies. At most, he assumed that the school either wanted or required Mr Lester to live in Greengates, but he did not know of the use to which the premises were in part being put. I am satisfied that he neither consented to nor acquesced in the use of Greengates for such purposes; he did not know of such use.

In the result, the judge declared that the tenancy was not one to which Part II of the Act applies and dismissed the trustees' application for a new tenancy. The trustees now appeal from his judgment essentially on the grounds that he went wrong on the third and fourth issues. The landlord has put in a respondent's notice contending that the judge's decision should be affirmed on additional grounds. Those grounds are in substance that the judge erred in his conclusions on the first two issues. The third issue, relating to breach, was the one first argued before us and I propose to deal with it first.

If the trustees were carrying on business in all or any part of the property, were they doing so "in breach of a prohibition (however expressed) of use for business purposes which subsists under the terms of the tenancy and extends to the whole of the property"?

Mr Jay, for the landlord, pointed out the words of section 23(4) of the Act quoted above are widely drafted. As his first submission in this context he contended that they pose an inescapable dilemma for the trustees. If (contrary to the landlord's contention) the judge was right in holding that the trustees, through persons standing in their shoes, occupied the property "for the purposes of a business carried on by them or for those and other purposes" within the meaning of section 23(1), then (it was submitted) it must inevitably follow that the trustees were using the property in breach of clause 2(X) of the lease which obliged the tenant:

> to use the demised premises for the purposes of a private residence in single occupation only.

A covenant in these terms, in Mr Jay's submission, by necessary implication prohibits use of the property for the purposes of a business. He referred us in this context to *German* v *Chapman* (1877) 7 ChD 271, in which the Court of Appeal held that the erection of a building to be used for the education and lodging of 100 girls in connection with a charitable institution for the daughters of missionaries, supported by voluntary contributions, was a breach of a covenant that "no house or other building to be erected or built upon the land shall be used or occupied otherwise than as and for a private residence only, and not for any purpose of trade". There is, in Mr Jay's submission, a "symmetry" between section 23(1) and section 23(4) which leads inevitably to the conclusion that the trustees in the present case must fall foul of section 23(4) if they can bring themselves within section 23(1).

I cannot, for my part, accept that this neat symmetry exists. As Mr Ferris, for the trustees, pointed out, a tenant does not have to satisfy the court that he is actually carrying on a business *on the premises* in order to bring himself within section 23(1). All he has to show is that he is occupying the premises "for the purposes of a business carried on by him or for those and other purposes".

Though they may perhaps be rare, it is possible to conceive cases where a tenant occupies premises "for the purposes of a business carried on by him or for those and other purposes", within the tests explained in *Chapman* v *Freeman* (*supra*) without actually carrying on a business on the premises. To postulate a hypothetical case, let it be supposed that a hotelier employs a chef at a very high wage, but for exceptionally long hours, and requires him, as a condition of his employment, to live in a cottage situated very near the hotel, of which cottage the hotelier is tenant, because this is the only practical way to ensure that the chef will be able to perform properly his onerous duties in the hotel. Applying the *Chapman* v *Freeman* test, I see no reason why the hotelier-tenant should not be able to assert that the chef is occupying the cottage for the purposes of his business. On the other hand, I would see no grounds for saying that that hotelier is carrying on his business in the cottage, because the chef himself would be using the premises solely as his own private residence and would not be performing any of his chef's duties in the cottage.

Mr Ferris, rightly alive to this distinction, more than once submitted that, on the facts of the present case, Mr Lester's occupation of the property was "necessary for the better performance of duties *discharged elsewhere*" (emphasis supplied). The case, he submitted, is for example analogous to that of a shopkeeper who lives in a flat above his shop, because this is necessary for the better performance of his duties in the shop itself; an arrangement of this kind does not involve the carrying on of any business in the flat itself. In his submission, it was the judge's conclusion that *Cheryl Investments* v *Saldanha* required that the availability of Mr Lester to react to emergencies outside normal school hours had to qualify as a "significant purpose" of his employment, which led him to conclude that the breach involved in use of the property for this purpose was "a significant breach".

I take this last point and regard the sentence in the judgment which refers to "significant breach" as adding nothing material to the judge's essential reasoning. Nevertheless, it does not destroy it and I am unable to accept the shopkeeper analogy, and other similar analogies suggested by Mr Ferris, as applicable on the facts of the present case. The insuperable obstacle facing the trustees in this context is the judge's finding that "the property was used for the purposes of rendering Mr Lester available to hear the alarm and to receive messages and from the nearest point available to the school to react promptly". This finding is, I think, incapable of challenge.

The job description of 1980, and the letter of 1988, make it clear that Mr Lester had to hold himself available during non-working hours. Mr Cotes, the bursar of Kent College, said in the course of his oral evidence-in-chief: "When the caretaker is off duty, he is only off duty with the approval of the school after adequate notice. He is on call seven days and seven nights a week."

Under cross-examination, Mr Cotes said: "The school requires Mr Lester to be on call." These two citations by themselves show how far removed the present case is from the hypothetical example of the hotelier and chef to which it is referred above. The essential difference is that in the present case Mr Lester is using the property partly for the purpose of carrying out duties of his employment (as caretaker), which have to be performed *in the property*, that is to say by continuously being on call and making himself available — the alarm and the telephone being mechanisms available to enable him to receive and respond to the calls. On his finding of fact quoted in the preceding paragraph, the judge was, in my judgment, right in holding that a breach of clause 2(X) had been and was being committed.

Has the landlord acquiesced in the breach?

As was pointed out by this court in *Bell* v *Alfred Franks & Bartlett Co Ltd* [1980] 1 WLR 340, in the context of section 23(4) of the Act the word "consent" is inserted in plain antithesis to "acquaintance". "Acquiescence" in the relevant breach on the part of an immediate landlord, such as the landlord in the present case, will suffice to render section 23(3) inapplicable.

Mr Ferris pointed out that, in the present case, the landlord has accepted rent for the entire contractual term of the lease and thereafter during the statutory continuation. He has never complained of the use of the property by letter or action. His solicitors treated the tenancy as a "business tenancy" by serving the section 25 notice, and indeed going so far as to say in the note to that notice that the landlord would not oppose an application for a new tenancy under Part II of the Act.

I accept that if, before the proceedings began, the landlord had the requisite knowledge, there could scarcely be a more obvious case of acquiescence. Knowledge, however, is a prerequisite of acquiescence. In the words of Shaw LJ in the *Bell* case (*supra*) (at p346):

> What is meant by acquiescence? It may involve no more than a merely passive attitude, doing nothing at all. It requires as an essential factor that there was knowledge of what was acquiesced in.

It is, I think, clear that in the present case the landlord could not be said to have acquiesced in the continuing breach of clause 2(X) of the lease *unless he knew the facts which gave rise to the breach*. If the foregoing analysis of the legal situation is correct, the fact which gave rise to the breach was the continuing use of the property by the school's caretaker, in part, for the purpose of being available on call during ordinary non-working hours.

Early in his judgment, the judge found the following facts. The landlord was aware of Mr Lester's occupation of Greengates shortly after July 1977. The landlord is a prison officer currently serving in HM prison at Rochester and residing in his own home there. However, having some of his family in Canterbury, he frequently visits this city and passes the property. He has met Mr Lester about three times. He believed him to be one of the school maintenance staff and assumed that the school would either want or require him to live at the property. Later in his judgment, when he came to answer the third question, he made the further findings quoted above.

The judge did not find that the landlord knew that Mr Lester was employed as the school's caretaker or that he was using the property in part for the purpose of being available on call. On the contrary, he found that the landlord thought that Mr Lester was one of the school's maintenance staff and that he did not know of the use to which the premises were in part being put.

Mr Ferris invited us to overrule these findings. He drew our attention in particular to a letter dated October 19 1981 written by Mr Cotes, the school bursar, to the landlord, inviting him to consider the possibility of extending the lease of the property. In one sentence this letter referred to "our resident caretaker" (*sic*), though in the next sentence, somewhat confusingly, it referred to our resident "caterer" (*sic*). Mr Ferris also relied on a letter dated June 28 1988, written by the trustees' surveyors to the landlord's surveyor, in which reference was made to "alternative arrangements to house the caretaker and his family". He also referred us to the judge's notes of the landlord's oral evidence and to letters written by the landlord's solicitors in which initially it was implicitly accepted that the tenancy was subject to the Act. In all the circumstances, in his submission, the landlord must have known of the use to which the property was being put.

I see the force of Mr Ferris' submissions on this point. It is, perhaps, at first sight surprising that the landlord had not, during the course of eight years or so, acquired a closer knowledge of Mr Lester's job and duties than the judge found. Nevertheless, the judge had the great advantage of seeing and hearing the landlord (as well as Mr Cotes) give evidence. The landlord's oral evidence, as noted by the judge, was that "I did not believe Mr Lester was a caretaker. I thought he was more of the maintenance staff". The letter of October 19 1981 was specifically put to the landlord in cross-examination. The judge had all the relevant correspondence before him. I see no sufficient grounds on which this court would be justified in disturbing the judge's findings of fact on this point; he was entitled to accept the landlord's evidence, and did accept it.

Accordingly, in my judgment, the landlord has not acquiesced in the breach.

Did the trustees occupy the property, and if so did they occupy it for the purposes of a business carried on by them and for those and other purposes?

If the answers to the third and fourth questions above are correct, it must follow that the trustees' appeal must be dismissed and that it is not strictly necessary to consider the issues raised by the landlord's respondent's notice. Nevertheless, in this judgment it has so far been assumed in favour of the trustees that the judge was right in deciding that they are in occupation of the property. Since this issue has been argued before us and is of some general interest, I will make a few *obiter* observations on it. As has already appeared, the judge held that the board of management were "beneficiaries" within the provisions of section 41(1) of the Act in relation to the trustees' interest under the lease. He treated the words "beneficiaries under the trust" in the context of section 41 as including persons whose interests under the relevant trust are such as to give them the right as against the trustees to occupy the relevant property or are such that the trustees may properly let them into possession. This was the route by which he held that the trustees, rather than the board of management, were in occupation of the property.

The judge's interpretation of the phrase "beneficiaries under the trust" appears to have been derived from an explanation of its meaning given in very similar terms by Sir Raymond Evershed MR in *Frish Ltd* v *Barclays Bank Ltd* [1955] 2 QB 541 at pp548-549. However, in the immediately succeeding sentence, Sir Raymond continued:

> In other words, the words "beneficiaries under the trust or any of them" in that context must be limited to persons occupying by virtue of their interest as beneficiaries.

Similarly, (at p554 *ibid*) Jenkins LJ said:

> I think that the interest of the beneficiary must be an interest under the trust on the strength of which he intends to occupy.

If, as I am disposed to think, the phrase "beneficiaries under the trust" in section 41(1) is limited to persons occupying by virtue of their interest as beneficiaries, I find it difficult to see how the board of management can be said to fit the description, bearing in mind that, under the trust as constituted, they are not intended to take any personal benefit. I incline therefore to the view that Mr Jay is right in submitting that in truth section 41 has no application whatsoever on the facts of this case, and that it is the board of management, through Mr Lester, who are in occupation of the property.

In *Groveside Properties Ltd* v *Westminster Medical School* (1983) 47 P&CR 507, which was followed by Scott J in *Linden* v *Department of Health and Social Security* [1986] 1 All ER 691, the Court of Appeal held that a medical school, which was a body corporate with the object of carrying on Westminster Medical School for the purposes of education of students of medicine or research, was in occupation of a flat for the purpose of section 23(1) by virtue of the substantial degree of control which it exercised over it. On the facts of that case occupation was established without any reliance on section 41. I think it unlikely, however, that on the facts of this case the trustees could point to a sufficient degree of control over the property to bring themselves within the ambit of that decision. The interposition of the board of management, which has the power to give directions to the trustees, renders this a special case.

I, accordingly, incline to the view that the trustees have not established that they are in occupation of the property. If this were right, their appeal would have to be dismissed on this ground, if no other. However, the operation or otherwise of section 23 of the Act in cases such as this where the tenanted property is held by trustees on trust not for individuals but for public charitable purposes, may give rise to knotty problems on which I think it unnecessary to elaborate in this judgment.

Conclusion

As things are, I prefer to base my decision on the third and fourth issues dealt with by the judge on which, in my judgment, his conclusions cannot be disturbed. For these reasons, I would dismiss this appeal.

Rose and **Glidewell LJJ** agreed and did not add anything.

Appeal dismissed with costs.

Court of Appeal

February 9 1993

(Before Sir Thomas BINGHAM MR, Lord Justice McCOWAN and Lord Justice HIRST)

WARD-LEE v LINEHAN

Estates Gazette May 22 1993

[1993] 20 EG 125

Landlord and Tenant Act 1954 — Failure of county court to serve originating application for new tenancy on landlord — Whether tenant entitled to leave to serve the originating application out of time

The appellant, who held a seven-year lease of ground-floor premises of a shop in Edenbridge, Kent, which expired on March 31 1992, made a request to the respondent landlord on April 22 1991 for a new tenancy under section 26 of the Landlord and Tenant Act 1954; the respondent replied, making no objection to the grant of a new tenancy but suggesting a higher rent. On July 4 1991 the appellant's solicitors sent a form of originating application for a new tenancy under Part II of the 1954 Act to East Grinstead County Court, with a copy for the landlord and a request for the issue of a summons. The county court stamped one copy of the originating application, assigned a cause number to the proceedings and returned a stamped copy to the appellant's solicitors. In accordance with CCR 1981, Ord 7, r10(1)(b) it was the duty of the county court to serve the proceedings on the respondent by first-class post; the county court did not serve the proceedings on the landlord by post or in any other way. On September 30 1991 the landlord's solicitors wrote to the tenant's solicitors inquiring as to whether proceedings for renewal had been issued and by a letter dated October 1 1991 the tenant's solicitors confirmed that proceedings had been issued. On November 29 1991 the landlord's solicitors stated that they had not received the originating application and contended that the tenant had lost her right of statutory renewal because of the expiration of the time-limit. On January 9 1992 the tenant's solicitors issued an application seeking leave to serve the originating application out of time, which the landlord opposed. On February 26 1992 Judge Lovegrove QC dismissed the tenant's appeal from the decision of the district judge dismissing her application for leave to serve out of time. In his judgment the judge accepted that he had a discretion to extend time, but decided he would not grant an extension because of the prejudice to the landlord and the two-month delay of the tenant in pursuing the question of service after the landlord's solicitors' letter of September 30 1991. The tenant appealed.

Held: **The appeal was allowed. Even where an originating application is not served within two months of the date of issue and application to extend is not made within four months of the date of issue, the county court has jurisdiction on proper grounds being shown to extend time for service and to treat the failure to extend and serve as an irregularity. There was a good reason and a satisfactory explanation on behalf of the tenant for the failure to apply for extension before the validity of the proceedings expired. The following factors were relevant in exercising the discretion to extend: that until January 1992 the landlord had raised no objection to a new tenancy; that the difference over rent was unlikely to prove an insuperable stumbling block; that but for the court's error the application would have been duly served; that the tenant did not learn of the error in time to cure it; and that the delay in service has of itself caused the landlord no prejudice, since the landlord's change of circumstances, whereby she wanted possession of the premises for herself, would have occurred even if she had been served in time. The judge's exercise of discretion was demonstrably wrong in that he had given little or no weight to these considerations.**

The following cases are referred to in this report.

Baxendale (Robert) Ltd v *Davstone (Holdings) Ltd* [1982] 1 WLR 1385; [1982] 3 All ER 496; (1983) 45 P&CR 127; (1982) 264 EG 713, CA

Bernstein v *Jackson* [1982] 1 WLR 1082; [1982] 2 All ER 806

Kleinwort Benson Ltd v *Barbrak Ltd* [1987] AC 597; [1987] 2 WLR 1053; [1987] 2 All ER 289; [1987] 2 Lloyd's Rep 1, HL

Leal v *Dunlop Bio-Processes International Ltd* [1984] 1 WLR 874; [1984] 2 All ER 207, CA

Lewis v *Wolking Properties Ltd* [1978] 1 WLR 403; [1978] 1 All ER 427; (1978) 36 P&CR 140; (1977) 245 EG 655, CA

This is an appeal from the decision of Judge Lovegrove QC, who on February 26 1992 had dismissed an appeal from a decision of the district judge dismissing an application by the tenant, Beryl Elizabeth Louise Ward-Lee, for leave to serve an originating application seeking a new tenancy out of time on the landlord, Collette Linehan.

Timothy Fancourt (instructed by Stephens & Scown, of Truro) appeared for the appellant; Ben Patten (instructed by Lufflum & Ainsworth, of Sutton) represented the respondent.

Giving the judgment of the court, **Sir Thomas Bingham MR**: This appeal raises a question of some practical importance on the powers of the county court to extend time for the service of proceedings. We give the judgment of the court.

The applicant was the tenant of ground-floor premises at Edenbridge, Kent, where she ran an antique shop. She held the premises under a seven-year lease, which was due to expire on March 31 1992. The respondent was her landlord. We shall refer to the parties as "the tenant" and "the landlord".

On April 22 1991 the tenant made to the landlord a request in writing for a new tenancy under section 26 of the Landlord and Tenant Act 1954. She proposed a term of 12 years and an annual rent of £2,750, with four-yearly reviews.

On May 15 1991 the landlord's solicitors acknowledged receipt of this notice. They stated that the landlord had no objection to the grant of a new tenancy, but they suggested a rent of £7,500.

On July 4 1991 the tenant's solicitors sent a form of originating application for a new tenancy under Part II of the 1954 Act to East Grinstead County Court, with a copy for the landlord and a request for the issue of a summons. The county court stamped one copy of the originating application, assigned a cause number to the proceedings and returned a stamped copy to the tenant's solicitors. The county court also issued, stamped, numbered and sent to the tenant's solicitors a summons addressed to the landlord, although in accordance with usual practice no date for a hearing was fixed. All this was done on July 5 1991.

The tenant's solicitors did not indicate any wish that they or the tenant should serve the summons or the originating application on the landlord personally. It was accordingly the duty of the county court to serve the proceedings on the landlord by first-class post at the address stated in the request for the summons (CCR 1981, Ord 7, r10(1)(b), (2)). The county court did not serve the proceedings on the landlord by post or in any other way. Nor, in sending the stamped, dated and numbered documents to the tenant's solicitors, did the county court indicate that the landlord had not been served. The natural inference was that the county court had served the landlord in accordance with the rules. That was what the tenant's solicitors assumed.

Under the county court rules, an answer to the tenant's application should have been filed by the landlord within 14 days of service (Ord 43, r 2(1), Ord 9, r 18(3)). In this case no answer was forthcoming because the landlord had not been served. The absence of an answer might have alerted the tenant's solicitors to the fact that the landlord had not been served. It did not do so. This was not at all surprising in the circumstances. The existing tenancy still had some eight months or so to go. The landlord had not objected to the grant of a new tenancy or to any of the proposed terms save the rent. There was nothing to suggest that agreement could not be reached on rent, as it usually is. The tenant was obliged to make application to the court in order to preserve her right to claim a new tenancy (1954 Act, sections 24(1), 29(3)). But if matters followed their usual course the application would never come to a hearing, and lack of activity on the landlord's part would not have appeared at all significant. The tenant's solicitors were not at fault in failing to suspect that the landlord had not been served and it was not argued that they were.

The silence was broken on September 30 1991 when the landlord's solicitors wrote to the tenant's solicitors as follows:

> Further to our correspondence last May we enquire whether your client has issued proceedings for renewal of her tenancy alternatively whether it is her intention to vacate the property next April.
>
> Please let us know either way as our client wishes to make her arrangements for re-letting if applicable. If you have issued proceedings please effect service upon us as Miss Lineham's solicitors.

The tenant's solicitors were clearly somewhat surprised to receive this letter. They replied by return of post in a letter dated October 1:

> We thank you for your letter of the 30th ultimo. Proceedings were issued on the 4th July last and we quoted your client's name c/o your office as the name and address of the respondent. A sealed copy of the application was received from the East Grinstead County Court on the 9th July.
>
> We are surprised that you have not heard from the court because we sent them a copy of the Originating Application for service. The Case Number allocated by the Court is 9101767.

The tenant's solicitors received no answer to that letter for nearly two months, during which they neither wrote again nor made inquiry of the landlord's solicitors or the county court to discover what (if anything) had happened. But on November 29 the landlord's solicitors wrote again:

> In reply to your letter of 1st October last we did not receive the Originating Application and nor were we authorised to do so. In consequence we believe that your client is now out of time and that she has lost her right of statutory renewal.
>
> Our authority to accept service dated 30th September last is hereby revoked.

At this point the tenant's solicitors did make inquiry of the county court. They were told that a note in the court records suggested that the summons and application had been sent to them for service, but this had not been done and they did not (until now) know that the court's understanding was otherwise. In early December 1991 the tenant's solicitors, following a suggestion made by the court, asked the landlord's solicitors to consent to late service, but the landlord's solicitors declined to do so.

On January 9 1992 the tenant's solicitors issued an application seeking leave to serve the originating application out of time. The landlord opposed. In an affidavit sworn on her behalf her solicitor said that her financial and personal circumstances had changed substantially since May 1991 when she had given her counternotice: she now sought the property for herself and would be prejudiced if the tenant's application were allowed. The district judge dismissed the application on January 21 1992. The tenant appealed. For purposes of the appeal the landlord made an affidavit in which she deposed:

> Since giving the Landlord's Counter Notice my financial position has deteriorated due to the loss of my husband's financial support. He has been in a Psychiatric Hospital as a result of an attempted suicide. He is now in a Rehabilitation Centre but I do not know whether he will be fit for employment again or whether we will resume co-habitation. I have a 16-year old daughter who is studying for GCSE who I am supporting. In the circumstances I need to obtain the maximum income possible from the said shop and I believe this will be achieved as a business proprietor rather than as a Landlord.

Judge Lovegrove QC dismissed the tenant's appeal on February 26 1992. In the course of his judgment the learned judge accepted that he had a discretion to extend time and that he had to balance the parties' competing interests, but decided that he should not grant an extension because of the prejudice to the landlord and (it would seem) the two-month delay of the tenant in pursuing the question of service after the landlord's solicitors' letter of September 30 1991.

Against that judgment the tenant now appeals with leave granted by this court. The issues are whether the county court had a discretion to extend time and, if so, whether it was a discretion which could and should have been exercised on the facts of this case. Before the judge the existence of some discretion appears to have been assumed. On appeal to this court the existence of such a discretion was at first challenged. But in argument counsel for the landlord did not contend that there was no discretion, only that the court's discretion could not properly have been exercised in the tenant's favour on the facts here.

It is first necessary to review the relevant County Court Rules. In doing so we make reference to the corresponding rules of the Supreme Court since it is not suggested that there is intended, in any relevant respect, to be a dichotomy of practice between the two courts.

Ord 7, r 20 of the County Court Rules provides:

> 20. (1) The time within which a summons may be served shall, unless extended under the following provisions of this rule, be limited
>
> (a) where leave to serve the summons out of England and Wales is required under Ord 8, r 2, to a period of 6 months;
>
> (b) in any other case to a period of 4 months, beginning with the date of issue of the summons.
>
> (2) subject to paragraph (3) the court may extend the period of service of a summons from time to time for such period, not exceeding 4 months at any one time, beginning with the day next following that on which it would otherwise expire, as the court may specify, if an application for extension is made before that day or such later day (if any) as the court may allow.
>
> (3) Where the court is satisfied on an application under paragraph (2) that, despite the making of all reasonable efforts, it may not be possible to serve the summons within 4 months, the court may, if it thinks fit, extend the period of service for such period, not exceeding 12 months, as the court may specify.

Ord 43, r 6(3) of the CCR substitutes a period of two months for the periods of four months and six months in the case of originating applications under section 24 of the 1954 Act. These rules correspond with RSC Ord 6, r 8 and Ord 97, r 6(3). On the facts here the two-month period for service expired on September 4 1991, at which time the tenant and her solicitors did not appreciate that the county court had fallen down on its duty to serve the proceedings on the landlord.

Ord 13, r 4 of the CCR provides:

> 4. (1) Except as otherwise provided, the period within which a person is required or authorised by these rules or by any judgment, order or direction to do any act in any proceedings may be extended or abridged by consent of all the parties or by the court on the application of any party.
>
> (2) Any such period may be extended by the court although the application for extension is not made until after the expiration of the period.

This corresponds in effect to RSC Ord 3, r 5.

Lastly, we should refer to Ord 37, r5 of the CCR, which (so far as relevant) provides:

> 5. (1) Where there has been a failure to comply with any requirement of these rules, the failure shall be treated as an irregularity and shall not nullify the proceedings, but the court may set aside the proceedings wholly or in part or exercise its powers under these rules to allow any such amendments and to give such directions as it thinks fit.
>
> (2) No application to set aside any proceedings for irregularity shall be granted unless made within a reasonable time, nor if the party applying has taken any step in the proceedings after knowledge of the irregularity.

This broadly corresponds to RSC Ord 2, r 1.

The effect of RSC Ord 6, r8 has recently been the subject of authoritative exposition by the House of Lords in *Kleinwort Benson Ltd* v *Barbrak Ltd* [1987] AC 597. It was there held that Ord 6, r 8(2) should be read as subject to an implied term that the power to extend the validity of a writ should be exercised only for good reason, a somewhat less stringent test than that of "exceptional circumstances" previously favoured. Where the application for extension is made after a writ has ceased to be valid and after the relevant limitation period has expired, an applicant has not only to show good reason for the court to exercise its power to extend but also to give a satisfactory explanation for his failure to apply for extension before the validity of the writ expired. We have no doubt that this reasoning should be applied by analogy to Ord 7, r20 of the CCR.

This rule does not, however, solve the tenant's problems, for even if she were able to show good reason and give a satisfactory explanation, the court's power to extend in a case under section 24 of the 1954 Act is limited to a period not exceeding two months. The primary two-month period expired on September 4 1991. An extension of two months under Ord 7, r20(2) would have carried her only to November 3 1991. By that date she had neither served the landlord nor applied to the court.

A somewhat similar case came before this court in *Lewis* v *Wolking Properties Ltd* [1978] 1 WLR 403*. At that time the period provided both for service and for renewal of an originating application of this kind was one month. The tenants applied to the county court on August 23 1976. The court tried to serve the landlords, but could not do so because the landlords had left the address given for service. The tenants' solicitors accordingly asked the landlords' solicitors for the landlords' correct address. The landlords' solicitors replied on September 17, six days before expiry of the one-month period allowed for service, giving no address for the landlords but agreeing to accept service if the summons had been timeously issued. The tenants' solicitors did not accept that invitation to serve the landlords' solicitors. Instead, they repeated their request for the landlords' address on September 21. The address was given on

*Editor's note: Also reported at (1977) 245 EG 655.

A September 29, by which time the month had expired. On October 11 1976 the landlords' solicitors took the point that service was out of time unless application for an extension had already been made. At that stage the tenants' solicitors applied to the county court for an extension, which was granted and, as the Court of Appeal held, rightly granted.

At the time of that case the CCR were somewhat different. The rule which then corresponded to what is now Ord 7, r20 contained two provisions which have now disappeared. First, the subrule which permitted extensions to be granted was subject to the proviso:

> That the time shall not be extended for any period unless the application is made within the currency of the last preceding period.

B Second, the rule provided:

> Where the summons has not been served within the time allowed for service by this rule, the action shall be struck out.

The landlords in *Lewis* accepted that if there was a discretion to extend time on application made after expiry of the initial one-month period the case was a proper one for exercise of the discretion in the tenants' favour, but they argued that, if service was not effected and application for extension not made within the initial one-month period, the court had no choice but to strike out the action.

The Court of Appeal did not accept that argument. It construed the rule which then corresponded to Ord 13, r 4 as giving a broader power to extend than the then equivalent of Ord 7, r 20. There was C accordingly a discretion to extend time, although a discretion to be sparingly exercised. Under the rules as they then stood, Ord 13, r 5 (now Ord 13, r 4) opened with the words "Subject to the provisions of these rules . . .". The court held that that qualification was intended to apply to rules (of which an example was given) where words plainly inconsistent with Ord 13, r 5 were used, and the rule in question used no such inconsistent words. The present Ord 13, r 4 opens, "Except as otherwise provided . . .", but that does not, in our view, mean anything different from "Subject to the provisions of these rules", unless it makes even clearer the need for specific inconsistency.

The tenant in the present case naturally relies very strongly on this authority. If the court was there able to save the tenants' proceedings, although their solicitors knew of the failure to serve within the period D for service and could have served the landlords' solicitors in time, she argues that the same result should follow here.

In *Robert Baxendale Ltd* v *Davstone (Holdings) Ltd* [1982] 1 WLR 1385* there were two cases before the Court of Appeal. In each the tenants had failed, within the period allowed for service (by this time two months), either to serve or to apply for an extension of time to serve. In one case the tenants had obtained an order extending time, in the other not. The rule corresponding to Ord 7, r 20 was in the same form as in *Lewis*, above, and the Court of Appeal relied on that authority in holding that, while a discretion to extend time existed, it was a discretion to be sparingly exercised and not to be exercised where no justification could be shown for failing to serve in time.

E The authority on which the landlord principally relies is *Bernstein* v *Jackson* [1982] 1 WLR 1082. In that case a district registrar made an order for substituted service of a writ the validity of which had expired over two years before, which had never been renewed and which he did not renew. On the defendant's application to set aside a default judgment entered against her, the judge held that the writ was not valid for service since it had not been renewed under RSC Ord 6, r8, but that this was an irregularity which could (and should) be cured under RSC Ord 2, r1. The Court of Appeal took a different view. Dunn LJ said at p1089D:

> I do not think that the judge could have extended the writ under RSC Ord 6, r8 and, that being the case, I think that there is great force in his [counsel's] submission that RSC Ord 6, r8 provides a compendious code for the extension F and renewal of writs, and that it is not the type of irregularity which was envisaged could be dealt with by the provisions of RSC Ord 2, r1. But, assuming that it could, in my view this was such a fundamental defect in the proceedings that the judge should not have exercised his discretion to make an order under RSC Ord 2 r1.

Slade LJ agreed at p1089M:

> If a specific application to renew the writ had been before him [the judge], I do not think that he could have properly extended its validity for the reasons already given by Dunn LJ. Correspondingly, I do not think that the plaintiffs can be in a better position than they would have been if such a specific application had been [made] before the judge.

This reasoning suggests, on our understanding, that retrospective G renewal for the necessary period was not permissible under Ord 6, r 8 and that the failure to renew was not a defect which could be cured under Ord 2, r 1. It is, however, noteworthy that no reference was apparently made to RSC Ord 3, r 5 and *Lewis* was not cited.

The Court of Appeal had occasion to reconsider *Bernstein* v *Jackson* in *Leal* v *Dunlop Bio-Processes International Ltd* [1984] 1 WLR 874. In that case the plaintiff suffered personal injuries on May 27 1978. He issued a writ at the end of the relevant limitation period on May 20 1981. On May 18 1982, the penultimate day of the writ's validity, without the leave of the court and without any order of renewal, the writ was served on the defendants out of the jurisdiction. The defendants applied to set aside service and dismiss the action on June 10 1982. The plaintiff applied on July 19 1982 for renewal of the H writ under RSC Ord 6, r 8, and for leave to serve out of the jurisdiction. Before the district registrar the plaintiff in effect succeeded. But he failed on the defendants' appeal to Neill J, who held (applying *Bernstein* v *Jackson*) that he could not and in any event should not grant the plaintiff the relief he sought.

The Court of Appeal upheld Neill J's exercise of discretion, but differed on the question of jurisdiction. In a leading judgment with which May LJ and, subject to one reservation, Slade LJ agreed, Stephenson LJ said of the court's judgment in *Bernstein* v *Jackson* at p879C:

> If Dunn LJ meant to say that the judge had no power to extend the validity of the writ under Ord 6, r8(2), even if an application to extend had been made, I respectfully consider that he was ignoring the terms of Ord 6, r8(2); if he J meant to say that the judge had no power to treat the failure to extend the writ as an irregularity under Ord 2, r1, again I have to disagree with him . . .

Slade LJ expressly agreed that Ord 2, r1 gave the court jurisdiction to treat the failure to extend the writ as an irregularity and to waive it accordingly. Emphasis was, however, laid on the special circumstances needed to justify renewal after expiry of a limitation period and it was largely for that reason that the judge's exercise of discretion was upheld.

In this case also it does not appear that *Lewis* was cited or reference made to RSC Ord 3, r5. Such reference could have strengthened only the conclusion which the court reached on the jurisdictional question.

K This review of the authorities and the rules leads us to conclude that, even where an originating application of this type is not served within two months of the date of issue and application to extend is not made within four months of the date of issue, the county court has jurisdiction on proper grounds being shown: (1) to extend the time for service under Ord 13, r4 of the CCR; and (2) to treat the failure to extend and serve as an irregularity and order the action to proceed under Ord 37, r5.

Counsel for the landlord did not press a contrary argument. Although the opening words of Ord 13, r4 differ from those considered in *Lewis*, we have already indicated that, in our view, the difference, if significant at all, strengthens the tenant's argument.

L It was argued for the landlord that, although the court had a discretion to extend time in this case, it was one which could properly be exercised only by refusing an extension. This was because Ord 7, r20 of the CCR, like RSC Ord 6, r8 (and Ord 11, r1), were of a special character such that it was impermissible to override them. We prefer to say nothing about Ord 11, r1, which is not in issue here and raises special questions. So far as Ord 7, r20 and Ord 6, r8 are concerned, the submission is, in our judgment, unsound. It is not consistent with *Lewis, Robert Baxendale* or *Leal*, and it is not supported by *Bernstein* as explained in *Leal*. It is not supported by the wide terms of RSC Ord 2, r1 and Ord 3, r5 and CCR Ord 13, r4 and Ord 37, r5. And it is not, in our view, consonant with the requirements of justice in some cases. M It is one thing to say that good grounds must be shown to justify a certain exercise of discretion; it is quite another to emasculate a discretion by holding that it can be exercised only one way.

We have already accepted that a party (such as the tenant) seeking an extension of time for service after the validity of the proceedings has expired and after expiry of any relevant limitation period must show good reason for the court to exercise its power to extend and also give a satisfactory explanation for the failure to apply for extension before the validity of the proceedings expired. We have no doubt that the tenant discharged these burdens. She took the correct steps to issue her originating application in good time. It was entirely the fault of the county court that the proceedings were not served. She was reasonably unaware of this until after September 4 1991 when the

---

*Editor's note: Also reported at (1982) 264 EG 713.

A period for service expired. It would be most unjust if she paid the penalty for the court's error.

That conclusion, we think, opens the door to consideration of more general matters relevant to the exercise of discretion. The landlord's solicitors were, at best, remiss in failing to respond to the tenant's solicitors' letter of October 1 until November 29, but those solicitors were even more remiss in failing to pursue the question of service during that period, and even when they learned the facts from the county court in early December they did not apply to the court until a month later. Time-limits such as these are intended to be short; it is incumbent on parties to comply with them; and if extensions were granted at all readily the time-limits would very quickly become a
B dead letter. That is why the discretion to extend must be sparingly exercised.

It is, on the other hand, relevant that:

(1) the landlord until January 1992 raised no objection to a new tenancy;

(2) the difference over rent was unlikely to prove an insuperable stumbling block;

(3) but for the court's error the application would have been duly served;

(4) the tenant did not learn of the error in time to cure it; and

(5) the delay in service has of itself caused the landlord no prejudice, since her change of circumstances would have occurred even if she had been served in time.

C In these circumstances there were, in our opinion, strong grounds for granting the extension sought.

This conclusion would not entitle this court to substitute its exercise of discretion for that of the judge unless it were satisfied that the judge's exercise of discretion was demonstrably wrong. We think it was. He appears to have given little or no weight to considerations (1), (2), (3) and (4) above; and he held that there was prejudice to the landlord without recognising that this was not caused by the tenant's delay.

We therefore allow the appeal, set aside the judge's order and grant the tenant such extension as she requires to effect service of the proceedings.

D *Appeal allowed with costs.*

Court of Appeal

December 21 1992

(Before Lord Justice NOURSE and Lord Justice RUSSELL)

FRENCH v COMMERCIAL UNION LIFE ASSURANCE CO PLC AND ANOTHER

Estates Gazette June 19 1993

E **[1993] 24 EG 115**

Landlord and Tenant Act 1954 — Determination of new rent and interim rent — Whether county court judge erred in determining rents

In 1981 the appellant took an assignment of a lease of unit 11, Kingfisher Walk, in the Kingfisher Shopping Centre, Redditch, for a term of 14 years from May 2 1973. Since 1985 the landlords have been the respondents. By a notice given under section 25 of the 1954 Act they duly terminated the tenancy at the expiration of the contractual term on May 2 1987. The annual rent payable during the last seven years of
F the contractual term had been £10,475, a figure agreed for rent review purposes between the tenant and the then landlords, Redditch Development Corporation, in September 1981. The appellant is a retail fruiterer. The Kingfisher Centre is a large development of shops and supermarkets with adjoining car parks. Unit 11 comprises a shop arranged on ground, first and basement levels. The interim rent fell to be assessed on May 2 1987, at which date the premises adjoining unit 11 on its west side, units 12 and 13, were a supermarket occupied by J Sainsbury plc. In Redditch County Court Judge Roy Ward QC determined the interim rent at £28,500 and the rent under the new tenancy, the grant of which the respondents did not oppose, at £36,098. The appellant appealed against the G
determination of the rents contending: (1) in regard to the interim rent that the valuation in terms of zone A of £51.50 per sq ft was too much and that the deduction of 3.275% of the interim market rent was too small a cushion; and (2) in relation to the rent under the new tenancy, the judge recognised the fact that, in a time of recession, retailers are not ready and willing to take leases of inferior sites, but did not, in fact, make anywhere near a sufficient reduction in that respect, and did not deal adequately with a transaction involving the unit originally occupied by J Sainsbury plc.

Held: **The appeal was allowed. The court had all the written** H
material and the judge's notes of evidence, although it could not differ from any finding of the judge which was based on demeanour or reliability of the witnesses; in this type of case such findings are rare and the appellate court is in as good a position as the trial judge to assess the merits of the expert evidence and make findings on it accordingly. The judge was wrong to place reliance on two comparables while at the same time disregarding a further comparable. He was also wrong to pay no regard to the known fact in May 1987 that J Sainsbury plc were to move out of their unit On May 2 1987 the value of unit 11 was £51.50 per sq ft ITZA and, after the 10% agreed reduction for the tenancy from year to year, the unit should be J
valued at £45.90 per sq ft ITZA, resulting in an interim market rent of £26,195. A cushion of 10%, to have regard to the old rent, should be made, leaving an interim rent of £23,576. In relation to the rent under the new tenancy, the judge had accepted that there should be some reduction in recognition of the fact that, in a time of recession, retailers are not ready and willing to take leases of inferior sites, but he did not, in fact, make anywhere near a sufficient reduction in that respect. On the evidence before the court, the unit should be valued in November 1991 at the rate of £56 per sq ft ITZA, resulting in a new rent of £31,858. K

The following cases are referred to in this report.

Charles Follett Ltd v *Cabtell Investments Ltd* (1987) 55 P&CR 36; [1987] 2 EGLR 88; (1987) 283 EG 195, CA

English Exporters (London) Ltd v *Eldonwall Ltd* [1973] Ch 415; [1973] 2 WLR 435; [1973] 1 All ER 726; (1972) 25 P&CR 379; [1973] EGD 439; 225 EG 255 & 433

This was an appeal by the applicant, Mr Anthony John French, from orders made by Judge Roy Ward QC in Redditch County Court whereby he determined an interim rent and rent under a new tenancy of premises in the Kingfisher Shopping Centre at Redditch owned by the responents, Commercial Union Life Assurance Co plc and Shell L
Pensions Trust Ltd.

John West (instructed by Wallis Robinson & Morgan, of Solihull) appeared for the appellant; David Smith (instructed by Stones Porter & Co) represented the respondents.

Giving the first judgment, NOURSE LJ said: This is an appeal by a tenant against the amounts of an interim rent and a new rent determined by the court on his application for a new tenancy of business premises under section 24(1) of the Landlord and Tenant Act 1954.

The tenant is Mr Anthony John French. In 1981 he took an assignment of a lease of unit 11, Kingfisher Walk, in the Kingfisher Shopping Centre, Redditch, for a term of 14 years from May 2 1973. M
The landlords since 1985 have been the respondents, Commercial Union Assurance Co plc and Shell Pensions Trust Ltd. By a notice given under section 25 of the 1954 Act they duly terminated the tenancy at the expiration of the contractual term on May 2 1987. The annual rent payable during the last seven years of the contractual term had been £10,475, a figure agreed for rent review purposes between the tenant and the then landlords, Redditch Development Corporation, in September 1981.

The tenant, who has seven shops in the West Midlands, is a retail fruiterer trading under the name or style of "Fresha Fruits". The Kingfisher Centre is a large development of shops and supermarkets with adjoining car parks, the whole of which is owned and managed

A by the landlords. Kingfisher Walk runs through the middle of the centre from Worcester Square in the east, the units being numbered consecutively from east to west on the north side and then back again along the south side. At about its middle point Kingfisher Walk widens into an area about 53 ft x 53 ft known as Clock Court. Unit 11, which comprises a shop arranged on ground, first and basement levels, is on the north side of Clock Court at about its middle point. On the determination of the contractual term on May 2 1987, the date at which the interim rent falls to be assessed, the premises adjoining unit 11 on its west side, ie units 12 and 13, were a supermarket occupied by J Sainsbury plc.

The tenant's application for a new tenancy was heard by Judge
B Roy Ward QC in Redditch County Court in September and November 1991. The landlords did not oppose the grant of a new tenancy, which it was agreed should be for a term of five years. There was also agreement on all other matters except the interim rent, the new rent and the tenant's claim that the new lease should incorporate an additional term. In his judgment delivered on November 18 Judge Ward refused to include the additional term. He fixed the interim rent at £28,500 and the new rent at £36,098. The tenant has not appealed against the judge's refusal to include the additional term, but he claims that both the interim rent and the new rent are too high.

The judge had the evidence of expert valuers on each side. Each expert expressed his views fully in writing and then gave extensive oral evidence as well. We have all the written material and the judge's
C notes of evidence. Not having seen and heard the witnesses give their evidence, we could not differ from any finding of the judge which was based on demeanour or reliability. But in this type of case such findings are rare. Broadly speaking, the appellate court is in as good a position as the trial judge to assess the merits of the expert evidence and make findings on it accordingly.

Interim rent

Although the judge dealt first with the new rent, in this court Mr West, for the tenant, started with the interim rent. I find it convenient to follow the same course. The court's power to determine an interim rent is given to it by section 24A of the 1954 Act, which was added by section 3(1) of the Law of Property Act
D 1969. Section 24A is in these terms:

> (1) The landlord of a tenancy to which this Part of this Act applies may —
> (*a*) if he has given notice under section 25 of this Act to terminate the tenancy; or
> (*b*) if the tenant has made a request for a new tenancy in accordance with section 26 of this Act;
> apply to the court to determine a rent which it would be reasonable for the tenant to pay while the tenancy continues by virtue of section 24 of this Act, and the court may determine a rent accordingly.
>
> (2) A rent determined in proceedings under this section shall be deemed to be the rent payable under the tenancy from the date on which the proceedings were commenced or the date specified in the landlord's notice or the tenant's request, whichever is the later.
>
> (3) In determining a rent under this section the court shall have regard to the
> E rent payable under the terms of the tenancy, but otherwise subsections (1) and (2) of section 34 of this Act shall apply to the determination as they would apply to the determination of a rent under that section if a new tenancy from year to year of the whole of the property comprised in the tenancy were granted to the tenant by order of the court.

The effect of section 34(1) and (2) of the Act is that the rent payable under a tenancy granted by order of the court shall, in default of agreement, be such as may be determined by the court to be that at which, having regard to the terms of the tenancy (other than those relating to rent), the holding might reasonably be expected to be let in the open market by a willing lessor, certain standard matters being disregarded.

Although the determination of an interim rent is a matter for the
F discretion of the court, the tenant has throughout accepted that one should be determined in this case. It is also agreed that the rent will run from May 2 1987 (being the date specified for the termination of the tenancy in the landlords' notice under section 25) until the expiration of three months beginning with the date on which the application is finally disposed of: see section 64(1) of the Act. Thus the interim rent will run for another three months from today, making a period of nearly six years in all. As to the amount of the interim rent, the parties are agreed that under section 24A (2) and (3) the court must assess the section 34 market rent for a tenancy from year to year commencing on May 2 1987 ("the interim market rent" and then decide what, if any, consequences are to flow from the obligation to "have regard to" the old rent of £10,475. The first part
of this exercise is one of valuation depending on the evidence of G
experts; the second is a matter for the judgment of the court.

Ignoring at present those parts of the premises not falling in "zone A", the method of valuation adopted by the experts on both sides was to make a valuation per square foot of floorspace in terms of zone A ("ITZA"), as if for a term of years, and then to discount it by 10% in order to tailor it to a tenancy from year to year. The judge arrived at a valuation ITZA of £57.30 per sq ft which, after deducting 10%, he rounded down to £51.50. For an agreed zone A area of 530 sq ft, that produced a figure of £27,295, to which he added £2,170 for the first floor and basement, making a total interim market rent of £29,465.

In carrying out the second part of the exercise, the judge took the increase of £19,000 over the old rent of £10,475 and, for the period
from May 2 1987 to November 18 1991, multiplied it by four and a H
half. Having thus calculated that the tenant would have to find some £85,500 over and above the old rent, at p11H of the transcript he continued:

> To some extent clearly he should, as a shrewd businessman, have contemplated an increase and made provision against it of a considerable part of that sum. Nevertheless, it seems to me to give rise to such a substantial immediate demand on his resources that he should be cushioned to some extent against the shock. Having regard to the fact that he should, it seems to me, have contemplated an increase in the region of £17,000 a year on his own advisers' calculations, it would be appropriate to allow a cushion of a further 5% of the capital sum. Doing that in round terms it would reduce the appropriate interim rent to £28,500 pa.

So the judge deducted 5%, not of the interim market rent of J
£29,465 but of the £85,500 extra which the tenant would have to find over the period of four and a half years. That produced a deduction of £4,275 which, averaged over the four and a half years, was the equivalent of an annual reduction of £950, thus producing an interim rent of £28,515, which the judge rounded down to £28,500. Viewed as a percentage of the interim market rent, the reduction was only 3.275%.

The first point to be made about the judge's interim rent is that the parties are agreed that the figure of £2,170 for the first floor and the basement was wrong, the right figure being £1,868. Mr Smith, for the landlords, therefore accepts that the interim rent ought to be reduced to £28,230, say £28,250. The tenant's outstanding complaints in
regard to the interim rent are twofold: first, that the valuation ITZA K
of £51.50 per sq ft was too much; second, that the deduction of 3.275% of the interim market rent was too little.

Under the first of these complaints, Mr West, having pointed out that the judge's figure for the interim market rent (£29,465) was higher than that put forward by the landlords' expert, Mr John Tinley [ARICS] (£29,348), made three main submissions. First, he submitted that the only two comparables considered by the judge were not, as the judge thought, the best; that he ought to have considered three different and better comparables; and that the calculation by which he applied the comparison was in any event arbitrary and unsound. Second, Mr West submitted that the judge
wrongly disregarded what has been called "the Sainsbury factor". L
Third, he submitted that the judge wrongly disregarded what has been called "the refurbishment factor".

The judge thought that the best comparables were units 5 and 7. Unit 5 was let on June 24 1987 for a term of 20 years at an agreed rent ITZA of £63.25 per sq ft. Unit 7 was the subject of an agreement as to an interim rent, as at February 1987, but before any discount for an annual tenancy, of £61.30 per sq ft ITZA. On the face of them, those dates were near in time to May 2 1987. However, Mr West argued that the rent for unit 5 was agreed nearly two months later at a time of rising values and that the interim rent for unit 7, although assessed as at February 1987, was not agreed until more than two years later. His more substantial complaint was that units 5 and 7 were not truly
comparable, being situated some way to the east of unit 11 and not in M
Clock Court, unit 5 being next door to Boots. Moreover, he criticised the judge's method of discounting the values on the basis of £1 per sq ft for every unit between units 5 and 7 respectively and unit 11, for which purpose he pointed to the judge's own description of it as arbitrary. Thus unit 5, being six units away, was discounted from £63.25 to £57.25 and unit 7, being four units away, was discounted from £61.30 to £57.30. It was thus that the judge arrived at his figure of £57.30 already referred to.

Mr West submitted that the better comparables were units 10, 29, and 26. Unit 10, immediately to the east of unit 11 and also in Clock Court, was let in November 1985, 18 months earlier, at £36 per sq ft ITZA. Unit 29, almost opposite unit 11 on the other side of Clock

A Court, was let from December 25 1986 at £49 per sq ft ITZA, with the first three months rent free. In his oral evidence Mr Tinley said that agreement as to that rent might have been reached later. Earlier he had described the next-door unit, unit 30, which is immediately opposite unit 11, as "the mirror image" of unit 11. Unit 26, which is in the western part of Kingfisher Walk and on the other side to unit 11, was let in April 1987 at £44 per sq ft ITZA. Mr West was prepared to accept that neither unit 10 nor unit 26 was especially helpful, the date of the former being too early and the latter, like units 5 and 7, being too far away. But he placed great reliance on unit 29, which was close both in date and location.

Sainsbury moved out of units 12 and 13 in July 1989. It is agreed that in May 1987 it was generally known that they were going to B move. The tenant's expert, Mr Richard Hardwick [FRICS] said that, since the move was anticipated, any tenant from year to year would have expected a discount rather greater than the normal 10%. The judge did not act on that view. He said, at p5D:

> The unfortunate applicant in this case was badly affected, sustaining a 40% loss in turnover which has never recovered. However, the applicant knew at the time of his assignment that Sainsbury were going to leave at some time in the future and that it was their national policy to move to purpose-built sites, and he contemplated at that time a drop in his own profits, although not appreciating that it would be as severe as he had not contemplated. I do not think that he can here complain that the eventual effect was worse than he foresaw. It might have proved to be better. I do not think that this is a factor which can properly influence my judgment either on the new rent or as to the C interim rent. It was a matter entirely for the commercial judgment of the applicant.

Mr West criticised that passage on two grounds. First, he submitted that the judge had failed to understand the true nature and effect of the Sainsbury factor, which was simply that a prospective tenant in May 1987 would have required a discount on account of the damage to his turnover, whatever it might have been beforehand, that would be suffered when Sainsbury moved out. Second, Mr West said that the judge's misunderstanding of the Sainsbury factor was illustrated by his mistaken belief that it could have an effect not only on the interim rent but also on the new rent. It was never suggested that it could have an effect on the new rent.

D The refurbishment factor was said to be of a similar nature and effect. In May 1987 it was known that there was going to be a major refurbishment of the centre, in particular around Clock Court. Again Mr Hardwick maintained that that prospect would also have depreciated the amount which a prospective tenant was prepared to pay at that time.

In my view, there is substance in some of Mr West's criticisms of the judge's approach to the assessment of the interim market rent. I think that he was wrong to place reliance on units 5 and 7 and to apply the arbitrary £1 per unit discount, while at the same time disregarding unit 29. I think that he was also wrong to pay no regard to the Sainsbury factor. The valuation of unit 29 some five months earlier at £49 per sq ft ITZA, which would itself have been influenced by the E Sainsbury factor and, if it had any influence, by the refurbishment factor, cannot be disregarded. Doing the best I can on the evidence available to the judge, I would value unit 11 on May 2 1987 at £51 per sq ft ITZA. It follows that, after the 10% agreed reduction for the tenancy from year to year, unit 11 should be valued at £45.90 per sq ft ITZA. If that valuation is applied to the 530 sq ft in zone A and the agreed £1,868 for the first floor and the basement is added on, the result is an interim market rent of £26,195.

The question then is whether any deduction ought to be made as a result of the statutory obligation to have regard to the old rent of £10,475. That is a question which has caused difficulty in other cases: see, for example, *English Exporters (London) Ltd* v *Eldonwall* [1973] F Ch 415*. However, both sides have accepted as a correct statement of Parliament's intention in requiring regard to be had to the old rent the following passage in my judgment (with which Fox LJ and Sir Denys Buckley agreed) in *Charles Follett Ltd* v *Cabtell Investments Ltd* (1987) 55 P&CR 36†, at p42:

> By 1969 it had been demonstrated that a tenant, in times of inflation, could readily spin out the steps prescribed by the 1954 Act and the rules of court, so as unfairly to prolong the continuation of the old rent under section 24. The defeat of such practices was the primary legislative purpose of section 24A. At the same time it was recognised that, while inflation benefits the tenant during the currency of a lease at an uninflated rent, it exposes him to an inordinate shock if its consequences are visited on him in full directly the lease has G determined. The legislative purpose of the requirement that regard should be had to the old rent was, where appropriate, to cushion the tenant against that shock.

The facts of *Charles Follett Ltd* v *Cabtell Investments Ltd* were recognised by this court to have been exceptional. The old rent agreed on a rent review not much more than seven years beforehand had been £13,500 and the interim market rent had been assessed by the judge at £80,000. The judge had then reduced that figure by 50%, arriving at an interim rent of £40,000. It was thought by this court to be an exceptional case if only for the reason that a reduction of 50% necessarily produced an interim rent nearer to the old rent than the interim market rent. But we decided no more than that the judge did H not err in principle or exercise his discretion in a manner which was plainly wrong.

Mr West has relied on the following passage in my judgment at p44:

> Certainly I would agree that to make a 10% reduction, a mere £8,000, would be paying no more than lip service to the old rent or, if you prefer it, not providing the plaintiff with any real cushion at all.

That does not mean that a 10% reduction or even less would not, in other cases, be appropriate. In that case the stark fact was that a reduction of only 10% would still have left the tenant paying an interim rent of more than five times the rent which had been agreed not much more than seven years beforehand. The judge was entitled J to take the view that that would have been an inordinate shock to him and that the cushion must be appreciably larger than 10%. On the other hand, I would very much doubt that a cushion as large as 50% could be sustained in any but the rarest of cases.

In the present case the difference between the old rent (£10,475) and the proposed interim market rent (£26,195) is £15,720. Thus the interim market rent is almost exactly two and a half times the old rent.

Ought there to be a cushion in this case and, if so, what should it be? Mr Smith has not suggested that there should be no cushion. He says that the judge's approach was correct. As I have said, the judge calculated his cushion by reference to the aggregate amount of the difference between the old rent and the interim market rent over the four and a half year period between May 1987 and November 1991. K Although I believe that it is usual to assess the amount of the cushion by taking a percentage of the interim market rent, I do not say that the judge's method of calculation was not open to him. However, it is preferable to express the deduction as a percentage of the interim market rent. On that footing the cushion was 3.275%.

In arriving at his figure, the judge, in the passage from his judgment already quoted, seems to have placed reliance on the fact that the tenant should, on his own advisers' calculations, have contemplated an increased rent in the region of £17,000 a year. Inferentially the judge suggested that the £19,000 increase awarded ought not, therefore, to have come as a shock to the tenant. But that is not the shock with which the statute is concerned. It is the shock of having to pay a greatly increased rent, irrespective of whether the L tenant has been forewarned of it or not. I therefore think that the judge's calculation of the cushion proceeded on an error of principle. In any event, a percentage of no more than 3.275 was, in all the circumstances of the case, far too little.

That means that we in this court must exercise an original discretion as to the amount of the cushion. What figure should we adopt? One of the most significant features of the case is that the interim rent will be payable for a period of nearly six years. In other words, the landlords will have been kept out of the new rent for a period very much longer than is either usual or desirable. Surely the obligation to have regard to the old rent must include a regard for the period after the end of the contractual term during which the old rent M has continued. We must therefore take into account the cushion which the tenant has already received in not having to pay anything more than the old rent until November 1991, when we were told that he had, by agreement, started to pay an increased amount. Without speculating as to what might have been appropriate in any other case, or in this case had the circumstances been different, I think that we should fix the cushion at 10%. That means that £2,619 should be deducted from the interim market rent of £26,195, leaving an interim rent of £23,576.

The new rent

I turn to consider the new rent, which fell to be assessed as at the date of the hearing in November 1991. Before dealing with

*Editor's note: See also [1973] 225 EG 255 & 433.
†Editor's note: See also [1987] 2 EGLR 88.

comparables, the judge considered a number of special and general factors affecting unit 11. He started with its location. Although he did not think that the location, close to a corner, would make a significant difference to a retail fruiterer, he accepted that for many other retailers it would make a less desirable site, that the field of potential lessees would be reduced and certainly that they would not be prepared to pay the same rent as they would for better sites. Having said that in times of a buoyant market that might not matter, he continued, at p3H:

> In a time of recession, however, it might make a significant difference and *pro tanto* substantially affect the balance of supply and demand. I do not regard it as pure coincidence in this case that two of the corner shops, numbers 12 and 31, have long been unlet. It is, in my judgment, [a] clear indication that retailers are not ready and willing to take leases of those premises at any rate at the rents which the landlords are presently demanding. There should accordingly be some reduction in recognition of that factor.

Next, the judge referred to a pedestrian flow count commissioned by the tenant's advisers and carried out on a Thursday in October 1988. Although in broad terms he accepted that the findings of that count, which showed that many pedestrians passed by unit 11 without going in, were evidence of the general pedestrian flow, he remained sceptical as to the proper inference to be drawn from it. He was certainly not prepared to accept Mr Hardwick's suggestion that the rent should be discounted by 20%, which he described as arbitrary.

Having then dealt with the Sainsbury factor, which I repeat was immaterial to the new rent, the judge turned to the impact of rates and service charges, as to which he rejected the submissions of Mr West on behalf of the tenant. Finally among the special and general factors affecting unit 11, the judge considered the impact of the recession. Having said that he must limit himself to the overall effect of the evidence in the case, he continued, at p6H:

> That evidence is to the effect that business is in recession, that shops are standing empty, that tenants of some shops are offering incentives by way of reverse premiums to anybody willing to take leases off their hands. Of that I am quite satisfied and there is, in my judgment, no useful purpose to be served in examining all the specific particulars of individual cases.

The judge then considered other individual sites, starting with the Sainsbury site and the Milletts Camping site. After Sainsbury had moved out in the autumn of 1989, their premises were empty for about 12 months. Later the major part of the premises was divided into two, part being let to Peacocks and part to Argos. That left unit 12, a small part in the shape of a dogleg immediately adjoining unit 11 to the west, which, shortly before the hearing below, was agreed to be let from September 29 1991 to Milletts at a rent of £70 per sq ft ITZA. The judge accepted that the Argos and Peacock leases were on far more favourable terms than the leases of smaller units, but on reflection he thought that he should discount that fact, on the ground that tenants such as they would be likely to have been treated by the landlord as loss leaders. He also found that he did not gain any assistance from the Milletts transaction. Although Mr West has argued to the contrary, I see no reason for differing from the judge over the leases to Argos and Peacocks. On the other hand, the Milletts transaction merits careful consideration and I will come back to it in due course.

The judge then considered units 32 and 30. I agree with him that unit 32 was of no positive assistance. Unit 30 is a more problematical case. That was let to Hogg Robinson in March 1990 at £65.76 per sq ft ITZA. As I have said, Mr Tinley described it as the mirror image of unit 11, being directly opposite it on the other side of Clock Court. Accordingly, on the face of it, the rental valuation of unit 30 in March 1990 ought to have been of some assistance in making a rental valuation of unit 11 in November 1991. However, the judge was not happy about the evidence as to this unit, thinking it proper to infer that there was some sort of a package deal which he could not assess. He therefore eliminated unit 30 from his consideration. At the same time he observed that unit 31 still remained unlet despite having been advertised by the landlords for nearly two years.

At p9A the judge turned to the effect of the recession generally, reciting the experts' views:

> But Mr Hardwick's view was that the property market reached its peak in September 1989, that it is still declining and that market prices at the moment are some 15% to 20% off the market at its peak. Mr Tinley's evidence is not very different. He accepts that the peak was in autumn 1989. He thinks the downward trend has now ceased and has done for some three or four months, and he puts the drop from peak times at some 10%.

The judge concluded that his proper approach ought to be to fix the rent by reference to comparables, making a further reduction, first, in respect of the inferior location of the tenant's site as opposed to prime sites and, second, in recognition of the recession and the consequent downturn in the market. Having referred to the evidence before him as to the 34 individual shops in Kingfisher Walk, he was prepared to accept that, by and large, those of the same size were originally let at much the same rents. He then said that if he excluded consideration of leases or rent reviews more than 18 months before November 1991 he was left with units 9, 15 and 28. At p10C, he continued:

> That would be an inadequate sample, it seems to me, and I propose to cast the net wider and I propose to have regard to other Zone A assessments as indicated on the plan . . . Doing my best to make a proper allowance for the slump in property prices and for a differential to affect the diminution in value attributable to its position in the square, I arrive at a Zone A assessment per square foot ITZA of £64.

The judge then applied that valuation to the 530 sq ft in zone A and added on £2,178 for the first floor and the basement, arriving at a new rent of £36,098.

In this court Mr West has advanced a number of criticisms of the judge's approach to the assessment of the new rent, both in regard to the special and general factors affecting unit 11 and in regard to the comparables. He complains in particular of the judge's omission to state to which other zone A assessments he had had regard. For my part, I do not see that criticism as being of any substance. Looking at the plan for myself, I would think it to be a matter of no great difficulty to see what other units the judge took into account.

In my view, Mr West's criticisms are, in two important and related respects, valid. First, I think that he is correct in saying that, although at pp3H and 6H the judge had accepted that there should be some reduction in recognition of the fact that, in a time of recession, retailers are not ready and willing to take leases of inferior sites (he instanced units 12 and 31 as having long been unlet), he did not, in fact, make anywhere near a sufficient reduction in that respect.

Second, there is the important matter of the Milletts' site. As I have said, unit 12 was agreed to be let to Milletts from September 29 1991 at a rent of £70 per sq ft ITZA. On the face of it, that letting of immediately adjoining premises of much the same size so soon before the hearing below ought to have had a decisive effect on the valuation of unit 11. No doubt the landlords, who put it forward as a comparable, hoped that it would. However, the matter was by no means as simple as that.

For some years Milletts had occupied a site at the north-eastern corner of Worcester Square, near the principal entrance to the Kingfisher Centre. After a rent review in 1987 they found that the rent there had reached a level at which their profits had become marginal. In 1990 they instructed Mr D C Haynes [FRICS], the senior partner in Kingstons, chartered surveyors, to dispose of the premises. He gave evidence in the court below. I take up the story in the words of the judge, at p7F-G:

> As a result of negotiations they have agreed to move to unit 12 which has now been empty for some two years. That agreement is by way of a package deal. The [landlords] are to pay Milletts a capital sum by way of premium of £62,500 for surrender of their present site. That has advantages to the [landlords] in that they will recover a presently untenanted 4,000 sq ft of first floor space, plus an ability to redevelop that site with its adjacent premises just outside the shopping centre which also are currently standing empty.
>
> Coupled with that payment of a capital sum is an agreement to rent to Milletts unit 12 at a rent of £42,000 pa, but with a rent-free period of nine months. I do not propose to examine the various permutations (which seem to me to be endless) of the consequences of that package deal, but I am bound to say that like Mr Haynes I find it impossible to accept that there was not in some substantial degree an incentive by the [landlords] given to Milletts to enter into a tenancy of 12 Kingfisher Walk; and that incentive is calculated to distort the apparent market rents available. On the other hand, I cannot find it a useful exercise to show what market rent it really represents. That depends entirely on how one splits up the capital sum and the rent finally agreed.

It is significant that the judge agreed with Mr Haynes that the landlords must have given Milletts a substantial incentive to enter into a tenancy of unit 12. Of course the judge was right to imply that it is very difficult to say with what result in terms of open market rental valuation the package should be broken down. But some attempt must be made. Mr Haynes himself suggested that it might have resulted in a valuation of £42.41 or even £32.81 per sq ft ITZA. For my part, I am unable to go anywhere near so far as Mr Haynes, not least because Mr Hardwick's own valuation was £51 per sq ft

A ITZA. The real value of the Milletts transaction as viewed by the judge is that it was a particular example of the general proposition, accepted by him, that in a time of recession retailers are not ready and willing to take leases of inferior sites without some substantial incentive, an incentive which, in terms of section 34, can be only by way of a substantial reduction in rent. As I have said, I do not think that the judge made anywhere near a sufficient reduction in that respect.

Doing the best I can on the evidence available to the judge, I would value unit 11 in November 1991 at £56 per sq ft ITZA. If that valuation is applied to the 530 sq ft in zone A and the judge's figure of B £2,178 for the first floor, and the basement (to which no objection has been made) is added on, the result is a new rent of £31,858.

In summary, I would reduce the interim rent determined by the judge from £28,500 to £23,576 and the new rent from £36,098 to £31,858. I would allow the appeal accordingly.

RUSSELL J agreed and did not add anything.

Appeal allowed with costs.

LANDLORD AND TENANT

RENT ACTS

A **Court of Appeal**

October 21 1992

(Before Lord Justice BUTLER-SLOSS and Lord Justice Simon BROWN)

GRAY v BROWN

Estates Gazette February 20 1993

[1993] 07 EG 97

Rent Act 1977 — Whether tenancy agreement a restricted B ***contract within section 21 of the Rent Act 1977***

By an oral agreement made in March 1978 for the exclusive occupation of a room at 40 Maidstone Road, London N11, the appellant tenant was to share the bathroom, lavatory and kitchen. The respondent landlord has always lived elsewhere. The respondent served a notice to quit on the appellant dated November 8 1990 and commenced proceedings in July 1991 seeking possession of the room. The appellant did not appear at the hearing, although he sent three friends to seek an adjournment on the ground that he was seriously affected by agoraphobia. The judge refused the adjournment, found that the agreement was a restricted contract within section 21 of the C Rent Act 1977 and made an order for possession. The appellant appealed contending, *inter alia*, that he held a tenancy protected by the Rent Act 1977.

Held: **The appeal was allowed. The terms of the 1978 agreement did not disclose any reference to the possibility of the landlord himself residing in the property. The words of the landlord's affidavit setting out the terms of the oral agreement were not sufficiently specific to make it clear that the landlord had reserved to himself the right to be one of the possible occupiers of the premises. To come within section 21, the terms must include clear words reserving the right to the** D **landlord to live on the premises sharing the accommodation with the tenant. It is sufficient for the landlord, who does not live on the premises, to reserve to himself the express right to share accommodation with the tenant, and it is not necessary that he should actually exercise that right, nor have a clear intention to do so, so long as the possibility of moving into the premises is genuinely within his contemplation at the time of the tenancy agreement. But on the evidence found in this case that was not the position. The appellant was a protected tenant under section 22 of the Rent Act 1977.**

The following cases are referred to in this report.

E *A G Securities* v *Vaughan; Antoniades* v *Villiers* [1990] 1 AC 417; [1988] 3 WLR 1205; [1988] 3 All ER 1058; (1989) 57 P&CR 17; [1988] 2 EGLR 78; [1988] 47 EG 193, HL

Stanley v *Compton* [1951] 1 All ER 859, CA

This was an appeal by the appellant, Mr Brown, the tenant of a room at 40 Maidstone Road, London N11, against a decision of Judge Stockdale at Barnet County Court on October 14 1991 making an order of possession upon a claim for possession by the respondent Mr Gray.

Simon Birks (instructed by Caines, of Enfield) appeared for the appellant; David Matthias (instructed by Attwater & Liell, of Loughton) represented the respondent. F

Giving judgment, **BUTLER-SLOSS LJ** said: This is an appeal by a tenant from an order for possession of 40 Maidstone Road, London N11, made by Judge Stockdale at Barnet County Court on October 14 1991.

By an oral agreement made in March 1978 for exclusive occupation of the second-floor rear room as a bedsitter, on a weekly tenancy of £15, the tenant was to share the bathroom, lavatory and kitchen. The landlord has always lived elsewhere.

The background to this case is that the tenant and his brother (and at one time his mother) lived in the property — a house with seven rooms owned by the landlord. At one stage the brother shared with G the tenant and in 1965 the brother himself became the tenant of a first-floor room, with a written agreement entered into in December 1985, which the court has not seen.

The landlord decided that he wanted to demolish the house and rebuild. He obtained planning permission. He issued proceedings for possession in respect of the first-floor room occupied by the brother. The hearing was before Judge Stockdale and an order for possession was made on May 15 1990.

At the 1990 hearing the judge did not have the brother in court, although he was legally represented. The judge commented that both brothers suffered from problems and that the tenant in the present appeal was said to be suffering from agoraphobia. H

In the written agreement, according to the judgment, the landlord agreed that the tenant had exclusive possession to the first-floor room with a right to share the kitchen and bathroom. The agreement gave the landlord the right to choose the other occupiers of the house, who would share the shared rooms (the words being "... whosoever I choose ... ") which, as the judge put it, gave the landlord the widest possible scope, even to be included himself. The judge held that the terms of the 1985 agreement came within the provisions of section 21 of the Rent Act 1977. It was therefore a restricted contract and the brother did not have a protected tenancy.

The landlord then brought proceedings to obtain possession of the second-floor room occupied by the tenant. The notice to quit was dated November 8 1990. Proceedings were issued in July 1991. The J tenant sent a home-made defence, claiming that he was a protected tenant. The case was due to be heard on August 20, but on the application of the landlord and agreement of the tenant, it was adjourned to October 14.

On that occasion the tenant did not attend, but sent three friends to ask for an adjournment and he provided letters from his doctors to explain that he was seriously affected by agoraphobia and had not left the house for several years.

The judge refused the adjournment. He obviously had a clear recollection of the unusual features of the previous case involving the brother and the judge relied upon that evidence in coming to his decision. It appears that no application was made under the Civil K Evidence Act 1968 to admit the evidence.

The landlord did not give any further evidence at this hearing and the judge made an order for possession against the tenant.

In the amended notice of appeal and skeleton argument, counsel for the tenant, Mr Birks, raises three issues: the refusal of the judge to adjourn, and the reliance exclusively upon the evidence of the previous proceedings, which seems to have been done by a somewhat rough-and-ready procedure. It is not necessary to deal with either of these points which in isolation would not, in my view, have been sufficient to allow the appeal, since the third issue is of substance and, if successful, renders the other two points unnecessary.

A The third issue is whether the tenancy granted to the tenant by the landlord falls within the provisions of either sections 21 or 22 of the Rent Act 1977.

Section 21 (which has subsequently been repealed) is headed: "Tenant sharing accommodation with landlord" and provides:

> Where under any contract —
> (*a*) a tenant has the exclusive occupation of any accommodation, and
> (*b*) the terms on which he holds the accommodation include the use of other accommodation in common with his landlord or in common with his landlord and other persons, and
> (*c*) by reason only of the circumstances mentioned in paragraph (*b*) above, or by reason of those circumstances and the operation of Section 12 of this Act, the accommodation referred to in paragraph (*a*) above is not a
> B dwelling-house let on a protected tenancy, the contract is a restricted contract notwithstanding that the rent does not include payment for the use of furniture or for services.

That is to be contrasted with section 22, headed "Tenant sharing accommodation with persons other than landlord" which provides:

> (1) Where a tenant has the exclusive occupation of any accommodation ("the separate accommodation") and —
> (*a*) the terms as between the tenant and his landlord on which he holds the separate accommodation include the use of other accommodation ("the shared accommodation") in common with another person or other persons, not being or including the landlord, and
> (*b*) by reason only of the circumstances mentioned in paragraph (*a*) above, the separate accommodation would not, apart from, this section, be a dwelling-house let on or subject to a protected or statutory tenancy, the
> C separate accommodation shall be deemed to be a dwelling-house let on a protected tenancy or, as the case may be, subject to a statutory tenancy and the following provisions of this section shall have effect.

It is not necessary, for the purposes of this judgment, to set out any further part of section 22.

There is a paucity of evidence about the oral agreement in 1978. We accepted, as additional evidence, an affidavit sworn by the landlord on October 16 1992, para 3 of which reads:

> I gave evidence to the learned Judge in the previous action. I have been shown by my solicitors a copy of his notes of evidence from the case which include his note of my evidence . . . I gave evidence that John Brown had been let into occupation of the property on the same terms as in the case of his
> D brother, David Brown, ie that he could share the property with other people chosen by myself. I had expressly discussed this term with John Brown, and it was understood between us that this was the arrangement as there are, in all, seven bedrooms in the house.

The question is whether those words are sufficiently wide and sufficiently clear to include the landlord. If they are, it would appear that section 21 would apply; if they are not, then section 22 would apply.

We have a note of the argument put forward by those representing the tenant, together with a note of the judge's reply:

> Judge: It is hard to see what John Brown's defence is. I cannot see that John Brown can have any stronger case (than David Brown). It is not his house and he can have no defence. The landlord does not have to be there.

E Then at p 73:

> Judge: . . . John Brown does not have a protected tenancy. John Brown was misinformed (that he had a protected tenancy).

In the note of his judgment (which is the subject of this appeal) the judge said:

> I made an order for possession last year when I found that there was an arrangement for sharing and that David Brown was not protected.

Then a little later he said:

> But the question remains, does he have a defence? He has filed a defence. This is based on a misapprehension. It seems to me that the landlord must share. It was a deciding point in David Brown's case . . . there is no defence provided in
> F the defence.

The judge was clearly relying on his previous decision which had not been appealed. In the previous judgment he found that the landlord had never resided there.

I refer next to where the judge said in the David Brown judgment:

> Although the agreement says it relates to occupation of the house, it is agreed it is to be read as referring to a particular room with shared rights to the living rooms. According to the agreement these were to be shared with other occupiers chosen by the landlord, and this gives to the landlord the widest possible scope — even to be included himself, according to a case concerning a similar agreement in the House of Lords . . .

— I interpolate to say it has not been possible to find that case.

At p 89 he said:

> The landlord may choose himself to be one of the occupiers. But must he move G
> in? The early authorities show it is not necessary, so long as he can, before the issue of the proceedings . . . The landlord can become a person entitled to do the sharing, unless the contract bars that or says only named persons may share. The landlord does not have to actually move in to come within section 21.

Although not entirely clear, it appears the judge was relying upon the decision of this court in *Stanley* v *Compton* [1951] 1 All ER 859. In that case the landlord let the flat to the tenant on a protected tenancy, and the tenant let a room in the flat to the subtenant, with the sharing of the kitchen. The landlord obtained possession against the tenant and claimed that the subtenant was a trespasser. It was held that the subtenant became the tenant of the landlord on the same terms as he had been of the tenant, and could not be in a better H
position than he would have been with the tenant. He had no protection against the tenant, who could have claimed under the previous section analogous to section 21, and consequently no protection against the landlord.

Section 8(1) is the previous section analogous to section 22. On the facts of that case the landlord was exercising the rights of the tenant even though he himself had not exercised the right to share the kitchen. More particularly, the subtenant was unable to establish that he was sharing with someone other than the landlord, so as to come within the provisions of section 8(1). This decision is not, in my judgment, authority for the very wide proposition advanced and relied upon by the judge. There appears to be no other decision on J
this point.

Section 21 appears to have been passed to protect the interests of a landlord who was, or might be, sharing accommodation with a tenant or tenants, who were consequently not protected. On the other hand, tenants sharing accommodation, not including the landlord, were protected. It would therefore be of importance for a tenant to be aware of the terms of the agreement and whether he was a protected tenant.

The terms of the 1978 agreement as set out in the affidavit of the landlord do not disclose any reference to the possibility of the landlord himself residing in the property. The words are not sufficiently specific to make it clear that the landlord is reserving to himself the right to be one of the possible occupiers of the premises. I K
do not believe, myself, that a tenant ought to be expected to interpret those words as including the landlord himself. To come within section 21 the terms must include clear words to reserve the right to the landlord to live on the premises and share the accommodation with the tenant. This 1978 agreement, in my view, brings the tenant into the ambit of section 22, and not section 21.

Mr Birks has also argued that an express reservation of the right to live in the property and share accommodation cannot bring a tenancy agreement within the scope of section 21 unless, at the time of the tenancy agreement, he clearly intends at some stage to come to live on the premises. In my view, it is sufficient for the landlord, who does not live on the premises, to reserve to himself the express right to share other accommodation with the tenant. It is not necessary that L
he should actually exercise that right, nor have a clear intention to do so, so long as the possibility of moving into the premises is genuinely within his contemplation at the time of the tenancy agreement — that is to say, there is a real prospect at some future date of his doing so. That real prospect is to be contrasted with a colourable device designed to exclude the operation of the Rent Acts.

In *AG Securities* v *Vaughan* [1990] 1 AC 417*, Lord Templeman was, at p462, considering a reservation by a landlord of a power to share the flat with the tenant to be exercised either by himself or those he authorised. Lord Templeman laid stress upon a genuine intention to exercise the powers contrasted with the reservation in question, the only purpose of which was to deprive the tenant of the protection of M
the Rent Acts. I do not understand Lord Templeman's reference to "genuine intention" to mean that the landlord must actually have decided to share accommodation; rather he must be able to demonstrate that there was a real prospect that this might happen.

I would allow this appeal and hold that the tenant is a protected tenant under the provisions of section 22 of the Rent Act 1977 and set aside the possession order.

Simon Brown LJ agreed and did not add anything.

Appeal allowed with costs. Leave to appeal to the House of Lords refused.

---

*Editor's note: Also reported at [1988] 2 EGLR 78.

LANDLORD AND TENANT

RENT REVIEW

A

Chancery Division

July 23 1992

(Before Mr MANN QC, sitting as a deputy judge of the division)

LONDON & LEEDS ESTATES LTD v PARIBAS LTD

Estates Gazette January 30 1993

[1993] 04 EG 119

Landlord and tenant — Rent review — Construction — Meaning of "fit for immediate occupation and use"

B By a lease dated August 19 1986 the respondent tenant holds from the applicant landlord a term of 25 years from August 8 1986 of 27-37 Wigmore Street, London W1; the lease contains provisions for a rent review on every fifth anniversary of the term. If the rent at any review date is reviewed, to the "rack rental market value of the demised premises", clause 6(b) of the lease directs certain assumptions are to be made including the following: "(1) That the demised premises are fit for immediate occupation and use in a state of good repair and condition and that all fitting out and other tenant's works required by such willing tenant have already been completed". C The premises had been let to the tenant in the state described as a "developer's finish" and the tenant incurred the expense of fitting them out before it could use them as offices. In respect of the rent review for August 8 1991, a number of questions of construction arose between the parties. The tenant argued that the valuation approach required by clause 6(b)(1) was that on the review date the premises are fitted out and completed but that work has been carried out by the hypothetical tenant at his own expense (with the hypothetical landlord's consent) prior to the review date and prior, therefore, to the agreed letting. The landlord contended for a valuation approach that at the review date the D fitting-out works have been completed, but without any assumption as to who carried out the work.

Held: **The correct approach to questions of construction is to seek to discover the parties' intention from the written language which they have used. It is only when their language lends itself to alternative constructions that it is proper to depart from the literal by choosing the reasonable in preference to the perverse. Under the lease it is the demised premises which must be valued and these include works provided by the tenant (other than tenant's fixtures). Clause 6(b)(1) is completely neutral as to who carried out the tenant's requirements and it follows that the subclause similarly E reflects the possibility that the outgoing tenant's expenditure might be of no value to the landlord as an inducement to the hypothetical willing tenant, so it is equally implicit that such expenditure should be disregarded. Accordingly, the language of the rent review clause requires no assumption to be made as to who paid for the hypothetical fitting-out works required by the hypothetical willing tenant and, by parity of reasoning, requires no assumption to be made as to what such (if any) works required by the hypothetical willing tenant might be. It is sufficient for the valuer to know only that any such works have been carried out and that such works will F enable such tenant immediately to occupy and use the demised premises as authorised by the lease. The hypothetical willing tenant is not to be assumed to have suffered any financial burden by reason of being allowed to fit out the demised premises during a permissive period of occupation prior to the date of review.**

The following cases are referred to in this report.

Cornwall Coast Country Club v *Cardgrange Ltd* [1987] 1 EGLR 146; (1987) 282 EG 1664

G *Iceland Frozen Foods plc* v *Starlight Investments Ltd* [1992] 1 EGLR 126; [1992] 07 EG 117

Orchid Lodge (UK) Ltd v *Extel Computing Ltd* [1991] 2 EGLR 116; [1991] 32 EG 57

Ponsford v *HMS Aerosols Ltd* [1979] AC 63; [1978] 3 WLR 241; [1978] 2 All ER 837; (1978) 38 P&CR 270; [1978] EGD 137; 247 EG 1171, HL

Pontsarn Investments Ltd v *Kansallis-Osake-Pankki* [1992] 1 EGLR 148; [1992] 22 EG 103

This was an application by way of originating summons pursuant to section 2(1) of the Arbitration Act 1979 for the determination of questions of construction of a lease dated August 19 1986 and made between the applicant landlord, London & Leeds Estates Ltd, and the respondent tenant, Paribas Ltd.

H Michael Barnes QC (instructed by Titmus Sainer & Webb) appeared for the applicant; Kim Lewison QC (instructed by Linklaters & Paines) represented the respondent.

Giving judgment, **MR MANN QC** said: This is an application pursuant to section 2 (1) of the Arbitration Act 1979 whereby the parties to a reference seek with the consent of an arbitrator the determination by the court of questions arising upon the construction of a lease. The questions have been agreed by the parties and set out in the originating summons by which the application is brought before the court. For reasons which I need not recite, the application is somewhat out of time under RSC Ord 73, r 5(3), but since neither party makes any complaint about it I will give leave extending the time within which it may be made. Accordingly, it is J now regularly before me.

The lease, made on August 19 1986 between London & Leeds Estates Ltd and Paribas Ltd (then named "Banque Paribas Capital Markets Ltd") and another for a term of 25 years from August 8 1986 of 27-37 Wigmore Street, London W1, reserves a substantial rent which, by clause 6, is reviewable every fifth anniversary of August 8 1986 (defined in the lease as a "Date of Review"). The rent review clause (so far as material and with appropriate truncation) provides:

> 6(a) With effect from 8th August 1991 ("the Date of Review") the yearly rent for the time being payable hereunder shall be reviewed and from and after each such date of review the yearly rent shall be such an amount ("the Amount of the Revised Rent") equal to whichever is the greater of:
>
> K 6(a)(i) the yearly rent payable immediately prior to such Date of Review; and
>
> 6(a)(ii) the rack rental market value of the demised premises at such Date of Review
>
> 6(b) The rack rental market value of the demised premises at each Date of Review shall be such an amount as may be agreed between the Landlord and the Tenant or determined in accordance with sub-clause (c) of this clause as representing the best open market rent at which the demised premises might if available reasonably be expected to be let as at such Date of Review for a term having the unexpired residue of this term or of 15 years whichever shall be the longer with rent reviews every 5 years at the relevant Dates of Review as between a willing landlord and a willing tenant with vacant possession without taking a fine or premium and subject to the provisions of this lease (other than the term and the amount of rent but including these provisions for

A rent reviews) and upon the following assumptions (if not a fact in each case) namely

6(b)(1) That the demised premises are fit for immediate occupation and use and in a state of good repair and condition and that all fitting out and other tenant's works required by such willing tenant have already been completed

6(b)(2) That no work has been carried out on the demised premises by the tenant its sub-tenants or predecessors in title during the term which has diminished the rental value of the demised premises

6(b)(3)-(6) . . .

but taking no account of:

6(b)(i) any goodwill attributable to the demised premises by reason of any trade or business carried on therein by the tenant or any lawful underlessee

6(b)(ii) any increase in rental value of the demised premises attributable to the existence at the relevant Date of Review of any improvements to the B demised premises or any part of them carried out (with consent where required) otherwise than in pursuance of an obligation to the landlord . . . during the term or during any period of occupation prior to the term arising out of any agreement to grant the term

6(b)(iii) any effect on rent of the fact that the tenant . . . may have been in occupation of the demised premises; and

6(b)(iv) . . .

6(b)(v) any effect on rent of the absence of any rent free period or contribution towards fitting out costs or other inducement which might then be the practice in the open market to make or allow tenants on a new letting with vacant possession

I need not cite clause 6(c), which sets out the procedure for the review and reference to an arbitrator. It is sufficient to know that the parties elected that the new rent be determined by Mr GR Stanton C FRICS of John D Wood and that questions of construction having arisen he has given his consent to this application.

Rent review clauses ought in principle to give rise to few problems given conveyancers' familiarity with them. Such clauses are *prima facie* intended simply to compensate landlords for changes in the value of money, which depreciate historic rents and distort changes in the real value of property. On review, landlords ought, in principle, to receive the current market rental value of the building demised, less the rental value of improvements lawfully made and for which tenants have paid. The formula evolved by conveyancers to carry this principle into effect has been adopted by statute and is embodied in section 34 of the Landlord and Tenant Act 1954. The D formula requires the current market rental value to be assessed by reference to an assumed market place and hypothetical willing landlord and tenant who will eventually agree upon the terms of a new lease. Personal attributes, such as goodwill, of the tenant in occupation including the fact of occupation are disregarded as are effects on rent of the tenant's improvements lawfully carried out otherwise than pursuant to an obligation to the landlord.

It is a regrettable fact, though, that this utopian ideal is rarely achieved in practice, and there is now a considerable body of authority which amply illustrates the organic properties of this branch of the law of landlord and tenant.

The questions on this application spring out of practices which, though general, are of comparatively recent origin. These days office E buildings are seldom finished or refurbished fit for occupation and use by tenants. Developers are disinclined to accept the financial risk of doing so since tenants' tastes and requirements differ widely. It is often better by far to offer a building to the market in a state often described as "developer's finish" so that a tenant can fit out the building. But in the tenants' market which now prevails it is common also to offer a prospective tenant a financial inducement to take a lease. Examples of such inducements are rent-free periods (rent remissions), reverse premiums and the like, which directly or indirectly contribute towards the tenant's costs of fitting out. Draftsmen try to provide for the rental effect of these and other exigencies so that one or other of the parties shall not receive a double benefit or, correspondingly, a double detriment on review. They do F not always succeed.

Although rent review clauses introduce the concept of the willing hypothetical landlord and tenant, it does not follow that it is appropriate to require valuers to value a hypothetical building. It is axiomatic that there has to be some physical state or *quantum* for the valuer to value if the task of valuation is not to be rendered inordinately difficult or artificial; so hypothesis gives way to actuality except to the extent that it is appropriate in order to do justice between the parties. The balance of justice is struck by introducing specific "assumptions" and "disregards" in addition to the general assumptions and disregards usually necessary to implant the hypothetical market place. These specific assumptions and disregards therefore signpost the additional departures from actuality which the parties agree should enable the valuer to do justice G between them.

Even so, it is important to bear in mind the cardinal principle of construction that if parties fail to reflect in their rent review clause what might otherwise be supposed to be reasonable intentions, the court will enforce their language notwithstanding: *Ponsford* v *HMS Aerosols Ltd* [1979] AC 63.

From the general, I turn now to the particular — 27-37 Wigmore Street is an office building with 11 floors, including a basement and ground floor. The applicant landlord let it to the respondent tenant in a state of "developer's finish". The tenant had to incur the expense of fitting out the demised premises before it could use them as offices. The language of the rent review clause allows the tenant no credit for H this, but subclause 6(b)(1) is said to admit of three possible valuation approaches, one of which might do, namely: (1) at the review date, the demised premises are physically in a state of "developer's finish"; and (2) the premises are fitted out and completed at the review date but that work has been carried out by the hypothetical willing tenant at his own expense (with the hypothetical landlord's consent) prior to the review date and prior, therefore, to the agreed letting.

Approach (2), therefore, predicates:

(a) that consent for the tenant's works has been given some months prior to August 8 1991, and

(b) that at the review date, on the one hand the hypothetical willing tenant does not have any period of time within which to fit out, but on J the other hand he is treated as having paid for it because he has done the work himself. This is the approach for which the tenant argues in the present case.

The landlord argues for the third approach, which is:

(3) at the review date, the fitting-out works have been completed as in (2), but there is no super-added assumption as to who has carried out the work.

It is common ground that the first approach is inapposite. The rental effects of these approaches have been canvassed in correspondence between the expert valuers appointed by the parties respectively. I shall allude to these as necessary later in this judgment but as the questions which they raise are agreed upon there is no need for me to refer to them specifically or at length. K

The questions are:

(1) Whether it is to be assumed that the fitting out and other tenants, works required by the willing lessee ("the hypothetical fitting out works") have been carried out at the expense of the willing lessee;

(2) If the answer to question (1) is in the affirmative whether it is to be assumed that the hypothetical fitting out works have been carried out (a) before or (b) after or (c) simultaneously with agreement on the rent;

(3) If the answer to (1) is affirmative whether it is to be assumed that the willing lessee has been permitted a period of occupation to enable the hypothetical fitting out works to be completed;

(4) If the answer to question (1) is in the negative whether the rack rental market value of the premises is to be assessed on the assumption that the landlord is entitled to the increase (if any) in the rack rental market value L attributable to the hypothetical fitting out works;

(5) If the answer to question (4) is in the affirmative whether the increase is to be diminished if and to the extent that the hypothetical fitting out works are found by the arbitrator to be the same as the actual fitting out works (or other work of improvement) carried out by the Respondent.

There is a sixth question, namely:

(6) Where (a) the practice in the open market at the review date is that a rent free period or contribution towards fitting out costs or other inducement would be made or allowed to a new tenant of the demised premises, and (b) the absence of such benefit being available to the willing lessee would reduce the rent below what it would otherwise be, whether that reduction is to be left out of account in determining the rack rental market value of the demised premises. M

While not ignoring relevant surrounding circumstances such as I have described, the correct approach to questions of construction is to seek to discover the parties' intention from the written language which they have used. It is only when their language lends itself to alternative constructions that it is proper to depart from the literal by choosing the reasonable in preference to the perverse.

By clause 1(2)(f)(viii) of the lease, the demised premises include

all fixtures (other than trade or tenant's fixtures) . . .

Thus, the premises which are to be valued *exclude* the actual tenant's fixtures and fittings in the legal sense of fixtures and fittings which are severable from the building of which the demised premises form part *and include* others which the tenant has provided but which because

A of the degree and/or method of annexation have accrued to and become unseverable from it so as to fall to be treated either as landlord's fixtures and fittings or simply as part of the building (I shall call these "tenant's residual works").

So this is the physical state or *quantum* of the demised premises which (leaving aside the result of work of a type within subclause 6(b)(ii)) clause 6(b) *prima facie* requires to be valued. However, subclause 6(b)(1) now requires it to be assumed that:

> all fitting out and other tenant's works required by such willing tenant have already been completed

which inferentially predicates that the tenant's residual works could just as well suit the hypothetical willing tenant's requirements as not,
B so as possibly to be of no value at all either to such tenant or, it must follow, to the hypothetical landlord. The subclause is completely neutral as to who carried out the tenant's requirements.

It must follow that the subclause similarly refelcts the possibility that the outgoing tenant's expenditure might be of no value to the landlord as an inducement to the hypothetical willing tenant, so it is equally implicit that such expenditure should be disregarded. By contrast, subclause 6(b)(ii) prohibits any account being taken of any increase in rental value attributable to expenditure by the tenant of a type within that subclause, that is, a qualifying improvement which, inferentially, benefits the landlord's reversion.

Accordingly, the language of the rent review clause requires no assumption to be made as to who paid for the hypothetical fitting-out
C works required by the hypothetical willing tenant, and, by parity of reasoning, requires no assumption to be made as to what such (if any) works required by the hypothetical willing tenant might be. It is sufficient for the valuer to know only that any such works have been carried out and that such works will enable such tenant immediately, that is at the date of review, to occupy *and* use the demised premises, and in this context "use" means as authorised by the lease.

It is no less trite in the specialised field of landlord and tenant law than it is under the general law that a term is not to be implied as a term of a written agreement except where it is necessary to give effect to the parties' plain and obvious intention.

The curiosity in this case is that both parties contend that their opposing contentions require an implication to be made as to which
D of the hypothetical willing landlord and tenant carried out the hypothetical fitting-out works. Thus, on the one hand, the tenant contends that because the landlord's argument is that none of the fitting-out works required by the hypothetical willing tenant have been carried out by the tenant it is necessary to assume that the landlord carried out such works, which is contrary to the fact and would result in an injustice to the tenant. This is because in such circumstances the valuer must assume that the hypothetical tenant has suffered from no financial burden in fitting out the demised premises. The landlord will, therefore, benefit from a rental uplift unjustified by any expenditure on its part.

The landlord, on the other hand, contends that the tenant's
E argument that the hypothetical willing tenant has been allowed into occupation before the date of review, in order to carry out the works, necessitates introducing additional words into subclause 6(b)(1).

In my judgment, these divergent possibilities amply illustrate the reason for the legal principle. The court cannot possibly know whether the parties would have agreed upon a term appropriate to provide for what they each perceive the problem to be and, for my part, I do not perceive there to be a problem for which the parties have not in any event adequately provided.

The construction for which the landlord contends results in it getting the benefit of any effect on the rent of it being assumed that the hypothetical willing tenant can immediately occupy *and* use the demised premises. It does not follow that it will unjustly get the
F benefit of works for which the actual tenant has paid. Indeed, by inference, they have agreed that such works shall be treated as of, at most, nominal benefit to the landlord's reversion, since had they deemed otherwise subclause 6(b)(ii) would, certainly should, have prohibited any account being taken of any increase of rental value or rental effect attributable to these works also being in existence at the date of review.

Moreover, in a case where a landlord has offered and the tenant has accepted an inducement to take the lease, such landlord has, *ex hypothesi*, directly or indirectly contributed to the costs of such works, even notionally paid for them, at the inception of the lease; so there is plainly no justification for the tenant receiving the benefit of an assumption which is neither express nor necessarily consequential on subclause 6(b)(i). It is *a fortiori* where the tenant has not received an inducement. G

There is therefore no warrant for the court to presume that the parties intended differently. Accordingly, no consequential assumption is necessary either way. In particular, the hypothetical willing tenant is not to be assumed to have suffered from any financial burden by reason of being allowed to fit out the demised premises during a permissive period of occupation prior to the date of review.

If further support were needed for this construction, it can be found in the language which the parties have incorporated in the rent review clause, for if subclause 6(b)(1), in particular the specific directions as to fitness for immediate use and fitting-out works to the H
hypothetical willing tenant's requirements having already been completed, were absent, evidence of market practice would warrant the valuer assuming that the hypothetical willing tenant might have to fit out notwithstanding the presence of subclause 6(b)(v) which directs, *inter alia*, that no account shall be taken of any effect on rent of the absence of any market practice of offering predicated types of inducements. Subclause 6(b)(1) therefore serves the specific purpose of preventing the tenant from praying in aid a different type of inducement, namely that of offering to a hypothetical willing tenant the option of taking up occupation prior to the lease in order to carry out fitting-out works in so far as this type of inducement is not subsumed within the language of subclause 6(b)(v). Alternatively, it underpins subclause 6(b)(v), should it embrace a market practice J
comparable to that which the tenant's valuation approach predicates, though I know not whether such exists.

There is, I would add, an inherent conflict in the tenant's approach to valuation in any event because the hypothetical market place assumes not only that the parties will come to terms at the date of review but also that the new term will commence at such date, which appears to me logically to favour a construction which precludes any assumption of a prelet occupation period for fitting out or any other purpose.

Counsel for the tenant, Mr Lewison QC, submitted that the landlord's construction will produce an injustice because on review the landlord will receive an income return on a capital injection which the landlord never made. For the reasons which I have stated, I do not K
agree.

Counsel for the tenant also submitted that the court leans against construing a rent review clause in a sense which requires the valuer to decide for himself how to value. I am not sure that I understand why the landlord's construction places an inappropriate or unusual burden on the valuer which should influence the court to an alternative construction. But where, as in this case, the language admits only one construction without the implication of an additional or consequential term, it would, in my judgment, be singularly inappropriate and wrong in principle for the court to do so. I might add that it would be presumptuous of the court to make any assumptions as to the competence of valuers highly experienced L
in their field of specialisation to resolve what is pre-eminently a problem of valuation.

Accordingly, clause 6(b)(1), construed literally, requires the valuer to assume:

> (a) that the hypothetical willing tenant can take up occupation at the review date, and
>
> (b) that the hypothetical willing tenant can actually use the demised premises from such date for the purposes permitted by the lease.

It does not require or let in a super-added assumption as to what the fitting-out work consisted of or who did it.

A number of authorities have been cited in argument, including *Cornwall Coast Country Club* v *Cardgrange Ltd* [1987] 1 EGLR 146*, *Orchid Lodge (UK) Ltd* v *Extel Computing Ltd* [1991] 2 EGLR M
116†, *Iceland Frozen Foods plc* v *Starlight Investments Ltd* [1992] 07 EG 117‡ and *Pontsarn Investments Ltd* v *Kansallis-Osake-Pankki* [1992] 22 EG 103§.

Cornwall was cited for Scott J's clarification that hypotheses can properly be used as a basis for valuation only when they are expressly directed to be taken into account: see pp 149K-M, 15OA-D, 152H-T. With respect, that is obviously correct. The valuer has to enter and leave the dreamland created for him by the same gate through which

*Editor's note: Also reported at (1987) 282 EG 1664.
†Editor's note: Also reported at [1991] 32 EG 57.
‡Editor's note: Also reported at [1992] 1 EGLR 126.
§Editor's note: Also reported at [1992] 1 EGLR 148.

A he entered using the key the draftsman has given him; there is no secret garden for him to browse in of which he alone is aware and to which he alone has the key, and there is no reason in law for the court to cut one for him.

In *Orchid* the Court of Appeal had to decide what physical state or *quantum* the valuer had to value having regard to the language of a proviso in a licence permitting assignment and change of user. The proviso was in these terms:

> Whilst it is authorised to use the Demised Premises for the purposes authorised in the Schedule hereto and the Assignee does in fact occupy the Demised Premises for those purposes the rent payable pursuant to the terms of the Lease will be reviewed in accordance with the provisions for review contained therein on the basis of the authorised use at the relevant review date
> B (the Assignee hereby acknowledging that the Demised Premises are fit for use and occupation therefor) assuming a willing Lessor and a willing Lessee but disregarding any improvements carried out to the Demised Premises by the Tenant otherwise than in pursuance of an obligation to the Landlord

Nearly nine months later, the landlord licensed the assignee to make extensive alterations to the demised premises.

Delivering the principal judgment, Dillon LJ held that the relevant state or *quantum* was that which existed at the date of the license to assign, that is, before the permitted alterations took place. This was because of the assignee's acknowledgement that the demised premises were then fit for use and occupation as authorised by the schedule which was inimical to an alternative construction that would have required the valuer to value some hypothetical building
C fit for use as authorised, but which had never existed and never would. Conversely, it did not matter that the valuer would have to value something which did not exist at the date of review because the assignee's improvements had to be disregarded.

The case is of relevance as yet further authority for the correctness of the approach adopted and analysed by Scott J in *Cornwall*.

Iceland, where a review clause required a valuer to assume premises,

> remain in existence and are ready for immediate occupation and use

embodies the same reasoning. I shall need to refer to certain of it later in this judgment.

In *Pontsarn*, where the assumption was,

> D vacant but fit for immediate occupation and use

Paul Baker QC, sitting as a deputy judge of the High Court, stated, *obiter*, that the reasoning in *Iceland* was fatal to a landlord's contention that:

> The expert is required to assume that the premises are fitted out ready for occupation by a hypothetical tenant. The building is not in a shell state but fitted out so that the tenant can just walk in and start business. It does not matter who paid for it. This has the consequence that there is to be no remission of rent to allow for fitting out to be done. But there is to be an uplift of rent because the hypothetical tenant does not have the financial burden of paying a capital sum to get in

He preferred the tenant's contention, which involved making a
E distinction between premises "fitted out" and premises "fit for immediate use". The tenant argued:

> the expression "fit for immediate use" . . . meant that the building was free from defects and ready for the tenant to go in, fit it out for his business and commence trading. A building is fit for occupation when it is ready to be occupied for fitting-out purposes. Ready for occupation does not mean ready for a tenant to go in and start trading immediately. If this were correct, it would mean that a tenant could negotiate for a reduction of rent over the fitting-out period . . .

As the tenant's argument illustrates, there is possibly a difference between the assumption in that case and an assumption that the demised premises have already been fitted out to the tenant's requirements; the latter being a concept which, so to speak, oversteps
F a condition that the premises are fit for occupation.

Paul Baker's reasoning was that the landlord's construction would give the landlord more than compensation for changes in the value of money, that is, rent for something which the landlord had not provided. He was thus adopting a purposive approach to construction; one which is irreproachable on general principles if the parties' choice of language warrants it. However, fatal to the landlord's contention was a passage in the judgment of Dillon LJ in *Iceland*, where, at p 110, he said,

> But it seems to me to make nonsense of the rent review clause as drawn and of the aim that the rent on review should be fixed so as to bear as close a resemblance to reality as possible, if it is not possible to charge rent for actual improvements but is possible to charge the actual tenant with rent for hypothetical improvements which have never been carried out.

But Dillon LJ was addressing himself to a contention that though the G
premises were to be taken as being in their actual condition for the time being, even so the valuer must value on the basis that the hypothetical willing tenant would be able to carry out works amounting to improvements themselves dependent upon the fulfilment of a number of hypotheses none of which were predicated by the assumption let alone language elsewhere in the lease.

In the present case, the assumption necessarily involves taking into account the rental effect of the existence of hypothetical fitting-out works, possibly including improvements, even though the actual tenant's improvements within subclause 6(b)(ii) are required to be left out of account. It is neither necessary nor permissible to make any additional assumption as to what that work might have been or who H
might have paid for it. It is simply irrelevant.

I shall therefore answer the questions in the originating summons as follows:

(1) No; (2) and (3) are not applicable; (4) Yes; (5) No.

It was common ground between Mr Barnes QC for the landlord and Mr Lewison QC for the tenant that the answer to question (6) is "Yes".

I am indebted to counsel for their submissions.

Declarations accordingly.

J

Chancery Division

July 27 1992

(Before Mr Justice HOFFMANN)

LAURA INVESTMENT CO LTD v HAVERING LONDON BOROUGH COUNCIL (NO 2)

Estates Gazette February 27 1993

[1993] 08 EG 120

Landlord and tenant — Rent review — Whether rent to be assessed on assumption that premises are let with vacant possession or subject to underleases existing at date of review K

By a lease dated June 14 1972 a predecessor in title to the plaintiffs demised to the defendants 5 acres of undeveloped land at Langley Close, Havering, for a term of 62 years. By clause 3(b) of the lease the defendants covenanted to underlet the demised premises in separate plots. By clause 3(h)(iii) the defendants covenanted not to underlet without written consent, such consent not to be unreasonably withheld. By clause 3(i) the lease recited that consent was given less a few
days to H P Borey & Co Ltd. Further underleases of other L
plots were granted to other tenants in 1977, 1979 and 1981. In *Laura Investment Co Ltd* v *Havering London Borough Council* [1992] 1 EGLR 155 Hoffmann J declared that the rent at review should be assessed on the assumption that buildings erected on the land since the grant of the lease were included in the hypothetical demise. Following an adjournment of the originating summons, and by an amendment, the defendants contended that the rent should be assessed on the assumption that the premises are let subject to the tenancies which exist at the date of the rent review.

***Held*: Declaration in favour of the defendants. It was clear** M
from the lease that underlettings were contemplated, indeed mandated, and the parties intended the rent to be calculated on an assumption of reality, assuming the underlettings were in place. The landlords' power to refuse consent to an underletting unless they were being unreasonable gave them control over the terms of the future underleases such as to enable them to protect their position on a rent review from arbitrary damage. Unless the rent review provision is construed to assume that the premises are in fact underlet, the defendants were bound to be putting themselves in a position in which, on a rent review, they would be paying to the

landlords a higher rent then they would be receiving from their tenants.

The following cases are referred to in this report.

Avon County Council v *Alliance Property Co Ltd* (1981) 258 EG 1181
Forte & Co Ltd v *General Accident Life Assurance Ltd* [1986] 2 EGLR 115; (1986) 279 EG 1227
Laura Investment Co Ltd v *Havering London Borough Council* [1992] 1 EGLR 155; [1992] 24 EG 136
Scottish & Newcastle Breweries plc v *Sir Richard Sutton's Settled Estates* [1985] 2 EGLR 130; (1985) 276 EG 77

This was an adjourned originating summons seeking a declaration as to the proper construction of a lease held by the defendants, Havering London Borough Council, from the plaintiffs, Laura Investment Co Ltd.

Michael Barnes QC (instructed by Binks Stern) appeared for the plaintiffs; Nicholas Patten QC and Timothy Fancourt (instructed by the solicitor to Havering London Borough Council) represented the defendants.

Giving judgment, **HOFFMANN J** said: This is the adjourned hearing of an originating summons on the construction of a rent review clause. The first question raised by the plaintiffs was whether the rent should be assessed on the assumption that buildings erected on the land since the grant of the lease were included in the hypothetical demise. On March 13 I decided this question in the affirmative and gave a judgment which is reported at [1992] 1 EGLR 155*.

I must now answer the second question, which has been raised by amendment on behalf of the defendants. This is whether the rent should be assessed on the assumption that the premises are let with vacant possession or subject to the tenancies which exist at the date of the rent review.

As appears from my earlier judgment, the lease was granted on June 14 1972 by the Greater London Council, the defendants, to Havering London Borough Council. It was a demise for 62 years of about 5 acres of undeveloped land. The GLC wanted the land developed for industrial purposes and, by clause 3(b), the tenants covenanted:

as speedily as possible, subject to the provisions of subclauses 3(h), 3(n) and 3(s) hereof, to underlet the demised premises in separate plots and to cause to be erected on each of such plots factory or warehousing premises together with associated premises.

Clause 3(h)(iii) was a covenant not to underlet without written consent, not to be unreasonably withheld. At the time of the execution of the lease, the defendants had already entered into a building lease agreement with a company called H P Borey & Co Ltd, by which they were to grant in consideration of the erection of a building on one plot an underlease for the whole 62-year term, less a few days.

Clause 3(i) of the lease recited that consent to that underletting had been given. The underletting to H P Borey & Co Ltd was completed by the grant of an underlease dated March 23 1973. The rent no doubt reflected the fact that the tenants had been obliged to erect the building.

There were 15-year rent reviews which required the rent to be assessed on the assumption that the premises were a bare site. Further underleases of other plots were granted to other tenants in 1977, 1979 and 1981. For present purposes the terms were not materially different.

The rent review clause does not say that the hypothetical letting is to be with vacant possession. It is remarkably terse and simply speaks of "the market rental value" at the rent review date. The only expressed assumption is that all covenants and conditions in the lease have been duly observed and performed. It was the absence of any express assumption that improvements made by the tenants were to be disregarded which led to my decision on the first question.

The landlords here say that, in the absence of express words, there is a presumption that the hypothetical letting is to be with vacant possession. I was referred to three cases in which judges have had to consider whether such an assumption should be made or not. In chronological order they were: *Avon County Council* v *Alliance Property Co Ltd* (1981) 258 EG 1181, a decision of Judge Mervyn Davies QC, sitting as a High Court judge; *Scottish & Newcastle Breweries plc* v *Sir Richard Sutton's Settled Estates* [1985] 2 EGLR 130*, a decision of Judge Paul Baker QC, sitting as a High Court judge; and *Forte & Co Ltd* v *General Accident Life Assurance Ltd* [1986] 2 EGLR 115†.

In the second of those cases Judge Baker accepted the submission of Mr Barnes that vacant possession valuation

is likely to be the intention unless it is expressed to the contrary.

The learned judge added:

Were it otherwise the landlord would be in the position on a review of rent of being in the hands of the tenant if the tenant decides what terms he is to arrange with the subtenant.

Likewise in the *Forte* case, Peter Gibson J said at p116J:

... I accept that the ascertainment of an open market rental of premises which were let with vacant possession would ordinarily involve ... an assumption [of vacant possession] . . .

In a case where the premises are, in fact, underlet at the rent review date, such a presumption would, on the face of it, conflict with what has been called the presumption of reality, which is that one normally assumes for the purposes of the hypothetical letting that circumstances are what they are in real life.

It can be said, however, that the reason which Judge Baker gave for a presumption of vacant possession, namely that otherwise it would be entirely within the control of the tenant to determine the rental value of the land by agreeing such terms as he pleased on an underletting, is such a powerful one that it would inevitably rebut any presumption of reality.

Having said that, I do not think that it matters a great deal whether in this case one starts with a presumption in favour of vacant possession or a presumption of reality. All the cases emphasise that the question must turn upon the terms of the individual lease, and a presumption in favour of vacant possession is somewhat weakened if the terms of the lease protect the landlord against being made subject to the arbitrary acts of the tenant in the terms on which he underlets.

In the *Forte* case Peter Gibson J regarded the fact that the premises were already underlet at the time of the grant of the lease for a term, which extended beyond the review date, as an indication that the parties intended the review rent to be assessed on the assumption that that underletting continued to exist. He said at p116J:

. . . it seems to me that different considerations apply when part of the premises is let already subject to a tenancy in favour of a third party, the term of which will continue beyond the review date. It is not a necessary part of the hypothesis of an open market letting that such a tenancy should be ignored, and in my judgment reality should not be departed from more than is necessary to give effect to the assumptions required by the rent review clause.

What indication do we get from the terms and surrounding circumstances of this lease? First, it is clear that underlettings were contemplated, indeed mandated, by the terms of the lease. The terms of one such underletting had already been agreed.

Second, the landlords' power to refuse consent to an underletting unless they were being unreasonable gave them control over the terms of future underleases such as to enable them to protect their position on a rent review from arbitrary damage inflicted by the terms on which the tenants chose to underlet. It would, in my judgment, have been reasonable for a landlord to object to the grant of a lease at a lower rent in return for a premium if, upon the true construction of this lease, the review rent had to be assessed on the assumption that the premises were underlet. This would have been true both of an original underlease and any subsequent underlease granted on the expiry or surrender of an earlier one. On the other hand, it would not have been open to the landlords to object to a lower rent in consideration of the erection of buildings, which the lease itself contemplated.

The next significant factor is that the rent review clause, as I decided on the last occasion, required the rent to be assessed on the assumption that the demised premises included the buildings which the undertenants were going to erect. On the other hand, it could not have been contemplated by the parties that the undertenants would be paying a rent which was calculated on the assumption that there had been demised to them the buildings for which they themselves have paid. It was clear from the terms of the underlease which the landlords had already approved that no such assumption could be made.

Unless, therefore, the rent review is construed to assume that the premises are in fact underlet, the defendants were bound to be

*Editor's note: Also reported at [1992] 24 EG 136.

*Editor's note: Also reported at (1985) 276 EG 77.
†Editor's note: Also reported at (1986) 279 EG 1227.

A putting themselves in a position in which, on a rent review, they would be paying to the landlords a higher rent than they were receiving from their tenants.

This would be an uncommercial thing to do. It might be the consequence of the express language of the lease or the clear presumption of construction such as I applied in the first question on the summons. In a question such as this, however, where one has to deduce the intention of the parties from all the circumstances, including their agreement that the buildings are presumed to be included in the demise, I think that one is free to assume that they intended a fair and commercial result rather than an unfair one.

I do not think that it is possible to draw a hard and fast line between cases such as *Forte* in which there is an existing underletting at the
B time of the grant, and cases in which the underletting is contemplated or even agreed, but not yet actually completed.

In my judgment, the terms of the underlettings contemplated by this lease, when related to the rent review provision of this lease, point to the conclusion that the parties intended the rent to be calculated on an assumption of reality, that is to say, assuming the existing underlettings to be in place. I shall declare accordingly.

Declarations accordingly.

C

Chancery Division

November 27 1992

(Before Mr Justice ALDOUS)

CITY OFFICES PLC v BRYANSTON INSURANCE CO LTD

CITY OFFICES PLC v ALLIANZ CORNHILL INTERNATIONAL INSURANCE CO LTD

Estates Gazette March 20 1993

[1993] 11 EG 129

D

Landlord and tenant — Rent review — Construction — Effect of disregard of any notional rent-free concession or fitting-out period

City Offices plc* v *Bryanston Insurance Co Ltd

By a lease dated July 24 1988 the respondent tenant, Bryanston Insurance Co Ltd, holds from the applicant landlord a term of 25 years of premises at 27 Leadenhall Street, London EC3; the lease contains provisions for rent review every five years. In providing for the determination of the rent at review, clause (1)(b) of the fourth schedule to the
E lease provided that the current market rental value shall be the best yearly rent which would reasonably be expected to be payable after the expiration of any rent-free concession or fitting-out period which might be given. The clause contained a number of assumptions and disregards. The fifth assumption was that at the review date the premises are suitable and available for immediate occupation and use. The fourth disregard was of any notional rent-free concession or fitting-out period for which allowance would or might be given to the tenant if the demised premises were let in the open market with vacant possession

By an interim award, Mr Geoffrey Perkins FRICS, as
F arbitrator, determined the rent payable from the first rent review at £479,678. In making his award, the arbitrator had the benefit of a written opinion of Mr D M Barnes QC to the effect that no reduction was to be made to the rent for the absence of any rent-free period for fitting out in the hypothetical lease; the rent was to be determined as the rent which would have been paid in the absence of any further rent-free period or other inducement to the hypothetical tenant to take the subject lease. The arbitrator concluded that a 12-month rent-free period would have been given, of which three months would be required for fitting out; nine months was equivalent to 15% and he deducted this percentage from the open market rent of £564,327 by way of allowance for this in G arriving at his award. The landlord appealed, contending that the rent to be determined under the lease was the current rack-rental value after the expiry of any rent-free period.

City Offices plc* v *Allianz Cornhill International Insurance Co Ltd

By a lease dated July 10 1986 the defendant tenant holds from the plaintiff landlord a term of the fourth floor in 27 Leadenhall Street, London EC3; the provisions for review being substantially the same as those in the lease held by Bryanston Insurance Co Ltd. The plaintiff sought a H declaration as to the meaning of provisions similar to those in issue in *City Offices plc* v *Bryanston Insurance Co Ltd*.

***Held*: *City Offices plc* v *Bryanston Insurance Co Ltd*. The appeal was dismissed. The rent has to be the best yearly rent payable after the expiry of any rent-free period whether provided as a fitting-out period or not. The periods referred to in the fourth disregard have to be disregarded for all purposes in the rent determination, the period of payment and the amount of payment. In arriving at the rent of £564,327 the arbitrator concluded that in the hypothetical negotiation the hypothetical landlord had conceded nine months' free rent and a further three months for fitting out. Those notional J periods have to be disregarded. Necessarily the disregarding of the notional free periods results in an adjustment of the rent. Upon the literal meaning of the lease the approach of the arbitrator was correct.**

***City Offices plc* v *Allianz Cornhill International Insurance Co Ltd*. The difference between the two leases was not material to the point of construction. The fourth disregards in the *Allianz* lease must therefore be construed as contended for by the defendant.**

The following case is referred to in this report. K

Basingstoke and Deane Borough Council v *Host Group Ltd* [1988] 1 WLR 348; [1988] 1 All ER 824 (1988) 56 P&CR 31; [1987] 2 EGLR 147; 284 EG 1587, CA

In *City Offices plc* v *Bryanston Insurance Co Ltd* there was an appeal by the applicant, City Offices plc, against an interim award of an arbitrator determining the rent at review in a lease held by the respondent, Bryanston Insurance Co Ltd. In *City Offices plc* v *Allianz Cornhill International Insurance Co Ltd* the plaintiff, City Offices plc, sought a determination by originating summons of the construction of a lease held by the defendant, Allianz Cornhill International Insurance Co Ltd.

Terence Cullen QC and Nicholas Dowding (instructed by Herbert L Smith) appeared for City Offices plc; Hazel Williamson QC (instructed by Stephenson Harwood) appeared for Bryanston Insurance Co Ltd; David Neuberger QC (instructed by Beachcroft Stanleys) represented Allianz Cornhill International Insurance Co Ltd.

Giving judgment, **ALDOUS J** said: First I will give judgment on the appeal by City Offices plc against the interim award dated March 21 1992 of Mr Geoffrey Perkins FRICS sitting as an arbitrator. He determined that the rent payable under the rent review provisions of a lease dated July 24 1988 should be £479,678. I will then go on and decide the construction summons between City Offices and Allianz Cornhill Insurance Co Ltd. M

The parties to the July 1988 lease were the landlords, City Offices plc, and Bryanston Insurance Co Ltd, the tenants. It demised to Bryanston the ground, upper mezzanine, first floor and basement of the building known as 27 Leadenhall Street, London EC3, for 25 years at a yearly rent of £459,571 or such greater yearly rent as shall be agreed or determined in accordance with the provisions of the fourth schedule of the lease. The rent was to be paid quarterly in advance, but the first payment was deferred to August 14 1986.

Part I of the fourth schedule to the lease contained provisions for rent reviews. Clause (1)(a) provided that for the first five years the yearly rent should be £459,571. The relevant parts of clause (1)(b) are as follows:

> For the next five years of the term and for each subsequent period of five years of the term (hereinafter called the review period) such rent as shall have been agreed between the Landlord and the Tenant or determined as hereinafter provided to be the current market rack rental value of the demised premises at the review date and for this purpose current market rack rental value shall mean the best yearly rent which would reasonably be expected to be payable in respect of the demised premises after the expiry of any rent free concession or fitting out period which might be given to the Tenant if a letting of the demised premises were negotiated in the open market between a willing landlord and a willing tenant upon a letting on the review date . . .

The clause continues:

> . . . on the assumptions if not in each case of fact that: . . .

I will not read the first four assumptions, but the fifth assumption is in this form:

> (5) that the demised premises are suitable and available for immediate occupation and use and are fully carpeted and that the demised premises incorporate on each floor boxes appropriate and suitable for servicing the demised premises as first class offices (judged in the light of standards prevailing at the relevant review date) . . .

Thereafter, there are set out five matters to be disregarded. The fourth is in this form:

> (iv) any notional rent free rent concession or fitting out period for which allowance would or might be given to the Tenant if the demised premises were let in the open market with vacant possession.

The schedule also provided that the rent should be payable on the days and in the manner provided in clause 2 of the lease.

The schedule goes on to lay down the procedure that should be followed for the rent review. As the parties could not agree what the new rent should be, Mr Perkins was appointed as the arbitrator and he resolved a number of issues between the parties and gave his conclusion in an interim award. He concluded, having had the advantage of the written opinion of Mr D M Barnes QC, that:

> No reduction is to be made to the rent for the absence of any rent free period for fitting out in the hypothetical Lease, but the rent on review is to be determined as the rent which would have been paid in the absence of any further rent free period or other inducement to the hypothetical Tenant to take the subject Lease.

Upon the evidence he concluded that at the review date the hypothetical landlord would have given to the hypothetical tenant a rent-free inducement of not less than 12 months to take the hypothetical lease, but that three months of that would have been for fitting out. He went on to conclude that a nine-month inducement period was equivalent to a reduction in rent of 15% over the review period. He valued the market rent for the premises at £564,327.60 pa. Thus, if a 15% allowance were made for the inducement period, the rent payable came to £479,678.46 pa. He therefore held that the rent should be the £479,678 to which I have referred.

The applicant, the landlords, had contended that any inducement period that might have been offered should be ignored. Upon the assumption that that construction was right, the arbitrator concluded that the appropriate rent was £564,327 pa.

The dispute before me is whether the rent should be assessed taking into account the value of an inducement period.

The applicant submitted, both before the arbitrator and before me, that, upon the proper construction of the fourth schedule to the lease, the rent was to be determined as the current market rack-rental value after the expiry of any rent-free concession. Thus, if the hypothetical landlord would have agreed to a nine-month rent-free period as an inducement to take the lease and thereafter a rent of, say, £40 per sq ft, the rent would be £40 per sq ft payable from the review date. The respondent submitted that the arbitrator was right and that the rent should be assessed on the basis that no rent-free period was given as an inducement. Thus, on the example I have taken, the hypothetical landlord would seek and obtain, say, £35 per sq ft payable from the rent review date.

Clause (1)(b) of the fourth schedule provides that the rent for the next five years shall be the current-market rack-rental value of the premises at the review date. For the purpose of the lease, that is defined as the best yearly rent payable after the expiry of any rent-free concession or fitting-out period, which might be given to the tenant in open market conditions.

The reason for choosing the period after any rent-free concession or fitting-out period may well be to ensure that, at the start of the period to be considered, rent would be payable. If the review date had been chosen, it could have been argued by the tenant that at that date no rent was payable because the hypothetical landlord would have agreed with the hypothetical tenant that no rent should be paid for, say, three months for fitting-out or for, say, nine months as an inducement. Thus, the best rent would be less than the current rent.

The rent review clause also provides that the best yearly rent shall be determined upon five assumptions, one of which is that the premises are suitable and available for occupation. Thus, if it be assumed in the hypothetical negotiation that the landlord would grant one year free as an inducement, the best rent referred to in clause (1)(b) would be calculated after the one year had elapsed, but upon the assumption that the premises would be ready for occupation. Thus, no fitting-out period would be needed and the tenant could not claim an allowance for the cost of fitting out.

The clause also has five disregards. I have read the fourth, which is the crucial clause so far as the dispute before me is concerned. It lays down that any notional rent-free, rent-concession or fitting-out period, if the premises were let on the open market, must be disregarded. Thus, the best yearly rent payable from the review date has to be determined upon the assumption that no rent-free or other like period would or might be given.

The fourth disregard when read with the opening part of clause (1)(b) charts a somewhat tortuous path, but I believe that its effect is clear. That effect is best explained by adopting the figures found by the arbitrator, namely that the market rent would be about £560,000 pa, but that a fitting-out period of three months would be required and a further nine months, rent-free period. Thus, the hypothetical tenant would pay about £560,000 pa for four years of the five-year lease.

The opening parts of clause (1)(b) state that the rent payable from the review date shall be the best yearly rent after any rent-free concession or fitting-out period has expired. That rent would be about £560,000 pa. Assumption (5) requires the arbitrator to assume that the premises are fitted out. Thus, the hypothetical tenant cannot seek a reduction of that rent for the three months that would be necessary, nor could the parties pray in aid the cost of fitting out. Disregard (iv) requires the arbitrator to disregard any notional rent-free, rent-concession or fitting-out period. To arrive at the figure of £560,000 pa the arbitrator, using comparables, concluded that in the hypothetical negotiation the hypothetical landlord had conceded nine months free of rent as an inducement and a further three months for fitting out. Those notional periods have, according to the fourth disregard, to be disregarded. Necessarily the result of disregarding the notional free periods results in an adjustment of the rent. In this case it means a reduction of 15% to about £480,000. By that route the draftsman artificially fixed the date at which the rent was to be valued, in this case the review date plus one year, but said that the valuation had to be made on the assumption that the premises were fitted out and no rent-free period or the like had been, or would be, given. Thus, the hypothetical landlord and tenant must be assumed to agree that the rent was payable yearly for each of the five years and was the best rent obtainable upon that assumption.

There was before me considerable argument as to the phrase "after the expiry of any rent-free concession or fitting-out period". Both parties pointed out that a fitting-out period was a rent-free period. On behalf of the respondent, it was submitted that the phrase should be read as a whole and, if so, referred to a rent-free period granted to cover a period for fitting out and to that alone. That appears to have been the meaning given to the phrase in the opinion of Mr Barnes. I believe that submission to be wrong. In my view, it violates the language used, which is clear. The rent has to be the best yearly rent payable after the expiry of any rent-free period whether provided as a fitting-out period or not.

There was also argument as to the meaning to be given to the words "rent-free concession", because the parties appeared to accept that a fitting-out period was a rent-free period which the applicant believed to be a concessionary period. Of course, a fitting-out period involves a rent-free concession, but the lease is its own dictionary. It uses the words "any rent-free concession" as meaning something different to a fitting-out period. That does not create a difficulty even though the draftsman could have used a composite phrase to include all rent-free periods.

I have construed the opening words of clause (1)(b) in the way which the applicant submitted that they should be construed. However, I have construed the fourth disregard in a manner that results in the whole clause having the meaning for which the respondent sought. The applicant submitted that such a construction was wrong and that all that the fourth disregard did was to make it

A plain that the rent was to be paid as from the first day of the five-year period. The fourth disregard may have the effect for which the applicant contends, but it is not limited to that effect. The periods referred to in the fourth disregard have to be disregarded for all purposes of the rent determination. Thus, it is relevant to both the period of payment and the amount of payment.

I have tried to construe the lease purposefully. In so doing I believe that my approach and the result accord with the judgment of Nicholls LJ in *Basingstoke and Deane Borough Council* v *Host Group Ltd* [1988] 1 WLR 348. At p353 he said:

> The question raised on this appeal is one of construction of a rent review clause in a lease. In answering that question it is axiomatic that what the court is seeking to identify and declare is the intention of the parties to the lease B expressed in that clause. Thus, like all points of construction, the meaning of this rent review clause depends on the particular language used interpreted having regard to the context provided by the whole document and the matrix of the material surrounding circumstances. We recognise, therefore, that the particular language used will always be of paramount importance. Nonetheless it is proper and only sensible, when construing a rent review clause, to have in mind what normally is the commercial purpose of such a clause.
>
> That purpose has been referred to in several recent cases, and it is not in doubt. Sir Nicolas Browne-Wilkinson V-C expressed it in these terms in *British Gas Corporation* v *Universities Superannuation Scheme Ltd* [1986] 1 WLR 398, 401:
>
> "There is really no dispute that the general purpose of a provision for rent C review is to enable the landlord to obtain from time to time the market rental which the premises would command if let on the same terms on the open market at the review dates. The purpose is to reflect the changes in the value of money and real increases in the value of the property during a long term."
>
> To the same effect Dillon LJ said in *Equity & Law Life Assurance Society Plc* v *Bodfield Ltd* [1987] 1 EGLR 124, 125:
>
> "There is no doubt that the general object of a rent review clause, which provides that the rent cannot be reduced on a review, is to provide the landlord with some measure of relief where, by increases in property values or falls in the real value of money in an inflationary period, a fixed rent has become out of date and unduly favourable to the tenant. The exact measure of relief depends on the true construction of the particular rent review clause".

Upon the conclusions of fact arrived at by the arbitrator, the market rent of the premises which could have been obtained, on the D assumption that it was fitted out, was about £480,000 pa. The submission of the applicant was that the tenant should pay about £560,000 pa even though such a rent could not have been obtained in the open market. Thus, the landlord would obtain, upon the landlord's submissions, a windfall profit with resulting loss to the tenant not only in the amount of rent payable but also in the cost of any assignment of the lease.

Nicholls LJ in the *Basingstoke and Deane* case went on to hold that, save in cases where a rent review clause expressly or by necessary implication stated the contrary, the parties are to be taken as having intended the letting to be on the same terms, the reason being that accorded with the general intention underlying the inclusion of rent E review clauses.

Similar reasoning is applicable to the facts of this case. Construction of the lease so that the tenant paid £560,000 pa for five years, as sought by the applicant, would give to the applicant rental income, which did not reflect the changes in property values and the value of money. That is reflected by a rent of about £560,000 for four years, amounting to about £480,000 for five years as awarded by the arbitrator. I conclude that, upon the literal meaning of the lease, the approach of the arbitrator was correct. Further, that meaning coincides with the presumed intention of the parties. I therefore dismiss the appeal.

In the construction summons, City Offices plc is the plaintiff and Allianz Cornhill International Insurance Co Ltd is the defendant. By F that summons they seek construction of the rent review provisions of a lease dated July 10 1986 relating to the fourth floor of 27 Leadenhall Street, London EC3.

It is therefore not surprising that the Allianz lease is very similar to the Bryanston lease which I have already construed. Further, it is not surprising that the matters in dispute concern the construction of the rent review clause in clause (1)(b) of the fourth schedule.

The opening words of clause (1)(b) differ slightly to the words used in the Bryanston lease. It provides that the reviewed rent should be:

> . . . the current market rack rental value of the demised premises at the review date and for this purpose current market rack rental value shall mean the best yearly rent which would reasonably be expected to become payable in respect of the demised premises after the expiry of a rent free period of such length as would be negotiated in the open market between a willing landlord and a G willing tenant upon a letting on the review date . . .

The difference between the two leases is that in the Allianz lease it is required that the rent be the best yearly rent "after the expiry of a rent free period of such length as would be negotiated in the open market between a willing landlord and a willing tenant upon a letting on the review date"; whereas the Bryanston lease required the best yearly rent to be ascertained "after the expiry of any rent free concession or fitting out period which might be given to the Tenant if a letting of the demised premises were negotiated in the open market between a willing landlord and a willing tenant upon a letting on the review date". In both leases the assumptions and disregards are the same.

The difference between the Allianz and Bryanston leases is not H material to the point of construction I have to decide. Having regard to the way I have construed the fourth disregard in the Bryanston lease, the same conclusion must be reached for similar reasons to those given, namely the defendants' view as to the effect of the lease is correct. Therefore, I will make an appropriate declaration to that effect.

Declaration accordingly certificate and leave to appeal granted under section 1(7)(b) of the Arbitration Act 1979.

Chancery Division J

November 16 1992

(Before Judge COLYER QC, sitting as a judge of the division)

PRUDENTIAL PROPERTY SERVICES LTD v CAPITAL LAND HOLDINGS LTD

Estates Gazette April 17 1993

[1993] 15 EG 147

Landlord and tenant — Rent review — Whether a letter constituted a valid counternotice to landlord's trigger notice K

By a lease dated May 21 1982 A C Frost & Co Ltd, a nominee of the plaintiff, Prudential Property Services Ltd, holds a term from the defendant, Capital Land Holdings Ltd, of 20 years from May 21 1982, of 117 High Street, Chesham, Buckingham. Where the landlord desires to implement any of the four-yearly rent reviews, the lease provides that the landlord serves at least three months before any rent review date a notice upon the tenant specifying the rent required. Where the parties fail to agree the rent within a month of the landlord's notice, then by para 2(c) of the fourth schedule to the lease, "it shall be determined at the election of the tenant L by counter-notice in writing to the landlord, not later than one month after the landlord's said notice (time to be of the essence hereof), by an independent surveyor . . . ". Following the service of a notice by the defendant on February 14 1990 specifying a rent of £17,500 pa for the review date on May 21 1990, a letter was sent by the plaintiff and received by the defendant on March 6 1990 acknowledging the defendant's letter and stating that the plaintiff's client was not agreeable to the rent proposal and adding "in accordance with clause 2 of the lease please acknowledge this letter on the landlord's behalf as formal notice of our disagreement to your rent proposal". After the expiration of the one-month time-limit M the defendant asserted that the plaintiff had failed to serve an effective counternotice within the time-limit and the rent for the four years from May 21 1990 was therefore that specified in the defendant's letter of February 14 1990. The plaintiff sought a declaration that the letter of March 6 1990 was an effective counternotice served within the time-limit.

***Held*: The letter was an effective counternotice and a declaration was granted accordingly. It is sufficient that, to be an effective counternotice, a letter makes clear that the tenant is exercising the relevant election; a tenant may do that either by indicating that the letter is a counternotice under the**

A **relevant clause or by spelling out the consequences which he seeks to achieve. The reference to clause 2 in the letter can have been meant only as para 2 of the fourth schedule to the lease. The letter must be read in the context not just of the lease but in the context of the letter to which it is responding. The letter was in response to the landlord's trigger notice and referred to para 2 of the fourth schedule, mistakenly described as clause 2 of the lease; that paragraph makes provision for the service of the counternotice making an election for the determination of the rent by an independent surveyor.**

B The following cases are referred to in this report.

Amalgamated Estates Ltd v *Joystretch Manufacturing Ltd* [1981] EGD 84; (1980) 257 EG 489, CA
Barrett Estate Services Ltd v *David Graig (Retail) Ltd* [1991] 2 EGLR 123; [1991] 36 EG 155
Bellinger v *South London Stationers* (1979) 252 EG 699
Edlingham Ltd v *MFI Furniture Centres Ltd* (1981) 259 EG 421
Glofield Properties Ltd v *Morley* [1988] 1 EGLR 113; [1988] 02 EG 62
Nunes v *Davies Laing & Dick Ltd* (1985) 51 P&CR 310; [1986] 1 EGLR 106; 277 EG 416
Oldschool v *Johns* (1980) 256 EG 381
Sheridan v *Blaircourt Investments Ltd* [1984] EGD 176; (1984) 270 EG 1290
United Scientific Holdings Ltd v *Burnley Borough Council* [1978] AC 904; [1977] 2 WLR 806; [1977] 2 All ER 62; (1977) 33 P&CR 220; [1977] EGD
C 195; 243 EG 43 & 127, HL

This was an application by the plaintiff, Prudential Property Services Ltd, by an originating summons to have determined the validity of a counternotice served under the provisions of a lease dated May 21 1982 held by A C Frost & Co Ltd from the defendant, Capital Land Holdings Ltd.

Jonathan Steiner (instructed by Masons) appeared for the plaintiff; Jeremy Carey (instructed by Roman Coleman & Co) represented the defendant.

Giving judgment, **JUDGE COLYER** said: I proceed this morning to give judgment in relation to matter no 1990 P7150, being that
D originating summons whereby originally Prudential Property Services Ltd, but now the substituted plaintiff Woolwich Property Services Ltd, seeks against Capital Land Holdings Ltd the determination of the court of the question whether a letter dated March 5 1990 sent by Prudential Property Services Ltd, which is described in the originating summons as being "the tenant at that time of a lease dated May 21 1982 — the benefit of which lease is now assigned to the plaintiff" — and the originating summons continues "and who is now the tenant of the said lease", constituted a valid counternotice pursuant to clause 2 of the fourth schedule of the same lease, of premises known as 117 High Street, Chesham, Buckinghamshire. The originating summons was issued as long ago as July 20 1990; it was followed five days later by a second originating
E summons, which I shall deal with after I have concluded this matter, whereunder the tenant seeks an extension of time for service of such counternotice pursuant to section 27 of the Arbitration Act 1950. I need say nothing about that for the moment.

I would immediately observe that in fact Prudential Property Services were not the tenant under the lease. The legal term of years was vested in a company, A C Frost & Co Ltd, a nominee, it is said in the evidence, for Prudential Property Services Ltd, so that in any event the declaration which is sought even after amendment is entirely inappropriate. It is said that the letter, though sent by Prudential Property Services Ltd, was sent on behalf of the nominee company and that is not disputed by the landlords.

F I turn then, first, to consider the facts. This lease was granted by one Keith Depty and one Robin Westly on May 21 1982, to tenants Chiltern HiFi (Chesham) Ltd. There were sureties to the lease, but that does not matter. The term was for 20 years from May 21 1982, and there were provisions for rent review at four-yearly intervals; thus the initial rent of £4,500 per year was to apply only until May 2 1986.

The terms of the lease are irrelevant to the present dispute, other than the stipulations contained within the rent review clause, to which I now turn. That is contained in the fourth schedule, and by part (b) of para 1 of that schedule the rent review dates are designated. In consequence, May 21 1986, May 21 1990, May 21 1994 and May 21 1998 are rent review dates. I pause there to say that there is a drafting slip there, in that the draftsman obviously became confused between the concept of a rent review period and the initial G period of the lease whereunder for four years the initial rent of £4,500 was to be paid. Nothing turns on that incongruity in the reddendum to the lease in the present dispute.

The review is an upwards-only review, and it is stipulated that the reviewed rents payable by the tenant during the first review period, and the remainder thereof — that must mean the other rent review periods — shall be determined in the manner following — that is to say, it shall be whichever shall be the higher of the rents then current in the open market — and again the word "rent" has got lost, yet another indication of the sloppy drafting of this lease — and the open market — I interpose the word "rent" of the demised premises for the appropriate review period.

That is then defined — market rental — and nothing turns upon H that. Then by para 2 of the same schedule we have the machinery for the determination of the figure. It is to be determined as follows and I quote:

> (a) It shall be such sum as shall be specified in a notice in writing by the landlord to the tenant in the case of the first review period at the expiration of the period of three years and six months after the commencement of the term hereby granted; and in the case of the remaining review periods at any time before the expiration of three months before commencement of the relevant review period.

I pause there to observe that upon the first rent review date, being May 21 1986, the rent was revised without problems to an annual rent of £9,000; and that after that date and before the second review date J the residue of the term became vested in A C Frost & Co Ltd, and the reversion upon the lease in the defendants.

The review timetable therefore required that by February 20 — being three months before the relevant review date — the landlords should specify in a notice in writing the rent which they required. The lease then continues:

> (b) as shall within one month after such notice be agreed between the parties in writing, in substitution for the same sum

— that is the sum in the landlord's notice —

> or, (c) it shall be determined at the election of the tenant by counternotice in writing to the landlord, not later than one month after the landlord's said notice (time to be of the essence hereof), by an independent surveyor appointed for that purpose by the parties jointly in writing, or upon their K failure to agree upon such appointment, within one month after the date of the said counternotice, then by an independent surveyor appointed for that purpose by the President for the time being of the Royal Institution of Chartered Surveyors; and every determination shall be made insofar as not inconsistent with the provisions of the Arbitration Act, 1950, or any statutory modification or re-enactment for the time being in force; and shall be subject to further provisions of the next succeeding sub-clause.

The next succeeding subclause deals with the problem of any shortfall of rent which results from the determination after the due date of the revised figure. No thing turns upon that in the present proceedings.

Before leaving the quotation from para 2 of that fourth schedule I would observe that, fitting the schedule to the facts of this case, the L notice by the landlord, as I have pointed out, had to be served by February 20. As will be seen in a moment, that notice was given and the letter giving it is not impugned as a valid notice on February 14 1990; thus it was in time. As a result, the counternotice had to be given by March 14 1990 and in relation to that date time is expressed to be of the essence. Once the landlord's notice had been served, if a counternotice were served then a further month is stipulated by the lease, and time is not made of the essence in relation to that one-month period for the agreement by the parties upon their choice of independent surveyor. Put another way, whereas the first stipulation that the counternotice must be served within a month involves a deadline for the serving of the counternotice, the month that is M stipulated for the agreement upon the identity of an independent surveyor does not involve a deadline at all; it merely enables the application to be made to the President once that one-month period has elapsed and once it is apparent that there has not within that period been any agreement. It is my view, on the terms of this clause, that either party to the lease could after that month apply to the President for such appointment.

In the landlord's notice of February 14 1990 the figure of £17,500 pa was stipulated. The notice was contained in a letter from the landlords' solicitors, John Broomfield & Co, consultant surveyors and valuers. They wrote their letter, dating it February 14. It has been assumed that all letters arrived in the ordinary course of post. Again, nothing turns in this case upon whether we should therefore regard

A the notice as given on February 15 or treat it as given on the date that it bears, although, in my view, it should be treated as given on the date it arrives. Because nothing turns on it, it is easier to treat the documents as they are dated.

The letter was addressed to Mr Brook of Prudential Property Services at the address and it read as follows:

> Dear Sir
> 117 High Street, Chesham
> Following your letter of fourth January we write firstly to thank you for the arrangements made for us to inspect the accommodation held on lease by assignment at the above address by A C Frost & Co Ltd

I pause and interject that the landlords therefore had no doubt who B the tenant was, and were writing to Prudential Property Services as the agent of that tenant. The letter continues:

> You will know that we are instructed by the freeholder to settle with the company the rent to be paid as from the forthcoming review date. Please be advised now that, having considered carefully the size and location of the premises, and the terms of the lease as made on 21st May 1982 and subsequently assigned to your company, that we assess the open market rental value for the third period of the lease to be in the sum of £17,400

— which was then repeated in parenthesis in words —

> per annum exclusive, and that this is the rental which we ask you to confirm for payment from 20th May 1990. For the avoidance of doubt, we confirm that this communication represents our client's formal notice in writing as per the provisions of clause 2 of the said lease.

C I interject again that the reference to "clause 2" could be a reference only to para 2 of the fourth schedule. There is in the lease an operative clause 2, but it is singularly not that which the writer of the letter had in mind, for clause 2 of the lease merely reads:

> The tenant hereby covenants the landlord in the terms set out in the fifth schedule,

and all the covenants of the tenant are contained in that fifth schedule. So there is an obvious drafting slip or mislabelling of the relevant clause; but it is so obvious that no one could conceivably have been misled by it. Anyone reading the landlord's letter knew that the provisions of the fourth schedule, and para 2 of it, were now being invoked by the landlords; in short, knew that the landlords' D letter was the landlords' trigger notice for the rent review.

The landlords' agent's letter continues:

> In due course the lease and counterpart will need to be endorsed with a note of the review rent, and our clients will prepare a standard form of memorandum for signature by the parties. Please kindly acknowledge safe receipt of our communication. We trust that with the local evidence available our own assessment can be agreed by your goodselves.
> Yours faithfully,

The response to that letter came not from the person — H Brook Esq — to whom it had been addressed, but from a colleague of his within Prudential Property Services. It was undated, but appears to have been despatched on March 5 and received on March 6. If, therefore, it be a counternotice, then it was within the one-month E period; and it is common ground that there is no other document which was sent within that period. This letter, therefore, is the only possible counternotice.

The letter is addressed to John Broomfield & Co at their address. It takes up their reference on the previous letter, gives the writer's reference. It is sent also by recorded delivery, like the trigger notice, and it reads thus:

> Dear Sir
> Re 117 High Street, Chesham.
> I hereby acknowledge receipt of your letter of 14th February 1990 incorporating rent proposal of £17,500 per annum for the review effective 20th May 1990.

The syntax then goes a little awry, but it is important that the letter F be quoted verbatim and in full.

> My client is not agreeable to settlement of rental proposal on this basis, and in accordance with clause 2 of the lease please acknowledge this letter on the landlords' behalf as formal notice of our disagreement to your rent proposal. Our rent proposal is £9,000 per annum. I will be inspecting the property in the next few days and will contact you to commence the negotiation on this review.
> Yours faithfully

It is signed John Gay, for and on behalf of Prudential Commercial Property Services. They, on any view, were acting as agent for the tenant, although they were not in fact the tenant. If the notice is a valid counternotice, then it is not disputed by the landlords that it was duly served within the time provided and on behalf of A C Frost & Co Ltd. It appears, however, that Mr Gay and the solicitors then acting G became a little confused, and launched the proceedings in the name of Prudential Property Services — a matter which may have implications as to costs or as to the section 27 proceedings, but which I can ignore in relation to the construction issue which now faces the court.

The letter did not provoke a response from the landlords' surveyors and was followed by a further letter on March 21, which read as follows:

> I refer to recent correspondence in respect of this matter. I note that you have proposed a figure of £17,500 per annum in respect of the forthcoming rent review effective from 20th May 1990. I should be grateful if you would provide me with a breakdown of your floor areas together with a note of any H comparable evidence you may have in support of your proposal in order that I may give the matter due consideration. Yours faithfully

and it is signed by another surveyor within Prudential Commercial Property Services — a Mr Anthony Doyle. That letter was very promptly replied to — indeed by return of post — and the letter was faxed before being posted on March 23 1990. The response sent read:

> We write to thank you for your letter of 21st and for the enquiry which you make. We note all you say, and once you have undertaken your inspection of the property we will gladly exchange floor area notes with you. Please look at the terms of the lease and the subsequent correspondence. Our view is that the lessee is estopped from arguing that the review rent should be less than the amount of £17,500 per annum exclusive as sent out in the rent notice sent on 14th February last. We make this observation having regard to the provisions J of clause 2(c). Having looked at our file we can see that we have not received a counter-notice containing the request to have the rent determined by an independent surveyor. Please let us hear further from you. In terms of evidence we think you will find that your own tenancy of 3 White Hills has been the subject of the equivalent of the zone A rent of some £22 per square foot ever since September 1988. We await to hear from you.

I need not read the subsequent correspondence, but it is sufficient to say that on March 23 1990 the landlords are therefore taking the point that no valid counternotice has been given. It is their first response to the tenant's letter of March 6, and of course the one-month time-limit on which time is made of the essence for the giving of that counternotice had by then passed, for that time-limit would have expired on March 14 or 15, depending on the date of arrival of K the trigger notice dated February 14. I do not know whether the non-acknowledgment of the tenant's earlier letter was deliberate, and whether the landlords' agents were sitting there waiting for, as they saw it, the trap to fall; or whether it was inadvertent; or whether it was simply a minor discourtesy on their part in not acknowledging the tenant's notice of March 6.

Certain it is that, in my view, whether they replied to it or not, that could not make the notice into a counternotice in the absence of a letter from them saying, "We don't think your letter was a counter-notice but we will treat it as such", which singularly, of course, did not occur here. So one cannot draw inferences as to the effect of the counternotice from the fact that they did not reply. It is, however, of L some relevance — not huge relevance, but some relevance — that they were asked with such particularity to acknowledge the letter on the landlords' behalf, "as [a] formal notice".

There has been a plethora of decisions on the effect of counternotices in the context of rent review clauses. My attention in this case has been drawn to no less than eight decided cases in relation to counternotices. May I say, first of all, that I think it is of great importance in this type of situation that one remembers at all times that the task of the court is to construe a document — that is the lease — and then, in the light of that decision on the construction of the document, to conclude whether in the facts of a particular case the communication made by that particular tenant does or does not satisfy the requirements of the instrument which has just been M construed. It is quite wrong, therefore, in my view, slavishly to take decided cases and to say: "Well, here was a letter written in identical terms in relation to an identical clause. Without more ado that is conclusive; the court must come to the same conclusion in this case as it came to in the other case." In this manner a seductive jurisprudence is being constructed about the phraseology of rent review clauses, and even that of letters given in response to them. I have to say that I think this is entirely misguided. We must bear in mind that unless they are landmark decisions which lay down principles of law — and a classic example that instantly springs to mind would be *United Scientific Holdings Ltd* v *Burnley Borough Council* [1978] AC 904 — all such decisions will be either pure construction of documents or pure decisions on facts of particular cases, or a combination of both.

So "precedent" is not wholly compelling or conclusive, although it may provide illustration or inspiration for arguments as to the instruction and effect of leases.

Here I am assisted immensely by the other cases, but I do not consider that I am following or not following particular decisions. I have to deduce from the cases the principles of law with which to approach my task of construction, and indeed my assessment of the facts, and then apply them to the facts of this case and the lease in this case; and the facts in this case really mean the letter which I have quoted of March 5.

I turn then to those cases. Although not the earliest case on counternotices, the decision in *Amalgamated Estates Ltd* v *Joystretch Manufacturing Ltd* (1980) 257 EG 489 is a decision of the Court of Appeal. It does indicate the approach to be taken in what I term the "counter-notice situation". The Court of Appeal, consisting of Naughton, Templeman and O'Connor LJJ, rejected in that case a contention by the tenant that a sufficient counternotice had there been given. I need briefly to turn to the facts of that case and the context in which that notice fell to be served. Under the lease in question — and I quote from the third paragraph of the relevant proviso in the lease:

> The open market rental value shall be determined in manner following; that is to say (a) specified in notice in writing signed by or on behalf of the Landlord and posted by recorded delivery post in a pre-paid envelope addressed to the Tenant at the demised premises at any time provided that if a notice is posted after the review date the open market rental shall not be payable until the second quarter day following the posting of said notice

— and I can then omit words —

> or (b) agreed between the parties before the expiration of three months immediately after the date of posting such notice

— and I omit words again. Then in a separate subparagraph a further possibility:

> determined at the election of the Tenant (to be made by counter-notice in writing served by the Tenant on the Landlord not later than the expiration of the said three months)

— so it is the same three months that are referred to above —

> by an independent surveyor appointed for the purpose by the parties jointly in writing or upon their failure to agree, upon such appointment within one month immediately after the date of service of the said counter-notice, then by an independent surveyor appointed for that purpose on the application of either party alone by the President for the time being of the Royal Institution of Chartered Surveyors

— and I again omit words.

The clause therefore had a very close generic relationship or similarity to the fourth schedule of the lease of the premises, which I am now considering. In relation to that clause in *Amalgamated Estates Ltd* v *Joystretch Manufacturing Ltd* other problems of construction arose which the court had to consider and which do not arise here. It is sufficient to say that a trigger notice was given by letter by the landlord and that it was responded to in these terms:

> We are in receipt of your letter of December 8th 1978 and the contents of same.
>
> We cannot agree with your rent increase for, as you know, they are building a new Centre which is going to push Regent Street right out of position. There are shops very close to our own that are being offered for around £4,000 per annum. Perhaps therefore you should explain how you arrive at £10,000.

(I observe that, happily, that was not Regent Street, London, but Regent Street, Rugby, that was being referred to.)

The landlords did not reply to that letter, and subsequently contended successfully before the Court of Appeal that the letter was not a sufficient counternotice. If one applies any form of analysis to that letter, brief as it was, it is difficult, indeed, to get out of it anything other than the critical statement, "We cannot agree with your rent increase". There was then some special pleading as to the local trading and property development situation, and a request for information — therefore nothing beyond an intimation that, "We do not agree" was given by that letter. Nevertheless, it was contended by the tenant that that was a sufficient counternotice.

In considering what such a counternotice had to do in the context of a lease which did not specify any form of counternotice other than that it be in writing — which again is the identical context to the present lease — Templeman LJ, at the top right-hand column of p493, observes:

> It is true that no magic formula is required but, in my judgment, the tenant must make it clear to the landlord that he proposes to have the rent decided by arbitration in accordance with the provisions of the lease. The landlord and tenant may reach a compromise before arbitration has been reached, but, nevertheless, on the plain words of the lease it seems to me that the tenant is directed to serve a counternotice making it clear to the landlord that he proposes to go to arbitration if that is his intention. I cannot spell that out of the tenant's letter, which simply said he did not agree with the landlord's £10,000 and asked for an explanation.

and his lordship observes that similar views had been expressed by Golding J in *Bellinger* v *South London Stationers* (1979) 252 EG 699, and Mr Michael Wheeler, sitting as a deputy judge of this division, in *Oldschool* v *Johns* (1980) 256 EG 381. His lordship was therefore impliedly approving both of those two decisions, in each of which the landlord had indeed succeeded in defeating the suggestion that a counternotice had been served.

I would turn for a moment to the *Oldschool* case, because that does contain some helpful observations; and also one observation with which I must respectfully register a slight disagreement. In *Oldschool* v *Johns* (1980) 256 EG 381 Mr Michael Wheeler QC had to consider a similar problem, a similar clause. He observes — I read from p385, left-hand column:

> I think that the notice must either be in a form which clearly says it is a counternotice under clause 5(2)(c) or it must clearly indicate to the landlord that the tenant is exercising his election to have the rent fixed independently.

That is clearly part of the *ratio decidendi*; it is clearly something which was looked upon with approval by the Court of Appeal in *Amalgamated Estates* v *Joystretch*; and I would respectfully adopt that approach; that is to say, that in order to be a counternotice the document must either state, "This is a counter-notice under [the relevant clause]"; *or*, if it does not do that, then it must state clearly that the tenant is exercising his election to have whatever remedy the clause contemplates, ie reference to independent expert, reference to independent surveyor acting as arbitrator, or as may be.

So one can do it either by shorthand, simply saying, "This is a counter-notice under clause whatsoever", or by longhand, by a document which explains what you want.

On the right-hand column of p385 of the same report an observation is summarised from counsel's argument in the other earlier case to which I referred, *Bellinger* v *South London Stationers Ltd*. Goulding J quoted counsel there:

> So [says counsel] I must ask myself whether anyone receiving the letter . . . must have realised at once that the tenant was requiring arbitration; or, putting it another way in the judge's later words, that the recipient would be in no doubt as to what the tenant was up to.

If (which I do not believe to be the case) the learned judges in *Bellinger* or in *Oldschool* were adopting that submission and argument and were therefore holding that one must — and I quote — "realise at once" upon reading the letter what it meant, I would respectfully demur.

In the light of the subsequent decisions to which I shall refer and in the light of the observations of Templeman LJ and the other lord justices in *Amalgamated Estates* v *Joystretch* it is, in my view, sufficient that to be an effective counternotice a letter makes it clear that the tenant is exercising the relevant election; and the tenant may do that *either by* indicating that the letter is a counternotice under the relevant clause or by spelling out the consequences which he seeks to achieve.

Before leaving the other cases, I would, however, make nodding reference to certain of the authorities. I was referred, in addition to those to which I have referred, to *Edlingham Ltd* v *MFI Furniture Centres Ltd* (1981) 259 EG 421. There, the clause had a similar provision for trigger notice and then for the rent to be "determined (at the election of the Tenant such election to be made by counternotice in writing served by the Tenant upon the Landlord not later than the expiration of the said three months) by an independent surveyor". That stipulation and, indeed, all stipulations in the clause were made of the essence.

The tenant responded to the landlord's trigger notice in these terms.

> I refer to your letter dated February 7

— That is the trigger notice —

> and return herewith one copy of the notice duly acknowledged.
>
> Will you please accept this letter as counternotice to the effect that we consider that the rent of £50,000 is excessive and will appreciate it if you will kindly forward to us comparables on which you have based this figure.

The learned judge, McNeill J, did not consider that that was a valid counternotice and, having quoted from several cases, including the *Oldschool* case, to which I have referred, commented:

A I do not think it is necessary for the purposes of this judgment to express a view as to whether or not a counternotice, to be effective, must expressly refer to the relevant clause of the lease which is in point. My inclination would be to say that that degree of formality is probably unnecessary. But the notice must identify the form of relief which the tenant is seeking and must do so unequivocally.

With respect to the learned judge, he is therefore disagreeing with Mr Wheeler in the *Oldschool* case, which he says he is following, Mr Wheeler having indicated that the election can be either by, if by expressing making your letter, a counternotice or by, as I put it a moment ago, doing it longhand and spelling out the relief that you need. McNeill J is limiting the valid counternotice to a letter which does the second thing, which spells out the relief. I quote again:

B . . . the notice must identify the form of relief which the tenant is seeking and must do so unequivocally. If a notice may be read as an invitation to negotiate or as an election to arbitrate it seems to me to be a defective notice.

In other words, if a notice can be read in the alternative, as either an invitation to negotiate or as an election to arbitrate, he would see it as ambiguous. He continues a little later:

It is not a question, as Templeman LJ put it, of a "magic formula", but in his words "the tenant must make it clear to the landlord that he proposes to have the rent decided by arbitration in accordance with provisions of the lease".

His lordship did not consider that it had been made clear or, rather, as he put it, it was not unequivocal. In *Sheridan* v *Blaircourt Investments Ltd* Nicholls J also used in similar context the phrase
C "unequivocal", holding that it was not possible to construe the wording in four successive letters as an unequivocal intimation to the landlord's advisers that the tenant was requiring ascertainment of the substituted rent to be referred to a referee.

In view of the very different content of those letters to the letter which I have to consider, I do not find that decision of any real help. There, the letters were expressly contemplating a later reference — all but parading the point that those letters were not exercising the relevant election.

If there is — and I doubt that there is — but if there is any conflict on the case as to whether one should look for a "clear" or whether one should look for an "unequivocal" election (words which in this context I am bound to say I do not find it easy to see producing
D different results) then, in my view, that conflict is laid to rest by two things.

First, by the fairly obvious proposition that *Amalgamated Estates* v *Joystretch* is a decision of the Court of Appeal, and none of the other decisions to which I refer is, and I am bound to follow that decision. Second, by the fact that in *Nunes* v *Davies Laing & Dick Ltd* [1986] 1 EGLR 106, the then Vice-Chancellor, Sir Nicolas Browne-Wilkinson, considered the entire problem of counternotices in some detail. I will omit the full quotation of the clause in the lease before him, but I need to quote the notice and the counternotice. The trigger notice read:

On behalf of the present Lessors we hereby give you notice, pursuant to
E Clause 2 of the Lease that the market rent from the review date

— which he then specified —

shall be the annual sum of £23,000.

That provoked a response which I am bound to observe was expressed rather as a crossproposal or crossnotice, rather than a counternotice, for it read:

Dear Sirs,

I am instructed by the tenants to give you hereby formal notice that the open market rental is £12,000 per annum and call on you under the terms of the above Lease to agree this. Please confirm that this is accepted as due notice.

Faithfully

F Then it was signed.

The problem was summed up succinctly and most helpfully by the Vice-Chancellor, who pointed out that there are, in reality, two questions: (1) what in law is required to constitute a valid election under the relevant clause; and (2) whether the letter in question did so. His lordship then considered the earlier cases. Under his heading: "What are the legal requirements of a valid notice?" (and of course he is considering only a valid notice under that lease, not the lease now before the court, but his observations are of immense assistance and provide persuasive argument and reasons). His lordship observed:

In the reported cases this question has not been addressed separately. It has been treated as rolled up in the question of the construction of the alleged counternotice. But, in my judgment, the authorities do disclose that the law G
requires the alleged counternotice to have a certain degree of clarity in its phraseology. It is not merely a question of the judge construing the document; the question is whether the document brings home to the mind of the ordinary landlord the fact that the tenant is exercising his rights under paragraph (c).

His lordship then observed that the authorities disclose two possible tests: "a clear" indication or that the counternotice must be "unequivocal" and that there may be a difference between what is clear and what is unequivocal, and his lordship observes:

In my judgment, the test is that applied by the Court of Appeal in the *Amalgamated Estates* case, namely that the counternotice should be in terms which are sufficiently clear to bring home to the ordinary landlord that the tenant is purporting to exercise his right under . . .

And I interpose there: "under the relevant clause of the relevant H
lease". Applying that test, his lordship held that what I termed the crossnotice was a sufficient counternotice under that clause of that lease.

The tide at that point seems historically to have turned in the tenant's favour; *Nunes* is followed, in the sense of being chronologically succeeded, by two later cases to which my attention was drawn: *Glofield Properties Ltd* v *Morley* [1988] 2 EG 62; [1988] 1 EGLR 113 (a decision of Hutchinson J) and *Barrett Estate Services Ltd* v *David Greig (Retail) Ltd* [1991] 2 EGLR 123 each of which were decided in favour of the tenant.

I should say I derive no assistance from *Barrett*, because in *Barratt* the clause was sufficiently different as to avoid the very problems J
which one faces in the instant case. There, all that the counternotice needed to indicate was that the tenant did not agree the figure, which is a very different task for a counternotice to achieve than that required by the present clause.

In *Glofield*, presented with similar requirements for trigger notice and counternotice, the landlords having served a trigger notice, the tenants' agents responded:

We act for the tenant of the above premises. We have been passed a copy of your letter . . .

— being the trigger notice —

. . . in which you proposed a rent of £9,750 per annum at the forthcoming rent review. Please accept this letter as formal objection and counternotice. We would suggest an early meeting to discuss the matter in detail. K

It was contended by the landlords in that case that this was not an unequivocal election, that it was, indeed, suggesting a meeting and discussion. His lordship relied extensively both on the *Amalgamated Estates* decision and on the *Nunes* decision and respectfully demurred to the decision of McNeill J in *Edlingham*, if and in so far as it conflicted. Faced with a possible conflict between decisions at first instance, he followed the most recent of those decisions, that in *Nunes*, given that it had considered the earlier decision. *A fortiori*, I, too, should follow the decision in the *Nunes* case.

After considering the way in which the clause worked and applying the test suggested by the *Amalgamated Estates* case, his lordship concluded that that letter was an effectual counternotice. That, in my L
view, rather vindicates the observation made by Mr Wheeler in the *Oldschool* case, that I have already adopted, that the election can be either by expressly describing one's letter as a counternotice under the relevant clause or, alternatively, by doing it longhand. If we look at the letter sent in the *Glofield* case again, it is clear that that did not in longhand, as I put it, express an election to go to reference or arbitration; it merely gave formal objection, suggested a meeting to discuss the matter in detail, but nevertheless used the magic words "counternotice":

Please accept this letter as formal objection *and counternotice*.

I add emphasis which was not in the actual document to the words "and counternotice".

My attention has been drawn also to the existence of a body of law M
on the construction of notices to quit. It is sufficient to say that that was not pressed upon me. I expressed the view in argument, and I express it even more strongly now, that those decisions do not assist the court at all in construing notices in the context of trigger notices and counternotices relevant to rent review.

I turn now to the facts of this case and consider whether this letter of March 5 does or does not satisfy the requirement of this lease in relation to the giving of a counternotice and making an election. I observe in the first place that it is wrong to take the letter in total isolation. One must read it in the context not just of the lease but in the context of the letter to which it is responding. I return again to the provisions of the lease. It is clear that under para 2(c) of the schedule:

A It shall be determined at the election of the tenant by counter-notice by an independent surveyor . . .

It does not say there: "by a counternotice which spells out that which is required" — that is reference to an independent surveyor and appointment of a President if the parties cannot agree upon him, which the remainder of the subparagraph of the schedule envisage. It does not say the counternotice must spell all this out; it merely says "by counter-notice in writing".

So the requirements are that there be a counternotice, that it be in writing and that it be within time. The letter was in writing of course, it was in time. Was it a counternotice? I hold, following the earlier decisions, that it would be a counternotice if either it expressed itself B to be just that or, alternatively, if, without referring to the clause number or perhaps even without using the word "counternotice", as I earlier put it, the draftsman had in longhand spelled out just what he required, thus describing the consequences of a counternotice. Either method would be a valid election.

Mr Gay did not spell it out "in longhand"; and, on behalf of the landlords, it is urged upon me that he also did not express his letter to be a counternotice. I am bound to say that I think that he did. I think that any reasonable landlord, reading this letter, although he might not instantly have known that this was an election, would immediately have consulted his lease and have known that the relevant notice had been served upon him.

In the first place the letter was responding to the trigger notice. It C was served with the same degree of solemnity and particularity as the trigger notice by recorded delivery. It was served by a writer who clearly was anxious that the letter get to its addressee in time and who not merely sent it by recorded delivery but asked for acknowledgment:

Please acknowledge this letter [on the landlord's behalf] as formal notice . . .

I pause there. If the letter had stopped at that point and the next words "of our disagreement to your rent proposal" had been omitted, I doubt that this case would have come to court. It is the inclusion of those words which, if one reads the letter in isolation and without reference to the trigger notice itself and the clause in the lease itself, introduce what I regard as a specious ambiguity to the D document.

If, however, one reads the letter in context, and John Broomfield & Co, the authors of the trigger notice, hardly could read it other than in the context of their own earlier letter, then any momentary doubt that they had as to the effect of this letter could and should have been quickly, if not instantly, resolved were they to turn to the lease.

I slip into the language of the first person and pretend for a moment to be the hypothetical recipient of the letter dated March 6. If that hypothetical person said to himself: "Dear me, what is this letter trying to do? It says here: Please acknowledge this letter on the landlord's behalf as formal notice of our disagreement to your rent proposal', and he is asking me to do that in accordance with clause 2 of the lease. So he thinks he is taking a step under clause 2."

E I leave the language of the first person and I turn to clause 2 and, re-examining clause 2, one finds that there is but one notice which a tenant can serve under it. It is a counternotice under para 2 in fact. (The parties, of course, have been calling it clause 2, but each of them is clearly referring to the same stipulation.) There is but one notice which the tenant can serve, ie the counternotice envisaged by clause 2(c). That being the case, although I think that the addition of the words "of our disagreement to your rent proposal" is certainly an unhappy phrase and an inappropriate phrase to add, nevertheless this letter is proffered, to the person who received it as "formal notice under clause 2", ie under para 2, which must be para 2(c), and it must be intended to stop the time-limit running.

F Anyone who gets a recorded delivery letter in which great care is taken to request an acknowledgement is reasonably put on notice that the sender of the letter realises that he is meeting a time-limit and I think those reflections, too, should have taken the recipient of this letter fairly quickly beyond the doubt which he might momentarily have had. I do not consider, therefore, that this letter fails to be what I would term the clear giving of a counternotice under the relevant clause of the lease. That being so, although not a model of drafting, it satisfies the legal requirements of this particular lease and is a valid election. In those circumstances the answer to the question posed to the court is, in my view, better expressed in these terms:

It is declared that a letter dated 5th March 1990 and sent by Prudential Property Services Ltd on behalf of A C Frost Ltd, being the tenant at that time of a lease dated 21st May 1982, the benefit of which lease is now assigned to the plaintiff, who is now the tenant under that lease, did constitute valid counter-notice pursuant to clause 2 of the fourth schedule of the said lease of premises G known as 117 High Street, Chesham in Buckinghamshire.

I now consider the ramifications which that has as to costs and I need in that context to consider the complication that there are the other proceedings commenced five days later in which an extension of time under section 27 of the Arbitration Act 1950 was sought.

I have expressed the view that the chances of success in the section 27 proceedings were not great, partly by reason of the fact that I take the view that any application for extension of time under section 27 must be pursued with such diligence and be launched with such immediacy as is appropriate to the particular time-limit of the arbitration clause which is in issue. Here, there was a one-month time-limit directed to securing the rapid ascertainment of the revised H rent of the premises and there have at times been very significant delays in these proceedings.

Those pauses or delays, of course, do not affect in any way the problem of construction, but they do affect the prospects of success and even more the position as to costs in the section 27 proceeding. I therefore invite brief submissions from counsel as to costs, knowing as I do, because they have told me, that there is a single brief fee and I think that their submissions as to costs should be made globally in relation to the costs of both matters. Mr Steinert, since you have won the preceding that I have decided, I invite you to address me first.

Declaration accordingly.

J

Chancery Division

December 1 1992

(Before Mr T MORISON QC, sitting as a deputy judge of the division)

LONDON GENERAL HOLDINGS LTD v KINGSTON UPON THAMES NOMINEES LTD

Estates Gazette May 22 1993

[1993] 20 EG 128 K

Landlord and tenant — Rent review — Construction: whether review procedure exhausted by rent review date

By a reversionary lease dated August 24 1976, the plaintiff holds a term for 35 years from May 19 1981 of a six-storey building in Kingston upon Thames; the defendant acquired the reversion in July 1991. On January 7 1991 the defendant's predecessor in title served a notice requiring the rent to be reviewed at the review date (May 19 1991); by the review date the parties had not agreed a rent nor had a surveyor been appointed to determine it in accordance with the provisions of L clause 6 of the lease. Clause 6(c) provided that: "If at any date of review the Landlord and the Tenant shall be unable to agree on the amount of the rack rental market value as aforesaid then the same shall be decided by a Surveyor (who shall act as an expert and not as an arbitrator) to be agreed upon by the parties hereto or . . .". The plaintiff contended that, under the rent review provisions in clause 6, if by May 19 1991 a new rent had not been agreed nor decided on by a surveyor then the rent review provisions under clause 6(a)-(c) were exhausted in relation to that review date, although the landlord could serve a fresh notice under clause 6(d), but in this event the new rent would take effect only from, at the earliest, one month after M such notice was served. The defendant contended that once a notice was served under clause 6(a) the review was under way and was not exhausted by a failure to agree the rent or have it determined by a surveyor by May 19 1991; regard must be had to the opening words of clause 6(c).

***Held*: The originating summons was dismissed. The plain meaning of "If at any date of review" in clause 6(c) is that it was envisaged that the parties would try to agree terms right up to the date of valuation (the date of review) and only thereafter would the surveyor proceed to decide the rent. The landlord remained entitled to pursue the rent review.**

A No cases are referred to in this report.

This was a hearing of an originating summons of the plaintiff, London General Holdings Ltd, to determine the construction of a lease of premises which it held in Kingston upon Thames the reversion of which was owned by the defendant, Kingston upon Thames Nominees Ltd.

Kim Lewison QC (instructed by Paisner & Co) appeared for the plaintiff; Nicholas Dowding (instructed by Rabin Leacock Lipman) represented the defendant.

Giving judgment, **MR T MORRISON QC** said: By an originating summons dated November 6 1991 the plaintiff ("the tenant")
B claimed a declaration against the defendant ("the landlord") that their rent had not been reviewed with effect from May 19 1991.

The summons raises a short point of construction of the rent review provisions contained in a reversionary lease made between the parties' predecessors in title and dated August 24 1976. The lease is of a six-storey office block in Kingston upon Thames Royal London Borough Council and runs for a period of 35 years from May 19 1981.

Clause 6 contains the rent review provisions, the relevant parts of which I now set out:

> 6. PROVIDED ALWAYS AND IT IS HEREBY FURTHER AGREED that:
>
> (a) At the expiration of the fifth year and each period of five years thereafter of the term hereby granted (the time in each case being computed from the
> C date of the commencement of the term and the date of expiration of each such period being hereinafter referred to as the "date of review") the Landlord shall have the right to review the yearly rent first hereinbefore reserved and for the time being payable hereunder on giving not more than twelve nor less than three months' notice in writing prior to each date of review of its intention so to do and if the Landlord shall give to the Tenant a notice as aforesaid then from and after each date of review the yearly rent first hereinbefore reserved shall be increased to an amount which shall represent the rack rental market value of the demised premises at each date of review; . . .
>
> (b) The rack rental market value of the demised premises at each date of review shall be such amount as may be agreed between Landlord and Tenant or determined in accordance with subclause (c) of this Clause . . .
>
> (c) If at any date of review the Landlord and the Tenant shall be unable to agree on the amount of the rack rental market value as aforesaid then the same shall be decided by a Surveyor (who shall act as an expert and not as an
> D arbitrator) to be agreed upon by the parties hereto or in the event of failure so to agree within a period of two months from the date of receipt by the Tenant of the notice referred to in subclause (a) hereof by a surveyor to be nominated by the President for the time being of the Royal Institution of Chartered Surveyors . . .
>
> (d) Notwithstanding subclause (a) of this Clause if the yearly rent hereinbefore reserved and for the time being payable hereunder shall not have been reviewed at a date of review then the Landlord shall at any time thereafter have the right to review the said rent once only before the next review date upon giving to the Tenant not less than one month's notice in writing of its intention so to do and if the Landlord shall give the Tenant a notice as aforesaid then from and after the date therein specified (which shall be a date not earlier than one month from the date of service of the said notice) the yearly rent first hereinbefore reserved shall be increased to an amount
> E which shall represent the rack rental market value of the demised premises at the date specified in the said notice and the amount thereof shall be agreed between the Landlord and the Tenant or determined in accordance with the provisions of subclause (c) of this Clause;
>
> (e) Notwithstanding the decision of the expert hereinbefore referred to in no event shall the rent payable by the Tenant to the Landlord after each date of review be less than the rent payable by the Tenant to the Landlord immediately before such date of review;

The relevant facts are as follows. Immediately prior to May 19 1991, accepted by both parties as the "date of review" within the meaning of clause 6 of the lease, the annual rent of the demised premises was £333,333.33 pa. On January 7 1991 the then landlord served a rent review notice on the tenant, informing him that:

> F the landlord requires the rent reserved by the Lease to be reviewed at the expiration of the 10th year of the term granted by the Lease.

The notice did not mention a new proposed rent. There were, apparently, informal discussions between the parties and by letter dated April 22 1991, the tenant wrote to the then landlord saying that the tenant was anxious to have the matter settled and stated that:

> We had expected to receive your quotation prior to February 18th 1991

— that is, three months before the review —

> in line with lease terms.

Confirmation was sought that if the new rent could not be agreed before the review date the landlord would not levy interest on any outstanding rent which might be agreed to be paid after the review date. The then landlord made no reply to this letter. Accordingly, the G
tenant wrote again on May 10 1991 proposing a new rent of £400,000 pa and drawing attention to the landlord's lack of initiative in getting the review under way. Again there was no reply and again, on May 22 1991, the tenant wrote complaining and suggesting that the landlord's silence could be taken as an acceptance of the new rent proposed by the tenant. The landlord's response, the following day, was simply to reject the proposals in the tenant's earlier letter and reserving their right to review the rent. On June 3 1991 the tenant wrote yet again suggesting that the question of the new rent might have to be referred to arbitration. Sometime in July the then landlord assigned its interest in the demised premises to the defendant, who, through agents, proposed a new rent of £1,150,000 pa.

The parties then argued, in correspondence, the points which have H
been fully argued before me. The only other relevant fact is that the landlord, without prejudice to his contentions in relation to the validity of the earlier notice, served a notice under clause 6(d) on about November 8 1991 reviewing the rent as from December 11 1991.

The case was well and succinctly argued on both sides. I reserved judgment at the end of the argument, which lasted less than one and a half hours.

The rival contentions of the parties may be summarised as follows:

For the tenant

1. Clause 6 gives the landlord a right to have the rent reviewed. He J
may exercise that right by giving notice (in relation to the particular review date in question) at any time between May 19 1990 (not more than 12 months before the review date) and February 19 1991 (not less than three months before the review date). The notice required is a notice of the landlord's intention to review the rent (clause 6(a): "notice in writing prior to each date of review of its intention so to do").

2. Clause 6(b) contemplated that, following service of the notice of intention to review, the parties would seek to agree the new rent: (clause 6(b): "shall be such amount as may be agreed between Landlord and Tenant"). The parties had two months in which to agree the rent and, where appropriate, the nomination of a surveyor to act as an expert or, in default, to have an expert appointed by the K
president of the RICS (clause 6(c): "the same shall be decided by a Surveyor . . . to be agreed upon by the parties hereto or in the event of failure so to agree within a period of two months from the date of the receipt . . . of the notice . . . "). The earliest a surveyor could be appointed would be July 19 1990 and the latest April 19 1991.

3. If by May 19 1991 a new rent had neither been agreed nor decided upon by a surveyor then the rent review provisions in clause 6(a)-(c) inclusive were exhausted in relation to that review date. The landlord could, however, serve a fresh notice under clause 6(d), but in this event the new rent could take effect only from, at the earliest, one month after such a notice was served. The words of clause 6 which lead to this conclusion are to be found in subclause (d): L

> If the yearly rent . . . shall not have been reviewed at a date of review then the Landlord shall at any time thereafter have the right to review the said rent once only . . . upon giving to the Tenant not less than one month's notice in writing of its intention so to do

Counsel for the tenant stressed the contrast between, on the one hand, a notice under subclause (a) which is a notice of the landlord's intention to review the rent and, on the other, a review of the rent as contemplated by subclause (d) ("shall not have been reviewed"). Thus, it was argued, the parties must have accepted that the notice under subclause (a) was not the review itself. The review means a reconsideration and the reaching of a conclusion. If no new rent had been concluded "at a date of review" then a notice under (d) was permissible. M

Hence the force of the words "Notwithstanding subclause (a)" at the beginning of subclause (d). It was argued that even if, contrary to his prime contention, the notice under subclause (a) could be said to be a review, on the facts no new rent was proposed by the landlord until after the date of review. Therefore, on the facts, the rent was not, and could not have been, reviewed at the date of review, namely May 19 1991.

4. Counsel submitted that his construction gave effect to the language used by the draftsman, gave weight to each part of the clause, represented a perfectly sensible bargain in commercial terms (in particular he stressed the need for certainty so that a tenant could plan his financial affairs) and provided an inducement to the

A landlord to act timeously. The later the landlord left serving notice under subclause (a) the more risk he ran that the process would not have been completed by the review date. Here, the landlord had left serving his notice until near the end of the period and then failed to do anything about the review itself, leaving the tenant in a state of limbo, as it were.

For the landlord

1. The argument presented on the tenant's behalf was impractical and failed to give effect to the opening words of subclause (c) ("If at any date of review"). What was contemplated by clause 6 was a notice under subclause (a) followed by a period of negotiation:

> If at [the] date of review
>
> B — namely, May 19 1991 —
>
> the Landlord and the Tenant shall be unable to agree on the amount of the [new rent] . . . then the same shall be decided by a Surveyor . . .

If the rent review contemplated by subclauses (a)-(c) had to be completed by the date of review, as the tenant argued, the first time that the surveyor could be asked to decide would be the last time he could do so, apart from a *scintilla temporis*. The parties cannot have intended such a bizarre result. Further, the valuation date for the new rent was the date of review. It was unlikely that the parties intended a surveyor to be asked to decide the new rent much, if at all, before that date.

2. Service of a notice under subclause (a) is a condition precedent to C the landlord's right to review the rent. The right to review is crystallised by the service of such a notice; service of the notice triggers the review itself. It is not merely a notice of intention to review. Once served, the process is inexorably under way. Clause 6(a) refers to the landlord having the "right to review". The words in subclause (d), "shall not have been reviewed", are referring not to completion of the whole process but to the review process itself. The parties might have chosen to use a different expression (eg "if the yearly rent . . . shall not have been varied" — see p3 of the lease — "(subject to variation)") if it had intended the consequence contended for on behalf of the tenant.

3. On the tenant's argument the landlord would be put into a D difficult position if a surveyor was appointed and he did not proceed to his decision with due dispatch or became ill. Further, in the real world, the parties cannot have contemplated that, were a surveyor to be instructed, he could or would complete his task without receiving representations from the parties of valuation, with each putting forward their own comparables, although there was nothing in this lease requiring the parties to make such representations. None the less that is the way in which the clause would probably operate in practice. That would all take time and it was unlikely that the process could be completed by the date of review unless it was started a long time before. The tenant's argument effectively rendered the process in clause 6(a)-(c) unworkable.

4. Further, the tenant's argument in favour of certainty and speed E would not work in relation to clause 6(d). There, it would not matter how long the expert took to reach his decision. The landlord's right of review would not be lost once a proper notice had been served under subclause (d), however much time had elapsed.

I am of the view that the arguments on behalf of the landlord are to be preferred to those on behalf of the tenant. In essence, I think that the tenant tries to make too much of the words in subclause (d) on which his argument is rested. The effect of his submission is that the words "shall not have been reviewed at a date of review" mean "shall not have been agreed, or in default of agreement, decided by a Surveyor appointed in accordance with subclause (c)". I do not think the words bear that interpretation. In my view, subclauses (a) and (d) contemplate that the landlord reviews the rent by serving the F appropriate notice. In this sense what subclause (d) does is to provide that, if the landlord has not reviewed the rent by serving a notice (of his intention to do so) within subclause (a), he may review the rent in accordance with subclause (d), namely by serving a one-month notice. In other words, a notice under (a) both "reviews the rent" and serves notice of the landlord's intention to do so.

In my judgment, some support for this conclusion may be found in the fact that subclause (d) begins with the words: "Notwithstanding subclause (a) of this Clause." It is to be noted that there is no express reference to subclause (c); subclause (a) is largely concerned with the serving of notice, while subclause (c) is concerned with fixing the amount by agreement or decision of a surveyor. If the meaning contended for on the tenant's behalf had been correct I would have expected clause 6(d) to have said "Notwithstanding the provisions of G subclauses (a) and (c) of this Clause".

It seems to me that counsel for the landlord is right when he contends that the tenant's argument gives no meaning to the words at the beginning of subclause (c) "If at any date of review". In his reply, counsel for the tenant suggested that this phrase was not temporal and for the word "at" one might read "in relation to". I do not agree. The plain meaning of these words, in my judgment, is that it was envisaged that the parties would try to agree terms right up to the date of valuation (the date of review) and only thereafter would the surveyor proceed to decide the rent. Of course, there would be nothing to prevent the parties from getting the surveyor to decide the rent issue in advance of the valuation date, but I accept that, in the normal course of events, it would not be commercially sensible for H that process to be carried out too much in advance of the review date, since rents might fluctuate in a relatively short space of time.

Therefore, I accept that the tenant's construction would be unworkable in practice and would be capable of bizarre consequences. In short, for the reasons which he gave (and which I have set out above) counsel for the landlord seems to me to be right and the tenant's argument wrong.

It follows, therefore, that I refuse to make the declaration sought and dismiss this originating summons. Costs will follow the event.

It is only right that I should pay tribute to the arguments on both sides. For a short time I was entertained to advocacy of a high order.

Originating summons dismissed. J

Court of Appeal

February 9 1993

(Before Lord Justice Ralph GIBSON, Lord Justice LEGGATT and Lord Justice HOFFMANN)

HOLICATER LTD v GRANDRED LTD

Estates Gazette June 12 1993 K
[1993] 23 EG 129

Landlord and tenant — Rent review — Time of the essence — Construction of clause providing for appointment of expert

By a lease, restaurant premises in Great Yarmouth were demised to the appellant tenant for a term commencing on November 15 1979 and ending on March 31 2006. The lease provided for a rent review on April 1 1981 and every fifth anniversary of that date. The rent review clause provided that, if the parties could not agree the rent by not later than three months prior to the rent review date, such rent was to be determined by an expert to be appointed by the President of L the Royal Institution of Chartered Surveyors on the application of either party "made at any time before the said next review date or the expiration of the term as the case may be (time being of the essence) . . . ". In respect of the rent review due on April 1 1991 the respondent landlord did not attempt to agree the rent or apply for the appointment of an expert until after that date, claiming that it had until March 31 1996 to apply. The appellant appealed from the decision of Judge Weeks (sitting as a judge of the High Court), who had decided that the respondent was entitled to have the rent review. M

Held: **The appeal was dismissed. "The expiration of the term" means the expiration of whatever turns out to be the last rent review period; then it follows that in that period the landlord has until its end to apply for the appointment of the expert. By parity of reasoning, the landlord must have been intended to have until the end of each earlier rent review period, ie until the date of the next one. The "next" rent review means the date immediately following the end of each rent review period other than the last. The word "said" has to be rejected as meaningless.**

No cases are referred to in this report.

A This is an appeal from the decision of Judge Weeks, sitting as a judge of the High Court, who had made a determination on the construction of a rent review clause on the application of the appellant, Holicater Ltd, to which the respondent was Grandred Ltd.

Anthony de Freitas (instructed by Dunne & Crome, of Lowestoft) appeared for the appellant; John Furber (instructed by Davis Hanson) represented the respondent.

Giving the first judgment at the invitation of Ralph Gibson LJ, **Hoffmann LJ** said: This appeal raises a short point of construction on a rent review clause in the lease of a restaurant in Great Yarmouth. The demise is said in clause 3 to be for a term commencing on November 15 1979 and ending on March 31 2006. The initial yearly B rent was £15,000 subject to review in accordance with clause 4(2) of which the material part reads as follows:

> The rent shall be reviewed as at the First day of April One thousand nine hundred and eighty one and on every fifth anniversary of that date each of which dates and anniversaries are hereinafter called "the review date" and after such review date the rent shall be the greater of
>
> (a) the rent hereinbefore reserved
>
> (b) the rent payable following any previous rent review
>
> (c) . . .
>
> Unless the Landlord and Tenant shall agree by not later than three months prior to the review date the market rent payable from a review date such market rent shall be determined . . . by a person acting as an expert and not as an Arbitrator nominated in default of agreement between the Landlord and the Tenant by the President for the time being of the Royal Institution of C Chartered Surveyors on the application of either the Landlord or the Tenant made at any time before the said next review date or the expiration of the term as the case may be (time being of the essence) . . . and in the meantime the Tenant shall continue to pay rent at the same rate as before the review date and any increase in rent following the review date together with interest thereon from the review date shall be paid in one sum on the quarter day next following its determination in respect of the period from the review date until such quarter day . . .

The dispute is over the meaning of the words "the said next review date or the expiration of the term". The tenant says that it means the commencement of the period for which the review rent is to operate. The landlord says it means the end of that period. The issue arises because the landlord did not attempt to apply for the appointment of D a surveyor or even to agree the new rent until after April 1 1991. The tenant says he is too late. The landlord says he has until March 31 1996.

The term "review date" is defined to mean each of the series of dates on which the rent is reviewed, being the fifth anniversary of April 1 1979. In the timetable for determining the new rent the term "review date" standing by itself is used to mean the date in respect of which the rent is to be determined. Thus the right to appoint a surveyor arises unless the rent has been agreed "not later than three months prior to the review date". But the application to appoint a surveyor can be made "at any time before the said next review date". The difficulty arises from the combined use of the words "said" and "next". If the draftsman meant the date in respect of which the rent E had to be determined, ie the same review date as he had mentioned in connection with the three-month deadline, he could simply have said "the said review date" or even "the review date". On the other hand, if he meant the review date next following that in respect of which the rent had to be determined, the word "said" is out of place, because nothing has previously been said about that date. The landlord therefore says that, to make sense of the language, it is necessary to discard either "said" or "next".

The judge chose to discard "said" and what most influenced him in adopting that construction were the words "or the expiration of the term as the case may be". It seemed to him that those words were used to allow for the fact that in the last rent review period of the term there would be no "next review date". The reference to the F expiration of the term showed that the landlord was intended to be able to apply for the appointment of a surveyor at any time during the period in which the reviewed rent would apply.

Mr de Freitas, for the tenant, submits that one does not need to discard either "said" or "next". The said review date was the next review date because the application for the surveyor had to precede it. On the same reasoning, however, the term "next review date" would also have been appropriate in relation to the three-month deadline. To call the one "the review date" and the other "the next review date" would, if not for the use of the word "said", suggest that some distinction was intended.

Mr de Freitas then has to explain the reference to the expiration of the term. He says that the expiration of the term could itself be a review date because, despite the terms of clause 3, the lease would not G necessarily expire on March 31 2006. Clause 1(11) defines "the term" as meaning "the term of years hereby granted and any statutory continuation thereof". As this was a business tenancy, the provisions for statutory continuation which the parties are most likely to have had in mind are those of the Landlord and Tenant Act 1954. Section 24(1) of that Act provides that a business tenancy shall not come to an end unless it is terminated in accordance with the provisions of the Act. These require the service of a notice to terminate in statutory form by the landlord or the tenant, following which the tenant may apply for a new tenancy. If such an application is made, the tenancy cannot terminate until three months after the application has been finally disposed of. The term as defined may therefore continue for an indefinite period after March 31 2006. H

The expression "review date" is defined in clause 4(2) to mean, without limitation of time, every fifth anniversary of April 1 1981. It is not expressly confined to anniversaries occurring before March 31 2006. Since clause 4(2) determines the rent reserved by the lease, there is no reason why the term "review date" should not include anniversaries which occur after the contractual period has expired and while the lease is being continued by statute. The judge thought not, but I am inclined to agree with Mr de Freitas that April 1 2006 could be a review date and so, for that matter, could April 1 2011. But this means that, if Mr de Freitas is right on the meaning of "said next review date", there was no need to add the words "or the expiration of the term" in clause 4(2). The term "review date" would have J included all the review dates during the statutory period of extension. In fact, on Mr de Freitas' explanation, the reference to the expiration of the term is not quite accurate, because the contractual term would expire on March 31 2006 and the review date would be the next day.

Furthermore, if one has regard to the definition in clause 1(11), the "expiration of the term" in clause 4(2) does not necessarily mean the contractual expiry date. It means that date or any subsequent date on which the term as statutorily extended comes to an end. On no view could that be a rent review date. The judge thought that Mr de Freitas' argument for having rent review dates after the end of the contractual term placed too great a weight on the definition of the term in clause 1(11). I think, on the contrary, that it gives that definition too little effect. The argument assumes that "the K expiration of the term" in clause 4(2) means March 31 2006, whereas, in my judgment, it must mean the expiration of the term as defined.

It must, therefore, follow that, even if one rejects the judge's view that there could be no rent reviews after the end of the contractual term, his reasoning remains in essence correct. If "the expiration of the term" means the expiration of whatever turns out to be the last rent review period, then it follows that in that period the landlord has until its end to apply for the appointment of a surveyor. By parity of reasoning, the landlord must have been intended to have until the end of each earlier rent review period, ie until the date of the commencement of the next one. Mr de Freitas was driven in the end to submit that perhaps the draftsman had intended to distinguish L between the earlier rent review periods and the last one of all, but, in my judgment, this would be completely irrational. The "next" rent review date must therefore mean the date immediately following the end of each rent review period other than the last. It follows that the word "said" has to be rejected as meaningless. This is perhaps a lesson for everyone because the word seldom serves any purpose and this case shows that its indiscriminate use can confuse what would otherwise be clear.

In my judgment, the judge was right and the appeal should be dismissed.

Leggatt and **Ralph Gibson LJJ** agreed and did not add anything. M

Appeal dismissed with costs.

NEGLIGENCE

A Court of Appeal

July 31 1992

(Before Lord Justice PARKER, Lord Justice NOLAN and Lord Justice KENNEDY)

EHMLER AND ANOTHER v HALL

Estates Gazette January 16 1993

[1993] 02 EG 115

Negligence — Damaged building — Lease temporarily relieves tenant of obligation to pay rent — Whether landlord
B *entitled to recover loss of rents from tortfeasor*

On December 6 1990 a van driven by the appellant defendant left the road and crashed into a car showroom in Springfield Road, Tottenham, owned by the respondent plaintiffs. The showrooms were rendered unusable for a little over eight weeks. At the time they were let to Mr Dagistan under a lease dated February 13 1990 for a period of nine years from October 30 1989. By clause 3 of the lease the tenant covenanted to pay the reserved rent of £26,000 pa and the sum incurred by the landlord in maintaining insurance against the "insured risks". The insured risks included damage from
C impact by road vehicles. By clause 6(3) of the lease the tenant was relieved of his obligation to pay the reserved rent so long as the premises were unfit for use by reason of an insured risk. The damage caused by the appellant brought clause 6(3) into operation and the respondents suffered from a loss of rent in the sum of £4,305.03. The respondents brought proceedings and obtained a judgment for this sum in the county court. The appellant appealed, contending that the respondents had no right of recovery of this sum by reason of the decision of the court in *Rust* v *Victoria Graving Dock Co*.

Held: **The appeal was dismissed. *Rust* v *Victoria Graving Dock Co* can be distinguished. In that case the circumstances**
D **and nature of the disallowed claim were wholly different from those in the present appeal, there the landlord was entitled to receive a ground rent for the term of the lease and there was no permanent damage to his reversion. The general rule, cited in the *Rust* case, that a reversioner can sue only in respect of damage which will continue to affect the property when the reversion falls in, was applicable to the facts of that case but is not of universal application. The court in the *Rust* case upheld the right of the landlord to recover loss of actual rent; the respondents' claim in the present case was stronger and clearer and on this point the case could not be distinguished. Because**
E **the respondents' building suffered physical damage the claim could not be disallowed as a claim for purely economic loss. Nor could it be disallowed as being for damages which the appellant could not have foreseen; damage to the building could cause financial loss to the owner or tenant and it was not necessary that the appellant should foresee the precise terms of the lease.**

The following cases are referred to in this report.

Hughes v *Lord Advocate* [1963] AC 837; [1963] 2 WLR 779; [1963] 1 All ER 705; 1963 SC (HL) 31; 1963 SLT 150, HL

Rust v *Victoria Graving Dock Co* (1887) 36 Ch D 113 F

This was an appeal from a decision of Mr Recorder Goudie QC, sitting in Edmonton County Court, who had allowed a claim by the respondents, Laurence James Ehmler and Adrian James Harris, for damages in negligence against the appellant, Steven John Hall.

Timothy Briden (instructed by B K J Lewis) appeared for the appellant; Nicholas Dean (instructed by Sharman & Trethewy, of Bedford) represented the respondents.

Giving the first judgment at the invitation of Parker LJ, **Nolan** LJ said: On December 6 1990 a van which was being driven by the appellant defendant down Springfield Road, Tottenham, left the
road and crashed into a car showroom owned by the plaintiffs. The G
showroom premises were thereby rendered unusable for a period of a little more than eight weeks.

At the time they were occupied by a Mr Dagistan under a lease which had been granted to him by the plaintiffs on February 13 1990. The lease was for a period of nine years running from October 30 1989. By clause 3 of the lease the tenant was obliged to pay, first, an annual rent of £26,000, subject to review, and, second, by way of further or additional rent the sum or sums which the landlord might expend in maintaining insurance against the "insured risks". By clause 1(4)(a) the insured risks were defined to include risks in respect of damage from impact by road vehicles. Clause 6(3) provided, among other things, that if the demised premises were damaged
during the period of the lease by any of the insured risks, so as to be H
unfit for use, the rent first secured — the £26,000 pa — should cease to be payable by the tenant until the premises were reinstated.

The damage caused by the defendant's vehicle brought clause 6(3) into operation and, accordingly, Mr Dagistan was relieved of the rental obligation during the period for which the premises were unusable. The plaintiffs thus suffered from a loss of rent which has been agreed in the remarkably precise sum of £4,305.03. The plaintiffs claimed this sum (together with two small additions in respect of expenses) from the defendant by way of damages for negligence. No claim was made in respect of the cost of repairing the premises. We were told that this cost had been met by the plaintiffs'
insurers, and had been covered by an agreement between them and J
the defendant's insurers. The defendant admitted that the damage to the plaintiffs' premises had been caused by his negligence, and the plaintiffs' claim for damages in the amount mentioned was upheld by Mr Recorder Goudie QC in Edmonton County Court.

At first sight, it seems a straightforward case. The defendant now appeals, however, contending as he did before Mr Recorder Goudie that landlords such as the plaintiffs have no right of recovery for loss of rent. Mr Briden, on his behalf, accepts that if the terms of the lease had been such that the tenant was obliged to continue to pay rent during the period when the premises were out of use the defendant would have had to reimburse the tenant for the unproductive expenditure thus incurred, but he maintains that the converse does
not hold good so far as the landlords are concerned. For that K
proposition he relies upon the decision of this court in *Rust* v *Victoria Graving Dock Co* (1887) 36 Ch D 113. The facts of that case are summarised in the headnote, which so far as is relevant reads as follows:

> Owing to the negligence of the Defendants a building estate belonging to the Plaintiff was overflowed by a flood. Part of the land was covered with houses (A) which were in the Plaintiff's possession. Another part was covered with houses (B) erected by builders under building leases . . . The amount of damages to which the Plaintiff was entitled was referred to a special referee. In respect of (A) the referee allowed as damages (1) the expense of repairing the houses, and the rent during the period of repairs; . . . As to (B), he found that

A there was no injury which would last to the end of the leases, but he allowed a sum for depreciation of the selling value of the landlord's interest, in consequence of the houses being worth less to let.

The reference to "rent" at (A)(1) connotes the rent which would have been payable if the properties were let.

The award of the special referee was upheld by Chitty J in the Chancery Division. There was no appeal against his decision so far as the damages under heading (A)(1) were concerned, but the Court of Appeal reversed his decision in respect of the damages under heading (B). Cotton LJ said, at p 129 of the report:

> Now one principle to be borne in mind is that damages can only be granted for injuries which are the direct result of the act of omission or commission complained of. Then it must also be borne in mind that where there are B divided interests in land the amount of damages to be paid by the Defendants must not be increased in consequence of that subdivision of interests. All that they ought to be required to pay is damages for the injury caused directly to the whole estate, and if it is cut up into divers interests there is the difficulty of ascertaining, when the parties are not all before the Court, what is to be awarded to the owners of the different interests into which the fee simple is divided, a difficulty which in the present case is considerable.
>
> There is also another principle which must not be forgotten. It is an undoubted general rule that a reversioner or other person who has not an interest in possession in the land cannot recover any damages unless it is shewn that the injury to the land is of a permanent character, and will be injurious to the land when his estate comes into possession. Mr Jeune gives for this the reason (which I think probably is the sound one) that the tenant can bring his action and recover damages for any injury caused by the act complained of to C his estate or interest. Then the reversioner can bring an action only for the damages which will continue after the tenant's interest has come to an end and his reversion has become an estate in possession.

Cotton LJ went on to note that in respect of the houses which were in the plaintiffs' possession the special referee had given the sum required to repair the damage caused by the overflow and had also allowed the rent which would have been produced by the houses if let during the period of repair. Turning to the claim under heading (B) he said, at p131:

> The Plaintiff's counsel did not dispute the general rule, that a reversioner can only sue in respect of permanent injury which will continue to affect the property when his interest comes into possession; but, as I understand it, the Plaintiff's case was put this way. When a man buys land for the purpose of D letting it out on building leases the usual course is, that when some of the houses are built and the ground rent is secured, he sells the reversion, or, according to a common expression, sells the ground rent, in order to raise money to go on with the speculation. It is said that, this being so, any depreciation in the selling value of the ground rents is a ground for giving damages to the reversioner. In my opinion that is erroneous. His interest is a reversion, that is to say, his estate in the land is a reversion, and on the general rule he cannot get any damage for any wrongful act of the Defendants unless the damage is one which will endure and be continuing when the reversion becomes an estate in possession.

At p134 of the report Lindley LJ, in a concurring judgment, said that he did not propose to do more than make one or two observations upon two points, and continued at p135:

E

> The first is one of some importance, and that is the question what is the proper method of proceeding in cases of this description. Now I take it that the problem to be solved is simply to find out the measure in money of the damage done to the houses and land by the flood. That is the thing to be got at. Having got at that, you have to consider how that sum ought to be apportioned among the persons interested in the property damaged, ie, in the present case, how much of that sum ought to be awarded to the Plaintiff in respect of his interests, and how much to the tenants who are not before the court. Unless that principle is kept steadily in view there is great danger of error. If a skilled person is asked to value the damage done to each particular interest in the land without giving him a warning to keep the above principle in mind, there which be great danger of his assessing a number of amounts which in the aggregate will far exceed the total damage done, and it appears to me that that is F precisely what has been done here. The great importance of bearing the right principle in mind has induced me to make these remarks.
>
> The only other point to which I will allude is the diminution of ground rents, ie, the diminution in value of the Plaintiff's reversionary interest. That appears to me to be a wholly inadmissible item. It is established by the report of Mr Clifton and also by what he told us yesterday in Court that no damage has been done to this property of so permanent a nature that the reversion when it falls in will be worth one farthing less. If that is so, how can the reversioner get anything at all in the way of damages? When we were asked to give him something for diminution in the amount for which his reversion would now sell, we should be departing entirely from the principle upon which the Courts have always proceeded in determining the amount of damages to be given to the owner of a reversion for injury to it. The principle is perfectly well settled and will be found laid down in *Simpson* v *Savage*; in *Mumford* v *Oxford, Worcester and Wolverhampton Railway Company*, and in *Baxter* v *Taylor**. Upon looking at those cases it will be found that the Court has never G given what is in substance a loss of profit.

Pausing there, it is clear that the "loss of profit" to which Lindley LJ referred in that case was the temporary loss of unrealised profit which was claimed under heading (B). It was unrealised because the plaintiff had not sold the land at its diminished value nor, apparently, had he any intention of doing so. The loss was temporary because no permanent damage had been done to the property: when the reversion fell in it would be worth not a penny less than it was before the flood. It is hardly surprising, therefore, that the damages claimed under heading (B) were disallowed. But that carries the present case no further because the circumstances and the nature of the claim under heading (B) are wholly different from the circumstances and claim for loss of rent in the present case. So far as the houses H comprised in heading (B) were concerned, Mr Rust was the lessor of a building lease, which presumably ran for a long time, and during the currency of which he would be entitled only to a ground rent. He would be unable to charge any rent for the use of the buildings until the reversion fell in. In the present case, the plaintiffs are the lessors of a completed building which, prior to the accident, was bringing them in a rack-rent. Their simple claim for loss of rent is wholly different from the claim of Mr Rust under heading (B) for temporary and unrealised loss of value in the land. Mr Rust apparently continued to receive his ground rent in respect of the heading (B) properties and, therefore, he did not and could not make any claim for loss of rent. The court simply did not consider such a claim. In J these circumstances I do not see how its decision upon the heading (B) aspect of the case could possibly be regarded as authority for the proposition that a landlord can never make a claim in tort for loss of rent. The general rule, cited in the *Rust* case, that a reversioner can sue only in respect of damage which will continue to affect the property when the reversion falls in, was applicable to the facts of that case but is not of universal applicability.

What the court did consider, and uphold in that case, was Mr Rust's claim under heading (A)(1) for the lost rent which would have been produced by the houses if they had been let during the period of repair. I am, for my part, unable to see any distinction in principle between this claim, and the plaintiffs' claim in the present case for the loss of actual rent, which, but for the defendant's negligence, would K have been produced by the premises under an actual letting during the period of repair. The plaintiffs' claim in the present case seems to me to be stronger and clearer than that of Mr Rust in respect of heading (A)(1).

I am unable, therefore, to find any support for Mr Briden's argument in the *Rust* decision. In saying that, I do not overlook the passages in the judgments about the need to ensure that the defendant is protected from overlapping claims by persons having separate interests in the same premises, but no such problem arises in the present case. On the contrary, if the defendant is right, the fact that the plaintiffs and the lessee both have interests in the premises under the lease of February 13 1990 prevents him, by virtue of clause 6(3) of L the lease, from having to pay damages to either. That would be a strange and unjust state of affairs.

Mr Briden advanced two further arguments. One was that, because the damage to the premises would have been repaired long before the end of the nine-year lease, this was a case in which the plaintiffs had suffered no physical loss, but were claiming for a pure economic loss. This argument, however, seems to me again to involve transplanting some of the statements from the judgment in the *Rust* case and applying them to wholly different circumstances. The plaintiffs, unlike Mr Rust in respect of heading (B), suffered from physical damage to their building, which was an income-producing asset. The fact that the cost of repairs to the building was covered by insurance is neither here nor there. The loss of rent arising out of the M damage to the building was consequential upon that damage: it was not pure economic loss.

Finally, Mr Briden suggested that, although the defendant could reasonably be expected to foresee that the showroom was being put to commercial use and that accordingly he might have to compensate the tenant for loss of profits whose computation would include any rent which the tenant might have to pay during the period of repair, the defendant could not have been expected to foresee the effects of clause 6(3), this being an unusual clause in a lease. Therefore, he said, the loss to the plaintiffs lay beyond the limits of reasonable

*Editor's note: Reported at (1843-60) All ER Rep 748; (1856) 156 ER 1107 and (1832) 110 ER 382 respectively.

A foreseeability and was too remote to be recovered. In my judgment, however, this suggestion falls foul of the principle stated by Lord Reid in *Hughes* v *Lord Advocate* [1963] AC 837, at p845, that a negligent defendant "can only escape liability if the damage can be regarded as differing in kind from what was foreseeable". Thus, it is clear that the defendant could not be expected to foresee the precise terms of the lease, but he could reasonably be expected to foresee that his negligence would cause physical damage and consequential financial loss to the owner or tenant of the premises. That was enough, in my judgment, to make him liable in damages to whichever of those parties, under the contractual arrangements between them, in fact suffered from the particular loss for which damages are
B claimed.

For these reasons, which are broadly in line with those of the learned recorder, I would dismiss this appeal.

Kennedy and **Parker LJ** agreed and did not add anything.

Appeal dismissed with costs. Application for leave to appeal to the House of Lords refused.

Queen's Bench Division

C February 24 1992

(Before Mr Justice Douglas BROWN)

F G WHITLEY & SONS CO LTD v THOMAS BICKERTON AND OTHERS

Estates Gazette February 20 1993

[1993] 07 EG 100

Negligence — Surveyors giving planning advice — Whether defendants liable in negligence — Whether period of limitation expired before issue of writ

D On November 15 1973 the Secretary of State granted full conditional planning permission for the extraction of silica, stone and sand from land at Moel Findeg, Maeshafn, Mold, Denbighshire. On May 20 1976 the plaintiff, F G Whitley & Sons Co Ltd, purchased the site with the benefit of that planning permission and on July 1 1976 it instructed the first defendants, Cooke Leathes & Bickerton, chartered surveyors, to prepare and submit to the county council an application for the approval of the detailed matters contained in the conditions to the planning permission. On July 20 1977 the first defendants submitted detailed plans and specifications
E with a view to seeking agreement as to a scheme of working pursuant to one of the conditions. The county council delayed dealing with the submitted proposals until a meeting on October 25 1978. On that day, against the advice of the county planning officer, the subcommittee resolved that the scheme proposed "be not approved". On November 10 1978 the second defendants, Robert Moore & Sons, solicitors for the plaintiff, requested the county council that either the matter be submitted to the Secretary of State or that an extension of the time-limit be agreed. By letter of November 15 the county council refused to ask the Secretary of State to settle the scheme and rejected any course of action which would involve
F the planning committee considering a further application by the plaintiff before November 30. The county council also stated that if works were started before November 30 without a scheme having been settled by the Secretary of State then enforcement action might ensue. On November 21 1978 the second defendants wrote to the Secretary of State asking him to determine the submitted schemes of working. On November 22 1978 the plaintiff terminated the first defendants' retainer and instructions. On November 28 1978 the plaintiff commenced the development on the site by setting out boundaries to an access road, stripping topsoil and vegetation off the site and excavating silica, sand and overburden. On November 30 1978 Charles F Jones & Co, G chartered surveyors, purported to appeal to the Secretary of State against the non-determination by the county council of the working schemes and a related planning application. On December 1 1978 the second defendants also wrote to the Secretary of State asking him to determine the submitted schemes of working. By letter dated May 10 1982 the Secretary of State determined the scheme without prejudice to the question of the validity of the planning permission and its survival after November 30 1978. On May 5 1983 the plaintiff commenced further quarrying operations at the site and on December 2 1983 the county council purported to serve a valid H enforcement notice in relation to the works on the basis that they had been undertaken without planning permission and that the 1973 planning permission had lapsed on November 30 1978. The plaintiff appealed against the enforcement notice, and further appealed to the High Court against the Secretary of State's decision upholding the enforcement notice. It was agreed that the outcome of a further appeal to the Court of Appeal did not affect the preliminary issue to be decided in the present action. In March 1984 the third defendants, Walker Smith & Way, were retained by the plaintiff as solicitors and that retainer was terminated on November 23 1984. The writ in the action was issued on November 28 1984. At the trial of a J preliminary issue as to when the plaintiff's cause of action arose against the first and third defendants.

***Held*: It was admitted that the first defendants owed the plaintiff both a contractual and tortious duty of care and the particulars relied on were that the first defendants failed to appeal to the Secretary of State against the county council's non-determination of the scheme in sufficient time for the Secretary of State to determine it for work to start before November 30 1978, and in failing to apply to the county council for the removal or modification of condition 11 in sufficient time for an extension to be granted. Although it** K **mattered not when the breach of contract occurred, because any claim in contract was statute-barred, it was essential to the process of arriving at the date of actual damage in tort to see when the breach was because the breach of contract and the breach of the duty of care happened at the same time. The breach of contractual duty occurred on December 1 1977 and at the same time the first defendants were in breach of their duty of care. This date was appropriate having regard to the implacable opposition of the county council to mining development on the site, the right of the holder of planning permission, within a reasonable time of submission of the** L **scheme to the council for approval, to ask the Secretary of State to determine the scheme and the comprehensive conditions as to the details of the scheme which had to be considered. The cause of action in tort accrued when actual damage was sustained by the plaintiff, which, in the present case sustained damage at and from the time of the breach. From the end of November 1977 anyone valuing the site would know that it was probable that the scheme would not be determined by the Secretary of State in time for development to begin by the end of November 1978. Accordingly, the cause of action in tort arose on December 1 1977.**

M

The following cases are referred to in this report.

Baker v *Ollard & Bentley* (1982) 126 SJ 593

Bell v *Peter Browne & Co* [1990] 2 QB 495; [1990] 3 WLR 510; [1990] 3 All ER 124, CA

Forster v *Outred & Co* [1982] 1 WLR 86; [1982] 2 All ER 753, CA

Moore (D W) & Co Ltd v *Ferrier* [1988] 1 WLR 267; [1988] 1 All ER 400, CA

Whitley (FG) & Sons v *Secretary of State for Wales* (1989) 60 P&CR 185; [1990] 2 PLR 44; [1990] JPL 678

This was a trial of a preliminary issue arising in an action by the plaintiffs, F G Whitley & Sons Co Ltd, against the first to third defendants, Cooke Leathes & Bickerton, Roberts Moore & Son and

Walker Smith & Way respectively, for damages alleging negligence in contract and in tort.

Robert Griffiths and Paul Stinchcombe (instructed by Clement Jones & Co, of Holywell) appeared for the plaintiffs; Nicholas Nardecchia (instructed by Barlow Lyde & Gilbert) represented the first defendants; Richard Clegg QC (instructed by Weightman Rutherfords, of Liverpool) represented the third defendants; the second defendants did not appear and were not represented.

Giving judgment, **Douglas Brown J** said: In this action the plaintiffs, F G Whitley & Sons Co Ltd (FGW), who are public works and civil engineering contractors, sue Cooke Leathes & Bickerton, chartered surveyors, as first defendants (CLB), Roberts Moore & Son (RMS) as second defendants and Walker Smith & Way (WSW), as third defendants. The second and third defendants are solicitors. In each case the defendants are sued for professional negligence both in contract and in tort. The proceedings arise out of the employment of CLB to advise and assist FGW in and about the implementation of a planning permission granted in 1973 to extract silica, stone and sand from land owned by FGW. The action was begun by a writ issued on November 28 1984 against the first and second defendants and amended on February 15 1989 to bring in the third defendants. On May 4 1989 Mars-Jones J ordered that the following matter be tried as a preliminary issue: did FGW's cause or causes of action against CLB arise at:

(a) October 25 1978 — when the county council refused to approve the scheme of working submitted by first defendants; or

(b) November 22 1978 — when FGW dismissed CLB; or

(c) November 30 1978 — when (if at all) the planning permission of November 15 1973 expired; or

(d) some other date?

The plaintiff appealed to the Court of Appeal and on October 23 1990 that appeal was dismissed. The preliminary issue to be tried was varied by the consent of all parties by an order made by District Judge Harrison on May 20 1991, by which it was ordered:

1. That WSW have leave to be heard at the trial of the preliminary issues to limitation.

2. That such trial of the preliminary issues to limitation raised by CLB should extend to and include the limitation issue raised by WSW. Such preliminary issues to be heard together at the same time.

RMS have taken no part in the trial of the preliminary issue.

I have heard no oral evidence, but by agreement the material I have considered consists of: (1) the pleadings; (2) a statement of facts agreed between FGW and CLB; (3) a bundle of documents agreed between the same parties; (4) an affidavit in support of the application for the trial of the preliminary issue by Mr Allan Speechley [FRICS], partner in CLB; and (5) an affidavit from Mr Graham Bolton, chartered town planner and development consultant on behalf of WSW.

All three defendants by their original defence admitted that the duty of care alleged by FGW arose in contract and in tort. Before the hearing began before me, I gave leave to WSW to amend their defence to withdraw that admission in so far as it related to potential tortious liability. That considerably increased the scope of the trial of the preliminary issue but, by agreement, any argument as to whether and, if so, when a cause of action in tort arose in relation to WSW was deferred until after the trial of the preliminary issue as envisaged by Mars-Jones J and District Judge Harrison's orders.

Before dealing with the issues and counsel's submissions I refer to the history and factual background to this litigation. In 1968 the then owner of land at Moel Findeg, Maeshafn, near Mold, Denbighshire, applied to Denbighshire County Council for permission to extract silica, stone and sand. The application was refused on the grounds of harm to the landscape and the amenities of local residents. An appeal to the Secretary of State for Wales was unsuccessful after a public inquiry where there was considerable opposition from local residents, but at that time no objection from the local authority. Almost immediately, in April 1971, a second planning application was submitted, which was refused on amenity grounds identical to those given for the first refusal. In March 1972 another public inquiry was held and, on this occasion, the local planning authority joined with local residents in opposing the appeal. In August 1972 the Secretary of State dismissed the appeal, but suggested that a fresh application might be submitted in relation to a reduced area. It was that suggestion which led to a further application, the result of which was the permission which lies at the heart of this litigation. Because of continued local objections, the Secretary of State called in the application and dealt with the matter by way of written representation.

On November 15 1973 the Secretary of State granted full conditional planning permission for the extraction of silica, stone and sand from the Moel Findeg site. It is material to consider the nature of a number of the conditions:

> 2. No working shall take place except in accordance with a scheme to be agreed with the local planning authority or, failing agreement, as shall be determined by the Secretary of State and scheme shall among other matters include provision for:
>
> (a) the order, direction depth and method of working, including the arrangement for and times of blasting operations where they may be necessary;
>
> (b) the preservation of the ridge along the western boundary of the site;
>
> (c) the stripping and storing of top soil and overburden separately until disposal as required by condition 3 below;
>
> (d) the location and height of any heaps of topsoil, overburden or other excavated materials;
>
> (e) adequate measures for cleansing the wheels of vehicles leaving the site.
>
> 3. Progressive restoration of the site shall take place in accordance with a scheme to be agreed with the local planning authority or, in default of agreement, to be determined by the Secretary of State, such scheme to be agreed or determined before working takes place, and the scheme shall, among other matters, include provision for:
>
> (a) the even spreading and disposal into the excavation of the overburden and topsoil stripped and stored in accordance with condition 2;
>
> (b) the grading of the finished slopes of the excavation;
>
> (c) the final levels of the restored land.
>
> . . .
>
> 5. The access road to the working area shall be constructed and the junction of the access road with the county road shall be laid out in accordance with a scheme to be agreed with the local planning authority or, in default of agreement, to be determined by the Secretary of State.
>
> . . .
>
> 11. The development hereby permitted shall be begun on or before 30 November 1978.

On May 20 1976 FGW purchased the site with the benefit of that planning permission. On July 1 1976, FGW instructed CLB to act as surveyors on their behalf in preparing and submitting to Clwyd County Council successors to Denbighshire County Council (the county council) an application for the approval of the detailed matters contained in conditions 2, 3 and 5. CLB carried out the work and entered into discussions with the county council and, on July 20 1977, submitted to the planning authority detailed plans and specifications with a view to seeking their agreement to them as a scheme of working pursuant to condition 2.

Thereafter, despite the efforts of CLB though meetings and correspondence, the county council delayed in dealing with the proposal submitted, from the date of submission until April 1978 and then the planning authority merely sought further information. CLB provided that further information in July 1978, but the matter did not go before the county planning subcommittee until September 13 1978 when it was deferred by that committee to a further meeting on September 21 when again it was deferred to a meeting on October 25 1978. On that day, against the advice of the county planning officer, the members of the planning subcommittee resolved that the scheme proposed by FGW "be not approved".

On November 2 1978 Mr David Charles [FRICS], another surveyor acting for FGW, asked the Secretary of State by letter to take action to determine the application or begin that process by November 30, when by condition 11 the life of the permission expired. On November 10 1978 RMS wrote to the county council suggesting that either the matter be submitted before the Secretary of State or that an appropriate alternative course of action be taken and an extension of the time-limit of November 30 be agreed. The county council, by a letter of November 15, refused to ask the Secretary of State to settle the scheme and rejected any course of action which would involve the planning committee considering a further application by FGW before November 30 1978. The county council also stated that if works were started before November 30 without a scheme having been settled by the Secretary of State, then enforcement action might ensue. On November 21 1978 RMS also wrote to the Secretary of State asking for him to determine the submitted schemes of working. On November 22 1978 CLB's retainer and instructions were terminated.

On November 28 1978 FGW in actual or purported compliance with condition 11 commenced development on the site by:

(a) setting out approximate boundaries to the access road to the site;

(b) stripping topsoil and vegetation off the site;

(c) excavating silica, sand and overburden from the site and removing it by lorry.

On November 30 1978 Charles F Jones & Co, chartered surveyors who replaced CLB, purported to appeal to the Secretary of State against the non-determination by the county council of the working schemes and a related planning application. Last, on December 1 1978, RMS also wrote to the Secretary of State asking for him to determine the submitted schemes of working.

The Secretary of State began the process of determining the scheme on June 13 1980, having in the meantime engaged in prolonged correspondence with the parties on the subject of the survival of the planning consent after November 30 1978. After extensive consultation, written representations and a visit to the site by an inspector, the Secretary of State determined the scheme by letter dated May 10 1982, without prejudice to the question of the validity of the planning permission.

On May 5 1983 FGW commenced further quarrying operations at the site and, on December 2 1983, the county council purported to serve a valid enforcement notice in relation to the works commenced in September on the basis that the same had been undertaken without planning permission. It was asserted that the 1973 planning permission lapsed on November 30 1978. FGW appealed against the enforcement notice to the Secretary of State and after an inquiry on October 26 1988 the Secretary of State dismissed FGW's appeals, upholding the enforcement notice. On October 18 1989, FGW's appeal to the High Court under section 246 of the Town and Country Planning Act 1971 [section 289 of the Town and Country Planning Act 1990] against the Secretary of State's decision was successful and the matter was ordered to be remitted to the Secretary of State for determination and rehearing. The decision in that matter of Sir Frank Layfield, QC (sitting as a deputy judge of the Queen's Bench Division) is reported at *F G Whitley & Sons* v *Secretary of State for Wales* [1990] JPL 678. An appeal to the Court of Appeal by the county council from the deputy judge's order is due to be heard shortly. It is agreed at the bar that the ultimate decision in that litigation does not affect the question I have to decide, although it would clearly be material to any assessment of damages that takes place. In March 1984 WSW were retained by FGW. That retainer was terminated on November 23 1984.

On those facts Mr Nardecchia, for CLB, accepted that any claim under the contract is statute-barred. Any contractual cause of action must have arisen before CLB's services were dispensed with on November 22 1978, more than six years before the issue of writ on November 28 1984.

The cause of action in tort arises upon actual damage being sustained by the plaintiff. He contends that FGW have suffered actual damage by or on October 25 1978 at the latest when the county council refused to approve CLB's submitted scheme. By that date there was no time left for the Secretary of State to consider and determine the scheme in time for development to take place by November 30. Without conceding that, on the facts, it was remotely possible, even if I took the view that the county council might have been amenable to requests to reconsider the matter, the letter of November 15 firmly shut the door and in the five weeks remaining it was inconceivable that the Secretary of State could carry out the extensive process of determination on any reasonable and lawful basis.

Mr Robert Griffiths, for the plaintiffs, contends for November 30 1978, the date of expiry of the permission, as the earliest date on which the plaintiffs sustained any damage.

Mr Richard Clegg QC, for the third defendants, made submissions supporting Mr Griffiths in relation to the date of November 30 1978 being the date for the accrual of the plaintiffs' damage. That was after the date on which WSW's retainer was terminated. Against the event that I rejected Mr Griffiths' argument and was considering either of the dates canvassed by Mr Nicholas Nardecchia, for the first defendant, Mr Clegg had an alternative solution based on the view contained in Mr Bolton's affidavit, unchallenged by any other evidence, that to have any chance of securing a determination from the Secretary of State by November 30 1978, a scheme would have to have been submitted to the Secretary of State at the latest by the end of November 1977 and, as it was not so submitted, that was the date that the plaintiffs sustained damage and that was the date his cause of action therefore accrued.

Turning to these contentions in more detail, Mr Nardecchia submitted that the ascertainment of actual damage was largely a question of fact. The researches of all counsel had not discovered a similar case but there was help from authorities which provided useful analogies. The principle he contends for is that it is frequently the case on the authorities that actual damage will arise before actual financial loss accrues. The cases mainly involved solicitors but the principles equally applied to chartered surveyors and other professional men. The main authorities were all decided in the Court of Appeal and are: *Baker* v *Ollard & Bentley* (unreported May 12 1982 Court of Appeal (Civil Division) transcript no 155 of 1982); *Forster* v *Outred & Co* [1982] 1 WLR 86; *Moore (DW) & Co Ltd* v *Ferrier* [1988] 1 WLR 267; and *Bell* v *Peter Browne & Co* [1990] 3 WLR 510.

Of the cases he cited I refer to *Bell* v *Peter Browne & Co* to illustrate Mr Nardecchia's argument. The facts of that case are conveniently stated by Nicholls LJ at p513:

> This appeal raises some short points concerning the date of accrual of the plaintiff's causes of action for breach of contract and in tort, arising out of his solicitors alleged negligence. The facts as pleaded lie in a very small compass. In October 1977 Mr Barry Bell, the plaintiff, consulted the defendant firm regarding the breakdown of his marriage. One of the points he discussed with his solicitors concerned what was to happen to the matrimonial home . . . The house was registered in the joint names of the plaintiff and his wife . . . It was worth about £12,000, and subject to a mortgage for about £8,000. So, if the house had then been sold, the plaintiff would have received about £2,000; that is about one-sixth of the gross proceeds. The plaintiff was agreeable to the house not being sold for the time being, but he was to receive one-sixth of whatever might be the gross proceeds when a sale did take place. Meanwhile the house would be transferred into the sole name of the wife. This would help him to obtain a mortgage to buy other accommodation for himself. His continuing interest in [the matrimonial home] would be protected by a trust deed or a mortgage.
>
> These arrangements were agreed by the plaintiff and his wife and their respective solicitors. On 1 September 1978 the plaintiff executed a transfer of the house into his wife's sole name. But no steps were taken by the plaintiff's solicitors to protect his one-sixth share of the proceeds of sale. No declaration of trust or mortgage was prepared or executed. No caution was registered at the Land Registry. The parties were divorced in the following year.
>
> In December 1986 the plaintiff learned from his former wife that she had sold the property in July 1986 for £33,000, that is, almost eight years after the property had been transferred into her sole name. She had spent all the proceeds. Thus the plaintiff had lost his one-sixth interest in the proceeds of sale.
>
> So the plaintiff brought this action against his former solicitors for damages for professional negligence. The writ was issued on 20 August 1987. The sole issue arising on the appeal was whether the action was bound to fail because it was statute-barred.

On p515, dealing with the claim in tort, the lord justice posed the question: when did the plaintiff first sustain damage by reason of his solicitors' negligence? He distinguished between two failures: (a) to see that the parties' agreement was recorded formally in a suitable declaration of trust or other instrument; and (b) to protect the plaintiffs' interest in the house or the proceeds of sale by lodging a caution. As to failure (a) Nicholls LJ said at p515:

> . . . clearly the damage, such as it may have been, was sustained when the transfer was executed and handed over. At that point the plaintiff parted with title to the house, and he became subject to the practical inconveniences which might flow from his not having his wife's signature on a formal document.

Accordingly, Nicholls LJ rejected the plaintiff's argument that the damage occurred only when the former wife sold the unencumbered property and spent the proceeds. The plaintiff had suffered prejudice when the transaction was completed without his having the protection of a formal document and he continued at p516:

> The extent of that prejudice depended on the attitude adopted thereafter by his former wife. All that we know is that, according to the pleadings and the plaintiff's affidavit evidence, when she sold the house she disposed of all the proceeds and did not account to her former husband for his agreed one-sixth share. But the uncertainty surrounding her future intentions goes only to the quantum of the loss the plaintiff sustained when the transfer was executed without him having the same degree of protection as would have been provided by a formal document. Likewise in the decision in the Court of Appeal in *Baker* v *Ollard & Bentley* . . . cited in *DW Moore & Co Ltd* v *Ferrier* . . . There the plaintiff acquired a share in property rather than as she ought to have received, the security of a long lease of one floor of the property. The amount of her loss depended on the attitude of her co-owners. But, even so,

the damage was held to be suffered by the plaintiff at the time of the conveyance, when she received her precarious interest.

Nicholls LJ went on to deal with failure (b) in relation to the failure to make an appropriate entry in the land register. That was clearly different from failure (a), because it was within the plaintiff's own power to remedy the failure so long as the house continued to belong to his former wife. Nicholls LJ held that that was not a material difference and continued at p516G:

> I am unable to accept that remediability puts failure (b) on the other side of the line from failure (a). The solicitors' breach of duty in 1978 was remediable by the plaintiff, but that was only possible after he had become aware that there had been a breach of duty. Apart from any other consideration, to treat the plaintiff's ability to remedy the breach himself without the concurrence of his wife as a ground of distinction between this case and cases such as *Baker* v *Ollard & Bentley* . . . would be to disregard the unlikelihood in practice of the plaintiff ever being in the position to remedy the breach.

Mr Nardecchia accepted that on November 30 1978 the plaintiff sustained damage because no scheme had been determined by that date and the land ceased to have the clear benefit of an unexpired planning permission. But, he says, that was further damage and that actual damage had been sustained before that date.

If there was negligence, and that had to be assumed, arguably that had occurred in late 1977; it is even arguable that the date might be some time in 1976, that argument being based on the actual period taken by the Secretary of State to determine the scheme, namely two years. Land with a full planning permission has an enhanced value by reason of that permission, but if there are conditions which are difficult of performance then the value is reduced. When it is impossible to comply with the conditions, the enhanced value disappears and if that is brought about by negligence actual damage is suffered On the material before me that had happened by October 25 1978 at the latest. In the autumn of 1978, there were two avenues of approach to FGW.

(1) The avenue through the county council which they attempted to go down. However, it is plain from the letter of November 15 1978 that the county council refused to co-operate in waiving the time-limit and completely refused to co-operate in reconvening a special committee meeting before November 30 1978 to consider or reconsider the working scheme. Whatever the views of the county planning officer, the members of the council were taking a hard line and it was inconceivable that they would change their minds.

(2) The route via the Secretary of State; it was, as Mr Nardecchia put it, crystal clear to anyone experienced in town and country planning matters that the plaintiff was not going to get a decision from the Secretary of State on a matter of such complexity in the five-week period from October 25 1978 to November 30 1978. Mr Nardecchia submitted finally that the nearer one got to November 30, the more hopeless the plaintiffs' case became. Even if the Secretary of State had been sent the documents and plans with the clearest and fullest explanation of the scheme, it is inconceivable that any determination would have been made by the expiry of the permission. As it happens, as a matter of fact — and the documents establish this — the Secretary of State was not put in a position to make any determination because he was not fully furnished with the necessary documentation. Neither the letter from Mr Charles nor from RMS enclosed any details of the proposed scheme and without those the Secretary of State could not begin to consider the matter. Even on November 30 the new surveyors' application does not seem to have been accompanied by full details of the scheme. Mr Griffiths made three principal submissions:

(1) A greater prospect of harm is not harm. Notional damage or damage arising on a contingency is not damage. There is, he said, no reported decision supporting the proposition that where no damage is caused, but, where it is inevitable that damage will be caused in the future, the cause of action arises at an earlier date. In the cases relied on by Mr Nardecchia, each case the tort was complete at the date of, eg the execution of the document or of the negligent advice. In *Forster* v *Outred* the plaintiff could have sued immediately after the execution of the mortgage deed encumbered with a legal charge. She was subjected to a liability which might mature into a future loss. Actual damage accrued at the outset notwithstanding that she did not become liable for the repayment of the loan until a demand was made some years later. Her cause of action against her solicitors was complete. In the other cases cited, the cause of action was complete with accrued damage and Mr Griffiths referred to part of the judgment of Bingham LJ in *D W Moore & Co Ltd* v *Ferrier* [1988] 1 WLR 267. In that case solicitors advised and drafted documents for directors of a company to deal with the situation, *inter alia*, which might arise if a fellow director and employee left the company. One of the documents drafted was a covenant against competition after he left. At just below "G" Bingham LJ said:

> On the plaintiff's case, which for the purposes of this issue may be assumed to be wholly correct, the covenants against competition were intended and said by the defendants, to be effective but were in truth wholly ineffective. It seems to me clear beyond argument that from the moment of executing each agreement the plaintiffs suffered damage because instead of receiving a potentially valuable chose in action they received one that was valueless.

By contrast Mr Griffiths said the tort in the instant case was not complete until November 30 1978. The cause of action did not accrue before then because events up to that date could have extinguished the cause of action. The county council, for example, might have changed their mind and heeded the advice of the county planning officer. The scheme might have been approved or the condition varied to extend the expiry date. The Secretary of State might have moved with considerable speed and urgency and determined the scheme. If it was theoretically possible that those events could happen, then actual damage could not occur before November 30. There is, he said, no such thing as a provisional cause of action which lasts only until some new event has happened which extinguishes it. Once a cause of action accrues, it survives until extinguished by the Limitation Act or dealt with by judgment.

(2) By contrast with Mr Nardecchia's approach, which is arbitrary and imprecise, Mr Griffiths' test for the accrual of damage is based on a precise ascertainable date. The statutory framework of limitation is based on a precise ascertainable date so that prospective plaintiffs and defendants know where they stand. Mr Nardecchia's argument did not lead to any one date, eg he canvasses a date prior to October 25, October 25 itself and November 15. Why not, said Mr Griffiths, November 16 or any time up to the 59th minute of the 11th hour before the permission runs out? That cannot be the right way to decide the date of accrual of a cause of action following a claim for loss of a planning permission. In the end, he relied on *Forster* v *Outred* and the other cases and said the damage here was similar to the damage in those cases. The status of the land up to November 30 was land with planning permission; that status changed on that date and there was at that point actual damage.

(3) Finally, in answer to the main argument advanced by Mr Nardecchia, that damage had occurred earlier because of the recognisable inevitability of loss of the planning permission, Mr Griffiths referred to the way the plaintiffs' claim was pleaded against the first defendant. It was alleged in para 27 of the amended statement of claim that FGW had suffered and would suffer loss and damage particularised in this way.

> (A) If 1973 planning permission lapsed and no planning permission is granted upon appeal:
> (i) Diminution in the value of the Moel Findeg site because of the loss of planning permission for mineral extraction, or alternatively the loss of profits therefrom;
> (ii) particular loss arising from FGW's inability to supply particular contracts from is own quarry;
> (iii) professional fees and other expenses incurred in making any subsequent planning applications and pursuing them to appeal; alternatively the same are claimed as reasonable expenses incurred in an endeavour to mitigate loss.
> (B) If 1973 permission lapsed and planning permission is granted on appeal:
> (i) deferment of profits from quarrying between December 1978 and the date quarrying commenced pursuant to such permission.
> (ii) As above
> (iii) As above

After mentioning that it was possible for him to argue that actual damage did not accrue until December 2 1983 when the second enforcement notice was served by the county council, Mr Griffiths stood by November 30 as the earliest date. His argument ran in this way — in the first place the court must identify what damage was suffered. Damage in this case consisted of the loss of the right to implement the planning permission. The loss of the right cannot be any less than the loss of the legal right to implement the planning permission. That is a different matter from the essentially practical question of when it became inevitable that planning permission could not be implemented before it lapsed. The planning permission conferred a right on the conferee to do something within a prescribed period of time and that right was lost only when that period expired. That is that date of accrual of the damage. A diminution in the value of the land before November 30 would be a different claim. It is not

A the claim raised by the plaintiff in the pleading. The claim as pleaded is not based on the inevitability that damage would be caused but on actual damage. He made submissions, which I need not refer to in detail, based on the correspondence as to the possibility of agreement as to the scheme by the county council or determination of the Secretary of State before the expiry date. None of the facts precluded the possibility that either the county council or the Secretary of State could have approved the scheme before that date. Between November 15 and 30 there was a possibility of determination by the minister, there was no fixed statutory procedure which bound the Secretary of State — what had to take place was "quasi arbitration". From the date of Mr Charles' letter of November 2, there was even
B more time before the planning permission ran out for the minister to act. Mr Nardecchia's approach of looking at what happened after the permission expired was misconceived. The two-year period between June 1980 and May 1982 was no guide as to what might have happened in the autumn of 1978 if the minister had acted as a matter of urgency.

The advantage of November 30 1978 is that it was a fixed date, no other investigation was required and the time of accrual of the cause of action should not depend on assessments of dates of risk because certainty was needed.

As I have indicated, Mr Clegg did not seek to persuade me that Mr Griffiths was wrong but in case he was, Mr Clegg in the course of his cogent and succinct submissions argued that the evidence was
C overwhelming and pointing in only one direction. There came a point in time about a year before the permission expired when it would not have been possible for the Secretary of State to determine the submitted scheme and accordingly the damage accrued at that time, leaving a precise date to be arrived at by me.

In deciding this matter I return to the order for the preliminary issue, as amended, which defines the issue in this way: when did the plaintiffs' cause or causes of action against the first and third defendants arise?

As between FGW and CLB it is admitted that CLB owed FGW both the contractual and tortious duty of care. The allegation of breach of contract and negligence, which I must assume are well
D founded, are pleaded in para 26 of the amended statement of claim. No distinction is made in relation to the breach of contract or the breach of common law duty of care, and the particulars relied on are common to both. CLB were in breach of contract and in breach of their duty of care:

(1) By failing to appeal to the Secretary of State against the county council's non-determination of the scheme, in sufficient time for the Secretary of State to determine it for work to start before November 30 1978. There is also a complaint as to the lack of advice on taking that step;

(2) In failing to apply to the county council for the removal or modification of condition 11 in sufficient time for an extension to be
E granted and similarly a complaint is made in relation to the lack of advice in that regard.

The parties have concentrated on accrual of damage as the starting point for tortious liability. However, I have to determine when the cause of action in contract arose. It is not only an inescapable duty under the terms of the order but it is an essential starting point for any consideration as to when actual damage was caused. It may not matter greatly to the parties because whenever the breach of contract occurred any claim in contract must be statute-barred, but it is in my view essential to the process of arriving at the date of actual damage in tort to see when the breach was. The breach of contract and the breach of the duty of care happened at the same time.

The breach of contract could not be before July 20 1977 when CLB
F submitted their details of the scheme proposed to the county council. There is no easy identifiable date, such as the date of execution of a mortgage deed or the date when a negligently drawn agreement is entered into. The difficulty is caused by the allegation, which I must assume to be correct, that CLB did not act "in sufficient time". What period amounts to sufficient time? It is not a matter altogether easy to determine on the trial of a preliminary issue and I have some residual doubts as to whether it is an exercise that the court should be engaged on. Nevertheless, I have to decide that in order to ascertain the date of the breach on the material I have. In arriving at the view I have formed I have taken into account three principal matters:

(1) The planning history, the most obvious feature of which was the implacable opposition of first Denbighshire County Council and then Clwyd County Council to mining development in this G
prominent rural site;

(2) The right given to the holder of the planning permission, within a reasonable time of submission of the scheme to the county council for approval, to ask the Secretary of State to determine the scheme. The time-limits provided for by the Town and Country Planning Act 1990 were not in operation, of course;

(3) The comprehensive conditions as to the details of the scheme which had to be considered and a considerable amount of work involved in assessing and appraising the proposed scheme.

In all these circumstances, the longer the Secretary of State had to consider the matter, the more likely it was that he would determine the scheme to enable work to start by the end of November 1978. I H
have the advantage of knowing how long it did take the Secretary of State to determine this scheme. In effect, he began on June 13 1980 and finished by making his determination on May 11 1982, having carried out a lengthy consultation process with interested parties. I cannot say that it would have taken the length of time before November 30 because the assumed breach of contract carries also the assumption that there would have been sufficient time between the submission to the county council on July 20 1977 and the end of November 1978. FGW and CLB provided no evidence on the question of sufficient time. I have, however, been referred to Mr Bolton's affidavit which contains this opinion evidence:

> the first defendant should have appealed by December 1977 at the latest, otherwise the Secretary of State would not in all probability have had time to J
> determine this proposal by 30th November 1978.

On the material before me I hold that the breach of contractual duty occurred on December 1 1977 and therefore the cause of action in contract accrued at that date. At the same time CLB were in breach of their duty of care to the plaintiffs, or so I must assume. The next question which arises is: when did the cause of action in tort accrue? It is common ground that that would be when actual damage was sustained by the plaintiffs — whether the plaintiffs knew of the damage or not.

I am grateful to counsel for their analysis of the three reported cases, *Forster* v *Outred*; *D W Moore* v *Ferrier* and *Bell* v *Peter Browne & Co*. Those cases establish or restate, it seems to me, a clear line or authority that: (1) the question is a question of fact in each case; (2) if K
at the time of the defendants' negligence the plaintiff suffers some actual damage it matters not that at a later stage events might occur which would increase that damage. The cause of action accrues at the outset at the time of negligence if actual damage is caused as opposed to presumed or assumed damage. In all those cases and in *Baker* v *Ollard & Bentley* there is an easily ascertainable date for the breach and for the damage. The fact that that date is more difficult to ascertain in the instant case does not alter the general principle and the court's duty is to ascertain the date if it can. Mr Nardecchia was correct when he submitted that FGW had suffered actual damage by the latest at October 25 1978. It was plain to anyone with appropriate knowledge and skill, such as a valuer, that the planning permission L
was going to lapse or would be put at risk in legal proceedings and that the owner of the land would be open to action from the local planning authority. As Mr Nardecchia put it, the FGW's interest in the land was encumbered by insecurity, and that had inevitably an effect on the value of the interest and caused damage — real damage as opposed to assumed damage, although at that stage perhaps not immediately quantifiable. The tort was complete and the plaintiffs could sue. I am assisted in this context by another passage from the judgment of Bingham LJ in *D W Moore & Co* v *Ferrier* [1988] 1 WLR 267, where at C he says at p280:

> The matter may be tested. It is common ground, on the assumption that the plaintiff's pleaded case is correct, that the defendants were in breach of contract when they negligently advised and settled documents in 1971 and M
> 1975. A cause of action then arose. Suppose, per impossibile, that the plaintiffs had sued at once and before the later difficulties with Mr Fenton (the departing director) arose. They would have been bound to succeed. If of the opinion that the plaintiffs had suffered no damage, the judge would have awarded them nominal damages. But it seems to me plain that the judge would not have done that on these facts. He would have assessed as best he could on the available evidence the loss fairly and reasonably flowing in the usual course of things from the defendants' breach of contract, reaching a figure that might have been large or small but would not have been nominal. I think that the plaintiffs were in argument inclined to accept that. If, in a contractual claim for negligence, the Court would have awarded other than nominal damages, I do not see how it can be said that an action in tort based on the same negligence would have been bound to fail for want of any damage as an essential ingredient of the cause of action.

A I do not think that there is, on straightforward application of familiar principles, any escape from these conclusions. I am not therefore surprised to find that other Courts have also reached the same conclusions.

And he then referred to *Forster* v *Outred & Co*; *Baker* v *Ollard & Bentley* and two first instance decisions and continued:

It is pointed out, necessarily correctly, that the facts of those cases are different from those of the present, and in some the damage suffered by the claimant at the time of the execution of the agreement in question is perhaps more obvious than in the present case. Each of the cases is, however, in my view inconsistent in principle with the plaintiffs' contention in this case that they suffered no damage in 1971 and 1975.

Mr Griffiths urged upon me that the task of ascertaining the date B on Mr Nardecchia's basis was so difficult that the court should not embark on it, but decide the preliminary issue on the basis of the one certain date of damage, ie the date the planning permission on the face of it expired.

Although the reported cases are all concerned with easily ascertainable dates, in *Bell* v *Peter Browne & Co* Beldam LJ envisaged on the facts of that case the cause of action accruing at a time which had to be assessed by a judge. At p523 he said:

I consider therefore that the plaintiff's cause of action arose when he parted with the property

— an identifiable date —

or at the latest at the time when the careful solicitor would have effected C registration either of the charge or of a caution.

The alternative clearly envisages judicial appraisal of evidence as to when the prudent solicitor should have acted. In the instant case, I have unchallenged evidence as to when the prudent chartered surveyor should have acted.

Mr Griffiths' attractive argument was that the cause of action could not have arisen before November 30 because events might overtake it and, in effect, "cancel" the cause of action. That is because if the cause of action arose before that date, the determination of the scheme by the Secretary of State or a change of mind and heart by the county council would bring it to an end. It would not have that effect but would reduce perhaps to a nominal amount the damages recoverable. If Mr Griffiths is wrong, the D question remains: when was actual damage caused? I can understand why Mr Nardecchia concentrated upon October 25 as being the date up to which it was possible the county council would approve the scheme as the county planning officer recommended they should. Thereafter it was obvious, if it was not obvious before, that the elected councillors were unmovable in their opposition to mineral extraction on this site; and equally obvious that the time remaining available would not permit the Secretary of State to determine the scheme. However, the Secretary of State had had insufficient time from a much earlier date. I am persuaded by Mr Clegg that the plaintiff sustained damage at and from the time of the breach. In the early years of the planning permission any valuer assessing the value E of the site would do so in the knowledge of the nature of the permission and the condition and with the knowledge that time was passing, but from the end of November 1977 anyone valuing the site would know that it was probable that the scheme would not be determined by the Secretary of State in time for development to begin by the end of November 1978.

On the facts of this case, where clearly in my view on the contractual claim for negligence more than nominal damages would have been awarded, I do not see that it can be said that the action in tort based on the same negligence would have been bound to fail for want of any damage as an essential ingredient of the cause of action.

I bear in mind that FGW's claim in negligence is based not only on the failure to apply to the minister for a determination but also in a F failure to apply to the county council for an extension of the life of the permission. This second limb did not receive from counsel the same close attention as the first and no separate arguments were addressed upon it. However, it seems to me that if it was imprudent and negligent not to apply to the Secretary of State at a point a year away from expiry, it was equally imprudent and negligent not to ask for an extension of the life of the permission at the same time. I must again assume that it was negligent not to make this application and the statement of claim makes the same allegation that it was not done in sufficient time. It falls then for me to determine first, in relation to breach of contract when that breach took place, and I would hold it took place at the same time as the breach in relation to application or appeal to the minister. Any valuer would ask the obvious question: have you applied for an extension? And a negative answer to that G question would inevitably have some effect on the value of FGW's interest. What effect that would have on the assessment of damages is difficult to say and it would depend upon valuation evidence but it would be more than nominal and, accordingly, I answer the question raised in the preliminary issue in relation to the cause of action in tort in this way, that it arose on December 1 1977.

I have not dealt with the limitation issue as between FGW and WSW as envisaged by the district judge's order. No submissions were made by counsel on this and Mr Clegg was content to act within the first part of the order which gave him leave to be heard on the issue between FGW and CLB. As my determination of that preliminary issue removes any possibility of liability on the part of WSW, no need arises for any consideration of the limitation period in the H FGW/WSW litigation.

Queen's Bench Division

January 25 1993

(Before Judge RICE, sitting as a judge of the division)

PRIVATE BANK & TRUST CO LTD v S (UK) LTD

Estates Gazette March 6 1993

[1993] 09 EG 112 J

Negligence — Valuation of property for redevelopment — Whether plaintiff relied on valuation — Whether defendant negligent

By a letter dated December 5 1988 a Mr Carter, of the defendant firm, was instructed by Mr Thomas Brent, on behalf of Rentcrest Ltd, to provide a valuation of 2-4 Hoxton Square, London N1. By a further letter of December 21 1988 Mr Carter was instructed by Chancery Financial Services Ltd to provide a valuation of the property for mortgage purposes and for an assessment of the demand and the rental value K having regard to the development of the site. Mr Carter valued the property for redevelopment to a mainly B1 with some B2 uses at £1.7m. He allowed building costs of £100 per sq ft, giving £4.065m, and a rental value of £23.48 per sq ft on the basis of an investment yield of 8%. Planning permission was granted for the redevelopment of the property in March 1989 and Mr Carter confirmed his valuation in the light of this permission. Mr Carter provided a further valuation in November 1989 estimating building costs of £79.34 per sq ft, a rental value of £25 per sq ft and a yield of 9% giving a residual value of £1.75m. In April 1990 Mr Carter advised Chancery Financial Services in writing that due to a change in the L property market the value of the property was then between £1.35m and £1.45m. In June 1990 the defendant consented to the plaintiff being provided with the defendant's valuations and confirmed them. Because the valuation exceeded the plaintiff's minimum figure of £1.25m, it made a loan to Rentcrest on which Rentcrest later defaulted. The plaintiff claimed damages against the defendant alleging that: (1) it owed it a duty of care; (2) the defendant was negligent; and (3) the plaintiff relied on the defendant's representations of value and was induced to loan money on the strength of the valuation. M

***Held*: Judgment was given for the defendant. On the evidence the plaintiff relied on the defendant's valuation and the valuation was the cause of the plaintiff's loss. The defendant owed the plaintiff a duty of care. In making his residual valuation Mr Carter was entitled to a permissible margin of error of 15% either side of his bracket of £1.35m to £1.45m. His valuation was within this margin and, accordingly, he was not negligent.**

The following cases are referred to in this report.

Corisand Investments Ltd v *Druce & Co* [1978] EGD 769; (1978) 248 EG 315, 407, 504

A *Singer & Friedlander Ltd* v *John D Wood & Co* [1977] EGD 569; (1977) 243 EG 212 & 295

This was a claim for damages in negligence by the plaintiff, Private Bank & Trust Co Ltd, against the defendant, S (UK) Ltd, formerly Sallmanns (UK) Ltd, arising out of a valuation of June 1990.

Adrian Brunner (instructed by Lloyd Cooper) appeared for the plaintiff; Nigel Pitt (instructed by William Davies Meltzer) represented the defendant.

Giving judgment, **JUDGE RICE** said: In 1988 Mr Murray Sweet and an architect, Thomas Brent, formed a company known as Rentcrest Ltd for the purpose of acquiring a property at 2-4 Hoxton Square.
B The site area, which was approximately 7,750 sq ft with a frontage of 62 ft and a depth of 121 ft, was the subject of planning application and development plans and was sold to the company for £1,250,000.

In December 1988 Thomas Brent, on behalf of Rentcrest, verbally instructed a Mr Carter, who was currently employed by Sallmanns (UK) Ltd, to carry out a valuation of the property. These instructions were confirmed by letter from Rentcrest dated December 5 1988. At the time of his appointment he was 59 years old, was not a chartered surveyor, but had been employed throughout his life in the property business. Mr Carter was told that Rentcrest intended to construct a new office building on the site to top-level B1 user requirements.

On December 21 1988 Mr Carter received a letter from Chancery Financial Services Ltd informing them that they were providing a
C bridging loan to Rentcrest Ltd and that they would be relying on the valuation he was to prepare. Chancery Financial Services Ltd set out the terms of Mr Carter's instructions as follows:

> The open market valuation for mortgage purposes of: The property in its present condition at the date of your inspection with the benefit of:
> (i) Existing planning permission and existing lettings and vacant possession

(I interpose there to say that, in fact, there were no lettings or vacant possession, and indeed I should say at the outset that the development here was site development and that all the existing buildings were to be demolished.) The letter goes on:

> (ii) Existing planning permission but with the expectation of planning permission being granted for the scheme as envisaged.
> D (2) Your opinion as to the likelihood of planning permission being granted.

Then it goes on to say,

> We have asked for advice on the demand and current rental value for the type of accommodation now provided and as a development site the demand and rental value for the type of accommodation to be provided in the completed development.

On December 23 1988 the defendants, Sallmanns, through Mr Carter, sent Chancery Financial Services their first report. The first page of the report reads:

> The open market value of the freehold interest for mortgage purposes of: i The property in its present condition at the date of our inspection with the benefit of:
> E (i) Existing planning permission, but with the expectation of planning permission being granted for the scheme envisaged.

The second paragraph reads:

> (ii) Planning permission for approximately 33,000 sq ft of net usable B1 Offices and partial B2 industrial space.

Then (iii), the third paragraph, reads:

> (iii) With the benefit of full Bl planning permission for development of offices, providing a net lettable floor space of between 35,000 and 36,000 sq ft

At p9 of the report he gives his valuations. First, he gives a valuation in respect of a letting area wholly B1, 35,000 sq ft to 36,000 sq ft, as £1,870,000. I will not go into further details, because in fact
F planning permission was granted in respect of his second alternative, namely:

> With the benefit of planning permission being granted for a building of similar size to the Schedule under Appendix 2, ie a lettable floor space of about 33,000 sq ft of which up to 20% was to be used as B2 Industrial, with the remainder as B1 office accommodation, we are of the opinion that the value of the freehold interest, subject to and with the benefit of such a consent is in the order of £1,700,000.

He then considers a third alternative. He says that he believes that it is "a reasonable assumption that planning permission will be granted for a mixed-use building, with accommodation similar to that outlined in the schedule under Appendix 2", that is, the use to which I have referred in detail. He then envisages a possible extension of planning permission, namely:

> Due to the improving nature of the area, it is also possible that an unrestricted G
> B1 office planning permission, as outlined [above], would be granted, but it may be necessary to appeal if the initial application is refused.

The report makes clear that Mr Carter had been supplied with a set of six plans prepared by Thomas Brent, which showed a building designed as B1 office space containing a basement, ground floor and six upper floors which Mr Carter calculated provided a floor area of approximately 37,900 sq ft. The general specification of the proposed new building was to be to high-level B1 office user, including, *inter alia*, fan-coil air conditioning, carpets and high-speed lifts.

If one looks at his appraisal one sees that the building cost is anticipated there as £4,065,000, that is, the square footage, and he
allows the building costs at £100 per sq ft. He estimates the rental H
value on the 33,000 sq ft, which he takes as being B1 development area, at £23.48, and this is on the basis of an investment yield of 8%. The total gross development value was given as being £9,688,500. This was also on the basis that 20% was to be used as B1 industrial. He estimated, as I have said, that the residual site value would be £1,700,000 on planning permission being given in the terms that he envisaged in his report, and he sets out how he arrives at that figure.

The report having been sent to Chancery Financial Services on December 23 1988 at their request, a copy was sent by Mr Carter to Rentcrest on January 18 1989. On March 21 1989 Hackney London Borough Council granted conditional planning consent, and that was for the erection of a building on six main floors plus basement J
and the B1 area was to be 2,888m2 (I stress this area is in metres) on six main floors instead of seven. The height of the building was reduced by one floor as a result of objections raised by Heritage. There was also going to be 454m2 of B2 industrial use, and that was to provide for a loading and unloading bay and eight car-parking spaces in the basement.

Mr Carter, having been informed of the planning consent, informed Chancery Bank in April 1989 that there was no reason to alter his valuation of the residual site value of £1,700,000. That perhaps is not surprising because, of course, his valuation was anticipating the planning permission that, in fact, was granted.

On November 16 1989 Mr Carter did a further valuation or
appraisal of the site. This was on the basis of outlined planning K
consent and also a letter from Davis Langdon & Everest estimating the building costs. This estimate of the building costs was for an area of 28,442 sq ft at £79.34 per sq ft. He estimated on this occasion that the rental value per sq ft, on the basis of an estimated rental value of £25 per sq ft, the residual site value would be increased from his earlier figure of £1,700,000. He increased it to £1,750,000. This was on a 9% yield.

A copy of this later valuation was sent by Mr Carter to Chancery Securities Financial Services on November 3 1989. One notes that, on the second page of the valuation, the estimated rental value has been increased from £23.48, in his first report, to £25, and the yield was
increased from 8% to 9% in the second valuation. This valuation was L
subsequently submitted to Rentcrest.

On April 12 1990 Mr Carter wrote to Chancery again following conversations with them. On the second page of that letter he wrote as follows:

> You are well aware that since our initial report there has been a considerable downturn in market demand for this type of development situation, largely due to the increase in interest rates and the resultant high cost of funding. The effect of this is that developers seeking sites for such purposes are limited and likely to adopt a more cautious approach, with allowance for a longer term in respect of commencement and/or fruition of any scheme they may embark upon. At the same time, because of market conditions, tenant demand for the type of buildings envisaged for the site has weakened.
> Conversely, the majority of property owners are not selling, but holding M
> their investments, where possible, until a relaxation of interest rates, and improvement in market conditions. It is thus reasonable to comment that the only purchasers in the market at the present time, for a development site of this description are bargain hunters, seeking a heavily discounted price.
> Bearing all the above in mind, and subject to there being good and marketable title, free from any unusual, onerous or hampering restrictions, and that the information with which we have been provided being correct, we are of the opinion that the present value, on the basis set out in our "terms of reference" is in the region of between £1,350,000/£1,450,000.

How he arrived at these figures is shown at p119 of the bundle preceding the report or letter, which I have just read out. His workings there show this. He took the valuation of November 30 1989 (I remind myself that it was £1,750,000) and then he proceeded as follows. It is in his own manuscript and consists of a few lines: "Value as at April 1990", and then he takes £1,750,000 and then he says underneath that, "*Less* carrying cost of interest 1 year @ 16%", and he deducts therefrom £280,000, which brings it down to £1,450,000. That of course was the higher figure which he gave. Then, "or", and he takes again his November valuation of £1,750,000 and then marks it down, "*Less* interest of 18 months" (18 months is a longer period, the previous one being one year) "at 16% — £422,000", and he subtracts that which gives him the final figure of £1,330,000 which he rounds up to £1,350,000. That was the basis for the figures which he gave in his letter of April 12.

The plaintiffs became involved in 2-4 Hoxton Square through being approached by Christopher Humber and Murray Sweet. The latter, as well as being a director in Rentcrest was, along with Mr Humber, a director of Marketbase & Humber Contract Interiors Ltd. Marketbase proposed in 1990 to acquire a share in the capital of Rentcrest and to extend an interest-free loan to that company. In order to do this Marketbase approached the plaintiffs, and on July 3 1990 it was agreed that they would advance the sum of £780,000 to Rentcrest for one year against the security of the first legal charge over the site in Hoxton Square, plus the plaintiffs also required guarantees from Mr Humber and Mr Sweet each to the sum of £65,000, which would enable Rentcrest to repay Chancery Securities the money they had borrowed from them.

According to Mr Hanna, who is and was a director of banking at the plaintiff company and acted on their behalf, it was a condition precedent to the offer that the plaintiffs received a satisfactory valuation prepared for the benefit of the plaintiffs in respect of the property indicating a minimum market value of £1,250,000. According to Mr Hanna, the plaintiffs were not prepared to lend anything against the security of the property if the valuation was less than that sum.

On June 27 the defendants agreed that their earlier reports, namely December 13 1988 and November 16 1989, should be sent to the plaintiffs and agreed that their letter dated June 21 1990 addressed to Marketbase, which reiterated the views contained in the report of April 14 1990, should be shown to the plaintiffs. On June 27 1990 the defendants wrote to the plaintiffs as follows:

> At the request of your clients, Marketbase, we confirm our consent for the reproduction to you of the report to them of 21st June in respect of the above property, a copy of which is attached hereto and it may be regarded as being addressed to your bank.

Mr Hanna of the plaintiffs stated that, as the defendants' valuation exceeded the bank's minimum figure of £1,250,000, the loan was made. Rentcrest defaulted in payment of the principal sum plus interest and the value of the property and the guarantees are insufficient to meet the mortgage of the plaintiffs.

The plaintiffs sue the defendants on the basis that they were negligent in making the representation that in June 1990 the property was worth between £1,350,000 and £1,450,000 and thereby inducing the plaintiffs to advance money and are therefore liable for the loss and damage suffered by them. It follows, therefore, that the plaintiffs have to prove: first, that the defendants were under a duty to the plaintiffs to take care in the making of the representations; second, that the defendants were negligent in making the representations; and, third, that the plaintiffs relied on the representations and were thereby induced to advance money on the strength of the valuation.

Reliance and causation are capable of easy resolution in this case and accordingly I shall deal with that first. It was Mr Hanna, as one of the two directors responsible for the generation of new business and administration of banking business, who advised the executive committee of the plaintiffs to advance the money. He knew the property had been purchased in about 1988 for £1,250,000 and that the company Rentcrest had been formed solely for that purpose. As is clear from the liability bundle he had no mandate to proceed if the valuation was less than £1,250,000. Further, even if the site was valued at £1,250,000 or more, the plaintiff bank was prepared to advance only 62.4% of that sum and, in addition, required personal guarantees from both Mr Humber and Mr Sweet. He conceded in cross-examination that, although he was happy to do business with Mr Humber and Mr Sweet and regarded them as potential clients, he had no authority to recommend a loan unless the valuation exceeded £1,250,000 and that he relied on the defendants' valuation of the site's worth being between £1,350,000 and £1,450,000.

When pressed on this matter he said he could not say whether the committee would have advanced 62.4% of, say, £1,000,000 if the site had been so valued, but for a number of reasons he would not have recommended the loan and he was the person who was responsible for putting the proposal to the committee.

I am quite satisfied, both from Mr Hanna's evidence and the document to which I have just referred that this or any loan would never have been made but for the defendants' valuation. I am also satisfied that, as this loan was part of a package involving guarantees and paying off Chancery and the bank's practice of not making small loans, if the valuation had been below £1,250,000 they would not have advanced a lesser sum, or indeed any money at all. In the premises the plaintiffs have satisfied me that when they advanced the money in July 1990, they did so in reliance on the defendants' valuation which they believed to be true.

To succeed the plaintiffs have to show that the defendants owed them a duty to take care when making the representations and that they were negligent. It is not seriously disputed in this case that the defendants owed the plaintiffs a duty of care. They have discharged that burden.

In the absence of any express term to the contrary, the standard required of a surveyor is that of an ordinary skilled man. Nor is this standard lessened if the valuation is carried out in difficult circumstances, as where the market is showing signs of collapsing into a depression. However, in order to determine whether this standard has been complied with, the court must have regard to the circumstances and knowledge current at the time of the survey. The court must be careful to guard against hindsight and determine the surveyor's skill and competence on the knowledge and information available to him at the time of the survey.

In *Corisand Investments Ltd* v *Druce & Co* (1978) 248 EG 315, Watkins J listed the matters of principle and fact to which the valuer should have regard, though in that case it should be borne in mind that the valuation concerned a hotel, whereas in this case it is limited to the site. So many of the matters to which he referred do not apply.

Therefore, the valuer of a site is not concerned with any goodwill which may be attached to the building, but he should, by inspection and inquiry, determine the site's value at the time (which in this case was June 1990). Having inspected the site, it is for the surveyor to determine its value. It was agreed by the plaintiffs and the defendants that there were no close comparables, that is, in Hoxton Square itself, and that the proper way was to establish the gross development value and then to make the appropriate deductions in order to determine the residual site value.

In *Singer & Friedlander Ltd* v *John D Wood & Co* (1977) 243 EG 212 & 295, Watkins J held:

> The valuation of land by trained and competent and careful professional men is a task which rarely, if ever, admits of precise conclusion. Often beyond certain well-founded facts so many imponderables confront the valuer that he is obliged to proceed on the basis of assumptions. Therefore, he cannot be faulted for achieving a result which does not admit of some degree of error. Thus, two able and experienced men, each confronted with the same task, might come to different conclusions without anyone being justified in saying that either of them has lacked competence and reasonable care, still less integrity, in doing his work.
>
> The permissible margin of error is said . . . to be generally 10% either side of a figure which can be said to be the right figure . . . In exceptional circumstances the permissible margin could be extended to about 15%.

Mr Pitt has urged on behalf of his clients, the defendants, that Mr Carter was faced with a difficult task, in that he was asked to give a site valuation against a falling market. Therefore, the defendants should be given further latitude and licence by being allowed a bracket around the figure he has given and then be allowed a further 15% around the bracket.

I reject that approach in this case because Mr Carter's valuation did not give a single figure, but put it within the bracket of £1,350,000 to £1,450,000 and in effect to extend that bracket further would be to allow the defendants too much licence. However, I think the defendants are entitled to a variation of up to 15% either side of the bracket. I am fortified in this view because it was endorsed by both Mr Eric Shapiro [FRICS] and Mr Ralph Last [FRICS], the expert surveyors.

Mr Carter, the defendants' surveyor responsible for producing the valuations, is now 63 years old and resident in Australia, and his statement dated December 24 1992 has been tendered under the Civil Evidence Act. However, in order to assist the court in determining whether Mr Carter was negligent in making his valuations, the plaintiffs and defendants have enlisted the services of two qualified

A surveyors, namely Mr Last and Mr Shapiro respectively. Mr Last has prepared two reports: the first is dated December 4 1992 and the second, which attempts to reply to the issues raised by Mr Shapiro in his report, is dated January 5 1993.

As one might expect, all three surveyors are in agreement that the residual site value should be determined by the gross development value. In order to highlight the area of disagreement between the surveyors, the plaintiffs have produced a schedule (which I shall refer to as P1), the first two columns of which show the gross development value given to the site by the defendants on December 13 1988 and November 16 1989. The third and fourth columns show the gross development value given to the site by Mr Last. The fifth and sixth B columns show the gross development values given to the site by Mr Shapiro.

Both the expert witnesses gave a second valuation on the basis of criticisms raised by their counterpart. I shall not, obviously, in the course of this judgment go through all the points raised by Mr Last and Mr Shapiro, but only those which assist the court, in my view, in determining whether the defendants' valuation of June 1990 was negligent.

Mr Last's major criticism of Mr Carter's June report is that it departs from the proper way to carry out a valuation of the site. In December 1988 and November 1989 he carried out such a valuation in order to determine the residual value, and that is shown in columns 1 and 2 of P1. What he did in April 1990 was to discount the November C 1989 site valuation by deducting a carrying cost of one year to 18 months in the hope that the property market would have picked up within that period of time. Mr Last is of the view that a competent surveyor, when expressly asked to provide a valuation of a site for a bank, should have carried out a new development appraisal incorporating current information available at the time, including building costs, rental values, interest rates and yield. Such an appraisal would reflect the current market conditions and would take into account void periods.

He prays in aid graphs and figures, which appear at pp 69 and 76 of the expert bundle, which show that there was a growing increase of office accommodation available in the second-half of 1989 and in the first-half of 1990. This increase was particularly apparent in the D outskirts of the fringe where he places Hoxton Square, removed as it was from the main thoroughfares such as Old Street or City Road.

A further criticism is that the defendants increased the rental value for the development between 1988 and November 1989 from £23.48 to £25 per sq ft which was out of step with market trends. By June 1990 rentals per square foot for this type of development had fallen below the 1989 rental values. He does not quarrel with the investment yield of 9%, which the defendants ascribed to the site in 1989; indeed, he would put it somewhat lower, between 8.5% and 8.75%, but higher than Mr Shapiro's yield of 8%.

Another criticism of the defendants' valuation is that, though the November 1989 valuation provides for an increase of rent, the E proposed building cost was dropped from £100 per sq ft in the December 1988 valuation to £79.34, whereas it would be logical to expect the converse to be the case. As has been seen, the December 1988 valuation provided that the building should be constructed to a very high standard in order to command a rent of £23.48 per sq ft, whereas in November 1989 not only had the rent been increased to £25 per sq ft and building costs reduced, but on August 1 1989 Thomas Brent Associates had produced the building estimates and provide for a shell standard only. There is no provision for air conditioning, which seemingly has to be provided by the tenant. Mr Last is of the view that to attract tenants to this site, given its location, carpeting and suspended ceiling and further electrical work should have been provided along with the ventilation to the car park and a F sprinkler system.

In their December 13 1988 report the defendants state that the plans show that there are areas on a number of the floors where natural daylighting will be restricted. This has caused them concern, as this would not only affect the liability but also the marketability of the resultant investment. They further state that they had assumed that such problems can be resolved, but, as Mr Last points out in his report, there was no provision for such resolution. He is of the view that any valuation should take account of the poorly lit areas by discounting the rent in respect of them.

He allows for a lesser rent of £12.50 per sq ft in the basement and one of £17.50 per sq ft on the ground, first, second and third floors in respect of these poorly lit areas. Mr Last also criticises Mr Carter for failing to make proper allowance for the void periods and, as a result, G he believes that by his failures, to which I have referred, this has caused him to give a false figure; and that had he done a proper valuation in June 1990 then the residual site value would, on any view, be less than £500,000, as is shown in his own gross development valuations, which appear at (3) and (4) of P1. Certainly, he says, it would have been well below the plaintiffs' minimum of £1,250,000.

Mr Shapiro in his reports and evidence seeks to refute Mr Last's allegation of negligence. In his report he accepts that he, like Mr Last, would have carried out a further development appraisal, though when he came to give evidence he thought that on the information available to him at the time Mr Carter's deduction of a carrying cost of a year to 18 months in order to determine the current site value was H a shrewd and acceptable alternative. He was doubtless influenced in this view by the fact that his development appraisal for the site for June 1990 gives a residual site value not far removed from Mr Carter's lower figure. Accordingly, he contends that, even if the defendants were negligent in not carrying out a development appraisal, nevertheless the method he adopted came up with the correct figure and that the plaintiffs did not suffer any loss as a result of this failure.

There is a fundamental difference in approach between Mr Last and Mr Shapiro on the valuation of the site in respect of which planning permission has been obtained. Mr Last's valuation of the site is on the basis of the existing planning permission, though he concedes that it is open to the valuer to state in his valuation report J that the site has considerable potential and that, if there was an appeal from the existing order or a further application, the value of the site might be considerably increased; in other words, he is in favour in such circumstances of giving two figures, one based on the existing planning permission and the second the value of the site if a successful appeal were lodged. He said that this would be his approach in the example which was put to him in cross-examination where planning permission had been granted for dog kennels on a site in central London. Mr Shapiro felt that one figure should be given in the valuation, and that this should be on the site's potential and not limited by any existing planning permission.

I think Mr Last's approach in this respect is the proper one, since, though the potential of the site must be borne in mind and possible K provision made for it, to value it on the basis of a successful appeal or a change in planning approval introduces an element of speculation and uncertainty, particularly in this case where there was no evidence of an appeal or a further application. However, in my view, the point is an academic one, since the defendants' valuation of November 16 1989, though it mentions the possibility of a further planning application, is based on the March 1989 planning consent which provided for mixed B1 and B2 user.

Mr Last attacks Mr Carter's valuation on November 16 1989 as being based on wrong floorspace which, in his valuation is given as 28,750 sq ft. This is based on B1 floorspace of 26,160 sq ft net and B2 floorspace of 2,590 sq ft net. Under cross-examination Mr Last L conceded that he should have taken gross measurements for the B2 floorspace, but had made allowance for this by allowing a greater rent, namely £10 per sq ft, than would otherwise have been the case. For obvious reasons I was unimpressed by this explanation, which I regarded as a clumsy attempt by him to cover up a mistake. If one takes the B1 and B2 areas as being those given at p47 of the experts' bundle, the floorspace is B1 26,160 sq ft, B2 4,882 sq ft, making a total of 31,042 sq ft which is greater than he allowed for in November 1989. In the premises Mr Carter's valuation of November 1989, so far as the floorspace is concerned, is more accurate than that of Mr Last.

The chief criticism that can be made against Mr Carter's valuation of November 16 1989 is that he increased the rent from £23.48 to £25, and at the same time allowed for reduced building costs from £100 per M sq ft to £79.34 per sq ft. However, one can see the basis for his valuation if one looks at his residual approach. There he started by giving the site a value of £2,345,650, but, because he thought this was too great, one can see from the sensitivity analysis that he decided to reduce the site value to £1,761,437 by giving a high yield of 9% which is greater than Mr Last's of 8.5% and Mr Shapiro's of 8%. Thus it could be said that he has sought to compensate the higher rent and lower building costs by giving a higher yield. Mr Shapiro supports Mr Carter's valuation by giving an estimated rental of £25 per sq ft for the entire B1 floorspace, which is shown at D4, whereas Mr Last distinguishes between B1 floorspace where the lighting is good by giving it an estimated rental value of £22.50 and floorspace where the lighting is bad is given on an estimated rental value of £17.50.

This means that the overall B1 estimated rent in Mr Last's valuation is £19.29 instead of the £25 allowed for by both Mr Carter and Mr Shapiro. This, along with the fact that he has given B2 floorspace as 2,590 sq ft instead of 4,882 sq ft, helps to explain the difference between his valuation and that of Mr Shapiro.

Mr Shapiro is of the view that Mr Last was wrong to allow a discount in respect of the poorly lit floorspace, since the proposed development was going to be an open shell and a tenant was to partition it as he saw fit. In such a case Mr Shapiro states that he has never seen a building having its floorspace divided in the manner done by Mr Last in his valuation. He further states that all open-plan offices rely on artificial lighting and the amount of natural light admitted to the building is a comparatively minor matter in determining the rent. I think Mr Shapiro is right on this matter and that Mr Last was wrong in introducing two bands of B1 accommodation and arrived at too low an estimated rental value.

The main difference between the two surveyors is the state of the property market in June 1990. In 1990 Mr Last was of the view that there was no demand for development sites in the Hoxton Square area, and he seeks to support this opinion by producing an article of Savills which would have been available to Mr Carter at the time. Figure (4) of that article shows a sharp increase in vacancies in the City and City fringe for 1989. The same article states that the rental levels have declined in central London, and at p74 of the experts' bundle there is a further graph giving office completions and availability in 1989.

According to Mr Last the boom years for property of the middle and late 1980s had come to an end and Mr Carter should have appreciated that the bottom had dropped out of the market. Mr Last believes that Mr Carter not only should have, but did, appreciate this fact, and he cites p2 of his report dated April 12 1990. In particular he refers to these passages. I have referred to part of this before, but I will read it again because it shows Mr Carter's state of mind at the time:

> You are well aware that since our initial report there has been a considerable downturn in market demand for this type of development situation, largely due to the increase in interest rates and the resultant high cost of funding. The effect of this is that developers seeking sites for such purposes are limited and likely to adopt a more cautious approach, with allowance for a longer term in respect of commencement and/or fruition of any scheme they may embark upon.

Then the next paragraph reads:

> Conversely, the majority of property owners are not selling, but holding their investments, where possible, until a relaxation of interest rates, and improvement in market conditions. It is thus reasonable to comment that the only purchasers in the market at the present time, for a development site of this description are bargain hunters, seeking a heavily discounted price.

It is said by Mr Last on behalf of the plaintiffs that in the above passage Mr Carter was giving the right warning signals based on a market which was in rapid decline, but that this was not reflected in the valuation that he gave between £1,350,000 and £1,450,000.

Mr Shapiro, for the defendants, contends that in 1989 and 1990 the Government was seeking to deflate the economy by increasing the rate of borrowing and this had the effect of bringing the rate of expansion of the previous years to a halt. This had a corresponding effect of decreasing the demand for office development, but in the first-half of 1990, indeed up until and including June 1990, his view was that the property market had slowed down, but not dried up. Mr Shapiro said that this did not happen until August 1990 at the time Iraq invaded Kuwait when, as it were, overnight there was no longer an office development market, not only in the fringe areas, but also in the city itself.

In support of this contention he cited his own experience, admittedly in February 1990, in respect of a site development at 5-7 Leonard Street, which is in the fringe area, though nearer to the City than Hoxton Square. His valuation of these premises, which was a development site of 5,582 sq ft, is set out at D5. There he allowed £80 per sq ft building costs with an estimated rental value of £25 per sq ft and a yield of 8%, giving a residual site value of £421,755. If one compares this with D4, which is his valuation of the Hoxton Square site in June 1990 (D4 is an extrapolation from the valuation which appears in his first report), he allowed building costs of £90 per sq ft, which is £10 up on the Leonard Street site, a similar estimated rental value of £25 per sq ft, but a yield of 8.25%, which is 0.25% up on the Leonard Street site and gave a site value of £425,000 which was realised in April 1990.

As perhaps is not unusual in cases of this kind, there is a conflict between the surveyors. I find that both have fallen into error. I find that Mr Last fell into error in basing his valuation on a site area of 28,750 sq ft, whereas it should have been 31,042 sq ft. I think he was also wrong, for the reasons advanced by Mr Shapiro, in, first, making a distinction for letting purposes between well-lit and poorly lit B1 areas and, second, in making the substantial reduction which appears at p101. This has resulted in him arriving at an estimated rental value which is too low. I think Mr Shapiro is wrong in supporting, albeit halfheartedly, the method Mr Carter adopted in giving the plaintiffs the June valuation, namely by taking the April valuation and then arriving at a residual site valuation by deducting a carrying cost. As Mr Shapiro under cross-examination accepted, he should have done a fresh valuation. But, had he done so, then according to Mr Shapiro he would have produced D4.

Further, I prefer the approach of Mr Last to that of Mr Shapiro, namely that any valuation should be based on the criteria that existed at the time of the valuation and that, therefore, it should take into account the existing planning consent. To base one's valuation on the possibility of an appeal and more favourable planning permission being obtained introduces a speculative element which is bound to affect its accuracy. I much prefer the approach of Mr Last, namely that the valuation should be based on any planning consent and then as in the extreme case put in cross-examination, a second figure given in the event of a successful appeal.

However, though in his initial report of December 23 1988 he refers to the possibility of a successful appeal, Mr Carter bases his valuations on the existing planning consents as has Mr Shapiro when he produced D4. Therefore, though I think Mr Shapiro is wrong in this respect, his views have not affected the gross development values that he has produced. Equally, though I find that Mr Carter was wrong in the criteria that he applied in making the June 1990 valuation, the result he has achieved was within the 15% of Mr Shapiro's valuation at D4; in other words, as Mr Shapiro said, his wrong approach, if it was wrong, did not, in effect, make any difference.

Perhaps the greatest divergence, however, between Mr Last and Mr Shapiro was the state of the property market in June 1990. There is no doubt that they, along with Mr Carter, were aware that the property boom was over and that the country was experiencing a treasury-inspired contraction. Mr Shapiro contends that it was not until August that it was clear that, instead of a slowing down of the economy and its consequent effect on property development, what the country was heading for was a recession with a catastrophic effect on sites awaiting development. Both sides have produced articles and graphs which they pray in aid to support their respective arguments. On the balance I prefer the view of Mr Shapiro and indeed that of Mr Carter, namely that in June 1990 though the market was going through a difficult period on the information then available nobody could predict the decline that was to follow in the second half of the year. I am fortified in this view by the site value which was achieved for Leonard Street in April 1990. Though the site was nearer the City and the sale took place in April, nevertheless a site of 5,582 sq ft, of which 1,015 sq ft was basement, achieved a price in excess of £450,000. One might have expected a site of 31,042 sq ft containing 26,160 sq ft of B1 accommodation — being therefore nearly six times as large — would have a residual site value of at least £1,250,000.

In the circumstances, therefore, though I am quite satisfied that the plaintiffs made the loan on the strength of the defendants' valuation and suffered the losses which have been agreed between the parties, subject of course to liability, I do not find that the defendants, through Mr Carter, were negligent in making a valuation. In the premises there must be judgment for the defendants.

Judgment for the defendants.

A

Chancery Division

September 29 1992

(Before Mr Andrew BLACKETT-ORD QC, sitting as a deputy judge of the division)

MACEY v DEBENHAM TEWSON & CHINNOCKS

Estates Gazette March 27 1993

[1993] 12 EG 127

Negligence — Valuation report of surveyors — Whether
B *surveyors negligent*

In 1985 Mr Verney, of the defendant firm of surveyors, Debenham Tewson & Chinnocks, was instructed by the plaintiff, Mr Macey, to advise in connection with the plaintiff's interest in purchasing freehold retail investments with long reversions to full rental values. In the spring of 1986 Mr Verney advised the plaintiff to purchase a property in the High Street, Maidenhead, for £1.026m. This property was let under two leases one of which was for a term to June 23 2016 without review. By a report dated July 10 1986, Mr Verney confirmed his advice to purchase and set out his analysis of yields, including a net equivalent yield of 4.42%. The
C purchase was completed on September 17 1986. The plaintiff brought proceedings alleging that the defendants had been negligent in the advice given in July 1986 and that at the date the value of the property was only £530,000 or, if "hope value" should be included, £633,750. The defendants denied negligence, their expert valuing the property in 1986 at £993,266, rounding it up to £1m, a figure close to the £1.026m.

***Held*: The action was dismissed. Mr Verney's report could be criticised in that he did not make clear that his net equivalent yield rate of 4.42% was arrived at by taking a rack-rent yield on comparables and adding an uplift to the**
D **reversionary element. But his figure for rack-rent yield was not without foundation; there were comparables in Maidenhead to support his chosen yields. The valuation approach to the plaintiff's expert would not necessarily produce the right answer. The plaintiff failed to show on a balance of probabilities that Mr Verney's report, and the calculations and deductions on which it was based, were such as could not have been arrived at if he had shown reasonable care and skill.**

No cases are referred to in this report.

E This was a claim for damages by the plaintiff, Mr Macey, arising out of a valuation report prepared by Mr Tom Verney ARICS, of the defendant firm, Debenham Tewson & Chinnocks.

John Powell QC (instructed by Macey & Co) appeared for the plaintiff; Jonathan Gaunt QC (instructed by Cameron Markby Hewitt) represented the defendants.

Giving judgment, **MR ANDREW BLACKETT-ORD** said: The plaintiff in this action is a solicitor practising in Blackheath under the style of Macey & Co. In about 1983, it seems, he started investing in commercial property. He had income in excess of his immediate needs. He was aged in his 30s and contemplated that, when in his 50s, he would want to work under less pressure. One can sympathise with
F him. He was primarily interested in freehold retail properties in good locations let at relatively low rents under leases with something like 15 years unexpired, to meet these objectives. At this time the maximum income tax rate, one must remember, was 60% and the maximum capital gains tax rate was 30%.

By buying these freehold reversions he aimed at capital profit, as well as increased income when the leases fell in and the properties could be let at rack-rents. He thought that he could buy a good property currently producing a low income cheaper than a rack-rented one and he was prepared to wait for the returns. At the latest by 1985 he was employing the defendant firm (since incorporated) to find him suitable properties and his contact was one Tom Verney [ARICS]. In his statement he gives the impression that he met Tom Verney rather later than this, but I accept that this may have been a G misunderstanding as to what was required to go in the statement.

In this action he sues the firm for negligence for advising him to pay too much for a property, 88-90-90A High Street, Maidenhead, which he bought for £1.026m in mid-1986. In para 19 of the amended statement of claim he alleges that the value of the freehold, subject to the subsisting leases, was only £530,000 and says that that is the price which he should have been advised to pay. In the alternative, if the price should have included what has been referred to as "hope value" (which I will return to) he says that the value was £633,750. These latter are the figures which his expert, Mr Edwin Rhodes [ARICS] of Weatherall Green & Smith, puts forward and he now says that the latter figure is what should have been paid, it being accepted that H hope value would have been included in the price.

I will drop the "90A" for convenience as I only mentioned it to indicate that the evidence is that the site is capable of forming, perhaps once did form, three retail units fronting the High Street. No 88-90 is an unusual proposition in several ways. All the experts agreed that it is difficult to value. It occupies what in 1986 was probably the best retail site in Maidenhead. It is in the part of the High Street which, in the current jargon, had then been pedestrianised. Directly opposite, on the other side of the road, the High Street is joined by a road called Nicholson's Wharf (also pedestrianised), which had then been recently refurbished as a shopping centre and which formed a natural channel from the principal car park in the town to the High Street. Prospective shoppers coming from the car park down J Nicholson's Wharf would see Woolworths directly opposite. And there was a rear frontage for no 90 (and not, I think, no 88) to West Street with vehicular access. As can be seen from the map, which appears in various volumes, behind the High Street and fronting West Street the Woolworths' store extends westwards and it apparently owned the freehold of this part. The whole site, 88-90 and the Woolworths freehold, was occupied by one purpose-built store. No 88, with about one-third of the frontage to the High Street, had an agreed ground-floor sales area of about 1,430 sq ft, was held under a lease expiring on June 23 2016 under which the current rent was £2,500 pa. This was subject to review, upwards only, to the current market level in June 1993, 2002 and 2011, ie initially nine-yearly periods. No 90 had a ground-floor sales area of about 7,500 sq ft and K was also let until June 23 2016, but the current rent was only £2,150 and was not subject to review.

It is agreed that each property was valued for rent purposes in 1986 by the zoning method at a ground-floor rental value for zone A of £45 a sq ft. It is also agreed that the rack-rent values (including storage space, which I need not go into) in 1986 were about £27,500 for no 88 and £88,000 for no 90, about £115,500 in all. What capital value that would represent is not agreed, but, of course, under the leases it was only the rental value of no 88 which would be realised before 2016 and, even for that, in 1986 there were seven years to wait.

In the case of a long lease the strength of the covenant of the original lessee is considered to be of importance, although, of course, L less so where the rent is low. It is not disputed that the covenant of Woolworths, which was the original lessee in each case, was strong.

In the spring of 1986 the property was being marketed by a Mr George Tindley [FRICS] of Savills, who gave evidence and produced his working papers showing how it was that he came to seek offers in the region of £1m. He was a disinterested and honest witness and the dispute is not about his veracity but the details of his valuation, as will appear. He mentioned the property first to a few associates, including Tom Verney + who passed brief details to the plaintiff in his letter of April 2 1986. After describing it he said:

> The vendors are seeking offers in the region of £1m for their freehold interest, which they calculate will show a purchaser reversionary yields of 3% in 1993, 11.2% in 2016 and an equivalent yield of 4.45%. M

The main question in the case is whether this figure of 4.45% can be justified. Mr Verney is dead and his side of the story can be obtained only from his correspondence and working papers and, to some extent, from the evidence of his superior Mr Justin Roberts [FRICS], who gave evidence.

Mr Verney wrote again on April 10, after various telephone calls to the plaintiff, telling him that Savills proposed to test the market further before committing themselves to accepting £1m and advising him not to put in a bid of £1m at that stage, but either to "come in strongly over the asking price with a time limit for acceptance" or wait until the position clarified. He set out the result of such investigation as he had been able to carry out into the rental values in the High Street, which broadly showed an increase over the previous couple of years and in his view indicated that £45 for zone A was not an unrealistic rental value for nos 88-90.

The plaintiff's expert, Mr Rhodes of Weatherall Green & Smith, advanced the view in evidence that the rental value should be reduced because of the size of the building, making it a "store" (as indeed I have called it) rather than a "shop", but £45 was eventually agreed. The reference to the size of the site emphasises the different requirements of different parts of the market. The evidence now shows that, while a large site may put off some buyers interested in liquidity, it can attract long-term holders like charities whose interest is largely in acquiring a valuable site and who are more interested, therefore, in the value of the land. The plaintiff was happy with the advice given to date (see his letter of April 11 1986) though he wanted to pay less than £1m if he could secure the property for less.

On pp6 and 7 of vol 3 is apparently a specimen of the letter which Mr Tindley sent out to about 50 agents and clients to test the market, asking for offers in the region of £1m. He listed the following attractions to justify his suggested equivalent yield of 4.45%:

1. The substantial marriage value.
2. The rarity of prime long-dated retail reversionary investments.
3. The exceptionally good location.
4. The obvious tax advantages to high-tax-bracket individuals.

All the experts agree about 2, 3 and 4. The tax advantage, 4, would of course be lost if the property was bought with borrowed money and interest exceeded the income against which it could be set. In the present case the plaintiff borrowed more than half the price but in his letter to Mr Verney of May 29 1986 he says:

> Please do not be concerned if the income does not cover interest upon a loan of half a million. The income from other properties which are not subject to loans exceeds any interest that would be due to the bank and the bank are already fully satisfied with the position.

I find that Mr Verney was justified in assuming that the purchase would have substantial tax advantages for the plaintiff, that the plaintiff had reason to think this and that it was one of his reasons for seeking reversionary property. I do not mean that the plaintiff should have been advised to pay more than the property was worth, only that the open-market value depends upon who is in the market.

A peculiarity of this property was that an income which might go a long way towards meeting borrowing charges was expected in 1993 from no 88, although the principal capital gain would not be realised until 2016.

I must say a word about marriage value. When a property is let at less than the rack-rent the leasehold interest has a value. Broadly, the longer this state of affairs is expected to continue the greater that value is and the smaller the value of the freehold. But if in such circumstances the freehold and leasehold interests are separately valued it is found that the sum of those values is not as great as that of the unencumbered freehold. The difference is known among valuers as "the marriage value". It may be realised during the term of the lease if the tenant acquires the reversion or the reversioner acquires the lease. In either event, when this is done on commercial terms the marriage value will be divided between them, the proportion being a matter for negotiation.

When, as in the present case, a freehold is sold subject to a long lease the open-market value includes a small proportion, say 10%, of the marriage value because in the open market there will always be hopeful prospective purchasers of the freehold, apart from the tenant, who are prepared to pay something over the odds for the chance of realising some part of the marriage value during the term. This is known as "hope value". In this case the experts are agreed that it formed an intrinsic part of the open-market value I understand that the plaintiff now appreciates this, although it is clear from the later correspondence (for instance, his letter of November 15 1990) and, indeed, from the statement of claim that he did not appreciate it then. It is the addition of a figure of £103,750 for hope value which gives the alternative open-market value of £633,750 in the statement of claim.

Mr Rhodes agreed that a purchaser not prepared to pay hope value in a case like the present would be offering less than the open-market value of the property and would be unlikely to be successful. There has been some evidence that Woolworths are not interested in buying the freehold, but views change and there is still, in my judgment, an element of hope value in the open-market value of the property. The marriage value of course reduces as the term nears its end and the hope value with it, but it is not lost to the freeholder, who is compensated by the increased value of the freehold as the date for vacant possession approaches.

Reverting then to the narrative, Mr Tindley's informal tender process produced four bids, as follows:

(1) Mr Macey, the plaintiff, £1,026,000.
(2) Robert Fleming Properties Ltd, £906,000.
(3) Syndicate Administration Ltd, £855,000.
(4) Hatton Properties Ltd, £650,000.

It was suggested by Mr Rhodes that the under bids may not have been firm and that, as they were all made by limited companies, this puts them in a different category to that of the plaintiff, but there is no evidence of that. On the contrary, the evidence was that well-known property dealers or owners are slow to resile from bids in case they lose credibility in the market. Moreover, Robert Fleming Properties Ltd is referred to as bidding for "a client" and Syndicate Administration Ltd was said by Mr Tindley to be the nominee of a number of wealthy individuals. So I think that the under bids must be treated as comparable.

The Robert Fleming bid was £120,000 less than that of the plaintiff, but over £250,000 more than Mr Rhodes's upper valuation. The plaintiff's bid was accepted. Contracts were exchanged on July 30 1986 and completion took place on September 17.

Between April and June the plaintiff negotiated to buy 77 Clarence Street, Kingston on Thames, which was let at a relatively low rent with a reversion nine years ahead. He tried unsuccessfully to get the vendor to postpone the falling in of the reversion to 19 years by way of a sale-and-lease-back. In early June he offered £1.2 m for the property, subject to the existing lease, but the offer was not accepted. In my view Kingston was intended as an alternative to Maidenhead if the latter fell through. It may be some confirmation of the sort of money that the plaintiff was prepared to lay out at this time and the type of property that he was seeking.

With the recent further evidence which I will mention shortly, it also, I think, indicates the number of balls which he was trying to keep in the air at the same time, because it now appears that he was also interested in a property in Norwich and sought to persuade a Mr Hill of Drivers Jonas to put no 90 on the market to release funds to pay for Norwich. This was not later than early July 1986, long before contracts were exchanged, but Mr Hill advised against marketing no 90 before he had bought it and the proposal was postponed.

At some stage (it is not clear to me precisely when) it was understood that Mr Verney would supply a report. The first reference to it in the correspondence is in Mr Verney's letter to the plaintiff of May 15 1986. After confirming that the offer of £1,026,000 had been accepted subject to contract he continued:

> We are accordingly in the course of preparing our report in respect of the property and we understand that you would propose to use this report for the purposes of securing an advance by way of a commercial mortgage from Lloyds Bank. We would therefore propose to address this report both to yourself and Lloyds Bank. Within this report we would make comment as to the amount of advance which we as a firm would consider appropriate in respect of this freehold, subject to the tenancies.
>
> Our initial reaction is that the report will state that the property forms a reasonable security in today's market for a loan of up to half a million pounds. We will, however, have to point out to your bank that the income currently emanating from the investment is unlikely to cover the interest payable upon a loan of this amount. This situation is unusual but is occasioned by the very reversionary nature of the investment.
>
> I would advise you that our report will contain a structural survey and a fire insurance valuation.

What may be called the plaintiff's instructions are contained in his letter of June 6, part of which reflects the letter of May 29, to which I have already referred. He says:

> Your report should contain a structural report, as previously mentioned, and full details of the current rental value as assessed by yourself, together with a breakdown of your calculations. Your report should also contain your valuation of the investment and the valuation for insurance purposes. The report should please contain full details of the matters relating to the alterations made by the tenant and the effect of such alterations on the rent review. There is no need to trouble too much with regard to Lloyds Bank. They do not need a separate valuation, nor a valuation for loan purposes. They will be satisfied with a copy of your report.

For completeness I can mention Lloyds Bank's letter of July 2 1986 to the plaintiff requiring a professional valuation on an investment basis from Debenham Tewson & Chinnocks to be addressed to Lloyds Bank plc as one of the terms of the proposed loan.

The report mentioned in the correspondence was prepared by Mr Verney and is dated July 10 1986. It calls itself a "Report and

A Recommendation". It is addressed to the plaintiff and it is accepted that he relied on it in making his purchase. It says twice that it is not a valuation but it contains a recommendation to purchase for the sum of £1,026,000m, which, in my view, meant that Mr Verney valued the property at least at that figure. The opening paragraphs describe the area, the property and the leases to which it was subject and are not controversial. Para 10 deals with the rental value and is also now agreed. Para 11 is in these terms:

> 11. *Capital and return* We set out below our analysis of the purchase price based upon our estimated rental value as set out above.

Then follow figures in columns which it is not convenient to read out. The rental value, current and deferred, is valued on an
B equivalent yield of 4.42% and gives a gross purchase price of £1,055,013. Then it says "*Less* 2.75% costs, £29,013; net purchase price £1,026,000". Continuing the quotation:

> Our analysis of the purchase price produces net yields of .44% initial, 2.81% reversionary in 1993 and 10.94% reversionary in 2016 and a net equivalent yield of 4.42%

In para 12, headed "*Recommendations*" Mr Verney summarised some of the advantages of the property and continued:

> An equivalent yield of 4.42% is, we consider, an attractive yield for this property. We understand that a transaction took place at Nos 63-67 High Street where the freehold was sold earlier this year for a figure representing an equivalent yield in the order of 4%. In addition, we understand No 68 High Street has also recently been the subject of a freehold sale at a price reflecting
> C a yield in the order of 4.4%. Both these properties, however, are held on normal institutional lease terms unlike Nos 88/90 High Street.
>
> In conclusion, and having regard to your current investment requirements, we recommend the purchase of the freehold interest in Nos 88/90 High Street subject to the lease upon the terms described, at the price agreed of £1,026,000.

Surprisingly, in view of the terms of the plaintiff's instructions, the report disclaims liability for any party wishing to extend mortgage finance, but this is not a matter which I need go into.

It appeared from the evidence given by the plaintiff last June that he was content with his purchase until the question of value arose again, in the summer of 1990, when he wished to raise further finance and asked the defendant firm what it would charge for a further
D valuation. Mr Verney had ceased to act (I do not know if he had already died) so the plaintiff was sent a draft valuation dated August 7 1990 which, according to Mr Roberts who prepared it, was intended as a discussion document and set out four different figures based on different assumptions. I need mention only the first two:

> 1. Open market value assuming no allowance for any additional value attributable to potential offer from the tenant or prospective investor, £860,000.
>
> 2. Open market value assuming allowance by prospective purchaser other than tenant for special tenant offer, £1,300,000.

The plaintiff was, understandably, confused. He had expected the property to have doubled in value because, as he thought, rents had
E doubled in the four years since he had bought. This inference was not justified because by 1990 the market was falling instead of rising and yields were therefore rising.

He thought that the figure of £860,000 was reached on the same basis as the price he had paid. It is now agreed that here also he was wrong because he did not appreciate that (as is agreed by the expert witnesses) the market value included hope value. He now appreciates that it is the second figure, £1.3m, which is comparable. Debenhams were somewhat dismissive, so he consulted Weatherall Green, who produced the figures which I have quoted from the statement of claim.

I reserved judgment in the case in July to give myself time to consider the expert evidence and arranged to give judgment
F yesterday. But earlier this month I granted an application for the defendant firm to call further evidence. The core of this came from the files of Jones Lang Wootton, the firm of Mr Jeremy Waters [FRICS], the defendant's expert. It derived from his partner, Mr Robbin Broadhurst [FRICS], who had been acting in his capacity as surveyor to the Drapers Company, which is probably why it did not come to light earlier. There is also evidence from Drivers Jonas of a valuation exercise in 1988.

It will be recalled that contracts were exchanged on July 30 1986. The new evidence establishes that early in August, immediately after exchange, Mr Macey formally instructed Drivers Jonas, through Mr Hill (mentioned above), to seek a purchaser urgently for no 90 at a price of £645,000. Mr Macey was a friend of a partner in Drivers Jonas and, as I have mentioned, a subsale had been mooted for some G
time to enable him to buy the Norwich property. Mr Verney seems not to have been brought in. No 90 was the part of the property to which, in reaching his total of £633,000-odd, Mr Rhodes attributed a value of £145,000 only plus a share of the marriage value (much the major share, certainly) but giving a total of about £240,000 in all.

The marketing efforts were not at first successful. Mr Hill circulated 30-odd likely agents or purchasers, but it was probably not the best time of year for the exercise and many would know that the property had very recently been on the market and had presumably not then been interested. Mr Hill said that, as he was selling part only, it was a very different animal and he tried to take care to establish this, but maybe to some extent his efforts were seen through and the H
marketing may have been affected by what someone referred to as the "stale bun" syndrome.

The best offer received in August was £350,000 though later Mr Duncan Preston [ARICS] of Jones Lang Wootton, acting in this case for the Buccleugh Estates, wrote rather vaguely on October 17 to his client of it possibly being a "very rational decision" to buy at £580,000 if it suited their tax planning. But in fact it did not and so that proceeded no further. Still, the figure of £350,000 was £100,000 over Mr Rhodes's figure and it was an opening shot by Hillier Baker for a possible purchaser and probably they did not pitch it too high.

Most prospective purchasers, however, who were approached thought a reversion of 30 years without review was either too long or too expensive, or both. But, just as it was an unusual property, so the J
Drapers Company could be said to be an unusual purchaser as a charity able and willing to take a longer view than most.

In November 1986 Mr Broadhurst heard of the property and in December he recommended it to the Drapers Company, at a price of £575,000 for no 90. They were very interested and, having confirmed with Drivers Jonas that this figure would be acceptable to Mr Macey, Mr Broadhurst wrote a formal report in mid-December for consideration by the governing body of the company. Mr Macey then withdrew, but the relevance of the evidence is that Mr Broadhurst in his report recommended a purchase at the price I have mentioned to show an equivalent yield of 4.4%. In my judgment, this provides strong support for the views which have been expressed on the K
defendant's side.

Neither Drivers Jonas nor Mr Rhodes seem to have considered the possibility of a non-tax-paying purchaser. I do not accept Mr Powell's argument that Mr Broadhurst's view should be disregarded as based on inadequate research. It is, of course, only a view. All valuations are only a view, but it is a view of a very experienced surveyor basing himself on research carried out on his instructions. His clients did not have the normal requirements of a property investor seeking profit in the foreseeable future. To a large extent they were interested in the long-term value of the site, which was very substantial although for a considerable time ahead it would also be probably very illiquid. They were not interested in a quick turnover.

Mr Macey, when questioned about this new evidence, said that he L
did not recall approaching Drivers Jonas and that the advice to sell no 90 in August came from Mr Verney at a time, presumably before exchange, when Mr Macey was expressing a wish to withdraw from the transaction because the lot size was too big and the reversion too distant. He certainly said, in cross-examination in June, that in July 1986 he was having cold feet. But he said that Mr Verney persuaded him to go on because to withdraw at that stage would damage his (I think Mr Verney's) credibility. Possibly I suppose the credibility of both of them. He said that his ultimate decision was to retain no 90 until something cheaper turned up, when he would sell no 90 to pay for it. But it now seems that this is not what happened and I cannot accept his evidence in this respect. It is noticeable that no publicity was given to his purchase and I consider it probable that this was M
because he realised that it would be detrimental to his proposed subsale and was not done to please Mr Verney.

There is no dispute as to the law concerning claims against valuers for negligence. It is conveniently summarised in *Jackson & Powell on Professional Negligence*, 3rd ed, p219:

> The onus of proof rests "fairly and squarely" upon the plaintiff in an action for professional negligence against a surveyor. It is not enough to prove that he was wrong. Negligence must be distinctly proved. It may even be difficult to prove he was wrong by showing that the defendant's valuation was significantly higher or lower than another surveyor's valuation or the price actually achieved or that his estimate of individual items making up the calculation was on the high side. It must be shown that his valuation and

A calculations were such that they could not have been arrived at by the exercise of reasonable care and skill.

The first question is whether the plaintiff has proved on the balance of probabilities that Mr Verney did not exercise such reasonable care and skill. Did he apply the correct principles taking into account the relevant factors and ignoring others? So far as this is uncertain, the court should have regard to a second question, namely whether the figure for the valuation of the property was right or, at least, within an acceptable margin of error. This element is particularly relevant by way of a check as Mr Verney is not here to explain his thought processes. Finally, if he seems not to have shown reasonable care and skill, what loss, if any, has the plaintiff suffered? B This, in my view, relates to the value of the property in 1986 and not in 1990 or today.

In the summer of 1986 the property market generally was not quite as strong or, one could say, as speculative as in the previous autumn but I am satisfied that nevertheless it was to be described as "very strong". Of course it was not the same all over the country. Maidenhead, in the then affluent Thames Valley, was regarded by those in touch with the market as a place where both rental and capital growth were to be expected. Demand for property in the best area, which included the High Street, was considerable. The difference between the valuers in this case arises purely, first, from the different rack-rent values which they attached to the property and, second, from the different uplift which they apply to allow for C the terms of the leases. It is agreed that this is a question of judgment. The capital valuations are based on the yield figures.

Mr Rhodes' valuation (the appendix to his first report) was based on a valuation process which he agreed with Mr Jonathon Gaunt QC, for the defendant, in cross-examination. He assessed the estimated rental value to an amount not unlike that of Mr Verney and then identified a rack-rental yield of 4.5% as appropriate to each part of the property. This he agreed was a matter of judgment. He then added uplift of 0.75% for no 88 and 3% for no 90 on account of the presence or absence of rent reviews in the leases. Taking the resulting yields of 5.25% and 7.5% and allowing for the periods until the respective rents could be increased, he valued the income streams at a total of £530,000. He then valued the occupational leases and D calculated the marriage value at £995,000. He then divided the marriage value between the freehold and leasehold interests in proportion to the value that he had reached for them and took 25% of the freeholder's share as hope value. This process also was largely subjective. Adding the hope value of just over £100,000 to £530,000 gave him the £633,750 as the open market value of the freehold of the whole. On this figure he worked out an equivalent yield of 5.72%. £633,750 is, of course, below all the four bids which Savills got for the property in April.

Mr Rhodes' figures are impressively precise; too precise, Mr Roberts and the defendants' expert, Mr Waters of Jones Lang Wootton, would say. Valuation is not an exact science like E mathematics. An element of judgment is involved. As well as the marriage value calculations the rack-rental yield of 4.5% and the figures for uplift of 0.75% and 3% are also matters of judgment. The former particularly is reached against comparables. Different views can and have been expressed as to the value and interpretation of the relevant comparable evidence but these judgments are the foundation of the calculation and, if they are wrong, the edifice built on them collapses.

Mr Rhodes's figures are not accepted by the defendants or by Mr Waters, who suggests a rack-rental yield for the property in mid-1986 to be about 3.75% and applies an uplift of 0.75% for the whole. Mr Waters had much the same comparables to consider as Mr Rhodes but he interpreted them differently, thinking that a prospective F purchaser would see more prospect of growth in the value of the Woolworths site taken as a whole than Mr Rhodes was able to agree. As regards hope value, it was generally accepted that this was about £100,000, though the valuers reached this destination by very different routes. This lends support for the defendant's view that there was more than one way of approaching this admittedly difficult valuation as against the view of Mr Rhodes that his method was the only right one.

In reaching his asking price of about £1m Mr Tindley of Savills took a current rack-rent yield of 4.25% for each part of the property. He then uplifted it by 0.75% in respect of no 88, but made no uplift at all in respect of no 90, his reason being that he considered that the prospect of a low-income yield for 30 years and consequential capital G gain would attract higher-rate tax payers and that it had an increasing scarcity value. Mr Rhodes and Mr Waters confirmed the attraction and the scarcity, but still thought an uplift necessary.

Mr Tindley can perhaps claim to have been justified by events. His overall equivalent yield came out at 4.45%. Although Mr Rhodes disagreed with Mr Tindley's approach he would not go so far as to say that it was incompetent. He had to agree that in respect of another property at Bridgwater, which Savills sold at the same time, Mr Tindley's judgment of the asking price was also proved correct.

Mr Rhodes did not claim to have had any particular feel for the market in Maidenhead in 1985 or 1986. He formed his judgment of equivalent yields on comparables and various indices to which he referred in his evidence, but he insisted that the two rental streams H had to be valued separately because it would not be impossible to sell one without the other. However, he agreed in cross-examination that in addition to the initial yields and the reversionary yields a purchaser would require to know the overall equivalent yield. All these figures were given by Mr Verney in the 1986 report. Mr Rhodes accepted the analysis in para 11 of the 1986 report but thought that more explanation should have been given in para 12, where the equivalent yield is discussed, to explain how the valuation figure was arrived at; particularly, I think, that the question of uplift should have been expressly dealt with.

At first sight this criticism seems justified. Mr Verney could have made it clear that his equivalent yield figure of 4.42% was arrived at J by taking a rack-rent yield on comparables and adding an uplift to the reversionary element. Why he did not do so I think appears from Mr Roberts' evidence. The latter was behind the 1986 report as Mr Verney's superior. He indicated that he is a market man and not a valuer. In mid-1986 he was very familiar with the market for retail shops in Maidenhead. He knew that, whatever the indices might say and whether or not they described Maidenhead as "a prime town", the retail property market there was very strong though not quite as strong as six months earlier.

I accept that indices are based on averages. They show trends but not the peculiarities of any particular property and so cannot be the basis of a valuation. Mr Rhodes placed too much weight on them. In mid-1986 properties in the middle of Maidenhead were changing K hands at a prime or rack-rent yield of under 4%. Optimism was in the air. Woolworths, a household name, occupied probably the best site in the town. It was an original lessee and its covenant inspired confidence. Mr Roberts took the rack-rent yield of nos 88-90 as 3.75% and from his market knowledge gave an uplift of 0.5 or 0.75% only. When the rental income figures were put through a computer the plaintiff's proposed offer of £1,026,000 showed an equivalent yield of 4.42%.

His figure for the rack-rent yield cannot be said to have been without foundation. He had available as comparables 68 High Street (Contessa) showing an equivalent yield of 4.36% in August 1985 and 63-67 (McDonald's) showing 4.16% at March 1986. Both these L figures are agreed. They were based on rack-rent lettings of properties which, for various reasons, Mr Roberts considered inferior and as justifying a figure of about 3.75% for Woolworths.

He took the view, and Mr Waters agreed, that to sell the property in its two constituent parts would not realise its full potential. Mr Rhodes did not fully accept this but agreed that there was a potential for devaluation. Of course we know that Mr Macey did try to sell no 90 separately, but it would seem from the evidence that, had he succeeded in doing so, he would have been left with no 88 which, although carrying a prospect of an earlier income increase, had various structural difficulties attached to it such as an absence of a staircase and, I think, no rear access, so he would, by selling no 90, M have devalued no 88.

Mr Roberts had no doubt that with an equivalent yield of 4.42% the property was worth the price and was happy to advise the plaintiff to go ahead. He made no marriage-value calculation, but went straight to the market price of the property as a whole. Mr Verney's thoughts seem to have been similar when he prepared the 1986 report, para 12, and I do not consider that its format is open to serious criticism.

There was an issue of fact between the plaintiff and Mr Roberts as to whether in 1990 Mr Roberts told the plaintiff on the telephone that yields on reversionary property were about 3% higher than on rack-rented property. I do not accept that Mr Roberts said this in any context relevant to the 1986 transaction. I have already said that the

A plaintiff was confused by the 1990 draft report and this may have been part of the confusion.

In his statement and oral evidence the plaintiff sought to emphasise his inexperience in property dealing and his reliance on the defendant firm. I thought these matters were over-emphasised and that his evidence was tainted by his irritation at the rather derogatory way in which the defendants dealt with his queries in 1990. In 1986 he was swept away by the excitement of the property market and probably paid more than he could afford, but this is not the same as saying that he paid too much. His bid was higher than the others, but was not excessive for someone keen to succeed. The defendants' expert, Mr Waters, valued the property at £993,266 in 1986,
B rounding it up to £1m, and showing an equivalent yield of 4.42%.

Both he and Mr Rhodes are men of considerable experience but I accept Mr Waters' view that the valuation method adopted by Mr Rhodes would not necessarily produce the right answer in the present case. Mr Waters' evidence, supporting Mr Roberts, was that the market would regard the property as one and that the value of the whole would be greater than that of the two constituent parts plus hope value. In the fluid market of 1986 a somewhat broad-brush approach would be likely to give the truer result.

In some cases it is possible to fix on the correct figure and judge whether or not the defendant's figure falls within an acceptable margin of error or bracket. I do not find this approach helpful in the present case because the nature of the property and the state of the
C market led to competent people taking widely different views.

Mr Waters' view is supported by Mr Broadhurst and to some extent by mysterious papers from a Drivers Jonas file dating from 1988 and attaching a value (it is not clear for what purpose) of £765,000 to no 90 and £1,465,000 to the whole property. It was clearly a formal exercise in valuation and I think they must have had instructions, though Mr Macey denies giving them. The figures show a capital gain of 25% pa over 21 months.

All this is not strictly relevant to the defendants' 1986 valuation but D
is not inconsistent with it and it throws further doubt, in my judgment, on Mr Macey's credibility, or at least his memory. I do not have to decide the value the property unless I come to the question of assessment of damages.

My initial task is to determine on the evidence and on the balance of probabilities whether Mr Verney's report in 1986, and the calculations and deductions on which it was based, were such as could not have been arrived at if he had shown reasonable care and skill. The plaintiff has failed to satisfy me on this point. If I am wrong and Mr Verney's calculations are less precise than the plaintiff was entitled to expect then still the basic figure of £1m was, in my judgment, on the evidence justified. The extra £26,000 was offered E
by the plaintiff after advice, but of his own volition, to secure the property and did not relate to any negligent conduct on the part of the defendant firm. The action is accordingly dismissed.

Following an application by the defendants for costs on an indemnity basis **Mr Andrew Blackett-Ord** continued: It seems to me that this is not one of those very special cases in which indemnity costs should be ordered under the new rule. One has to remember that one is dealing, in effect, with an order on the old "common fund" basis, which was a good deal more generous than the "party and party" basis. I think the authorities to which I have been referred show clearly that the standard basis is very much the norm even in cases of this kind where the parties are not on very good terms and F
there are various matters of complaint.

In my judgment, therefore, the proper order for costs (and this includes the application to hear further evidence) is that the plaintiff should pay the costs on the standard basis.

Action dismissed.

RATING

A
House of Lords

December 10 1992

(Before Lord TEMPLEMAN, Lord OLIVER OF AYLMERTON, Lord GOFF OF CHIEVELEY, Lord MUSTILL and Lord SLYNN OF HADLEY)

FARMER (VO) AND OTHERS v BUXTED POULTRY LTD

Estates Gazette March 20 1993

[1993] 11 EG 124

B

Rates and rating — Exemption — Appellant owning and occupying 67 farms situated between a quarter of a mile and 120 miles from poultry processing factory and provender mill which it owned and occupied — Whether provender mill and poultry processing factory an "agricultural building" within the meaning of section 26 of the General Rate Act 1967

The appellant company owned and occupied a provender mill and a poultry processing factory at Dalton, Thirsk, North Yorkshire; it also owned and occupied 67 poultry breeding and rearing farms situated between a quarter of a mile and a 120 miles from the subject hereditaments. The mill produced C and supplied to the 67 farms pelleted feed for consumption by the poultry and also supplied 6% to 8% of its total production to farms occupied by independent turkey rearers. The Court of Appeal rejected the finding by the Lands Tribunal that the mill was both occupied together with the farms and used solely in connection with them; there was no appeal from that decision. The poultry processing factory received the substantial production of poultry from 48 of the farms; it was processed at the factory and either frozen or sold fresh to depots. The sole question before the House of Lords was whether the factory could be said to have been "occupied D together with" buildings used for the keeping or breeding of livestock so as to satisfy the requirements of section 2(1)(*b*) of the Rating Act 1971, which extends the definition of "agricultural buildings" in section 26 of the General Rate Act 1967 for the purposes of derating.

Held: **The appeal was dismissed. For a building to be "occupied together with" another for the purposes of the 1967 Act, they must be in the same occupation and the activities carried on in both must be jointly controlled or managed. To satisfy each of these necessary conditions separately or together is not sufficient to establish that one building is "occupied together with" another for rating E purposes. Nor is there any geographical test which gives a conclusive answer. It is not sufficient to ask generally whether the building or buildings and land in question are all part of the same business enterprise. What is necessary is to show that the two buildings, or as the case may be the buildings and agricultural land, are occupied together so as to form in a real sense a single agricultural unit. Contiguity or propinquity may go far to show that they are. In the present case there were 48 farms with their broiler houses and each broiler house must be surrounded by at least 2 ha of land to qualify. They are kept separate and distinct, in part, in order to prevent or reduce the F spread of disease. Yet it is an inescapable finding that they are separate and distinct farms and are to be treated as such for rating purposes since it has not been suggested that any two or more of the broiler houses are, in reality, run as a single unit. It is impossible to say that each broiler house is occupied together with all the other broiler houses as one unit, or that the factory is occupied as one unit together with all farms, some of which are 100 miles away. Applying the test as to whether the several buildings are worked together as one agricultural unit, and have regard to their physical separation, as part of this test, the tribunal could not possibly conclude G that the 48 farms are "occupied together with" the factory for the purposes of the Act. The appeal should be dismissed and the appellant's processing factory and premises should be reinstated in the valuation list at the agreed appropriate rating valuation of £20,150.**

The following cases are referred to in this report.

Eastwood (W & J B) Ltd v *Herrod (VO)* [1971] AC 160; [1970] 2 WLR 775; [1970] 1 All ER 774; [1970] RA 63, HL

Handley (VO) v *Bernard Matthews plc* [1989] 1 EGLR 215; [1989] 14 EG 67; [1988] RA 222

Hilleshog Sugar Beet Breeding Co v *Wilkes (VO)* (1971) 17 RRC 275; [1971] RA 275 H

Midlothian Assessor v *Buccleuch Estates Ltd* [1962] RA 257; 1962 SC 453

Taylor (Maurice E) (Merchants) Ltd v *Commissioner of Valuation* [1981] NI 236

This was an appeal by Buxted Poultry Ltd from a decision of the Court of Appeal (*sub nom, Farmer (VO)* v *Hambleton District Council*) [1992] 1 EGLR 179 allowing an appeal from a decision of the Lands Tribunal (C R Mallett FRICS and T Hoyes FRICS) given on November 24 1989 dismissing appeals against a decision of the North Yorkshire Local Valuation Court following the appellant's proposals to delete a provender mill and poultry processing factory at Dalton, Thirsk, North Yorkshire, from the valuation list on the J ground that they were agricultural buildings.

Peter Curry QC and Malcolm Davis-White (instructed by Stephens & Scown) appeared for the appellant, Buxted Poultry Ltd; Christopher Cochrane QC and Simon Bird (instructed by Sharpe Pritchard, for the solicitor to Hambleton District Council) represented the first respondents; the valuation officer did not appear and was not represented.

In his speech, **LORD SLYNN OF HADLEY** said: The underlying question at issue in this appeal is whether a poultry processing factory owned, occupied and operated by the appellant is an "agricultural building" within the meaning of section 26 of the General Rate Act 1967 and sections 1 and 2 of the Rating Act 1971, so as to be exempt K from rates.

The local valuation court (to whom an application was made bv the appellant contesting the decision of the second respondent) and the Lands Tribunal held that it was an "agricultural building" and was exempt. The Court of Appeal held that the Lands Tribunal had misdirected itself in law and remitted the question to the Lands Tribunal [1992] 1 WLR 330.

By section 26 of the General Rate Act 1967 no agricultural land or agricultural buildings are liable to be rated or included in any valuation list. By subsection 4:

In this section, the expression "agricultural buildings" —

(*a*) means buildings (other than dwellings) occupied together with agricultural land or being or forming part of a market garden, and in either case used solely in connection with agricultural operations thereon; and

(*b*) includes a building which is used solely in connection with agricultural operations carried on on agricultural land and which is occupied either —

(i) by the occupiers of all that land; or

(ii) by individuals who are appointed by the said occupiers for the time being to manage . . .

By section 1 of the Rating Act 1971, the definition of "agricultural buildings" in section 26 of the 1967 Act is extended for purposes of derating. The extensions include the following sections:

Livestock Buildings

2. —(1) Subject to subsections (2) to (4) of this section, each of the following is an agricultural building by virtue of this section —

(*a*) any building used for the keeping or breeding of livestock; and

(*b*) any building (other than a dwelling) which is occupied together with one or more buildings falling within paragraph (*a*) above and is used in connection with the operations carried on in that building or those buildings.

(2) A building used as mentioned in subsection (1)(*a*) of this section is not an agricultural building by virtue of this section unless either —

(*a*) it is solely so used; or

(*b*) it is occupied together with agricultural land (as defined in the principal section) and used also in connection with agricultural operations on that land, and that other use together with the use mentioned in subsection (1)(*a*) of this section is its sole use.

(3) A building occupied and used as mentioned in subsection (1)(*b*) of this section is not an agricultural building by virtue of this section unless either —

(*a*) it is solely so used; or

(*b*) it is occupied also together with agricultural land (as defined in the principal section) and used also in connection with agricultural operations on that land and that other use together with the use mentioned in subsection (1)(*b*) of this section is its sole use.

(4) A building is not an agricultural building by virtue of this area section unless it is surrounded by or contiguous to an of agricultural land (as defined in the principal section) which amounts to not less than [2 hectares]; . . .

But for the purpose of deciding whether a building is so surrounded by or contiguous to an area of agricultural land certain items are to be disregarded.

The Lands Tribunal found that the objects of the appellant company included the breeding, rearing, slaughtering, processing, packing, selling and distributing of turkeys and chickens for human consumption. For this purpose it owned and occupied 67 poultry breeding and rearing farms situated between a quarter of a mile and 120 miles from the hereditaments in question at Dalton, Thirsk, North Yorkshire, where the appellant owned and occupied a provender mill and premises and a separate processing factory and premises.

The mill produced and supplied to the 67 farms pelletted feed for consumption by the poultry, but it also supplied 6% to 8% of its total production (amounting to between 162 and 216 tonnes per week) to farms occupied by independent turkey rearers. The Lands Tribunal accepted that the provender mill was both occupied together with the farms and used solely in connection with them. The Court of Appeal rejected both of these contentions. The 6% to 8% supplied to other farmers prevented the supply from being the sole use in connection with the appellant's farms and even if the *de minimis* rule applied (which as a matter of construction the Court of Appeal found that it did not) this quantity was not *de minimis*. There has been no appeal to your lordships' House from that decision.

The factory received the substantial production (though not the entire production) of poultry from 48 of the farms, again scattered over a very large area. Poultry was not received from other farms. The poultry was processed at the factory and either frozen or sold fresh to depots. The Court of Appeal, like the Lands Tribunal, accepted that the factory itself was used solely in connection with the operations carried on at the 48 broiler-rearing houses on the farms. The Court of Appeal, however, regarded the Lands Tribunal's decision that the ancillary buildings (a sales office, shop and maintenance workshop) were so linked to the factory that they could be regarded as solely used for the same purposes, as one of fact and degree which could not be set aside. There is no appeal from that part of the Court of Appeal's decision.

The sole question, therefore, is whether the factory can be said to have been "occupied together with" buildings used for the keeping or breeding of livestock so as to satisfy the requirements of section 2(1)(*b*) of the 1971 Act.

The appellant's case is that the factory was occupied together with those buildings since they were both occupied by the same person, they were both so occupied during the same period and they both took part in one continuous process of rearing, slaughtering and preparing poultry for sale so that a test of functional unity was satisfied. There is no requirement that the relevant building should be contiguous or adjacent to the farms. The absence of contiguity in the present case was not found by the Lands Tribunal to prevent the factory being "occupied together with" the broiler houses. That is a question of fact and degree which cannot be interfered with on appeal.

The Lands Tribunal in the present case found that the distance of the factory from the farms did not appear to weaken the functional link between the factory and the farms, nor prevent the company from occupying those distant farms "together with" their buildings at Dalton. The members concluded:

For these reasons we do not think that a geographical test has merit in these cases and we find that the provender mill and the factory are, as a matter of fact, "occupied together with" the farms in the occupation of the company.

In the Court of Appeal Glidewell LJ, with whom Sir John Megaw agreed, considered that not to treat the geographical test of propinquity as a relevant consideration was an error of law. Purchas LJ was of the same view and considered that the Lands Tribunal misdirected themselves in so far as they ignored the geographical relationship between the factory and the agricultural building in coming to the conclusion that it was "occupied together with" those buildings.

Glidewell LJ was of the opinion, with which the other members of the Court of Appeal agreed at p339 F-G, that:

a building can be said to be occupied together with another building used for the breeding or keeping of livestock, provided, first, that they have a single occupier; secondly, that the activities carried on in both are jointly controlled or managed; and, thirdly, that the physical communication between the two buildings is, by reason either of physical nearness or some other factor, so close and convenient that they can properly be regarded as being occupied as parts of the same enterprise.

The phrase "occupied together with" has been considered in a number of cases arising either under the 1967 Act or under the 1971 Act and different approaches have been adopted. Thus in *Hilleshog Sugar Beet Breeding Co Ltd* v *Wilkes (VO)* [1971] RA 275, the question was whether a plant research centre was occupied together with a number of widely scattered plots of land, some of which were up to 40 miles away from the centre. There, men and machines were sent out from the centre to work on the plots of land in order to plant out seed beds with a view to producing an improved type of beet. Sir Michael Rowe QC, President of the Lands Tribunal, quoted at p284 the proposition in *Ryde on Rating*, 12th ed (1968), p289, that "occupied together with" seems to imply a functional rather than a geographical connection. He accepted that the centre was occupied together with the plots of land.

In *Handley (VO)* v *Bernard Matthews plc* [1988] RA 222*, the hereditament was used for the production of pelletted feed for turkeys on 29 farms owned and occupied by the company and situated between 9 and 74 miles from the hereditament.

Mr V G Wellings QC concluded at p229:

On the facts, I find that the appeal hereditament at all material times was and is occupied together with the livestock buildings on all the 29 farms. I am not impressed by any geographical test, at all events in the present case. If it is desired to ascribe a specific meaning to the words "occupied together with", it appears to me that the most likely meaning is "occupied at the same time as": see Stroud, *Judicial Dictionary*, 5th ed (1986).

In *Maurice E Taylor (Merchants) Ltd* v *Commissioner of Valuation* [1981] NI 236, a company carried on business as a shipper of seed and ware potatoes. Some of the land on which the potatoes were grown was 10 to 20 miles from the buildings in question, which consisted of a shed, where the potatoes were graded and bagged, and a store. Gibson LJ at p248 could see:

no principle of rating law which would suggest that, once it has been proved that the land and buildings are in common occupation, their physical relationship should determine whether the buildings are to be treated as agricultural or not. I would accept the view stated in *Ryde on Rating* (13th ed) at p300 that "occupied together with" seems to imply a functional rather than a geographical connection.

*Editor's note: Also reported at [1989] 1 EGLR 215.

A Lord Lowry LCJ at p251 rejected the proposition that the buildings and the land must be contiguous or adjacent and at p252 he said:

> In any case there is really no reason why the words "together with" should connote more than the occupation of buildings and land by the same person, since the need for a functional connection is ensured by the words "and used solely in connection with" etc.

Finally, on the English cases reference must be made to *W & J B Eastwood Ltd* v *Herrod (VO)* [1971] AC 160 where, again, the rearing, killing and preparing of poultry for sale were at issue. The case turned on whether the buildings in question were used "solely in connection with the agricultural operations on the land", it being B conceded that all the buildings were occupied together with the 1,150 acres of land. Viscount Dilhorne expressed surprise that that should be so. He said at p180:

> But for this concession I do not think I should have found it easy to conclude that the packing station in Gainsborough nine miles or so away was occupied together with the agricultural land in the sense in which those words are used in the definition, and it may be that I would have had difficulty in coming to that conclusion in relation to the five layer houses at Norton Brisney some six miles away and some of the other buildings.
>
> In its context "occupied together with agricultural land" may connote more than common ownership. My impression on reading the definition of "agricultural buildings" is that it was an attempt by the draftsman to define a farm in statutory language and that it was intended to include buildings used and occupied together with the land for the purpose of farming the land, not C buildings far distant and not used in connection with an operation on the land, even though owned by the same person.

In *Midlothian Assessor* v *Buccleuch Estates Ltd* [1962] RA 257 the Lands Valuation Appeal Court in Scotland was concerned with woodlands, comprising several scattered areas of land and sawmills on separate sites. The majority decision was that sawmills other than one sawmill contiguous to woodlands were not part of agricultural lands. They were not "buildings thereon". Both Lord Sorn, who dissented, and Lord Kilbrandon approached the matter on the basis that the relevant question was what was the unit or *unum quid* for rating purposes? Lord Kilbrandon said at p268:

> It has never yet been admitted that you can have a unit of valuation consisting D of widely scattered heritable subjects connected only by some functional or commercial nexus, and I do not see why it should be. I do not think one is being merely old fashioned or obscurantist in insisting, in the conception of *unum quid*, on a fairly close physical relationship between what might be considered as parts of a commercial unit; one is, after all, attempting to value not a business but heritable subjects, and it may be that the precedents, which all insist on such a physical relationship, indicate a determination to preserve that essential distinction.

I agree with Glidewell LJ that for one building to be "occupied together with" another for the purposes of this Act they must be in the same occupation and the activities carried on in both must be jointly controlled or managed. I also consider that the buildings must be so occupied and the activities so controlled and managed at the E same time. These are necessary conditions to be satisfied, but to satisfy each of them separately or together is not sufficient to establish that one building is "occupied together with" another for rating purposes. Nor is there any geographical test which gives a conclusive answer — though the distance between the buildings is a relevant consideration, as the Court of Appeal held.

It is not, however, sufficient to ask generally whether the buildings F or buildings and land in question are all part of the same business enterprise. What it is necessary to show is that the two buildings or, as the case may be, the buildings and agricultural land, are occupied together so as to form in a real sense a single agricultural unit. Contiguity or propinquity may go far to show that they are. Thus farm buildings surrounded by land which is farmed with other land nearby though not contiguous or even land in another neighbouring village may well as a matter of fact be found to be "occupied together with" each other. On the other hand, separation may indicate that they are not and the greater the distance the less likely they are to be one agricultural unit.

In view of the extension in the 1971 Act to derate further hereditaments, it is not right now to ask whether the two premises G constituted one "farm" in the ordinary sense, but Viscount Dilhorne in the passage quoted above, in my view, indicates in the right direction. Though I consider that the actual decision in the case is to be treated as one on its special facts and the correctness of which may in any event be debatable, the sense of "togetherness" referred to by Sir Michael Rowe in *Hilleshog* perhaps equally shows that the important question is whether the two buildings or the buildings and land are worked together so as to form one agricultural unit.

The Rating and Valuation (Apportionment) Act 1928 (c44), which was the forerunner of the 1967 Act, gives a similar indication since it distinguishes "agricultural hereditaments" from other hereditaments and defines "agricultural buildings" as meaning H "buildings (other than dwelling-houses) occupied together with agricultural land or being or forming part of a market garden, and in either case used solely in connection with agricultural operations thereon". That, again, gives the impression of an agricultural unit.

In the present case there are 48 farms with their broiler houses and each broiler house must be surrounded by at least 2 ha of land to qualify. They are kept separate and distinct, in part, in order to prevent or reduce the spread of disease. Yet it is an inescapable finding that they are separate and distinct farms and are to be treated as such for rating purposes since it has not been suggested that any two or more of the broiler houses are, in reality, run as a single unit. It seems to me that it is quite impossible on the findings of the Lands Tribunal to say that each farm or broiler house is occupied together J with all the other broiler houses as one unit or that the factory is occupied as one unit together with all the farms, some of which are 100 miles away.

Applying the test as to whether the several buildings are worked together as one agricultural unit, and having regard to their physical separation, as part of this test, it seems to me that the Lands Tribunal could not possibly conclude that the 48 farms are "occupied together with" the factory for the purposes of the Act.

I take the view, therefore, that the appeal should be dismissed, that there is no point in referring the matter back to the Lands Tribunal and that it be ordered that the appellant's processing factory and premises at Dalton, Thirsk, North Yorkshire, should be reinstated in K the valuation list, it being agreed that the appropriate rating valuation is £20,150.

Lords Templeman, Oliver of Aylmerton, Geoff of Chievely and **Mustill** agreed and did not add anything.

REAL PROPERTY AND CONVEYANCING

A
Chancery Division

November 22 1991
(Before Mr Justice FERRIS)

SURREY COUNTY COUNCIL AND ANOTHER v BREDERO HOMES LTD

Estates Gazette January 23 1993
[1993] 03 EG 112

Development agreement — Breach of covenant to develop in B *accordance with agreement — Whether defendant developer liable for breach of covenant — If defendant liable whether plaintiffs entitled to nominal damages only or to substantial damages equivalent to such sum as might have been obtained had modification of the covenants been agreed to*

In 1980 the plaintiffs, Surrey County Council and Mole Valley District Council, were respectively the registered proprietors of two adjoining parcels of land at the Ridgeway, Fetcham, Leatherhead, Surrey, the total area of which was 12.33 acres. Following the issue of a development brief inviting prospective developers to make offers for the site, a contract was entered into dated November 26 1980 between C the plaintiffs, as vendors, and the defendant, Bredero Homes Ltd, as purchaser. The contract was completed by two transfers dated January 2 1981 each containing covenants in identical terms by the defendant with the respective councils "to commence the development of the land hereby transferred in accordance with the planning permission issued by Mole Valley District Council reference number MO/80/1214 and dated 11th December 1980 within six months from the date hereof and thereafter to diligently pursue the development of the land hereby transferred to its completion complying with the said planning permission and D the scheme approved by the transferor and the County Council of Surrey for the development of the land hereby transferred and adjoining land acquired . . . by a transfer of even date herewith". Planning permission MO/80/1214 provided for the development of the site with 72 detached bungalows and houses in six different designs. Following an application by the defendant, planning permission was granted on June 29 1983 under reference MO/83/0368. The effect of this new permission was to raise the total number of dwellings to be built on the site from 72 to 77. In due course the defendant completed the development of the last 3.64 acres of E the site in accordance with the second planning permission and not in accordance with the original permission. The plaintiffs contended that there had been a breach of the respective covenants in the two transfers and that the defendant had carried out a more advantageous development; although the plaintiffs neither desired nor sought any injunctive relief to compel the defendant to develop the estate in accordance with the covenants, they sought damages equal to the payment that might have been extracted from the defendant in return for agreed modifications of its covenants in such a way as to permit the development which has in fact been carried out. The defendant admitted the breach of F covenant and contended that the modified development did not make the development more valuable to it, but if that were wrong it denied that the plaintiffs were entitled to recover anything more than nominal damages.

Held: **At common law the plaintiffs are entitled to recover *only* nominal damages. Contrary to the general principle on which damages for breach of contract are to be assessed, what the plaintiffs were seeking to do was not to restore themselves to the position in which they would have been if the covenants had been performed but were seeking to obtain a share of the profit which the defendant was said to have made by reason of G the covenants not being performed. This was contrary to the general principle that, in assessing damages for breach of contract, the court is concerned with the plaintiffs' loss and not with the defendant's profit. The decisions in *Wrotham Park Estate Co Ltd* v *Parkside Homes Ltd* and *Johnson* v *Agnew* did not prescribe a new rule for the assessment of damages at common law which is to be used, in appropriate cases, as an alternative to the rule stated in *Robinson v Harman*. The decision in *Wrotham Park* was a decision on the award of damages under Lord Cairns' Act and nothing else and had not been transmitted into a decision on common law damages by virtue of the observations of Lord Wilberforce in H *Johnson* v *Agnew*.**

Substantial damages could not be awarded in equity for the breach of the covenants. The defendant had disposed of all the houses on the site before the writ in the action was issued and could not therefore have complied with any mandatory injunction requiring the site to be redeveloped in accordance with the original planning permission. The injunctive relief sought by the plaintiffs upon amendment at the invitation of the court was not such an order as the court would ever be prepared to make. There never was a case for an injunction or for specific performance at the date of the writ and it J inevitably followed that there was no jurisdiction to award damages in equity in lieu of the equitable relief.

The following cases are referred to in this report.

Bain v *Fothergill* (1874) LR 7 HL 158, HL
Bracewell v *Appleby* [1975] Ch 408; [1975] 2 WLR 282; [1975] 1 All ER 993; [1976] EGD 190; (1974) 237 EG 731
Carr-Saunders v *Dick McNeil Associates Ltd* [1986] 1 WLR 922; [1986] 2 All ER 888; (1986) 53 P&CR 14; [1986] 2 EGLR 181; 279 EG 1359
Ferguson v *Wilson* (1866) 2 Ch App 77
Hipgrave v *Case* (1885) 28 ChD 356
Johnson v *Agnew* [1980] AC 367; [1979] 2 WLR 487; [1979] 1 All ER 883; (1979) 38 P&CR 424; [1979] EGD 969; 251 EG 1167, HL K
Penarth Dock Engineering Co Properties Ltd v *Pounds* [1963] 1 Lloyd's Rep 359
Proctor v *Bayley* (1889) 42 ChD 390
Radford v *De Froberville* [1977] 1 WLR 1262; [1978] 1 All ER 33; (1977) 35 P&CR 316
Robinson v *Harman* (1848) 1 Exch 850, [1843-60] All ER Rep 383, 18 LJ Ex 202, 13 LTOS 141
Sotiros Shipping Inc v *Sameiet Solholt ("The Solholt")* [1981] 2 Lloyd's Rep 574; [1953] 1 LLoyd's Rep 605, CA
Strand Electric and Engineering Co Ltd v *Brisford Entertainments Ltd* [1952] 2 QB 246; [1952] 1 All ER 796, CA
Stoke-on-Trent City Council v *W&J Wass Ltd* [1988] 1 WLR 1406; [1988] 3 All ER 394; (1988) 87 LGR 129, CA

Tito v *Waddell (No 2)* [1977] Ch 106; [1977] 2 WLR 496; [1977] 3 All ER 129
Watson, Laidlaw & Co Ltd v *Pott, Cassels & Williamson* 1914 SC HL 18; [1914] 31 RPC 104
Whitwham v *Westminster Brymbo Coal & Coke Co* [1896] 2 Ch 538
Wrotham Park Estate Co Ltd v *Parkside Homes Ltd* [1974] 1 WLR 798; [1974] 2 All ER 321; (1973) 27 P&CR 296

This was a trial of two issues as to whether the defendant, Bredero Homes Ltd, was liable to the plaintiffs, Surrey County Council and Mole Valley District Council, for breach of covenant and if so liable as to whether the plaintiffs were entitled to substantial rather than nominal damages arising out of the breach of covenant in two transfers dated January 2 1981.

Sir William Goodhart QC and Bernard Weatherill (instructed by Howell-Jones & Partners, of Kingston upon Thames) appeared for the plaintiffs; Colin Rimer QC and Nicholas Peacock (instructed by Turner Kenneth Brown) represented the defendant.

Giving judgment, **Ferris J** said: This action raises an interesting point on the measure of damages applicable to a breach of a covenant to develop land in a particular way.

The facts are fairly straightforward and are not in dispute. In 1980 Surrey County Council and Mole Valley District Council were respectively the registered proprietors with absolute title of two adjoining parcels of land lying to the west of a highway known as The Ridgeway, Fetcham, Leatherhead, Surrey. The total area of these two parcels was some 12.33 acres. The land had orig,inally been acquired by the councils, or their predecessors, for road purposes but by 1980 it was no longer required for these purposes and the two councils decided to act together in offering the entire site for development as a housing estate. To this end a development brief was prepared inviting prospective developers to make offers for the site. The brief stipulated that offers:

> shall take the form of a financial offer for the land linked with a submitted design for the layout and buildings. The successful offer will be the one reflecting, in the Councils view, the best balance between the financial offer and the submitted design.

The offer which found favour with the councils was an offer from the defendant, Bredero Homes Ltd ("Bredero"), the financial terms of which included a purchase price of £1.52m. This offer led to a contract dated November 26 1980 between the two councils, as vendors, and Bredero, as purchaser. This contract was completed by two transfers both dated January 2 1981, one being made by the district council and the other by the county council. In the transfer made by the district council, Bredero covenanted with the council:

> To commence the development of the land hereby transferred in accordance with the planning permission issued by Mole Valley District Council reference number MO/80/1214 and dated 11th December 1980 within six months from the date hereof and thereafter to diligently pursue the development of the land hereby transferred to its completion complying with the said planning permission and the scheme approved by the transferor and the County Council of Surrey for the development of the land hereby transferred and adjoining land acquired by the transferee from the said County Council by a transfer of even date herewith.

The transfer made by the county council contained a covenant by Bredero in identical terms, but with additional words expressing the covenant to be entered into in pursuance of section 5 of the Surrey County Council Act 1958. These additional words, taken with section 5, are said to make positive covenants enforceable against successors in title of Bredero, but nothing turns on this for present purposes.

The scheme referred to in the two covenants has not been identified in this action and it can be ignored. The planning permission MO/80/1214 was granted pursuant to the application of Bredero. It provided for the development of the land with 72 detached bungalows and houses in six different designs. The layout of the estate and the numbers and positions of the bungalows and houses of each design were specified in the application and in drawings referred to in the planning permission.

Bredero duly commenced the development of the estate in accordance with the planning permission referred to. In the course of the development Bredero applied for and obtained a number of further planning permissions which modified permission 1214 referred to in the covenant. Most of these permissions made insubstantial variations in permission 1214. It is not suggested that development in accordance with these insubstantial variations gave rise to any claim for damages for breach of covenant.

In December 1982, however, it came to the notice of the district council that Bredero was contemplating a further variation in the development which, if carried out, would be more advantageous to Bredero than the development provided for by permission 1214. On December 9 1982 the clerk and chief executive of the district council wrote to Bredero in the following terms:

> I understand from the Planning Officer that you are considering making an application for planning permission to develop the remainder of the site to a greater density than that permitted under planning permission MO/80/1214 dated 11th December, 1980. Whilst the Council's Planning Committee will, of course, consider any planning application on its merits, I would point out that the undertaking of a different development to that permitted under the 1980 planning permission is at variance with the covenant contained in clause 2 of the transfer of the land to your Company.
>
> If planning permission is granted, I assume you will be making formal application to the Council's Lands Officer who would be pleased to discuss the variation of the covenant.

No reply to that letter appears to have been received. Bredero did, however, apply for a fresh planning permission in respect of 3.64 acres of the estate. The remainder of the estate had, it seems, by this time been developed in accordance with permission 1214, as varied by the immaterial modifications, which I have mentioned. This fresh planning permission was granted on June 29 1983 under reference MO/83/0368. The effect of this new permission was to raise the total number of dwellings to be built on the estate from 72 to 77, the main changes being a reduction in the number of four-bedroom houses and an increase in the number of three-bedroom houses, the size and arrangement of plots being amended, but the layout of roads and verges remaining unchanged. Subsequently Bredero completed the development of the last 3.64 acres of the estate in accordance with the fresh planning permission 0368 and not in accordance with planning permission 1214.

In subsequent correspondence and negotiations the district council acted on behalf of themself and the county council. The two councils took the view that there had been a breach of the covenants on the part of Bredero contained in the two transfers. Neither council desired or sought any injunctive relief to compel Bredero to develop the estate in accordance with its covenants. Neither council claims to have suffered any damage to any adjoining land owned by them. They both took the view that in developing the last 3.64 acres in accordance with planning permission 0368, Bredero had carried out, in breach of covenant, a more advantageous development than that which was permitted by planning permission 1214 and the covenants contained in the transfers. They have sought, and now seek, damages equal to the payment that might have been extracted from Bredero in return for agreed modifications of its covenants in such a way as to permit the development which has in fact been carried out. The councils do not allege that they have suffered any other loss.

As appears from the events as I have described them, Bredero never sought an agreed modification of its covenants. It accepts that it is in breach of covenant. It denies that the modifications in the development provided for by permission 0368 made the development more valuable to it and it says that it would not have been prepared to pay for a modification of its covenants. But, even if it were wrong in these matters, it denies that the councils are entitled to recover anything more than nominal damages.

The parties having taken up their stand on this last issue as a matter of principle, the two councils commenced this action by writ issued on March 25 1988. Bredero is the only defendant to the action. It is accepted that by the date of the writ Bredero had disposed of all the houses on the estate to third parties. None of these third parties has been joined as a defendant. The claim of the two councils against Bredero as set out in the statement of claim endorsed on the writ was exclusively for damages and for interest pursuant to section 35A of the Supreme Court Act 1987.

By an order of Master Munrow made on December 18 1990, it was ordered that the following issues be tried at the trial of the action before any question or issue as to the amount of any damages recoverable by the two councils is tried, namely:

> (a) whether the Defendant is liable to the Plaintiffs for breach of covenant as alleged in paragraph 8 of the Statement of Claim;
>
> (b) whether if the Defendant is so liable to the Plaintiffs, the Plaintiffs are entitled to damages assessed on the basis pleaded in the second sentence of paragraph 9 of the Statement of Claim;

It is these issues which are before me for trial. Having regard to Bredero's admission that it is in breach of covenant, the question which I have to consider is whether the two councils are entitled (a) to nominal damages only, which is what Bredero contends; or (b) to

A substantial damages equivalent to the sum that the two councils could have obtained if Bredero had sought a modification of its covenants entitling it to develop in accordance with planning consent 0368 instead of in accordance with planning consent 1214. If I decide the matter in favour of substantial damages the actual quantification of those damages will be referred to an inquiry.

It was common ground between the parties that the usual principle on which damages for breach of contract are to be assessed is that stated in these terms by Parke B in *Robinson* v *Harman* (1848) 1 Exch 850 at p855:

> The rule of the common law is, that where a party sustains a loss by reason of a breach of contract, he is, so far as money can do it, to be placed in the same situation, with respect to damages, as if the contract had been performed.

B This rule has been repeatedly approved and applied by the House of Lords and by lower courts. Mr Colin Rimer QC, on behalf, of Bredero took me to a number of cases where this had been done.

Sir William Goodhart QC, on behalf of the two councils, did not dispute that this rule embodies the normal basis on which damages for breach of contract are to be assessed. He submitted, however, that the rule does not represent the only measure of such damages. As an example of a case where another measure has been adopted he cited *Radford* v *De Froberville* [1977] 1 WLR 1262, where the owner of one parcel of land whose neighbour was in breach of a covenant to build a wall on the boundary between the two plots recovered the cost of erecting the requisite wall on his own land and was not limited to C recovery of the diminution in the investment value of his property caused by the absence of a wall. Sir William also referred to the type of case discussed in *Chitty on Contracts*, 26th ed, vol 1, at paras 1802-1803, where in certain cases of breach of contract the injured party can recover the amount of his own expenditure incurred in performance of his own obligations in anticipation that the party in breach would perform his obligations too.

The authority, said to represent a departure from the general rule, on which Sir William primarily relied was *Wrotham Park Estate Co Ltd* v *Parkside Homes Ltd* [1974] 1 WLR 798. In that case a developer acquired land which was affected by a covenant not to develop the land for building purposes except in strict accordance with a layout plan to be first submitted to and approved in writing by the vendor or D its surveyor, such plan to indicate thereon the roads, sewers and drains to be constructed. The plaintiff was the successor in title of the original covenantee and the developer was the successor in title of the original covenantor. Although there was no privity of contract between them the covenant was enforceable by the plaintiff against the developer in equity. The developer developed part of the land with a housing estate without submitting a layout plan to the plaintiff. Subsequently the plaintiff commenced proceedings against the developer and the purchasers of the houses, who were also bound by the covenant, seeking, *inter alia*, a mandatory injunction for the demolition of the buildings erected in breach of covenant. On this claim the judge, Brightman J, concluded as follows:

> E I cannot close my eyes to the fact that the houses now exist. It would, in my opinion, be an unpardonable waste of much needed houses to direct that they now be pulled down and I have never had a moment's doubt during the hearing of this case that such an order ought to be refused. No damage of a financial nature has been done to the plaintiffs by the breach of the lay-out stipulation. The plaintiffs' use of the Wrotham Park Estate has not been and will not be impeded. It is totally unnecessary to demolish the houses in order to preserve the integrity of the restrictive covenants imposed on the rest of area 14. Without hesitation I decline to grant a mandatory injunction. But the fact that these houses will remain does not spell out a charter entitling others to despoil adjacent areas of land in breach of valid restrictions imposed by the conveyances. A developer who tries that course may be in for a rude awakening.

F Brightman J then turned to the question of damages. He said:

> I am able under the jurisdiction which originated with the Chancery Amendment Act 1858 (Lord Cairns' Act) to award damages in substitution for an injunction. But before I discuss quantum of damages I must answer the question, damages against whom? For I apprehend that I am not able to award damages against a defendant except as a substitute for an injunction that I could have granted against the same defendant. Parkside has parted with the ownership of the allotment site except the roads. I am able to grant a mandatory injunction to compel Parkside to remove the roads, at any rate once the houses have gone. I can, therefore, award damages against Parkside in substitution for that injunction. I am not able to award damages against Parkside as a substitute for an injunction to remove the houses because I cannot order Parkside to demolish them. But I can award damages against each of the purchasers in substitution for any mandatory injunction that I could have granted against the same purchaser.

It is to be observed that the jurisdiction originally conferred by G Lord Cairns' Act is now conferred by section 50 of the Supreme Court Act 1981. At p 812 Brightman J came to consider the question of damages. He said:

> The basic rule in contract is to measure damages by that sum of money which will put the plaintiff in the same position as he would have been in if the contract had not been broken. From that basis the defendants argue that the damages are nil or purely nominal, because the value of the Wrotham Park Estate as the plaintiffs concede is not diminished by one farthing in consequence of the construction of a road and the erection of 14 houses on the allotment site. If, therefore, the defendants submit, I refuse an injunction I ought to award no damages in lieu. That would seem, on the face of it, a result of questionable fairness on the facts of this case. Had the offending development been the erection of an advertisement hoarding in defiance of H protest and writ, I apprehend (assuming my conclusions on other points to be correct) that the court would not have hesitated to grant a mandatory injunction for its removal. If, for social and economic reasons, the court does not see fit in the exercise of its discretion, to order demolition of the 14 houses, is it just that the plaintiffs should receive no compensation and that the defendants should be left in undisturbed possession of the fruits of their wrongdoing? Common sense would seem to demand a negative answer to this question.

He considered a number of cases, not involving breach of contract, where compensation for various forms of wrongful conduct had been assessed by reference to what would have been a reasonable price for a licence to do what had been done. He stated his conclusion at p 815 as follows: J

> In the present case I am faced with the problem what damages ought to be awarded to the plaintiffs in the place of mandatory injunctions which would have restored the plaintiffs' rights. If the plaintiffs are merely given a nominal sum, or no sum, in substitution for injunctions, it seems to me that justice will manifestly not have been done. As I have said, the general rule would be to measure damages by reference to that sum which would place the plaintiffs in the same position as if the covenant had not been broken. Parkside and the individual purchasers could have avoided breaking the covenant in two ways. One course would have been not to develop the allotment site. The other course would have been for Parkside to have sought from the plaintiffs a relaxation of the covenant. On the facts of this particular case the plaintiffs, rightly conscious of their obligations towards existing residents, would clearly not have granted any relaxation, but for present purposes I must assume that it could have been induced to do so. In my judgment a just substitute for a K mandatory injunction would be such a sum of money as might reasonably have been demanded by the plaintiffs from Parkside as a quid pro quo for relaxing the covenant.

He went on to quantify this sum by reference to the facts of the case before him. Sir William Goodhart submitted that the court should adopt a similar approach in this present case. He accepted that he had not sought any form of injunction. That was relief which the councils had never wanted and had never sought. Sir William said it should not make any difference to the court's jurisdiction that the plaintiffs were not seeking relief which they did not want and which, if they had asked for it, would have been likely to be refused on the same grounds as those upon which injunctive relief was refused in the *Wrotham Park* case. Sir William also made it clear that his claim was for L damages at common law, rather than in equity. The present action is between the original parties to the relevant covenants and the claim is a simple common law claim. In *Wrotham Park* the only remedy available to the plaintiff was equitable relief, whether by way of injunction or damages, because the covenant was enforceable against the defendants only in equity, not at common law. Sir William submitted that this circumstance explained why Brightman J expressed himself to be exercising the equitable jurisdiction to award damages in lieu of an injunction. He had no power to award damages at common law. In this case, however, the court does have that power. In this connection I was referred to *Johnson* v *Agnew* [1980] AC 367* where, at p 400, Lord Wilberforce, with whom the rest of their lordships agreed, dealt with the question of damages. He posed M two questions, the first of which was whether Lord Cairns' Act provided a different measure of damages from the common law. He answered this question as follows:

> Since the decision of this House, by majority, in *Leeds Industrial Co-operative Society Ltd* v *Slack* [1924] AC 851 it is clear that the jurisdiction to award damages in accordance with section 2 of Lord Cairns' Act (accepted by the House as surviving the repeal of the Act) may arise in some cases in which damages could not be recovered at common law: examples of this would be damages in lieu of a quia timet injunction and damages for breach of a restrictive covenant to which the plaintiff was not a party. To this extent the Act created a power to award damages which did not exist before at common

*Editor's note: Also reported at (1979) 251 EG 1167.

> A law. But apart from these and similar cases where damages could not be claimed at all at common law, there is sound authority for the proposition that the Act does not provide for the assessment of damages on any new basis. The wording of section 2 "may be assessed in such manner as the court shall direct" does not so suggest, but clearly refers only to procedure.

He then referred to certain authorities and concluded:

> On the balance of these authorities and also on principle, I find in the Act no warrant for the court awarding damages differently from common law damages.

In the light of what was said by Lord Wilberforce in these two passages Sir William submitted that the *Wrotham Park* case is now to be regarded as authority on the assessment of damages generally, not B merely on the assessment of damages in equity.

Mr Rimer, on behalf of Bredero, submitted that the damages sought by the two councils are not recoverable in English law. He based himself firmly upon the general rule, stated in *Robinson* v *Harman*, in terms which I have already read and repeatedly affirmed in courts of all levels, including the highest. This, he submitted, was the only true measure of the damages to be awarded for breach of contract in English law. If, in accordance with what Lord Wilberforce said in *Johnson* v *Agnew*, the measure of damages in equity is the same as that at common law in cases when the equity and common law provide a remedy in damages, then *Wrotham Park* may be wrongly decided. He submitted that it is not necessary for me to consider in the present case whether or not this is so because C *Wrotham Park* was a case where the only remedy was in equity and it was also a case where a claim for an injunction had been made and persisted in and where the court had accepted that it had jurisdiction to grant an injunction. None of these features is present in this case. If *Wrotham Park* is correct when applied to its own circumstances, it cannot be elevated into a general authority on the assessment of damages at common law by reliance upon what Lord Wilberforce said in *Johnson* v *Agnew*, where *Wrotham Park* was not cited.

Mr Rimer's submission was that, in accordance with the principle stated in *Robinson* v *Harman*, the position of the two councils has to be considered by reference to what it would have been if the covenant had been performed. Making this comparison the two councils have suffered no loss. No property of theirs has been diminished in value. D It cannot be said that Bredero's breach of covenant has destroyed the value of the covenants. If the two councils had sought equitable relief by way of injunction or specific performance at the appropriate time they would have been likely to obtain it. Mr Rimer submitted that the covenants are positive covenants and that specific performance rather than an injunction would have been the appropriate relief. I was not convinced by this, but it makes no difference in principle to his submissions. If and so far as the value of the covenants, which the council obtained, has been diminished, this has occurred because of the deliberate decision of the councils not to seek specific performance or an injunction. It is not, itself, the result of Bredero's breach of covenant.

E What the councils are now seeking to do, said Mr Rimer, is not to restore themselves to the position in which they would have been if the covenants had been performed. They are seeking to obtain a share of the profit which Bredero is said to have made by reason of the covenants not being performed. This is contrary to the general principle that, in assessing damages for breach of contract, the court is concerned with the plaintiffs' loss and not with the defendants' profit, the latter being wholly irrelevant: see *Sotiros Shipping Inc* v *Sameiet Solholt:The Solholt* [1983] 1 Lloyd's Rep 605 at p 608.

In my judgment, Mr Rimer's submissions must be correct unless it can be said that the *Wrotham Park* case, taken in conjunction with *Johnson* v *Agnew*, has prescribed a new rule for the assessment of damages at common law which is to be used, in appropriate cases, as F an alternative to the rule stated in *Robinson* v *Harman*. It clearly cannot be regarded as a subrule, derived from *Robinson* v *Harman*, and Sir William Goodhart did not submit that it could.

In giving his decision in *Wrotham Park*, Brightman J drew an analogy with certain earlier authorities. The first was *Whitwham* v *Westminster Brymbo Coal & Coke Co*[1896] 2 Ch 538, where trespass in the form of unauthorised tipping of colliery waste was remedied by an award of damages equivalent to a licence fee rather than an amount equal to the diminution in the capital value of the land. The next case was *Watson Laidlaw & Co Ltd* v *Pott Cassels & Williamson* [1914] 31 RPC 104, when a party infringing the plaintiff's patent was ordered to pay a sum equivalent to a licence fee. A similar result was reached in *Strand Electric & Engineering Co Ltd* v *Brisford Entertainments Ltd* [1952] 2 QB 246, where a defendant G had detained and used for his own profit chattels owned by the plaintiff for the purpose of hiring them out.

In none of these cases was the plaintiff claiming damages for breach of contract. In *Penarth Dock Engineering Co Properties Ltd* v *Pounds* [1963] 1 Lloyd's Rep 359, where a similar principle was applied, it seems that there was a breach of contract, but there was also a trespass and the Court of Appeal applied the decision in *Whitwham*'s case. There have been a few cases since 1974 in which *Wrotham Park* has been applied, but of these *Bracewell* v *Appleby* [1975] Ch 408 was a trespass case and *Carr-Saunders* v *Dick McNeil Associates Ltd* [1986] 1 WLR 922* concerned a nuisance consisting of obstruction to the plaintiff's right of light. In none of these cases, therefore, did the measure of damage applied compete with such a H well-established principle as that stated in *Robinson* v *Harman* or, so far as I can see, conflict in any way with the usual principles for the assessment of damages in cases of the particular type in question. In the Canadian case of *Arbutus Park Estates Ltd* v *Fuller* (1976) 74 DLR, 3rd series, 257, *Wrotham Park* was applied in the assessment of damages for breach of a restrictive covenant, but that was a case in which a mandatory injunction was sought, and refused only on discretionary grounds and the award of damages was made under Lord Cairns' Act, which is apparently applicable in British Columbia. The *Arbutus Park* case appears, therefore, to be on all fours with *Wrotham Park*.

There are, however, two cases in which the *Wrotham Park* J decision has been discussed, which I must consider in greater detail. The first of these is *Tito* v *Waddell (No 2)* [1977] Ch 106. The material part of the judgment in that case is that in which Sir Robert Megarry V-C discussed the remedy of the plaintiffs for breach by the British Phosphate Commissioners of their covenant to replant with trees and shrubs land from which phosphate had been extracted. The primary contention of counsel for the plaintiffs was that the damages would be the cost of doing the work of replanting; his secondary contention was that there should be awarded by way of damages a sum equivalent to that which the commissioners would have paid in order to be relieved from their obligation to replant. It was accepted that this sum would inevitably be less than the cost of replanting. On the primary submission, Sir Robert Megarry discussed at some length the K consideration which he considered to be relevant in order to determine whether the damages should be the cost of replanting or the diminution in the market value of the plaintiffs' land caused by failure to replant. On the secondary submission he referred to *Wrotham Park* and *Bracewell* v *Appleby* and continued (at pp 335-6):

> I find great difficulty in seeing how these cases help [counsel for the plaintiffs]. If the plaintiff has the right to prevent some act being done without his consent, and the defendant does the act without seeking that consent, the plaintiff has suffered a loss in that the defendant has taken without paying for it something for which the plaintiff could have required payment, namely, the right to do the act. The court therefore makes the defendant pay what he ought to have paid the plaintiff, for that is what the plaintiff has lost. The L basis of computation is not, it will be observed, in any way directly related to wasted expenditure or other loss that the defendant is escaping by reason of an injunction being refused: it is the loss that the plaintiff has suffered by the defendant not having observed the obligation to obtain the plaintiff's consent. Where the obligation is contractual, that loss is the loss caused to the plaintiff by the breach of contract. In the present case, the loss caused to the plaintiffs by the British Phosphate Commissioners' failure to replant is the diminution in the value of their land resulting from that failure, or if it is established that the land would be replanted, the cost of replanting. In the latter case, no doubt, the British Phosphate Commissioners might well have been willing to pay something to be released from their obligation to replant, though that something would probably be rather less than the total estimated cost of replanting. But the point is that not unless the British Phosphate Commissioners would be liable to replant or pay damages equal to the cost of M replanting would there be any liability from which the British Phosphate Commissioners would seek release on the basis of paying a sum equal to the discounted cost of replanting. If Mr Macdonald establishes that liability, he does not need his less favourable secondary contention: if Mr Macdonald fails to establish that liability, there is no foundation on which to base his secondary contention. Of course, until it has been determined whether or not some burden exists, the person who would be subject to that burden may always be willing to pay something to be relieved of the risk: but I do not think that this can affect the measure of damages in the case which determines that the burden does exist. In any case, the two authorities in question seem to me to be a long way away from a case where the issue is not one of invading the property rights of another without consent, but of breach of a contract to replant his land.

*Editor's note: Also reported at [1986] 2 EGLR 181.

Sir William Goodhart naturally relied upon the first of these paragraphs as recognising and restating as a general principle what had been decided in *Wrotham Park* and *Bracewell* v *Appleby*. He attached importance to the words "If the plaintiff has the right to prevent some act being done without his consent" as embracing both the type of case where, as in *Wrotham Park*, there is an express reference to the obtaining of consent in the words creating the obligation, and a case such as *Bracewell* v *Appleby*, where the need for consent arises only from the fact that without consent the defendants' act constitutes a wrong.

What was said by Sir Robert Megarry in the paragraph relied upon was, I think, *obiter* because he proceeded to deal with the quantification of damages on the basis of choosing between the cost of replanting and the diminution in value of the plaintiff's land. Moreover, it seems to me that the two councils in this case face difficulties not unlike those which Sir Robert Megarry discussed in the second of the two paragraphs which I have read. If one excludes the possibility that the councils might, if they had acted in time, have obtained relief by way of specific performance or an injunction, Bredero would not, in practice, have been likely to pay a significant sum in return for a modification of its obligations unless it is liable to pay substantial damages for breach of these obligations. The argument then appears to depend upon the assumption of the very thing which it is intended to prove. I hope it is not unfair to suggest that the argument amounts to saying that the court ought to award damages equal to the cost of purchasing a release because if a release were not purchased the court would award these very damages.

The second authority commenting upon *Wrotham Park* which requires discussion is *Stoke-on-Trent City Council* v *W&J Wass Ltd* [1988] 1 WLR 1406. In that case, so far as material, the city council claimed damages for the infringement by the defendant of the council's right to hold a statutory market. The council could not establish that they had suffered any quantifiable loss in the form of loss of stallage or other dues, but, relying on the principle that the levying of an unlawful same-day market within $6\frac{2}{3}$ miles of a franchise or statutory market is actionable at the suit of the market owner without proof of loss, they claimed damages calculated by reference to the licence fee which they could reasonably have required the defendant to pay in return for consent to operate its market up to the time that continuance was restricted by injunction. The defendant contended that, at most, only nominal damages could be recovered by the city council.

Nourse LJ approached the matter on the basis that the levying of an unlawful market is a tort, probably a nuisance rather than a trespass, and that under the general rules applicable to actions in tort the best that the city council could recover was nominal damages. But he recognised that there were exceptions to the general rules, which he proceeded to examine "in order to see whether a further exception ought to be made in this case": see p 1410. He then discussed the authorities which had been referred to by Brightman J in *Wrotham Park* and also some other cases of the same kind. As to the *Wrotham Park* case itself he said:

> [That] decision is in my opinion one which stands very much on its own. The conclusion of Brightman J may, I think, be more fully explained as follows. An injunction is frequently granted to enforce an express negative covenant, especially a restrictive covenant affecting land, without proof of loss to the plaintiff. Injunctions could therefore and would have been granted in that case but for the social and economic reasons against ordering the demolition of 14 houses. If injunctions had been granted, the loss to the defendant purchasers would have been enormous. If, on the other hand, injunctions were not granted and no damages were awarded, the purchasers would have been left in undisturbed possession of the correspondingly enormous fruits of their wrongdoing. Accordingly, if the plaintiffs had not been awarded substantial damages, justice manifestly would not have been done. If this analysis is correct, the practical result of the *Wrotham Park* decision was something akin to an award of exemplary damages for breach of contract, albeit that their amount bore no relation to the loss which would have been suffered by the defendant purchasers if they had had to demolish their houses. In saying this, I do not wish to suggest that that case was wrongly decided. Indeed, I regard the result as having been entirely appropriate and I see no reason why it should not serve as a precedent for other cases of the same kind. I merely wish to emphasise that it stands a long way away from the present problem and does not assist in its solution.

Nourse LJ then dealt with the other exceptions and came to the conclusion that none of them justified the making of a further exception in relation to markets.

Nicholls LJ recognised as an established principle, which he called "the user principle", that a person who wrongfully uses another's property without causing the latter any pecuniary loss will, in general, be liable to pay, as damages, a reasonable sum for the wrongful use he has made of the other's property. He treated the authorities relied upon by Brightman J in *Wrotham Park*, and certain other authorities, as examples of the application of the user principle. He declined to extend the user principle to the disturbance of market rights. In his main discussion of these matters he made no reference to the *Wrotham Park* case itself, but he mentioned that case at the very end of his judgment at p 1420, saying that it took the matter no further so far as the *Stoke on Trent* case itself was concerned. Mann LJ agreed with the judgments of both Nourse LJ and Nicholls LJ.

In my judgment, what emerges from the *Stoke on Trent* case which is potentially material to the present case is that Nourse LJ explained the *Wrotham Park* case as being an award of damages in lieu of an injunction which could and would have been granted but for social and economic reasons, expressly disclaimed any suggestion that the decision was wrong and stated that he saw no reason why it should not serve as a precedent "for other cases of the same kind". The other two members of the court cannot, I think, be said to have taken any different view of *Wrotham Park*.

Having now discussed *Wrotham Park* and the cases which have followed it at some length I have to say that, in my judgment, nothing is to be found in any of them which justifies the view that *Wrotham Park*, taken with *Johnson* v *Agnew*, has prescribed a new rule for the assessment of damages at common law. In my view, *Wrotham Park* was a decision on the award of damages under Lord Cairns' Act and nothing else and has not been transmuted into a decision on common law damages by virtue of the observations of Lord Wilberforce in *Johnson* v *Agnew*.

Moreover, I do not accept Sir William Goodhart's submission that damages at common law for breach of covenant are in other types of cases assessed in accordance with principles which differ from the rule stated in *Robinson* v *Harman*. As I have mentioned, Sir William cited *Radford* v *De Froberville* as being such a case. But, in my judgment, the issue in that case was not whether *Robinson* v *Harman* or some other principle was applicable, but whether the principle of *Robinson* v *Harman* required that the plaintiffs' loss is always to be assessed by reference to the diminution in value of his own land or, in an appropriate case, permitted that loss to be assessed by reference to a cost which the plaintiffs had incurred or would undoubtedly incur in making good the defendants' default. It seems to me that in deciding in favour of the second of these alternatives Oliver J was applying the rule in *Robinson* v *Harman*, not departing from it.

The second type of case in which, according to Sir William, the court applies a rule different from that in *Robinson* v *Harman* is that in which the plaintiff recovers by way of damages wasted expenditure which he incurred in reliance upon his expectation that the defendant would perform his part of the bargain. This type of case is discussed in *Chitty on Contracts*, vol 1 at paras 1802 and 1803. But it does not seem to me that this is truly a departure from the ordinary rule. If the injured party can show that he has suffered a loss of bargain he will usually recover that loss and he cannot recover in addition to it the expenditure which he himself has laid out in order to obtain the bargain. But if for any reason he cannot establish a loss or a gross profit on the transaction, or if he is prevented from recovering that profit by the rule in *Bain* v *Fothergill* (1874) LR 7 HL 158 (now, of course, abolished), he will be entitled to recover the amount of his own expenditure incurred in order to obtain that profit because, in all the circumstances, that expenditure will be the amount required to put him as nearly as possible into the position in which he would have been if the contract had been performed. (I assume, of course, that the amount of expenditure will not be such that it exceeds the potential gross profit on the transaction. Where it does, a different calculation will need to be made, as discussed in *Chitty*.)

The conclusion which I reach, therefore, on the claim for damages at common law advanced on behalf of the plaintiff is that substantial, as distinct from nominal, damages are not recoverable by the plaintiffs at common law.

During the course of argument there was debate as to whether the result might be different if damages could be awarded in equity instead of at common law. This led to discussion whether, on the pleadings as they stood at the beginning of the trial, a claim to damages in equity was open to the plaintiffs. Thereafter Sir William Goodhart applied for leave to amend the plaintiffs' prayer for relief by adding a claim for

A (1) An Order that the Defendant do use its best endeavours to take down and remove or to procure the taking down and removal from the Fetcham Gap South site in Fetcham Surrey of all or any buildings and boundary features erected thereat and which do not comply with the scheme of development permitted by Planning Permission MO/80/1214 and do as soon as practicable thereafter take such steps as shall be required in order fully to implement the said scheme.

and by amending the simple claim for "damages" so that it became a claim for "damages at common law or alternatively in equity".

In fairness to Sir William it is right to say that he made this application with a marked lack of enthusiasm and only in response to encouragement from the bench. Indeed, he even volunteered that, if leave to amend were granted, the plaintiffs would undertake not to
B pursue the claim for mandatory relief embodied in the new prayer. In effect, therefore, the claim for injunctive relief was advanced merely in case it was necessary to advance such a claim in order to obtain damages under Lord Cairns' Act, now section 50 of the Supreme Court Act 1981.

Mr Rimer resisted this amendment on the ground that the new claim was doomed to failure and that leave ought not to be given to make a futile amendment. He did not suggest that the amendment would give rise to any embarrassment to his clients or necessitate an adjournment or substantially prolong the trial. In the somewhat unusual circumstances I decided to give leave for the amendment to be made without expressing any view at that stage on Mr Rimer's contention that the new claim was doomed to failure. I did this
C because it seemed to me to be more convenient and satisfactory to deal with the claim for damages in equity in my substantive judgment. Accordingly, I now turn to this claim.

Section 50 of the Supreme Court Act 1981 provides that "where . . . the High Court has jurisdiction to entertain an application for an injunction or specific performance" it may award damages in addition to, or in substitution for, an injunction or specific performance. As a note in vol 2 of the Supreme Court Practice 1991 states, this section has substantially re-enacted section 2 of Lord Cairns' Act. But in 1866, not very long after that Act was passed, it was held by a Court of Appeal, of which Lord Cairns was himself a member, that the Act was not applicable if the plaintiff in a suit in
D equity had no equitable right at the time the proceedings were commenced: see *Ferguson* v *Wilson* (1866) 2 Ch App 77. Accordingly, where specific performance has become impossible because the land which is the subject matter of the contract has been sold to a third party, damages in lieu of specific performance are not available in equity either: see *Hipgrave* v *Case* (1885) 28 ChD 356. Correspondingly where there is no case for an injunction a court of equity cannot award damages in substitution for an injunction: see *Proctor* v *Bayley* (1889) 42 ChD 390.

In the present case, as I have already said, Bredero had disposed of all the houses on the estate before the writ in this action was issued. Bredero could not, therefore, comply with a mandatory injunction requiring the estate to be redeveloped in accordance with planning
E permission 1214 even if the court would otherwise have been prepared to grant an injunction. It was for this reason that Sir William Goodhart formulated the injunctive part of his amended relief in terms of a requirement that Bredero should use its best endeavours to procure a particular result. The only means whereby Bredero could in fact achieve this result would be by repurchasing demolishing and reconstructing the houses developed in accordance with planning permission 0368, which would depend upon such unlikely or uncertain factors as the concurrence of the present numbers of the houses and the financial resources of Bredero. In my judgment, the injunctive order sought as part of the amended relief is not such an order as the court would ever be prepared to make. In truth, therefore, there never was a case for an injunction or for
F specific performance at the date of the writ or probably for a substantial period of time before that. In fairness to Sir William Goodhart I record that his reluctance to make the amendment was at least in part attributable to his recognition that this is so.

In my judgment, it inevitably follows that there is no jurisdiction to award damages in equity. This is the basis on which Mr Rimer argued that the amended claim was futile because it was doomed to failure. I have to agree that this is the case.

In the result I hold that the plaintiffs cannot recover substantial, as distinct from nominal, damages either at common law or in equity. I reach this conclusion with a measure of regret because, like Brightman J in the *Wrotham Park* case, I find it unsatisfactory that, assuming that the defendant has profited from its breach of covenant, the plaintiffs will receive no compensation and the G
defendant will be left in undisturbed possession of the fruits of its wrongdoing. Had relief by way of injunction still been available, as it was in the *Wrotham Park* case, I am practically certain that I would have refused injunctive relief. But I am equally certain that I would have followed the course taken in *Wrotham Park* and directed an inquiry as to damages on the basis that they should be the amount which the two councils might reasonably have charged for a variation of the covenants. I would have left it to a higher court to consider, if Bredero had so contended, whether *Wrotham Park* was rightly decided. As it is, however, there is a vital distinction between this case and the *Wrotham Park* case and the course which was taken there is not open to me.

I invite submissions from counsel as to the precise answers which I H
ought to give in relation to the issues described in the master's order which is referred to earlier in this judgment and as to the disposal of the action as a whole.

Court of Appeal

December 18 1992

(Before Lord Justice DILLON, Lord Justice FARQUHARSON J
and Lord Justice Simon BROWN)

Chancery Divison

October 29 1992

(Before Sir Donald NICHOLLS V-C)

NORWICH UNION LIFE INSURANCE SOCIETY v P&O PROPERTY HOLDINGS LTD AND OTHERS

Estates Gazette April 3 1993

[1993] 13 EG 108 K

Expert — Dispute arising under a funding agreement referred to expert for determination — Whether injunction can be granted to restrain expert from proceeding with determination of dispute pending the decision of the court on question of construction

The plaintiff was providing funds for the development of a substantial shopping centre in Sutton, being carried out by the first to third defendants. The development was being controlled by a building contract and funding agreement. The building contract specified April 12 1992 as the date for completion. The architect issued a certified certifying May 19 L
1992 as the date for practical completion. A date of considerable importance to the parties is the date defined in the funding agreement as "the completion date". The plaintiff contended: (1) that the expression "completed" in the funding agreement means a higher standard than "practical completion" under the building contract, because there was outstanding at practical completion asubstantial snagging list; and (2) some of the work, contrary to the funding agreement, does not conform to design drawings. In accordance with the funding agreement the fourth defendant was nominated as an expert to determine, *inter alia*, these issues. In the Chancery Division the Vice-Chancellor M
dismissed the plaintiff's application for an interlocutory restraining the fourth defendant from proceeding pending the determination by the court of these issues. The plaintiff appealed.

***Held*: *Before the Vice-Chancellor*. The application for an interlocutory application was dismissed. The Vice-Chancellor held that as a matter of principle, the issue on which a determination by the court is being sought is either within the matters remitted to the expert for his decision or it is not; if it is, then the court will not intervene either before or after, unless there has been fraud or collusion.**

Court of Appeal. **The plaintiff's appeal was dismissed. Whether the development has been completed must essentially depend on the facts and not on any clear-cut issue of law. The nominated arbiter will have to consider in what respects it is said that the development has not been completed. That, and the identification of the design documents, are matters which are pre-eminently for the decision of the nominated arbiter and not for the court to consider in anticipation of the arbiter reaching his decision. The court does not have a discretion whether or not to grant a declaration in an advance of an expert's determination. The function of the expert is to make the decision and that is not the function of the court where the decision has been entrusted to the expert.**

The following cases are referred to in this report.

Campbell v *Edwards* [1976] 1 WLR 403; [1976] 1 All ER 785; [1976] 1 Lloyd's Rep 522, CA
Chelsea Man plc v *Vivat Holdings plc* unreported, August 24 1989
Cozens v *Brutus* [1973] AC 854; [1972] 3 WLR 521; [1972] 2 All ER 1297; (1972) Cr AppR 799, HL
Jones v *Sherwood Computer Services plc* [1992] 1 WLR 277; [1992] 2 All ER 1780, CA
Pontsarn Investments Ltd v *Kansalli-Osake-Pankki* [1992] 1 EGLR 148; [1992] 22 EG 103
PosTel Properties Ltd v *Greenwell* [1992] 2 EGLR 130; [1992] 47 EG 106
Royal Trust International Ltd v *Nordbanken* Unreported, October 13 1989

This was an appeal by the plaintiff, Norwich Union Life Insurance Society, from a decision of the Vice-Chancellor dismissing its application for an interlocutory injunction to restrain the fourth defendant, John H Reilly, a nominated expert, from proceeding with his determination of certain questions of construction of a funding agreement with the first to third defendants, P&O Property Holdings Ltd, Cannon Assurance Ltd and Maybrook Developments Ltd.

Hazel Williamson QC and Christopher Pymount (instructed by the solicitor to Norwich Union Life Insurance Society) appeared for the appellant; Michael Barnes QC and Marcus Tavener (instructed by Speechly Bircham) represented the respondents.

Chancery Division

Giving judgment, **Sir Donald Nicholls V-C** said: This is an application for an interlocutory injunction to restrain a nominated arbiter, as he is described, from proceeding with his determination of a dispute referred to him under a funding agreement pending the decision by the court of a question of interpretation of that agreement and of certain other questions.

The agreement relates to a substantial new shopping-centre development in the London Borough of Sutton. The developers are a consortium, which comprises the first three defendants. The development is being funded by the plaintiff society, the Norwich Union Life Insurance Society.

As one would expect, the building contract and the funding agreement are elaborate documents. The overall shape of the arrangements was that the developers would carry out and complete the development, a long headlease to be granted by the local authority would be vested in the society, the society would pay the developers the costs of the development up to a specified maximum of £108m or thereabouts, the developers would bear any cost of the development above the specified maximum and the society would pay the developers a sum calculated as a multiplier of the rents achieved from letting the component parts of the development above a certain sum.

Work began in August 1989. The building contract specified April 12 1992 as the date for completion of all the building works. On May 19 the architects issued certificates of practical completion for the 11th and last section of the works and, accordingly, for the entirety of the building works. The centre was opened to the public two days later, on May 21. Seventy per cent of the centre was let and occupied, but many of the shop units were not. Because of the current economic and letting circumstances, both the society and the developers expect to suffer substantial financial loss.

The financial provisions in the funding agreement are complex. All I need say is that a date of considerable financial importance, in the circumstances as they have turned out, is the date defined in the funding agreement as "the completion date". It is in the financial interests of the society that this date should be later rather than sooner. Conversely, it is in the developers' financial interests that the date should be May 19 which, as I have said, was the date of the certificate of practical completion of the building works.

Clause 6 of the funding agreement is concerned with how the completion date is to be ascertained. The procedure is this. First, when the development has been practically completed in accordance with the building contract the architects are to issue a certificate of practical completion. Before that certificate is issued the society and its consultants have to be given prior notice by the architects of their intention to inspect the development in anticipation of certifying the date of practical completion. The date of practical completion is defined, unexceptionally, as "the date specified in the certificate the architect shall issue to the effect that the development has been practically completed in accordance with the building contract subject, nevertheless, to any list of outstanding works". The society's architect is then to be permitted to accompany the architects on the inspection, and the society's architect may make representations to the architects concerning the intended issue of the practical completion certificate and its contents and any list of outstanding works annexed to it. After the issue of the certificate the society's architect is not entitled to raise matters on which he could reasonably be expected to have made representations before the issue of the certificate.

Clause 6(8) provides for the consequence if the society's architect makes no representations. So far as material, it provides that in that event "the certificate of practical completion shall be deemed to have been accepted without qualification and, in the case of the last such certificate, the date specified in such certificate shall be deemed to have been accepted as the completion date".

Clause 6(9) deals with the situation which has arisen here, where the society's architect did make representations and, indeed, the representations covered very many matters, before the practical completion certificate was issued. Subclause (9) reads:

> If the Society's architect represents against the issue of a certificate of practical completion, and the effect is that there is a dispute as to whether the date of practical completion can also properly be regarded as the completion date, the architects and the Society's architect shall endeavour to reach agreement concerning the completion date, but if they fail to do so by 20 working days after the date of practical completion, the matter shall be referred for determination to the nominated arbiter or, if he is unable to act, to an independent and appropriate person acting as an expert, who shall be nominated jointly or, on failure to agree, at the instance of either party by the President of the Royal Institution of Chartered Surveyors, and the completion date shall be that so agreed or that so determined.

The completion date is defined as:

> The date established, pursuant to clause 6, as being the date on which the development has been completed in accordance with the design documents.

The nominated arbiter is defined as:

> A person agreed between the Society and the developers as an independent and fit person to act under clause 6 to determine a dispute concerning the completion date, which person shall be agreed upon by a date six months prior to the anticipated completion date or, failing such agreement, by such date shall be appointed under the provisions of clause 17.

There are essentially two matters on which, by the originating summons, the society seeks a ruling from the court before the nominated arbiter proceeds further. First, as to the meaning of "completed" in the definition I have read of the completion date. The society contends that this is not the same standard as that applied in determining whether the development has been practically completed. A substantial snagging list was scheduled to the final certificate of practical completion. The society's case is that "completed" in relation to the completion date envisages a higher standard than practical completion. It envisages completion of all the snagging items. The nominated arbiter needs to know which standard he is to apply when determining whether in fact the development has reached the requisite standard.

The second matter raised by the society's originating summons concerns identification of the "design documents". The "design documents" are defined as certain specified drawings and other documents, or any approved variations thereto and any other document intended to define or specify the development or the method or timing of its completion agreed between the society and the developers. Clause 5 of the funding agreement provides for variations to the design drawings to be sent to the society's consultants for consideration and that the developers shall not implement any proposed variation to which the society's consultants shall object in writing within 10 days.

A The society's case is that not only is much of the work not completed. In addition, the work does not conform to the design drawings. In this connection it is said that one of the issues which arises is whether certain documents which were seen by the society's consultants have become design drawings. The society's consultants, it is said, saw some documents but did not object thereto because they did not realise, and they were not told, that the documents were being submitted to them for approva under clause 5. Further, the society does not at present know what documents are being relied on by the developers as constituting design drawings in respect of some of the works.

So the second issue raised by the originating summons, in essence, B concerns identification of the design drawings. Until this is settled, the nominated arbiter does not know what precisely is the development with whose completion he is concerned.

In short, the developers' case is that both these matters are within the remit of the nominated arbiter and there is no occasion or, indeed, jurisdiction for the court to intervene and give a ruling on these matters which the parties have, by their contract, agreed should be determined by the nominated arbiter. A nominated arbiter, Mr John Reilly, is in office and has been so since December 1991. As intended, since then he has been familiarising himself with the development and its progress. He is ready to proceed with his work, that is to say, to take the necessary steps leading in due course to his determination.

In my view, the key question on this application is whether the C society is entitled to have the court, rather than Mr Reilly, determine the matters raised by the originating summons. If yes, then an injunction will follow, subject always to questions of the so-called balance of convenience. If not, not. Whether the society is so entitled in turn depends on what, as a question of interpretation of the funding agreement, is the ambit of the matters entrusted by the parties to the nominated arbiter for determination by him. This accords with the approach enunciated by the Court of Appeal in the leading case of *Jones* v *Sherwood Computer Services plc* [1992] 1 WLR 277. Dillon LJ at p287 said this regarding the principle:

> The first step must be to see what the parties have agreed to remit to the expert, this being, as Lord Denning MR said in *Campbell* v *Edwards* [1976] 1 WLR 403, 407G, "a matter of contract".

D If, on the proper construction of the agreement, the question sought to be put to the court is an issue within the scope of the matters remitted to the nominated arbiter, then (by definition) the parties have chosen, for better or for worse, to have that question determined by a tribunal other than the court.

In answering that question of interpretation, I must apply the ordinary principles of construction. I must have regard to the language used and interpret that in the context of the funding agreement and its provisions as a whole. I must have in mind the commercial background to the agreement.

Miss Williamson submitted that the parties are not to be taken to have intended to submit to the nominated arbiter jurisdiction to E determine a question of the proper interpretation of the funding agreement such as that raised by the first question, as I have described it, in the originating summons. Authority to the expert to determine an issue of fact does not necessarily carry with it authority to determine, at any rate to the exclusion of the court, questions of interpretation which are prerequisite to determining the issue of fact. Here there are indicia that the parties did not intend the nominated arbiter to have such a wide authority. First, the parties cannot have intended that an architect, with the professional expertise of an architect, should be the tribunal to determine, to the exclusion of the court, what is the yardstick of completion which the nominated arbiter should apply. That is a question of law. Second, there are other provisions in the agreement which support this. There are other F provisions in the funding agreement which are geared to the development, or parts of it, being "completed" in accordance with the design drawings: see, for example, clause 21(e). A dispute over these provisions would be referred to arbitration under clause 17, and in any such arbitration the decision of the arbitrator is expressed to be "final and conclusive and binding on the parties and without appeal, save on questions of law". Further, as to identifying the design drawings, this may involve complex issues of fact and questions of law or questions of mixed fact and law, and there may be questions of estoppel. Here also a dispute over the identity of the design drawings could arise in a different context, such as clause 5(4), para (b).

I am unable to accept these submissions on the proper construction of this agreement. Under clause 6(9) the parties have, in short, agreed that a third party (the nominated arbiter) shall determine whether the G completion date has arrived. Parties to a contract such as this enter into a clause such as clause 6(9) with the object of obtaining a speedy and conclusive determination on the matter in dispute by the tribunal they have chosen. They are not readily to be taken to have intended that any necessary prerequisite to that determination, which raises a question of law, is to be outside the matter so remitted. On the contrary, they are unlikely to have intended that fine and nice distinctions were to be drawn between factual matters which fall within the expert's remit and questions of law or questions or mixed law and fact which do not.

In the present case I am unable to accept that the remit to Mr Reilly does not embrace the two matters raised in essence by the originating H summons. In order to decide whether the development has been completed in accordance with the design drawings, inevitably and obviously the nominated arbiter will have to identify what are the design drawings. He will also have to consider whether the works have reached the standard which the word "completed" would reasonably be understood as bearing in the context of the definition in this agreement. The parties must have intended that the nominated arbiter should determine these matters as well as looking at the physical state of the works.

Far from this being unlikely, it seems to me that identifying what are the design drawings in accordance with which the development was to be completed was typical of the questions the nominated arbiter would be expected to answer if a dispute arose on whether a J particular document was or was not a design document. Answering this question may give rise to issues of fact or mixed fact and law, but architects and surveyors not infrequently are faced with such issues in the course of references of this nature. Likewise as to the meaning of the word "completed". What the parties are to be taken to have intended by that word in clause 6(9) and in the definition of "the completion date" is, as a question of construction, a question of law as distinct from a question of fact. But questions of interpretation of documents frequently arise in the course of resolution of rent review disputes by surveyors or building contract disputes by architects. I am therefore not persuaded that the parties to the funding agreement should be taken to have excluded these questions from the scope of K the matters to be decided by the nominated arbiter if, as has happened, answering such questions becomes necessary in the course of determining what was the completion date.

Accordingly, since the matters raised by the summons have been remitted to the nominated arbiter, it is not for the court to intervene and take away from him the determination of these questions.

On this question of interpretation, each agreement must depend on its own terms, read in its own context. Comparing one case and one document with another gives, at best, very limited assistance. I mention only one case in this connection to which I was referred, namely *Re Davstone Estates Ltd's Leases* [1969] 2 Ch 378. The question which arose there concerned the remit of a landlord's surveyor when issuing a certificate as to the amount due by the tenant L by way of service charges, the sum claimed being in respect of expenses incurred in making good certain defects in the building. Ungoed-Thomas J held that the question whether those defects fell within the provisions of the relevant clause was not a matter for decision by the surveyor. That, it seems to me, was a very different case and, as the judge himself pointed out, that was a case where there was a reference to a surveyor appointed by one side only.

Miss Williamson sought to draw a distinction between the position before the expert makes his determination and the position afterwards. Afterwards, the court will not intervene, partly because it will often be impossible to know what is the basis on which the expert has made his award, and partly because, by proceeding with the hearing and arguing the issue in front of the expert, the parties M have impliedly clothed him with authority to determine that issue. By their conduct and whatever they might say, they have elected to have the matter determined by him. Beforehand, however, those considerations do not apply.

I do not think that a distinction between before and after drawn in this way is satisfactory. This analysis does not get to the heart of the matter. As a matter of principle, the issue on which a determination by the court is being sought is either within the matters remitted to the expert for his decision or it is not. If it is, then the court will not intervene either before or after, unless there has been fraud or collusion.

I was referred to the decision of the Court of Appeal in *Chelsea*

A *Man plc* v *Vivat Holdings plc*, given on August 24 1989 (unreported). Suffice it to say, certain of the observations made in the course of the judgments in that case must be read in the light of the fact that that decision ante-dated the leading case of *Jones* v *Sherwood*, which has set out clearly the modern approach to these questions.

I was also referred to the unreported decision of Mr Timothy Lloyd QC, sitting as a deputy judge of the Chancery Divison, in *PosTel Properties Ltd* v *Greenwell**, given on July 17 1992. There the judge had before him an originating summons raising questions on the proper interpretation of a rent review clause. Certain assumptions had to be made by the independent surveyor and a dispute had arisen over the meaning of the assumptions. The judge refused to strike out the summons and proceeded to determine those questions. As I read B that case, counsel for the respondents did not contend that the jurisdiction of the court to entertain those questions had been excluded by the provisions of the lease. In that case it seems, therefore, not to have been argued that on the proper interpretation of the lease the questions raised by the summons were within the remit of the surveyor and, therefore, not open to determination by the court. The matter seems to have proceeded on the basis that the court had a discretion. That being the basis on which the case proceeded, it does not assist in the present case, nor does it assist on any question of general principle.

Finally, reference was made to a dictum of Judge Paul Baker QC, sitting as a judge of the High Court, in *Pontsarn Investments Ltd* v C *Kansallis-Osake-Pankki* [1992] 1 EGLR 148†. At p152 the learned judge made an observation on which Miss Williamson placed some reliance. In that case the judge was concerned to address, and in his judgment he rejected, a submission that, if an expert went wrong in law, he exceeded his jurisdiction and his determination was a nullity. The judge rejected a suggested analogy with inferior courts and tribunals. He rejected a submission that the principles applied by the High Court in exercising control over inferior courts and tribunals should be applied to the determinations of experts, even though they operated in the field of private law. The judge gave several reasons for rejecting this submission. He gave, as one of his reasons:

> Second, if it is thought that some question of law requires resolution, either party can seek a declaration from the court.

D He stated that as a general proposition concerning a possible course available in the case of references to an expert. As I read the learned judge's judgment, that was a very general comment, and I do not understand it to have been directed at the point now under consideration.

Miss Williamson also submitted there was a distinction to be drawn between a case where the arbitrator or expert has jurisdiction to decide a matter and a case where the arbitrator or expert has been given *exclusive* jurisdiction to decide a matter. Only in the latter case will the court decline to intervene in advance of a determination. The difficulty I have with this submission is that in the normal way, when agreeing to refer a matter to a third party for his decision, parties E must intend that he and he alone should decide the matter. It is, I suppose, theoretically possible to have a case where the parties intend that the expert's jurisdiction shall cover certain matters, but so that this should not exclude the court's jurisdiction to decide some of those matters as well. That would be a very strange and unusual arrangement. Clear language would be required before a court could safely conclude that the parties intended the expert's jurisdiction to be of this non-exclusive character. That is not this case.

For these reasons my conclusion is that the questions raised by the originating summons are matters which fall to be determined by Mr Reilly. There is, therefore, no occasion for the court to grant an injunction as sought. I shall, therefore, dismiss this application.

I add two footnotes. First, the nominated arbiter is concerned to F decide only one question: has the development been completed in accordance with the design drawings and, if so, on what date? He will not decide other matters; for example, whether the developers are in breach of obligations regarding some of the reserved matters. His decision will be confined to the one question before him, unless the parties otherwise agree.

Second, Miss Williamson voiced some concern that the developers might wish to pursue a line that the development has been completed even though design documents are not produced relating to all the work done. She expressed anxiety that the developers might wish to say that the existence of work done outside the design drawings does G not preclude the development from being completed in accordance with the design drawings. Again, it seems to me this is an issue which falls within the remit of the nominated arbiter. If such an argument is advanced it will fall to be evaluated by Mr Reilly in the course of his determination whether the works of demolition and clearance and construction in accordance with the design drawings have been completed and, if so, on what date they were completed.

Motion dismissed with costs.

Court of Appeal

Giving first judgment, **Dillon LJ** said: This is an expedited appeal by the plaintiffs in the proceedings, the Norwich Union Life Insurance Society, against an order of Sir Donald Nicholls V-C made H on October 29 1992, whereby he dismissed an application by the plaintiffs for an interlocutory injunction. The injunction sought was an injunction which would restrain the fourth defendant in the proceedings, Mr John Reilly, as nominated arbiter under a written agreement dated July 27 1989, and made between the plaintiffs of the first part, the first, second and third defendants of the second part, and certain parties as guarantors of the third part, from making or purporting to make or proceeding to make any determination of any issue between the parties or of any other matter in any case arising out of that agreement until after the trial of the action or further order in the meantime.

The agreement of July 27 1989 is referred to as a funding J agreement. It was an agreement whereby the plaintiffs, Norwich Union, agreed to provide the first three defendants, who were developers, with the funds for carrying out a development of a modern shopping centre with multi-storey car parking and other facilities at St Nicholas Place in the London Borough of Sutton. The developers had an agreement for a lease of the site from Sutton Corporation and were going to have it constructed for them by contractors under a building agreement. This agreement was the funding agreement to provide the finance for the development.

The nominated arbiter was appointed pursuant to the funding agreement and the particular point that he had to decide was what was the completion date of the development for the purposes of the funding agreement. That arose under clause 6 of the funding K agreement to which I will come. But it is pertinent to note first certain provisions from the definitions.

First, the nominated arbiter is defined as meaning:

> A person agreed between the Society [ie the plaintiffs] and the developers [ie the first three defendants] as an independent and fit person to act under clause 6 to determine a dispute concerning the completion date, which person shall be agreed upon by a date six months prior to the anticipated completion date or, failing such agreement by such date, shall be appointed under the provisions of clause 17 hereof.

Clause 17 is the arbitration clause and it provided for disputes, other than questions of law, to be referred:

> . . . to an independent and fit person who [should] either be agreed upon L between the parties . . . or failing such agreement [should] be nominated on the application of either of the parties . . . by the President for the time being of The Royal Institution of Chartered Surveyors . . . [It was provided that] the decision of such person [should] be final conclusive and binding on the parties and without appeal (save on questions of law).

It was further provided, in relation to the nominated arbitrator in clause 6(9) of the funding agreement, that if the nominated arbiter was unable to act, the matter under that clause was to be referred "to an independent and appropriate person acting as an expert who [should] be nominated jointly or, on failure to agree, at the instance of either party by the President of the Royal Institution of Chartered Surveyors, and the completion date shall be that so agreed or that so determined".

M Since, as it were, an alternative nominated arbiter appointed under the final part of clause 6(9) is expressly stated to be acting as an expert, it is accepted that the original nominated arbiter, Mr Reilly, was intended to act as an expert and not as an arbitrator. That then is the procedure for the appointment of the nominated arbiter.

The completion date is defined in the definition clause as meaning:

> . . . the date established, pursuant to clause 6, as being the date on which the development has been completed in accordance with the design documents.

I have to come back to further definitions.

The scheme of clause 6 which is headed "Practical Completion" is as follows. First, under subclause (1) it is said that:

> Upon the Development being practically completed in accordance with the terms of the Building Contract the Developers shall procure the Architects

*Editor's note: Reported at [1992] 2 EGLR 130.
†Editor's note: Also reported at [1992] 22 EG 103.

A — and that is the developers' architects, not the Society's architects —

> shall issue a certificate of Practical Completion which shall not be unduly delayed.

In the present case the certificate of practical completion was issued on May 19 1992, but it had a substantial snagging list attached to it.

Subclause (2) of clause 6 then provides that:

> The Developers shall procure that the Architects give to the Society [the plaintiffs] and to the Society's Consultants [ie their architects] separate prior notices in writing not less than fourteen days before the date on which the Architects intend to carry out their inspection of the Development in
> B anticipation of certifying the Date of Practical Completion or Practical Completion of any part of the Development.

Here we are concerned with the final certificate.

> (3) The Developers shall procure that the Society's Architect is permitted to inspect the Development following such notice as is referred to in sub-clause (2) above and to accompany the Architects on their said inspection of the Development in anticipation of certification.
>
> (4) On such inspections the Society's Architect may make representations to the Architects concerning the intended issue of the Architects' said certificate and the contents hereof and any list of outstanding works to be annexed thereto (which representations shall have regard to the definition of the Completion Date hereinbefore appearing) and the Society shall procure that as soon as practicable following such inspections the Society's Architect shall deliver his representations to the Architects in writing.

C

Then there is provision for the developers to "supply the Society and the Society's Architect with a verified copy of the . . . certificate and any documents annexed thereto".

Under subclause (6):

> The Society's Architect [is not to be] entitled to make representations following the issue of the Architect's . . . Certificate as to matters in respect of which he could reasonably have been expected to make representations at or following his previous inspections but failed to do so.
>
> (8) Unless the Society's Architect makes representations against the issue of any Certificate of Practical Completion (or Practical Completion Certificate in respect of part of the Development) within 10 working days after receipt by him of a copy of such certificate it shall be deemed to have been accepted
> D without qualification and (in the case of the last such certificate) the date specified in such certificate shall be deemed to have been accepted as the Completion Date.

That is not what happened.

Then the crucial clause is subclause (9) of clause 6:

> If the Society's architect represents against the issue of a certificate of practical completion [or practical completion certificate in respect of part of the development] and the effect is that there is a dispute as to whether the Date of Practical Completion can also properly be regarded as the completion date, the Architects and the Society's Architect shall endeavour to reach agreement concerning the completion date, but if they fail so to do by 20 working days after the date of practical completion the matter shall be referred for determination to the nominated arbiter

E

— with the alternative if he was unable to act, which I have already mentioned.

There is thus a reference of the matter of the completion date to the nominated arbiter. The two elements in the definition of the completion date are the date on which the development has been completed in accordance with the design documents. I stress completed and design documents. In the present case the plaintiff society raises by the originating summons two issues. (1) What is the standard of completion to be applied by the fourth defendant as nominated arbiter in determining whether the development has been completed in accordance with the design documents? (2) What are, by virtue of the funding agreement and in the events which have
F happened, the actual design documents upon the basis of which the nominated arbiter should make his determination?

The plaintiffs claim to have those two issues decided by the court and sought by their motion the injunction which I have indicated, which would prevent the nominated arbiter from determining the issues until after the trial of the action or further order in the meantime. The Vice-Chancellor rejected the motion for interlocutory relief on the ground that the issues lay within the remit of the nominated arbiter and were therefore issues which it was for him to decide and not for the courts to decide.

I should treat them separately. The first concerns the word "completed" in the definition of the completion date. I find it helpful in that context to refer to a passage in the speech of Lord Reid in the case of *Cozens* v *Brutus* [1973] AC 854 at p861 between C and D, where he said: G

> The meaning of an ordinary word of the English language is not a question of law.

He then goes on to say:

> The proper construction of a statute is a question of law. If the context shows that a word is used in an unusual sense the court will determine in other words what that unusual sense is.

Then he says:

> It is for the tribunal which decides the case to consider, not as law but as fact, whether in the whole circumstances the words of the statute do or do not as a matter of ordinary usage of the English language cover or apply to the facts which have been proved. If it is alleged that the tribunal has reached a wrong H
> decision then there can be a question of law but only of a limited character. The question would normally be whether their decision was unreasonable in the sense that no tribunal acquainted with the ordinary use of language could reasonably reach that decision.

Then he points to the difficulties which would arise if there was a different approach. In the present case, whether the development has been completed must essentially depend on the facts and not on any clear-cut issue of law. The nominated arbiter will have to consider in what respects it is said that the development has not been completed and how significant they are and so forth. They are all matters which are rolled into the question that he has to consider, whether at the stage in which he is considering it the development has been completed. Those are pre-eminently, in my judgment, matters for the J
decision of the nominated arbiter and not for the court to consider in anticipation of the arbiter reaching his decision. It is to the arbiter and not to the court that the issue has been remitted.

Then there is the question of the design documents. That term is defined in the definition clause as meaning:

> (i) the specification and the drawings and other documents details of which appear in part one of the First Schedule hereto copies of which have been signed for the purpose of identification by the parties hereto or (to the exclusion of the document or plan prior to its variation) any variations of the same subsequently approved pursuant to the provisions of Clause 5 hereof and (ii) any other document intended to define or specify the Development or the method or timing of its completion subsequently agreed between the Society and the Developers for the purpose of the operation of this agreement.

K

Obviously there is no difficulty about the documents specified in part one of the first schedule. Ordinarily there would be no difficulty about any subsequent agreement.

But there is however the reference to approval pursuant to the provisions of clause 5, and I should refer to subclause (10) of clause 5 which is concerned with approvals. Clause 5 in this rather lengthy agreement sets out various covenants by the developers with the society.

> (10)(i) To procure to be carried out such variation expansion refinement and (if necessary) supplementation of the Design Documents as shall be necessary in either case to render them suitable and sufficient at the relevant time (i) for annexation to the Building Contract and (ii) for use in the course of the construction of the Development and (iii) to render them compliant at any L
> time with the terms or conditions of the Planning Permission and any other consents to be obtained pursuant to sub-clause (1) of . . . clause 5
>
> (ii) The Developers shall procure that every proposed amendment or addition to the Design Documents is referred to the Society's Consultants for consideration as soon as practicable after such amendment or addition has been issued by the Professional Consultant to the Developers for approval
>
> (iii) The Developers shall not implement any such proposed amendment or addition to which the Society or the Society's Consultants shall have expressed an objection in writing within 10 working days until such objection is dealt with to the reasonable satisfaction of the Society
>
> (iv) The Society will not unreasonably withhold its consent to the substitution of materials of equal quality and performance

I do not think I need read any further.

So the sort of questions that may arise in respect of the design M
documents are whether there has or has not been some document constituting a subsequent agreement between the parties or a variation which has been approved, as opposed to one which has not been approved, pursuant to the provisions of clause 5. Those again are questions dependent on the facts. It would be necessary to consider not any bare point of law but what the circumstances are which are said to amount to a subsequent agreement or are said to have led to approval having been granted under clause 5(10), or conversely to approval not having been sought or not having been granted under that clause.

In my judgment, those questions are entirely within the remit of the nominated arbiter since they are matters which he must necessarily

A decide as he proceeds to amass and marshall the design documents for the purposes of determining whether the development has been completed in accordance with the design documents.

We were referred to a number of authorities by Miss Williamson QC, for the appellant. One is the case of *Pontsarn Investments Ltd* v *Kansallis-Osake-Pankki* [1992] 1 EGLR 148, decided by Judge Paul Baker QC, sitting as a judge of the High Court. At p151F the judge said:

> What is remitted to the determination of the expert is the determination of the open market rent as defined in clause 1.2 of the second schedule. The definition involves the making of certain assumptions and disregarding certain effects. The task may involve interpreting the language used, for example the reconciliation between the direction to assume in fixing the rent
> B that the premises are fit for immediate occupation and use, and the direction to disregard any effect on that same rent of improvements carried out by the tenant after the commencement of the term but before the review date.
>
> Mr Oddy indeed saw as one of the matters referred to him the application of the assumption that the premises were to be regarded as fit for immediate occupation and use. He noted the different interpretations of the landlord and tenant, backed up by the opinions of counsel in comparable cases, and reached his determination accepting the arguments of the tenant. This resulted in no increase of the rent to take account of any notional fitting out at the expense of the landlord, or at all events not at the expense of the tenant, or any decrease of rent to take account of any rent-free period to allow for the time which the tenant might take to fit out the premises at his expense.
>
> The resolution of this question of interpretation is, in my judgment, one which the expert necessarily had to undertake along the way to determine the
> C open market rent as defined in the second schedule. Accordingly, as I see it, the question was within the remit of the expert and the answer of Mr Oddy is final and binding and not subject to review.

Precisely the same applies, in my judgment, to the determination what the design documents are in the nominated arbiter's progress to determining the completion date.

In his judgment in the court below the Vice-Chancellor [see p109] said:

> Under clause 6(9) the parties have, in short, agreed that a third party (the nominated arbiter) shall determine whether the completion date has arrived. Parties to a contract such as this enter into a clause such as clause 6(9) with the object of obtaining a speedy and conclusive determination on the matter in dispute by the tribunal they have chosen. They are not readily to be taken to
> D have intended that any necessary prerequisite to that determination, which raises a question of law, is to be outside the matter so remitted. On the contrary, they are unlikely to have intended that fine and nice distinctions were to be drawn between factual matters which fall within the expert's remit and questions of law or questions of mixed law and fact which do not.
>
> In the present case I am unable to accept that the remit to Mr Reilly does not embrace the two matters raised in essence by the originating summons. In order to decide whether the development has been completed in accordance with the design drawings, inevitably and obviously the nominated arbiter will have to identify what are the design drawings. He will also have to consider whether the works have reached the standard which the word "completed" would reasonably be understood as bearing in the context of the definition in this agreement. The parties must have intended that the nominated arbiter should determine these matters as well as looking at the physical state of the
> E works.

I respectfully and entirely agree with the Vice-Chancellor in what he there said for the reasons he has given.

We were referred also to a decision of Hoffmann J in a case of *Royal Trust International Ltd* v *Nordbanken* decided on October 13 1989, but unreported. At p6F of the transcript he said:

> I do not think it is right that the court has no jurisdiction to make declarations in advance of an expert's determination except with the consent of the parties.

He considered that the court had a discretion whether or not to grant such declarations and to stay the proceedings, if necessary, pending the making of such declarations.

But with all respect I do not agree. The function of the expert is to
F make the decision and that is not the function of the court where the decision has been entrusted to the expert. It is otherwise if both parties agree — as they often do — to get a ruling from the court to determine the basis on which an expert is to proceed, and if it is practical to assist the court will do so. But here there is no such agreement.

I would therefore, for the reasons that I have given, dismiss this appeal.

Farquharson and **Simon Brown LJJ** agreed and did not add anything.

Appeal dismissed with costs; application for leave to appeal to the House of Lords refused.

Court of Appeal G

October 19 1992

(Before Lord Justice NOURSE and Lord Justice FARQUARSON)

ROBINS AND ANOTHER v TUPMAN AND ANOTHER

Estates Gazette April 17 1993

[1993] 15 EG 145

Nuisance — Whether the plaintiffs' obstruction of easement of drainage the cause of the nuisance complained of

In 1978 the common owner of "The Chantry", East H
Village, Crediton, Devon, conveyed the house, garden and northern part of some further land on the other side of a public road, including the septic tank to which the house drained, to the defendants. The conveyance did not contain any express grant of an easement of drainage through drains from the septic tank to a soakaway on the southern part of the further land. In 1979 the common owner conveyed the southern part of the further land (which became known as "Applewood") to the plaintiffs. On purchasing the land, the plaintiffs commenced building a bungalow and garage and, on severing the drain from the defendants' septic tank, requested them to cease using the system. Faced with the plaintiffs' building J
works which obstructed their drainage, the defendants constructed an alternative system in 1982 and sought the costs of the same from the plaintiffs. At the trial of the action Judge Neville awarded the defendants the full amount of their claim of £2,358.69. From 1984 the plaintiffs complained of problems to their land arising from the operation of the new system and brought the present proceedings alleging nuisance. In Exeter County Court Judge Sir Jonathan Clarke dismissed the plaintiffs' claim. He found that the old sewage system worked satisfactorily before 1979 and that the steps taken by the defendants in 1982 to provide the alternative system K
represented a reasonable attempt to solve the problem. The plaintiffs' appealed.

***Held*: The appeal was dismissed. It was the destruction of the original drainage system by the plaintiffs which was responsible for the matters complained of in the present action. In taking the steps they took the defendants had mitigated their damage and had acted reasonably. There was a sufficient chain of causation between the plaintiffs' actions in obstructing the original drainage and the matters of which they now complained.**

The following case is referred to in this report. L

Wheeldon v *Burrows* (1879) 12 ChD 31

This was an appeal by the plaintiffs, Geoffrey Russell Robins and Margaret Anne Robins, from the decision of Judge Sir Jonathan Clarke in Exeter County Court, when he dismissed their claim alleging nuisance on the part of the defendants, William Arthur Tupman and Jean Tupman, in the use of their property.

James Townsend (instructed by Trump & Partners, of Bristol) appeared for the appellants/plaintiffs; Michael Brabin (instructed by Stones, of Exeter) represented the respondents/defendants.

Giving the first judgment, **Nourse LJ** said: This is an appeal from M
a decision of Judge Sir Jonathan Clarke given in Exeter County Court on May 1 1991 in a long-running and most regrettable dispute between owners of adjoining properties at East Village, Crediton, Devon.

Until 1978 the properties were in joint ownership. There was a house at the northern end called "The Chantry" standing to the north-west of a crossroads. It had a garden and further land to the south-west of the crossroads, on the other side of the east/west road, stretching down to the south. The general fall of the garden and further land is to the south-west. At that time sewage and other effluent from the house passed down a pipe running (with manholes at each end) under the north/south road across the crossroads until it reached a position to the east of the garden. There it ran westward at right angles to the road into a septic tank situated in the garden, from which there was in turn a pipe running southwest into a soakaway placed on the lower part of the land.

In 1978 the common owner, a Miss Blenman, conveyed the house, the garden and the northern part of the further land, including the septic tank but not the soakaway, to the defendants, Mr William Arthur Tupman and Mrs Jean Tupman. Although the conveyance did not contain an express grant of an easement of drainage into and through the soakaway on the southern part of the further land, it is clear that the conveyance operated to include such a grant by virtue of the rule in *Wheeldon* v *Burrows* (1879) 12 ChD 31. In 1979 Miss Blenman conveyed the southern part of the further land (now known as "Applewood") to the plaintiffs, Mr Geoffrey Russell Robins and Mrs Margaret Anne Robins. That conveyance was not expressed to be subject to the easement of drainage and we are told that it was not disclosed to the Robins' solicitors on preliminary inquiries. However, being a legal easement, it was enforceable against the Applewood land irrespective of notice.

In his judgment, Judge Clarke said of the easement:

> It is important to note in this connection that the easement was not confined to the use of the pipe and to discharge into that soakaway. In the nature of things, the liquid entering the soakaway would then do just that — soak away. In other words, it would then percolate, as water will, through the land, finding its own level elsewhere. Thus, liquid from "The Chantry" must, after leaving the soakaway, have spread through what is now "Applewood" land.

The Robins purchased their land in order to build a bungalow and, to the north, a garage. Although they knew beforehand of the existence of the easement and, in general terms, the line of the pipe and the approximate position of the soakaway, they went ahead with their plans. I take up the story in the words of Judge Clarke:

> Mr Robins excavated without regard, or without sufficient regard, for "The Chantry" drain pipe and in or about August 1979 Mr Robins cut through that drain pipe. Sewage effluent flowed out. What was his reaction? It is, I am afraid, indicative of the plaintiffs attitude to this whole matter — not "I am sorry" — no. They told the Tupmans to stop using the system. The effect in mid-August of being deprived of a bath and other facilities of that kind can be imagined.
>
> To put it shortly, the plaintiffs went on with their building work. They provided a temporary drainage system which, as Judge Neville found, failed and, as he said, was bound to fail within a short time.

Judge Neville's part in the story will be explained presently. I should point out that, although the temporary drainage system failed, that was not until about 17 months later. I return to the story as told by Judge Clarke:

> By the end of 1981, therefore, the Tupmans were facing a real problem. Their sewage was not escaping properly. It was backing up. The Robins would do nothing to deal adequately with the problem, which they and they alone had caused. They were indeed denying liability at that time and so the Tupmans had to do something for themselves. What they did, was to take steps to alleviate the problem. Their easement had been denied, a large garage block, consisting of three garages and a workshop, had been built, which in length was practically as long as the Tupmans entire southern boundary. That building has substantial footings and effectively acts as a dam or barrier to all liquid, whether rain drainage or any other kind of drainage, coming down the fall of the land from the north. So, acting as I find on advice, the Tupmans constructed an elaborate system of drainage designed to operate within the limited confines of their garden. What they did was to construct a system by which effluent entered a septic tank where it was allowed to settle. Thereafter, liquid is taken into a pump chamber from whence it is pumped uphill to the north into a third chamber. From this chamber the effluent is discharged by gravity into a series of land drain pipes laid throughout the garden. By and large this works adequately, but there have been problems.

It appears from that passage that the judge may have thought that a new septic tank was installed as part of the new system. That was not, in fact, the case. The existing septic tank was incorporated into the new system.

The new system was constructed by Tupmans in the spring 1982. About a year earlier, in April 1981, they had commenced an action for damages against the Robins in Exeter County Court. Having constructed the new system, the Tupmans sought to charge the Robins for the costs of its construction and ancillary costs by way of damages for disturbance of their easement. The Robins' defence was that the problem could and should have been solved more cheaply, but that the Tupmans had refused to discuss it with them and had gone ahead and incurred excessive costs.

At the trial of the action on June 23 1982 Judge Neville awarded the Tupmans damages in the full amount of their claim, namely £2,358.69, including nearly £260 for clearing the septic tank on 13 occasions between August 30 1979 and March 12 1982, 11 of them in the last 10 months of that period. That shows the scale which the problem had by then assumed from the Tupmans' point of view and it no doubt accounts for their decision to construct the new system. In giving judgment, Judge Neville said:

> I do not think the new system is an improvement, indeed in a way it will be a nuisance because it is under their garden now and involves an electric pump, which may go wrong.

I would add that not only was it under the Tupmans' garden but it was, if Judge Clarke's description of the operation of the system was correct, also to a lesser extent under the ground of Applewood. That, then, was the position when Judge Neville gave his decision in June 1982.

From 1984 onwards the Robins began to complain about problems on their land arising from the operation of the new system. It is these complaints which form the basis of the present action started by the Robins in February 1988. First, they complained that on occasions the sewage from the new system had backed up into the more southerly of the two manholes in the road and had overflowed. Effluent had run down the road causing an unpleasant sight and smell; on occasions it had run from the road into the Robins' driveway. Judge Clarke had no doubt that that had, on occasions, happened. Second, the Robins complained that sewage effluent from the Tupmans' system was seeping through on to their land and was causing ponding in an area due west of their garage. Ponding occurred when the water table was high. The judge had no doubt in finding that it consisted largely of rainwater, but rainwater which on occasions could well be contaminated to a slight degree from the effluent discharged into the new system which, owing to the fall of the land, would in times of heavy rain and a high water table mingle with the rainwater percolating through the soil and finding its way downhill to a natural low point. He also noted a subsidiary complaint under the same head, namely that on occasions, notably in 1984, vegetables grown in the Applewood garden shrivelled and died as if poisoned. He said that was, happily, a rare occurrence. He also pointed out that earlier in 1991 the Tupmans had carried out remedial action at their own significant expense to cure both the overflowing from the manhole and the ponding on the Applewood land.

Judge Clarke made two further important findings. First, he found that the old sewage-dispersal system had worked satisfactorily before 1979. He said that Judge Neville had also found that as a fact. Our note of the latter's judgment does not disclose an express finding to that effect. But I agree with Mr Brabin, for the Tupmans, that it is implicit in the passage I have quoted from his judgment. Second, Mr Townsend, for the Robins, accepts that Judge Clarke effectively found that the steps taken by the Tupmans in 1982 represented a reasonable attempt to solve the problem for the future.

The essence of Judge Clarke's decision was expressed thus:

> I have heard much expert evidence. I am satisfied that while the system installed by the Tupmans is in general terms adequate, it does at times of heavy rain and a high water table, cause excess effluent to soak into the "Applewood" land. The building of the deep physical barrier constituted by the "Applewood" garage block effectively causes all liquid seepage from "The Chantry", be that rainwater or a mixture of rainwater and excess sewage effluent, to be diverted from its original seepage course to the west and so sometimes to puddle at 9.015.
>
> The whole problem stems from two things. First, the destruction of the original drainage system and second, the construction of that garage barrier.
>
> To the extent that the work of alleviation carried out by the Tupmans in 1982 and subsequently is inadequate, that is, as I find, not negligence nor an actionable nuisance for which the original tortfeasors have any claim at law.

Both sides are agreed that the Robins' claims in this action lie essentially in nuisance. The main question debated in this court has been that of causation. Mr Townsend very properly accepts that if, as Judge Clarke held, the problems which have occurred since 1984 were caused by the wrongs done by the Robins in 1979, their appeal, however it is put, cannot succeed. If they caused the nuisance, it cannot have been caused by the Tupmans.

Mr Townsend submits that by no stretch of the imagination can it be said that the post-1984 problems were caused by the Robins' wrongdoing in 1979. He says that the present problems arise entirely because of the Tupmans' choice of the new system in 1982. At that time, he submits, the Tupmans had a choice. They could, in assertion of their easement, have insisted that the old system be reinstated. They could, had they wished, have applied for an interlocutory prohibitory injunction in August 1979; and they could also have claimed a mandatory injunction at the trial in 1982. At that stage they

A would have had a good case for a mandatory injunction. They chose not to press that case. Instead, they chose to install a system which has proved inadequate. Because it has proved inadequate, they are themselves liable in nuisance to the Robins.

Mr Townsend also questions the judge's finding that the problem was caused not only by the destruction of the original drainage system but also by the construction of the garage barrier. The footings of the garage, which was built over the soakaway, descend 4ft 6ins into the ground. Mr Townsend maintains that there was no evidence before the judge on which he could have made his finding as to the effect of those footings. Although I am not satisfied that there was no such evidence before the judge, it does not seem to me that that is a point of importance. I say that because if it was the B destruction of the original drainage system which was responsible for the matters of which the Robins now complain, that would be enough to defeat their cause of action in nuisance.

I reject Mr Townsend's submissions. In my view, there was, as the judge held, a sufficient chain of causation. I agree with Mr Brabin that the proposition that the Tupmans had a realistic choice in 1979 cannot be sustained. As I have said, there was a temporary system which appears to have worked for some 17 months. Even at that stage it might have been a strong thing for the court to grant an injunction restraining further building. But whether that be right or wrong, I think it highly unlikely that a mandatory injunction would have been granted at trial, bearing in mind that during the 17-month period the C Robins' had gone ahead and finished building the garage.

So what, then, could the Tupmans have done and what should they have done? Mr Brabin submits that they did all that they had to do in order to minimise the amount of the claim against them. I agree. In technical terms they mitigated their damage. They constructed a new system, which Judge Neville decided that they had been justified in constructing, a decision which has been confirmed by the implied finding of Judge Clarke that they acted reasonably at the time. It is an extraordinary proposition to say now that the Tupmans ought, in 1982, to have asserted their easement and thus to have caused the Robins far greater expense and inconvenience than they were caused by the course which the Tupmans in fact took. It does not lie in the mouths of the Robins to say that the Tupmans ought to have taken D some other course.

Alternatively, it is suggested that the Tupmans should have installed a more effective system. On the evidence before the judge the only practicable alternative would have been to put in a cesspit. But that would have involved the Tupmans in very great expense. So, too, would the boring of a hole deep into the ground, another suggested means of disposing of the effluent. In my view, the Tupmans were not obliged to adopt either of those courses.

It is true, as the judge found, that the new system has not been adequate for all occasions. I am bound to say that the matters complained of, although naturally of very great irritation and concern to the Robins, did not seem to have been as bad as they might have been. Whether that be right or wrong, the Tupmans having E acted reasonably in taking the steps which they did take, the matters of which the Robins now complain cannot have been caused by the Tupmans. In so far as they have been caused by them in the literal sense they have inevitably been brought about by the actions of the Robins in 1979 and the steps which the Tupmans were then forced to take in order to remedy the situation. For these reasons I think that Judge Clarke's decision on the question of causation was correct, and that is enough to dispose of this appeal.

I must mention two further points. First, Mr Brabin developed a very interesting argument to the effect that the original easement of drainage over the Robins' land had never been abandoned by the Tupmans. He submitted that, if the Tupmans were to fail on the F question of causation, they could nevertheless rely on their easement as a justification for everything they had done since. The argument raised some interesting questions. Does the right to deviate apply to an easement of drainage as it applies to an easement of way? Did the Tupmans increase the burden of the easement or alter the dominant tenement in such a way as to destroy it? And so on. Those are interesting questions. Since it is unnecessary to decide them, I prefer not to do so.

Second, the judge, having held that the easement had not been abandoned, went on to say:

> Even if I am wrong and the easement on the facts has been abandoned, then I would hold *ex turpi causa oritur non actio* that the plaintiffs cannot in this court be heard to complain that their own wrongdoing has set off a chain of events which now causes them a legal nuisance.

If there is any part of the judgment of the judge with which I would G venture a respectful disagreement, it is that. It seems that he relied on the maxim *ex turpi causa oritur non actio* as support for his decision on the question of causation. His observations have caused some confusion in this court. His decision could be, and was, soundly based on causation alone. It was neither necessary nor appropriate to invoke the maxim as well.

For the reasons I have given I think that Judge Clarke arrived at an entirely correct decision. I would dismiss the Robins' appeal accordingly.

Agreeing, **Farquharson LJ** said: It was in September 1979 that the first plaintiff severed the soakaway pipe of the defendants' sewage system in the course of digging the footings for the plaintiffs' H new garage.

The first plaintiff then constructed a new soakaway close to the boundary of the defendants' land. This operated satisfactorily for some 17 months, but in the end proved to be inadequate to contain the effluent passing through it. It was in those circumstances that the defendants, under expert advice, constructed a new disposal system within the confines of The Chantry garden in 1982. This new system proved to be successful in the main in containing the effluent from The Chantry, but there had been some percolation on to the plaintiffs' land at Applewood, both from The Chantry garden and by way of a manhole in the road which borders the two properties.

The question which the court has been considering today is whether the defendants should be held responsible for the damage J caused to the plaintiffs by the seepage of effluent from The Chantry garden on to Applewood or, on the other hand, whether the plaintiffs must be held responsible by reason of the first plaintiff's act in fracturing the soakaway pipe.

Mr Townsend, for the plaintiffs, argues that there is a straightforward question of causation. While he accepts that the problem began with the first plaintiff breaking the soakaway pipe, the plaintiffs cannot be responsible for the defective nature of the new system about which they were not consulted. Mr Townsend submits that in August 1979 it would have been open to the defendants to have applied for an interlocutory injunction to restrain the plaintiffs from interfering with their sewage system. Such an K application, he says, would have been likely to succeed had the defendants elected to do so. Instead, the defendants commissioned the installation of an entirely new system to be installed within the confines of their own property. By so doing, he argues, not only did the defendants abandon their easement of effluent disposal through the original pipe and soakaway but they also exonerated the plaintiffs from any further responsibility. Accordingly, it is said, any fresh discharge of effluent from The Chantry to another part of the plaintiffs' garden is a nuisance for which the defendants are liable in damages.

It would be a curious result, in my judgment, if that were the effect of the relevant law. As the disposal of their effluent appeared to be satisfactory following the immediate repairs carried out by the L plaintiffs there was no need, and indeed little likelihood of success, for the defendants to apply for an injunction. The crucial question was, what should be done when those repairs proved ineffective? By this time the garage, with its deep footings, was completed. The original soakaway was destroyed. What, then, were the defendants to do? Their duty was to act reasonably and to mitigate their damage as best they could. In the interest of all parties the defendants sought a solution which would solve the problem without dismantling the garage. That was how the new disposal system came to be installed and, to a considerable extent, it appears to have worked.

Such shortcomings as remained can hardly be ascribed to the defendants when the judge concluded that they had acted M reasonably. In my judgment, the plaintiffs' original act of destruction remained the effective cause of any shortcomings for the new system. I, too, would dismiss the appeal on that ground, and in those circumstances I agree it is unnecessary to consider the question of whether the easement had been abandoned by the defendants' installation of the new system, or whether the easement on a proper construction permitted the defendants to discharge effluent over parts of the plaintiffs' garden other than through the original soakaway, nor whether the suit fails by the application of the maxim *ex turpi causa oritur non actio*, as the learned judge considered.

Appeal dismissed.

A

Chancery Division

October 16 1992

(Before Mr M R REID QC, sitting as a deputy judge of the division)

LEHMANN AND ANOTHER v HERMAN

Estates Gazette April 24 1993

[1993] 16 EG 124

London Building Acts (Amendment) Act 1939 — Whether notice given by one of two joint tenants of leasehold interest, purportedly under section 47 of the Act, valid

B

The plaintiffs, Mr and Mrs Lehmann, are the owner-occupiers of 35 Reddington Road, London NW3. The defendant and his wife, Mr and Mrs Herman, own a long lease of the ground-floor flat in the adjoining property, 33 Reddington Road; the freehold of which is owned by Mrs Herman's grandmother, Mrs Goodman. The division between the back gardens of nos 33 and 35 is marked, *inter alia*, by a true party wall. In February 1991 Mrs Goodman's agents, Clifford Tee & Gale, informed the plaintiffs that works were proposed at no 33 and, by a letter dated October 13 1991, notice under section 47 of the London Building Acts

C (Amendment) Act 1939 was purportedly given to the first plaintiff on behalf of the defendant. By a letter dated December 23 1991 the plaintiffs' solicitors informed the defendant that, by reason of the notice having been served solely on his behalf, and because the defendant and his wife were the joint tenants of the lease of their flat, the notice was invalid: the letter invited the defendant to procure that his wife confirm that she was bound by the notice. The defendant contended that the service of a notice by one of two joint tenants was valid.

Held: **The notice was invalid. The defendant and his wife**

D **would constitute the "building owner" under the 1939 Act for the purpose of serving a notice under section 47 or exercising any rights under section 46. The definition of "owner" in the London Building Act 1930, section 5, includes "every person in possession . . . or in the occupation of any land" and "in the occupation" must be construed restrictively. Where there are joint tenants occupying the premises, it cannot be said that either alone is in *the* occupation of the premises; both are in the occupation and both must serve a notice under section 47.**

The following cases are referred to in this report.

E *Crosby* v*Alhambra Co Ltd* [1907] 1 Ch 295

Featherstone v *Staples* [1986] 1 WLR 861; [1986] 2 All ER 461; [1986] 1 EGLR 6; (1986) 278 EG 867, CA

Hammersmith and Fulham London Borough Council v *Monk* [1992] 1 AC 478;[1991] 3 WLR 1144; [1992] 1 EGLR 65; [1992] 09 EG 135, HL

Newman v *Keedwell* (1977) 35 P&CR 393; 244 EG 469

Solomons v *Gertzenstein (R) Ltd* [1954] 2 QB 243; [1954] 3 WLR 317; [1954] 2 All ER 625, CA

This was a hearing of an originating summons of the plaintiffs, Mr and Mrs Lehmann, to determine the validity of a notice under section 47 of the London Building Acts (Amendment) Act 1939 served on behalf of the defendant, Mr Herman.

Benjamin Levy (instructed by Adlers) appeared for the plaintiffs;

F Michael Bowsher (instructed by Eppel & Co) represented the defendant.

Giving judgment, **MR REID QC** said: This is a dispute between neighbours. The issue is whether a party wall notice under section 47 of the London Building Acts (Amendment) Act 1939 served on behalf of the defendant on the first plaintiff is valid. The defendant says that it is valid; the plaintiffs say that it is not. They say that to be valid it should have been given on behalf of the defendant and his wife and not on behalf of the defendant alone.

The plaintiffs, Mr and Mrs Lehmann, are the owners and occupiers of 35 Reddington Road, London NW3. Mr Herman, the defendant, and his wife live at 33 Reddington Road. They have a long lease of the ground-floor flat there. The lease includes the part of the back garden which abuts no 35. The freehold of no 33 is owned by G Mrs Herman's grandmother, Mrs Goodman. The division between the back gardens of nos 33 and 35 is marked not by one but by two walls. One wall is entirely on no 33's land, the other is, on the material before me, a true party wall.

It is evident from the correspondence that relations between the Lehmanns and the Hermans have been strained for some time, but the matters giving rise to these proceedings began in February 1991, when managing agents acting for Mrs Goodman wrote a letter addressed to the freeholder of no 35 indicating that works would be done to a rear retaining wall and pathway at no 33. There followed correspondence between Mr Lehmann and his representatives on the one hand and Mrs Goodman and her representatives and Mr Herman H and his representatives on the other. The correspondence was lengthy and involved. For the purposes of the present proceedings I need say only that it led eventually to Clifford Tee & Gale, as surveyors and architects for Mr Herman, serving the document, the validity of which is the issue in these proceedings. The notice was sent under cover of a letter dated October 13 1991 and is in these terms:

To: R J Lehmann Esq of 35 Reddington Road, NW3.

I, I Herman, of [then no address is given] as owner of 33 Reddington Road, London, NW3 which adjoins your premises known as 35 Reddington Road, London, NW3, hereby serve you with notice that in accordance with my/our rights under S46(1) and with reference to the party fence wall separating the above premises, it is intended to carry out the works detailed below after the expiration of one month from service of this notice. The proposed works are: J

The reconstruction of a new retaining wall including special foundations running parallel with the rear wall of no 33 to replace the existing defective retaining wall and to rebuild part of the defective boundary wall of no 33 which runs parallel with and touches the party fence wall. It is intended to commence works as soon a notice has run, earlier by agreement. Under S49 if you do not consent to the works within fourteen days you are deemed to have dissented and a difference is deemed to have arisen. In such case S55 of the Act requires that both parties should concur in the appointment of a surveyor or should each appoint one surveyor and in those circumstances I/we would appoint Mr C J Graham, of Clifford Tee & Gale . . .

— and the address is given. It is then signed on behalf of Mr Herman and dated as I have indicated. With it was enclosed a form of acknowledgement, the relevant parts of which read: K

I/we, R J Lehmann Esq, 35 Reddington Road, London NW3 having received the notice served by I Herman in respect of 33 Reddington Road, London NW3 which joins with my/our premises known as 35 Reddington Road in relation to which the works proposed under S46(1) . . .

— then there is provision for either consent or dissent.

Mr Lehmann passed the letter and its enclosure to his solicitors. They replied on behalf of both Mr and Mrs Lehmann raising a number of points. This led to further detailed correspondence between the parties' advisers. I do not need for the purposes of these proceedings to go through it, save to say that the Lehmanns appointed a surveyor pursuant to the Act, but the appointment was expressed to be under protest and without prejudice to their L contentions as to the validity of the notice.

The culmination of the correspondence was a letter dated December 23 1991 from the plaintiffs' solicitors, in these terms, sent to Mr Herman:

We act on behalf of Mr and Mrs Ralph Lehmann, your neighbours at 35 Reddington Road, with regard to the above.

— The above being 33 and 35 Reddington Road, London NW3, party wall matters —

We hereby formally put you on notice that the Party Structure Notice dated 15th October 1991 served on Mr Lehmann appears to have been served solely on your behalf. We should inform you that in view of the fact that you are joint owner of Flat A, No 33 Reddington Road, with Laurel Susan Herman, M your wife, and in view of the provisions of the London Building Act 1930 and London Building Acts (Amendment Act) 1939 the above said notice is not a valid notice for the purposes referred to therein. Accordingly, our clients require you to procure and your wife formally to undertake to treat herself as bound by the said notice as well as yourself. Moreover our clients require written confirmation of the same. You should be left in no doubt that should you fail to provide such evidence within seven days of the date of this letter or seek to prevaricate, our clients will forthwith commence proceedings to have the said notice declared invalid. Yours faithfully . . .

— and signed by the solicitors. The invitation to Mrs Herman to treat herself as bound by the notice was not taken up and these proceedings then followed.

The basis upon which Mr Herman acted and the reasons why Mrs

A Herman did not join in the notice or take up the suggestion that she should undertake to treat herself as bound by it, appear from two paragraphs in the affidavit sworn by Mr Herman. They read as follows:

> I am advised and verily believe that as a matter of law service by one of two joint tenants or tenants in common is sufficient just as service upon one of two joint tenants will suffice. In the circumstances I respectfully submit that service by me, a joint tenant together with my wife, of a notice upon the first plaintiff, a joint tenant together with the second plaintiff, is sufficient. In any event, however, I am the person who proposes to carry out the works. I appointed the party wall surveyor, Mr Graham, I have appointed Tant Building Management Limited to design and supervise the works, I shall be entering into the building contracts in my own name, I shall pay for the works
> B by cheques drawn on my account. My wife has had no involvement in any of this and it is not intended that she will become involved, save perhaps to the extent of offering mugs of tea to the building contractors from time to time. I respectfully submit that even if the assertion in paragraph 15 above is wrong my wife cannot be construed as the building owner.

Those, then are the facts.

The first of the relevant statutory provisions is section 46 of the 1939 Act. It begins:

> (1) Where lands of different owners adjoin and at the line of junction said lands are built on or a boundary wall being a party fence wall or the external wall of a building has been erected, the building owner shall have the following rights: —
> C (*a*) A right to make good underpin thicken or repair or demolish and rebuild a party structure or party fence wall in any case where such work is necessary on account of defect or want of repair of the party structure or party fence wall.

Section 47(1) then sets out requirements that have to be fulfilled before the rights given by section 46 can be exercised. It provides:

> Before exercising any right conferred on him by section 46 (Rights of owners of adjoining lands where junction line built on) of this Act a building owner shall serve on the adjoining owner notice in writing (in this Act referred to as "a party structure notice") stating the nature and particulars of the proposed work the time at which it will be begun and those particulars shall where the building owner proposes to construct special foundations including plan sections and details of construction of the special foundations with
> D reasonable particulars of the loads to be carried thereby.

In section 48 there is provision for a counternotice to be served on the "building owner" which in context means the building owner who served the party structure notice. There are then further provisions and section 55 provides for the settlement of differences by a surveyor or surveyors, for appeals from their decisions and for the award of the costs. Section 56 relates to the apportionment of expenses as between the building owners and the adjoining owner. It begins:

> The following provisions shall apply with respect to the apportionment of expenses as between the building owner and the adjoining owner: —
> (*a*) Expenses incurred in the exercise of the rights conferred by paragraph
> E (*a*) subsection (1) of section 46 (Rights of owners of adjoining lands where junction wall built on) of this Act shall be defrayed by the building owner and the adjoining owner in due proportion regard being had to the use which the two owners respectively make or may make of a party structure or party fence wall;

Section 57 enables an adjoining owner to obtain security for expenses from a building owner.

Under section 1 of the Act it is necessary to look at section 5 of the London Building Act 1930 for the definitions of "adjoining owner", "building owner" and "owner". The definition section is prefaced by words: "In this Act, save as is otherwise expressly provided therein and unless the context otherwise requires, the following expressions have the meanings hereby respectively assigned to them, that is to say: . . .

F

— and then the definitions follow. The respective definitions are as follows:

> "Adjoining owner" and "adjoining occupier" respectively mean any owner and any occupier of land buildings storeys or rooms adjoining those of the building owner; . . .
> "Building owner" means such one of the owners of adjoining land as is desirous of building or such one of the owners of buildings storeys or rooms separated from one another by a party wall or party structure as does or is desirous of doing work affecting that party wall or party structure;

"Owner" is then defined:

> "Owner" includes every person in possession or receipt either of the whole or of any part of the rent or profits of any land or tenement or in the occupation of any land or tenement otherwise than as a tenant from year to year or for any G less term or as a tenant at will;

As is clear from those definitions, for the purposes of the 1939 Act there can be more than one building owner in respect of a building. It is also common ground that Mr and Mrs Herman together could constitute the building owner for the purpose of serving a notice under section 47 or exercising rights under section 46.

Mr Bowsher, for the defendants, went further. He contended that Mr Herman alone could be a building owner for the purpose of section 47. Basing himself on the statutory material that I have cited he submitted that the Act was not intended to raise complex issues of property law. It was, he said, intended that any one of the owners of an interest in the property should be able to serve a notice under H section 47(1) of the Act if he were desirous of doing works to which section 46 applied and there was nothing in the Act which required all joint tenants to join in serving a notice. He submitted that each joint tenant was, by himself, a building owner within the meaning of the Act and so capable of serving a notice under section 47(1), independently of any other joint tenant. Mr and Mrs Herman were both occupying their flat and therefore each alone is a building owner for the purposes of the section. He argued that only the owner who desires to carry out work affecting a party structure is required to serve a party structure notice.

In support of his submission he referred to *Crosby* v *Alhambra Co Ltd* [1907] 1 Ch 295, a case on the London Building Act 1894. In that case the issue was as to the meaning of the term "adjoining owner" in J the 1894 Act. Neville J said, at p299:

> I hold, therefore, that the true interpretation of the two sub-sections is that all the persons coming within the definition of "owner" in sub-s 29 must be served, except in the case where several persons hold, as tenants in common or as joint tenants, some particular interest in the land, in which case service on one of such tenants in common or joint tenants would be sufficient.

This, said Mr Bowsher, showed by analogy that one of two joint tenants could serve a notice under section 47(1) of the 1939 Act.

Mr Levy on behalf of the plaintiffs submitted that the context and purpose of the London Building Acts (Amendment) Act 1939 required that any notice served by joint tenants be served by or on behalf of all of them. He pointed out that it would be unsatisfactory K if one joint owner could serve a notice which created new rights and obligations between the parties to the notice and that the objection to a notice served by only one joint tenant could not be regarded as a mere technicality. He suggested that if Mr Bowsher's submission were right, where there were joint tenants, one of whom was impecunious, that one could serve the notice and the adjoining owner would effectively be left without remedy in respect of any award of costs under section 55 and perhaps would also have difficulties in recovering expenses, though he accepted that the securities provisions of section 57 provided some protection in that respect. He also submitted that it might be very difficult to operate the expenses provision of section 56(1) properly where one of two joint tenants could be the building owner. L

He drew a distinction between the service of a notice on one of two joint tenants as in the *Alhambra* case and the service of a notice creating rights by only one joint tenant. In support of his argument he cited *Newman* v *Keedwell* (1977) 35 P&CR 393 and referred to *Featherstone* v *Staples* [1986] 1 WLR 861 and *Hammersmith & Fulham Borough Council* v *Monk* [1992] 1 AC 478 . There would, he submitted, be difficulties about any ensuing arbitration which would bind only one of two joint tenants. In his submission it would be unreasonable to expect the Act to be formulated so as to enable one of two or more joint tenants to take the benefit and bear the burden of the notice and for the others to say "we shall not be bound". The words in section 5 of the 1930 Act "such one of the owners of adjoining land as is desirous" must therefore be construed as M meaning "such owner of any layer of ownership of adjoining land as is desirous". By "layer of ownership" he explained that he meant, for example, that the freehold was one layer, a lease another layer and an underlease a third layer.

In my judgment, it would be surprising if the statute were intended to have the effect for which Mr Bowsher contends. Simply in practical terms it would be very odd if the statute provided for one of two joint owners to deal with an adjoining owner without the other joint owner being involved. In real property law terms the concept of one joint owner being able to deal with the property without the other being party to the transaction has been foreign to English law since the 1925 property legislation. The fact that notice could validly be

A given to one of two joint tenants so as to be effective as notice to both and to bind both, does not carry the matter any further forward. Does the definition of building owner then compel such conclusion? In considering that, of course I bear in mind the opening words of section 5 "unless the context otherwise requires", and I take the view that even if Mr Herman alone could otherwise be a "building owner" within the definition, the context here requires otherwise. But could he anyway be described as a "building owner"? In order for a person to be a building owner for the purposes of the Act, he must be an "owner". The definition of "owner" is not exhaustive, it begins "owner includes". It then continues:

> . . . every person in possession or receipt either of the whole or of any part of
> B the rent or profits of any land . . .

It is not suggested that Mr Herman alone could be regarded as being in possession of his flat, nor that he is in receipt of any rents or profits. In any event this phrase itself must be read as being limited in some sense. It does not include an agent who is in receipt of rents: see *Solomons* v *R Gertzenstein Ltd* [1954] 2 QB 243. Similarly, in the ordinary case, neither joint tenant of premises let at a rent could be said, when taken alone, to be in receipt of any part of the rent. Technically the whole rent would be payable to them jointly as trustees (even if received by one as agent for both) and the money, so far as available for distribution paid out as trust income, distributable to each beneficiary according to his entitlement.

C The remaining words of the definition are "or in the occupation of any land or tenement otherwise than as a tenant from year to year or for any less term or as a tenant at will". Here "in the occupation" must be construed restrictively. This is clear from the following words; the Act cannot be intended to be construed so that a tenant for 11 months certain is not an owner, but a mere licensee is. The clue lies in the words "the occupation". The Act is concerned with persons who are in *the* occupation, not persons who are occupying. Where, as in the present case, there are joint tenants occupying the premises, it cannot be said that either alone is in *the* occupation of the premises. Nor can the definition of owner properly be stretched as to include one of the occupiers on the basis that the definition is not exhaustive.

It follows that, in my judgment, a notice under section 47 served by
D or on behalf of only one of two joint tenants is not served by a building owner within the meaning of the section and is invalid. I, therefore, hold that the plaintiffs are entitled to succeed in these unfortunate proceedings.

Court of Appeal

November 3 1992

Before Lord Justice GLIDEWELL and Lord Justice Simon BROWN)

E

WOODHOUSE v CONSOLIDATED PROPERTY CORPORATION LTD

Estates Gazette May 15 1993

[1993] 19 EG 134

Party wall — Whether third surveyor appointed under London Building Acts (Amendment) Act 1939 had jurisdiction to make award determining responsibility for collapse of party wall

Ord 14 summons — Whether condition that defendants pay
F ***into court £350,000 before leave to defend proper exercise of court's discretion***

The plaintiff was the owner of 61 Endell Street, Covent Garden, London WC2, where he carried on a timber merchant's business, and on which stood a timber shed. The defendants were the owners of the adjoining premises, a former mission hall, the basement floor of which lay some 2.3m below the ground level of the timber yard. In 1990 the defendants commenced building works to the mission hall, including lowering the basement floor. On January 14 1991 the defendant gave the plaintiff a notice under section 47 of the London Building Acts (Amendment) Act 1939 of their intention to carry out work to the wall of the mission hall on G the boundary of the properties. On January 25 1991, before the expiration of the 14 days provided by the Act for the giving of consent, the wall of the mission hall on the boundary collapsed and seriously damaged the timber-storage shed. Following the appointment by the parties of surveyors under the 1939 Act, the two surveyors appointed a third surveyor. The plaintiff commenced the present action by the issue of the writ on December 10 1991 claiming damages for nuisance and trespass; the damages included £430,000 for the cost of rebuilding the timber shed and loss of profits continuing at £2,660 per month. By an award of December 10 1991 the third H surveyor determined that the prime cause of the collapse was the defendants' work for which they were responsible. On June 15 1992 the plaintiff issued a summons for summary judgment under RSC Ord 14 and on June 27 1992 the plaintiff sought leave to amend the statement of claim with a pleading that by the third surveyor's award the defendants were estopped from denying that they were responsible for the collapse of the wall. At the hearing of the summons Judge Hordern QC (sitting as a judge of the High Court) gave leave to amend the statement of claim and granted the defendants leave to defend on condition that £350,000 be paid into court within 14 days. J The defendant appealed against both orders contending that if the statement of claim were to be amended there should be an adjournment of the Ord 14 summons.

Held: **The appeal was dismissed. Where an amendment to a statement of claim is allowed, then formally the plaintiff should have taken out a fresh summons under Ord 14 and in the circumstances an adjournment should have been granted. However, as the defendants had had sufficient time to consider the amendments, and had produced additional affidavit evidence, the court would formally dismiss the defendants' appeal against the amendment and hear the appeal against the judge's further order as if a fresh Ord 14 K summons had been issued.**

Under the 1939 Act the third surveyor had no jurisdiction to make the award he purported to make because section 55 of the Act, which provides for the resolution of disputes between adjoining owners, is limited to the resolution of disputes between adjoining owners as to whether one of them shall be permitted under the Act to carry out works the subject of a notice under section 47, and if so, the terms and conditions. Section 55 of the Act does not permit or authorise the surveyor appointed under the Act to determine other disputes arising between the parties. Having regard to the other causes of L action relied on by the plaintiff, the judge was correct in giving leave to defend and in saying that he could almost give judgment for the plaintiff in justifying the condition for a payment into court. Having regard to the likely total claim of the plaintiff, the sum ordered as a condition for giving leave to defend cannot be said to be wrong.

The following cases are referred to in this report.

Dalton v *Angus* (1881) 6 App Cas 740; [1881] All ER Rep1; 50 LJQB 689; 44 LT 844; 46 JP 132; 30 WR 191, HL

Ladd v *Marshall* [1954] 1 WLR 1489; [1954] 3 All ER 745, CA

Wringe v *Cohen* [1940] 1 KB 229 M

This was an appeal by the defendants, and a cross-appeal by the plaintiff, from orders of Judge Hordern QC (sitting as a judge of the High Court) who, on the hearing the plaintiff's summons seeking leave to amend his statement of claim and for summary judgment under RSC Ord 14, had granted the plaintiff leave to amend and granted the defendants leave to defend on condition that they first paid into court £350,000.

Bruce Mauleverer QC and Mark Raeside (instructed by Beachcroft Stanleys) appeared for the appellant/defendants; Stephen Hockman QC and Adrian Jack (instructed by Barrett & Co, of Reading) represented the respondent/plaintiff.

Giving judgment, **Glidewell LJ** said: The plaintiff, Mr Woodhouse, owns and occupies 61 Endell Street, Covent Garden, London WC2, and a yard behind that building. At those premises he carried on business as a timber merchant, trading as Latchfords (Endell Street). The business is long-established. Mr Woodhouse himself has been in the business since 1953, at first as an employee, then as a partner and latterly as sole proprietor.

The defendants, the appellants in this court, are the owners of 68a Neal Street, Covent Garden. This property is a former mission hall, approached only by a narrow pedestrian passage from Neal Street, which runs parallel to Endell Street. The north-easterly wall of the mission hall lies immediately adjacent to the south-westerly boundary of the plaintiff's timber yard. In the timber yard, immediately adjacent to the wall of the mission hall, there was a storage shed for timber. This structure consisted of five brick pillars, three adjoining the wall of the mission hall, two at the north-easterly side, with a monopitch roof at the same height as the roof of the mission hall.

The mission hall had ground and first floors and a basement. The floor of the basement lay some distance — one report says approximately 2.3m — below the ground level of the timber yard. Thus, in practice, the wall of the mission hall, below the ground floor of that structure, acted as a retaining wall for the timber yard and the soil beneath it.

In the autumn of 1990 the defendants commenced works on the mission hall, which included lowering the floor of the basement. There is a dispute as to what stage those works had reached, and the condition of the building, at the relevant date.

On January 14 1991 the defendants served on the plaintiff a notice under the London Building Acts (Amendment) Act 1939, section 47, of their intention to carry out works to the wall of the mission hall on the boundary of the properties. It is agreed that the wall was a party wall within the definition in section 44 of the 1939 Act. The works referred to in the notice included "underpinning and reinforced concrete retaining wall". In accordance with the Act, the notice provided that if within 14 days the plaintiff did not consent to the works, he was deemed to have dissented. In that case a difference would have arisen between the parties under the relevant part of the 1939 Act. The notice said that it was intended to commence works earlier than the expiration of the 14 days by agreement. As I have already said, the works had commenced some time previously and according to the defendants the underpinning was substantially complete.

On January 25 1991, ie before the expiration of the 14 days under the notice, the wall of the mission hall adjoining the timber yard collapsed. Initially there was a suggestion that the collapse was towards the mission hall, but it now appears that the experts are agreed that above ground level the collapse was into the timber yard. In collapsing, the wall seriously damaged the timber-storage shed.

Both the plaintiff and the defendants appointed surveyors to act for them in the dispute which had arisen under the Act of 1939. The plaintiff appointed Mr Paul Cackett [BSc] FRICS, and the defendants, Mr Andrew Halstead [BSc] ARICS. They, in turn, appointed a third surveyor to resolve the dispute, Mr Anthony Poole, [FRICS ACIArb]. He signed an award on December 10 1991. One of the issues in this appeal concerns the validity of that award, to which I shall have to return.

The action

The plaintiff issued his writ on December 10 1991. It claimed damages against the defendants for nuisance, trespass to land, trespass to goods and negligence. The statement of claim was served on February 5 1992. In its original form it made no reference to the award of Mr Poole under the 1939 Act. In the particulars of special damage, the major item was the cost of rebuilding, which was put at some £403,000. Damages were also claimed for loss of profit of the business and additional expenses at the rate of £2,660 per month, together with a number of smaller items.

The defendants filed their defence on April 14 1992. On June 15 1992 the plaintiff's solicitors took out a summons for judgment under RSC Ord 14, and for an order for an interim payment, and for speedy trial of the action. The date for the hearing of these applications was fixed before a judge for July 30 1992. On July 27 1992 the plaintiff's solicitors took out a summons for leave to amend the statement of claim, which was served on the defendants' solicitors on July 28 1992. The defendants also made an application for an order striking out parts of an affidavit sworn by Mr Woodhouse in support of his application for judgment under Ord 14.

The applications were heard by Judge Hordern QC, sitting as a High Court judge, on July 30 1992. He ordered that the plaintiff should have leave to amend his statement of claim, with consequent leave to the defendants to amend their defence. He refused an application by counsel for the defendants to adjourn the hearing of the application for summary judgment under Ord 14. On that application he granted the defendants leave to defend on condition that within 14 days they paid into court the sum of £350,000. He made no order on the defendants' application with regard to Mr Woodhouse's affidavit.

The defendants now appeal to this court against both orders made by the judge. By a subsequent order of this court, compliance with the condition requiring payment into court has been stayed, pending the hearing of this appeal.

The appeal against the order for amendment of the statement of claim

The major amendment was the addition of four paragraphs following para 10. The first three of these paragraphs read:

> 10a. By an award dated 10 December 1991, lawfully made by Anthony Poole, the third surveyor duly appointed pursuant to the London Building Acts (Amendment) Act 1939 to resolve a difference arising between the plaintiff and the defendant herein, namely the responsibility for the cost of rebuilding the wall and certain related fees, it was determined:
>
> (a) that the prime cause of the collapse pleaded in paragraph 6 hereof was work being carried out by the defendant; and
>
> (b) that the defendant was responsible for the collapse of the wall as pleaded in paragraph 6 hereof; and
>
> (c) that the defendants should pay the costs of the plaintiff's surveyor and of the third surveyor in the said reference.
>
> 10b. No appeal has been made from the said award.
>
> 10c. In the premises the defendant is estopped *per rem judicatam* from denying that it is responsible for the said collapse.

Para 10d sets out the additional costs which it is alleged have not been satisfied by the defendants.

The amendments also include an allegation that the collapse of the wall was caused by the defendants "failing timeously to service Notices under the London Building Acts".

At the hearing before Judge Hordern, the defendants did not oppose the amendment as such, but sought an adjournment if that amendment were to be granted. It was argued on their behalf that formally the plaintiff's solicitors, after amending their statement of claim if leave were granted, should take out a fresh summons under Ord 14, thus giving the defendants the opportunity to file evidence in reply to the allegations in the amended statement of claim. Counsel for the defendants argued that if the judge did not grant an adjournment, he should consider the Ord 14 application only on the material before him which related to the statement of claim in its original form.

In my judgment, these submissions by counsel for the defendants were correct. The judge apparently took the view that in seeking an adjournment the defendants were merely trying to delay the resolution of the dispute. But the plaintiff, by his amendment, was making a major change in the basis of his claim. I agree that if such amendment were to be permitted (and there was no good reason why it should not have been) then formally the plaintiff should have taken out a fresh summons under Ord 14. No doubt it was permissible to waive this formality by treating the original summons as applying to the amended statement of claim, but if that were to be done, the defendants should have been granted an adjournment and the opportunity to file further evidence. I appreciate that the judge's decision in this respect was one made in the exercise of his discretion, but, in my view, he exercised that discretion on a wholly wrong basis.

This raised the question, what should we now do in this court in relation to this appeal? By now the defendants and their advisers have had ample time to consider the amended statement of claim. They have filed an amended defence on August 24 1992. Moreover, the defendants seek leave to adduce further evidence at the hearing of this appeal, namely a further affidavit from Mr B G McCarthy, the defendants' solicitor, sworn on October 5 1992, to which is exhibited a further report from Mr A G Antrobus. Mr Antrobus, whose qualifications are not stated in his report, is a member or employee of the Brian Clancy Partnership, chartered engineers and civil and structural engineers.

After some discussion Mr Bruce Mauleverer QC, for the

appellants, agreed that provided we gave leave for the additional evidence to be adduced, there would now be no disadvantage to his clients in the amendment of the statement of claim, and thus the main appeal could proceed as if the Ord 14 summons had been issued after the amendment of pleadings.

However, Mr Stephen Hockman QC for the plaintiff/respondent, then submitted that in considering whether to admit the further evidence, we were obliged to apply the principles derived from *Ladd* v *Marshall* [1954] 1 WLR 1489. In particular, the further report from Mr Antrobus, which it is sought to adduce, could have been available before the hearing before Judge Hordern and thus should not be admitted in evidence.

RSC Ord 59, r 10(2) provides:

The Court of Appeal shall have power to receive further evidence on questions of fact, either by oral examination in court, by affidavit . . . but, in the case of an appeal from a judgment after trial or hearing of any cause or matter on the merits, no such further evidence (other than evidence as to matters which have occurred after the date of the trial or hearing) shall be admitted except on special grounds.

It is commonplace that the principles in *Ladd* v *Marshall* define the special grounds for admitting such further evidence.

The essential question therefore is, the judge having granted the defendants leave to appeal on condition of a payment into court, has there been a trial or hearing of the action on the merits? The current edition of the *Supreme Court Practice* (1993 ed) contains the following note, numbered 59/10/8:

It was held in *Langdale* v *Danby* [1982] 1 WLR 1123; [1982] 3 All ER 129, HL that a judgment given under O14 (or O86) was a judgment given after a hearing on the merits; it follows that the *Ladd* v *Marshall* conditions apply where there is an application to adduce further evidence in an appeal against such a judgment (*Lodge Green Limited* v *Leitch* (*trading as Manx Electronics*), November 2 1982 (unrep), CA and *K/S/A/S Oil Transport* v *Saudi Research and Development Corporation Ltd* [1984] 1 Lloyd's Rep 5, CA). But an order refusing summary judgment or granting conditional leave to defend is not an order made after a hearing on the merits (see below).

In *Weller* v *Dunbar*, January 27 1984, CA (unrep) it was held that an order setting aside a default judgment is not a decision after a "hearing on the merits" (even though the merits are taken into account in deciding whether to set aside the default judgment), because the hearing on the merits in the shape of the trial of the action is yet to come. It is submitted that the same reasoning applies where an application for summary judgment is refused, or conditional leave to defend is granted, and therefore the *Ladd* v *Marshall* conditions do not have to be satisfied in such cases. . .

I have set out this reasoning because I entirely agree with it and I need add no words of my own. It follows that the defendants do not need to show "special grounds" for the admission of the fresh evidence in this case and the *Ladd* v *Marshall* principles do not apply.

It was for these reasons that at this stage in the hearing we announced that we would formally dismiss the appeal against the judge's order allowing the amendment to the pleadings, and would proceed to hear the main appeal as if the Ord 14 summons had been taken out after those amendments had been completed and the further evidence had been filed.

The issues

The original statement of claim in essence alleged three causes of action. These were:

(i) negligence;

(ii) simple nuisance;

(iii) breach of an easement of support, for which the right of action is also in nuisance.

The amended statement of claim pleaded that, as a result of Mr Poole's award, the defendants were estopped from denying that they were responsible for the collapse of their wall. Clearly if this plea were established, the plaintiff would be entitled to judgment on the Ord 14 summons. I therefore propose to consider this plea as a fourth issue.

I will therefore consider each of these issues in turn.

(i) *Negligence*

Proof that the defendants, or those for whose acts they were responsible when negligent, depends upon proof of facts which are in issue between the parties. Mr Hockman accepts that this as a cause of action cannot be the basis of the judgment under Ord 14.

(ii) *Simple nuisance*

This cause of action is based upon the decision of this court in *Wringe* v *Cohen* [1940] 1 KB 229.

At the commencement of his judgment, Judge Hordern said:

In my view, without anything more, the plaintiff is clearly entitled to damages, for unless the wall fell from some act of a stranger, act of the plaintiff himself or some insidious cause which the defendants could not know about, then I think that Mr Hockman is right on the authority of *Wringe* v *Cohen* [1940] 1 KB 229 and, indeed, if I may say so without disrespect to the learned judge who decided that case, I would have thought also on pure matters of first principle the plaintiff would obviously have been entitled to succeed.

Wringe v *Cohen*, like the present case, was an action between the owners of two adjoining properties. The plaintiff was the owner of a shop which was substantially damaged when the gable end of the adjoining house, owned by the defendant, collapsed on to it during a storm. There was evidence that the gable wall had been in a state of disrepair for some years. The house was let to a tenant. There was an issue as to whether the defendant, as owner not in occupation, was liable for the damage caused by the collapse of the gable wall in the absence of proof that he knew or ought to have known of its state of disrepair.

Atkinson J, giving the judgment of the court, said at p233:

In our judgment if, owing to want of repair, premises on a highway become dangerous and, therefore, a nuisance, and a passer-by or an adjoining owner suffers damage by their collapse, the occupier, or the owner if he has undertaken the duty of repair, is answerable whether he knew or ought to have known of the danger or not. The undertaking to repair gives the owner control of the premises, and a right of access thereto for the purpose of maintaining them in a safe condition. On the other hand, if the nuisance is created, not by want of repair, but, for example, by the act of a trespasser, or by a secret and unobservable operation of nature, such as a subsidence under or near the foundations of the premises, neither an occupier nor an owner responsible for repair is answerable, unless with knowledge or means of knowledge he allows the danger to continue. In such a case he has in no sense caused the nuisance by any act or breach of duty.

Although Atkinson J referred to the liability of an owner or occupier of premises on a highway, in that case it appears to have been irrelevant to the decision that the defendant's property was on a highway. Thus the decision appears to support the proposition that, if the wall of the mission hall collapsed due to want of repair, the defendants are liable in nuisance without proof that the defendants knew or should have known of the dangerous condition of the wall, unless it can be shown that the collapse was due either to the plaintiff's own act, to the act of a stranger or to some latent defect.

The decision in *Wringe* v *Cohen*, if it does establish the proposition I have just set out, has been much criticised. Mr Mauleverer submits that this proposition is not good law, in the sense that it is not supported by, or consistent with, other authorities.

Mr Hockman has saved us from deciding this interesting but difficult point by conceding that, as the defendants plead that in this case the collapse of their wall was caused by the plaintiff's own act, in order to secure a judgement under Ord 14 he has in practice to defeat that defence by proving what he requires to prove in order to establish nuisance by breach of an easement of support. It is therefore not necessary for us to consider the cause of action is simple nuisance further.

It is convenient to consider the next issue.

(iv) *Estoppel by reason of the award of Mr Poole under the London Building Acts (Amendment) Act 1939*

Under section 49 of the 1939 Act if an owner of land on whom a party structure notice under section 47 has been served does not consent in writing to the carrying out of the work specified in the notice, he is deemed to have dissented from the notice and "a difference shall be deemed to have arisen between the parties". Section 55 of the Act then contains a detailed procedure for settlement of such a difference. If, as here, each party appoints a surveyor and they do not agree, they must appoint a third surveyor.

So far as material, section 55(*i*) provides:

The agreed surveyor . . . shall settle by award any matter which before the commencement of any work to which a Notice under this Part of this Act relates or from time to time during the continuance of such work may be in dispute between the building owner and the adjoining owner.

Section 55(*k*) provides:

The award may determine the right to execute and the time and manner of executing any work and generally any other matter arising out of or incidental to the difference.

In a letter to Mr Poole, dated April 23 1991, the plaintiff's surveyor, Mr Cackett, said:

The dispute concerns the responsibility of costs for reconstructing the party

A wall between the two properties, the associated costs arising from the collapse of the party wall, ie damages, surveyors' and engineers' fees, consequential losses and the reconstruction costs for Latchford's including all fees etc.

Mr Poole accepted the responsibility of determining the dispute so defined and in his award he provided:

> 1. The building owner will pay the proper cost of rebuilding the party wall . . .
>
> 2. The building owner will pay the proper cost of reconstructing those parts of the (plaintiff's) premises which were damaged or collapsed as a result of the collapse of the party wall.

Mr Mauleverer submits that Mr Poole had no jurisdiction to make such an award under the scheme of the 1939 Act. If that is correct, he further submits that the award would be of no effect and thus could
B not estop his clients in any way.

Although subpara (*i*) of section 55 requires the surveyor to settle by his award "any matter which . . . from time to time during the continuance of such work may be in dispute . . .", this must, in my view, be read in its context. The context in particular includes the provisions of subpara (*k*), which commences: "The award may determine the right to execute and the time and manner of executing any work . . .". In my judgment, the provisions of section 55 relate only to the resolution of differences between adjoining owners as to whether one of them shall be permitted under the Act to carry out works, the subject of a section 47 notice, and if so, the terms and conditions under which he is permitted to carry out such work. A
C matter which arises during the carrying out of the works, about which there is a dispute, must therefore be a matter which relates to the consent for the works to be carried out, eg whether the building owner is complying with a particular requirement in the consent. Section 55 does not permit or authorise the surveyor appointed under this part of the 1939 Act to determine other disputes arising between the parties.

It follows, in my judgment, that under the 1939 Act Mr Poole had no jurisdiction to make the award which he purported to make. Mr Poole's award in this action is evidence, but no more. That is how the judge treated it in his judgment.

(iii) *Breach of the easement of support*

D In para 5 of his statement of claim (which was not altered in the amendment) the plaintiff pleads as follows:

> The plaintiff and his predecessors in title have for upwards of forty years, alternatively twenty years, before 1 January 1987 and/or before 25 January 1991, enjoyed support from the wall and in the premises by virtue of the Prescription Act 1837 . . . the plaintiff at all material times has enjoyed an easement of support from the wall.

By their amended defence, the defendants complain that this paragraph does not properly plead the nature of the easement of support which is claimed. It has been made clear during the hearing (if it was not clear before) that what the plaintiff claims is that he was entitled to support from the lower part of the defendants' wall for the land adjoining the wall forming part of his timber yard, and for the
E storage shed upon it, in its use as a shed for storing timber. In my view the pleading sufficiently covers this case, though I agree it could be more clearly set out in the statement of claim.

Apart from not admitting that such an easement exists, the defence in essence is that the collapse of the wall was caused, not by any failure on the defendants' part to continue to support the plaintiff's land and building, but as a result of the plaintiff bringing timber on his land which surcharged the land.

Mr Mauleverer submits that, as a matter of law, an easement of support if it has been created is for the support of the land and the building, but not for chattels, ie the timber on the land and in the building. In my judgment this is not a correct way to approach the
F issue. If an easement has been acquired by prescription, so that a plaintiff is entitled to enjoy support for his land and building from his neighbour's adjoining land or building, that right is for support for the land and building as they have been used during the period of prescription. Only if it can be shown that the nature of the use has materially altered can it be said that the easement does not benefit the plaintiff. We have not been referred to any authority in which this question has been raised directly, but, in my view, the proposition I have sought to set out is consistent with and follows the principles to be derived from the leading authority, *Dalton* v *Angus* (1881) 6 App Cas 740.

I consider next Mr Mauleverer's submission that the judge adopted a wrong approach when considering whether he should make leave to defend conditional upon the defendants' bringing a substantial sum G
into court.

It is a commonplace that if, on an application by a plaintiff under Ord 14 for summary judgment, the defendant persuades the court by his evidence that there is a triable issue, he should be given leave to defend. However, the courts are increasingly willing to make use of the power to grant such leave only subject to a condition requiring the defendant to pay money into court. The authorities on the grant of conditional leave are set out in some detail in the note in the *Supreme Court Practice*, 14/3-4/15, and I need not refer to them further.

In the decided cases, different expressions have been used to describe the circumstances in which a court is justified in imposing such a condition. The expressions most frequently used are that the court "is prepared very nearly to give judgment for the plaintiff", or H
that the defence can properly be described as "shadowy". I do not detect any difference of substance between these expressions. Mr Mauleverer submits, and I agree, that for a defence properly to be described as "shadowy", the court must have doubts about the defendants' good faith in advancing it. But the less likely a defence is to succeed, the easier it is for a court to infer that the defendant is seeking leave to defend for some ulterior reason, of which delay is the most obvious. Such a reason obviously casts doubt on the defendant's good faith.

In his judgment, the judge said:

> It seems to me that it is impossible for me to say on the affidavit evidence without actually trying this matter on reading experts' evidence, which I am J
> certainly not anxious to do, that there is no triable issue.

We should not, in my view, disagree with that conclusion. It follows that I would dismiss the plaintiff's cross-appeal.

The judge also said:

> All those considerations lead me to think that it matters not how one puts it, though I think it is probably wrong to say the defence is shadowy because it is actually quite clear. It is just that its chances of success are less than ample, shall we say.

When, in the authorities, judges have in the past referred to a defence being "shadowy", they have clearly, in my view, meant that it had little substance and perhaps that for this reason it was difficult to discern whether there really was a defence. They have not been referring to the clarity with which the defence was, or was not, K
pleaded. Thus in this sentence of his judgment I believe the judge was not using the word "shadowy" in the sense in which it was used in the authorities.

Mr Mauleverer also submits that where a defendant has shown that he will be well able to meet any judgment which might be given, a condition of payment into court should not be imposed on granting leave to defend. For my part I am not prepared to accept, without further argument, that when a judge is prepared "very nearly to give judgment for the plaintiff", he should nevertheless not impose a condition on granting leave to defend to a defendant who proves that he has ample resources to meet a judgment, while being willing to impose such a condition on a defendant of lesser means. However, in L
this case it is not necessary to decide whether this submission is correct, because on the material before us the defendants have not shown, to the necessary high standard, that they are at present well able to meet any judgment which they may have to meet.

Since I agree that the judge was correct in granting leave to defend, it is clearly inappropriate for me to rehearse the evidence as to the reasons why the wall collapsed. It suffices to say that, having considered that evidence, I am satisfied that the judge was right to say that he could "almost give judgment for the plaintiff". Accordingly, I also think that he was justified in imposing a requirement that the defendants should make a substantial payment into court as a precondition to the grant of such leave.

I note that the order actually made by the judge was: M

> If the defendant within 14 days pays £350,000 into court he may defend the action as to the whole of the plaintiff's claim.

It is apparent that he intended that if the defendants failed to comply with the condition, they should not have leave to defend. However, in my view, the order should also say in terms that if the defendants fail to comply with the condition, within the specified time, the plaintiff should be entitled to sign judgment for damages to be assessed.

Quantum

This leaves only the question which is raised by the appeal as to whether the amount of £350,000, which the judge required to be paid

A into court, is excessive. The plaintiff's total claim, including a continuing claim for loss of profits of the business and interest, comes to approximately £525,000. The defendants' main ground of objection to the amount ordered to be paid into court is that the evaluation of the rebuilding cost which forms the major item in the plaintiff's claim, namely some £343,000, is on their own pleading and evidence the cost of modernisation of the major part of their premises, not simply the cost of rebuilding the demolished building. However, the judge heard this argument and in the end decided to order payment in of a sum amounting to two-thirds of the plaintiff's total claim. I cannot say, on the material before us, that he was wrong so to order.

B *Conclusion*

I would therefore dismiss this appeal, and affirm the order made by the judge, save that:

(a) the time for compliance with the provision should be altered to enable the defendants now to comply with the condition; and

(b) there should be added a provision entitling the plaintiff to judgment for damages to be assessed if the defendants do not comply with the condition in the time required by the order.

Simon Brown LJ agreed and did not add anything.

Appeal dismissed. Cross-appeal dismissed. Leave to appeal to the House of Lords refused.

C

Court of Appeal

December 21 1992

(Before Lord Justice RALPH GIBSON, Lord Justice BELDAM and Sir John MEGAW)

WILSON AND ANOTHER v MARTIN'S EXORS

Estates Gazette June 19 1993

[1992] 24 EG 119

D

Adverse possession — Whether acts of squatter constituted dispossession of owner

In 1941 a Mr Martin purchased all or part of the Isenhurst Estate, Heathfield, Sussex. Part of the land was Sunset Farm. In 1947 Mr Martin, retaining the rest of the land, sold Sunset Farm to a Mr Crouch who, in 1961, sold the farm to Mr John Howitt. Mr Howitt remained in possession of Sunset Farm until 1985 when he sold it to the plaintiffs. Throughout that time Mr Martin was in occupation of the estate land until his death in 1981, and thereafter, the defendants, his executors,
E were in occupation. The land owned by the executors includes OS no 1770 the disputed area, a piece of woodland of about 2 acres-plus. On May 21 1989 the defendants went on to the disputed land and fenced it. In upholding the plaintiffs' claims for a declaration and damages for trespass, Judge Norrie in Eastbourne County Court found that the following acts of Mr Howitt constituted dispossession of Mr Martin within the meaning of the Limitation Act 1939: walking the boundary in 1962, cutting chestnut in 1961 and 1962 to mend fences, clearing fallen timber for firewood in 1963, repairing a wire fence in 1963-64 and cutting trees for sale in 1979. The judge was satisfied that Mr Howitt had always regarded the disputed
F land as his land and when he sold it to the plaintiffs he conveyed "such rights title and interest as he as transferor may have acquired by adverse possession". On appeal it was contended on behalf of the defendants that the judge was wrong in law in declining to apply the presumption laid down in *Smirk* v *Lyndale Developments Ltd* and, second, that there was no sufficient evidence on which the judge could find that Mr Howitt was at all material times in adverse possession of the disputed land.

***Held*: The appeal was allowed. It is irrelevant for the purposes of the Limitation Act 1939, in the absence of concealed fraud, that the true owner is ignorant that he has been dispossessed. There was no evidence in this case that Mr** G **Martin ever knew, or in probability must have known, of Mr Howitt's claim to the disputed land. The acts found by the judge to have been done by Mr Howitt before 1979 failed to amount in law to adverse possession. The repairing of an existing fence was insufficient to "make it perfectly plain to the world at large that Mr Howitt intended to exclude Mr Martin as best he could". The acts of cutting chestnut for the purposes of repairing fences were of no significance whatsoever and the act of taking fallen timber for firewood was of no greater significance. The concept of adverse possession is that of a landowner who fails to take legal action** H **to enforce a cause of action to recover possession of land wrongly taken by a trespasser. The acts of Mr Howitt which the judge found to have been done by him did not make it plain to the world at large that he has been intending to exclude Mr Martin from the 2 acres-plus of his woodland.**

The following cases are referred to in this report.

Boosey v *Davis* (1988) 55 P&CR 83, CA

Buckinghamshire County Council v *Moran* [1989] 3 WLR 152; [1989] 2 All ER 225, CA

Powell v *McFarlane* (1979) 38 P&CR 452

Smirk v *Lyndale Developments Ltd* [1975] Ch 317; [1975] 2 WLR 495; [1975] 1 All ER 690; (1974) 29 P&CR 407, CA J

This was an appeal by the defendants, the executors of the will of Frederick Harry Martin deceased, against the order of Judge Norrie in Eastbourne County Court on October 3 1991 that the plaintiffs, Mr and Mrs Wilson, were the owners of a disputed piece of land annexed to the boundary of Sunset Farm, Heathfield, Sussex.

Patrick Lawrence (instructed by Cripps Harries Hall, of Uckfield) appeared for the appellant defendants; Timothy Sisley (instructed by Pegram Heron, of Heathfield) represented the respondent plaintiffs.

Giving judgment, **Ralph Gibson LJ** said: This is an appeal by the defendants, who are the executors of F H Martin (deceased) in an action in Eastbourne County Court, against the order of Judge Norrie of October 3 1991. K

By that order the judge in substance held that the plaintiffs, Mr and Mrs Wilson, were the owners of a disputed piece of land next to the boundary of Sunset Farm, near Heathfield, Sussex, in which title had been gained by adverse possession. The defendants contend in the appeal that the decision of Judge Norrie cannot be sustained in law on the evidence.

When the dispute arose the defendants on May 21 1989 went on to the disputed land and fenced it. Applying the provisions of the Limitation Act 1939 as amended, the central question — the crucial question as the judge saw it and as has not been in issue — was whether Mr Howitt, the predecessor-in-title of the plaintiffs, was in adverse possession of the disputed land on May 21 1977, 12 years L before.

The facts

The story starts in 1941 when Mr Martin purchased all or part of the Isenhurst Estate, near Heathfield, Sussex. Part of that land was Sunset Farm. In 1947 Mr Martin, retaining the rest of his land, sold Sunset Farm to a Mr Crouch, and in 1961 Mr Crouch sold Sunset Farm to a Mr John Howitt. For a time Mr Howitt's father had an interest in it, but that does not matter. This case is concerned with the actions of Mr Howitt with reference to the disputed land and the consequences in law of those actions. Sunset Farm is to the west of land retained by Mr Martin and near the disputed land the boundary runs north and a bit east. M

Mr Howitt remained in possession of Sunset Farm until 1985 when he sold the farm to the plaintiffs, Mr and Mrs Wilson. Throughout that time Mr Martin was in occupation of the estate land until his death in 1981. Thereafter his executors have been in occupation.

The dispute is over a piece of woodland of about 2.5 acres. It is common ground that Mr Martin was the freehold owner of the disputed land. His land included OS nos 1770, which was woodland. The western boundary of Mr Martin's land at that point ran along the edge of the woodland where the woodland met open grazing or parkland part of Sunset Farm including OS no 1776 and 1775. There was at all material times a fence along that boundary, which boundary followed an irregular line. The disputed land consists of a

A piece of land cut out of Mr Martin's woodland by means of a straight line running about NNE, which starts from a point on the field fence, the undisputed boundary, cuts off a roughly triangular piece of dense woodland and rejoins the undisputed boundary again at a point in the field fence between woodland and pasture.

The acts of Mr Howitt, which according to the ruling of Judge Norrie constituted dispossession of Mr Martin by Mr Howitt, began after Mr Howitt and his father bought Sunset Farm from Mr Crouch in 1961. Mr Crouch did not own the disputed land. He had not bought it and it was not conveyed to him. The acts of Mr Howitt may be listed from the judgment of Judge Norrie as follows:

(a) he walked the eastern boundary in 1962;

B (b) he cut chestnut from the disputed land in 1961 and 1962 to mend fences;

(c) he cleared fallen timber from the disputed land after 1963 for firewood and nobody objected;

(d) in the winter of 1963-64 he repaired — it must be emphasised that he repaired and did not erect — a wire fence along the eastern boundary of the disputed land, ie along the line enclosing the disputed land as described above;

(e) in 1979 he cut trees for sale from the disputed land. That was an act done outside the relevant period of time, but the judge regarded it as having some evidential value.

As to the subjective belief of Mr Howitt, the judge was satisfied that he always regarded the disputed land as his land, as did the
C plaintiffs when they purchased Sunset Farm.

When Mr Howitt came to sell Sunset Farm he said that he was told by his agent by reference to the deeds that he had no paper title to the disputed land. When he sold it to the plaintiffs he conveyed, so far as concerned the disputed land, "such rights title and interest as he as transferor may have acquired by adverse possession".

Mr Howitt did not at any time assert to Mr Martin, or to anyone acting on behalf of Mr Martin, that Mr Howitt had taken possession of the disputed land or claimed to exclude Mr Martin from it. Mr Martin had no estate manager or agent upon his land. Mr Howitt saw him only on rare occasions on the estate. There was no evidence that Mr Martin or any agent before him had ever seen Mr Howitt doing any of the acts said to have constituted the taking of possession by Mr
D Howitt. At most it was suggested that that might have happened.

Mr Howitt had an arrangement with Mr Martin whereby Mr Howitt used some fields on Mr Martin's land for grazing. The judge referred to this as a "complication on the facts" and described it as follows:

> The tenancy [of the fields] was agreed on the basis of payment of a small amount of rent and that Mr Howitt would repair the fences and ensure that the stock would not trespass. This agreement continued until 1978 or 1979 and the fields were vacated in 1980. Mr Howitt said that Mr Martin was a sympathetic landlord and that if he had a bad year then Mr Martin would propose that Mr Howitt maintain the drive of the Estate in lieu of payment of rent.

The amount of the rent was £30, increasing eventually to £100 pa.
E Mr Howitt accepted in evidence that, under his arrangement with Mr Martin, there was no barring of him from any area on the estate, but he was interested only in the fields and some buildings.

Upon the case made for the plaintiffs, on the evidence of Mr Howitt and of Mr Wilson, the eastern boundary of the disputed land was claimed as a straight line running as described above from one point on the undisputed boundary to another: see the plan attached to the amended particulars of claim. Mr Howitt, upon walking the eastern boundary of his newly acquired farm in 1962, must presumably have walked along that line; and that presumably was the line of the fence which he repaired in 1962-63. Mr Hewitt's evidence was that the fence which he repaired was intact from 1970 to 1972 but was, as he put it, possibly in disrepair by 1984. The judge accepted Mr
F Hewitt's evidence and further held that "on the balance of probabilities the fences were intact in 1977".

On November 20 1990 there was a site meeting for inspection of the disputed land and of such signs as there were of the fence. The meeting was attended by the two surveyors advising the plaintiffs and the defendants; and Mr Wilson, the first plaintiff, and Mr Dutnall, one of the defendants, were also there. Such traces of a fence as there were were charted on a drawing. There was nothing to show a line of fence along the line claimed on the plan attached to the particulars of claim nor a particular place where the line began at the field fence at the southern end of the line. There were traces of wire and of the fixing of wire to posts and some fallen posts or spiles and traces of wire in growing trees, some to the east of that line and some not only to the west of the line but also on the line of the paper-title boundary G
fence. The agreed report of the meeting and the drawing were before the judge as part of the evidence. The surveyors were not called.

The hearing was conducted over two days in February 1991. According to the judge's note, the case as opened for the plaintiffs was that from 1962 Mr Howitt had exercised rights of ownership over the land appropriate to the nature of the land and also "fenced off land along the eastern boundary — he undertook the fencing after 1963 — a post and wire fence along the eastern border of Sunset Farm, and when he came to the land he fenced to the east and therefore enclosed that land". Therefore, Mr Martin's title, it was said, was extinguished in 1973 or 1975.

The explanation for the fact that there was no visible fence on inspection was given that Mr Wilson, who had started the business of H
organising war games on his farmland in which the participants shoot at each other with pellets and dye, had removed Mr Hewitt's fence because it was dangerous.

The witnesses called for the plaintiffs included Mr Howitt and Mr Wilson.

The submissions to Judge Norrie for the defendants by Mr Lawrence included the following:

(i) it had not been proved that Mr Howitt had taken possession of the land so as to exclude Mr Martin;

(ii) on the evidence the fence, if it was repaired by Mr Howitt, must have disappeared or had been in very dilapidated condition by 1970 and in obvious contrast to the well-maintained fence along the J
original boundary line;

(iii) Mr Howitt had permission to go over the disputed land;

(iv) even if Mr Howitt was not on the disputed land by licence his so-called acts of possession did not make it clear that he was dispossessing the true owner. In particular, he had not erected a fence or enclosed the land, but had merely repaired an existing fence;

(v) Reliance was placed on the presumption applied in *Smirk* v *Lyndale Developments Ltd* [1975] Ch 317 that, where a tenant takes possession of a piece of his landlord's land, the land taken is to be treated as part of his holding under the tenancy.

Reference was made to *Buckinghamshire County Council* v *Moran* [1989] 2 All ER 225, but not to the decision of Slade J (as he then was) in *Powell* v *McFarlane* (1979) 38 P&CR 452 where the K
principle of law upon which Mr Lawrence clearly based his submissions was given very clear expression.

It is convenient to state at this point what I regard as the principle upon which this appeal is to be decided. At p472 Slade J said:

> An owner or other person with the right to possession of land will be readily assumed to have the requisite intention to possess, unless the contrary is clearly proved. This, in my judgment, is why the slightest acts done by or on behalf of an owner in possession will be found to negative discontinuance of possession. The position, however, is quite different from a case where the question is whether a trespasser has acquired possession. In such a situation the courts will, in my judgment, require clear and affirmative evidence that the trespasser, claiming that he has acquired possession, not only had the requisite intention to possess, but made such intention clear to the world. If L
> his acts are open to more than one interpretation and he has not made it perfectly plain to the world at large by his actions or words that he has intended to exclude the owner as best he can, the courts will treat him as not having had the requisite *animus possidendi* and consequently as not having dispossessed the owner.

In *Buckinghamshire County Council* v *Moran*, at p240, Nourse LJ said that Slade J's judgment accurately stated the law in all material respects and, for my part, I respectfully agree.

Judgment was reserved by Judge Norrie in February and not given until October 4 1991. We do not know what may have been the cause of that delay. It is unexplained. It must be referred to as a period of delay, which is to be regretted. The judge reached her conclusions by the following steps. M

Although the plaintiffs took possession of the property in 1985 there was no complaint or assertion of title by the defendants until March 1988, when the defendants queried the grant of possessory title with the Land Registry; and it is not without significance that this coincided with an application for planning permission for change of use of part of the land by the plaintiffs, ie for the activity of war games. As to the law, after reference to the provisions of the Limitation Act, the judge stated that the plaintiffs must establish that Mr Howitt had dispossessed or ousted Mr Martin as the paper owner of the disputed land and, accordingly, that Mr Howitt not only had actual possession but also adverse possession within the meaning of the Act. Adverse possession means possession inconsistent with the

A title of the true owner by a person who has done acts which show that he is seeking to dispossess the true owner.

The judge referred to a visit made by herself with the parties' representatives before the hearing and to the agreed report of the joint inspection on November 20 1990. She commented only that she had been able to see the posts and barbed wire and other features referred to in the agreed report, but did not identify anything as having any particular significance.

Having considered all the evidence and stated the acts of Mr Howitt which I have listed earlier in this judgment, the judge observed:

(a) That it was "natural that the fence along the perimeter of the field on the western edge of the land in dispute was more substantial B than the inner (*sic*) eastern fence for that was a stock fence and had to be more robust to prevent the animals from straying into the area of dense and overgrown woodland which comprised the disputed land". I add that it is not clear to me why the eastern fence was referred to as "inner". To Mr Howitt it was his outer fence at that point.

(b) Acts sufficient to show that a trespasser is seeking to dispossess the owner must depend on the nature of the land. This was overgrown woodland which was fenced by Mr Howitt and from which he took timber. The sale of timber for pulp from the disputed land in 1979 could not, by itself, be sufficient (because within 12 years of 1989) but it was of evidential significance because it demonstrated the C continuing use and occupation of the area.

(c) The western fence of the disputed land (along the paper boundary) was a stock fence which needed to be more secure to contain cattle whereas the eastern boundary simply marked the boundary and Mr Martin, coming across it, must have realised it was such. It cannot, said the judge, be argued that the now dilapidated eastern fence was a fence internal to the estate for there is no evidence to support that proposition and it does not run alongside any footpath.

(d) The circumstances of the case were quite different from those in the case of *Smirk*. The leasing of a few fields for pasture on another part of Mr Martin's estate could not change the true relationship between Mr Howitt and Mr Martin, which was that of adjoining D freeholders. Finally, the judge expressed her conclusion as follows:

> I am satisfied that Mr Howitt had exclusive use [of the disputed land] of the triangular piece of woodland at the relevant time in 1977 and that his earlier fencing of it and use of it amounted to adverse possession. The fact that Mr Howitt was allowed to go onto other parts of the estate is not relevant as his right to do so was with permission and to enable him to reach the fields, which were not adjacent to the land in dispute, which he tenanted whereas his going onto the disputed land was not subject to any such granting of permission.
>
> Having considered all the evidence, the surveyors report and my inspection of the site I find that Mr Howitt took possession of the disputed land in 1963 when he fenced and that he used the woodland in the only way he could thereafter using it and treating it as his own and that possession was adverse and continued after the sale of Sunset Farm to the plaintiffs in 1985 and until E this action was brought.

The notice of appeal of October 31 1991 asserted, first, that the judge was wrong in law in declining to apply the presumption laid down in *Smirk* v *Lyndale Developments Ltd* and, second, that there was no sufficient evidence upon which the learned judge could find that Mr Howitt was at any material time in adverse possession of the disputed land. The plaintiffs served no respondents' notice and relied upon the reasons given by the judge. Each member of this court, on reading the papers, formed the provisional view that the appeal appeared to turn upon the second ground of appeal, and Mr Lawrence was asked to defer the making of his submissions upon the presumption in *Smirk's* case until after we had heard Mr Sisley upon the issue of adverse possession. In the result we did not find it F necessary to hear argument directed to the presumption applied in *Smirk's* case.

Mr Lawrence in substance repeated in this court the submissions made by him to the learned judge, but supported by the citation of additional authorities. His submissions were based upon acceptance of the judge's primary finding of fact, save for the judge's finding that the fence along the eastern side of the disputed land was intact in the mid-1970s. For that finding he asserted there was no evidence to support it. Mr Howitt said the fence was intact by 1970-72 and that Mr Martin had sold timber up to the Isenhurst line (which I take to be a reference to the fence), but by 1985 the fence was in disrepair; and that the fence in heavy woodland would last 10 to 15 years. By 1985 the condition of the fence as described by Mr Wilson in evidence was that it was "dilapidated fencing along the approximate disputed G boundary" with "no fencing at the south west" and he had removed the fence along the eastern side of the disputed land because it was old, decrepit fencing.

Next it was contended that, whether or not the judge's finding that the fence was intact in the mid-1970s is to be upheld, the acts which the judge found to have been proved were not sufficient to constitute adverse possession: see *per* Nourse LJ in *Boosey* v *Davis* (1988) 55 P&CR 83 at p86. Mr Lawrence relied in particular on the fact that Mr Howitt did not enclose the disputed land by fencing it. He did no more than to repair an existing fence, which was internal to the Isenhurst Estate. Further, the presence of Mr Howitt on the disputed land would not have been seen by Mr Martin as calling for particular H notice or comment. He had, as he accepted, permission to go where he chose to go upon the estate.

Mr Sisley contended that the facts found were indeed sufficient in law. By the work done to the fence by Mr Howitt in 1963 the true owner had indeed been, as he put it, "set at defiance". Thereafter, Mr Howitt had made such use of the land as was appropriate to the nature of the land. The possessory title was complete by 1977 whatever happened to the fence thereafter. Mr Howitt could not reasonably be expected to have done more to make it plain to the world at large that he intended to exclude Mr Martin if more were necessary for him to do because, as the judge found, Mr Howitt believed that he was the true owner of the disputed land. The acts done on the disputed land were sufficient even if the repair of the J fence be ignored. The fact that the traces of the repaired fence, as observed by the surveyors and described in the agreed report, did not support the line asserted on the plan attached to the amended particulars of claim by reference to which the judge's order was made was, it was said, irrelevant. The defendants had not contended for any other line and had, in effect, relied upon the discrepancies as reasons why the judge should not accept the accuracy of recollection of Mr Howitt, but that attack on the evidence of Mr Howitt had failed.

Conclusion

I have no doubt that the order of the learned judge cannot be sustained in law. I accept Mr Lawrence's main submission. K

I have sympathy with Mr Sisley's rueful comment that Mr Howitt had no occasion from his point of view, if there was need as Mr Sisley denied that there was, to do more in order to make it perfectly plain to the world at large by his actions or words that he was intending to exclude Mr Martin from the small section of Mr Martin's extensive woodlands because, according to Mr Howitt, Mr Howitt believed that he already had full title to it. Mr Howitt, however, had only himself to blame. The boundary of the land which he and his father acquired in 1961 was perfectly clear on the deeds. The judge noted that Mr Howitt "walked the eastern boundary in 1962". That, I am sure, is a very sensible thing for the purchaser of land to do so as to make certain where on the land are the boundaries of the land L acquired. It is, however, a useless exercise if the new owner does not have the plan of the paper title with him because without the plan he will discover no more than what he thinks the boundary to be. Mr Howitt, according to him, did not take that obvious step until he decided to sell in 1985. If he had done that in 1962 he surely would not have laid claim to 2.5 acres of his neighbour's woodland.

It is irrelevant for the purposes of the Limitation Act 1939, in the absence of concealed fraud, that the true owner is ignorant that he has been dispossessed: see *per* Slade J in *Powell* v *McFarlane* at p480. He added:

> In view of the drastic results of a change of possession, however, a person seeking to dispossess an owner must, in my judgment, at least make his intentions sufficiently clear so that the owner, if present at the land, would M clearly appreciate that the claimant is not merely a persistent trespasser, but is actually seeking to dispossess him.

There is no evidence in this case that Mr Martin ever knew, or in probability must have known, of Mr Howitt's claim to the disputed land. The cutting of timber in 1979 might, I think, have qualified as an act of constituting adverse possession, but we have no findings as to the nature or extent of what was done. If what had gone before could constitute adverse possession the cutting of timber was of evidential significance, as the judge said, demonstrating the continuing use and occupation of the area. The acts found by the judge to have been done by Mr Howitt before 1979, however, fail in my judgment to amount in law to adverse possession.

The act of repairing an existing fence in the circumstances of this case seems to me to be obviously insufficient to "make it perfectly plain to the world at large that Mr Howitt intended to exclude Mr Martin as best he could". The act was not done so as to be seen by Mr Martin. Mr Howitt described what was done as "replacing some wire and some posts" in November/December 1963. Later, he said, we "managed to fence right through". If Mr Martin had found the fence to be in better repair than he had previously observed it to be — and why should he? — I can see no reason why he should interpret its improved condition as a claim of Mr Howitt to exclude him.

There was no evidence from either side as to why the fence which Mr Howitt repaired was where it was. In its place it was, without question, internal to Mr Martin's estate and I do not understand why the judge should think that evidence was necessary from the defendants to support that proposition. The strands of wire were on Mr Martin's land. If it is necessary to speculate after Mr Martin's death as to why the strands of wire were there, and I do not think it is, one answer could be that he wished to separate one section of his woodland from another for some purpose of cutting or other aspect of management, perhaps once seen as of some importance and later abandoned.

The acts of cutting chestnut for the purposes of repairing fences are, I think, of no significance whatever. If Mr Martin had seen Mr Howitt do it he would have supposed, being a sympathetic landlord as Mr Howitt acknowledged him to be, that Mr Howitt was not stealing wood but taking wood to repair fences as Mr Howitt had agreed to do. The act of taking fallen timber for firewood is of no greater significance. If Mr Martin had observed the action he would have supposed, and I do not doubt accurately, that Mr Howitt believed that Mr Martin would not object.

It would, I think, be plainly unjust if the law required this court to uphold the decision of Judge Norrie. The concept of adverse possession is that of a landowner who fails to take legal action to enforce a cause of action to recover possession of land wrongly taken by a trespasser. The acts of Mr Howitt which the judge found to have been done by him did not, in my judgment, make it plain to the world at large that he was intending to exclude Mr Martin from the 2.5-acre piece of his woodland.

I do not know why the judge thought it of significance that the assertion of title by the defendants did not come until March 1988 when Mr Wilson made application for planning permission. There was no evidence that before that date the claims of Mr Howitt or of the plaintiffs to the disputed land were known to the defendants.

I would, therefore, allow this appeal. I would wish to hear counsel as to the terms of the judgment of the court on the counterclaim and as to any directions which may be necessary for the determination of the damage claim of the defendants on their counterclaim.

Beldam LJ and **Sir John Megaw** agreed and did not add anything.

Appeal allowed with costs; defendants to have costs in the court below.

For further cases on this subject see p 197

TOWN AND COUNTRY PLANNING

A
House of Lords

January 21 1993

(Before Lord TEMPLEMAN, Lord ACKNER, Lord OLIVER OF AYLMERTON, Lord MUSTILL and Lord WOOLF)

CANTERBURY CITY COUNCIL v COLLEY AND ANOTHER

Estates Gazette March 20 1993

[1993] 11 EG 126

B *Town and country planning — Revocation of planning permission — Compensation — Whether assumption that planning permission would be granted for development of the land of any class specified in Schedule 8 to the Town and Country Planning Act 1971 applied*

On November 16 1961 Kent County Council granted an outline planning permission for "the demolition of house and erection of new dwelling" relating to property at Whitstable. Following that permission, in September 1963 the house originally standing on the property was demolished, but no new dwelling was ever in fact erected. On January 31 1986 the appellants purchased the site at a price of £14,500 and C thereafter there ensued a lengthy dispute between the appellants and the respondents, Canterbury City Council, as the local planning authority for the area, relating to the continuing validity of the 1961 planning permission. That dispute was finally resolved in the appellants' favour. On November 12 1987 the respondents notified the appellants that they had made an order revoking the 1961 permission and, following submission to the Secretary of State for confirmation and a public local inquiry, the order was confirmed on August 17 1989. The appellants made a claim for compensation under section 164 of the 1971 Act and on D March 27 1991 the Lands Tribunal made an interim award of £108,626.84 consisting of £106,759 depreciation in value of the land and £1,866.84 for abortive design expenditure. The member of the tribunal valued the land with the benefit of the 1961 permission at £115,000; without the permission, and disregarding the assumption required to be made by section 164(4) of the 1971 Act, the valuation was £8,250. On the basis of making the statutory assumption, however, the member of the tribunal found that the post-revocation value of the land was £70,000, thus reducing the compensation on this footing to £45,000. In his decision the member said: "that in revocation order cases section 164 will only operate as E Parliament intended it to operate if the development resulting from the Schedule 8 assumption, although a hypothetical one, is not that which also corresponds with the subject of the associated revocation order". This could be achieved by notionally adding at the end of subsection (4) of section 164 words such as "unless such planning permission is the subject of revocation order proceedings". The Court of Appeal, which had allowed the respondents' appeal from the tribunal, gave leave to appeal.

Held: **The appeal was dismissed. The provisions of section 164(4) of the 1971 Act are, in terms, mandatory and there is no F escape from the proposition that they require to be applied even in the case postulated by the notional permission to be assumed to be the very permission which has in fact been revoked. The section is, in terms, postulating an artificial assumption and dealing with a situation where no planning permission is, in fact, in existence or likely to be in existence. The legislative history, starting with the Town and Country Planning Act 1947, and taking into account the substantial amendments made by the Town and Country Planning Act 1954, underlines the impossibility of escaping the conclusion that the clear purpose of the legislature was to limit the G amount of compensation payable for depreciation owing to a revocation of planning permission by fixing a base value for the land in all cases on the footing that planning permission for any class of Schedule 8 development would be granted. That must include, however arbitrary it may seem, even development of that very class which was the subject of the revoked permission. It may well be that the particular circumstances of the revoked permission being the very permission comprehended in the statutory assumption was not one which the legislature foresaw as unlikely to occur.**

The following cases are referred to in this report. H

Central Control Board (Liquor Traffic) v *Cannon Brewery Co Ltd* [1919] AC 744; (1919) 17 LGR 569, HL

Stock v *Frank Jones (Tipton) Ltd* [1978] 1 WLR 231; [1983] 1 All ER 948; [1978] ICR 347, HL

This was an appeal by the appellants, Mark Colley and Janine Colley, from a decision of the Court of Appeal which had allowed an appeal by the respondents, Canterbury City Council, from a decision of the Lands Tribunal determining the amount of compensation payable by the respondents to the appellants following the revocation of a planning permission granted in 1961.

David Keene QC and Barry Payton (instructed by Hempsons) J appeared for the appellants; Roger Henderson QC and Edward Cousins (instructed by Sharpe Pritchard, as agents for Canterbury City Council) represented the respondents.

In his speech, **LORD OLIVER OF AYLMERTON** said: Section 45 of the Town and Country Planning Act 1971 (re-enacting provisions which first appeared in section 21 of the Town and Country Planning Act of 1947) empowers a local planning authority, subject to certain conditions and subject, in the absence of consent from persons affected, to confirmation by the Secretary of State for the Environment, to order the revocation or modification of a planning permission which has been granted. Where such an order (other than one made with the consent of the person affected) has been made and K confirmed, a person interested in the land who has incurred expenditure rendered abortive by the order or who has otherwise sustained loss directly attributable to the revocation or modification is entitled, under section 164 of the Act, to claim and receive compensation for the expenditure or loss so incurred or sustained. Subsection (4) of section 164, however, contains a provision limiting the amount of any loss capable of being claimed for depreciation of the value of an interest in the land affected by the revocation or modification. It is in these terms:

In calculating, for the purposes of this section, the amount of any loss or damage consisting of depreciation of the value of an interest in land, it shall be assumed that planning permission would be granted for development of the land of any class specified in Schedule 8 to this Act.

Schedule 8 (which is headed "development not constituting new development") is primarily related to the provisions of section 169 of the Act, which applies not to a revocation of an existing planning permission but to the refusal by the Secretary of State, either on an appeal or on a reference, of an application for planning permission. If the application is for development of a class specified in Part II of Schedule 8, that section confers a right to compensation. Part I of the Schedule relates to development not ranking under section 169 for compensation for a refusal for planning permission and it embraces, for relevant purposes, the rebuilding of buildings in existence on January 1 1948 or coming into existence thereafter so long as the cubic content of the original building is not exceeded by certain defined limits.

The present appeal, which is brought with the leave of the Court of Appeal, relates to a property at Whitstable known as Marley House for which, subject to conditions and detailed approval, an outline planning permission had been granted on November 15 1961 by Kent County Council, then the local planning authority, for "the demolition of house and erection of new dwelling". Following that permission, the house originally standing on the property was demolished in September 1963, but no new dwelling was ever, in fact, erected. On January 31 1986 the appellants purchased the site at a price of £14,500 and there thereafter ensued a lengthy dispute between the appellants and the respondents, now the planning authority for the area, relating to the continuing validity of the 1961 planning permission. That dispute was finally resolved in the appellants' favour. In the interim the appellants had made a new application for the erection of a house within the same curtilage as the original Marley House, although not in precisely the same position. That application was refused, but in June 1987 the respondents told the appellants that the original permission was regarded as valid and that they would consider proposals for implementing it. The appellants sought to follow this up and expended moneys in abortive design expenditure. That sum is not now in dispute. On November 12 1987, however, the respondents notified the appellants that they had made an order revoking the 1961 permission and had submitted it to the Secretary of State for confirmation. Following a public inquiry the order was confirmed on August 17 1989.

The appellants then made a claim for compensation under section 164 which was referred to the Lands Tribunal. On March 27 1991 the tribunal made an interim award of £108,626.84 consisting of £106,750 depreciation in value of the land and £1,876.84 abortive design expenditure. The only question currently in issue is the correctness of the member's assessment of the depreciation in value of the land at the sum of £106,750 which was arrived at by, in effect, disregarding the assumption required to be made by section 164(4). The member valued the land with the benefit of the 1961 permission at £115,000. Without that permission, and disregarding the statutory assumption, the valuation was £8,250. On the basis of making the statutory assumption, however, the member found that the post-revocation value of the land was £70,000, thus reducing the compensation on this footing to £45,000.

In reaching the conclusion that he did in apparent defiance of the express requirement of section 164(4) the member was much influenced by the fact that the development to which the revoked permission related was, in substance at least, the very permission which the subsection required to be assumed to be still subsisting. As it happened, the evidence satisfied him that the 1961 permission would have enabled the appellants to build a considerably larger house than the original Marley House, with the result that the actual post-revocation value of the site was greater than that which resulted from the simple application of the statutory assumption. The statutory assumption would, however, result, on any analysis, in the appellants being deprived of a substantial part of their real loss by the attribution of a notional planning permission which was, in fact, incapable of implementation. Indeed, as Mr Keene QC has forcefully argued, if one supposes a planning permission covering merely the erection of a building of the same size and on the same site as the building demolished, the effect of the statutory assumption is to eliminate any compensation at all for the undoubted depreciation of the land value occasioned by the revocation. The member's view was that this could not possibly have been the intention of Parliament in enacting a provision designed to provide compensation.

It seems to me

— he said —

that in revocation order cases section 164 will only operate as Parliament intended it to operate if the development resulting from the Schedule 8 assumption, although a hypothetical one, is not that which also corresponds with the subject of the associated revocation order. This could be achieved by notionally adding at the end of subsection (4) of section 164 words such as "unless such planning permission is the subject of revocation order proceedings".

From this decision the respondents appealed by way of cases stated to the Court of Appeal which, on March 20 1992, allowed the appeal but granted leave to the appellants to appeal to your lordships' House.

My lords, that the results of the application of section 164(4) can, in appropriate circumstances, be anomalous is beyond doubt. The respondents' argument involves, as Sir Donald Nicholls V-C pointed out in his judgment in the Court of Appeal:

> . . . as a necessary corollary, the proposition that in assessing compensation for the revocation of permission to demolish and rebuild a house, the valuer must assume that permission to rebuild the house would be granted. On its face the proposition seems bizarre.

Indeed, the result of the statutory assumption, where the holder of a permission to demolish and rebuild has actually carried out the demolition at the date when the permission is revoked, is to leave him without a house and with a site upon which he can neither rebuild nor effectively claim any compensation for its depreciation in value.

Nevertheless, the provisions of section 164(4) are, in terms, mandatory and I cannot, for my part, see any escape from the proposition that they require to be applied even in the case postulated of the notional permission to be assumed being the very permission that has, in fact, been revoked. It has been suggested that something turns on the use by the legislature of the word "assumed" and it is argued that you cannot "assume" that which, because it cannot in fact be implemented, is incapable of being rationally assumed. That, however, I find myself quite unable to accept. The section is, in terms, postulating an artificial assumption and dealing with a situation where no planning permission is, in fact, in existence or likely to be in existence.

More convincing, perhaps, is the appeal to the well-known canon enunciated by Lord Atkinson in *Central Control Board (Liquor Traffic)* v *Cannon Brewery Co Ltd* [1919] AC 744 at p752, that an intention to take away the property of a subject without giving him a legal right to compensation for the loss of it is not to be imputed to the legislature unless that intention is expressed in unequivocal terms. But the difficulty here is that, even assuming that there is room for the application of this principle in a section whose express purpose is to provide and, at the same time, to limit compensation, the terms of the section are, indeed, entirely unequivocal.

Then it is said that the provision, if applied strictly in accordance with its terms, leads to an absurdity. It is pointed out that, in the case of a refusal of planning permission for development specified in Part II of the Schedule 8, there is, in section 169, specific provision for compensation to be assessed on the full difference in the value of the land with the permission and the value of the land without it. It is, it is argued, absurd that where permission has been granted and is then revoked there should be a harsher rule and one which can, in certain circumstances, result in the subject receiving no compensation at all. Reliance is placed on the speech of Lord Simon of Glaisdale in *Stock* v *Frank Jones (Tipton) Ltd* [1978] 1 All ER 948. But that reference is, as it seems to me, fatal to the submission. Lord Simon said at p954:

> . . . a court would only be justified in departing from the plain words of the statute were it satisfied that: (1) there is a clear and gross balance of anomaly; (2) Parliament, the legislative promoters and the draftsman could not have envisaged such anomaly and could not have been prepared to accept it in the interest of a supervening legislative objective; (3) the anomaly can be obviated without detriment to such legislative objective; (4) the language of the statute is susceptible of the modification required to obviate the anomaly.

Even assuming that the anomaly, which arises only in very limited circumstances, is such as to satisfy the second of these conditions, it is really impossible, as Mr Keene was, I think, constrained to admit, to suggest any way in which the statutory language could be modified save by the addition, for which there is no context, of some such words as those used by the member of the Lands Tribunal. That involves more than a mere purposive construction. It involves substantially rewriting the section on the supposition that the legislature, had it thought about the particular case, would have

A expressed itself in substantially different terms from those which it in fact chose to use. Accepting, as I do, that the appellants have suffered hardship in being deprived of a substantial part of the value represented by the revoked permission, I cannot, for my part, regard this as a legitimate approach to construction of the statute.

In a clear and most helpful address, Mr Keene's final appeal was to a consideration of the legislative history in order to support a submission that a purposive construction of section 164 leads to the conclusion that subsection (4) cannot be applied in accordance with its literal terms where this would result in compensation being denied. The origins of the section and of Schedule 8 and their subsequent amendment and incorporation into a succession of consolidating statutes has been fully and carefully analysed and B described in the judgment of the vice-chancellor which, in the ordinary way, I would be content simply to adopt in full without repetition. In deference to Mr Keene's very careful and detailed submission, however, I will endeavour to describe the history in summary form. There was nothing revolutionary or novel about the provisions of sections 45 and 164 or those of Schedule 8. Substantially similar provisions were first enacted in the Town and Country Planning Act 1947 in sections 21 and 22 and Schedule 3. In particular, the assumption in section 164(4) formed an essential part of the compensation calculation and was provided for in section 22(7) of the Act of 1947. While, however, in that Act it served the purpose of providing a valuation base from which depreciation of C land value for compensation purposes was to be calculated, the rationale of that valuation base was quite different.

The philosophy behind the Act of 1947 was not only that development, whether it consisted of building operations or change of user, should be controlled but that the development value of land, over and above the value attributable to an artificially defined "existing use" of the land, should be taken into public ownership. Thus, when the Act came into force on the appointed day (July 1 1948) a landowner became entitled to develop his land in a way not classified as an existing use only upon payment to the public purse, in the shape of a central land board, of a development charge under section 69 of the Act which was to be calculated, in effect, by reference to the amount by which the value of the land was enhanced D by permission for the proposed development. The existing use for the purposes of these provisions was defined by Schedule 3 to the Act, Parts I and II of which were in substantially the same terms as Parts I and II of Schedule 8 to the 1971 Act, save that they were headed respectively "development included in existing use for purposes other than compensation under section 20" and "development included in existing use for all purposes". The significance of this distinction was that it was provided in section 20 (the forerunner of section 169 of the 1971 Act) that where, on an application for permission to carry out development of a class specified in Part II, that application, either on appeal or on a reference, was refused by the Secretary of State or was granted subject only to conditions, E compensation could be claimed from the local planning authority for the difference between the value which the land would have had the permission been granted or granted unconditionally and its value without the permission or with only the conditional permission. No provision, however, was made for compensation for refusal of permission to carry out development of a class specified in Part I of the Schedule and although this may seem arbitrary it could not, as Sir Donald Nicholls V-C remarked in the course of his judgment, have been accidental. The existing use provisions of Schedule 3 had a further relevance. Under Part VI of the Act there was established a compensation fund of an arbitrary sum of £300m to meet claims by landowners for loss of development value as a result of the Act, that is to say, the difference between the unrestricted value of land F immediately prior to July 1 1948 and its value subject to the restrictions imposed by the Act. Since development charges were to be levied only on development other than that comprised in the existing use provisions in Schedule 3, section 61 provided that the restricted value of the land for the purposes of this compensation was to be calculated on the assumption that planning permission would be granted for development of any class specified in this Schedule.

It is against this scheme that sections 21 and 22 of the Act of 1947 fall to be considered. Section 21 was, for material purposes, in substantially the same terms as section 45 of the Act of 1971, but the compensation provisions in section 22, albeit substantially in the same terms as those of section 164, contained a crucial difference in the form of a proviso that no compensation should be paid for depreciation in value unless (for relevant purposes) a development G charge had been paid in respect of the revoked permission. Subsection (7) provided for the assessment of compensation in accordance with the Acquisition of Land (Assessment of Compensation) Act 1919 and required it to be assumed (using words identical with those of section 164(4) of the 1971 Act) that planning permission would be granted for any class of development specified in Schedule 3. In the context of the scheme of the Act this was perfectly logical. Since, by definition, the development charge which formed the essential condition for payment of compensation became payable only on a development which was not within the existing use, it was both rational and necessary to provide in the calculation of the amount of depreciation a base value which assumed the existence of a permission for existing use development. H

So far so good. The difficulty arises in the next stage in the legislative history. The planning legislation was substantially amended in 1953 and 1954 when development charges were abolished. The Town and Country Planning Act 1954 contained, in section 38, provision for compensation for revocation of a planning permission similar to that contained in section 21 of the 1947 Act. It did this, however, not by enacting a new compensation permission, but by incorporating and applying section 22 of the 1947 Act, but deleting the proviso, which excluded compensation for depreciation in value when no development charge had been paid. The formula for the calculation of the amount of depreciation which was provided in section 22(7) was thus not only retained but was also expressly J incorporated by reference. This may be said to have been arbitrary and illogical because, with the abolition of the development charge, the reason for the statutory assumption in fixing a base value for calculation might be thought to have gone, so that it would have been logical simply to repeal that part of section 22(7) which included the statutory assumption. But this was not done and I cannot, for my part, regard the omission as having been accidental.

Where a statute contains a provision in words substantially identical with those used in a previous Act it may be possible to argue that the words of the later Act take on, from their context, a meaning different from that which they bore in the previous enactment. But where the later statute expressly incorporates and applies a section from an existing statute it is, as it seems to me, quite impossible to K attribute to it in its new application a meaning different from that which it bore in the statute from which it is taken. The section had a perfectly clear meaning in the context of the 1947 Act and clearly applied equally to the case where the permission revoked was one for a development specified in Schedule 3. When it was incorporated referentially in the Act of 1954 and repeated in terms, as it subsequently was in the consolidating Acts of 1962 and 1971, it cannot have changed its meaning. Thus the legislative history, far from providing support for the purposive construction which Mr Keene seeks to urge upon your lordships, seems to me to underline the impossibility of escaping the conclusion that the clear purpose of the legislature was to limit the amount of compensation payable for L depreciation owing to a revocation of planning permission by fixing a base value for the land in all cases on the footing that planning permission for any class of Schedule 8 development would be granted. That must include, however arbitrary it may seem, even development of that very class which was the subject-matter of the revoked permission.

The conclusion is not one which I embrace with any enthusiasm and it may well be that the particular circumstance of the revoked permission being the very permission comprehended in the statutory assumption was not one which the legislature foresaw as ever likely to occur. But, while this provides a sound reason for the hope expressed by the Vice-Chancellor that Parliament may look again at what he described as "an anachronistic relic", it cannot provide an avenue M for escape from the clear and express words of the section. I would dismiss the appeal.

Lord Templeman, Ackner, Mustill and **Woolf** agreed that the appeal should be dismissed, and did not add anything.

Chancery Division

October 21 1992

(Before Mr Justice Mervyn DAVIS)

HANDOLL AND OTHERS v WARNER GOODMAN & STREAT (A FIRM)

Estates Gazette March 27 1993

[1993] 12 EG 131

Town and country planning — Planning condition — Whether subject property subject to a condition that its occupation is limited to a person solely or mainly or last employed in the locality in agriculture

On May 31 1977 the second defendant, Mr Cook, submitted an application for planning permission for a proposed dwelling on land described as "West Fen Lane, Stickney, Boston, Lincs" on OS sheet 90/6 Plots 9 and 10. The location of the proposed dwelling inside the application form was stated as "Stickney OS XC:6. Fields 9 and 10 (part)". On October 27 1977 the fourth defendants, East Lindsey District Council, granted outline planning permission to erect a dwelling and garage in connection with agriculture. The location of the development was stated as OS sheet no 90.6, field no 10 pt, grid reference no 534450 358580. The permission included the words "The occupation of the dwelling shall be limited to a person solely or mainly employed, or last employed, in the locality in agriculture defined in Section 290 of the Town and Country Planning Act 1971, or in forestry or dependant of such a person residing with him (but including a widow or widower of such person)". On November 30 1977 an application was submitted to the fourth defendants in respect of the reserved matters accompanied by a drawing showing in a block plan the proposed dwelling located on enclosure 9 of the OS plan and immediately to the west of enclosure 10; approval to the reserved matters was given on December 2 1977 stating the location of the development as in the outline planning permission. As a result of the approval "The Paddocks" was constructed about 90 ft west of the location approved by the fourth defendants and wholly within enclosure 9. By a conveyance dated December 10 1988 the plaintiffs, Mr Stephen Colin Handoll, Mrs Margaret Charlotte Ginette Handoll and Mrs Muriel Hazel Suddick, acquired the property from the second and third defendants, Mr and Mrs S J Cook. On the trial of a preliminary issue in a consolidated action brought by the plaintiffs against the defendants, whether The Paddocks was subject to the planning condition imposed by the fourth defendants, the plaintiffs contended that the agricultural occupancy restriction did not blind The Paddocks because it was not a bungalow built pursuant to planning permission being built on land away from that in respect of which the planning permission was to operate.

***Held*: The departure from the planning permission was of a greater degree that the departure considered in *Kerrier District Council* v *Secretary of State for the Environment*. The Paddocks was built without permission. In that situation the permission ought not be treated as a nullity and that being so there was no reason why the occupancy restriction should not be enforceable.**

This was a trial of a preliminary issue arising in consolidated actions brought by the plaintiffs, Stephen Colin Handoll, Mrs Margaret Charlotte Ginette Handoll and Mrs Muriel Hazel Suddick, against the first defendants, Warner Goodman & Streat, the second and third defendants, Mr and Mrs S J Cook, and the fourth defendants, East Lindsey District Council, ordered by Master Murrow on October 16 1990.

Michael Michell (instructed by Tinn Thimbleby, of Spilsby) appeared for the plaintiffs; Graham Machin (instructed by Browne Jacobson, of Nottingham) represented the fourth defendants; the first, second and third defendants did not appear and were not represented.

The following cases are referred to in this report.

Kerrier District Council v *Secretary of State for the Environment* (1980) 41 P&CR 284; [1981] JPL 193, DC

R v *Basildon District Council, ex parte Martin Grant Homes Ltd* (1986) 53 P&CR 397

Giving judgment, **Mervyn Davies J** said: I have to decide a preliminary issue in a consolidated action. However, the issue is tried only as between the plaintiffs and the fourth defendants in the consolidated action. The other defendants have agreed, as I understand, to be bound by any finding on the preliminary issue. The three plaintiffs are Mr Stephen Colin Handoll, Mrs Margaret Charlotte Ginette Handoll and Mrs Muriel Hazel Suddick. The plaintiffs own 5.183 acres of land at Stickney, Lincolnshire, together with a bungalow and outbuildings built thereon called "The Paddocks". They acquired the property by a conveyance dated December 10 1984 in which Mr and Mrs S J Cook were the vendors. The first defendants in the consolidated action are Warner Goodman & Streat. They are the solicitors who acted for the plaintiffs when the property was bought. The second and third defendants are the Cooks. The fourth defendants are East Lindsey District Council. The trial of the preliminary issue was ordered by Master Munrow on October 16 1990. In brief, the question is whether the land that has been conveyed to the plaintiffs is subject to any restriction imposed by the district council whereby occupation of the land is limited to persons engaged in agriculture. The actual terms of the preliminary issue as set out in the master's order are as follows:

> Whether the property known as the piece or parcel of land situate in the parish of Stickney Lincolnshire being number 9 on the sheet Xc6(1904) on the Ordnance Survey Plan containing 5.183 acres or thereabouts together with the messuage or dwelling house and outbuildings erected thereon or some part thereof and known as The Paddocks, West Fen Lane, Stickney aforesaid is subject to any restriction imposed by the East Lindsey district council or otherwise whereby its occupation is limited to a person solely or mainly employed or last employed in the locality in agriculture as defined by Section 290 of the Town and Country Planning Act 1971 or in forestry or a dependant of such a person residing with him (but including a widow or widower of such person).

The Paddocks was built by the Cooks in 1978. On May 31 1977 Mr Cook, acting by his architect Mr Daubeny (since deceased), submitted an application for planning permission on the appropriate form. Exhibit DGB1 to the affidavit of Mr D G Brewster, sworn on February 7 1992, is a copy of the application. The application shows on its front page that permission or approval was sought for land described as "West Fen Lane, Stickney, Boston, Lincs" OS sheet 90/6 plots 9 and 10. Inside the application form one is required to state "Location of proposed dwelling". That is given as "Stickney OS Xc:6. Fields 9 and 10 (part)". Para 6 of the application form refers to a plan submitted with the application being Plan 770/5/77/B (hereinafter called "the application plan").

An outline planning permission was forthcoming from the district council. It is dated October 27 1977. It gives as particulars and location of the development submitted in these terms:

> Outline Application for permission to erect a dwelling and garage in connection with agriculture and for vehicular/pedestrian access at West End Lane, Stickney. Location of development: — OS Sheet No. 90.6, Field No 10pt Grid Reference No 534450 358580

It is to be noted that this permission refers to field no. 10pt, whereas the application referred to "Plots 9 and 10". The permission stated that the subsequent approval of the district planning authority was required as respects some such matters as siting, design landscaping etc. The permission also included the words:

> 3. The occupation of the dwelling shall be limited to a person solely or mainly employed, or last employed, in the locality in agriculture as defined in Section 290 of the Town and Country Planning Act 1971, or in forestry or dependant of such a person residing with him (but including a widow or widower of such person).

I need not refer to the other provisions of the permission.

On November 30 1977 an application was submitted to the district council as respects the matters which had been reserved in the outline planning permission. In this application the location of the land was simply referred to as West Fen Lane, Stickney, Boston, Lincs, grid reference 534450 358580. The application was accompanied by drawing 770/5/77/C — enclosure SJC3. The drawing showed the plans and elevations of the proposed bungalow and as well, in the bottom half of SJC3: (i) a block plan showing the bungalow located in enclosure 9 of the OS plan and immediately (as I see it) to the west of enclosure 10, and (ii) at the top of SJC3, a sketch which appears to show the land the subject of the application as being in part on enclosure 9, in part in enclosure 10 and in part in enclosure 11. Approval of the reserved matters was forthcoming on December 2 1977. The approval is part of exhibit "SPGF2". Therein the "location of development" is stated to be (as in the outline planning permission) "OS sheet No 90.6 Field No 10pt Grid Reference 534450 358580". The restriction as to agricultural occupancy is repeated.

With this December approval received it seems that the building of The Paddocks went ahead. It is now realised that the bungalow was not being placed on the location suggested in the application plan or by the application form or on the locations shown in SJC3. Evidence was given by a chartered surveyor, Mr Christopher Job [FRICS]. Mr Job produced "Plan 1". Plan 1 reproduces the 1904 OS map so as to show the immediate locality including enclosures 9, 10 and 11. The bungalow is, of course, not shown because it was not there in 1904. By reference to an enlargement — also on plan 1 — and by the use of transparent overlays Mr Job drew attention to some features referred to in the application plan and the other planning documents that I have mentioned:

(a) Overlay 2 shows that the thick-hatched square on the application plan is in part in enclosure 9, in part in enclosure 10 and in part in enclosure 11. Looking at the position of The Paddocks as shown on the enlargement one sees that The Paddocks has been built outside the thick-hatched square on the application form;

(b) Overlay 3 shows the grid reference already mentioned. It is in enclosure 9 about 48ft north of the (original) buildings shown on the 1904 ordnance survey plan and about 27ft from the nearest point of the boundary of enclosure 10;

(c) Overlay 4 shows the location of the bungalow as built and its location "on the Block Plan being part of the Drawing 770/5/77/B . . . ". I think that the reference to 770/5/77B is a mistaken reference for 770/5/77C because it is in the C plan that there is a "block plan". However that may be, it is clear from the overlay and from Mr Job's cross-examination that the bungalow is about 90ft somewhat westerly of the location as approved by the district council; and it is plain that, as the bungalow now stands, it is wholly within enclosure 9.

Mr Michell appeared for the plaintiffs. He submitted that the agricultural occupancy restriction is not binding on The Paddocks. The restriction would affect, he said, a bungalow built on land for which planning permission was given. But The Paddocks is not a bungalow built pursuant to any planning permission. It has been built on land away from that in respect of which the planning permission was to operate. He accepted, as I understand, that meant that The Paddocks was built in breach of planning control. Since the limitation period for enforcing planning control has passed, the bungalow may remain in being; but as such it is not, he said, affected by the occupancy restriction because that restriction is a term of a planning permission that has no relevance to The Paddocks. In other words, The Paddocks would have been subject to a restriction if built in the right place, but is not subject to a restriction now it has been built in the wrong place.

Mr Machin, appearing for the district council, submitted that the occupancy restriction binds The Paddocks. He drew attention to the crude and inaccurate features of the application plan leading to difficulties in ascertaining exactly what plot was intended by the district council as the approved plot for a new building. But, he said, even if the bungalow was built about 90ft to the west of the approved location one must take account of several considerations: (a) in the rural context and the 5.183 acres extent of enclosure 9 a removal of 90ft was not a substantial alteration of the planning approval; (b) such change of location was without any planning significance; (c) the bungalow built is exactly as to construction the same as that approved by the district council; (d) Mr Cook, when building the bungalow, proceeded on the footing that he was implementing the planning permission; and (e) the district council were, it seems, satisfied as to the building going up on the site that it now occupies: see the council's letter dated December 20 1978 in exhibit SJC3. In this connection see *R* v *Basildon District Council, ex parte Martin Grant Homes Ltd*, (1986) 53 P&CR 397.

In the light of such considerations Mr Machin said that the approach to the question of the enforceability of the restriction was indicated by *Kerrier District Council* v *Secretary of State for the Environment* (1980) 41 P&CR 284. That was a case where planning permission was given for the erection of a bungalow, but the bungalow was built otherwise than in accordance with the approved plan in that a basement (not shown on the plan) was included within the bungalow. The planning permission included an agricultural occupancy restriction. At p287 Lord Lane CJ said:

> It seems to us that there is another and perhaps equally important question. Can a planning authority enforce a condition subject to which planning permission has been granted even though the building as built does not comply with the approved plans?

And at p288 the Lord Lane says:

> We approach the question first as a matter of principle. The reasoning of the Secretary of State comprises four steps: (1) the dwelling-house actually built differed materially from that shown on the approved plan; (2) therefore, it was built without planning permission; (3) therefore, the planning permission was never implemented; (4) therefore, the condition never applied.
>
> Step (1) involves a finding of fact that does not appear anywhere in the inspector's report. It is the starting-point of the allegation of breach of natural justice, to which we shall return briefly later. Step (2) is well established law. Steps (3) and (4) involve, it seems to us, a *non sequitur*. When one speaks of something being done without permission, it may mean one of two things. It may mean that no permission has ever been granted at all, or it may mean that permission has been granted but that what has been done does not comply with that permission. The fact that something is done without permission in the latter sense does not mean that the permission must be treated as a nullity, and, if it is not a nullity, we see no reason why any condition subject to which the permission has been given should not bite.
>
> The error in the Secretary of State's reasoning may lie in his reference to the planning permission never having been "implemented". Implementation is not, as we understand it, a term of art in this connection. In one sense, the planning permission plainly was "implemented" since without the planning permission the house would never have been built at all. Having relied on the permission to build a house, it would seem strange that the occupiers should not be bound by the condition, particularly if it was by reason of their own default that the plans were not complied with. If the house had complied with the plans, the occupiers would have been bound by the condition. They can hardly be in a better position because the house did not comply with the plans. Two blacks do not make a white in this or any other branch of the law.

Mr Michell said that *Kerrier* was to be distinguished because there the departure from the terms of the planning permission was simply a matter affecting the interior parts of a building built on the very site for which planning permission was given, the departure being not of a very substantial nature. But here, he said, there was not only a removal from the approved site of the bungalow, but a removal of as much as 90ft. True it is that the departure here from the terms of the planning permission is of greater degree than the departure considered in *Kerrier*. But it seems to me that Lord Lane's words have as much force in the circumstances of the present case as they had in the *Kerrier case*. I accept that The Paddocks was built without permission. It was built without permission in the sense that a permission was granted, but that what was then done did not comply with that permission. In that situation I respectfully agree with Lord Lane that the permission ought not to be treated as a nullity. That being so there is no reason why the occupancy restriction should not be enforceable. Accordingly, I answer the preliminary issue in the affirmative. I have not been asked to consider any impact of the Land Charges Act 1925.

Declaration accordingly.

Court of Appeal

December 21 1992

(Before Lord Justice DILLON, Lord Justice FARQUHARSON and Lord Justice SIMON BROWN)

VAN DYCK AND ANOTHER v SECRETARY OF STATE FOR THE ENVIRONMENT AND ANOTHER
DONCASTER BOROUGH COUNCIL v SECRETARY OF STATE FOR THE ENVIRONMENT AND ANOTHER

Estates Gazette May 29 1993
[1993] 21 EG 112

Town and country planning — Breach of planning control consisting of change of use without planning permission of

A ***single dwellinghouse into two (or more) separate dwellinghouses — Whether the immunity from enforcement notice proceedings after four years provided by section 172(4)(c) applies to such subdivision***

Van Dyck v *Secretary of State for the Environment*

In 1982-83 2 Oakleigh Park Drive was divided into two flats without planning permission. In 1986 the appellants, unaware of the lack of planning permission, acquired a long lease of one of the flats. In 1989, on discovering the breach of planning control, the appellants applied to the second respondents, Southend-on-Sea Borough Council, for retrospective B planning permission. Their appeal against the second respondents' refusal of planning permission was dismissed by the inspector on the ground that the unauthorised conversion of the house into flats had resulted in more competition for the limited kerbside parking. The appellant's application to the High Court to quash that decision pursuant to section 288 of the Town and Country Planning Act 1990, on the ground that the inspector had failed to consider a material consideration, that the division into two flats was immune from enforcement action by reason of the four-year rule under section 172(4)(c), was dismissed by Sir Frank Layfield QC on October 2 1991.

C *Doncaster Borough Council* v *Secretary of State for the Environment*

In 1971 the second respondent subdivided 16 Markham Avenue, a two-storey house into two self-contained flats. On June 4 1990 the appellant council served on the second respondent an enforcement notice alleging as the breach of planning control "the making of a material change in the use of the land from a use for single family occupation to a use for the purpose of multiple paying occupation providing flat accommodation". The second respon- dent's appeal to the Secretary of State was allowed, the inspector accepting that D each flat was a separate dwellinghouse and therefore entitled to immunity under the four-year rule. The appellant council appealed that decision to the High Court under section 289 of the 1990 Act. On November 7 1991 Webster J held that the breach of planning control was not immune and allowed the statutory appeal.

The appeals before the Court of Appeal

The appellants in both appeals, including the Secretary of State in *Doncaster Borough Council* v *Secretary of State for the Environment*, submitted that the immunity from enforcement-notice proceedings after four years provided by E section 172(4)(c) (the four-year rule) applies equally to a breach of planning control consisting of the change of use without planning permission of a single dwellinghouse into two (or more) separate dwellinghouses as to the unpermitted use of a building as a single dwellinghouse.

***Held*: The appeals were allowed. The language of section 172(4)(c) of the 1990 Act which, construed in the context of sections 55(3)(a) (there is a breach of planning control if development has been carried out without the grant of planning permission required) and 336(1) ("building" includes any part of a building), is capable of encompassing F subdivision within its protection. If a whole house is divided into two flats, then it is perfectly possible to find within the terms of the section in respect of each new flat a change of use of part of the building (the original whole house) to use as a single dwelling-house. Given that construction, section 172(4)(c) is capable of being construed and applied so as to benefit all new separate residences after four years. The conditions in *Pepper* v *Hart* that Parliamentary materials may be referred to in construing a statute where: (a) legislation is ambiguous or obscure or leads to an absurdity; (b) the material relied upon consists of one or more statements by a minister or other promoter of the Bill together if necessary with such other parliamentary material as is necessary to G understand such statements and their effect; and (c) the statements relied upon are clear, did not apply with the result that the parliamentary material could not be relied upon in the determination of the appeal.**

The following cases are referred to in this report.

Backer v *Secretary of State for the Environment* (1981) 42 P&CR 98
Birmingham Corporation v *Mahib Ullah* [1964] 1 QB 178; [1963] 3 WLR 937; [1963] 3 All ER 668; (1963) 61 LGR 623; 15 P&CR 404; [1963] RVR 712, DC
Burdle v *Secretary of State for the Environment* [1972] 1 WLR 1207; [1972] 3 All ER 240; (1972) 70 LGR 511; 24 P&CR 174; [1972] EGD 678; 223 EG 1597, DC
Ealing Corporation v *Ryan* [1965] 2 QB 486; [1965] 2 WLR 223; [1965] 1 All H ER 137; 17 P&CR 15, DC
Impey v *Secretary of State for the Environment* (1984) 47 P&CR 157; DC
Okereke v *Brent London Borough Council* [1967] 1 QB 42; [1966] 2 WLR 169; [1966] 1 All ER 150, CA
Peacock Homes Ltd v *Secretary of State for the Environment* (1984) 83 LGR 686; 48 P&CR 20, CA
Pepper v *Hart* [1992] 3 WLR 1032
Rawlins v *Secretary of State for the Environment* [1990] 1 PLR 110; [1990] JPL 326; CA
Trevors Warehouses Ltd v *Secretary of State for the Environment* (1972) 23 P&CR 215, DC
Wood v *Secretary of State for the Environment* [1973] 1 WLR 707; [1973] 2 All ER 404; (1973) 25 P&CR 303, DC
Worthing Borough Council v *Secretary of State for the Environment* (1992) 63 P&CR 446; [1992] 2 PLR 40 J

In *Van Dyck* v *Secretary of State for the Environment*, this was an appeal by the appellants, Peter Louis Van Dyck and Janice Elizabeth Van Dyck, from a decision of Sir Frank Layfield QC, sitting as a deputy judge of the Queen's Bench Division, who on October 2 1991 dismissed their application under section 288 of the Town and Country Planning Act 1990 to quash the decision of the first respondent, the Secretary of State for the Environment, by his inspector, who had dismissed an appeal by the appellants from a decision of the second respondents, Southend on Sea Borough Council, to refuse retrospective planning permission of 2 Oakleigh Park Drive, Southend on Sea.

In *Doncaster Borough Council* v *Secretary of State for the* K *Environment* the appellants, Doncaster Borough Council, appealed from a decision of Webster J, who on November 1991 allowed the appellants' appeal from a decision of the first respondent, the Secretary of State for the Environment, by his inspector, who had quashed an enforcement notice issued by the appellants to the second respondent, Mr Dunnill, in respect of 16 Markham Avenue, Doncaster.

Martin Wood (instructed by Sharp Pritchard, for the solicitor to Doncaster Borough Council) appeared for Doncaster Borough Council; Andrew Newcomb (instructed by L Dawson & Co) appeared for Mr and Mrs Van Dyck; Christopher Katkowski (instructed by the Treasury Solicitor) represented the Secretary of L State for the Environment.

Giving judgment, **Simon Brown LJ** said: There are before the court two appeals which I shall call respectively the *Doncaster* appeal and the *Van Dyck* appeal. Both turn upon the same short point of law: the proper construction and application of section 172(4)(c) of the Town and Country Planning Act 1990 (now re-enacted in substantially similar form as section 171B(2) by section 4 of the Planning and Compensation Act 1991). More particularly, the issue raised by these appeals is: does the immunity from enforcement notice proceedings after four years provided by section 172(4)(c) (hereafter called the "four-year rule") apply equally to a breach of planning control consisting of the change of use without planning permission of a single dwellinghouse into two (or more) separate M dwellinghouses (hereafter called "subdivision") as to the unpermitted use of a building as a single dwellinghouse? The appellants (including the Secretary of State in the *Doncaster* appeal) submit that it does; Doncaster Borough Council submits not.

In late 1991 there were, in quick succession, three decisions on the point: two, the two now under appeal, holding that the four-year rule does not apply to the subdivision of a single dwellinghouse; the third — *Worthing Borough Council* v *Secretary of State for the Environment* (1992) 63 P&CR 446* (*Worthing*) — holding that it does.

*Editor's note: See also [1992] 2 PLR 40.

A Before considering the facts of these cases it is convenient first to set out the statutory framework within which the point of law arises and to give a brief indication of the rival contentions.

The relevant provisions of the 1990 Act are as follows:

> **55.**—(1) . . . "development", means . . . the making of any material change in the use of any buildings or other land.
>
> (3) For the avoidance of doubt it is hereby declared that for the purposes of this section —
>
> (*a*) the use as two or more separate dwellinghouses of any building previously used as a single dwellinghouse involves a material change in the use of the building and of each part thereof which is so used;"
>
> . . .
>
> **172.**—(1) Where—
>
> B (a) it appears to the local planning authority that there has been a breach of planning control after the end of 1963; and
>
> (b) the authority consider it expedient to do so . . . they may issue a notice (an "enforcement notice") requiring the breach to be remedied.
>
> (3) There is a breach of planning control —
>
> (a) if development has been carried out . . . without the grant of the planning permission required
>
> . . .
>
> (4) An enforcement notice which relates to a breach of planning control consisting in —
>
> . . .
>
> (c) the making without planning permission of a change of use of any building to use as a single dwellinghouse . . . may be issued only within the period of 4 years from the date of the breach.
>
> C . . .
>
> **336.**—(1) In this Act, except insofar as the context otherwise requires . . . 'building" includes any structure or erection, and any part of a building, as so defined
>
> . . .
>
> **174.**—(1) A person having an interest in the land to which an enforcement notice relates or a relevant occupier may appeal to the Secretary of State against the notice
>
> . . .
>
> (2) An appeal may be brought on any of the following grounds —
>
> . . .
>
> (d) in the case of a notice to which section 172(4) applies, that the period of four years from the date of the breach of planning control to which the notice relates had elapsed at the date when the notice was issued . . .

D The appellants' argument focuses particularly on sections 55(3)(a), 172(4)(c) and 336(1) and essentially amounts to this: when a single dwellinghouse is subdivided, say, into two flats, there is a breach of planning control consisting of a material change of use of each part of it — section 55(3)(a) so states. Having regard to section 336(1) (the interpretation section) section 172(4)(c) can and should be read as stating: "the making without planning permission of a change of use of any building *or any part of a building* [my emphasis] to use as a single dwellinghouse" may be issued only within four years of the breach. "Building" is plainly apt to include a dwellinghouse; likewise "any part of a building" is apt to include that part which becomes a flat following subdivision, and the flat itself becomes a E separate and accordingly a "single" dwellinghouse. Even assuming that it is theoretically possible to address the enforcement notice to the whole house as well as to each flat following its subdivision, the planning authority cannot thereby deprive the occupiers of the individual flats of the ground of appeal otherwise available to them under sections 172(4)(c) and 174(2)(d).

The respondents' contrary argument is essentially as follows: section 55(3)(a) exists merely so that no one may argue that the subdivision of a single dwellinghouse into separate dwellinghouses does not itself involve a material change of use. It casts no light on the proper construction and application of section 172(4)(c). For the purposes of this latter section it is wrong to judge whether there has been a material change of use to use as a single dwellinghouse by F reference to the change which the unauthorised development has brought about; rather, the occupiers acquire immunity under this provision only when each part of the building converted was itself properly to be regarded as a distinct planning unit before the development took place. That at least is how I understand the respondents' argument although I confess to having found it on occasion somewhat elusive.

I turn now to the decisions under appeal with which I propose to deal in chronological order, inserting between them such reference as is helpful to *Worthing*.

(1) *Van Dyck* v *Secretary of State for the Environment*

The facts

In 1982-83 2 Oakleigh Park Drive was divided into two flats. Planning permission should have been, but was not, obtained. In G 1986 the appellants, unaware of the lack of planning permission, acquired a long lease of one of the flats. Only in 1989 did they discover this breach of planning control and seek to cure it by an application to the council for retrospective planning permission. It was refused. They appealed to the Secretary of State. An inspector was appointed to determine the appeal. He concluded that there were two main issues, first, whether the conversion would adversely affect the character and appearance of the surrounding area, second, whether it would be detrimental to the safety and free flow of traffic on Oakleigh Park Drive. The first issue he resolved in the appellants' favour. As to the second, however, he expressed himself "convinced that the unauthorised conversion of the house into flats has resulted in more competition for the limited kerb-side parking, exacerbating H the congestion and so the potential hazards for drivers and pedestrians". He accordingly dismissed the appeal. That was in October 1990.

The appellants then applied to the High Court to quash that decision pursuant to section 288 of the 1990 Act. Essentially their case was that more than four years had elapsed since the breach of planning control in question and that it was therefore no longer possible for the planning authority to enforce against it. Nothing, therefore, was to be gained by refusing the planning permission sought. At the very least the inspector should have had regard to the fact that the breach could no longer be enforced against. Clearly that was a material consideration. Instead, he assumed the contrary, this J being evident from his regretful conclusion that the appellants "will have to bear the consequences and costs of restoring (the property) to a single house". In short, the appellants took before the court (although not, it seems, before the inspector) the point now under appeal.

The decision under appeal

The point failed. On October 2 1991 the statutory application was dismissed by Sir Frank Layfield QC sitting as a deputy judge of the Queen's Bench Division. The relevant part of his judgment reads as follows:

> Mr Jay, on behalf of the Secretary of State, the first respondent, concedes that if the applicants are correct in claiming that the conversion of each part of the K premises to a self-contained flat is immune from enforcement proceedings by the operation of section 172(4)(c) of the 1990 Act, then the Secretary of State has lost the case. It was only upon that section that (counsel) for the applicants, based his submission of immunity. Mr Jay pointed out, however, that the subject under review is the subdivision of a single dwelling into two dwellings. That is not the change to which section 172(4)(c) is addressed. The subsection applies where there is a breach of planning control which consists of the making of a change of use of any part of any building to use as a single dwellinghouse. The breach to which the subsection applies is, I consider, a quite distinctly different breach which consists in making a change of use of any dwelling into two dwellinghouses as the wording of section 55(3)(a) demonstrates. I find, therefore, that section 172(4)(c) does not itself confer immunity on the change of use made in 1983 at the latest. If I am correct in that reading, there was no obligation on the inspector to take account of an L inability of the council to take enforcement action.

The sentence starting "The breach to which the subsection applies" can perhaps usefully be fleshed out to clarify the meaning which I have no doubt Sir Frank intended:

> The breach to which the subsection [ie section 172(4)(c)] applies is, I consider, a quite distinctly different breach to [that alleged here] which consists in making a change of use of any dwelling into two dwellinghouses as the wording of section 55(3)(a) demonstrates. (The words in square brackets are those I have added for clarity.)

What in short I understand Sir Frank to have been saying was that whereas section 55(3)(a) is all about subdivision — change of use *from* use as a single dwellinghouse section 172(4)(c) deals with the converse situation — change of use to use as a single dwellinghouse. M

(2) *Worthing Borough Council* v *Secretary of State for the Environment*

The facts

A two-storey house in Worthing was converted into two flats, respectively a ground-floor flat and a first-floor flat. More than four years later Worthing Borough Council issued an enforcement notice alleging a breach of planning control consisting in the "unauthorised change of use of land from a single unit of residential accommodation into two units of accommodation", and requiring the occupiers, by way of the steps to be taken to remedy the breach, "to discontinue use of the land as two units of accommodation and to revert the use thereof to a single unit of residential accommodation''. The occupiers appealed to the Secretary of State and before the inspector appointed took the point now under appeal — contending that is, that they were entitled to immunity under the four-year rule. The inspector accepted the argument and quashed the enforcement notice. Worthing Borough Council then appealed to the High Court under section 289 of the 1990 Act.

The decision

This time it was held that the breach of planning control was immune. On November 5 1991 the planning authority's statutory appeal was dismissed by Sir Graham Eyre QC sitting as a deputy judge. His reasoning was largely, although not entirely, that now urged by the present appellants. He emphasised the importance of reading into section 172(4)(c) the definition of building from section 336 and continued:

> On the conversion of each self-contained flat each was a part of a building and the change of use of those parts was each to use as a single dwellinghouse . . . I note that Sir Frank Layfield appears not to have been taken to the definition of building in section 336 of the Act, which is fundamental to the construction of section 172(4)(c).

That would seem to be wrong: the passage I have already cited from Sir Frank's judgment appears clearly to recognise that the word ''building'' in section 172(4)(c) includes also reference to a part of a building.

Sir Graham then differed from Sir Frank as to the relevance of section 55(3)(a), expressing the view that it afforded no assistance to the proper construction of section 172(4)(c).

(3) *Doncaster Borough Council* v *Secretary of State for the Environment*

The facts

In 1971, Mr Dunnill, the second respondent to this appeal, subdivided 16 Markham Avenue, a two-storey house which he owned in Doncaster, into two self-contained flats, creating thereby, as in the *Worthing* case, essentially a ground-floor flat and a first-floor flat. These flats he thereafter let, mostly to young couples, so that they remained in continuous occupation. On June 4 1990 Doncaster Borough Council served upon Mr Dunnill an enforcement notice alleging as the breach of planning control ''the making of a material change in the use of the land from a use for single family occupation to a use for the purpose of multiple paying occupation providing flat accommodation'' and requiring Mr Dunnill ''to discontinue the use of the land for the purpose of multiple-paying occupation''.

Mr Dunnill appealed to the Secretary of State and, as in *Worthing*, successfully took the immunity point under section 172(4)(c). The inspector, while noting the particular allegation made by that enforcement notice, concluded that it was clearly referable to flat accommodation and further that the two flats ''are now separate planning units''. He accepted that each flat was a separate dwellinghouse and therefore entitled to immunity under the four-year rule. He accordingly quashed the enforcement notice. Doncaster Borough Council, like Worthing Borough Council, appealed under section 289.

The decision under appeal

This time it was held that the breach of planning control was *not* immune. On November 7 1991 Webster J allowed the planning authority's statutory appeal. He agreed with Sir Frank Layfield; disagreed with Sir Graham Eyre. Having noted the Secretary of State's arguments — substantially the same as those advanced before us — he continued as follows:

> There is, in my view, much force in many of those contentions. However, they overlook both the mischief aimed at by the relevant provisions and the obviously intended effect of section 172(4)(c). The mischief aimed at to the extent that it is not protected by section 172(4)(c) is that of converting one building, or part of it, whether or not it was a dwellinghouse before the conversion, into two or more dwellinghouses. The obviously intended effect of section 172(4)(c) is to give limited protection where a building, or part of it, is converted into one dwellinghouse.
>
> In this case, the enforcement notice related to the whole of one dwellinghouse which, in breach of planning control, had been converted not into one, but into two dwellinghouses. The enforcement of planning control was enforced against the house as a whole, the unit which had been developed, and not against each flat which had not existed before the conversion.
>
> The reply to Mr Havers' [counsel then appearing for the Secretary of State] submission that, in effect, the breaches were two breaches and not one, consisting of theconversions of two separate parts of the use into two separate flats, each of which conversions would, if viewed separately, have been protected by section 172(4)(c), is that the development actually carried out was not the conversion of any separate part of the house, but the conversion of the house as a whole. Moreover, it would, in my view, have been impossible to identify the relevant use for planning purposes of either of the two parts of the house, which were converted into the flats, before that conversion took place. Although the Act, as it applies to a building, is to be taken to apply to a part of a building, the definition section does not provide that when it applies to a dwellinghouse, it is to be taken to apply to a part of the dwellinghouse. I can think of no way of describing the use of either of the two parts of the house before conversion for planning purposes other than simply as, in each case, a part of the dwellinghouse.

It will be noted that *Van Dyck* came before the Secretary of State as a planning appeal; the *Worthing* and *Doncaster* cases as appeals against enforcement notices. This, I have no doubt, explains the difference in approach taken by the various inspectors. It was only the inspector in the planning section who adopted an approach inconsistent with that for which the Secretary of State now contends. That, however, is by the way; the important question is: which approach is right?

It seems to me necessary first to consider what, if any, assistance one gets from the judge-made concept of the planning unit. This, after all, is the concept underlying the respondents' arguments.

As is well known, the planning-unit concept evolved as a means of determining the most appropriate physical area against which to assess the materiality of change — to decide, therefore, whether in any particular case there has been a breach of planning control consisting of a ''material change in the use of any buildings or other land''.

The classic exposition of this doctrine is to be found in the judgment of Bridge J in *Burdle* v *Secretary of State for the Environment* [1972] 1 WLR 1207, from which the following short passage will for present purposes suffice:

> It may be a useful working rule to assume that the unit of occupation is the appropriate planning unit, unless and until some smaller unit can be recognised as the site of activities which amount in substance to a separate use both physically and functionally.

Counsel for the respondents drew our attention to *Wood* v *Secretary of State for the Environment* (1973) 25 P&CR 303 in which the divisional court was concerned with the varying uses of different parts of a smallholding which at all material times remained in the appellant's sole occupation. The particular question raised was whether it was right to treat a conservatory extension of the farmhouse as a separate planning unit. Giving the judgment of the Divisional Court, Lord Widgery CJ referred to a number of decisions in which ''. . . one finds a disinclination to cut up a holding into penny packets and treat them as separate planning units'', cited the passage from *Burdle* to which I have already referred, and continued:

> That authority seems to me to support the view, which I would think to be right without such assistance, that it can rarely if ever be right to dissect a single dwellinghouse and to regard one room in isolation as being an appropriate planning unit for present purposes.

Relying on that decision, Mr Wood submits that the court should be reluctant to conclude that the appropriate planning units in the case of a subdivided house are the flats resulting rather than the single dwellinghouse previously existing.

He further points to *Rawlins* v *Secretary of State for the Environment* [1990] 1 PLR 110 where this court upheld the planning authority's selection of an area greater than that against which they might have sought to enforce and greater certainly than that of the individual areas of occupation within it. The unsuccessful appellants there were travelling showmen occupying individual caravan plots. The enforcement notices were directed to the whole site and the breaches alleged included the use of the land ''as a caravan site, the construction of a road, the construction of hardstandings on the land, the use of the land for the parking, storage, maintenance and repair of motor vehicles, fairground rides, plant, equipment and machinery''. The court recognised the well-established principle that the question of what is the proper planning unit is essentially a matter of fact and degree and accepted that it was permissible in that case for the authority to have taken a large unit even though within it were individual plots in separate ownership.

Rawlins was relied upon also by Mr Katkowski for the Secretary of State, in particular the following passage from Butler Sloss LJ's judgment:

> . . . I do not see these appeals as setting aside the existing long-established practice of issuing and serving enforcement notices on a planning unit, which

A is, as Lord Widgery CJ said in [*Johnston* v *The Secretary of State for the Environment* (1974) 28 P&CR 424] "... the area occupied as a single holding by a single occupier" — using single occupier in the collective sense so that it would include two or more joint occupiers.

These ... cases have unusual if not exceptional features and characteristics ... which justify the less usual procedure adopted, but which do not appear to have arisen before and are therefore unlikely to arise frequently.

The burden of the appellant's submissions on the planning unit issue is that the cases of subdividing dwellinghouses, unlike *Rawlins*, are not exceptional so that the ordinary practice should apply; that is, enforcement action should be targeted at the unit of occupation, here each individual flat. For good measure, submits Mr Katkowski, the inspector in *Doncaster* in any event expressly found that the two flats B "are now separate planning units". In so far as the question is one of fact and degree, therefore, it was a question for him and he resolved it against the planning authority. Each individual unit should, therefore, he submits, be held entitled to the benefit of the four-year rule.

Having considered these competing submissions at some length I have reached the conclusion that the concept of the planning unit in fact affords no assistance to either side. As stated, the purpose of the concept — the only point in deciding upon the appropriate planning unit — is to decide whether there has been a material change of use. Here undoubtedly there has been — section 55(3)(a) so declares. There is accordingly no possible need to invoke the doctrine of the C planning unit to decide the question. Here, moreover, it was the very act of creating two separate units of occupation out of a single pre-existing unit which constituted the making of the material change of use. How inappropriate, therefore, that one should seek assistance from a doctrine substantially founded on units of occupation to illuminate the nature of the breach of planning control in question or to determine the appropriate target for enforcement. In short, none of the planning-unit cases to which I have referred has, I believe, any application whatever to the situation presently confronting this court.

I turn next to a brief consideration of the three judgments already given upon the point at issue.

1. *Van Dyck*

D I cannot accept Sir Frank Layfield's apparent view that because subdivision is within section 55(3)(a) it is not therefore within section 172(4)(c). Section 55(3)(a) is, one must note, only concerned to declare that a particular change of use is a breach of planning control. Section 162(4)(c) dictates simply for how long a particular change of use is vulnerable to enforcement. Beyond pointing out merely the first blush difference between the two provisions, Sir Frank seems to me to give no reasoning for his conclusion that: "... the subdivision of a single dwellinghouse into two dwellings ... is not the change to which section 172(4)(c) is addressed".

2. *Worthing*

Sir Graham Eyre reached the opposite conclusion from Sir Frank E Layfield, essentially by putting aside section 55(3)(a) and adopting a possible literal reading to section 172(4)(c) after building in the definition of "building".

3. *Doncaster*

Webster J's thoughtful reasoning is to be found in the long passage which I have already set out. There seem to me, however, fundamental difficulties within it. First, I have difficulty with the learned judge's suggestion that "the development actually carried out was not the conversion of any separate part of the house, but the conversion of the house as a whole". The development here in question, it must be remembered, was not the conversion (whatever precisely that may be) at all — it was rather the material change of use F occasioned by "the use as two ... separate dwellinghouses" of the building previously used as a single dwellinghouse: see section 55(3)(a). Obviously, whoever converts such a house (if by conversion is meant the physical creation of separate living facilities) initiates the situation in which the material change of use can then occur. But the actual breach of planning control commences only when separate dwellinghouses exist and, indeed — subject to only the decisions in *Impey* v *Secretary of State for the Environment* (1984) 47 P&CR 157 and *Backer* v *Secretary of State for the Environment* (1981) 42 P&CR 98 — only when these separate dwellinghouses actually come to be used as such.

Second, I cannot share the judge's view that "it would have been impossible to identify the relevant use for planning purposes of either of the two parts of the house, which were converted into the flats, G before that conversion took place". As it seems to me, section 55(3)(a) answers that too: the material change of use is declared to have taken place in regard to each part of the building; accordingly, the previous use of each part was that of a (different) dwellinghouse.

None of these three decisions, therefore, do I find compelling in their reasoning.

What additional arguments have been advanced before this court? Let me turn first to those which, for my part, I have found ultimately unhelpful in resolving this case.

Mr Katkowski drew our attention to the statutory provisions requiring enforcement notices to be served on all owners and occupiers (section 172(6)) and enabling any occupier, including even H a licensee, to appeal to the Secretary of State (section 174(1)(6)). So be it. This seems to me a slender basis for contending that the four-year rule must therefore apply to each part of the subdivided dwelling. Why should that be? of course I recognise that there are individual interests involved and equally obviously that there will often be acute problems of enforcement. Take a case where the owner has divided his house into two self-contained flats (or, indeed two separate vertical freeholds) and sold each to a different buyer. But even were the four-year rule to apply in such circumstances, the planning authority would still be entitled to enforce within that period and problems would still arise.

Next Mr Katkowski fixes upon the phrase "which relates to" in the J opening clause of section 172(4): "An enforcement notice which relates to a breach of planning control consisting in ... ". This phrase was considered with regard to the provisions of the 1971 Act (although not in the context of a para (c) breach) in *Peacock Homes Ltd* v *Secretary of State for the Environment* (1984) 48 P&CR 20 where Dillon LJ said:

The key words "which relates to" in the phrase "a condition or limitation which relates to the carrying out of such operations", are not words of art. They are words often used in the Act, as in other statutes, in a very general sense ... They may often mean little more than "which has to do with".

Let us then approach section 172(4) as if it read: "An enforcement notice which has to do with a breach of planning control consisting in ... ". That still leaves for decision to the critical question whether K para (c) is apt to apply to separate dwellings consequent upon subdivision.

Mr Newcombe submits first, that in the case of subdivision, enforcement action can be taken against only the subdivided parts and not against the original whole, alternatively that, even if it can be taken against the whole, it should be found immune after four years by virtue of the immunity of its parts.

On the first limb of his submission he prays in aid the decision of the divisional court in *Backer* v *Secretary of State for the Environment* (1981) 42 P&CR 98 in which it was held that "the question whether section 87(3)(c) (the predecessor to section 172(4)(c)) applies must be decided as a matter of reality on the facts L and cannot be artificially avoided or by-passed by the choice of words adopted by the planning authority in the enforcement notice" *per* Skinner J p101, finding some support for that view in the use of the phrase "which relates to". But that decision is to my mind of no direct value in the present case because it was there accepted on all sides that the breach would have fallen squarely within para (c) had the authority chosen only to enforce against it as a change of use rather than as a breach of condition. In the present case that is the very point at issue. Here, indeed, the invocation of substance and reality and the condemnation of artificiality is common to both sides! In short, *Backer* seems to me to beg rather than answer the question for decision in the present appeals. On the second limb of his argument Mr Newcombe refers us to *Trevors Warehouses Ltd* v M *Secretary of State for the Environment* (1972) 23 P&CR 215, a case illustrating the well-known *Mansi* principle. This is to the broad effect that steps required to be taken by an enforcement notice to remedy a breach of planning control should not extend to requiring the discontinuance of ancillary uses or the cutting down of uses to the extent to which, although unlawful, they are established and thus immune from enforcement action. Even assuming, the argument runs, that enforcement action can be issued against the whole building, the steps required to be taken cannot properly include anything which involves destroying the rights established after four years which attach to the constituent parts of the building. Once again, however, this argument seems to me to beg the central question as to whether section 172(4)(c) applies to cases of subdivision. If it does, the *Mansi* principle is not needed; if it does not, there are no established rights to preserve.

Mr Wood for the respondents submits that subdivision, as opposed to the creation, of a single dwelling creates particular difficulties of enforcement: it both gives rise to increased activity (creating, for instance, problems of parking and congestion as found in the *Van Dyck* appeal) and yet is difficult to monitor and detect. Parliament should not readily be thought to have intended to include such breaches of planning control within the four-year rule. These arguments, too, I find unconvincing. In the first place there appears a certain inconsistency in asserting on the one hand that the consequence of subdivision is all too evident and yet on the other hand that it is not easily detectable. And, second, it should not be beyond the wit of the planning authority's enforcement section to discover the creation of new separate dwellings whether by reference to the electoral register or indeed by counting door bells.

Mr Wood's central argument is that which I have already attempted to summarise: an argument depending upon the application of the planning-unit concept to the situation as it stood before the unauthorised development took place. For reasons I have already endeavoured to indicate, it is not an argument I feel able to accept.

I come back, therefore, to the language of section 172(4)(c) itself, language which, construed in the context of sections 55(3)(a) and 336(1), is, as already demonstrated, capable of encompassing subdivision within its protection. If a whole house is divided into two flats, then it is perfectly possible to find within the terms of the section in respect of each new flat a change of use of part of the building (the original whole house) to use as a single dwellinghouse. That, as stated, is the central thrust of the appellants' arguments. Given that possible construction why then should it not be adopted so as to provide immunity to the limited extent of the four-year rule for the new separate dwellings brought about by subdivision?

Mr Wood for Doncaster submits that such a construction involves a strained reading of section 172(4)(c), and that it would result in the application of this branch of the four-year rule (a branch created by the 1968 Act, which abolished the rule in the great majority of cases) in circumstances in which, until comparatively recently, no one appears to have thought that it applied.

The appellants, however, contend that the self-same considerations underlying the undisputed application of the four-year rule to the creation of a single dwellinghouse — namely Parliament's desire to protect long-established residential occupation — apply equally to the creation of more than one new residential unit. There is, they submit, neither sense nor logic in making a distinction between these categories; the rule should apply to the creation of all new separate dwellings — although not, of course, to conversions giving rise to multiple occupation with shared facilities, a different class of case altogether.

The appellants stress too the various anomalies which would result from the narrow construction contended for by the respondents. Take a three-storey office building, one floor of which is converted and becomes used as a dwellinghouse. Section 172(4)(c) would presumably apply: there has been a change of use of part of the building to use as a single dwellinghouse. Why should it be any different if two floors, or indeed all three, are converted? Would it be different again if the changes of use occurred at different times? If so, why? Instead of horizontal conversion of this nature, take instead a single office building converted vertically into one, two or three separate houses, and ask the same questions as before. Why, in short, should some of these developments be treated differently to others? The considerations for the new single owners will certainly be no different.

Although I do not pretend to have found this an easy problem to resolve, I have finally come to accept the appellants' central argument. Section 172(4)(c) is, I believe, capable of being construed and applied so as to benefit all new separate residences after four years. That being so, having regard to the broad policy which I conceive to underlie this provision, in my judgment, it should be thus construed and applied.

It follows that I, for my part, would allow both appeals. In the result I would remit matters to the Secretary of State in the case of *Van Dyck* for rehearing and determination by him. In the case of *Doncaster*, the decision of the Secretary of State should be restored.

Chapter II

Chapter I of this judgment was written and due to be handed down when along came *Pepper* v *Hart* [1992] 3 WLR 1032. The appeals were promptly relisted for further hearing and Chapter II is concerned with the arguments addressed to us following this important decision of the House of Lords.

First, I must set out the parliamentary material now placed before us as bearing upon the proper construction of section 172(4)(c). This appears in the *Hansard* report of the debate on May 22 1968 on the Town and Country Planning Bill, when the Minister of Housing and Local Government, Mr Nial MacDermott, was promoting by way of amendment the provision that eventually became section 172(4)(c). For the opposition, Mr Hugh Rossi was proposing that para(c) should instead read:

> The making without planning permission of a change of use of any building to residential use (rather than "to use as a single dwellinghouse").

The relevant passages in the debate are as follows: at column 664-665 the minister, explaining para(c)'s reference to a change of use to use as a single dwellinghouse said:

> 1. The reason for this is that changes of use to multi-occupation can have undesirable social as well as planning consequences. They come within the general mischief in that they are often difficult for the planning authority to detect and I understand that there have in practice been cases of this kind of development in which the 4 year rule has been a barrier, preventing the planning authorities from exercising the control which is patently desirable on social as well as planning grounds. This is the sort of abuse which can lead to an area's deterioration into a slum or near slum and I hope that the House will agree that this is a proper limitation.

2. At column 671 Mr Rossi said:

> There is no such class use in the Planning Acts as a single dwellinghouse. The class use is residential use . . . Is the Minister suggesting that there can be more concealment for multi-occupation than for single occupation?

3. The minister responded:

> I should have thought that it is very clear that if a building is used for residential purposes it is fairly readily observable, but one does not know how many people are living there.

4. At column 671-672 Mr Rossi continued:

> I concede at once that multi-occupation, particularly in our running twilight areas, is a great social evil which must be cured, but I suggest that it must be cured by redevelopment plans by housing associations or local authorities. . . . Multi- occupation may be treated in other ways under other legislation. I believe that the recent White Paper has something to say about problems of multi-occupation and overcrowding.
>
> I ask that Minister to consider the case of a shop with storerooms over it, all housed in one building. If the shopkeeper decides to change the user of part of the premises for residential accommodation for one or two members of his staff, he will be in breach of this provision. He will not be protected by the four-year rule. They are housed within one building, he is creating residential accommodation, and yet he is not protected by it. I ask the Minister to consider the further ramifications that the restriction . . . might well have.

5. At column 675 the minister concluded:

> I have pointed out the social evil which will result. It is not something new. I am not inventing the idea of a single dwellinghouse as a separate class. Section 12(3) of the 1962 Act provides:
>
> "For the avoidance of doubt it is hereby declared that for the purposes of this section —
>
> (*a*) the use as two or more separate dwellinghouses of any building previously used as a single dwellinghouse involves a material change in the use of the building and of each part thereof which is so used;"

— That, of course, is the present section 5(3)(a) —

> It already is a breach of planning control to convert from a single dwellinghouse to a multiple dwellinghouse, and for the social reasons I have outlined I suggest that it would be wise not to subject such a breach to the constraint of the four-year rule.

Before turning to the argument I must refer to the speeches in *Pepper* v *Hart* and in particular the threefold test now held to govern the use of Parliamentary material of this kind. It appears in the speech of Lord Browne-Wilkinson (with which the majority of the other members of the House agreed) at p1061F as follows:

> . . . the exclusionary rule should be relaxed so as to permit reference to Parliamentary materials where (a) legislation is ambiguous or obscure, or leads to an absurdity; (b) the material relied upon consists of one or more statements by a minister or other promoter of the Bill together if necessary with such other Parliamentary material as is necessary to understand such statements and their effect; (c) the statements relied upon are clear.

As to condition (c), Lord Browne-Wilkinson had a little earlier, at p1060C, said:

> The purpose of looking at Hansard will not be to construe the words used by the minister but to give effect to the words used so long as they are clear.

Two other passages in the speeches I have found helpful in illuminating the proper approach to condition (c): *per* Lord Bridge of Harwich at p1039H:

> It should, in my opinion, only be in the rare cases where the very issue of interpretation which the courts are called on to resolve has been addressed in Parliamentary debate and where the promoter of the legislation has made a clear statement directed to that very issue, that reference to Hansard should be permitted.

Per Lord Oliver of Aylmerton at p1042H:

> . . . relaxation . . . can apply only where the expression of the legislative intention is genuinely ambiguous or obscure or where a literal or *prima facie* construction leads to a manifest absurdity and where the difficulty can be resolved by a clear statement directed to the matter in issue.

As to condition (a), I myself have not the least doubt that it is satisfied in the present case. True, the first submission of each of the three counsel appearing before us was that para(c) unambiguously surrenders the meaning for which he contends. But the fact that these contentions are inconsistent with each other and that various experienced judges at first instance have reached differing conclusions upon the point, quite apart from the very real doubts which I myself experienced in writing chapter I of this judgment, lead me to conclude that this provision is indeed properly to be characterised as both ambiguous and obscure.

Condition (b) is plainly satisfied, the statements relied upon being those of the minister promoting the relevant amendment to the Bill.

It is upon condition (c) that the argument has centred. Mr Wood for Doncaster argues that the report of the debate read as a whole clearly reveals that the minister was using the phrases "multi-occupation" and "multiple dwellinghouse" to refer to all forms of multiple occupations including use of premises as self-contained flats. That, he submits, not merely accords with the ordinary use of language but is perfectly apparent from the minister's concluding remarks, in which he quoted what is now section 55(3)(a) and immediately continued:

> It already is a breach of planning control to convert from a single dwellinghouse to a multiple dwellinghouse and . . . it would be wise not to subject *such* a breach to the constraint of the four-year rule.

[Counsel's emphasis.]

The minister was clearly regarding a section 55(3)(a) breach as falling outside the protection of the four-year rule. For good measure, Mr Wood points out that that response followed upon the earlier passage in which Mr Rossi, giving the illustration of the shopkeeper changing the use of part of shop premises to residential accommodation, made plain his own understanding — consistent with Doncaster's case — of the effect of the proposed legislation, an understanding which the minister's response conspicuously left undisturbed.

Mr Katkowski for the Secretary of State submits to the contrary that the terminology used by the minister — his many references to "multi- occupation" and his final reference to "multiple dwellinghouse" — taken together with the specific reasons which he gave for excluding certain types of residential use from the benefit of the four-year rule — first, the difficulty of detecting such change of use through not knowing how many people live in particular premises; second, the fear of areas deteriorating into slums — clearly indicate that the Government's concern in promoting what is now section 172(4)(c) was to ensure that the four-year rule applied not to multi-occupation as that term is generally understood, but only to separate and self-contained units of occupation.

Mr Newcombe, on behalf of the Van Dycks, advanced an interesting if somewhat difficult argument which emphasised the distinction between on the one hand *dwellinghouses* in multiple occupation (such as those considered in what at the date of this debate were the relatively recent decisions in *Birmingham Corporation* v *Habib Ullah* [1964] 1 QB 178 and *Ealing Corporation* v *Ryan* [1965] 2 QB 486) and on the other hand *houses* in multiple occupation such as had been under consideration in *Okereke* v *Brent London Borough Council* [1967] 1 QB 42. Mr Newcombe's central contention was that, in his final response, the minister was first, by quoting section 12(3) of the 1962 Act, meeting Mr Rossi's suggestion that he had invented the idea of a "single dwellinghouse", and then, turning to a quite separate matter, identifying multiple occupation (as opposed to subdivision) of dwellinghouses as the particular feature of residential user which the Government was intent on excluding from the protection of the four-year rule.

My conclusions upon the matter are these:

1. The minister was clearly under the impression that section 172(4)(c) would not avail breaches of planning control consisting of subdivision (section 55(3)(a) cases) or even changes of use of *part of* a building (Mr Rossi's example).

2. That, however, was no more than an assumption on the minister's part. Those were not the particular breaches of planning control which concerned him and whose exclusion from the four-year rule he was specifically addressing.

3. The minister's final remarks were necessarily *extempore* responses to various points raised by Mr Rossi during the debate. Although I reject Mr Newcombe's suggestion as unacceptably far-fetched — the suggestion that the minister was invoking section 12(3) simply to justify his proposed use of the term "single dwellinghouse" and, when referring to conversion "from a single dwellinghouse to a multiple dwellinghouse", was then intending to describe some quite different breach of planning control — I certainly do not find in this crucial passage the clarity for which *Pepper* v *Hart* requires us to search, still less a clear statement directed to the very matter in issue: the application or otherwise of the four-year rule not merely to multi-occupation generally but to subdivision into separate flats in particular.

For my part, therefore, I would hold the *Pepper* v *Hart* test to be unsatisfied here with the result that the parliamentary material now before us must be put aside and the appeals determined on the basis indicated in chapter I of this judgment.

FARQUHARSON and **DILLON LJJ** agreed and did not add anything.

Appeals allowed with costs here and below; legal aid taxation for the Van Dycks; as regards the matter of relief, Van Dyck's *case be remitted to the Secretary of State for rehearing and determination, and the decision of the Secretary of State in* Doncaster *be restored; application for leave to appeal to the House of Lords refused.*

For further cases on this subject see p 207

LANDS TRIBUNAL

COMPENSATION

VICKERS v DOVER DISTRICT COUNCIL
(Ref/89/1992)

November 30 1992

(V G WELLINGS QC, President)

Estates Gazette May 22 1993

[1993] 20 EG 132

Land Compensation Act 1973, Part I — Depreciation in the value of land — Preliminary issue — Whether the Road Regulation Act 1984 confers immunity from actions in nuisance within the meaning of section 1(6) of the 1973 Act so as to allow compensation to be paid

The claimant is the owner of a 124-year lease of a flat on the sixth floor of a purpose-built block of flats at 11 Gateway, Marine Parade, Dover, Kent. The compensation authority is the freeholder and mortgagee of the flat. The claim relates to the use of land (in the vicinity of the block) at Camden Crescent, Dover, as a public car park, the site of which was acquired by the compensating authority on January 10 1989 by private treaty. The relevant date for the purposes of section 15 of the 1973 Act was August 22 1989, when, exercising their statutory powers under section 32 of the Road Traffic Regulation Act 1984, the compensating authority appropriated the land for use as a public car park to make provision, *inter alia*, for a building known as "the White Cliffs Experience". For the purposes of the preliminary issue, it was assumed that the car park by its use caused nuisance to the claimant. The tribunal heard a preliminary issue of law under r49 of the Lands Tribunal Rules 1975 as follows: "Does the Road Traffic Regulation Act 1984, under which the Dover District Council . . . provided and now use [as] a public car park, either expressly or impliedly, confer immunity on the Council from actions in nuisance within the meaning of section 1(6) of the Land Compensation Act 1973 so as to allow compensation to be paid under the Act?". The claimant contended that the compensation authority had immunity from actions in nuisance and accordingly he was entitled to make a claim for compensation under the 1973 Act. Two propositions were advanced on behalf of the compensating authority: (1) where a statute has authorised the doing of a particular act, or the use of land in a particular way, which act or user will inevitably involve nuisance, any resulting harm is not actionable; and (2) a body is liable for a nuisance by them attributable to the exercise of a power conferred by statute even without negligence, if by statute they are either expressly made liable, or not exempted from liability for nuisance.

***Decision*: Section 82 of the 1984 Act is entirely permissive: as to whether the powers of providing car parks are exercised at all, as to where they are to be situated and as to the manner in which they are to be exercised; there is no inevitability as to nuisance — The powers of providing car parks must be exercised in strict conformity with private rights; accordingly, section 32 does not, either expressly or by implication, confer on the compensating authority immunity from actions for nuisance — The tribunal therefore has no power to award compensation to the claimant.**

The following cases are referred to in this report.

Allen v *Gulf Oil Refining Ltd* [1981] AC 1001; [1981] 2 WLR 188; [1981] 1 All ER 353; [1981] JPL 353, HL
Department of Transport v *North West Water Authority* [1984] AC 336; [1983] 3 WLR 707; [1983] 3 All ER 273, HL reversing [1983] 3 WLR 105; [1983] 1 All ER 892
Manchester Corporation v *Farnworth* [1930] AC 171
Marriage v *East Norfolk Rivers Catchment Board* [1950] 1 KB 284; [1949] 2 All ER 1021; [1950] 66 TLR (Pt1) 225, CA
Metropolitan Asylum District v *Hill* (1881) 6 App Cas 193; 50 LJQB 353; 44 LT 653

The claimant appeared in person; John Litton (instructed by the solicitor to Dover District Council) represented the compensating authority.

Giving his decision, **MR V G WELLINGS QC** said: This is an issue of law ordered to be disposed of at a preliminary hearing under r49 of the Lands Tribunal Rules 1975. The preliminary issue arises under a notice of reference in which Mr Vickers claims from Dover District Council compensation under section 1 of the Land Compensation Act 1973 for depreciation in the value of land by physical factors caused by the use of public works. Subsection (6) of that section provides that:

> Compensation shall not be payable under this Part of this Act in respect of the physical factors caused by the use of any public works other than a highway unless immunity from actions for nuisance in respect of that use is conferred (whether expressly or by implication) by an enactment relating to those works . . .

The preliminary issue is stated as follows:

> Does the Road Traffic Regulation Act 1984, under which the Dover District Council ("the Council") provided and now use [as] a public car park, either expressly or impliedly, confer immunity on the Council from actions in nuisance within the meaning of section 1(6) of the Land Compensation Act 1973 so as to allow compensation to be paid under the Act?

The council and Mr Vickers on this question take up positions which are contrary to those which they might ordinarily be expected to adopt. The council say that they have no relevant immunity from actions in nuisance and Mr Vickers says that they have. If there is immunity, the reference can proceed; if there is no immunity, it cannot, and the only remedy of Mr Vickers would be an action in nuisance against the council.

There is an agreed statement of facts, which I summarise as follows:

(1) The property in question is a flat on the sixth floor of a purpose-built block at 11 Gateway, Marine Parade, Dover, Kent CT16 1LS.

(2) The claimant's interest is under a lease for 124 years and 159 days from January 21 1983.

(3) The council is both freeholder and mortgagee of the flat.

(4) The claim relates to use of land at Camden Crescent, Dover (in the vicinity of the block), as a public car park.

(5) The site of the car park was acquired by the council on January 10 1989 by private treaty.

(6) The relevant date for the purposes of section 15 of the Land Compensation Act 1973 is August 22 1989.

(7) The site was previously used as the site of a licensed hotel.

(8) The statutory power under which the public car park is provided is contained in section 32 of the Road Traffic Regulation Act 1984.

(9) The date of the claim was April 10 1991.

(10) On November 19 1991 the council began consideration of the

claim and on November 25 1991 rejected it on the ground that section 32 of the Road Traffic Regulation Act 1984 does not, either expressly or impliedly, confer immunity on the council from actions in nuisance.

At the hearing, I was informed by Mr Litton, counsel for the council, that the site of the car park was not purchased for the purposes of providing a car park, but was purchased for the purpose of providing housing accommodation. However, the council, having constructed the building known as "the White Cliffs Experience", needed a public car park in its vicinity and accordingly exercised their power under section 32(2) of the 1984 Act to appropriate the site for use as a public car park. The car park so provided serves the White Cliffs Experience, but not exclusively. It is indeed a public car park. Section 32 of the Act of 1984 provides, *inter alia*, as follows:

> (1) Where for the purpose of relieving or preventing congestion of traffic it appears to a local authority to be necessary to provide within their area suitable parking places for vehicles, the local authority . . .
> (a) may provide off-street parking places (whether above or below ground and whether or not consisting of or including buildings) together with means of entrance to and egress from them, or
> (b) may by order authorise the use as a parking place of any part of a road within their area, not being a road the whole or part of the width of which is within Greater London.
> (2) A local authority may adapt for use as, or for providing means of entrance to or egress from, a parking place any land acquired or appropriated by them for the purposes of this section, and may, with the consent of the authority or person responsible for the maintenance of a road, adapt any part of it for the purpose of providing means of entrance to or egress from a parking place.

Mr Litton conceded for the purposes of the preliminary issue (only) that I could assume that the car park by its use causes nuisance to Mr Vickers.

Mr Litton said that there were two relevant propositions.

Proposition 1: "Where a statute has authorised the doing of a particular act, or the use of land in a particular way, which act or user will inevitably involve nuisance, any resulting harm is not actionable": see *Clerk and Lindsell on Torts* (16th ed), paras 24-81 and also: *Metropolitan Asylum District* v *Hill* (1881) 6 App Cas 193, at p203; *Manchester Corporation* v *Farnworth* [1930] AC 171, at p183; *Allen* v *Gulf Oil Refining Ltd* [1981] 1 All ER 353, at p356.

Proposition 2: "A body is liable for a nuisance by it attributable to the exercise of a power conferred by statute even without negligence, if by statute it is either expressly made liable, or not exempted from liability for nuisance": see *Department of Transport* v *North West Water Authority* [1983] 1 All ER 892, at p895, affirmed by the House of Lords [1983] 3 All ER 273 at p275 and:

(1) *Allen* v *Gulf Oil (supra)* at p356, where Lord Wilberforce stated:

> . . . immunity from action is withheld where the terms of the statute are permissive only, in which case the powers conferred must be exercised in strict conformity with private rights

(2) *Metropolitan Asylum District* v *Hill* at pp201, 214 and 215.

Mr Litton submitted that the 1984 Act does not authorise the council to do a particular act or use the land in a particular way. This is not a case where the nuisance complained of is the inevitable result from the operation of the car park. Mr Litton said that section 32 of the Act of 1984 did not require the council to provide the car park; they were simply given the power to provide it if they so wished. The power was permissive. The 1984 Act neither provides for liability of the council nor does it expressly exempt the council from liability from nuisance and the council are therefore under proposition 2 subject to liability if nuisance is proved. Accordingly, Mr Litton submitted that no compensation was payable under section 1 of the 1973 Act.

Mr Vickers agreed that there was no express immunity from actions in nuisance in the Road Traffic Regulation Act 1984. He said that the authorities relied upon by Mr Litton all involved private acts in the interpretation of which the *contra proferentem* rule applies. He said that he was not alleging negligence on the part of the council, but the very nature of a car park was such that it must cause nuisance to somebody. If, he said, a car park was put in a large field there would be no nuisance, but if it were placed next to a block of flats it was inevitable that there would be nuisance. Accordingly, if the local authority chose to place a car park in a situation where it would inevitably cause nuisance, the statute under which the car park was provided must be assumed to have authorised or permitted the council to cause such nuisance resulting in immunity. He referred, in addition to the relevant authorities, to textbooks on the law of tort.

In my opinion, that submission is defeated by the authorities and more particularly by the decision of the House of Lords in *Metropolitan Asylum District* v *Hill (supra)*. That case was concerned with the Metropolitan Poor Act 1867, a public general Act, which authorised the formation of districts and district asylums for the care and cure of sick and infirm poor and created corporations for that purpose. It gave authority to the Poor Law Board to issue directions to the corporations to enable them to purchase lands and erect buildings for the purposes of the Act. The plaintiffs alleged that the district board had erected a certain hospital near their properties for the reception of persons suffering from small-pox and other infectious and contagious disorders, which was a nuisance, and had carried on the hospital so as to be a nuisance. At the trial the judge left certain questions to the jury two of which were as follows:

> (1) Was the hospital a nuisance occasioning damage to the plaintiffs *per se*?
> (2) Was it a nuisance by reason of the patients coming to or going from the hospital?

The jury answered "yes" to both those questions. It was held by the House of Lords that the district board could not set up the statute nor the orders of the Poor Law Board under it as an answer to an action or to prevent an injunction issuing to restrain the board from continuing the nuisance. The speeches at the hearing were read *in extenso*, but it is sufficient to cite a passage from the speech of Lord Watson, at p213, where he said:

> Where the terms of the statute are not imperative, but permissive, when it is left to the discretion of the persons empowered to determine whether the general powers committed to them shall be put into execution or not, I think the fair inference is that the Legislature intended that discretion to be exercised in strict conformity with private rights, and did not intend to confer licence to permit nuisance in any place which might be selected for the purpose.

He added, at p215:

> These powers appear to me to be from first to last permissive and not imperative. Whether they shall be exercised at all, and, if so, to what extent and effect their exercise shall be carried, is left to the discretion of the Local Government Board.

See also *Manchester Corporation* v *Farnworth* [1930] AC 171 at p183 where Viscount Dunedin said:

> When Parliament has authorised a certain thing to be made or done in a certain place there can be no action for nuisance caused by the making or doing of that thing if the nuisance is the inevitable result of the making or doing so authorised. The onus of proving that the result is inevitable is on those who wish to escape liability for nuisance, but the criterion of inevitability is not what is theoretically possible but what is possible according to the state of scientific knowledge at the time, having also in view a certain commonsense appreciation, which cannot be rigidly defined, of practical feasibility in view of situation and of expense.

See further *per* Jenkins LJ in *Marriage* v *East Norfolk Rivers Catchment Board* [1950] 1 KB at p305 *et seq*.

I do not agree with Mr Vickers' further submission, namely that after 1973 it is permissible, in determining whether a statute expressly or by implication gives immunity from actions in nuisance, to have regard to the remedy provided by the Land Compensation Act 1973, section 1. If that were permissible, section 1(6) of the Act of 1973 would surely say so.

I have come to the conclusion that section 32 of the Act of 1984 is entirely permissive: as to whether the powers of providing car parks are exercised at all, as to where they are to be situated and as to the manner in which they are to be exercised; there is no inevitability as to nuisance. As Lord Watson and Lord Wilberforce have said, the powers of providing car parks must be exercised in strict conformity with private rights. My decision, therefore, is that section 32 of the Act of 1984 does not, either expressly or by implication, confer on the council immunity from actions for nuisance. The Lands Tribunal, therefore, has no power to award compensation in this case to Mr Vickers. I answer the preliminary issue accordingly.

In my opinion, my decision on the point of law substantially disposes of the proceedings and I order that the argument be treated as the hearing of the case and dismiss the reference. Both parties asked for an order for costs on the preliminary issue in the event of success on it. Each asked me to order a lump-sum payment to be made. I order that Mr Vickers pay to the council the sum of £500 as a contribution towards their costs of the reference (being approximately one-half of the costs claimed by the council).

For a further case on this subject see p 213

SUPPLEMENT

CASES NOT REPORTED IN "ESTATES GAZETTE"

REAL PROPERTY AND CONVEYANCING

A
County Court

July 30 1992

(Before Judge JACK QC, sitting in Weymouth County Court)

JAGGARD v SAWYER AND ANOTHER

Trespass — Whether injunction and/or damages appropriate remedy — Whether damages can be awarded for future acts of trespass — Determination of damages

In 1987 the defendants, Mr and Mrs Sawyer, bought 5
B Ashleigh Avenue, Maiden Newton, Dorset, a semi-detached house at the end of a cul-de-sac. In December 1988 the defendants bought an area of land adjoining no 5. Following the grant of planning permission in 1988, the defendants constructed a dwelling-house on the additional land, which became known as no 5A. By reason of correspondence with the county surveyor the defendants believed that Ashleigh Avenue was a public highway, but were aware of a restrictive covenant affecting no 5 that no part of the land which was unbuilt upon should be used otherwise than as a private garden. In June 1989 the plaintiff, the owner of 1 Ashleigh Avenue and of a part of the land over which the avenue ran,
C contended that the defendants did not have any rights of way over the avenue for the benefit of no 5A. The present proceedings were commenced by the plaintiff on August 10 1989 by which time the walls and the roof of no 5A were well advanced: the plaintiff did not apply for interlocutory relief. The defendants sold no 5, less the access drive to no 5A, and following the completion of no 5A moved into the house and lived there until December 1991, when they let it. At the trial it was accepted that Ashleigh Avenue was not a public highway. The plaintiff sought an injunction restraining the defendants from trespassing on the plaintiff's part of Ashley Avenue,
D alternatively damages in lieu of an injunction under section 50 of the Supreme Court Act 1981.

Held: **(1) There were two lines of authority in relation to trespass, one of which effectively removed the possibility of awarding damages in lieu of injunction where the trespass was either continuing or anticipated: see *Anchor Brewhouse Developments Ltd* v *Berkley House* and the second where the court has a discretion to award damages under section 50 of the Supreme Court Act 1981: see *Bracewell* v *Appleby*. The *Bracewell* case was indistinguishable on its facts and should be applied. It was appropriate to refuse to grant an injunction**
E **and to award damages in lieu having regard to the conduct of the plaintiff, the defendants, the failure of the plaintiff to apply for interlocutory relief, the nature of the trespass and of the relevant land. (2) For the same reasons damages in lieu of an injunction in respect of the breach of the restrictive covenant should be awarded. (3) Applying *Wrotham Park Estate Co* v *Parkside Homes Ltd* the figure for damages is to be assessed as if the negotiation for the right to use Ashleigh Avenue as access to no 5A were taking place in the first half of 1989 prior to the commencement of building. Having regard to a notional profit on no 5A of £22,000 and that the defendants would have been prepared to pay not less than** F
£5,000 to achieve access and not more than £7,500, damages were assessed as the mean of £6,250 divided by the number of residents in the road, namely £694.54.

The following cases are referred to in this report.

Anchor Brewhouse Developments v *Berkley House* (1987) 38 BLR 82
Bracewell v *Appleby* [1975] Ch 408; [1975] 2 WLR 282; [1975] 1 All ER 993; (1974) 29 P&CR 204
Charrington v *Simons & Co Ltd* [1971] 1 WLR 598; [1971] 2 All ER 588; (1971) 22 P&CR 558; [1971] EGD 432; 218 EG 1164, CA
Eardley v *Granville* (1876) 3 ChD 826
Goodson v *Richardson* (1874) 9 Ch App 221 G
Kelsen v *Imperial Tobacco Co (of Great Britain and Ireland) Ltd* [1957] 2 QB 334; [1957] 2 WLR 1007; [1957] 2 All ER 343
Leeds Industrial Co-operative Society v *Slack* [1924] AC 851
Patel v *W H Smith (Eziot) Ltd* [1987] 1 WLR 853; [1987] 2 All ER 569, CA
Shelfer v *City of London Electric Lighting Co Ltd* [1895] 1 ChD 287; [1891-4] All ER Rep 838, CA
Stokes v *Cambridge Corporation* (1961) 13 P&CR 77; [1961] EGD 207; 180 EG 839, LT
Swordheath Properties Ltd v *Tabet* [1979] 1 WLR 285; [1979] 1 All ER 240; (1978) 37 P&CR 327; [1979] EGD 330; 249 EG 439, CA
Trenberth (John) Ltd v *National Westminster Bank Ltd* (1979) 39 P&CR 104; 253 EG 151
Wrotham Park Estate Co Ltd v *Parkside Homes Ltd* [1974] 1 WLR 798; [1974] 2 All ER 321; (1973) 27 P&CR 296
Woollerton & Wilson Ltd v *Richard Costain Ltd* [1970] 1 WLR 411; [1970] 1 All ER 483 H

This was a claim by the plaintiff, Heather Jaggard, for an injunction and damages in lieu for trespass and breach of covenant against the defendants, Colin and Lorna Sawyer, in respect of premises at Ashleigh Avenue, Maiden Newton, Dorset.

Mark Treneer (instructed by Miln & Lyall, of Bridport) appeared for the plaintiff; Jeremy Griggs (instructed by Clarke Willmott & Clarke, of Crewkerne) represented the defendants.

Giving judgment, **JUDGE RAYMOND JACK QC** said: The plaintiff in this action, lives in the village of Maiden Newton, Dorset, at 1 Ashleigh Avenue. The defendants, Mr and Mrs Sawyer, used to live in no 5A, which they built. They hold the legal title to it. In Maiden J
Newton there is a public house with an unusual name, the Chalk & Cheese. In the context of this unhappy case it seems appropriate.

Ashleigh Avenue is a small cul-de-sac built as a private development and completed in 1959 or 1960. Bull Lane leads up from the main road to the railway and Ashleigh Avenue is off to the left towards the top. It is some 50 yds long and the roadway about 15 ft wide. Before the addition of no 5A there were 10 dwellings, four small-detached bungalows on each side and a semi-detached pair at the end facing towards Bull Lane. No 1, where Mrs Jaggard lives, is the first house on the left. In 1987 Mr and Mrs Sawyer bought no 5. It is the left-hand house of the semi-detached pair. In December 1988 they bought the back half of the land comprised by 13 Bull Lane, which gave them an additional area adjoining the left of no 5. On this K
land they built no 5A. For access they laid a driveway running through to Ashleigh Avenue in part over land originally attached to no 5. In February 1990 they sold no 5, less than part of the driveway, and less a strip of garden next to the left-hand boundary. This action is brought by Mrs Jaggard to restrain Mr and Mrs Sawyer from using that part of Ashleigh Avenue which she owns in front of her house as an access road to no 5A and to restrain their use of land originally attached to no 5 as a driveway for no 5A.

It was at first contended that Ashleigh Avenue had become a public highway, and section 31 of the Highways Act 1980 was relied upon. It was later accepted that there was no evidence before the court that

A this cul-de-sac had ever been used by the public as of right. I approach the case on the basis that Ashleigh Avenue is a private road.

When Ashleigh Avenue was being developed and the plots were being sold off, covenants were given and taken in the same terms in each case. It is accepted that these covenants benefit and bind between themselves the present owners of the original plots. Thus each such owner has a right of way against each of the others over that part of the road which those others own outside their houses, and each one is bound by restrictive covenants to which I will refer. It was not suggested that no 5A is entitled to any right of way. Conversely, the 5A land which has never been part of no 5 is free from any covenants.

B The questions which I have to determine are:

(a) Should the court grant an injunction to restrain Mr and Mrs Sawyer from trespassing on Mrs Jaggard's part of Ashleigh Avenue? Or, should the court award her damages in lieu of an injunction under the jurisdiction originally provided by Lord Cairns' Act and now by section 50 of the Supreme Court Act 1981, a jurisdiction made available to the county court by section 38 of the County Courts Act 1984 as amended by the Courts and Legal Services Act 1990?

(b) Is the use of part of the land originally attached to no 5 as a driveway for no 5A in breach of covenant? If it is, should there be an injunction, or should there be an award of damages in lieu?

(c) If damages are to be awarded in lieu of injunctions, what should the figure be?

C Before I answer these questions, I must set out the history in greater detail, thereby making my findings of fact relevant to the dispute. But I first record that it is clear that the wish of Mr and Mrs Sawyer to build has raised strong feelings, feelings which have become far from neighbourly. It is not necessary for me to refer to the evidence relating to this. In particular I need not refer to the evidence of Mrs Nicholas who lives next to 13 Bull Lane and therefore owns the garden which at its rear abuts no 5A. But I bear this aspect in mind in assessing the motives of the parties.

Mr and Mrs Sawyer purchased no 5 in May 1987. They were probably then the only young couple in Ashleigh Avenue. Both were keen to get on and to do well. Mr Sawyer's parents live in Cerne Abbas and hers in Dorchester. It was a convenient place for them to
D be. Their first child was born in August of that year. No 5 has two bedrooms. Mr Sawyer initially thought of building on. Then he had the idea of building a separate house in the garden of no 5. He applied for planning permission in April 1988. Mrs Jaggard got up a petition against it and all save one of the other residents of Ashleigh Avenue signed. The petition referred to their refusal of access over Ashleigh Avenue as a private road other than to existing neighbours. Permission was refused on the ground of plot size.

Mrs Collyer is the owner of 13 Bull Lane. She offered to Mr Sawyer to sell him the land at the rear of her property for £15,000, and he agreed to buy subject to his getting planning permission. That application was made in the latter part of 1988. By this time Mr and
E Mrs Sawyer were aware from discussion they had had with Mrs Jaggard, and perhaps others, that the grounds of objection were the covenants and the fact that the road was private. This was also set out in a letter dated November 8 1988 to them from solicitors, Williams & Co, instructed by Mrs Jaggard. Planning permission was granted and the land for no 5A was conveyed by Mrs Collyer on December 22 1988. Mr Sawyer went about arranging the building of a new bungalow on the no 5A plot. To avoid the problem of builders' traffic up Ashleigh Avenue he arranged a temporary access to the plot from the industrial site to its rear. He then went round the residents of Ashleigh Avenue. On hearing that the traffic would not be using their road many withdrew their objections or took a neutral stance.

When they bought the no 5A from Mrs Collyer, Mr and Mrs
F Sawyer were advised by Creech & Co. Following the purchase they were advised as to their position regarding access and were told that there might be a problem. Creech & Co wrote to the county surveyor. He replied on February 7 1989 that Ashleigh Avenue was a public highway in his view. Mr Sawyer also received a visit on his behalf from Mr Wilkie, who confirmed that view in plain terms. Mr Sawyer was aware of the covenants regarding the no 5 land and that he could not build on that land. He does not seem to have appreciated that there might be a problem in using part of the no 5 land as a driveway for no 5A. The council's view, that Ashleigh Avenue was a public highway, was hotly disputed by a number of residents, including Mrs Jaggard. The council confirmed that view in letters dated March 9, April 3 and May 11 1989. A year later, in a letter to Mrs Jaggard of July 27 1990, they took the opposite view. No reference was made to G
the change and no explanation was offered.

By April 1989 Mr and Mrs Sawyer were being advised by Clarke, Wilmott & Clarke, who wrote to Mrs Jaggard on April 5. In May Mr Sawyer went to see Mrs Jaggard and told her that he was going ahead and would start building soon. She and Mrs Smith (of no 7) went to see Humphries Kirk. They wrote to Mr Sawyer on June 13 1989 raising the questions of the covenant and of the private road. Building was then just starting. They asked him to discontinue and threatened an injunction. Clarke Wilmott replied by fax on June 15 stating that as there was no building on the no 5 land, there was no breach of covenant, and making a somewhat naive point as to the road, that there was no building on the road and so no trespass. Mr H
Sawyer's position as to the road was in fact that he was relying on the council's view that it was public. Humphries Kirk replied by fax the same day in terms which did not clarify the points at issue. Clarke Wilmott pointed that out in their letter of the next day. On June 21 Humphries Kirk wrote stating that Mrs Jaggard would take proceedings for a declaration as to the question of trespass and right of way. They did not comment further on the question of breach of covenant. They asked that work should cease and threatened an application for an injunction if it did not. Clarke Willmott replied on June 22 stating their view that any action would fail and they would seek instructions. On July 4 1989 Humphries Kirk wrote referring to the case of *Bracewell* v *Appleby* [1975] 1 All ER 993. That is a case with facts very close to those in the present case and where an J
injunction was refused. The letter stated that Mrs Jaggard would probably seek the alternative remedy to an injunction — damages in lieu. The file shows that the letter was copied to Mrs Jaggard and Mrs Smith. I would expect that copies of Humphries Kirk's earlier letters were also sent to them. The proceedings were commenced on August 10. The work had then been going on for two months: the walls and the roof were very well advanced. No interlocutory relief was applied for. In April 1990 Mrs Jaggard changed her solicitors. She told me that she has been pressing for an injunction to be applied for from April 1989, but Humphries Kirk had failed to act on her instructions. In view of the terms of the correspondence and the fact that at least the letter of July 4 was copied to her, I do not think that the position was as simple as she suggests. But in any event, as between the parties K
to the action I have to proceed on the basis that no application for an interlocutory injunction was made and that on July 4 it was indicated that damages might be acceptable.

Meanwhile, in that July a second son was born to Mrs Sawyer and building was proceeding. I find that it proceeded in an ordinary manner and that there was no attempt to rush the work to frustrate any legal proceedings. Mr Sawyer funded the work in part by borrowing £45,000 from a friend, Mr Gandi, who was a builder though not the builder he was using for the work. It was completed in December 1989. The total cost was £75,770 including the £15,000 for the land.

After no 5A was built and as the action continued, the atmosphere L
in Ashleigh Avenue was very unpleasant for Mr and Mrs Sawyer, particularly because of the attitude of Mrs Jaggard and Mrs Nicholas. As a result in December 1991 they moved out and let no 5A. The arrangement they made was an unusual one. I do not have copies of the documents but they have been seen, if only briefly, by counsel for Mrs Jaggard. It appears that Mr Gandi or his company owned a house at Horton. It is there that Mr and Mrs Sawyer are now living. They purchased it at a price of £159,000, raising a mortgage of £50,000 to do so. They paid £54,000 to Mr Gandi and executed a deed of trust in respect of a one-third interest in the Horton house in his favour. They also executed a deed of trust over no 5A in his favour, to secure the £30,000 which remained outstanding from his original loan of £45,000, and the remainder of the purchase price for the M
Horton house, namely some £50,000 or £54,000.

It was urged on me that Mr and Mrs Sawyer developed no 5A with the idea of making a quick profit and moving on. I do not accept that. They badly needed a bigger home, Maiden Newton was genuinely convenient for them. No doubt in the atmosphere of the property market of 1988 and 1989 they also hoped that they would do well financially out of the exercise, and it may be that they talked unwisely to some of the residents about that. Mr Sawyer told me that it was still his desire to live in no 5A. In view of what has happened I am more doubtful of that.

I reject that allegation that Mr Sawyer simply intended to go ahead regardless of the legal position. I find that he believed that the road

A was public in reliance on the county surveyor's firm view. I do not think that he appreciated the problem of the covenant and the driveway through the no 5 land. I think that he might have shown more care in the investigation of his position. I put that down to his inexperience in a complicated situation. At the important stage in 1989 he was receiving legal advice and it was not suggested, let alone established, that the advice was that he had no right to do what he intended.

Mrs Jaggard was asked her reasons for bringing the action and for now pressing the case for an injunction. She answered that she felt that Mr Sawyer was proceeding in defiance of the law and she wanted the law upheld. She was concerned about the additional traffic which no 5a brought to Ashleigh Avenue. I note here that there was no
B evidence suggesting that it was any more than the light traffic one would except from the addition of an 11th house. Mrs Jaggard was concerned also that Mr Sawyer would not be contributing to the maintenance of the road. He has always been willing to do so. I find that the reason which weighs with Mrs Jaggard is that Mr Sawyer should not be permitted to behave as she thinks that he has.

The evidence as to the maintenance of the road was very slight. It hardly appears to have received any. I find that the traffic during the building works which ignored the instruction to use the access from the industrial estate caused no damage to the road. The road looked to me as if it might require resurfacing in the not too distant future. But it may have looked like that for years.

C I should mention the fact that having obtained a temporary access for the duration of the building works from the industrial site above no 5A, Mr Sawyer asked the owner of the site about a permanent access. He received an offer of a 10-year access with no charge save legal fees. He did not take that further. He was not pursued as to this possibility. It received no mention in the submissions of either side. I therefore put it out of my mind and proceed on the basis that the only access to no 5A is via Ashleigh Avenue.

It is against the factual background which I have set out that I have to answer the first question, namely whether Mrs Jaggard should be granted an injunction against Mr and Mrs Sawyer prohibiting trespass on that part of Ashleigh Avenue owned by her, or whether they should be an award of damages in lieu. This involves an
D examination of a number of authorities. For it is submitted on behalf of Mrs Jaggard that in an action of this nature it is established by the authorities that an injunction should go and that damages in lieu are inappropriate. I will take the cases to which I need to refer in chronological order.

Shelfer v *City of London Electric Lighting Company* [1895] 1 Ch 287 is a leading authority on the application of Lord Cairns' Act of 1858. The headnote begins:

> *Lord Cairns' Act* (21 & 22 Vict c 27), in conferring upon Courts of Equity a jurisdiction to award damages instead of an injunction, has not altered the settled principles upon which those Courts interfered by way of injunction; and in cases of continuing actionable nuisance the jurisdiction so conferred ought only to be exercised under very exceptional circumstances.

E The Court of Appeal reversed the decision at first instance and granted an injunction to restrain the operation of heavy machinery which was causing a serious nuisance. I will quote from the judgment of Lindley LJ at p315:

> The jurisdiction to give damages instead of an injunction is in words given in all cases; but the use of the word "damages" has led to a doubt whether the Act applies to cases where no injury at all has yet been inflicted, but where injury is threatened only. Subject, however, to this doubt, there appears to be no limit to the jurisdiction. But in exercising the jurisdiction thus given attention ought to be paid to well settled principles; and ever since *Lord Cairns' Act* was passed the Court of Chancery has repudiated the notion that the Legislature intended to turn that Court into a tribunal for legalizing wrongful acts; or in other words, the Court has always protested against the
> F notion that it ought to allow a wrong to continue simply because the wrongdoer is able and willing to pay for the injury he may inflict.

and at p316:

> Without denying the jurisdiction to award damages instead of an injunction, even in cases of continuing actionable nuisances, such jurisdiction ought not to be exercised in such cases except under very exceptional circumstances. I will not attempt to specify them, or to lay down rules for the exercise of judicial discretion. It is sufficient to refer, by way of example, to trivial and occasional nuisances: cases in which a plaintiff has shewn that he only wants money; vexatious and oppressive cases; and cases where the plaintiff has so conducted himself as to render it unjust to give him more than pecuniary relief. In all such cases as these, and in all others where an action for damages is really an adequate remedy — as where the acts complained of are already finished — an injunction can be properly refused.

The judgment of A L Smith LJ contains the following well-known G passage at p322:

> Many Judges have stated, and I emphatically agree with them, that a person by committing a wrongful act (whether it be a public company for public purposes or a private individual) is not thereby entitled to ask the Court to sanction his doing so by purchasing his neighbour's rights, by assessing damages in that behalf, leaving his neighbour with the nuisance, or his lights dimmed, as the case may be.
>
> In such cases, the well-known rule is not to accede to the application, but to grant the injunction sought, for the plaintiff's legal right has been invaded, and he is *prima facie* entitled to an injunction.
>
> There are, however, cases in which this rule may be relaxed, and in which damages may be awarded in substitution for an injunction as authorised by this section.
>
> In any instance in which a case for an injunction has been made out, if the H
> plaintiff by his acts or laches has disentitled himself to an injunction the Court may award damages in its place. So again, whether the case be for a mandatory injunction or to restrain a continuing nuisance, the appropriate remedy may be damages in lieu of an injunction, assuming a case for an injunction to be made out.
>
> In my opinion, it may be stated as a good working rule that —
>
> (1) If the injury to the plaintiff's legal rights is small,
>
> (2) And is one which is capable of being estimated in money,
>
> (3) And is one which can be adequately compensated by a small money payment,
>
> (4) And the case is one in which it would be oppressive to the defendant to grant an injunction: —
>
> then damages in substitution for an injunction may be given
>
> There may also be cases in which, though the four above-mentioned J
> requirements exist, the defendant by his conduct, as, for instance, hurrying up his buildings so as if possible to avoid an injunction, or otherwise acting with a reckless disregard to the plaintiff's rights, has disentitled himself from asking that damages may be assessed in substitution for an injunction.
>
> It is impossible to lay down any rule as to what, under the differing circumstances of each case, constitutes either a small injury, or one that can be estimated in money payment, or an adequate compensation, or what would be oppressive to the defendant. This must be left to the good sense of the tribunal which deals with each case as it comes up for adjudication. For instance, an injury to the plaintiffs's legal right to light to a window in a cottage represented by £15 might well be held to be not small but considerable; whereas a similar injury to a warehouse or other large building represented by ten times that amount might be held to be inconsiderable. Each case must be decided upon its own facts; but to escape the rule it must be brought within the K
> exception.

In *Leeds Industrial Co-Operative Society Ltd* v *Slack* [1924] AC 851 the House of Lords held that Lord Cairns' Act conferred a jurisdiction to award damages in lieu of an injunction against a threatened injury. The claim was for an injunction to restrain the erection of buildings to block the plaintiff's windows. The trial judge held that the interference with the plaintiff's rights would be small and was capable of being estimated in money, and that damages would be an adequate compensation. But he held himself bound by authority to hold that he had no jurisdiction to award damages. The House of Lords held that there was jurisdiction even though no damage had yet occurred. In his speech Viscount Finlay stated at L
p860:

> It has been urged as an objection to the construction of Lord Cairns's Act above stated, that it would give the Chancery Courts power to legalise the commission of torts by any defendant who was able and willing to pay damages. The Courts have on more than one occasion expressed their determination to prevent any abuse of the Act in this direction.

The case was remitted to the Court of Appeal where it was held that on the facts it was appropriate for damages to be awarded: [1924] 2 Ch 475. Each judgment of the Court of Appeal referred to the good working rule proposed by A L Smith LJ in the *Shelfer* case.

With that I can come to a number of modern cases beginning with *Kelsen* v *Imperial Tobacco Co (of Great Britain and Ireland) Ltd* [1957] 2 All ER 343. This concerned an advertising sign, which M
trespassed into the space above the roof of a shop. On the question of injunction or damages McNair J cited passages from the judgment of A L Smith LJ in *Shelfer* and continued at p351:

> I have no doubt at all that in the present case I have a discretion, which I have to exercise judicially, to award damages in lieu of an injunction, if I see fit. It is true that the injury to the plaintiff's legal rights in this case is small. The sign in his air space does him no harm and does not diminish his enjoyment. I doubt whether it is a case in which one can estimate the damage in terms of money, because the damages, if estimated at all, would have to be stated to be nominal. I do not find, however, that it would be in any way oppressive to the defendants to grant an injunction. It is true that considerable expense, stated to be some £220, was incurred some seven years ago in erecting the sign, but I have no evidence at all whether the defendants have not had

A good value for that expenditure. I know that Messrs Gallagher are apparently prepared today to pay £75 a year for a smaller advertisement in a less prominent position. Furthermore, I think it is relevant in this connexion that the defendants throughout the case have been insisting on the right to display this advertisement as a matter of right. I think that that is a circumstance which the court is entitled to take into account in determining whether a small money payment with a declaration of right should be sufficient or whether an injunction should be granted. Cases in which an injunction has not been granted on the ground of hardship have, I believe, been mostly cases in which there has been some accidental invasion of the plaintiff's rights. I was referred by counsel for the plaintiff to *Goodson* v *Richardson* ((1874) 9 Ch App 221), where a strong Court of Appeal held that the mere fact that the invasion of the highway in that case did not cause any serious damage to the owner of the highway did not disentitle him from an injunction. I am further impressed by B the fact that, if I refuse to grant an injunction in this case, there is nothing to prevent the defendants from continuing to display the sign, and leaving it to the plaintiff to put forward a subsequent claim for damages in a further action. If I were to decide that an appropriate remedy would be a small money payment of nominal damages, I would be, in effect, saying that, although such implied licence, if any, as the defendants had has been determined, nevertheless, the defendants are entitled to continue to display their sign.

In my judgment, bearing in mind that both parties, in pursuance of what they claimed to be their business interests, have attempted to bring commercial pressure to bear one of the other, this is a proper case in which the court should direct that there should be a mandatory injunction that such portion of the sign as projects over the plaintiff's premises be removed within twenty-eight days.

I have cited the whole of this passage because it raises the difficulty C that if damages are awarded the trespass will continue and there will be the possibility of a further action. That difficulty will also arise where there is a nuisance which is continuing. In my view, the answer may be found as follows. If at some future time a plaintiff who has been awarded damages in lieu of an injunction were to try a second action, the action would fail unless the circumstances were sufficiently different from those of the first. For whether or not there was strictly an estoppel *per rem judicatam*, the court would exercise its discretion to refuse an injunction. A claim for damages would be met by the answer that the plaintiff had already received his compensation. For in awarding damages in the first action the court is looking to compensate the plaintiff for the threatened wrong, that is, the wrong which it is anticipated will occur, or continue, in the D future. It will award him such sum as will compensate him over the anticipated period of the wrong. That may be as far into the future as can be foreseen. If he has been awarded a sum on that basis, he cannot come forward at a later date and ask for more: he has been compensated. Although the basis on which it might be so was not considered in argument, it seems to me that such a plaintiff's successors in title should be in no better position. It also seems to me that by being awarded compensation such a plaintiff waives or gives up any right to self-help to prevent the continuing tort, such as the use of reasonable force to prevent trespass.

Woollerton & Wilson Ltd v *Richard Costain Ltd* [1970] 1 All ER 483 concerned a trespass into airspace by an overhead crane on a E building site. Stamp J stated at p465:

It is in my judgment well established that it is no answer to a claim for an injunction to restrain a trespass that the trespass does no harm to the plaintiff. Indeed the very fact that no harm is done is a reason for rather than against the granting of an injunction; for if there is no damage done the damages recovered in the action will be nominal and if the injunction is refused the result will be no more nor less than a licence to continue the tort of trespass in return for a nominal payment. Furthermore, the very fact that the plaintiffs are the owners of the property are —

"... that no interference of this kind can lawfully take place without [their] consent, and without a bargain with [them], gives [their] interest in this land, even in a pecuniary point of view, precisely the value which that power of veto upon its use creates, when such use is to any other person desirable and an object sought to be retained".

F (See the judgment of Lord Selborne LC in *Goodson* v *Richardson*) Sir George Jessel MR in *Eardley* v *Granville* remarked of the defendant in that case:

"... he is a mere trespasser, and he being a trespasser comes within the well-established doctrine of *Goodson* v *Richardson*, and *Rochdale Canal Co* v *King* where damages would be no compensation for a right to property, and the Plaintiffs are entitled to prohibit him by injunction. There may be little or no injury to the estate, but if they restrain him he will be glad to pay a way-leave".

Counsel for the defendant referred me to a number of cases of which he submitted was to modify the principle or rule so stated in *Eardley* v *Granville* and to justify the court in refusing an injunction in a case such as the present. These were cases in which the claim was based on nuisance not on trespass. The gist of an action for nuisance is damage. And since the tort of nuisance can only exist if there be damage, in an action for nuisance damages can be obtained which will be measured by the extent of the nuisance and the plaintiff G in such a case is not in a situation of a plaintiff in an action for trespass who may recover only nominal damages. In an action for nuisance the licence which the court by refusing an injunction may be said to give the defendant to continue the nuisance will be compensated by the damages which the plaintiff will receive on his claim for damages.

and he cited the good working rule from *Shelfer*. He referred to the judgment in *Kelsen*. He held that the plaintiffs were entitled to an injunction, but that its operation should be suspended for a period which would enable the work to be completed. The plaintiffs thus got neither an injunction nor compensation. The correctness of the suspension has since been raised by the Court of Appeal in *Charrington* v *Simons & Co Ltd* [1971] 2 All ER 588 at p582. If the court in *Kelsen* had felt able to award an appropriate sum by way of H damages, on my reading of the judgment it would have done so.

Damages were awarded in lieu of an injunction in *Wrotham Park Estate Co Ltd* v *Parkside Homes Ltd* [1974] 2 All ER 321. This concerned a breach of covenant attaching to all property of an estate, namely not to develop land save in accordance with an approved lay-out plan. A development company bought part of the land and commenced building. A writ was issued claiming an injunction against building in breach of the covenant and a mandatory injunction to pull down buildings erected. No interlocutory injunction was sought. It was held that damages should be awarded in lieu of an injunction, the sum being such sum as might reasonably have been demanded by the plaintiffs from the defendants as a quid pro quo for relaxing the covenants, namely 5% of the profits of the J commercial development. In determining not to grant an injunction Brightman J stated at p337:

Counsel for the plaintiffs submitted, and I accept, that it is no answer to a claim for a mandatory injunction that the plaintiffs, having issued proceedings, deliberately held their hand and did not see the assistance of the court for the purpose of preserving the status quo. On the other hand, it is, in my view, equally true that a plaintiff is not entitled "as of course" to have everything pulled down that was built after the issue of the writ. The erection of the houses whether one likes it or not, is a fait accompli and the houses are now the homes of people. I accept that this particular fait accompli is reversible and could be undone. But I cannot close my eyes to the fact that the houses now exist. K

The basis for the award of damages adopted by the court is supported by a number of cases, which were reviewed by Brightman J, in which trespass to land or chattels involving little or no actual damage was compensated for by awarding damages on the basis of what would have been a reasonable rent or hire for the subject-matter of the trespass. The most recent reported example appears to be *Swordheath Properties Ltd* v *Tabet* [1979] 1 All ER 240. In a second action, which was also before the court, but where building had not yet begun, Brightman J granted an injunction.

The approach of the court in *Wrotham Park* was followed in *Bracewell* v *Appleby* [1975] 1 All ER 993, a case very close on its facts to the present. The scene was a building estate in Farnham and involved a private road. The defendant owned a house which had a L right to use the private road. He built a second house partly on land which he had acquired and partly in the garden of his existing house, and went to live there. He intended to use the private road as access. The plaintiff was another resident. Interlocutory relief was applied for but refused because of delay, the house then being finished. In deciding at the trial that an injunction should not be granted Graham J stated at p997:

Having seen both Mr Wright and the defendant in the witness box there was clearly a clash of personalities and Mr Wright said they have not been on speaking terms since the start of this trouble. I have not much sympathy for the way in which either of them has conducted his affairs. As I say, like Pennycuick V-C, I would not have granted an interlocutory injunction, but M furthermore I would not grant an injunction now if otherwise satisfied that the plaintiffs were entitled to relief following the same principles as those enunciated by Brightman J in *Wrotham Park Estate Co Ltd* v *Parkside Homes Ltd*. Although an injunction here would be against trespassing on the plaintiffs' land and would not go to pulling the house down, it would in effect make the house, no 2A, uninhabitable and would put the plaintiffs into an unassailable bargaining position.

Woollerton was cited in argument, but was not referred to in the judgment. Damages were awarded on the basis of what would have been a fair price for the right of way.

In *John Trenberth Ltd* v *National Westminster Bank Ltd* (1979) 39 P&CR 104 the wrong was trespass by a scaffold erected for the repair of an adjacent building. Walton J held that because the damage was very slight that was the very reason for an injunction to be granted. He stated a p107:

> But it is perfectly clear that the actual damage, apart from any question of aggravation, caused by the mere trespass, both by oversailing the front of the plaintiffs' property and by actually resting upon the rear of the plaintiffs' property, is comparatively slight: so slight that if an action were brought for it, it would hardly command the smallest coin in the realm. But so far from that being a reason why an injunction should not be granted, it has been said in many of the cases to which Mr. Munby drew my attention that the fact that any damage would be trifling is the very reason why an injunction should be granted. People are not to infringe the property rights of others and then say, "And I am entitled to go on doing it because I am really doing you no tangible harm, and fivepence will amply compensate you for that harm".

On p108 he stated that there could have been no answer if an injunction had been applied for *quia timet*, and questioned how the plaintiffs could be in a better position because they had erected the scaffolding and commenced the trespass.

In *Patel* v *W H Smith (Eziot) Ltd* [1987] 2 All ER 569 the Court of Appeal had before it a case of trespass by parking in a yard. The first instance judge had refused an interlocutory injunction on the ground that there was a serious issue to be tried as to whether the parking constituted a trespass. The Court of Appeal held that there was clearly a trespass and granted an injunction. In the course of his judgment Balcombe LJ stated at p573:

> What, then, are the principles which a court should apply in a case of this type? It seems to me that, first, prima facie a landowner whose title is not an issue is entitled to an injunction to restrain trespass on his land whether or not the trespass harms him. In support of that proposition there are two comparatively recent cases at first instance. The first is *Woollerton & Wilson Ltd* v *Richard Costain Ltd* [1970] 1 All ER 483, [1970] 1 WLR 411, which was a reserved judgment of Stamp J, and from the report I note that *Behrens* v *Richards* [1905] 2 Ch 614 was cited in argument, though the judge does not refer to it in his judgment. Stamp J said ([1970] 1 All ER 483 at 485, [1970] 1 WLR 411 at 413):
>
> "It is in my judgment well established that it is no answer to a claim for an injunction to restrain a trespass that the trespass does no harm to the plaintiff. Indeed, the very fact that no harm is done is a reason for rather than against the granting of an injunction; for if there is no damage done the damage recovered in the action will be nominal and if the injunction is refused the result will be no more nor less than a licence to continue the tort to trespass in return for a nominal continue the tort to trespass in return for a nominal payment."
>
> To the like effect is the decision, also at first instance, of Walton J in *John Trenberth Ltd* v *National Westminster Bank Ltd* (1979) 39 P&CR 104, and I cite a short passage from the judgement (at 108) where the judge accepted a submission by counsel for the plaintiffs in that case that —
>
> "when one dealing with the direct physical invasion of a right of property by a trespass one is very close to the line of cases stemming from *Doherty* v *Allman* ((1878) 3 App Cas 709) which decide that an injunction to enforce a negative stipulation in a contract goes almost as of course. The parties, having agreed that something shall not be done, the court simply says that what the parties have agreed shall not be done."
>
> When the court resumed this morning, argument having been concluded, counsel for the defendants drew our attention to the case which I have mentioned of *Brehens* v *Richards* [1905] 2 Ch 614. This is a decision of Buckley J and the headnote states quite accurately that the effect of the decision was: "The Court may refuse to grant an injunction to restrain persons from trespassing on land if the landowner is not injured thereby." That was a case relating to exercise by the public of claimed rights of way over a path where it was clear there could have been no damage of any kind done to the landowner.
>
> If it came to a choice between these two lines of authority, for my part I think I would prefer the more recent authority, but I do not think it necessary to go to the extent of saying that this court must choose to follow one line of authority rather than another. In normal circumstances the prima facie test should be that indicated by Stamp J in the *Woollerton & Wilson Ltd* case, but there may be exceptional circumstances, such as those considered by the court in *Behrens* v *Richards*, when the court will not think it appropriate to grant an injunction.
>
> The two recent cases to which I have referred, *Woollerton & Sons Ltd* v *Richard Costain Ltd* and *John Trenberth Ltd* v *National Westminster Bank Ltd*, are also authority that the same principle, namely prima facie a landowner whose title is not an issue is entitled to an injunction to restrain trespass, applies where the claim is for an interlocutory injunction.

Once it was established that the parking in the yard was a trespass, the situation appears a relatively simple one. The court was not looking at a situation where it was suggested that the application of Lord Cairns' Act was appropriate.

Lastly, I come to *Anchor Brewhouse Developments Ltd* v *Berkeley House* (1987) 38 BLR 82. This was another case involving an overhead crane on a building site. Scott J cited at p95 in the report the passage from the judgment of Stamp J in *Woollerton* where *Eardley* v *Granville* and *Goodson* v *Richardson* are cited. Scott J continued at p96:

> Mr Moss pointed out, and it is common ground, that there is a flaw in that exposition of the law. The learned judge proceeded on the footing that nominal damages only could be recovered for the trespass. He was not referred by counsel to the line of authority enabling damages to be awarded to represent the sum that on a "willing grantorgrantee" basis the plaintiff might have charged the defendant for a licence to commit the trespass. Mr Moss fastened on that error in order to submit that Stamp J's conclusion that an injunction should be granted to restrain the trespass was unsound. I do not agree. Stamp J based himself upon firm early authority in the form of the decision of Lord Selborne in *Goodson* v *Richardson* and of Sir George Jessel in *Eardley* v *Granville*. Both these authorities justify the grant of injunctive relief to restrain trespass notwithstanding that the damage to the plaintiff may be trivial only. Lord Selbourne and Sir George Jessel adopted a robust Victorian approach which might, perhaps, find less sympathy now. The ownership of property entitled the owner to licence or refuse to licence the use of it by others. If he is asked to licence the use of it by others, he can charge whatever he chooses for the licence. The law will recognise and protect the monopoly that his ownership carries with it. That is the philosophy, as I read the two cases, underlying the judgments of Lord Selborne and of Sir George Jessel. The grant of the injunction by Stamp J was in accordance with the principle expressed in these two cases. It did not depend on the nominal damages point.

Scott J cited from the judgment of Balcombe LJ in *Patel* and concluded (at p99):

> The Court of Appeal's approach in the *Patel* case requires me, in my judgment, to proceed on the footing that, prima facie, the plaintiffs are entitled to an injunction to restrain the continuing trespass. Mr Moss, however, has put forward special circumstances which, he submits, justify withholding an injunction. He points out, perhaps rightly, that the possibility of special circumstances justifying the withholding of an injunction to restrain a continuing trespass was accepted by Balcombe LJ.

He held that there was no special circumstances and granted an injunction. In respect of *Bracewell* v *Appleby* he said as follows (p101):

> I find some difficulty with *Bracewell* v *Appleby* mainly because, as it seems to me, the learned judge regarded the damages he was awarding as a once and for all payment. But it was, as I see it, not within the power of the judge to produce that result. Whether or not an injunction were granted, the defendant's use of the right of way would, after the judgment as well as before, represent trespass unless and until he were granted a right of way. The judge could not by an award of damages put the defendant in the position of a person entitled to an easement of way. So assuming, which is not clear from the case, that there had not been some agreement by the plaintiffs to treat the damages as entitling the defendant to a right of way, the defendant's subsequent use of the private road would have constituted a continuing trespass. A succession of further actions for damages could have been brought. In those circumstances it seems to me very difficult to justify the withholding of the injunction. By withholding the injunction the court was allowing a legal wrong to continue unabated. Nonetheless Mr Moss is entitled to refer to the case as one in which an injunction was refused.

Although in his extemporary judgment Scott J did not refer to Lord Cairns' Act or to *Slack's* case, they must have been in his mind.

It was thus argued before me that there were two lines of authority in relation to trespass, one of which effectively removed the possibility of awarding damages in lieu of an injunction where the trespass was either continuing or anticipated. There does not appear to be a conflict. But I do not think that it is one which I have to resolve. In this area of the law particularly, every case turns on its own facts. The facts before me are a long way from those in any of the cases in the line, the last of which is *Anchor Brewhouse*. I consider that I should accept *Bracewell* v *Appleby* as establishing a discretion to award damages under section 50 of the Supreme Court Act 1981 in the circumstances before me: it is for that purpose indistinguishable on its facts. In doing so I do not consider that I am in any way running contrary to the judgments of the Court of Appeal in *Patel*. *Bracewell* v *Appleby* was not cited on the appeal and, as I have stated, the court was not concerned with the application of Lord Cairns' Act.

Before moving on I should refer to the difficulty of a continuing wrong and of further actions which was raised by Scott J in *Anchor Brewhouse*. I have suggested a possible answer to this earlier in my comments following my citation of *Kelsen* v *Imperial Tobacco Co*. It was suggested on behalf of Mrs Jaggard that there might be further difficulties. For the access to no 5A could be used to service a number of houses built behind it. I do not regard that as likely. But, in any event, the answer would be that the present proceedings would be no bar to proceedings against the owners or occupiers of any such houses. I have also considered the question whether there is a problem because it would be open to Mrs Jaggard to sue third parties who use her part of the roadway to go to no 5A. I do not think that there is a problem there. If the court has declined to grant her an injunction against Mr and Mrs Sawyer, it will decline to grant her an injunction against their visitors, and she could recover no damages because the compensation which this court would have awarded her in lieu of an injunction would cover use by such persons.

I conclude therefore that the facts of this case provide no bar to my following the approach of the court in the *Wrotham Park* case and *Bracewell* v *Appleby*.

How then is my discretion to be exercised? In all the circumstances which I have set out I think it appropriate not to grant an injunction in respect on trespass of the part of Ashleigh Avenue owned by Mrs Jaggard, but to award damages in lieu. I have particularly in mind the conduct of the plaintiff and of the defendants and their reasons for acting as they have, the failure of the plaintiff to apply for interlocutory relief, the particular nature of the trespass and of the relevant land, and the fact that if an injunction is granted no 5A will have no access. Although the wide discretion given by section 50 is not to be limited as if the working rule proposed by A L Smith LJ in *Shelfer* was part of the Act, in my view, each of the four points there referred to are satisfied.

The terms of the covenant, the breach of which is alleged by the use of part of the original no 5 land as a driveway to no 5A, are as follows:

> No house or building to be erected on any part of the said land shall be used . . . in any manner calculated or likely to be a nuisance or cause annoyance to the Vendors or adjoining owners or residents of the neighbourhood or in any manner otherwise than as a private residence only and no part of the said land which is unbuilt upon shall be used otherwise than as a private garden.

I tend to the view that the use of part of a garden as a drive for the purpose of the house which that garden serves would not be in breach of this covenant. Such a use is one to which a part of modern gardens are commonly put. In my judgment, however, the use of part of the land as an access driveway for another property is use other than as a private garden. This construction is supported in the particular circumstances by the fact that such a use would not have been in contemplation when the covenant was framed because Ashleigh Avenue was intended to serve only the original houses which had a right of way over it. I, therefore, hold that the use of the relevant land as a driveway is in breach of the covenant.

For the same reasons as have caused me to refuse Mrs Jaggard an injunction but to award her damages in lieu in respect of trespass, I exercise my discretion to award her damages in lieu of an injunction in respect of the breach of covenant. In the circumstances the answer has to be the same in each case. The position in respect of breach of covenant does not face the same difficulties which I have had to consider in respect of trespass.

Lastly, what damages should be awarded? Figures are provided by means of an agreed valuation report by Humberts dated May 13 1992.

It was pointed out on behalf of Mrs Jaggard that the courts have not in any reported case had to deal with a combination of trespass and breach of covenant. It was submitted that the plaintiff should receive a share in the increase in the value of no 5A which the defendants will achieve by means of the rejection of the plaintiff's case for an injunction. Counsel took the value of no 5A at June 1989 at £100,000 with access and deducted a land value with no açcess, to give a profit of £96,000. Alternatively, he submitted that the profit figure should be the cost of the house on the basis that the house without access is valueless. For reasons which will appear when I refer to the authorities I reject these submissions as being contrary to principle. I think that they are also unrealistic.

A second basis of assessment was put forward on the plaintiff's behalf, namely what the defendants would have had to pay to buy land to provide access. This was put at £15,000 based on the cost of the land bought from Mrs Collyer. Again I do not think that this is an appropriate assessment.

Lastly, there was submitted an assessment on the basis of what the defendants might reasonably have paid for a right of way and the release of the covenant. The valuation for the completed development as at June 1989 is £100,000. That gives a profit over cost of some £24,000. I point out that this ignores the inclusion as part of no 5A of a substantial part of the garden of no 5. The value of this is not included in the costs which have been deducted. It was suggested that $\frac{3}{5}$ of that profit should be taken as the figure, namely £14,400, or $\frac{2}{5}$ + 5% (for the covenant, I think) which gives £10,800. Dividing these figures by nine, for the nine residents who are relevant gives £1,600 and £1,200 respectively. It was submitted that there should also be an additional amount to reflect trespasses over the last three years.

It was submitted on behalf of Mr and Mrs Sawyer that I should take a share of the paper profit made on the development of no 5A. Taking the midway figure of the valuations for August 1989 and September 1990 gives £93,500, and a profit of £17,730 (again excluding the no 5 land). Dividing by nine gives £656.

I was referred to three authorities. In *Stokes* v *Cambridge Corporation* (1961) 13 P&CR 77 the Lands Tribunal held for the purpose of assessing the value of land for compulsory purchase that a developer would in the particular circumstances of the case have surrendered one-third of his hypothetical profit on the development in order to buy land to secure access to the development.

I have already referred to the basis on which damages were assessed in the *Wrotham Park Case* [1974] 2 All ER 321. At p341 Brightman J stated:

> As I have said, the general rule would be to measure damages by reference to that sum which would place the plaintiffs in the same position as if the covenant had not been broken. The defendant Parkside and the individual purchasers could have avoided breaking the covenant in two ways. One course would have been not to develop the allotment site. The other course would have been for Parkside to have sought from the plaintiffs a relaxation of the covenant. On the facts of this particular case the plaintiffs, rightly conscious of their obligations towards existing residents, would clearly not have granted any relaxation, but for present purposes I must assume that they would have been induced to do so. In my judgment a just substitute for a mandatory injunction would be such a sum of money as might reasonably have been demanded by the plaintiffs from Parkside as a quid pro quo for relaxing the covenant. The plaintiffs submitted that that sum should be a substantial proportion of the development value of the land. This is currently put at no less than £10,000 per plot, ie £140,000 on the assumption that the plots are undeveloped. Mr Parker gave evidence that a half or a third of the development value was commonly demanded by a landowner whose property stood in the way of a development. I do not agree with that approach to damages in this type of case. I bear in mind the following factors: (1) The layout covenant is not an asset which the estate owner ever contemplated he would have either the opportunity or the desire to turn to account. It has no commercial or even nuisance value. For it cannot be turned to account except to the detriment of the existing residents who are people the estate owner professes to protect. (2) The breach of covenant which has actually taken place is over a very small area and the impact of this particular breach on the Wrotham Park estate is insignificant. The validity of the covenant over the rest of area 14 is unaffected. I think that in a case such as the present a landowner faced with a request from a developer which, it must be assumed, he feels reluctantly obliged to grant, would have first asked the developer what profit he expected to make from his operations. With the benefit of foresight the developer would, in the present case, have said about £50,000, for that is the profit which Parkside concedes it made from the development. I think that the landowner would then reasonably have required a certain percentage of that anticipated profit as a price for the relaxation of the covenant, assuming, as I must, that he feels obliged to relax it. In assessing what would be a fair percentage I think that the court ought, on the particular facts of this case, to act with great moderation. For it is to be borne in mind that the plaintiffs were aware, before the action took place, that the land was being offered for sale as freehold building land for 13 houses, and the plaintiffs knew that they were not going to consent to any such development. The plaintiffs could have informed the urban district council of their attitude in advance of the action or could have given the like information to Parkside prior to completion of the contract for sale. In either event it seems highly unlikely that Parkside would have parted with its £90,000, at any rate unconditionally. I think that damages must be assessed in such a case on a basis which is fair and, in all the circumstances, in my judgment a sum equal to 5 per cent of Parkside's anticipated profit is the most that is fair. I accordingly aware the sum of £2,5000 in substitution for mandatory injunctions.

Like the reasoning of the assessment made in *Stokes* this passage emphasises that the particular circumstances must be examined to see what is appropriate. In *Bracewell* v *Appleby* [1975] 1 All ER 993 at p1,000 Graham J dealt with the question of damages in the following terms:

> In the present case, the plaintiffs, for amenity reasons, did not want an extra house built in the cul-de-sac and I think it is right to regard them also as "reluctant", just as Brightman J did in the case of the plaintiffs before him. On the other hand, in all the circumstances, I think that for the purpose of estimating damages they and the other servient owners in Hill Road, albeit reluctant, must be treated as being willing to accept a fair price for the right of way in question and must not be treated as if they were in the extremely powerful bargaining position which an interlocutory injunction would have given them if it had been obtained before the defendant started operations and incurred expense. Such is to my mind the penalty of standing by until the house is built.

On the evidence here the probable figure of notional profit which the defendant has made, being the difference between the overall cost of the new house and its present day value seems to be somewhere between £4,000 and £6,000 and I think it is fair to take £5,000 as about an accurate a figure as one can get. The circumstances here are very different from those in *Wrotham Park Estate Co Ltd* v *Parkside Homes Ltd* and I think that the proper approach is to endeavour to arrive at a fair figure which, on the assumption made, the parties would arrive at as one which the figure which, the plaintiffs would accept as compensating them for loss of amenity and increased user, and which at the same time, whilst making the blue land a viable building plot, would not be so high as to deter the defendant from building at all. The defendant was not a speculative builder and in fact wanted to live in, and does now live in, 2A Hill Road himself and I think he would have been prepared to pay what is relatively to his notional profit quite a large sum for the right of way in question and to achieve the building of his new home. This was a time of rising property values and I think he would have been prepared to pay £2,000 to get his right of way and if he had made such an offer, I think the other five owners in Hill Road ought also to have been prepared to accept it.

Much of that can be applied in the present case.

Had there been a negotiation here it would have been in reality between Mr and Mrs Sawyer and all the residents, some of whom did not ultimately object to the development. They would all of course have wanted Mr and Mrs Sawyer to pay their share of the upkeep of the road. Mr and Mrs Sawyer remain responsible for the small part of road which was attached to no 5, because they still own it. The others would, none the less, want at least compensation for the additional use of the road for whose maintenance they are responsible. I consider that the figure is to be assessed as if the negotiation were taking place in the first-half of 1989 prior to the commencement of building. Taking the figures in Humberts report, Mr and Mrs Sawyer were then looking at a notional profit of some £24,000 if the value of the land from no 5 is ignored. I do not consider that it is right to ignore it. For with a larger garden no 5 could obviously have been sold for more, and Mr and Mrs Sawyer have suffered a real detriment in that respect by including it in with no 5A. The land in question is in area approximately one-third to -half of the garden of no 5. I have no valuation of it. The point was not referred to. I only know that the price for the dimished no 5 was £60,000. In this situation I should take a very modest figure. I will take £2,000, which reduces the notional profit on no 5A to £22,000. The sale of no 5 is not otherwise to be taken into account because its value was not increased by anything relating to no 5A: it was simply diminished by the loss of part of its garden.

There is in this case an unreality in supposing that any hypothetical negotiations would have included a calculation of the hypothetical profit. The reality is that they would have been in terms of offer and demand in round figures. The profit calculation can be a guide. I think that Mr and Mrs Sawyer should have been prepared to pay not less than £5,000 to achieve access, that is to acquire a right of way and to have the covenant waived. I do not think that the "come-back" figure should have been more than £7,500. I propose to take the mean, £6,250. That gives £694.44 per resident. It is between a quarter and a third of the hypothetical profit, whether that is taken at £24,000 or £22,000. Mr and Mrs Sawyer were not buying land as is postulated where it is said that a developer will commonly give up a third of his profit to acquire access. Looking at the figure of £6,250 in the round against the circumstances of the case I consider it appropriate. On the basis on which I have assessed it, it would have been paid by the beginning of June 1989, and it will carry interest from that date. In that way Mrs Jaggard will be compensated for the trespass which has taken place to date.

May I lastly record my gratitude to counsel for their researches and for the manner in which they presented the material.

Privy Council

February 22 1993

(Before Lord Keith of KINKEL, Lord Jauncey of TULLICHETTLE, Lord Donaldson of LYMINGTON, Lord BROWNE-WILKINSON and Sir Christopher SLADE)

WORKERS TRUST & MERCHANT BANK LTD v DOJAP INVESTMENTS LTD

Contract for sale — Deposit — Whether deposit in excess of 10% is a penalty or whether such deposit can be forfeited by the vendor

On October 5 1989 the appellant, Workers Trust & Merchant Bank Ltd, as second mortgagee, sold premises at auction to the respondent, Dojap Investments Ltd at a price of $J11,500,000. Clause 4 of the contract provided for the payment of a deposit of 25%, and a sum of $J2,875,000 was duly paid by the respondent to the appellant's solicitors. The contract provided that the reminder of the purchase money should be paid within 14 days of the date of the auction whereupon the appellant were to execute a transfer of the property to the respondent and lodge such transfer for registration. Clause 15 of the contract provided that time should be of the essence of all time-limits contained in the contract. Clause 13 provided that if the purchaser should fail to observe or comply with any of the foregoing stipulations on his part his deposit shall be forfeited to the vendor who would be at liberty to resell the property. On the date fixed for completion (October 19 1989) the respondent's attorney sent to the appellant a letter of undertaking from the Jamaica Citizens Bank Ltd to pay the balance of the purchase price, subject to certain conditions. The appellant's attorney rejected this and gave the respondent 24 hours to provide a satisfactory undertaking. On the respondent failing to provide a satisfactory undertaking, on October 23 1989 the appellant wrote to the respondent rescinding the contract and purporting to forfeit the deposit. The respondent refused to accept this and on October 26 1989 tendered to the appellant the balance of the purchase money with interest. The appellant returned the cheque the next day. On November 24 1989 the respondent commenced proceedings claiming specific performance or alternatively relief from forfeiture of the deposit. On June 25 1990 Zacca CJ rejected the respondent's claim for specific performance and its claim for return of the deposit. On appeal, the Jamaican Court of Appeal held that the respondent was entitled to relief from forfeiture to the extent that the deposit exceeded 10% of the price, but did not award any interest on that sum. The appellant appealed against that decision; the respondent cross-appealed claiming that it should have been awarded relief against forfeiture as to the whole of the 25% deposit and should have been awarded interest.

Held: **The appeal was dismissed and the cross-appeal allowed. A reasonable deposit has always been regarded as a guaranteed of performance as well as a payment on account, and its forfeiture has never been regarded as a penalty in English law or common English usage. In order to be reasonable a true deposit must be objectively operating as "earnest money" and not as a penalty. The correct approach is to start from the position that, by long-continued usage the customary deposit has been 10% and a vendor who seeks to obtain a larger amount by way of forfeitable deposit must show special circumstances which justify such a deposit. Notwithstanding the practice in Jamaica of the vendor discharging the purchaser's liability to transfer tax, and therefore deposits in excess of 10% have become common to provide for this, there was no need in the present case for such a high deposit because the tax was not payable before**

A **completion. Since the 25% deposit was not a true deposit by way of earnest, the provision for its forfeiture was a plain penalty. As the appellant never stipulated for a reasonable deposit of 10%, the whole 25% deposit constituted an unreasonable sum and had to be repaid as a whole together with interest at 12% pa from the date of rescission until the date of actual payment. However, the appellant is entitled to deduct the amount of any damages it has suffered as a result of the respondent's failure to complete from the purported deposit of 25%.**

The following cases are referred to in this report.

B *Commissioner of Public Works* v *Hills* [1906] AC 368
Howe v *Smith* (1884) 27 ChD 89
Linggi Plantations Ltd v *Jagatheesan* (1972) 1 MLJ 89
Stockloser v *Johnson* [1954] 1 QB 476; [1954] 2 WLR 439; [1954] 1 All ER 639, CA

This was an appeal by Workers Trust & Merchant Bank Ltd, from a decision of the Court of Appeal of Jamaica which had allowed, in part, an appeal by Dojap Investments Ltd, from a decision of Zacca CJ in proceedings for specific performance and relief from forfeiture of a deposit.

Roald Henriques QC and David Batts, of the Jamaican Bar,
C (instructed by Philip Conway Thomas) appeared for the appellant; Richard Mahfood QC, of the Jamaican Bar, and James Guthrie (instructed by Charles Russell) represented the respondent.

In delivery the opinion of the Judicial Committee, **LORD BROWNE-WILKINSON** said: This case raises the question whether a deposit in excess of 10% paid under a contract for sale of land can be lawfully forfeited by the vendor in the event of a failure by the purchase to complete on the due date.

On October 5 1989 Workers Trust & Merchant Bank ("the bank") as second mortgagee sold certain premises at auction to Dojap Investments ("Dojap") at a price of $J11,500,000. Clause 4 of the contract provided for the payment of a deposit of 25% of the contract price and a deposit of $J2,875,000 was duly paid by Dojap to the
D bank's solicitors. The contract provided that the remainder of the purchase money should be paid within 14 days of the date of the auction whereupon the bank were to execute a transfer of the property to Dojap and lodge such transfer for registration. Clause 15 of the contract provided that time should be of the essence of all time-limits contained in the contract. Clause 13 of the contract provided as follows:

> 13. If the purchase shall fail to observe or comply with any of the foregoing stipulation on his part his deposit shall be forfeited to the vendor who shall be at liberty (without tendering any transfer) to re-sell the property either by public auction or private contract at such time and in such manner and subject to such conditions as the vendor may think fit and any deficiency in price which may result on and all charges costs and expenses attending a re-sale or
> E attempted re-sale, together or rendered useless by such default, shall be made good and paid by the defaulting purchaser at the present sale and be recoverable from him by the vendor as liquidated damages. Any increase of price on a re-sale shall belong to the vendor.

On the date fixed for completion (October 19 1989) Dojap's attorney, Mr Clough, sent to the bank a letter of undertaking from the Jamaica Citizens Bank to pay the balance of the purchase price, subject to certain conditions. The bank's attorney, Miss Eaton, rejected this and gave Dojap 24 hours to provide a satisfactory undertaking. Dojap attempted to do so on October 20 1989, but the bank again rejected it. On October 23, the bank wrote to Dojap rescinding the contract and purporting to forfeit the deposit. Dojap refused to accept this, and on October 26 1989 tendered to the bank
F the balance of the purchase price with interest. The bank returned the cheque the next day.

On November 24 1989 Dojap started proceedings claiming specific performance or alternatively relief from forfeiture of the deposit. The case was heard by Zacca C J before whom a number of different issues arose for decision on the claim for specific performance. These are no longer in issue before the board. The judge gave judgment on June 25 1990, rejecting Dojap's claim for specific performance and its claim for return of the deposit.

On November 12 1990 Dojap arranged to purchase the same piece of land from the first mortgagee, Jamaica Citizens Bank, as a result of which the claim to specific performance of the contract of the bank became largely academic. Dojap appealed to the Court of Appeal, but did not pursue its claim for specific performance. On Dojap's G
alternative claim for relief from forfeiture and the return of the deposit, the Court of Appeal (Rowe P, Forte J A and Downer J A) held that Dojap was entitled to relief from forfeiture to the extent that the deposit exceeded 10% of the price but did not award any interest on that sum. The bank appeals to the board against the decision of the Court of Appeal to give such relief against forfeiture. Dojap cross-appeals claiming that it should have been awarded relief against forfeiture as to the whole of the 25% deposit and should also have been awarded interest.

In general, a contractual provision which requires one party in the event of his breach of the contract to pay or forfeit a sum of money to the other party is unlawful as being a penalty, unless such provision H
can be justified as being a payment of liquidated damaged being a genuine pre-estimate of the loss which the innocent party will incur by reason of the breach. One exception to this general rule is the provision for the payment of a deposit by the purchaser on a contract for the sale of land. Ancient law has established that the forfeiture of such a deposit (customarily 10% of the contract price) does not fall within the general rule and can be validly forfeited even though the amount of the deposit bears no reference to the anticipated loss to the vendor flowing from the breach of contract.

This exception is anomalous and at least one textbook writer has been surprised that the courts of equity ever countenanced it: see *Farrand, Contract and Conveyancing* 4th ed p204. The special treatment afforded to such a deposit derives from the ancient custom J
of providing an earnest for the performance of a contract in the form giving either some physical token of earnest (such as a ring) or earnest money. The history of the law of deposits can be traced to the Roman law of *arra*, and possibly further back still: see *Howe* v *Smith* (1884) 27 ChD 89 *per* Fry LJ at pp101-2. Ever since the decision in *Howe* v *Smith*, the nature of such a deposit has been settled in English law. Even in the absence of express contractual provision, it is an earnest for the performance of the contract: in the event of completion of the contract the deposit is applicable towards payment of the purchase price; in the event of the purchaser's failure to complete in accordance with the terms of the contract, the deposit is forfeit, equity having no power to relieve against such forfeiture.

However, the special treatment afforded to deposits is plainly K
capable of being abused if the parties to a contract, by attaching the label "deposit" to any penalty, could escape the general rule which renders penalties unenforceable. There are two authorities which indicate that this cannot be done. In *Stockloser* v *Johnson* [1954] 1 QB 476, Denning LJ in considering the power of the court to relieve against forfeiture said, *obiter*, at p491:

> Again, suppose that a vendor of property, in lieu of the usual 10% deposit, stipulates for an initial payment of 50% of the price as a deposit and part payment; and later, when the purchaser fails to complete, the vendor resells the property at a profit and in addition claims to forfeit the 50% deposit. Surely the court will relieve again the forfeiture. The vendor cannot forestall this equity by describing an extravagant sum as a deposit, any more than he L
> can recover a penalty by calling it liquidated damages.

In *Linggi Plantations Ltd* v *Jagatheesan* (1972) 1 MLJ 89 Lord Hailsham delivered the judgment of the board which upheld the claim to forfeit a normal 10% deposit even though the vendor had in fact suffered no loss. He referred on a number of occasions to a requirement that the amount of a deposit should be "reasonable" and said at p94:

> It is also no doubt possible that in a particular contract the parties may use language normally appropriate to deposits properly so called and even to forfeiture which turn out on investigation to be purely colourable and that in such a case the real nature of the transaction might turn out to be the imposition of a penalty, by purporting to render forfeit something which is in truth part payment. This no doubt explains why in some cases the M
> irrecoverable nature of a deposit is qualified by the insertion of the adjective "reasonable" before the noun. But the truth is that a reasonable deposit has always been regarded as a guarantee of performance as well as a payment on account, and its forfeiture has never been regarded as a penalty in English law or common English usage.

In the view of their lordships these passages accurately reflect the law. It is not possible for the parties to attach the incidents of a deposit to the payment of a sum of money unless such sum is reasonable as earnest money. The question therefore is whether or not the deposit of 25% in this case was reasonable as being in line with the traditional concept of earnest money or was in truth a penalty intended to act *in terrorem*.

The Chief Justice tested the question of "reasonableness" by reference to the evidence before him that it was of common occurrence for banks in Jamaica selling property at auction to demand deposits of between 15% and 50%. He held that, since this was a common practice, it was reasonable. Like the Court of Appeal, their lordships are unable to accept this reasoning. In order to be reasonable a true deposit must be objectively operating as "earnest money" and not as a penalty. To allow the test of reasonableness to depend upon the practice of one class of vendor, which exercises considerable financial muscle, would be to allow them to evade the law against penalties by adopting practices of their own.

However, although their lordships are satisfied that the practice of a limited class of vendors cannot determine the reasonableness of a deposit, it is more difficult to define what the test should be. Since a true deposit may take effect as a penalty, albeit one permitted by law, it is hard to draw a line between a reasonable permissible amount of penalty and an unreasonable, impermissible penalty. In their lordships' view the correct approach is to start from the position that, without logic but by long continued usage both in the United Kingdom and formerly in Jamaica, the customary deposit has been 10%. A vendor who seeks to obtain a larger amount by way of forfeitable deposit must show special circumstances which justify such a deposit.

As their lordships understood from the submissions made in argument, formerly the normal practice in Jamaica was to require a deposit of 10%. This was changed by the introduction of a transfer tax by the Transfer Tax Act, 1971. Under that Act, a transfer tax of 7.5% is payable on a transfer of land on sale. Although the tax is ultimately payable by the transferor (section 3), under section 18 it is collected from the transferee, ie the purchaser. As from April 3 1984, any contract for the sale of land must contain a requirement for the payment of a deposit of at least 7.5% and the purchaser is required to pay this sum to the Commissioner of Stamp Duty and Transfer Tax: section 18(4). The purchaser is entitled to recover from the vendor the amount of the tax so paid either by way of deduction from the purchase price or by action: section 18(1).

Their lordships were told that in practice this statutory machinery is not followed. Since the tax has to be paid within 30 days of the date of contract (failing which interest is payable by the vendor), a vendor is concerned to see that the tax is paid promptly. Accordingly, what happens in practice is that the contractual deposit is increased to at least 17.5% and is paid by the purchaser to the vendor. The vendor then pays the tax. It is apparently this practice that has caused the departure from the previously customary deposit of 10%.

If the contract of sale in respect of which the transfer tax is payable is not in fact completed, there is no liability to pay the tax: if such tax has been paid and the contract goes off, the tax can be recovered by the vendor and, by virtue of section 16(1), any amount so refunded "shall be dealt with according to the rights of the parties to the contract (including any requirement of a deposit implied therein under sub-section (4) of section 18)".

Since in the present case completion was supposed to take place within 14 days of the contract, in the ordinary course completion would have taken place before the transfer tax was due. Accordingly, there was strictly no need in the present case for the bank to insist on the inclusion in the deposit of a sum equal to 7.5% of the contract price so as to be in pocket to pay the tax when it fell due. However, the provisions as to the transfer tax are relevant. First, the transfer tax provides the explanation for the departure from the customary 10% deposit which was previously contracted for in Jamaica. Second, it illustrates how unconscionable it would be for the vendor to forfeit the deposit to the extent of 7.5% in the ordinary case. Where the tax has in fact been paid by the vendor out of the deposit and then the sale goes off, the vendor would recover the tax from the Revenue and then put the money in his own pocket.

In the present case, the attorney for the bank in evidence sought to justify the amount of the 25% deposit in part by reference to the amount of the transfer tax which would have been payable *viz* \$J862,500. This evidence indicates that far from the amount of the deposit having been fixed upon as a reasonable amount of earnest, the amount was substantially influenced by fiscal considerations having nothing to do with encouragement to perform the contract.

For the rest, although the attorney for the bank gave evidence that the amount of the deposit was fixed in part because it was a sum set "to ensure that persons do not bid frivolously at the auction" she also sought to justify the amount of the deposit by reference to the payments that would have had to be made on completion, ie tax, stamp duty, auction costs and auctioneer's commission. She accepted that the amount of the deposit was far in excess of what would have been required to cover the maximum out of pocket expenses which would have attended completion.

Their lordships agree with the Court of Appeal that this evidence falls far short of showing that it was reasonable to stipulate for a forfeitable deposit of 25% of the purchase price or indeed any deposit in excess of 10%. As for the tax element, the board do not suggest that it would be unreasonable for a vendor to require advance payment of an amount sufficient to discharge the liability for transfer tax on or before completion. But it does not follow that such advance payment of tax should be capable of forfeiture if completion does not take place: such tax is either not in the event payable or is recoverable by the vendor. However, quite apart from the specific tax element in this case, there is in the view of the board no sufficient evidence to justify a deposit of 25% as being a true deposit.

The question therefore arises whether the court has jurisdiction to relieve against the express provision of the contract that the deposit of 25% was to be forfeited. Although there is no doubt that the court will not order the payment of a sum contracted for (but not yet paid) if satisfied that such sum is in reality a penalty, it was submitted that the court could not order, by way of relief, the repayment of sums already paid to the defendant in accordance with the terms of the contract which, on breach, the contract provided should be forfeit. The basis of this submission was the view expressed in a considered *obiter dictum* of Romer LJ in *Stockloser* v *Johnson (supra)*.

In that case there was a contract for the sale of quarry machinery to the plaintiff, the purchase price to be paid by instalments. The contract provided that in the event of a default in payment of the instalments, the vendor could retake the machinery and all instalments of the price previously paid should be forfeit. Pursuant to the contract, the plaintiff took possession and used the machinery but defaulted in payment of an instalment. The defendant forfeited the instalments already paid. In the action, the plaintiff sought to recover the instalments, alleging that their forfeiture was a penalty. The Court of Appeal unanimously held that the forfeiture did not constitute a penalty on the facts of that case but went on to express conflicting views, *obiter*, as to whether, if the forfeiture had been a penalty, the court had jurisdiction to order repayment. Somervell LJ and Denning LJ expressed the view that there was such jurisdiction. Romer LJ held that there was no general right in equity to mend the parties' bargain and that, even where there was jurisdiction to relieve from forfeiture, that could only be exercised by allowing a late completion to a party who was in default in performance but willing and able to carry out the terms of the contract belatedly.

Their lordships do not find it necessary to decide which of those two views is correct in a case where a party is seeking relief from forfeiture for breach of contract to pay a price by instalments, the party in default having been let into possession in the meantime. This is not such a case. In the view of their lordships, since the 25% deposit was not a true deposit by way of earnest, the provision for its forfeiture was a plain penalty. There is clear authority that in a case of a sum paid by one party to another under the contract as security for the performance of that contract, a provision for its forfeiture in the event of non-performance is a penalty from which the court will give relief by ordering repayment of the sum so paid, less any damage actually proved to have been suffered as a result of non-completion: *Commissioner of Public Works* v *Hills* [1906] AC 368. Accordingly, there is jurisdiction in the court to order repayment of the 25% deposit.

The Court of Appeal took a middle course by ordering the repayment of 15% out of the 25% deposit, leaving the bank with its normal 10% deposit which it was entitled to forfeit. Their lordships are unable to agree that this is the correct order. The bank has contracted for a deposit consisting of one globular sum, being 25% of the purchase price. If a deposit of 25% constitutes an unreasonable sum and is not therefore a true deposit, it must be repaid as a whole. The bank has never stipulated for a reasonable deposit of 10%: therefore it has no right to such a limited payment. If it cannot establish that the whole sum was truly a deposit, it has not contracted for a true deposit at all.

As to interest, Downer JA in the Court of Appeal was under the misapprehension that Dojap had never made any claim for interest: he indicated that if they had done so he would have awarded 12% interest (being the rate provided in clause 5 of the contract). It is clear

A that in written submissions headed "Reply to defendant's submissions" counsel for Dojap before the Court of Appeal did claim such interest. Dojap is therefore entitled to interest at 12% pa from the date of rescission until the date of actual payment.

Finally, it appears that the bank may have suffered some damage as a result of Dojap's failure to complete. If so, the bank is entitled to deduct the amount of such damages from the "deposit" of 25%. Such damage has not been quantified in the judgment below but appears to be small in amount. It would not be right to keep Dojap out of all its money to await the outcome of the necessary inquiry as to damages. The bank ought accordingly to make immediate repayment of a substantial amount of the deposit, leaving a fund out of which the bank's damages, if any, can be satisfied.

B Their lordships will therefore humbly advise Her Majesty that the appeal ought to be dismissed and the cross-appeal allowed, and that the order of the Court of Appeal should be varied so as to provide: C

(1) an inquiry as to the damage (if any) suffered by the bank by reason of Dojap's failure to complete the contract;

(2) an order that the bank forthwith repay to Dojap the sum of $J2,000,000 (being part of the deposit) together with interest as 12% pa from October 23 1989 (being the date of rescission) down to the date of actual payment;

(3) that the sum, if any, found due under the inquiry as to damages be deducted from the remainder of the deposit ($J875,000) and that the balance of the said sum of $J875,000 be paid to Dojap together with interest as aforesaid; and

(4) an order that the bank must pay Dojap's costs of the appeal to the Court of Appeal.

The bank must pay Dojap's costs before the lordships' board. D

For further cases on this subject see p 159

TOWN AND COUNTRY PLANNING

A **Queen's Bench Division**

January 28, 1993

(Before Mr Justice POPPLEWELL)

BATCHELOR v SECRETARY OF STATE FOR THE ENVIRONMENT AND ANOTHER

***Town and country planning — Tree preservation order — Public local inquiry — Whether conduct of inquiry biased — Whether inspector failed to have regard to documentary
B material provided by the applicant and whether he wrongly treated application as one to clear fell and failed to consider alternative basis of granting partial consent for felling***

Since 1983 the application. Shortly after his purchase Maidstone Borough Council imposed blanket provision tree preservation orders on all trees and woodland on the estate. The application then made an application to the council and to the Forestry Commission to fell all the woodland and trees. The council then confirmed all the provisional tree preservation orders and the applicant again sought permission for consent to fell. In December 1986 the council issued notices refusing felling permission and certifying that the
C refusal was in the interests of good forestry management. The applicant appealed to the Secretary of State for the Environment against the refusal. In February 1987 the council issued enforcement notices requiring replanting of trees, which had been felled in breach of the tree preservation orders. The applicant lodged notice of appeal against the enforcement notices. On August 7 1989 notice of the inquiry into the appeal to commence on September 14 was sent to the applicant's solicitors stating that the inspector would be assisted by an assessor, a Mrs C Davis BSc NDArb FArborA. At the hearing before the inspector the applicant's solicitors said his
D client would not be appearing nor would counsel or any witnesses be present on his behalf but that he had been instructed to read a statement by his client asking for an adjournment and then to leave it if the adjournment were not granted. That statement expressed the applicant's concern about the presence of the single assessor who it was alleged used to be a tree officer with a local authority and might therefore have only a limited background in forestry and agriculture; the applicant felt he would not receive a fair hearing because there was to be no agricultural expert. The inspector refused the application for an adjournment and recommended that the appeals be dismissed. On July 30 1991
E the Secretary of State accepted that recommendation and dismissed the applicant's appeals against the enforcement notices. The applicant appealed against that decision contending that the inspector should not have been assisted by the assessor because she had been a local government tree officer and there might be an appearance of bias; that the inspector had failed to have regard to the documentary material provided by the applicant which she should have treated as evidence at the inquiry; and that the inspector treated the application as an application to clear *all* the trees on the premises and did not consider the alternative basis of F
granting partial consent.

***Held*: The appeal was dismissed. Applying the test of bias in *Metropolitan Properties Co (FCG) Ltd* v *Lannon* no reasonably objective person could have the slightest suspicion that the tree assessor might favour the respondents at the expense of the applicant. Although the inspector had been given a statement of case on behalf of the applicant, he was perfectly entitled within his discretion in the absence of *evidence* put forward by the applicant and/or his advisers as to the precise nature and extent of the tree felling proposal to G
treat the application as one for *clear* felling. The fact that the statement of case and an accompanying proof of the applicant's expert supported partial felling did not of itself require the inspector to consider it. The inspector could not be criticised in the absence of a clear application as to the precise nature of the applicant's case set out in *evidence* before him not dealing with a claim for partial clearance. Because of the attitude taken by the applicant at the outset of the hearing, no oral evidence was given by or on behalf of him. The inspector had the statement of case, proofs of evidence of two experts and a summary of the facts and matters in support of the appeal. It is clear from the Town and Country Planning H
Inquiry Procedure Rules 1988 that in the absence of oral evidence given by a witness an inspector is not required to take any account of written representations or documents. There was nothing wrong in the inspector choosing not to treat the documentary materials as "evidence"; it was not necessary for him to set in the decision letter the applicant's case as contained in documents. The inspector was properly entitled to treat oral evidence given on behalf of the respondents which was tested in cross-examination and having greater weight when it conflicted with the written submissions of the applicant's witnesses which was not given on oath or tested in J
cross-examination.**

The following case is referred to in this report.

Metropolitan Properties Co (FGC) Ltd v *Lannon* [1969] 1 QB 577; [1968] 3 WLR 694; [1968] 3 All ER 304, CA

This was an application and appeal under the Town and Country Planning Act 1990 by the applicant, Hubert Dorrington Batchelor, from the decision of the first respondent, the Secretary of State for the Environment, to dismiss appeals by the applicant from decisions of the second respondents, Maidstone Borough Council, which had refused planning permission to fell trees and had issued enforcement K
notices requiring the replanting of trees which had been felled in breach of tree preservation orders made under the Town and Country Planning (Tree Preservation Order) Regulations 1969.

Camden Pratt QC (instructed by Hallet & Co, of Ashford) appeared for the applicant; David Holgate (instructed by the treasury solicitor) represented the respondents.

Giving judgment, **POPPLEWELL J** said: By an amended notice of motion and appeal this applicant seeks to quash or remit for rehearing a decision by the first respondent, the Secretary of State for the Environment, under the Town and Country Planning Acts 1971 and 1990 and the Town and Country Planning (Tree Preservation Order) Regulations 1969.

Since 1983 the applicant has been the owner of the Howe Court Estate, Hollingbourne, and has been in almost continuous contention with Maidstone Borough Council about the preservation of trees on those premises.

The chronology can be very shortly stated. Soon after the applicant purchased the premises the borough council imposed blanket provisional tree preservation orders on all trees and woodland on the estate. The applicant then made an application to the council and to the forestry commission to fell all the woodland and trees on the estate. The council confirmed all the provisional tree preservation orders. The applicant again sought permission from the council for consent to fell. In December 1986 the council issued notices refusing felling permission and certifying that the refusal was in the interests of good forestry management. The applicant then appealed to the Department of the Environment against a refusal by the council for permission to fell the trees and, in February 1987, the council issued enforcement notices requiring replanting of trees which had been felled in breach of the tree preservation orders. The applicant lodged notice of appeal against those enforcement notices.

In September 1989 there was a public inquiry and on February 9 1990 the inspector reported adversely to the applicant. On July 30 1991 the Secretary of State gave his decision, which again was adverse to the applicant. It is that report and decision which are the subject of these proceedings.

Both the application and the appeal raise the same point and can be dealt with together.

There are three separate issues upon which the applicant relies.

(1) That the inspector should not have been assisted by the assessor, Mrs Davis, because she had been a tree officer in the employment of Hillingdon London Borough Council at some time and therefore there was an appearance of bias at the inquiry.

(2) That the inspector failed to have regard to the documentary material provided by the applicant which he should have treated as evidence at the inquiry.

(3) That the inspector treated the application as an application to clear fell *all* the trees on the premises and did not consider the alternative basis of granting partial consent.

The respondents, who are the Secretary of State for the Environment and Maidstone Borough Council, resist this application and appeal and further contend that if I were minded to take the view that the inspector should have dealt with a question of partial consent there is no need to quash the decision because it is open in any event to the applicant to make a further application and in the exercise of my discretion I should not accede to the application.

The statutory framework is as follows. A tree preservation order can be made under section 198 of the Town and Country Planning Act 1990. All references are to the Town and Country Planning Act 1990, though there were similar provisions in the Town and Country Planning Act 1971. A local planning authority may have to pay compensation in respect of tree preservation orders if there is a refusal of a consent (section 203(a)). A planning authority may issue a certificate that the refusal was in the interests of good forestry management or the trees had outstanding amenity value. If they issue such a certificate it may be the subject of appeal, as it was in this case.

The regulations governing tree preservation orders are the Town and Country Planning (Tree Preservation Order) Regulations 1969 SI 1969 No 17). The first schedule to that regulation indicates that the trees shall be specified individually either by description of tree, description of area, description of group of trees or description of woodlands; in the instant case they were the first, third and fourth descriptions. An application for felling consent is governed by para 3 of the Schedule to these regulations:

> An application for consent made to the authority under Article 2 of this Order shall be in writing stating the reasons for making the application, and shall by reference if necessary to a plan specify the trees to which the application relates, and the operations for the carrying out of which consent is required.

It will be seen therefore that the applicant's application was required: (1) to be in writing; (2) to set out the reasons; (3) to specify the trees to which the application related; and (4) the operations.

There is no dispute in this case that the area involved a large number of trees running to some thousands.

Under article 4 of the same schedule the authority may grant such consent either unconditionally or subject to conditions or may refuse consent. By article 4(2) the authority is required to keep a register of all applications for consent containing information as to the nature of the application, the decision of the authority thereon and any compensation awarded in consequence of such a decision and any directions as to replanting of woodland and every such register is to be available for inspection by the public at all reasonable hours.

It is common ground that although there is no statutory requirement for the local planning authority to consult with interested parties in tree preservation order cases, it is as a matter of practice certainly in relation to a large area such as we are concerned with here so to do. This was an area of outstanding natural beauty on the edge of the Downs and it is unthinkable that the local authority would not consult with the local council and with any preservation society.

Article 5 gives the local authority the power to issue a certificate in respect of compensation.

The applicant is alleged to have removed trees subject to the tree preservation order and, under section 206 of the Town and Country Planning Act 1990, there was a duty on him to replace trees. Tree replacement notices had been issued in February 1987.

Thus the matters which the inspector was required to consider were, first, the appeals against the refusal of consent to fell. Second, appeals against the certificates relating to compensation and, third, appeals against the tree replacement notices.

I turn now to the individual complaints.

Bias

The way that it is put in the amended notice of motion is as follows:

> The First Respondent and/or his inspector acted contrary to the rules of the natural justice in that there was an appearance or the real likelihood of bias at the Inquiry. The Assessor [Mrs Davis] had been a tree officer in the employment of the London Borough of Hillingdon. The First Respondent or his appointed Inspector failed to notify the Applicant at all and/or adequately and/or within a reasonable period of time prior to the hearing of the said enquiry of the full identity of the Assessor and failed to notify or inform the Applicant of her qualifications or experience or of the matters upon which she would be advising the inspector. The said Assessor did not appear to have expertise in weighing the needs of agriculture against environmental and arboricultural factors. The First Respondent and/or his Inspector failed to accede to the Applicant's Application for an adjournment of the enquiry to enable a different Assessor to be appointed and proceed with the inquiry in the absence of the Applicant.

The inquiry was due to start on September 14 1989. On August 7 notice of the inquiry was sent to the applicant's solicitors clearly stating that the inspector would be assisted by Mrs C Davis BSc NDArb FArbor A. She was appointed by the Secretary of State to assist the inspector on arboriculture aspects. It will be seen that the matters complained of in the particulars are really as to the expertise of Mrs Davis. There is nothing in the point that the applicant did not know in good time who the assessor was going to be.

At the hearing before the inspector, Mr Mortimer, the applicant's solicitor said his client would not be appearing nor would counsel or any witnesses be present on his behalf, and Mr Mortimer had been instructed to read a statement by his client asking for an adjournment and then to leave if the adjournment was not granted. The statement which the inspector recorded relating to this aspect of the case went as follows:

> It looked to Mr Bachelor as if the inquiry would turn on issues bearing no resemblance to the realities of planning practice or country life. Mr Batchelor felt very strongly; although he was of some substance his resources were more limited than the council's. He considered that although the inquiry would be capable of deciding between farming needs and the look of the area that was apparently not to be the situation. He was concerned about the composition of the tribunal, in particular the presence of a single assessor. No disrespect to the assessor was intended but Mr Batchelor had recently learned that the assessor used to be a Tree Officer with a local authority and therefore might well have only a limited background in forestry and agriculture. There was no agricultural assessor. Mr Bachelor did not feel he would receive a fair hearing. He decided that rather than sitting through lengthy proceedings with the scales loaded against him he asked that the hearing be adjourned and reconvened with a forestry expert and agricultural expert as assessors. If not he wished to withdraw from any part in the process.

The council contended:

> The main point seemed to be that there was no agricultural assessor. There had been two pre-inquiry meetings and this point could have been put up long ago. . . . The agricultural case raised by Mr Bachelor in his grounds of appeal was straightforward and covered the sort of issues commonly dealt with by inspectors. The second point appeared to be a challenge for the qualifications and background of Mrs Davis, the assessor.

The inspector ruled:

> First, the appellant had known about the appointment of the assessor some months previously, since at least the July pre-inquiry meeting and any concern about this could therefore have been raised well before the inquiry. Secondly, those acting for the Secretary of State had decided that appointing an assessor to assist and advise me on arboricultural matters would provide an appropriate mix of expertise to consider the evidence.

It could well be contended that the application by Mr Mortimer was not an allegation of bias either actual or perceived, but was related to the alleged lack of expertise by Mrs Davis as an assessor; indeed the inspector in an affidavit before the court said:

> The main purpose of the application, as put by Mr Mortimer by reading out his client's statement, was not as claimed in the . . . amended Notice of Motion, to enable a different assessor to be appointed, but for additional assessors to be appointed including an agricultural expert.

It does not seem to me to matter whether an additional expert or not was a ground of contention. That only goes to removal of the alleged complaint. What is important is that it is by no means clear to me that the application was then being made on the basis of bias or alleged bias.

However, Mr Batchelor has deposed as follows:

> 12. It was of decisive importance to me that I felt that during the course of the Enquiry I would not receive a fair and impartial hearing and that the Inspector appointed by the Department of the Environment would neither have the appropriate experience and background to deal with the matters before him nor that he would be advised by an Assessor or Assessors with such a background.
>
> . . .
>
> 13. I do not question the integrity or the particular abilities of the Assessor appointed . . . I do however maintain that she was not the appropriate person to be appointed to sit upon the Enquiry or sit upon the Enquiry alone without an additional Assessor to bring about a balance in the forum. I have been advised and verily believe that the said Mrs Davis' experience prior to her joining the Department of the Environment was that of Tree Officer in the London (Urban) Borough of Hillingdon. In such circumstances it is my respectful submission that it would have been difficult for the said Mrs Davis to be and seen to be totally objective in evaluating my position as against that of the Second Respondent and this apart I simply cannot believe that Mrs Davis' experience in dealing with tree problems in an urban area could possibly equip her to consider and advise the Inspector . . . in relation to countryside and agricultural matters in forestry management.

The criticism of her expertise is now not seriously pursued. The way it is put by counsel is this. That Mrs Davis as a tree officer in Hillingdon London Borough of Hillingdon would have been responsible for enforcing tree preservation orders. She would have had an urban experience. Therefore, she was or might appear to be biased in favour of Maidstone Borough Council in their desire to enforce tree preservation orders and as there had been a constant battle between the applicant and the officers for the council responsible for tree preservation, to employ as an assessor someone who had been in that position before might well give the appearance of bias.

There was no evidence as to how long Mrs Davis had been with Hillingdon or how long ago she had left and had ceased to act. The argument was simply presented on the basis that as she had at one time and for an indeterminate period of time been a tree officer with Hillingdon she was an inappropriate person to sit as an assessor.

Mr Batchelor further deposed in his affidavit in these proceedings to an article in the *Horticulture Week* September 7 1990 about Mrs Davis under the heading "Arboriculture Defender of the tree realm". Some flavour of the journalese involved in this article can be gathered by the first sentence:

> When Caroline Davis was a tree officer for a London local authority she was always the one sent out to face little old ladies angry about tree preservation orders.

It was suggested that that article showed that Mrs Davis was a person enthusiastic about trees. In one part of the article this appears:

> She doesn't see an intrinsic conflict between her own department, the Department of the Environment, and the other government departments such as the department of transport. "You fight for your corner against other departments. They promote the need for roads and we promote the need for trees but there is no conflict between individuals."

That extract is relied on in support of the proposition that Mrs Davis was fighting her corner for trees and therefore would be likely to be fighting against the applicant in this dispute. It is unnecessary to cite any further part of the article; all it does is to show that Mrs Davis is an enthusiast about trees.

It is accepted at the Bar that the proper test in relation to a complaint of bias is that laid down by Lord Denning in *Metropolitan Properties Co (FGC) Ltd* v *Lannon* [1969] 1 QB 577 at p599D:

> . . . if right-minded persons would think that, in the circumstances, there was a real likelihood of bias on his part, then he should not sit.

Applying that test I have not the slightest doubt that this is a wholly unfounded allegation which the inspector was right to reject. No reasonably objective person could have the slightest suspicion that Mrs Davis might favour the respondents at the expense of the applicant and there is nothing in the article which could possibly assist the applicant.

I turn next to the third point because my conclusion on the third point has some bearing on the applicant's second point — consent to clear felling or partial felling.

It is the applicant's contention that properly considered his application was an application for partial felling of the trees on the estate and that the inspector's conclusion that no case was made out for clear felling did not address itself to the issue which was before him.

It is not in dispute that the inspector did indeed deal with the application as if it were an application for clear felling and rejected it and that he did not deal with the question of partial felling. The inquiry was of course not only concerned with felling but with the question of certificates in relation to individual trees and with the replacement of trees. The original application, which is dated May 12 1983, was made pursuant to the Forestry Act 1967 and sought a licence for clear felling of nearly 7,000 trees. In the notice of appeal to the Secretary of State for the Environment giving rise to the public inquiry under "Details of Application" this appears:

> 1. To appeal against refusal by the authority of permission to fell the trees subject to the orders wholly or in part.

The applicant contends that that was sufficient to enable the inspector to deal with the case put forward in documents by the applicant's expert that at any rate part of the site should be cleared of trees.

The applicant's expert Mr Helliwell had put forward a scheme which was in effect to clear 20% of the site. There had been discussions during the years between the applicant's advisers and the respondent's advisers and others to see whether some agreement could be arrived at and various plans were put forward though no agreement was in fact reached.

It is the applicant's contention that the inspector should have treated the combination of the notice of appeal and Mr Helliwell's scheme as an application by the applicant for *clear* felling of that part of the land to which Mr Helliwell's scheme related.

It is clear that council had come to the inquiry to deal with the applicant's application in relation to tree felling on the basis that it was an application for *clear* felling. It is clear that the notice of appeal if it were to be treated as an application did not comply with the requirements of the schedule to the Town and country Planning (Tree Preservation Order) Regulation 1969 because it did not give sufficient detail of the trees to which the application related. It was particularly important in a case of this sort where a very large number of trees were concerned that everybody should know precisely what trees were involved because it affected not only the application to fell but also the certificates which were also the subject of the appeal. Neither Mr Helliwell nor the applicant gave evidence before the inspector. The inspector had of course the statement of case and Mr Helliwell's documents, but they were not supported on oath by the witnesses. The inspector said:

> The appellant's statement of case gives the impression that the application was not really intended to be taken literally as proposing clear felling; but as stated in a recent judgment (*Titterall* v *Secretary of State for the Environment and Tunbridge Wells Borough Council* (1989)), "there is no room for the concept of the gist of an application. An application is in the form in which one finds it by reference to the form itself".
>
> 111. One of the reasons for all these difficulties is that the original applications are confusing and unclear . . . Normally I would have sought to clear up these points at the inquiry but the appellant's withdrawal made that impossible.
>
> 115. No case has been put forward for felling on this massive scale, other than the claims that woodland is a liability and that clearing of as much woodland as possible would be in the owner's financial interest. The statement of case submitted for the appellant makes no real attempt to support such a proposal, the appellant's own expert advisers appear to be against it and indeed suggest that the appellant accepts that the majority of woodland in Howe Court should remain. I have no hesitation in finding that permission for clear felling should be refused.

A 116. The next question which arises is whether or not it is possible for the Secretary of State to consider granting consent for the scheme which the appellant appears now to be proposing. The scheme is very significantly different from clear felling, to the extent that on the appellant's figures some 80% of the trees and woodland present in 1983 would remain. No application has ever been made for such a scheme.

117. If an attempt were to be made to grant a partial consent, or a series of partial consents, so as to achieve the outcome apparently now sought by the appellant, the result would be a permission entirely different from the application. The practical effect would be to bypass the normal application and consultation procedures. Having regard to court judgments on this topic, notably the Whatcroft, Clwyd County Council and Wessex Regional Health Authority cases referred to in pages 20527 of the Encyclopaedia of Planning Law, there can be no doubt in my mind that such an approach could not be
B validly adopted. I therefore see the representations about the appellant's "woodland management plan" as only being relevant to the appeals concerning the tree replacement notices where variation of the notices may be possible.

The Secretary of State said at para 7 of his decision letter:

In considering the Inspector's report the Secretary of State considered whether partial consent to felling could or should be granted. Contrary to the view held by the Inspector, the Secretary of State is satisfied that both he, and the local authority to whom an application is made, may grant partial consent when there is a single application which clearly covers a number of separate, albeit related, trees or where part, or parts, of the application can properly stand alone. The Secretary of State does not consider, however, that the facts of this application fall into either of the categories. The application, if varied as requested by the applicant, would not appear to relate in any way to the
C original application, other than its geographical location. It would, if so varied, constitute a new application. The appellant could, of course, have submitted a new application at any time, but he did not do so. In these circumstances the Secretary of State has decided that he will consider the appeals as they were submitted at the time of the application, and not as varied.

There was some discussion at the Bar as to the effect of the decisions to which the inspector referred and whether in planning matters if at the end of the day the application is substantially different from the original application it can be refused.

I propose to deal with the problem on a slightly different basis. It seems to me that it would have been possible for the inspector to have adjourned the matter; to have allowed interested parties to consider
D Mr Helliwell's report and for the inspectorate to treat the notice of appeal coupled with Mr Helliwell's report as being the application. He was in my judgment however not bound to do so; it was entirely a matter within his discretion. There is no obligation for consultation as there is in applications for planning permission but there are references (see for instance Circular 36/78) to consultation. Quite clearly there were interested parties who had indeed been consulted. The fact that they were present at the hearing was no doubt a matter to be considered by the inspector as well as the fact that there had been in the past attempts to reach some sort of agreement between the interested parties and the applicant.

In my judgment, the inspector was perfectly entitled within his
E discretion in the absence of *evidence* put forward by the applicant and/or his advisers as to the precise nature and extent of the tree felling proposal to treat the application as one for *clear* felling. The fact that the statement of case and Mr Helliwell's proof supported partial felling did not of itself require the inspector to consider it. The position which the inspector was left in appears in his decision letter as follows:

Before Mr Mortimer [solicitor acting for Mr Batchelor] left I invited submission on two points: . . . second, the legal situation arising from the fact that the proposals which were apparently now being pursued by the appellant were greatly different from clear felling, to the extent that the appellant's advisers would seemingly have given evidence against their client's appeals. Mr Mortimer explained that in view of his instructions and lack of any Counsel he did not feel able to comment in detail, although he drew attention
F to the Secretary of State's powers to grant partial permission on an application. . . The council considered that the appeals against the refusals of consent were concerned with the clear felling proposal.

In my judgment, the inspector can scarcely be criticised in the absence of a clear application as to the precise nature of the applicant's case set out in *evidence* before him in not dealing with a claim for partial clearance. His reasons for not doing so I have already set out. I agree with his conclusion and with the approach of the Secretary of State. In any event even if I did not agree I would not have found that the decision was either perverse or irrational.

I turn to the final submission by the applicant.

The treatment of the applicant's "evidence".

Because of the attitude taken by Mr Batchelor at the outset of the hearing, no oral evidence was given by or on behalf of the applicant. G
The inspector had before him the statement of case, the proofs of evidence of two experts and a summary of the facts and matters in support of the appeal. The powers of the inspector are governed by the Town and Country Planning Inquiry Procedure Rules 1988, rule 14 which gives the inspector wide powers and discretion as to how to conduct the inquiry. It is clear from those rules that as a general practice oral evidence is to be given. A witness may read a summary of his evidence but it will then be subject to cross-examination. By rule 14(10):

The inspector may take into account any written representation or evidence or any other document received by him from any person before an inquiry opens or during the inquiry provided that he discloses it at the inquiry. H

A "Statement of case" is defined by rule 2 as follows:

. . . a written statement which contains full particulars of the case which a person proposes to put forward at an inquiry and a list of any documents which that person intends to refer to or put in evidence.

It is clear from those rules that in the absence of oral evidence given by a witness an inspector is not required to take any account of written representations or documents. It is entirely a matter for his discretion. In the instant case having regard to his decision to treat the application as an application for clear felling, having regard to the evidence given on behalf of the respondents and no doubt to the written representations by the applicant's experts that clear felling was not a viable proposition, it is not in the least surprising that the inspector came to the conclusion that clear felling was not a practical J
proposition.

It is the applicant's complaint however that in his decision letter the inspector has not dealt with the applicant's case.

It has to be remembered that there were three questions to be dealt with at the inquiry. First, the question of clear felling. Second, the question of certificates and, third, the replacement notices.

The inspector said:

Case for the appellant.

No evidence was presented for the appellant at the inquiry and no submissions were made other than those reported above. The written "statement of case" and other papers submitted before the inquiry are attached at documents 4-7.

(The submissions referred to were submissions relating to Mrs K
Davis.)

That is criticised on the basis that the inspector should have treated the documents submitted as evidence. In my judgment, what the inspector was saying is no *oral* evidence was presented. In that he was right.

The inspector then set out the case for the local planning authority by reference to the proofs of evidence which had been submitted and spoken to. It is clear that the respondent's witnesses were challenged by the inspector and by the assessor by reference to points contained in the applicant's expert's reports and documents. From time to time in the decision letter he makes an observation about the report of the applicant's experts. When the inspector came to deal with the appeals L
against the tree replacement notices he also clearly had regard to the points made in the appellant's case.

I find nothing wrong at all in the inspector choosing, as he undoubtedly has, not to treat the documentary material as "evidence" in the case. It is clear that the respondent's witnesses who did give oral evidence had their evidence tested by reference to those documents. But the inspector was perfectly entitled under the rules within his discretion not to treat the documents as *evidence*. It was not necessary for him to set out in the decision letter the applicant's case as contained in documents. He had the documents, and read them and these were before the Secretary of State.

However criticism is made of this passage in the inspector's M
decision:

In assessing the strength of the case for each side it is necessary to take account of the weight which should be given to evidence from various sources. As a general principle, where there is doubt or conflict I consider that greater weight should be accorded to evidence which I heard first hand at the inquiry and was able to test by questioning than to written representations.

As a general principle I confess to no surprise at this proposition. It seems to me to set out a proper approach. The inspector was properly entitled to treat oral evidence given on behalf of the respondents which was tested in cross-examination as having greater weight where it conflicted with the written submissions of the applicant's witnesses which was not given on oath or tested in cross-examination. There is in my judgment nothing in this point. The inspector went on:

Although I have used the term "proof of evidence" to refer to various documents submitted for the appellant because that is how they were labelled, these documents did not of course form the basis of evidence given at the inquiry.

All the inspector was there saying was this did not constitute "evidence" because it was not given on oath. Again I find nothing contentious in that submission.

There is, in my judgment, nothing in the point that the inspector has failed adequately to deal with the applicant's case. So far as tree felling was concerned the suggestion of *clear* felling was not supported and therefore no further reference needed to be made. In so far as certificates were concerned the main argument on behalf of the applicant was considered on the basis that the areas in dispute were scrub. The assessor did not agree with that nor did the inspector. In so far as the appeals against the tree replacement notices were concerned the inspector dealt in some detail with the four grounds which were raised.

Accordingly, I find nothing in this aspect of the case to criticise the inspector.

Finally, I deal with the submission that if I were to take the view that the inspector should have dealt with an application for partial felling I should quash this decision and allow the matter to go back for further inquiry. There is nothing, as I understand it, to prevent the applicant from now making an application for partial felling. There is no dispute but that the application for *clear* felling was bound to fail and accordingly in respect of that the inspector was right. Why then should I quash his decision when what is really complained about is that he has not dealt with the application before him? That can readily be cured by the applicant making another application. However, it is pointed out on his behalf that there remains in existence the tree replacement notices. The applicant is in this position. He is required by those notices to replace a large number of trees. If he was successful in his application for partial felling that might well include areas to which the tree replacement notices applied, thus putting him in the position of replacing trees which as a result of a successful application for partial tree felling he could then cut down.

This position arises from his deliberate and unjustified (as I find) decision to withdraw from the proceedings. That was entirely characteristic of his whole behaviour since 1983, which included a refusal to allow the inspector to visit the site during the inquiry. I confess to little sympathy for the applicant in the situation in which he finds himself. There is no criminal sanction against failure to comply with the tree replacement notice. The enforcement is governed by section 207 of the Town and Country Planning Act 1990 which gives the local planning authority power to serve the notice. Section 209 provides:

> If, within the period specified in a notice under section 207(1) for compliance with it . . . any trees which are required to be planted by a notice under that section have not been planted, the local planning authority may —
> (a) enter the land and plant those trees, and
> (b) recover from the person who is then the owner of the land any expenses reasonably incurred by them in doing so.

The note to those provisions says:

> This section provides the only means ultimately of enforcing the duty under section 206 for the replacement of trees removed in contravention of a tree preservation order.

A good number of the trees to which these notices apply were either destroyed or damaged in the clearest contravention of tree preservation orders. Those who defy the law or act illegally can scarcely expect to have the discretion of the court exercised in their favour. Accordingly, this is not a case where even if I had found in the applicant's favour on the issue of partial consent I would have been disposed to exercise my discretion to quash the *whole* proceedings.

Accordingly, I shall dismiss the appeal and the application.

Appeal and application dismissed with costs. Leave to appeal refused.

For further cases on this subject see p 182

LANDS TRIBUNAL

COMPENSATION

SLOT v GUILDFORD BOROUGH COUNCIL
(Ref/107/1990)
May 27 1992

(M ST J HOPPER FRICS)

Compensation — Article 4 direction withdrawing GDO permitted development rights — Agricultural land — Determination of compensation

The claimant owned 15.07 acres of agricultural land fronting Green Lane, Ockham, Surrey. On January 12 1990 Guildford Borough Council made a direction under article 4 of the General Development Order 1988, which was confirmed by the Secretary of State on May 29 1990, relating to the withdrawal of permitted development rights for development under Class A, Part 2 of the second schedule. This class concerns the erection, construction, maintenance, improvement or alteration of a gate, fence, wall or other means of enclosure. On November 13 1990 the council refused planning permission to the claimant for the erection of a glasshouse, storage barn and five fences, the latter were to divide the land into six parcels. The claimant submitted a claim for compensation under section 164(1) of the Town and Country Planning Act 1971, which at the hearing was amended to £114,596.72 to include £94,600 for depreciation in the value of the land. The council contended that a planning application must relate solely to the development (erection of fences etc) in respect of planning permission was withdrawn by the direction; because that was not the case, section 165(1)(*c*) of the 1971 Act was not satisfied and no claim could be made. The council also argued that in assessing the "after value" of the land it had to be assumed that permission would be granted for development within Schedule 8 to the 1971 Act.

Decision: **The planning application complied with section 165(1) of the Town and Country Planning Act 1971 notwithstanding that it included other proposed development in addition to the development the subject of the article 4 direction. In assessing the "after" value of the land it must be assumed that planning permission would be granted for development within Schedule 8 of the 1971 Act. On the evidence the "before" value was £82,500 (after making an allowance for the nuisance value of a strip of land over which there was a slight risk with regard to access rights). The "after" value was £66,000 after allowing for a 20% reduction for the effect of the direction. Compensation was determined at £19,230 to include legal and other expenses.**

The following cases are referred to in this report.

Burlin v *Manchester City Council* (1976) 32 P&CR 115; 238 EG 891; [1976] RVR 119, LT
Carter v *Windsor & Maidenhead Royal Borough Council* [1988] 3 PLR 6; [1988] RVR 241
Colley v *Canterbury City Council* [1991] EGCS 60
Essex County Council v *Essex Incorporated Congregational Church Union* [1963] AC 808; [1963] 2 WLR 802; [1963] 1 All ER 326; (1963) 14 P&CR 237; [1963] RVR 151; [1963] EGD 555; 185 EG 453, HL
Stock v *Frank Jones (Tipton) Ltd* [1978] 1 WLR 231; [1978] 1 All ER 948; [1978] ICR 347, HL

Edward Denehan (instructed by Giles Smeath & Co, of Fleet) appeared for the claimant; Christopher Lewsley (instructed by the solicitor to Guildford Borough Council) represented the compensating authority.

Giving his decision, **MR M ST J HOPPER** said: This is a reference to determine the compensation, if any, payable under sections 164 and 165 of the Town and Country Planning Act 1971("the 1971 Act") following the refusal of a planning application on February 13 1990 ("the material date") relating to 15.07 acres of land fronting Green Lane, Ockham, Surrey, which formed part of Ridings Farm ("the farm"), Long Reach, Ockham.

Mr Edward Denehan of counsel, instructed by Giles Smeath & Co, of Fleet, appeared for the claimant and called Mr David Edward Goldsmith FRICS, a director of John D Wood & Co (Residential and Agricultural) Ltd, the claimant and her former husband, Mr Robert Paul Slot. Mr Christopher Lewsley of counsel, instructed by the solicitor to Guildford Borough Council ("the council"), appeared for the compensating authority. He called Mr John Charles Catt BSc (Est Man) FRICS, a director of South East Marts plc, within which company he was the director in charge of all rural property agency carried on under the name of Bannister & Wellers at Guildford. He also called Mr Andrew Stallon MRTPI, the principal planning officer (development control) responsible for the eastern area of the Borough of Guildford.

The farm lies in the parish of Ockham, in an attractively wooded area, within the metropolitan green belt. It lies a short distance to the south of the village centre of Ockham, about two miles to the east of the village of Ripley, about two miles to the south of the village of East Horsley, where there is a railway station on a line between London (Waterloo) and Guildford and about seven miles from Guildford. The A3 trunk road is about one mile distant and junction 10 on the M25 motorway is about four miles distant.

At the material date the farm comprised a bungalow, farmbuildings and a total land area of about 86 acres. To the north the farm was bounded by woodland and open agricultural land. To the east the farm had a long frontage to Ockham Road North, a classified road. Mr Catt said that he understood there was a strip of land between the boundary of the farm and the highway in separate and unknown ownership but I received no proof of this. To the south-east the farm bounded an unclassified road known as Green Lane, although a strip of land not owned by the claimant lay between the farm and the highway. To the south of the farm lay Even Wood and to the south-west there was a long frontage to a highway known as Long Reach from which the principal access to the farm was taken. The land, the subject of the refusal of planning permission ("the reference land"), comprised a field (OS parcel 0050) which lay in the south-easterly part of the farm with a frontage to Green Lane, although, as previously noted, there was an intervening strip of land not in the claimant's ownership. There was access from Green Lane at two points at approximately either end of the Green Lane frontage.

Mr Catt was instructed in July 1990 and although he visited the farm in the years prior to 1986 he first inspected the reference land after the material date on August 22 1990. At that time he described the land as being in a very bad condition, infested with couch, thistles and other weeds. He described the boundaries as marked by fences, live hedges and ditches, generally in poor condition and not stock-proof.

Mr Goldsmith did not inspect the reference land until November 3 1991, but he referred to a description of it contained in a report dated August 19 1991 addressed to the claimant by his firm. This described the reference land as situated about 150ft above sea level, predominantly flat, the soil being a clay loam classified as Grade 3 by the Ministry of Agriculture, Fisheries and Food (MAFF). The boundaries of the land were said to be clearly defined by hedges and fences in poor condition. Mr Goldsmith added that the land had reverted to grass and was in poor order at the time of his inspection.

The direction

On January 12 1990 the council made a direction under article 4 of the Town and Country Planning General Development Order 1988 (the GDO) in respect of the reference land and a further area of land lying to the east thereof with frontages to Ockham Road North and Green Lane. The total area of land involved was 20.31 acres. The direction was approved by the Secretary of State for the Environment on May 29 1990. The direction related to certain development permitted by the GDO under Schedule 2, Class A Part 2 thereof, namely the erection, construction, maintenance, improvement or alteration of a gate, fence, wall or other means of enclosure. I hereafter refer to these forms of development as "fencing etc".

In an accompanying statement of reasons for making the direction the council stated:

1. The area is rural in character in a prominent position and comprises a mixture of open fields and woodland. The land covered by the Direction is visible from roads and public rights of way in the area.
2. The land is situated within the Metropolitan Green Belt and is therefore subject to the policies for the control of development as set out in the Surrey Structure Plan 1989 and the Guildford Borough Local Plan 1987 which seek to preserve and enhance the rural character of such areas and restrict any inappropriate new development.
3. It has to come to the Council's attention that it is the intention of the owner of the land to dispose of the land in lots which would result in fragmentation and it is considered that the erection of fencing or other means of enclosure would constitute a serious threat to the outstanding visual amenities of the area.

The application

On November 8 1990 the claimant made an application for planning permission in respect of the reference land. She sought permission for the provision of a glasshouse and a storage barn and the erection of five fences comprising 4ft-high chestnut posts and three strands of wire. These were to run from south-east to north-west and divide the reference land into six separate parcels.

On November 13 1990 the council refused the application on the following grounds:

01. The proposed fencing would fragment the existing open natural land to such a degree and in such a regularised manner that it would be visually instrusive detracting from the appearance of this locality to an unacceptable level contrary to policy C17 of the Surrey Structure Plan 1989.
02. The proposed buildings are not from the information submitted with the application required for any essential agricultural purpose and therefore the proposals are contrary to established green belt policies contained within both the Guildford Borough Local Plan and the Surrey Structure Plan 1989.
03. The proposed buildings by virtue of their size, location and regularised siting represent a significant visual intrusion into this sensitive, rural area which would be detrimental to both the existing natural open landscape and to the residential amenities of properties nearby, all contrary to policy CE37 of the Guildford Borough Local Plan 1987 and policy C19 of the Surrey Structure Plan 1989.

The claim

The claimant submitted a claim for compensation in respect of this refusal under section 164(1) of the 1971 Act on May 3 1990. As amended at the hearing the claim was made up of the following items:

1.	Depreciation in value of freehold interest in reference land	£94,600
2.	Legal costs in obtaining advice about the effect of the direction (including VAT)	2,794.50
3.	Surveyors' fees in connection with the submission of the planning application of November 8 1990	300
4.	Interest paid upon a loan of £80,000 advanced to the claimant	16,902.22
	Total	£114,596.72

Claimant's evidence

The claimant said that prior to 1986 she had had a building company but had developed a heart disease and been advised to seek an easier life. Both she and her former husband had forebears in farming and had been on farms but had no direct previous experience of farming. She purchased the freehold interest in the farm from a Mr Maiklem in June 1986 for the sum of £255,000. The farm had previously been a dairy unit but the vendor retained the milk quota. She had hoped to obtain her own quota but was unable to do so.

She sought the advice of MAFF. Their Mr Hickman spent a considerable time considering alternative farming methods and his first suggestion had been sheep farming. She said the farm formed an "island" in the middle of a sizeable village and neighbouring residents had been used to coming on to her land to walk their dogs etc. She had to discourage this activity, which was inimical to sheep farming. The sheep farming proved not to be viable and involved her in financial loss. She consulted Mr Hickman again and he produced a studied report. He thought that sheep farming could be profitable, but she said he did not allow for interest on the capital which would have been required on the repair of barns, erection of gates, stock-proof fencing etc.

Mr Hickman and she had looked at other alternatives and in the following two years she attempted mixed farming with 40 acres laid to barley and the remainder cropped for hay/silage with a reduced number of sheep. Owing to hot summers and the sandy soil not retaining good water levels the barley crops failed.

She again consulted Mr Hickman and he suggested a diversification scheme under which there would be a livery yard and stables, for which planning permission would be required, with 40 acres for sheep and horse grazing and 40 acres set aside. Planning permission was granted for the stables and MAFF undertook to grant approximately £8,000 towards the total estimated cost of the scheme of approximately £80,000 conditional upon her obtaining 75% of the remaining funds needed to complete the scheme. 20 acres of land along Green Lane was out on a limb and not visible from the ley of the farm at Long Reach. Horses and cattle had been rustled in the area and their presence upon the 20 acres could represent a security risk. Mr Hickman advised the land could be sold subject to "set aside".

In or about June 1989 she had therefore instructed Mann & Co, estate agents of East Horsley. She wanted a quick sale and thought that they were the most likely to succeed in this. They were to offer 20 acres of land for sale as a whole or in sections at a guide price of £160,000. The area offered included the whole of the reference land and the remaining area which later became subject to the direction, with the exception of a small area comprising a pond and woodland immediately to the north-west of the junction of Ockham Road North and Green Lane.

The claimant gave evidence as to offers which had been received. First, Mr Stephen Brian Castle offered the sum of £6,000 subject to contract for one acre with a gate into Green Lane. He wanted the land to graze ponies. Second, Mr K Hayford agreed to purchase five acres for the sum of £32,500 (equivalent to £6,500 an acre). He wanted to erect a greenhouse and use the land to plant trees, she believed conifers.

By October 1989 she was not wholly satisfied with the efforts of Mann & Co and in that month she instructed Philip Charles, estate agents of Guildford, to offer the land for sale at £10,000 an acre with the possibility of division into plots of a quarter of an acre. However, on or about October 10 1989 Mr Slot noticed an article which appeared in the *Surrey Advertiser*. This reported that the council were proposing to take action in the form of making an article 4 direction in respect of the reference land. Mr Slot consulted the claimant's solicitor on her behalf and spoke to one of the council's planning officers on the telephone. He said that the planning officer told him that the effect of the proposed direction would be that planning permission would be required for all forms of construction, fencing, buildings for agricultural use etc.

On the claimant's behalf, Mr Slot prepared the application for planning permission, which was dated November 5 1989. He included a greenhouse in the proposed development in view of Mr Hayford's known requirement. The barn was not included to suit any particular potential purchaser's requirement but was included on Mr Slot's advice. The fencing was included in order to enable the claimant to sell the land in parcels to meet the needs of existing potential purchasers and any that might later appear.

The claimant explained that the application related only to the reference land because at the time the application was made the direction had not been formally made and she did not realise that it would extend to the remaining area of approximately five acres to the east. A later planning application had been made in respect of that area. In November 1989 she also considered instructing Bannister & Wellers, rural property agents of Guildford, Mr Catt's firm, as agents to sell her land. Their Mr Simon Victor Ellis ARICS visited the farm on November 23 1989 and on the following day he wrote to Mr Slot. He recorded that he had suggested that it would be a good idea

A to put in a planning application to facilitate the division of the reference land into four paddocks. At the same time he gave advice about offering for sale land to the north-east of the reference land fronting Ockham Road North. He was instructed to offer that land but does not appear to have been instructed in relation to the reference land.

The claimant was asked why she did not take up Mr Ellis' suggestion of making a planning application for three fences to divide the reference land into four parcels. She said she preferred to adhere to her existing application. Had that been permitted, two of the parcels taken together would have met Mr Hayford's requirement, but he might not have gone ahead and in that event it B would have been untidy to have been left with such a large parcel. Meanwhile, Mr Castle had seen the article in the *Surrey Advertiser* about the direction. This put him off his purchase. He was very disappointed but withdrew his offer.

Just before Christmas a Mr Stamper had agreed to buy half an acre in the south-west corner of the reference land for £3,500. He recorded this agreement in a letter dated December 24 1989 with which he enclosed a cheque for £350 as a deposit. On January 3 1990 Mr Stamper appears to have drawn a cheque in favour of the claimant for the balance of the purchase price, £3,150. The claimant said that Mr Stamper later saw a notice of the making of the direction displayed on the land. He showed it to her and asked for his money back. She repaid his £350. Meanwhile, the cheque for £3,150 appears to have C been presented but stopped by Mr Stamper before it had been honoured. On January 18 1990 Mr Stamper's solicitors wrote to the claimant's solicitor to say that Mr Stamper's letter of December 24 1989 did not constitute a legal contract for the sale of land. Mr Stamper had said he wanted the land for sheep. None appeared but he had erected a tent on the land.

At the end of 1989 the claimant concluded that, in the light of the direction, she would not be able to fund the development of the livery stables from the proceeds of the sale of the reference land. It was suggested to her that she should seek a bank loan. She sought a loan from Barclays Bank plc to enable her to carry out the development. A loan of £80,000, described by her branch bank manager as a "home improvement loan", was sanctioned by January 3 1990. This was D confirmed by a formal offer on January 8 1990 and the loan was drawn down on January 11 1990. Although the loan was described as a "home improvement loan" the bank well knew the main purpose was to fund the development of the livery stables, although the work did include the erection of a garage for her car.

After the refusal of the application Mr Hayford withdrew his offer. Subsequently there was a lot of interest but no further offers for the land other than an offer from a gypsy at £1,000 an acre for the whole of the land on the market.

The farm was now mainly used in connection with the livery stables. Part was let for grazing sheep and part for clay pigeon shooting. The farm was financially successful. As to the reference E land she had decided to retain it and use it for equestrian events. She said that at the time of the refusal of her application she never received any indication from the council that they would be prepared to consider favourably other planning applications in respect of the reference land. Nothing in the statement of reasons for making the direction or the reasons given for refusing her application indicated that a lesser division of the land would be acceptable.

The claimant had attended a meeting on May 22 1991 at which were present Mr Slot, her solicitor and Messrs Stallon and Lingard of the council as well as Mr Catt. She understood from Mr Lingard that the council were unwilling to pay compensation but would give favourable consideration to an application for fencing to divide the reference land into four parcels. By that time she had had to raise the F loan to fund the development of the livery stables and had decided to retain the reference land. She did not, therefore, make an application as suggested by Mr Lingard. This had been the first occasion on which she had any indication that the council might favourably consider an application for fencing to divide the land into four plots, although she had met Mr Catt and Mr Lingard on three previous occasions regarding her claim and there had been much correspondence with the council.

The claimant gave evidence about the position regarding access to the reference land from Green Lane. She said that after the completion of her purchase of the farm the vendor, Mr Maiklem, said he had retained ownership of the strip of land along the frontage to Green Lane. There were two existing access points where a ditch along Green Lane was crossed. G The water in the ditch was piped under a filling of chalk, rubble and hardcore. She had used these accesses every year since her purchase of the farm for cutting grass, moving sheep, growing barley for two years, and general agricultural uses on the reference land. She had taken legal advice. She was advised that the contract for sale incorporated the selling agent's particulars and that as these stated that the reference land had access on to Green Lane she was entitled to such access. When she herself came to offer the land for sale, she told potential purchasers that she had been gaining access from Green Lane to the reference land without hindrance. However, she had also told them that if there was an "overnight problem" she would provide alternative access.

The only action ever taken by Mr Maiklem was that he wrote a H letter to her saying she should not use the accesses. When Mr Goldsmith came to see her in August 1991 he advised her that it would be worthwhile to approach Mr Maiklem with a view to purchasing the strip of land. This she did and by the second day of the hearing she had completed the purchase of the strip for the sum of £3,000.

The claimant said that Mr Slot had charged £900 for his work in submitting the planning application. This was a discounted charge because of their relationship. She had paid the charge. Mr Slot had said that £300 of it was referable to the fencing works. When she first heard about the proposed direction, she had never heard of such a matter. She took legal advice about its effect on many occasions subsequently. Her solicitor's bill, amounting to £2,430 plus VAT in the amount of £364.50, was correct and she had paid it. J

Mr Goldsmith thought that there was a considerable demand in the locality for small enclosures of land from half an acre to four/five acres for grazing ponies used for recreation as well as use in connection with forestry, rose-growing, uses allied to agricultural use, particularly in connection with horses, fish farming and clay pigeon shooting. There was also a potential demand from those wishing to use land for a purpose unlikely to be permitted under planning control but who hoped that they might yet obtain planning permission or be able to carry on their proposed use without suffering enforcement action by the local planning authority. I refer to such persons subsequently as "chancers".

Mr Goldsmith said that the demand was demonstrated by the number of inquiries the claimant received when she offered the land K for sale in 1989. He did not accept the claimant's description of the farm as an "island" site, but roads did circle it and there were houses along Long Reach and a few large farms in the area. He was aware of the position regarding the strip of land along the Green Lane frontage. He had advised the claimant to purchase this strip to resolve the matter and had been pleased to hear that the purchase had now been completed. As at February 1990 he thought a purchaser would have investigated the circumstances of the strip of land and the claimant's past use of the accesses from Green Lane. He would consider alternative means of access over the remainder of the farm and the possibility of buying the strip of land fronting Green Lane. He would have made an allowance for the position regarding access L to Green Lane. He foresaw no great difficulty in terms of either planning control or construction in providing alternative access to the reference land if no access was available from Green Lane. In particular, he did not admit that the type of alternative access which he had in mind, which he did not think would necessarily have to be a hardened track, would require planning permission.

Mr Goldsmith foresaw no difficulty in supplying water to any purchasers of the reference land; if need be, the claimant could provide a supply over the remainder of the farm. He accepted that the land was not in good condition but that it would have been very easy to tidy up — half a day's work, in his view.

His valuations were not based on any specific comparable transactions but on his experience in relation to transactions in M similar land in the southern counties. He had dealt with sales of 200-300 acres in parcels of less than 20 acres in the last two years, at least six within 25 miles, although none had been in Surrey.

He cited prices which he had recently been bid for a number of parcels of land on the Dell Park Estate at Englefield Green. He did not rely on these as directly comparable transactions as he thought Dell Park Estate a more valuable location, but in order to show that values an acre were higher for small parcels than for larger parcels.

His valuations of the reference land had been made as at November 1991. He had made two valuations of the freehold interest in the reference land excluding any effect of the direction and the subsequent refusal of planning permission. I refer to these as his

A "before" valuations. There were two alternative "before" valuations. The first was on the assumption that there was unrestricted access from Green Lane. This amounted to £102,500 — equivalent to £6,800 an acre. His second "before" valuation was made on the assumption that the land had the benefit of an access other than from Green Lane. It was in the amount of £90,000 — equivalent to about £5,970 an acre. As I understood his evidence, his second valuation reflected the bid which a purchaser would be likely to make in the light of the actual circumstances regarding the strip of land fronting Green Lane, and that upon the assumption that access was not unrestricted he made a deduction of about 12% from his valuation on the assumption of unrestricted access.

Mr Goldsmith had understood that the material date for valuation B purposes was the date of the hearing. He now understood that it was February 13 1990. In his experience, the value of land such as the reference land was about 10% higher at the earlier date and at that date his "before" valuations became roundly £112,500 (£7,465 an acre) and £99,000 (£6,570 an acre) respectively.

He had experience of selling land subject to article 4 directions, although not in Surrey. He thought that the effect of the direction and the refusal of planning permission was that the reference land could not be further subdivided for the purposes of occupation or sale to meet the considerable demand for small parcels of land which he had identified. Potential purchasers would take legal advice as to the considerable restrictions imposed on their ability to occupy and C manage land subject to a direction.

In this particular case, having regard to the state of the land, it would not be capable of carrying livestock without the improvements of existing fences for their proper containment. Purchasers were also deterred by the prospect of making repeated planning applications and other things being equal would always prefer to buy other land not affected by a direction. The result of making directions was often that land became less well managed and thus land values were reduced.

His valuations of the freehold interest in the land after the making of the direction and the refusal of planning permission (which I refer to as his "after" valuations) were £15,000 (equivalent to £1,000 an acre), assuming unrestricted access from Green Lane, and £12,750 D (equivalent to £846 an acre) with the benefit of access other than from Green Lane. Again, these "after" valuations were as at the date of the hearing and as at February 13 1990 they would have been 10% higher, ie roundly £16,500 (equivalent to £1,095 an acre) and £14,000 (equivalent to £930 an acre).

He did not accept the council's premise that his "after" valuations should be made on the assumption that planning permission for development within Schedule 8 to the 1971 Act, which would include fencing, etc would be granted. This would make a nonsense of the law. However, upon the basis that the council's premise were correct, he thought that the effect of the direction on market forces would still depress the value of the land despite the assumption that planning E permission would be granted for fencing, etc for agricultural purposes. There was a question as to what type of fencing must be assumed to be permitted. However, he conceded that if the council's premise were correct his "after" valuations were too low.

He was asked to compare:

(a) land which had the benefit of GDO rights to put up fencing etc and
(b) land in respect of which it was known that planning permission for development within Schedule 8 to the 1971 Act would be granted.

He said that (a) would produce the highest value. Only limited operations could be assumed under (b). He had not taken into account the fact that under Schedule 8 it had to be assumed that F planning permission would be granted for agricultural buildings without limitation as to size. He had not anticipated purchasers wanting to put up buildings.

He noted that the council were suggesting that a planning application for fencing etc to divide the land into four parcels would receive favourable consideration. In the light of the wording of reason 01 in the refusal of planning permission, he thought this extraordinary. However, in the event that planning permission could be granted, which would facilitate subdivision into four or fewer plots, the diminution in value would not be so great. However, the effect of the direction would remain in force and would inhibit future management of the land. On the basis that there was undeniable written evidence that planning permission would be granted for subdivision into four plots, his "after" valuation was 50% less than G his "before" valuation, less an allowance of £1,000 for the likely cost of making a planning application.

All Mr Goldsmith's valuations were subject to a deduction in respect of the costs of sale.

Council's evidence

Mr Stallon referred to the direction and the claimant's planning application. He said that the construction of a track across the farm to provide access to the reference land would require planning permission. If vehicular access could not be obtained from Green Lane, alternative access would be required from Ockham Road North or Long Reach over intervening land. The creation of such an access would comprise engineering operations under the 1971 Act H and planning permission would not be granted by the GDO, unless the access was reasonably required for agricultural purposes within the agricultural unit which was constituted by the farm.

He thought that access for the reference land, if it were to be occupied separately, would not be for agricultural purposes within the farm and would require specific planning permission. He thought vehicular activity on any such access would be detrimental to the visual amenity and rural character of the area and he would not have been prepared to recommend approval. Further, the highway authority had expressed concern about access from Ockham Road North and Long Reach and there might be a highway objection to any application in respect of access thereform. J

As to access from Green Lane, he acknowledged that there were two existing accesses but said that additional accesses would require planning permission. If required for agricultural use, planning permission would be granted by the GDO, otherwise specific permission would be required. Upon receipt of any application for such a permission he would have to consult the highway authority and could not say whether it would be permitted.

He accepted that the refusal of the application which gave rise to the claim did not suggest that any lesser form of development might be permitted. The words of reason 01, which stated the proposed fencing would fragment the land "to such a degree" as to detract from the appearance of the locality to "an unacceptable level", suggested that a lesser amount of fencing might be permitted, but K although there was sometimes an "informative" in a notice of refusal of a planning application, suggesting a lesser form of development might be permitted, there was none such in this case.

In answer to Mr Denehan, he said he had been concerned with the claim but never told the claimant her claim was misconceived nor, prior to the Friday before the hearing commenced (November 8 1991), had anyone of the council told her that her claim was worthless. He had attended the meeting between the council's officers, Mr Catt and the claimant in May 1991. The claimant had not then been told that her claim was misconceived and worthless and there had been discussion as to the level of compensation with a view to compromise. L

He accepted that whereas officers might make recommendations as to decisions whether to grant planning permission or to institute enforcement proceedings, council members made decisions and did not always accept the officers' recommendations. There had been cases where enforcement action recommended by officers had not been taken as a result of members' decisions and that "chancers" sometimes "got away with it". However, in the case of the reference land, which was relatively open and visible with dwellings in close proximity, he thought council members would be likely to sanction enforcement action.

In relation to access from Ockham Road North, he told Mr Denehan that in expressing his view on his likely recommendation to refuse any application for planning permission, he had had regard to M the proposal contained in John D Wood & Co's report on August 19 1991, which spoke of an access track along the north side of the reference land and the adjoining parcel (OS parcel 0072), with length of about 275 yds. He had assumed a track with a hardcore or possibly tarmac surface and a relatively large number of vehicular movements. If there were such an application made, he would need to know the actual use or uses proposed to be served by the access and the likely generation of vehicular traffic. On visual impact his objection would be both to the entrances and to the track.

He was not saying that there could not be an acceptable form of access, although his judgment was that there could not. He did not say that any form of entrance would be offensive. He thought the appearance of a track and the attendant vehicular activity would be unacceptable. However, it was possible that a wise applicant could devise a scheme which would be acceptable. In his view, it was unlikely but he could not express a definitive opinion. It was also possible that what would be acceptable to him would not be acceptable to an applicant. Should planning permission be needed for access from Green Lane it would be much easier to obtain than access from Ockham Road North.

Mr Catt had 30 years' experience of agricultural professional work in the Home Counties, all but two of those years in Surrey and adjoining counties. Between 1977 and 1988 he had acted for Mr Maiklem, who owned the farm from 1970 until 1986. In 1986 his then firm, Weller Eggar, acted jointly with Savills in the sale of the farm. He said it was only an "island" site in as much as it was surrounded by roads but not by non-agricultural property.

He was advised that as the application sought permission for buildings as well as fences his refusal did not provide the necessary basis for an entitlement to compensation. However, he had been asked to consider the compensation payable as if an application had been made for permission simply to fence six enclosures and had been refused on February 13 1990. He was also advised that section 164(4) of the 1971 Act required it to be assumed that planning permission would be granted for development within Schedule 8 after the refusal of the claimant's application. On this basis the loss in value as a result of the notional refusal of permission to which he had addressed his mind was nil, because, despite the direction, it must be assumed that planning permission would be granted in respect of the erection of fences and buildings for agricultural purposes.

Mr Catt produced his Appendix E showing:

(a) activities which did and did not require specific planning permission prior to the making of the direction;
(b) activities actually requiring and not requiring specific planning permission after the making of the direction;
(c) activities for which it must be assumed planning permission would be granted under Schedule 8 after the refusal of the claimant's application and in each case activities which did not constitute development under the 1971 Act.

In case that tribunal held against the council as to the effect of section 164(4), he had been asked to assess the compensation payable having regard to the matters set out in (a) and (b) in his appendix E. His assessments were made as at February 13 1990. He had had regard to all relevant factors as at that date, including the condition of the property, the means of access and the provision of water supply. He made three "before" valuations on three different assumptions. He expressed these simply in terms of a value per acre. I have also expressed them in terms of the total value in rounded figures. He described his three alternative valuations as calculations A, B and C.

Calculation A was based on the assumption that there was no legal right of access from Green Lane or any legal right to lay a piped water supply from Green Lane, that planning permission would be required and was unlikely to be granted for the construction of a hardcore and scalping farm track to access the reference land over other parts of the farm. On this basis his valuation was £1,500 an acre, equivalent to a total of £22,600.

Calculation B was based on the same assumptions as to legal rights in relation to Green Lane but on the assumption that it would be possible to provide access from Ockham Road North to the reference land by way of a hardcore and scalping track and to provide a piped water supply over the remainder of the farm. On this basis his valuation was £3,500 an acre, making a total value of £52,750. This calculation actually commenced from a value of £4,500 an acre (equivalent to a total value of £67,800) subject to a deduction of £1,000 an acre (or £15,000 in total) to allow for the cost of constructing the track, the fact that the site of the track could no longer be used for agricultural purposes, the cost of laying piped water supply, all of which he said the vendor would have to meet prior to selling the reference land. In addition, his allowance took into account the fact that the access would be less convenient than access from Green Lane.

Calculation C was based on the assumption that it was legally possible to create a number of separate accesses from Green Lane and to connect a number of piped water supplies therefrom. On this basis his valuation was £6,000 an acre, equivalent to a total value of £90,400.

In support of his valuations he said there was a strong demand for small areas of land from two to 20 acres in Surrey at the material date. All such land was likely to have been subject to green belt and other restrictive planning policies. Prospective purchasers had many and varied uses in mind. Many of these were likely to be in breach of planning policies but this did not seem to be a major deterrent to purchasers, either because they were unaware of the policies or because they were "chancers". However, many sales foundered between acceptance and exchange of contracts. This was sometimes because offers were made based on a whim or a dream. Later, a more realistic attitude led to an examination of financial or planning problems. As with any sale, the transaction was more likely to fail if a ready excuse could be found, eg poor access, costs of reseeding or refencing, water supply problems, etc.

In support of his "before" valuation based on calculation C, Mr Catt produced details of two comparable transactions. The first comprised the sale of 26.89 acres of arable and accommodation land at Hardwick Lane, Chertsey. This was about 11 miles by road from the reference land. Its location was not quite so attractive as the reference land but the land had been in good heart with very good access. There had been a hay barn on the land of little value. The land was offered for sale in two lots but sold as a whole for £156,000, equivalent to £5,801 an acre. Contracts were exchanged on January 11 1990 and the sale was completed on February 1. The purchaser wanted the land for pony grazing.

The second transaction comprised the sale of 6.73 acres of land at Elstead, about 19 miles by road from the reference land in a very attractive location, closer to the village of Elstead than the reference land was to the village of Ockham. There was very good access. The land was sold for £45,000, equivalent to £6,686 an acre. Contracts were exchanged on December 18 1990, the sale being completed on December 31. Again the vendor wanted the land for pony grazing.

As to his "after" valuation, he was advised that the effect of section 164(4) of and Schedule 8 to the 1971 Act required him to assume that planning permission would be granted for development referred to in para 4 of Schedule 8. As far as fencing etc was concerned, it had therefore to be assumed that planning permission would be granted. For practical valuation purposes he thought this meant that the value of the land was not reduced by the direction and the refusal of planning permission. However, as I understood him, he did accept that the direction had a depressing effect on the value of the reference land because, despite the assumption that planning permission would be granted for fencing, etc for agricultural purposes, there was a necessity to incur the costs and delay involved in making the necessary applications. In answer to me, he quantified the reduction in value at 20%.

In relation to the erection of buildings, however, he said that the provisions of para 4 of Schedule 8 meant that in his "after" valuation he could assume that planning permission would be granted for the erection of buildings to a greater extent than that permitted by the GDO. He cited three particular matters:

(a) Under the GDO, any building to be erected had to be required for the purposes of agriculture within the agricultural unit on which it was to be sited. However, under Schedule 8 the building could be used for general farming operations.
(b) Under the GDO the size of the building which could be erected was not to exceed 465 m^2, whereas under Schedule 8 there was no limitation as to size, and
(c) Under the GDO there was a condition requiring that any building to be erected for the use of livestock was not to be within 400m of another property. There was no such restriction under Schedule 8.

He thought that the Schedule 8 assumptions, in so far as they required the assumption that planning permission would be granted for development beyond that permitted by the GDO, opened up the market to a larger number of purchasers and increased the value of the land. He therefore thought that in so far as the erection of buildings was concerned the reference land was more valuable with the benefit of the Schedule 8 assumptions than it was with the GDO rights. He thought the additional value was at least 20% and that it offset the 20% reduction in value which he had attributed to the effect of the direction. He thus said that his "after" valuations were the same as his "before" valuations and that there was no depreciation in the value of the claimant's interest in the reference land.

However, in case the tribunal did not accept that his "after" valuations should reflect the Schedule 8 assumptions, Mr Catt produced "after" valuations on the same alternative assumptions as

A to access from Green Lane as his "before" valuations. These were in terms of values an acre. I have added the resultant total values in terms of rounded figures. On the assumptions on which his calculation A was based, his valuation was £1,000 an acre, or £15,000 in total. On the assumptions on which his calculation B was based, his valuation was £1,500 an acre, or £22,500 in total. On the assumptions on which calculation C was based, his valuation was £2,500 an acre, or £37,500 in total. In his valuation for calculation B he had deducted £500 an acre or £7,500 in total from a valuation of £2,000 an acre, or £30,000 in total. This reduction was an allowance for the same items that he had allowed for in his "before" valuation for the purposes of calculation C, although the allowance was less in amount than in his "before" valuation.

B He thought that in estimating both the "before" and "after" values it must be assumed that a purchaser would have been very concerned about the risk of not being able to obtain access or water supply from Green Lane and also about the risk of not obtaining planning permission for an access track from Ockham Road North across the intervening land. He therefore concluded that his valuations for the purposes of calculation C did not produce a true measure of the depreciation in the value of the claimant's interest. He thought the true value lay somewhere between his valuations for the purposes of calculations A and B, and that the depreciation in value in the freehold interest of the reference land as at the material date was £20,000.

C In support of his "after" valuations he cited the case of about 15 acres of land at St Martha's Hill, Chilworth, about 10 miles from the reference land. He described it as being in a very attractive position but with poor access up a long track. The land was subject to an article 4 direction in a very similar form to that affecting the reference land. An offer of £33,000, equivalent to £2,200 an acre, was accepted. The purchaser had intended to use the land for flying model aircraft. He had satisfied himself that no planning permission would be required for his intended use, but the transaction was not completed because the purchaser discovered that the land was the subject of a restrictive covenant against the use of the land for other than agricultural purposes and he had been unable to persuade the owner of the benefited land to agree to a modification.

D Mr Catt agreed with Mr Goldsmith's evidence as to the potential uses of the reference land and thought that horse or pony grazing was the most likely use and that potential purchasers were looking for parcels with a minimum area of two acres, normally five acres. They had regard to the total price. They were not generally interested in parcels of a lesser value than £10,000. He did not agree with Mr Goldsmith that the existence of a direction meant that land would be less well managed. Although it might be annoying to have to obtain planning permission, he could not believe that maintenance of fences would not be permitted.

As to Mr Goldsmith's addition of 10% to his valuations made as at November 1991, he accepted that values might reasonably be taken E to have been 10% higher as at February 1990, but he did not accept that the reference land was worth as much as £7,480 an acre at the latter date. He did not think that Mr Goldsmith had made sufficient allowance for the problem related to access from Green Lane in either his "before" or "after" valuations and that the latter were too low.

In answer to Mr Denehan, he agreed that his firm had acted for Mr Maiklem in the sale of the farm to the claimant and that according to his firm's sale particulars the land was sold with the benefit of access from Green Lane.

He confirmed that when he was instructed by the council he was not told that the claimant's claim was invalid or misconceived in its entirety. He had proceeded on the basis that the claim was valid until very recently.

F As to access from Green Lane, he accepted that the claimant had never been inhibited from using it. However, he thought potential purchasers and their solicitors would have to form a view about the importance of a potential dispute. With reference to the actual potential purchasers they may have been satisfied as to access but he did not know whether the transactions had reached the stage at which solicitors had considered the matter. Generally any distracting feature would have an adverse effect on a purchaser's mind and, given an otherwise equally acceptable alternative property, a purchaser would be likely to choose to buy the alternative property. If he preferred the land the subject of the dispute for other reasons, he would discuss the matter with the vendor, look for alternative access but shy away from a legal dispute.

As to the condition of the land, he thought that it had needed G complete reseeding, more than half a day's work. In answer to me, he said he had allowed £200 an acre in both his "before" and his "after" valuations because of the state of the land and boundary fences. He agreed that there were potential non-agricultural uses for the land, such as garden centres, clay pigeon shooting etc, but planning authorities did not always take enforcement action if there were no complaints about unauthorised uses and that some potential purchasers were "chancers" who formed part of the market.

He agreed that if the claimant was told in May 1991 that she would be able to obtain planning permission for fencing to divide up the reference land it would be irrelevant to the "after" value at the material date.

He did not agree that the Schedule 8 assumptions were of value H only in relation to the erection of buildings. Some purchasers might erect a building larger than that permitted under the GDO and take a chance on using it for non-agricultural purposes. Generally he assumed that "chancers" were in the market in both his "before" and his "after" valuations, although in answer to me he said that "chancers" were likely to be less enthusiastic after the direction had been made, which was one of the reasons why he thought the direction led to a 20% fall in value. He agreed that the ability to put up fences for agricultural use was of no value to a potential purchaser who wished to use the land for non-agricultural purposes, except for "chancers". He agreed that in so far as the Schedule 8 assumptions exceeded the GDO rights in relation to the erection of buildings, they J were of no value to a purchaser who did not wish to erect buildings. He accepted that his report dated May 23 1991 submitted to the tribunal made no mention of the Schedule 8 assumptions and the report's conclusion that the diminution in value of the claimant's interest in the reference land was £6,750 was consequently mistaken.

He admitted that he had not taken into account in either of his calculations A or B, the prospect of purchasing the strip of land fronting Green Lane to resolve any difficulty regarding access, although he did not think a purchaser of a small parcel of the reference land would have considered buying the strip.

His valuations did not reflect the possibility that potential purchasers might think that if the vendor had used the accesses from Green Lane without physical or legal action being taken to stop such K use and the only evidence of difficulty was a few solicitors' letters, a potential purchaser might think there was no substantive dispute. Further alternative valuations might have been made taking account of potential purchasers taking this view and the possibility of purchasing this strip of land, but he had not made them.

He did not think that his effective 75% reduction between his values on the basis of calculations C and A was too great, having regard to the assumptions upon which they were based. He thought a potential purchaser faced with the problem of estimating the value of the reference land would consider valuations on the lines of his calculations A and B and be likely to agree a price somewhere between the two, but nearer the result of calculation B than L calculation A.

He accepted that two of the offers made for parts of the reference land reflected acreage rates above his acreage rate in his "before" valuation under calculation C. However, the land might have been sold as a whole, in which event his acreage rate was likely to have been less, or in up to six small parcels, in which event the market for such small parcels might have been depressed. Nevertheless, he agreed that in the case of the Chertsey land the highest price had been produced by a sale of the whole of the land. He knew of the gypsy's offer but did not take much account of it. It was a verbal offer, verbally refused. There was no evidence that the market was fully tested and he preferred the evidence of the Chilworth land.

M As to the other items of claim, Mr Catt thought the claim for interest paid was invalid. Had the claimant sold the reference land and expended the net proceeds on the development of the livery stables, she would have foregone the interest which she would otherwise have been able to earn on the proceeds. The loss would have been broadly the same as the actual cost of interest incurred.

Inspection

I inspected the reference land on December 5 1991, at which date it was down to grass. I find that at the material date the land had on it the remains of the previous year's barley crop, that there had been some growth of weeds, that the boundaries were marked by fences and hedges that were not stock-proof and that there were two

A ungated accesses from Green Lane, which crossed a ditch close to the east and west ends respectively of the frontage to Green Lane.

Council's case

On behalf of the council, Mr Lewsley said that the entitlement to compensation could arise only if the requirements of section 165(1) of the 1971 Act were met. These were:

(a) That planning permission for the development of land had been granted by a development order. It was accepted that planning permission for fencing, etc was granted by article 3, Schedule 2, Class A of the GDO.

(b) That permission was withdrawn by the issue of directions under powers contained in that behalf conferred by the GDO. It was B accepted that by a direction under such powers the permission granted by article 3, Schedule 2, Part 2, Class A was withdrawn.

(c) That an application in that behalf under Part III of the 1971 Act for planning permission for that development was refused. This meant that an application must be made and refused for development formerly permitted by the GDO. This was the correct meaning of section 165(1)(*c*). First, because it was the natural meaning of the words of section 165(1). Second, it was even clearer in the equivalent provisions of the Town and Country Planning Act 1990 (the 1990 Act), which in section 108 (1) stated:

> Where —
>
> C (a) planning permission granted by a development order is withdrawn (whether by the revocation or amendment of the order or by the issue of directions under powers conferred by the order); and
>
> (b) on an application made under Part III planning permission for development formerly permitted by that order is refused. . .
>
> section 107 shall apply . . .

The 1990 Act was a consolidating statute. *Maxwell on Interpretation of Statutes*, 12th ed, p 20, stated: "in the construction of a consolidating Act the presumption that Parliament does not intend to alter the existing law applies with particular force". Thus the 1990 Act confirmed that under the 1971 Act, section 165(1), the application and refusal must be for development formerly permitted by the order.

D It followed that the intention of the legislature was to give the local planning authority the opportunity of considering and granting an application. It was not open to a claimant to ensure a refusal and therefore obtain compensation by including in an application development not formerly permitted by the order and which the local planning authority were likely to find objectionable.

The claimant's application was not for permission for development formerly permitted by the GDO because it included the provision of two buildings and fences and such development because it included buildings was not formerly permitted by the GDO. This had a crucial bearing on entitlement to compensation because, since there having been no planning application in respect of development E formerly permitted by the GDO, it was not known whether the council would refuse it or not. It was clear from the documents and Mr Stallon's evidence that the planning committee had never considered an application for fencing without buildings and therefore it was not known, and it would be wrong to speculate about or attempt to assess, what the committee would have decided had such an application been made. It would be a committee decision and the outcome was simply not known.

On behalf of the claimant, Mr Denehan had suggested that the application and refusal of permission had two distinct elements, fences and buildings, each of which was separately refused. This was simply not so. The application was for one development as a whole and was considered, as it had to be, and refused on that basis. It was F clear from the reasons for refusal, two relating to buildings and one to fences, that the buildings were an important factor in the decision.

Mr Denehan also suggested that the fences would clearly have been refused because one reason for refusal related entirely to fences and the other two to buildings. Again this was simply not so. It was accepted that the committee were concerned about the impact of fences but they had never considered whether the first reason for refusal amounted on its own to a sound and convincing reason for refusal of six enclosures sufficient to overcome the presumption in favour of granting an application, having regard to the entitlement to compensation. He reiterated that it was simply not known whether the committee would have refused an application for fencing six enclosures without buildings.

Mr Denehan had drawn attention to the fact that this point was not G raised until a letter was delivered to the claimant's solicitor on Friday, November 8. It was accepted that this was so. It was also accepted that the point had not been taken in discussions and correspondence between the council and the claimant. The council's legal advisers had not carefully examined the matter until the week before the hearing commenced. When they did, the point was drawn to the claimant's attention as soon as was practicable. All this may be relevant to costs but it cannot give the tribunal jurisdiction to award compensation where there is no entitlement.

In support of his contention, Mr Lewsley cited the case of *Essex County Council* v *Essex Incorporated Congregational Church Union* [1963] 1 All ER 326. In that case the county council objected to H a blight notice but failed to include in their objection a contention that the claimant's interest was not such as to entitle him to serve the notice. The claimant appeared in the Lands Tribunal, argued the point, but did not object that the tribunal had no jurisdiction because the point had not been included in the objection to the blight notice. The point on jurisdiction seemed to have been taken and brushed aside in the Court of Appeal but then held to be correct in the House of Lords, where Lord Reid said at p 330(H):

> I need not consider whether this amounted to a consent to widening the reference to the tribunal, because in my judgment it is a fundamental principle that no consent can confer on a court or tribunal with limited statutory jurisdiction any power to act beyond that jurisdiction, or can estop the consenting party from subsequently maintaining that such court or tribunal J has acted without jurisdiction.

and Lord Morris at p 333(G):

> The tribunal could not assume a jurisdiction with which it would only be endowed if certain steps had been taken and certain conditions satisfied,

and, lastly, Lord Hodson at p 335(E):

> The consent of the parties cannot confer jurisdiction where none exists nor can any question of estoppel arise for estoppel cannot enlarge the jurisdiction of a court of limited jurisdiction.

It was regretted that the point was not drawn to attention earlier, but this could not give the tribunal jurisdiction to award compensation where, because the requirements of the statute had not been met, no entitlement existed.

K Mr Lewsley referred to the tribunal's decision in *Carter* v *Windsor and Maidenhead Royal Borough Council* [1988] RVR 241*. In that case it was submitted by the council that "the claimant must establish that the refused application for planning permission was in respect of development which would have benefited from the deemed permission under Class 6". This was satisfied, said the council, only if the agricultural unit was the entirety of the reference land. Therefore, the unit of valuation must be the entirety of the reference land. The claimants did not challenge the council's submission that the refused application must be for development formerly permitted by the order. The claimants put forward two contentions:

1. That the council had misunderstood the order and that in fact the application did fall within Class 6 even though it related to 15 L enclosures within the entirety of the reference land — in short, the claimants were saying that the words "requisite for use of the land for the purposes of agriculture" did not refer back to the words "agricultural unit" and so were not restricted to use for the purposes of agriculture within the same agricultural unit as before but included also use for the purposes of agriculture within different units in the future.
2. That it was too late for the claimants to take the point. The president agreed with those submissions, holding that there was no legal impediment to valuation by reference to the land being sold in 15 plots. It could not be known whether he would have upheld the claimants on the second point alone. Further, no M objection was or could be taken to jurisdiction. This was important because the rules of estoppel, waiver and acquiescence which made up the claimants' second point could not operate to confer jurisdiction.

In the instant case, there appeared to be no challenge to the view that planning permission was required for the buildings. The GDO granted permission for buildings on agricultural land comprised in an agricultural unit provided that the buildings were reasonably required within that unit. At the material date the agricultural unit, if any, was the farm. The buildings were not required for agriculture within that unit.

---

*Editor's note: Also reported at [1988] 3 PLR 6.

A He then turned to the Schedule 8 assumptions. Section 164(4) stated in the clearest possible terms that "it shall be assumed that planning permission will be granted for" development within Schedule 8. He referred to the case of *Burlin* v *Manchester City Council* [1976] RVR 119*, in which the tribunal had stated: "then the question arises whether the value of the land after revocation should be taken as the claimants say as its Schedule 8 value or whether it should be taken with all its potentialities in the market". After considering the submissions on that point, the tribunal concluded: "We think the better view is that the reference to Schedule 8 in section 164(4) is intended to be an exclusive definition of what assumption is to be made as to the planning permission which would be granted". The clear words of section 164(4) and Schedule 8 should be given B effect to even though they operated against the council in that case by requiring actual potentialities of the land after the revocation to be disregarded if they fell outside Schedule 8.

Referring to the interim decision of the tribunal in *Colley* v *Canterbury City Council* (REF/27/1990) dated March 27 1991, Mr Lewsley said this was a stark example of where section 164(4) and Schedule 8 operated against the claimant. He accepted that this was unfair but no more unfair than when, as is likely often to be the case, the actual potentialities of the land had to be ignored in the post-revocation world even if they existed and added to post-revocation value. For example, in this case section 164(4) and Schedule 8 prevent the council from showing that in the post-revocation world they C would have granted planning permission for fences for any purpose for five enclosures. In *Colley*'s case, in order to achieve what was regarded as a fair result the tribunal made an ad hoc amendment to the section. The difficulty with that approach was that different ad hoc amendments are going to be required in different cases according to the facts. There was no basis for an assumption that Parliament intended Schedule 8 to be given effect to when it operated in favour of a claimant as in *Burlin*'s case and avoided where it operated in favour of a council as in *Colley*'s case.

Mr Lewsley also referred to the case of *Stock* v *Frank Jones (Tipton) Ltd* [1978] 1 All ER 948 in which Lord Simon had said at p 954:

> a court would only be justified in departing from the plain words of the statute D were it satisfied that: (1) there is clear and gross balance of anomaly; (2) Parliament, the legislative promoters and the draftsmen could not have envisaged such anomaly and could not have been prepared to accept it in the interest of a supervening legislative objective; (3) the anomaly can be obviated without detriment to such legislative objective; (4) the language of the statute is susceptible of the modification required to obviate the anomaly.

The whole planning system was restrictive of landowners' rights, often without compensation, so one could not be sure about the second point; but the real problem was the fourth point. If Parliament did not intend section 164(4) and Schedule 8 to have effect, what did they intend? This was unknown. Certainly it would be wrong to disregard Schedule 8 and not substitute something else for it. The council could not be deprived of both the post-revocation E realities as in *Burlin*'s case and the Schedule 8 assumptions in *Colley*'s case. In the absence of a clear indication of what Parliament did intend, it would not be right to make a further ad hoc amendment to the Act.

The whole foundation of *Colley*'s case was unfairness, but were the Schedule 8 assumptions unfair in the instant case? They were unfair only if the planning permissions which were assumed to be granted would be refused in the post-revocation world. If the assumptions represented or approximated to reality there would be no unfairness. In the post-revocation world, if planning permission were refused by the Secretary of State on appeal or on a reference to him, in respect of development within the Schedule 8 assumptions, F the claimant could seek compensation under section 169 of the 1971 Act. This is an important distinction from *Colley*'s case, where, because the relevant assumption lay within Part I of the Schedule, no compensation was payable under section 169 on a refusal of permission. In the instant case the relevant assumption is in Part II of the Schedule so that compensation would be payable under section 169 if a refusal occurred in the future. It followed that the essential foundation of unfairness present in *Colley*'s case did not exist in the instant case. It would be fair to assess compensation having regard to the Schedule 8 assumptions because if those assumptions later turned out to be incorrect further compensation would be payable under section 169.

*Editor's note: Also reported at (1976) 238 EG 891.

As to the questions of conditions likely to be imposed upon G planning permissions assumed to be granted under Schedule 8, section 278(1) of and Schedule 18 to the 1971 Act stated expressly where and what conditions are to be assumed. As held by the tribunal in *Burlin*'s case, in the context of Schedule 8, the statutory assumptions exclude other assumptions. It followed that it would be wrong to seek to modify the effect of section 278(1) and Schedule 18 by importing conditions outside those contained therein. In any event, neither side had called evidence which would enable the tribunal to do so.

In reviewing the evidence Mr Lewsley referred to *Carter*'s case and cited the words of the president at p 246:

> I agree with the counsel for the claimant's submissions. In my judgment, there is no legal impediment to valuation of the reference land by reference to H the aggregate of values which would be achieved if sold in plots or valuations, quite simply, of the individual plots (which amounts in effect to the same thing).

Mr Catt had proceeded on the basis that sale in individual plots would produce the higher value, so, he thought, did Mr Goldsmith. There had been no evidence that a purchaser of the whole for later division would be in the market, but if he were he would not outbid individual plot purchasers, as he would have to take account of all the risks and delay in acquisition, division and sale, provide finance and look for profit.

As to the question of access from Green Lane, it was common ground that there was a dispute. It was mentioned to at least one potential purchaser. He referred to a letter written by his solicitor J (Conrathes) on behalf of Mr Castle on September 11 1989.

The claimant had made use of an existing access across the strip of land owned by Mr Maiklem and he had written to her denying her right to do so. The claimant had been sufficiently concerned about it to purchase the strip of land. The claimant said that in the dispute she was in a strong position, referring to the contract of sale and the reference to the estate agent's sale particulars. However, neither the transfer document nor the pre-contract inquiries had been produced.

A purchaser of the land would be put off by such a dispute, which he would see as involving risk and delay. He would consider buying the strip of land or a right across it but he would not want to get involved in negotiations unless there was a substantial discount in the sale price. He would also have to face the prospect of seeking K planning permission for a new access and would be unlikely to get it. An access track from Ockham Road North or Long Reach would also require planning permission and, on Mr Stallon's evidence, this was unlikely to be granted.

As to the "before" valuations, Mr Goldsmith's acreage rate of £7,480 was higher than any of the comparable transactions in the locality and the prices offered by potential purchasers. He urged acceptance of Mr Catt's acreage rate of £6,000 an acre in his calculation C, which may well be optimistic because he disregarded planning difficulties regarding access from Green Lane as well as assuming that rights of way existed across the strip of land owned by Mr Maiklem. Taking account of the access difficulty, he urged an L acceptance of the amount of Mr Catt's deductions in his calculations A and B. Certainly Mr Goldsmith's deductions for the access difficulty at about 10% could not be right.

As to the "after" value, Mr Goldsmith had not taken into account the Schedule 8 assumptions. As a result he had no relevant or useful "after" valuations. Mr Catt's valuations took into account the Schedule 8 assumptions. He was an experienced valuer. He had compared what was permissible under the GDO with what was assumed to be permissible under Schedule 8. Considerably more was possible under Schedule 8, principally:

(1) The construction of general farmbuildings for livestock and
(2) The construction of general farmbuildings without limitation as to size and without being restricted as to use with an individual M plot or to the whole of the reference land or indeed the farm as a whole or any particular farm at all. Fences for agricultural purposes were assumed to be permissible under Schedule 8 and Mr Catt reiterated that the difference between this position and the position under the GDO which permitted fencing for any purposes was negligible in valuation terms.

Mr Catt accepted that in the real world the effect of the imposition of an article 4 direction tends to reduce the value of land (leaving aside the Schedule 8 assumptions) by around 20%, but his judgment, after taking into account the statutory assumptions that Schedule 8 applies after the revocation, was that there was no diminution in value.

A With regard to "chancers", he submitted that such people were in the market but that their significance had been exaggerated by the claimant. The site was too open and close to houses to be a good proposition for "chancers". As soon as the claimant marketed the field the council had acted. The reference land was not a site where "chancers" would prosper in relation to the circumstances of either the "before" or the "after" valuations. They were of no real significance in valuation terms. Such "chancers" as there were would take a greater interest in the potential offered by the Schedule 8 assumptions which were to be taken into account in the "after" valuation.

As far as other items of claim were concerned, if the tribunal had jurisdiction the claim for £300 in respect of submitting the
B application and £2,430 (exclusive of VAT) in respect of solicitors' expenses were accepted. VAT should be excluded, as the claimant could have claimed it back.

The claim for interest on the loan was not accepted for reasons given by Mr Catt in evidence. The loan was used to pay for the development of commercial stables which were financially successful. The claimant accepted that she had value for the loan. As Mr Catt had put it, she could not have the land and the interest to be derived from the capital value of that land. Incidentally, the interest claim included repayments of capital. It was common ground that the tribunal had no jurisdiction to award interest on compensation.

In conclusion, Mr Lewsley said that if the application and refusal
C of planning permission did not satisfy requirements of section 165(1) compensation was nil. Even if the requirements of section 165(1) were satisfied, the loss in land value was nil because of the Schedule 8 assumptions. Lastly, if the tribunal held against the council on both of these points, compensation should be £22,730, ie loss in land value *per* Mr Catt £20,000 plus expenses of £2,730.

Claimant's case

Mr Denehan dealt first with the factual background to the point of law as to whether the claimant had a valid claim. He said that Mr Catt had been instructed by the council in July 1990 to settle the claim. He had admitted that his instructions were unqualified and that he was to proceed on the basis that the claimant had a valid claim. He
D compared this admission with Mr Lewsley's argument that it was not possible to know what the council's decision would have been upon a planning application restricted solely to fencing.

He said it was admitted by Messrs Catt and Stallon that negotiations had taken place between the claimant and the council with a view to settling the claim and the claimant was never informed that her claim was misconceived on any ground. The claimant had referred the matter to the tribunal and incurred substantial costs in so doing. The claimant first learnt that the council were to take the point of law as to the invalidity of the claim during the afternoon of the last working day before the commencement of the hearing. There seemed to have been some confusion as to the date but Friday, November 8 was the last working day before the hearing commenced.

E No explanation had been given by the council's witnesses as to why the point of law had not been raised previously. Mr Denehan said the point was bad but by failing to inform the claimant that it would be taken at the hearing the council had denied the claimant the opportunity of making a further protective planning application in the unlikely event of the point being held to be good. He said that the point of law involved could properly be called a preliminary point of law, that the tribunal had a specific procedure under rule 49 of the Lands Tribunal Rules which should have been used in this case as only if it failed was it necessary to prepare and deal with the substantive issues involved in the reference and further a member could have been appointed who was best suited to determine the
F point. The council had failed to apply under rule 49 and should not now be permitted to take the point of law having failed to comply with the rules.

Mr Denehan relied on the *Essex County Council* case to support him. In that case, although the county council had not mentioned a point on validity in their counternotice in response to a blight notice, the tribunal did consider the point as a preliminary issue, the claimant having argued the point but not objected to the procedure on the ground that the point was not mentioned in the counternotice. He distinguished the present case in as much as the council had not complied with the rules and sought a hearing of the point of law as a preliminary issue. In these circumstances the point could not be taken. The tribunal had jurisdiction and could use its discretion.

Mr Denehan was grateful to Mr Lewsley for conceding that the G
council's conduct might effect an issue on costs if the council were successful on the point of law.

On the point of law, he submitted that the claimant had satisfied the requirements of section 165(1) of the 1971 Act and was entitled to claim compensation. The policy behind the compensation scheme was straightforward. A claimant was entitled to compensation for loss suffered as a result of GDO rights being withdrawn in the interests of the community as a whole. It followed that if a claimant was in the same position after the rights were withdrawn as before, the policy criteria had not been satisfied; for example, if planning permission was granted for development formerly within the GDO rights there was no entitlement to compensation.

Three elements had to be satisfied in order to qualify for H
compensation:

(a) GDO rights had existed;
(b) they had been withdrawn; and
(c) there had been a failure to obtain permission for development within the scope of the withdrawn GDO rights.

The claimant satisfied these elements in as much as:

(a) she enjoyed the rights granted by class A of Part II of the GDO;
(b) these rights were withdrawn by the direction; and
(c) she was refused permission for operations previously permitted under those rights

and the claim was entirely consistent with the policy of the compensation scheme. It was also consistent with the relevant J
statutory provisions in section 165(1) of the 1971 Act.

By article 3(1) and class A of Part II of the GDO the claimant was granted planning permission for, *inter alia*, the erection of fencing etc. Accordingly section 165(1)(*a*) was satisfied.

The direction provided that the permission granted by article 3 of the GDO should not apply to the development described in the Schedule to the direction which was that within class A of Part II. Accordingly, section 165(1)(*b*) was satisfied.

The claimant had made an application under Part III of the 1971 Act for planning permission to carry out development within class A Part II. This had been refused and accordingly section 165(1)(*c*) was satisfied.

The planning application of November 1989 had admittedly K
contained two elements, the building and the fencing. It was clear that from the application that these were separate elements and treated as such by the council. This was clear from the reasons for refusal, the first of which was concerned only with the proposed fencing and the second and third of which were concerned only with the proposed buildings. It was entirely wrong to suggest that the council did not consider the application for permission to erect fencing on its planning merits separate from the merits of the erection of buildings.

There was no express requirement in section 165 that there be a single application for permission for the development affected by a direction. There was no need to imply such a requirement. All that L
was required was that there be a refusal of permission for the development, which there had been in this case. Mr Lewsley asked the tribunal to put its head in the sand but the tribunal should look at the direction and the reasons for refusal. The first reason stood alone. There was no cross-fertilisation between that reason and the other two.

The claimant relied upon the opening words of article 4, which provided that if the local planning authority were satisfied that permitted development should not be carried out unless permission was granted for it "on an application" the authority may give a direction. Again there was no express requirement that there must be a single application in respect of the land the subject of the direction concerning only the class of development for which permission was M
withdrawn by the direction. There must be "an application" and there had been in the instant case.

Referring to Mr Lewsley's submission, Mr Denehan said this was not a case where the claimant had included development other than that in respect of which GDO rights had been withdrawn in order to ensure a refusal of permission. In the instant case the claimant had thought permission would be required both for fencing and buildings. Had the council thought the fencing permissible they could have refused permission specifying only the reasons relating to the buildings.

The claimant further relied on *Carter*'s case. In that case the claimants had not made an application in respect of the whole of the reference land, but 15 applications in respect of 15 different parts. Counsel for the compensating authority had contended that the applications were not in respect of development which could have benefited from GDO rights under class 6 because the reference land had never been used as 15 plots. This argument had been rejected by the learned president of the tribunal.

The point of law taken by the council offended the express words of the statute, the policy of the compensation scheme and common sense and should be rejected.

Turning to the substantive claim, Mr Denehan dealt first with the "before" value of the reference land. He described this as being its value prior to the making of the direction. The valuation involved, *inter alia*:

(a) the right of access from Green Lane, and
(b) the prospects of obtaining planning permission for access via a track from Ockham Road North.

As to (a), the claimant's case was that at all times she had a legal right to gain access from Green Lane. The contract of sale by reference to the sale particulars provided for it and she had enjoyed *de facto* use of two accesses from Green Lane at all times since her acquisition of the reference land in July 1986. Mr Lewsley had said that evidence of the transfer or inquiries upon purchase had not been given but the contract specifically incorporated the sale particulars which mentioned the rights of access from Green Lane.

The claimant's use of the accesses had not been interfered with by Mr Maiklem either physically or by legal process notwithstanding that he knew of the claimant's use of the accesses. He had challenged her right to use them in a letter. She had taken legal advice, which was to the effect that she had a legal right to use the accesses. No one had suggested that that advice was incorrect. What was called a dispute over the right of access was in reality no dispute at all. The claimant's unchallenged evidence was that none of the potential purchasers withdrew because of it. Mr Lewsley had referred to a letter from Conrathes (Mr Castle's solicitors) but the claimant's evidence had been unchallenged.

The claimant had not been asked why she had recently bought the strip of land along Green Lane. The land might have some nuisance value but there was no evidence that Mr Maiklem would have bothered purchasers. The claimant's case was that no allowance should be made for this factor in the "before" valuation or, in the alternative, that the allowance should be no more than 12%.

Turning to the prospect of obtaining planning permission for an access track from Ockham Road North, Mr Denehan said that Mr Stallon's report should be rejected as partial, unbalanced and incorrect. In evidence, Mr Stallon had accepted that it was possible that planning permission would have been granted for an access track from Ockham Road North to serve the reference land. The tribunal was invited to find that, provided sensible and proper consultation had taken place between an applicant and the planning officers, planning permission would have been granted for the construction of an access track from Ockham Road North to the reference land. The need to obtain such permission was not a matter which would have significantly depressed the value of the reference land. Mr Stallon had conceded that it would be "a good bit easier" to obtain planning permission for an additional four accesses from Green Lane to the reference land. The tribunal was invited to find that there would have been no difficulty in obtaining planning permission for the creation of a number of access ways from Green Lane to the reference land.

The perceived difficulty of obtaining a water supply to the reference land had been overstated by Mr Catt. The tribunal was invited to accept the evidence of Mr Goldsmith on the point. This was consistent with the reality of the situation. A water supply would have been readily available from the farm and the cost of an overground supply minimal.

As to the evidence of value, Mr Goldsmith's valuation took into account the possibility of selling the reference land in plots. He relied for guidance on the fact that smaller plots on the Dell Park Estate had realised higher prices an acre than the larger plots. The best comparable evidence was provided by the offers made to the claimant. These were £6,000, £6,500 and £7,000 an acre respectively. Mr Goldsmith's higher acreage rate of about £7,480 could be justified as it accommodated the disposal of the land in small plots. Mr Catt's acreage rate of £6,000 was less than the average of the prices paid in the two comparable transactions upon which he relied, which were £6,330.50 an acre. The tribunal was invited to accept Mr Goldsmith's evidence.

As to taking account of the "dispute" about access from Green Lane, it should be assumed that access was available and planning permission would be readily granted for such an access. If a reduction were to be made on account of the "dispute", it should not exceed 12%, Mr Goldsmith's evidence being that a hypothetical purchaser would have had regard to the nature and extent of the alleged dispute and the possibility of purchasing Mr Maiklem's strip of land. The inconvenience and expense of this among other things is reflected in his 12% allowance.

Mr Catt's analysis in his calculations A and B was false. He conceded in cross-examination that a prospective purchaser would take into account the matters mentioned by Mr Goldsmith. His calculations A and B did not do so. In calculation B his deduction for the "dispute" was 25% (from £6,000 an acre to £4,500). This was too great. In calculation A Mr Catt made a deduction of 75% (from £6,000 an acre to £1,500 an acre) for the "dispute", assuming there was no prospect of an alternative access being provided. Mr Stallon's evidence did not support the latter premise. No deduction should be made to reflect the need to obtain planning permission for an alternative means of access because:

(a) an alternative access would not be needed as access to Green Lane was available, and
(b) planning permission would be granted for a suitable alternative access.

On the assumption that access from Green Lane was a legal certainty it would be wrong to make any more than a nominal deduction to accommodate the cost of creating an alternative access. Mr Goldsmith's evidence was that his 12% deduction accommodated the possibility of having to construct an alternative access.

As to the condition of the reference land at the material date, both valuers said it was not in good condition but only Mr Catt thought that this would have diminished its value. Mr Goldsmith thought that there was a part of the market which would not have been concerned with the condition and it was submitted that no deduction should be made to reflect the condition of the reference land.

In summary, the claimant's case on the "before" valuation was that:

(a) the value should be taken as being £7,481.75 an acre;
(b) if, alternatively, there was to be a deduction to accommodate the "dispute" as to access from Green Lane, it should not exceed 12% and the value should be £6,949 an acre;
(c) in the further alternative, if Mr Catt's figure of £6,000 an acre was accepted and an allowance was to be made to accommodate the access "dispute" the value should be £5,280 an acre.

Turning to the "after" valuation, the council's argument was that the assumption required by section 164(4) of the 1971 Act that planning permission would be granted for development within Schedule 8 meant that the loss in value was nil because, despite the direction, fences and buildings for agricultural purposes could be erected. The claimant submitted that section 164(4) did not apply in this case in the manner suggested by the council notwithstanding the words of section 164(4), in other words, the Schedule 8 assumptions did not apply to produce a nil loss in value of the reference land. The claimant relied upon the interim decision of the tribunal in *Colley*'s case.

The claimant adopted the submission of the claimants in that case as follows:

(a) where a statute takes away property rights without giving a right to compensation, such an intention will not be imputed to the statute unless it is in clear and unequivocal terms. The fact that compensation is not always payable for planning decisions is of no assistance.
(b) there is authority that an absurd result would not be intended by statute, so absurd that Mr Goldsmith could not get to grips with it,
(c) a court is justified in departing from the plain words of a statute where it is satisfied:
    (i) there is a clear and gross balance of anomaly as there is in this case,
    (ii) Parliament, the legislative promoters and the draftsmen could not have envisaged such an anomaly and could not have been prepared to accept it in the interests of a supervening legislative objective,
    (iii) the anomaly can be obviated without detriment to such legislative objective, and

(iv) the language of the statute is susceptible of the modification required to obviate the anomaly.

The policy of the statutory compensation scheme, put simply, is that a landowner is to receive compensation in respect of loss suffered as a result of the withdrawal of GDO rights. If the Schedule 8 assumptions apply in the way the council say, what is given by section 164(1) is taken away by section 164(4) and Schedule 8. If the council are correct the claimant loses her GDO rights without compensation being payable, an absurd result which Parliament cannot have intended. The policy of Part VIII of the 1971 Act would be frustrated and the claimant's remedy in section 164(1) destroyed.

Section 164(4) should be interpreted as the member of the tribunal did in *Colley*'s case, namely by adding the words "unless such planning permission is the subject of revocation order proceedings", in the instant case the reference being to an article 4 direction rather than a revocation order.

Mr Lewsley had questioned whether there was any hardship involved in literal application of section 164(4) in this case, having regard to the possiblity of obtaining compensation under section 169 at some future date. This was no answer. In *Colley*'s case some compensation would have been payable in any event. The tribunal should look at the fairness of the application of section 164(4) in this case. If the result was absurd, the reasoning of the interim decision in *Colley*'s case should apply.

In fact, the absurdity and injustice that would be caused in the instant case would be greater than that which would have been caused in *Colley*'s case. Whereas in that case some compensation would have been payable in any event, in the instant case, according to the council, none would be payable.

Burlin's case was distinguished on the same ground that it was distinguished in *Colley*'s case, namely that the application of the Schedule 8 assumptions did not produce an absurdity in *Burlin*'s case. It was irrelevant that *Colley*'s case concerned a revocation of planning permission. The decision was a general application in all cases where compensation was claimed under section 164. In any event, where the claim is based upon a withdrawal of GDO rights the Act requires it to be assumed that there has been a revocation of planning permission.

The Schedule 8 assumptions should not be applied in this case. Accordingly, it is not to be assumed that planning permission referred to in para 4 of Schedule 8 would be granted to the claimant. Even if the Schedule 8 assumptions were to apply, Mr Catt's initial approach ignored the fact that prior to the direction fences could be erected for any purpose under the GDO and the depressing effect of the direction.

In answer to the tribunal, he had said that he could not think of a non-agricultural use requiring fencing. Mr Goldsmith had not been given an opportunity to comment on this point, but the claimant said that the keeping and schooling of horses, recreational uses such as archery, clay pigeon shooting, hobby farming and the keeping of rare breeds (in or on an area of less than one acre), camping and caravan sites, all require fencing.

As to the depressing effect of the direction, Mr Catt had estimated it at 20%. This was not in his report and was not put to Mr Goldsmith. The estimate of 20% must be viewed against the events that happened. The claimant's unchallenged evidence was that Messrs Castle, Hayford and Stamper all withdrew offers because of the making of the direction. Mr Stamper's withdrawal was particularly significant because he had paid the purchase price and taken up possession. This suggested that 20% to reflect the presence of the direction was too low. Even applying the Schedule 8 assumptions, the claimant submitted that the depressing effect of the reference land was such as to reduce it to something approaching the £1,000 an acre that Mr Goldsmith suggested. This was the amount offered to the claimant after the direction was made.

The evidence of the Chilworth land was not compelling as it did not concern the reference land and in the light of the actual offers made for the reference land.

Although not in his report, Mr Catt expressed the view that the value to the hypothetical purchaser of the right to erect agricultural buildings of unlimited size would offset and even surpass the depressing effect of the direction. Mr Denehan submitted that Mr Catt's analysis was false and an attempt to justify an unsupportable position. It effectively limited the market for the reference land to those wishing to use the land for agricultural purposes. This was not the market. Mr Catt admitted that the market contained those wishing to use the land for other non-agricultural purposes. Mr Catt's response was that a hypothetical purchaser who did not wish to use the land for agricultural purposes would nevertheless pay the full price of the land in order to exploit the assumed planning permission to erect agricultural buildings and take a chance of using them for non-agricultural purposes. This was inconsistent with what Mr Lewsley had said in closing about "chancers". It was admitted that the market included "chancers", but there must be very few, if any, "chancers" prepared to incur the cost of erecting a large building when there was a risk of enforcement action, particularly if the land was open and visible as Mr Stallon had described the reference land.

It was submitted that, despite Mr Catt's efforts, the only conclusion was that notwithstanding the Schedule 8 assumptions there was a difference between the value of the reference land before and after the issue of the direction. The matter was academic, as the Schedule 8 assumptions should not be applied.

In conclusion, the claimant's case was that the direction had the effect of reducing the value of the reference land to £1,000 an acre. It followed that the compensation for diminution in value of the reference land should be £96,000 (£112,500 less £15,000 less costs of sale £1,500). Alternatively, if an allowance should be made because of Mr Maiklem's ownership of the strip of land along the Green Lane frontage it should be £84,750 (£99,000 less £12,750 less costs of sale £1,500). Lastly, if Mr Catt's base figure for his calculation C was accepted it should be £73,920 (£90,420 less £15,000 less costs of sale £1,500).

Turning to the other elements of the claim, section 164(1) (*b*) of the 1971 Act as applied by section 165 provided that loss directly attibutable to the issue of the direction was recoverable. But for the direction no planning application would have been made for fencing. The fees for making the application in so far as it related to fencing, £300, were payable.

As to interest on the mortgage advance, but for the making of the direction the advance would not have been taken and therefore no interest paid. Mr Catt's criticism of this item was not accepted. If the claimant received the compensation she sought she would not be in the position of having the benefit of the value of the reference land and the interest thereon. Compensation was sought in order to replace money the claimant had expended on interest payments. The repayments include repayment of capital, but the amount of capital repayment was negligible.

Lastly, evidence of the amount expended on legal costs was unchallenged. But for the making of the direction they would not have been incurred and fell squarely within section 164(1) (*b*) of the 1971 Act.

Point of law

The point of law turns on the construction of section 165(1) (*c*) of the 1971 Act. It was clearly not in the council's contemplation until shortly before the hearing and the claimant only heard of it immediately before the hearing. At the commencement of the hearing Mr Lewsley suggested that it might be convenient for the point to be taken as a preliminary issue, but at that stage, with both sides prepared for a hearing of the case in full, the claimant preferred to proceed and Mr Lewsley did not pursue his suggestion.

The claimant contended that rule 49 of the Lands Tribunal Rules specifically provided for the determination of points of law at a preliminary hearing, that the council failed to comply with this rule and that in these circumstances it was not open to the council to take the point. The council regretted that the point was not drawn to the attention of the claimant sooner but contended that this could not widen the tribunal's jurisdiction to award compensation beyond what was permitted by the statute.

Both parties referred to the *Essex County Council* case and *Carter*'s case. The claimant distinguished the *Essex County Council* case on the ground that the point which had been omitted from the county council's counternotice had nevertheless been taken as a preliminary point before the tribunal. The council relied upon the case as an authority derived from the House of Lords that "estoppel cannot enlarge the jurisdiction of a court of limited jurisdiction". In *Carter*'s case the learned president of the tribunal accepted a submission that a point of law could not be taken by the compensating authority because it was too late to do so. This was accepted as one of two submissions and the council say that it is impossible to say whether it would have succeeded on its own. In any event the point did not go to jurisdiction.

A Rule 49 of the Lands Tribunal Rules is not mandatory but permissive in as much as either party can make an application for the preliminary hearing of a point of law. It is then in the tribunal's discretion whether to agree to such a hearing.

In the instant case I accept that it would probably have been desirable for a preliminary hearing to be held and for a member of the tribunal learned in the law to determine the point of law. However, the authority of the tribunal to award compensation is limited by the provisions of section 165 of the 1971 Act and I accept that "estoppel cannot enlarge the jurisdiction of a court of limited jurisdiction" *per* Lord Hodson in the House of Lords in the *Essex County Council* case. I therefore think that I have to consider and decide the point of law.

B The point of law is essentially a short one. Section 165(1) of the 1971 Act provides:

> The provisions of this section shall have effect where —
> (*a*) planning permission for the development of land has been granted by a development order; and
> (*b*) that permission is withdrawn, whether by the revocation or amendment of the order or by the issue of directions under powers in that behalf conferred by the order; and
> (*c*) on an application made in that behalf under Part III of this Act, planning permission for that development is refused, or is granted subject to conditions other than those previously imposed by the development order.

The parties agree that the first two requirements of the subsection C are met by the claimant in this case. The conflict arises in relation to the third requirement.

The council contend that the natural meaning of the words of section 165(1) (*c*) is that to qualify an application must relate solely to development in respect of which planning permission was withdrawn by a direction. This was said to be made even clearer by reference to the corresponding provision of the Act of 1990, a consolidating Act which it must be presumed was not intended to alter the previously existing law. In support of their contention the council said that the intention of the legislature was that the local planning authority should have an opportunity to consider and grant an application for development previously within the applicant's GDO rights. In the D instant case the committee had had no such opportunity as they had to consider the application as a whole and no one could know what their decision would have been upon an application relating solely to fencing.

The claimant contended that the application made on November 8 1990 was a qualifying application containing two distinct elements, the merits of each of which the council had clearly considered, the reasons for refusal clearly showing that they rejected the proposed fencing. Neither section 165(1)(*c*) nor article 4 of the GDO contained any express requirement that a qualifying application should relate solely to development previously within the applicant's rights and no such requirement should be implied.

I was referred to *Carter*'s case, where the president of the tribunal E accepted that the refusal of 15 applications for 15 parts of the land the subject of a direction met the requirements of section 164(1)(*c*).

The issue can be stated briefly. Was the application one which sought planning permission for development in respect of which GDO rights had been withdrawn by the direction or, even more briefly, was the application an application for planning permission to erect fencing? If, as I think, the words of the statute should be given their ordinary meaning, I do not see how this question can be answered in the negative. The fact that the application was also an application for planning permission for other development does not seem to me to take it outside the words of section 165(1)(*c*), there being no requirement in the statute that, to qualify, an application F must relate solely to development previously permitted under the GDO.

I have also considered the wording of section 108 of the 1990 Act, but it does not affect my conclusion as to the meaning of section 165(1)(*c*) of the 1971 Act.

I elicit some support for the view I take as to the ordinary meaning of the relevant words from the fact that until a very late stage none of the council's officers or their advisers thought the application fell outside the wording of section 165(1)(*c*).

Mr Lewsley said that if the statute is construed in the claimant's favour it leaves it open to a potential claimant to ensure the refusal of an application by including in it development not previously permitted by the GDO but clearly likely to be objectionable to a local planning authority.

In the instant case there is no suggestion that the claimant sought to G ensure a refusal. Clearly the claimant included buildings in the application because, at the time it was submitted, she understood the direction would withdraw her GDO rights in respect of buildings. In the event, it did not.

The officer's report and recommendation to the planning committee on the application was not put in evidence, but on the evidence of the reasons for the making of the direction and the reasons given for refusing the application it is clear to me that the council were very concerned about the effect of fencing the reference land into separate parcels and I do not think it is reasonable to doubt that had the application related solely to the proposed fencing it would have been refused.

H I take the point that, had the council been minded to accept the fencing but not the buildings, they would still have had to refuse the application as a whole. However, they could have omitted the first reason for refusal and also informed the claimant that they would be willing to grant an application for the proposed fencing alone.

Given that the evidence in the instant case is that the existence of a direction can have a depressing effect on the value of land, this might still have left the council liable to a claim for some compensation, which they could have avoided had they had the opportunity of granting an application related solely to fencing. This may be thought an unintended result of the legislation, but it would have been a fair result in as much as the claimant would have been entitled to claim no more than her true loss and, as Mr Lewsley pointed out in J another context, this field of legislation does not always produce results which are entirely fair to both parties.

Lastly on the point of law, I note that neither valuer found any difficulty in applying the law on the basis of the view I have taken as to its meaning.

I find against the council on the point of law.

Schedule 8 assumptions

Before turning to the consideration of the amount of compensation payable I have to consider a second point of law. This relates to the application of section 164(4) of the 1971 Act to the "after" valuation of the reference land. The council said that this meant that in considering the "after" valuation it must be assumed K that planning permission would be granted for development within Schedule 8 to the 1971 Act, which included various agricultural operations including fencing etc.

During the hearing the claimant conceded that the council's view was correct. However, before counsel made their closing submissions I drew their attention to the interim decision of the tribunal in *Colley*'s case of which neither party had then heard. As a result, the claimant, with the council's agreement, withdrew her concession as to the council's view on section 164(4).

The provisions of section 165 apply the provisions of section 164 as if a planning permission granted by a development order:

(a) had been granted by the local planning authority under Part III of the Act, and L

(b) had been revoked or modified by an order under section 45.

Section 164 provides for compensation in respect of the revocation or modification of a planning permission. The provisions of section 164, which are relevant to the instant case, are as follows:

> (1) . . . where planning permission is revoked . . . then if . . . it is shown that a person interested in the land . . .
> (*a*) has incurred expenditure in carrying out work which is rendered abortive by the revocation . . . or
> (*b*) has otherwise sustained loss or damage which is directly attributable to the revocation . . . the local planning authority shall pay to that person compensation in respect of that expenditure, loss or damage.
> (2) For the purposes of this section, any expenditure incurred in the preparation of plans for the purposes of any work, or upon other similar matters preparatory thereto, shall be taken to be included in the expenditure incurred in carrying out that work . . . M
> (4) In calculating, for the purposes of this section, the amount of any loss or damage consisting of depreciation of the value of an interest in land, it shall be assumed that planning permission would be granted for development of the land of any class specified in Schedule 8 of this Act.

The council contend that subsection (4) means that in making the "after" valuation it must be assumed that planning permission would be granted upon an application being made in respect of any form of development specified in Schedule 8, one such form of development being agricultural operations including fencing etc.

A The claimant argued that the plain words of the statute produce an absurd result. They meant that in assessing compensation for a refusal of planning permission it must be assumed in the "after" valuation that that permission would nevertheless be granted. An analogous situation had arisen in *Colley*'s case and the claimant adopted the submissions of the claimants in that case which, in their interim decision, the tribunal had accepted.

However, as the council submitted, there is a difference between the two cases in as much as in *Colley*'s case the revoked planning permission was for development which lay within Part I of Schedule 8, whereas in the instant case development comprising the erection of fencing for agricultural purposes lay within Part II of Schedule 8.
B Where planning permission is refused for development within Part II and the refusal is sustained by the Secretary of State on appeal, there is a right to compensation for loss suffered under section 169 of the 1971 Act. This is not the case if the development lies within Part I. I think this is an important distinction and because there was a separate right to compensation for refusal of planning permission, in respect of the relevant Schedule 8 development in this case, I do not think it can be said that giving the words of section 164(4) their plain meaning would create a clear and gross balance of anomaly unintended by the legislature. I therefore conclude that in the "after" valuation it must be assumed that planning permission would be granted for development within Schedule 8.

"Before" valuation

C The three offers actually received by the claimant prior to the making of the direction totalled £42,000, a mathematical average of about £6,460 an acre. I think it is reasonable to conclude that in each case the offerers took into account the condition of the land and assumed that there would be no legal restraint against their taking access directly from Green Lane. None of the offers appear to have been conditional upon planning permission being obtained for any necessary new access from Green Lane, although in Mr Castle's offer he did specify the inclusion of "a gate into Green Lane" and Mr Stamper's offer related to a specific area of land which included the western-most of the existing accesses.

The comparable transactions referred to by Mr Catt produced
D prices an acre of £5,801 (26.89 acres at Chertsey) and £6,686 (6.73 acres at Elstead). I inspected both parcels. The evidence was that the Chertsey land had achieved a greater price than the aggregate of the prices which would have been achieved if it had been sold in two lots. My impression was that it was not as convenient for dividing into small parcels as the reference land. The Elstead land was less than seven acres in area and met the demands of the market for small parcels without division.

Mr Goldsmith's value of about £7,480 an acre is not supported by either the offers or the comparable transactions. Mr Catt's valuation for calculation C was effectively £6,200 an acre less £200 an acre to reflect the poor condition of the land, a net £6,000 an acre.

The condition of the land was apparently deteriorating during the
E period between the dates when the offers were made and the date of Mr Catt's first inspection in June 1990. As at the material date the condition might reasonably be assumed to be somewhat worse than when the offers were made and somewhat better than when Mr Catt made his inspection. Mr Goldsmith thought no allowance necessary for the poor condition of the land.

It seems to me that the reference land was particularly suitable for division into small parcels because it had a long frontage to Green Lane. In the light of the offers and the evidence of the comparable transactions, I think Mr Catt's value for calculation C was a little low and I adopt an acreage rate of £6,250, net costs of sale, to reflect the condition of the land at the material date.

There was a suggestion that there might be difficulty in obtaining
F planning permission for accesses from Green Lane. I am not persuaded by Mr Stallon's evidence that there would have been any real difficulty in obtaining such permission or permissions. Mr Catt apparently made no allowance for it in his valuations for calculation C.

This leaves the matter of the strip of land owned by Mr Maiklem. Although the claimant had been legally advised that she was entitled to use the accesses and had done so for several years without Mr Maiklem doing more than writing a letter challenging her right to do so, I think the strip of land had some nuisance value in relation to the reference land. On the evidence, although I think it is probable that planning permission would have been obtainable for some alternative access, it would inevitably have involved delay and expenditure. Additional expenditure might also have been incurred G
in providing a water supply over the remainder of the farm. On the other hand, there was always the possibility of purchasing the strip of land as the claimant ultimately did for £3,000 in November 1991.

I note that the price was agreed after the direction was made and the refusal of the planning application which gave rise to the claim. At that time the nuisance value of the reference land was probably less than it was prior to the direction being made as according to the evidence the market value of land such as the reference land had fallen somewhat.

Mr Catt thought that the price likely to be agreed for the reference land would fall somewhere between his valuations for calculations A and B, rather nearer the latter than the former. From his evidence it is H
impossible to say exactly what this price was, but his "before" valuations for calculations A and B produced roundly £22,600 and £52,750 respectively. The mid-point between these two valuations is £37,675, so on the basis of his evidence his estimate of the price likely to have been agreed was probably no more than £40,000, which is £50,400 or about 55% less than his valuation on the basis of his calculation C, ie on the basis of access and water supply being available from Green Lane. I think such a deduction is much too great to reflect such nuisance value as Mr Maiklem's strip of land represented and I prefer Mr Goldsmith's evidence that an allowance of 12% would have been sufficient.

I therefore find that the "before" value was as follows: J

	£
15.07 acres at £6,250	94,188
Less: for nuisance value of land owned by Mr Maiklem — 12%	11,302
Leaving	82,883
Say	£82,500

"After" valuation

The "after" value is to be estimated by reference to a hypothetical world in which it is to be assumed that planning permission would be granted for Schedule 8 development. On this basis Mr Catt said that the value of the land was no less after the making of the direction than K
it was before. He accepted that the imposition of the direction had a depressing effect on the value of land, but he said that this was at least offset by the effect of the Schedule 8 assumptions in as much as the forms of development within Schedule 8 exceeded those permitted under the GDO. In answer to me he quantified the depressing effect at 20% and it follows that he thought the additional value due to the fact that Schedule 8 included forms of development additional to those permitted under the GDO must be at least 20%.

As to the value of land, the subject of a direction similar to that made in respect of the reference land, Mr Catt cited a failed transaction at Chilworth, which would have involved a sale of about 15 acres of land at £2,200 an acre. However, this would have been a transaction in the real world, ie without the benefit of the Schedule 8 L
assumptions and Mr Catt referred to it only to support a valuation based upon the assumption that the council's contention regarding the Schedule 8 assumptions was not upheld.

Mr Goldsmith said that the value of land was depressed by the making of a direction. He said that if it were known that planning permission would be granted for fencing to divide the reference land into four or fewer plots he would still anticipate a loss of value of 50% of his "before" valuations. However, he did not express any opinion as to the loss in value if it had to be assumed that planning permission would be granted for fencing for agricultural purposes to divide the land into more than four plots, although it is fair to assume that his estimate of the loss in value would have been less than 50%.

Both valuers agree that the direction had a depressing effect on the M
value of the reference land even taking account of the Schedule 8 assumptions in so far as they relate to fencing for agricultural purposes. I have no difficulty in accepting their view. It is clear that the market demand for the reference land included a number of potential non-agricultural users and it could not be assumed that planning permission would be granted for fencing for the purpose of such users. Additionally, the need to make applications for proposed fencing etc for agricultural purposes would involve cost and delay even though it had to be assumed that planning permission would be granted in respect of them.

The only evidence quantifying the depressing effect of the direction taking account of the Schedule 8 assumption regarding

A fencing etc for agricultural purposes is Mr Catt's estimate of 20%, which I adopt.

Mr Catt offset the depressing effect of the direction because of the added potential of the reference land in the hypothetical world of the Schedule 8 assumptions in so far as these provided that planning permission would be granted for development not permitted under the GDO. He cited three specific matters:

(a) agricultural buildings not required in connection with the agricultural unit on which they were to be sited,
(b) an agricultural building in excess of 465 m^2 in floor area, and
(c) a building for use by livestock nearer than 400 m to another property.

B Mr Goldsmith had not anticipated potential purchasers wishing to erect buildings.

The evidence as to the uses which potential purchasers would wish to put land such as the reference land, the offers made for the reference land and the uses to which the land the subject of comparable transactions had been put did not suggest to me that the right to erect agricultural buildings in excess of 465 m^2 in floor area or agricultural buildings to be used other than in connection with the agricultural unit in which they were to be sited, or buildings for livestock, would have been of interest to potential purchasers. I am not persuaded that the Schedule 8 assumptions in so far as they include development not within GDO rights relating to buildings would have added to the "after" value of the reference land.

C I therefore find that the "after" value of the reference land was:

	£
"Before" value	82,500
Less: for effect of direction — 20%	16,500
"After" value	£66,000

and that the diminution in value of the claimant's interest in the reference land at the material date was £16,500 (£82,500 less £66,000).

On the basis that the claim was valid the parties agreed that the fee for submitting the application was properly claimable, likewise the solicitors' expenses. The council said that the VAT element should be

D excluded as the claimant could have recovered it. In his closing speech, Mr Denehan said that the VAT point was in dispute, but I can find no evidence which controverted the council's evidence that the VAT element was recoverable by the claimant and I find that the compensation should include the solicitors' expenses excluding the VAT element.

Lastly, there is the claim for interest on the loan the claimant took out to finance the development of the livery stables. Mr Catt said that had the claimant sold the reference land and expended the net E proceeds on the development of the livery stables she would have forgone the interest she would otherwise have been able to earn by investing the net proceeds elsewhere. That potential interest broadly equated the actual interest paid on the loan and the claim for interest was thus invalid. It might be said that the rate of interest the claimant could have earned by investing the net proceeds elsewhere was likely to have been less than the rate of interest she paid on the loan, but the claim was not put in this way. I therefore accept Mr Catt's evidence and reject the claim for interest.

I therefore find that the compensation payable by the council to the claimant is £19,230 made up as follows: F

	£
(a) dimunition in value of the claimant's interest in the reference land	16,500
(b) fee for submitting planning application	300
(c) solicitors' expenses for advice in connection with the direction and application	2,430
Total	£19,230

I received submissions as to costs. The claimant sought costs to an indemnity basis. The council requested that there should be no order as to costs. The claimant succeeded on the pricipal point of law. The council succeeded on the point of law related to the assumptions G required by Schedule 8 to the 1971 Act, but the claimant had conceded the correctness of the council's contention until the closing submission by counsel.

As to quantum, the amount of my award fell well short of the claim and was effectively based upon the council's evidence.

The only evidence of any offer having been made is on an offer made, without prejudice save as to costs, by the claimant between the hearing of the evidence and counsel's final submissions. The offer specified an amount well above the amount of my award. There is no record of council's having made any offer.

In view of the fact that the council introduced the principle point of law at a late stage, the claimant seeks costs on an indemnity basis. The H council were entitled to argue their point of law and, in the event, I do not think the late introduction of their argument on this prolonged the hearing significantly. I do not think the council should be prejudiced in terms of the basis of costs.

In all the circumstances I have come to the conclusion that the council should bear 75% of the claimant's costs on the High Court Scale on the standard basis, such costs to be taxed by the Registrar in the event of disagreement.

For a further case on this subject see p 193